Marking Time

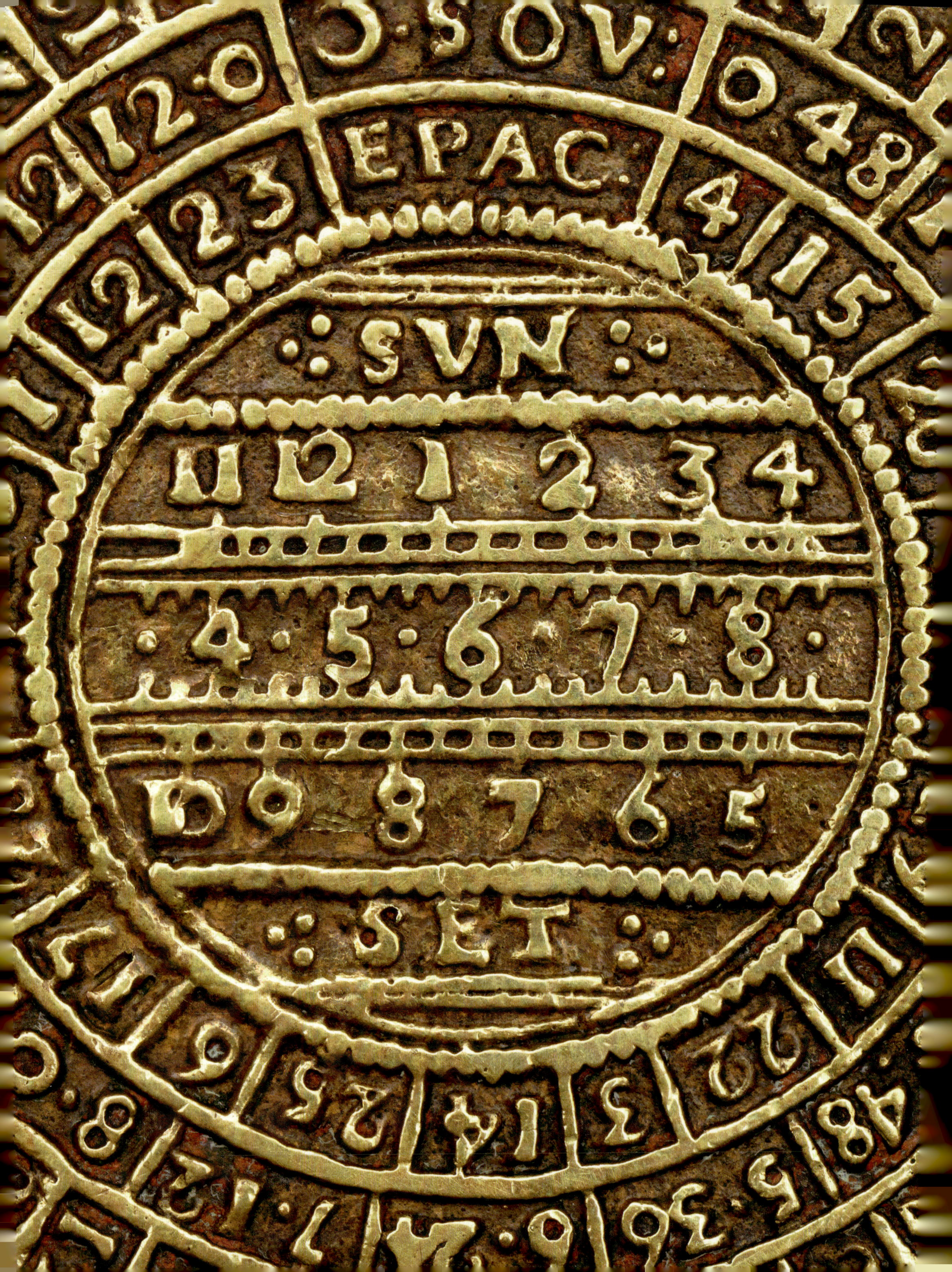
EPAC.
SVN
11 12 1 2 3 4
4 5 6 7 8
D 9 8 7 6 5
SET

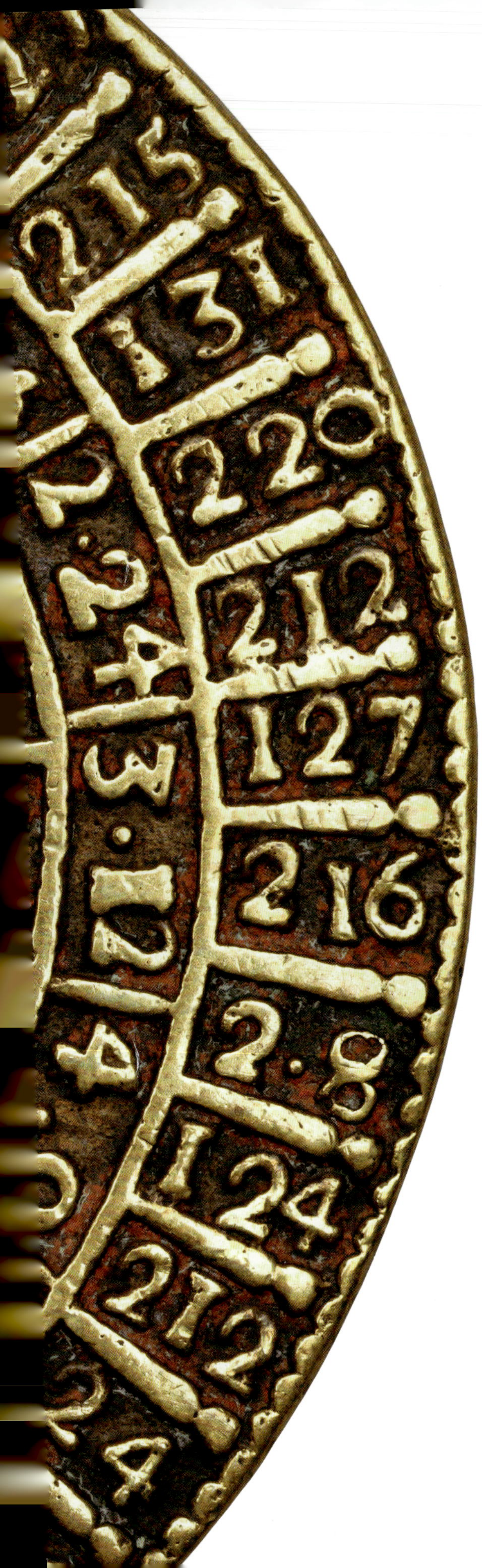

Marking Time

Objects, People, and Their Lives, 1500–1800

Edited by
EDWARD TOWN &
ANGELA MCSHANE

YALE CENTER FOR
BRITISH ART
NEW HAVEN

YALE UNIVERSITY PRESS
NEW HAVEN & LONDON

CONTENTS

Catalogue of Objects

Director's Foreword

MARKING TIME SURVEYS THE MATERIAL CULTURE OF THE BRITISH nation as it emerged as a dominant force on the global stage. This colonial expansion coincided with a horological revolution that improved the accuracy of clock time, igniting people's impulse to date the objects they owned and made. As the objects in *Marking Time* reveal, even small, everyday things—gifts, tools, keepsakes, furnishings—have their place in broader narratives of the past.

The idea that objects have agency has a long history at the Yale Center for British Art. The Center's first director, Jules Prown, was a pioneer of studies in material culture; his teaching legacy is evident in the three generations of scholars represented in this publication—a fitting testament in this, his ninetieth year. After Jules's retirement from the Yale University faculty, his mantle was assumed by Edward S. Cooke Jr., Charles F. Montgomery Professor of American Decorative Arts, who furnished one of the seven essays that set the stage for this volume's catalogue. Studies in time also preoccupy other members of Yale faculty, such as social historian Keith Wrightson, Randolph W. Townsend Jr., Professor of History, who provides the book's insightful introduction. Notions of time and space are also central to new scholarship, evidenced in essays by Justin M. Brown and Gavi Levy Haskell, doctoral candidates in the Department of the History of Art at Yale.

Given this long tradition, it is fitting that the Yale Center for British Art, which houses the collection of philanthropist Paul Mellon—the largest and most comprehensive collection of British art outside of the United Kingdom—would invite this collaboration with John H. Bryan (1936–2018). Mr. Bryan, whose collection inspired *Marking Time*, was a giant in American art philanthropy. He pursued numerous projects with purpose and profoundly ethical leadership. While Chair of the Board of Trustees of the Art Institute of Chicago from 1999 to 2006, Mr. Bryan oversaw the planning of architect Renzo Piano's Modern Wing. He was also the driving force behind Chicago's renowned Millennium Park. In addition, his fundraising talents made the public sculpture *Cloud Gate* by British artist Anish Kapoor a reality. During his time as Chief Executive Officer of the Sara Lee Corporation, Mr. Bryan launched the Sara Lee Millennium Gift, the distribution of fifty-two important works of art to forty museums worldwide—considered the single most significant gift to the arts in American history. He and his family have since made countless gifts to museums throughout the country.

Mr. Bryan's generous gift to the Yale Center for British Art was his support for *Marking Time*, through a grant from his Crab Tree Farm Foundation, named for Mr. Bryan's estate in Lake Bluff, Illinois. We extend our sincere gratitude to Mr. Bryan's son, John H. Bryan III, who orchestrated this gift.

The project was overseen by Edward Town, Head of Collections Information and Access at the Yale Center for British Art, with the assistance of Glenn Adamson, Senior Scholar. The publication was coedited by Town and contributor Angela McShane, Head of Research Development at the Wellcome Collection, London, who collaborated with Mr. Bryan and others on an earlier iteration of this book. The project was greatly enriched by our scholarly community at the Center, especially Maryam Ohadi-Hamadani, Postdoctoral Research Associate in the Research Department, and Jennifer Reynolds-Kaye, Educator, Academic Outreach. Jennifer and others—including Jenny Saunt—participated in a workshop the Center hosted at Crab Tree Farm in 2018. Jenny worked closely with Mr. Bryan and his son for many years, and she offers her personal reflection on Mr. Bryan in the afterword to this book.

That this complex volume has come together in such an attractive form is thanks to a fruitful partnership with the firm Lucia|Marquand and Design Director Ryan Polich. At the Center special thanks go to Nathan Flis, Head of Exhibitions and Publications, and a contributor to this volume; Deborah Cannarella, Development Editor and Publications Manager; Shaunee Cole, Publications Assistant; and Belene Day, Senior Administrative Assistant. They have stewarded this book through to completion, with the invaluable assistance of Kathleen Friello, Livia Tenzer, and Tina Henderson. John "Jack" Kyono (Yale University '20) and Paige Swanson (Yale University '20), student assistants in the Exhibitions and Publications department, assisted in the preparation of the catalogue, as did Ioann Popov (Yale University '21). We also thank Emmanuelle Delmas-Glass, Collections Data Manager, whose efforts allowed Edward Town to oversee this project without disruption to the department's endeavors. We also record our thanks to Amy Meyers, the Center's former director, who supported this project from its inception.

At Crab Tree Farm, Dru Muskovin fulfills a dizzying number of roles as on-site project manager, art handler, courier, conservator, researcher, liaison, and all-round indispensable person. This project could never have happened without her, and we are profoundly grateful. Besides serving as a haven for scholars, Crab Tree Farm is home to craftspeople who travel there to work alongside master craftsman Mike Jarvi. Jarvi's cabinetry showcases the signature finishes that he and Mr. Bryan developed, an expression of their shared love of handmade objects and natural materials. Other heroes at Crab Tree Farm include Tom Gleason, Alek Kasprazak, Eve Mangurten, and Nancy Novit. Special thanks also go to Jamie Stukenberg for his care in photographing every object in the book. The many additional experts, research assistants, and specialists who contributed to this project are recognized in Edward Town's preface on page 9 and in the editors' acknowledgments on page 516.

Finally, acknowledging that it is impossible to name everyone involved, I extend my sincere thanks to all who supported this project behind the scenes.

Courtney J. Martin
DIRECTOR, YALE CENTER FOR BRITISH ART

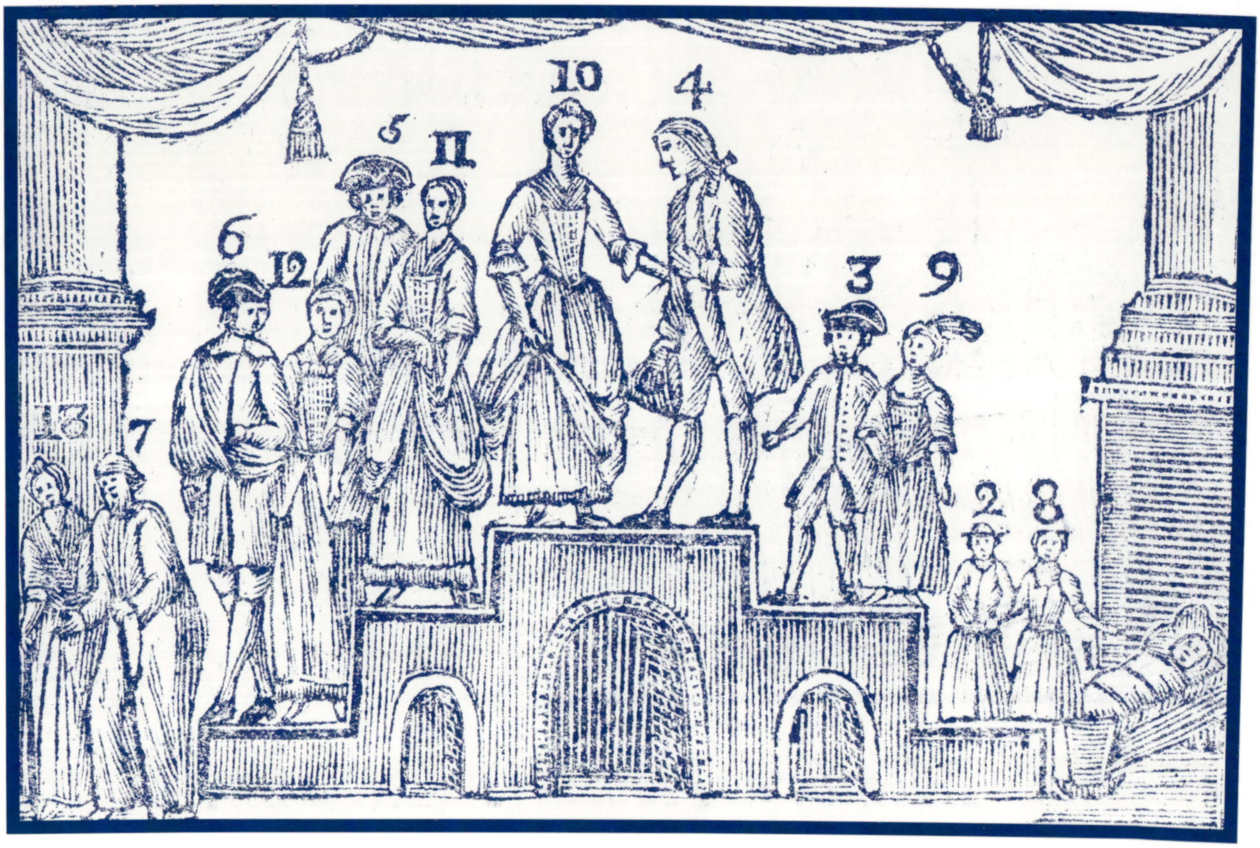

10
4
5
11
6
12
3
9
13
7
2
8

Preface

THE STORY OF THIS BOOK IS A COMPLEX NARRATIVE WITH A considerable cast of characters. It began more than half a century ago, when young John H. Bryan—the son of a Mississippi meat wholesaler—acquired a shipping carrier's worth of furniture while holidaying with his wife in Britain. The dubious quality of the furniture made this purchase an inauspicious start, but as months turned to years and years to decades, his collection grew into one of formidable size and reputation. Before long, experts from around the globe came to marvel both at the objects and at the magnificent setting that Mr. Bryan created for their study and enjoyment. Much like the objects on display, everything at his home at Crab Tree Farm in Illinois—from the finish on the furniture to the foliage around the architectural follies—was carefully selected, with great sensitivity for natural materials and craftsmanship.

Driving Mr. Bryan's relentless campaign of acquisition was his bold ambition to acquire at least one example of every type of object from early modern Britain—principally from England and Wales, but items from Ireland and Scotland were by no means discounted. Quixotic though this approach might seem, it did not discriminate based on status or style and therefore shaped a remarkably democratic collection. It became something of an encyclopedia of objects—some far from polite—that served as a means for Mr. Bryan and those he brought into his orbit to learn about the past. The encyclopedic quality of the Bryan Collection inspired the choice of woodcuts from the 1772 English edition of John Amos Comenius's popular *Orbis sensualium pictus… Or, A Nomenclature, and the Pictures, of All the Chief Things That Are in the World* as opening images for the prose sections of this book.

An innocuous comment by Dr. Tessa Murdoch of the Victoria and Albert Museum (V&A) during her visit to Crab Tree Farm in 2012 launched the project that became *Marking Time*. She asked the simple question, "Why were so many of the objects in the collection dated?" Mr. Bryan had no immediate answer. He, like many collectors, had simply considered a date on an object to be a desirable criterion for acquisition—but his thinking was soon to change.

Enter Dr. Angela McShane, a member of the Research Department at the V&A. With the assistance of several graduate students in the V&A/Royal College of Art master's program, Angela transformed Mr. Bryan's understanding of his own collection by contextualizing his objects within British social history. At this point, the "dated project" began in earnest.

The acquisition of dated objects became a daily pursuit, masterminded by Mr. Bryan but carried out by a network of experts and dealers who sought objects that met his criteria: they needed to have a date, to be of high quality, and—to use trade terminology—to be of particular "rarity," with earliest examples of their type always preferred. Running between auction houses, dealers, and libraries was Dr. Jenny Saunt, whose reflection on her time working with Mr. Bryan forms the afterword to this book. When objects arrived at Crab Tree Farm, they came into the care of Tom Gleason and Dru Muskovin, whose work laid the foundations upon which this catalogue is built. It was through their combined efforts that a significant portion of the material history of Britain came to reside at Crab Tree Farm.

In 2017, a second question changed the course of the project yet again. I had first met Mr. Bryan in 2014 during a visit to Crab Tree Farm to study his collection of Tudor and Stuart portraits with my colleague Jessica David, Senior Conservator of Paintings at the Yale Center for British Art. When I learned that Mr. Bryan was planning to produce a book about the dated project, I asked him whether he had ever considered that the project might also take the form of an exhibition. As the former Chair of the Board of Trustees of the Art Institute of Chicago and the de facto curator of his own collection, Mr. Bryan was well aware of the effort such a project might entail but he was nonetheless enthused. And so, with the support of Amy Meyers, then Director of the Yale Center for British Art, and under the guidance of Glenn Adamson, Senior Scholar at the Center, the book and exhibition *Marking Time* entered the planning stage. The project team took on two new members, Gavi Levy Haskell and Justin M. Brown, two PhD candidates from the History of Art Department at Yale University, and their respective expertise helped broaden the scope of the project. Contributing authors also include Edward S. Cooke, Jr., Charles F. Montgomery Professor of American Decorative Arts in the Department of the History of Art at Yale University, and Nathan Flis, the Center's Head of Exhibitions and Publications, whose research into British art focuses on the seventeenth century. With the social historian Keith Wrightson as an additional contributor to the publication, three generations of scholars came together to see this project through to completion. By early 2017, the research prepared by Dr. McShane and the graduate researchers was compiled as a manuscript, parts of which are incorporated into this publication. In the following year, the Center convened a meeting at Crab Tree Farm of additional contributors who reshaped the arc of the narrative. As a result, in this volume, seven essays exploring the temporal experience and material culture of early modern Britain introduce the catalogue of more than four hundred fifty objects from the Bryan Collection.

The Yale Center for British Art in New Haven, Connecticut, houses the largest collection of British art outside the United Kingdom. Herein lies one of the quirks in this story—which, if one were to apply an early modern sensibility, might be interpreted as cosmographically preordained: New Haven was established as a British Colony in the late 1630s on land inhabited by several Algonquian-speaking indigenous tribes. By tradition, the city served as a hiding place for three regicides who served as judges in the trial of Charles I and signed his death warrant in 1649: John Dixwell (1607–1689), Edward Whalley (ca. 1607–ca. 1675), and William Goffe (1605?–1679?). After the Restoration of Charles II to the throne in

1660, the men are believed to have hidden among the rocks now known as Judges Cave. As the final section of this book attests, the death of Charles I was arguably the single most significant event in the seventeenth century, lodged in the collective memory of those in the anglophone world well into the century that followed. The stones outside Judges Cave are today riddled with carved inscriptions made by pilgrims to this once important site in the nation's history, who display the same inclination as so many of the people represented in this book: to mark their names and their place in time, tethering their own lives and beliefs to the past. The names of Dixwell, Whalley, and Goffe are themselves inscribed into the very fabric of the city as the names of three roads that lead out of New Haven toward Judges Cave at West Rock. Today, neither New Haven nor Yale pays much mind to this aspect of the city's history. In Mr. Bryan's view, this lack of awareness would be understood as part of a broader antipathy toward early American history. It can also, however, be explained in relation to the university's increasingly uneasy relationship with its slaveholding benefactor Elihu Yale (1649–1721) and the city's reluctance to delve into a colonial history that is graphed onto three of its main streets.

The objects included in *Marking Time* make clear that major historic events—such as the execution of Charles I—formed promontories on the landscape of an era's collective memory, such that they came to serve as crucial temporal markers in individual lives. Further, these events had massive ramifications in the histories of Britain and America. The political and religious dissonances of the period, so evidently expressed in the objects in this book, formed the tenets of American identity and beliefs. For example, the candlestick dated 1709 (page 294) sustained conversation and debate in the Grecian, the coffeehouse where the Commonwealthmen gathered in secret nightly meetings to argue against the existence of a standing army and the tyranny of monarchs—ideas central to the ideologies of the American Revolution later in that century. Within this book, a small acknowledgement of the two countries' shared history is the choice of the typeface Caslon, developed by William and Elizabeth Caslon in London in 1722 and used on both sides of the Atlantic, not least for the printed version of the U.S. Declaration of Independence.

The notion that the small, the personal, and the everyday have direct relationships with the larger schema of historical events was inherent in Mr. Bryan's collecting and is writ large in the chronology at the back of this book, which lists the dated objects alongside a timeline of significant historical events. This section enables the reader easily to locate objects made in a specific time period. It also points to the interrelation of political and global histories and the lives and lived experiences of individuals who, even in their obscurity, were a part of these broader narratives.

Britain experienced significant social and economic change during the period from 1500 to 1800, which resulted in a significant shift in the occupational geography of the British Isles and a massive movement of people and communities across the globe. Yet, the maps that follow this preface illustrate that, for much of the period, most of Britain was still very much tethered to its rural and agricultural past. To be sure, the ports of Bristol and Liverpool have their place on the map, but the great northern cities of the Industrial Revolution are barely in evidence. Conversely, the great market town of Shrewsbury and its various dependencies in Shropshire—rich with the profits of the wool trade—still loom large. What the maps show is that the world of Crab Tree Farm—and by association the world of this book—is a world of the small-scale, the seasonal, the handmade, and also the global.

Mr. Bryan asked the question, "Why was it dated?" of every object in his collection, as did those charged with finding the answer. No single answer pertains to all of the objects, but over the course of the project it became apparent, through a survey of other collections, that the practice of dating objects in this demonstrative fashion was particular to the early modern period. This is not to suggest that people did not date objects previously or ceased doing so later; rather, the practice of inscribing dates grew steadily from the middle of the sixteenth century, reaching its apogee at about the time of the Restoration, and tailed off during the following century and sharply after 1750. This phenomenon was not unique to Britain during this period, however, as is evident in the material cultures of other Western nations, particularly in northern Europe, indicating that the practice was associated with Protestantism and anxieties provoked by the doctrine of predestination. In an attempt to glean God's plan and to learn of providential warnings and admonitions, Protestants scoured history, scripture, and astrological and cosmological sources for signs by which to chart their lives. These concerns combined with an increased awareness of the connections between time and the life cycle and a growing desire for self-knowledge, resulting in the proliferation of personalized almanacs and autobiographical writings. Each of these practices, like the dating of objects, served as a way of positioning the individual's existence in time.

Although not everyone shared the same experience or understanding of time, it governed people's lives—from the sounding of the bell in the morning to the ringing of curfew at night, from the moment of birth until the time of death. For this reason, the catalogue of this book is arranged according to the Seven Ages, the prevalent conception of the distinct phases of a life lived to its full extent. An illustration from *Orbis sensualium pictus* of the seven ages of a woman, entitled *Septem aetates hominis*, serves as the opening illustration for this preface, tracing the gradual progression in a life from birth to childhood to marriage to old age. Many objects in the Bryan Collection were made and dated to mark significant moments within this conceptual framework, often relating to the transition from one phase into the next, whether from apprenticeship to freedom or from singledom to marriage. The structure of the catalogue likewise opens out into thematic sections, allowing for a more expansive view of life in the early modern world. The final section of the catalogue, "Death & Legacy," illustrates how the lives of the dead sustained their presence in the world of the living. As such, this book eschews a linear arrangement of the objects in chronological order in favor of anachronic sequence. This ordering has the benefit of showing how people lived with and were persistently reminded of the past and the passage of time. It also mirrors the experience of encountering these objects at Crab Tree Farm, where the arrangement of Mr. Bryan's collection was in a constant state of flux, with no strict order of display, yet the date was always important. Accordingly, within the catalogue, the date or dates inscribed on the object are prominent in each catalogue entry. These dates do not necessarily relate to when the objects were made, however, and dates were often added to objects well after manufacture—both conditions making it impossible to create a strict chronology of objects in any case.

This collection of objects, spanning a broad chronological range, reveals how people's relationship with the concept of time and their personal experience of it underwent a series of revolutions during the early modern period, leading to the habitual dating of objects. The change in time consciousness can be attributed to a number of causes, but chief among them was the advent of affordable and widely available pendulum clocks and, soon after, the arrival of the pocket watch, which enabled accuracy to the minute. These timepieces provided individuals with the powerful, new experience of traveling through time and space. Indeed, a driving force in the pursuit of accurate timekeeping was the need to establish longitude when at sea. With accurate chronometry came the promise of maritime supremacy and the realization of Britain's colonial ambitions, which by the mid-eighteenth century were increasingly unchecked, leading Britain to wage war on its colonial competitors—principally France and Spain. Timekeeping devices also aided scientific and medical observation.

Many of the individuals who owned and commissioned objects featured in this book did so to celebrate worldly success derived from the profits of international trade, often founded upon the ensnarement and enslavement of others who then toiled unseen in lands "beyond the seas." This book includes numerous objects made from materials wrought great distances from Britain and brought to the British Isles and manufactured to satisfy the cultural habits created by the abundance of new commodities, such as tobacco, tea, and sugar—demonstrating beyond all doubt that there was indeed "slavery in small things." As Justin M. Brown argues in his poignant essay in this volume, for many, time was money, but for enslaved people, who generated wealth and capital, time was a stolen commodity. This fact, inherent in many of the objects in this book, is brought into stark focus by the ankle iron dated 1733 and inscribed with the family name Deverell, belonging to a Bristol surgeon and his son. Discussed in both Brown's essay (page 43) and an extended catalogue entry (page 384), the ankle iron demonstrates the grim affinity between temporal control and the subjugation of human dignity through enslavement. In fact, Britain's horological and colonial histories are uncannily concurrent: the increased ability to regulate and account for time placed Britain temporarily at the center of global affairs.

In Laurence Sterne's novel *Tristram Shandy*—first published in the so-called *annus mirabilis* of 1759, the year of the invention of the marine chronometer—the narrator's father, described as "one of the most regular men in every thing he did," punctiliously winds the house clock on the first Sunday night of every month. Through this routine, he gradually "brought in some of the other little family concernments," which one night results in Tristram Shandy's conception. Famously, in his attempt to tell his life story, the eponymous narrator struggles to adhere to a linear narrative, but in so doing he creates something that is truer to life and infinitely more human than traditional forms of storytelling. It is hoped that *Marking Time*, much like Sterne's great novel—in allowing for many lives, many voices, and many objects and ideas—will convey the richness, variety, and complexity of the early modern experience and expose a disorderliness often tidied away in grand narratives or focused histories.

What is constant is that time wastes too fast. Even though Mr. Bryan lived into his eighties, his passing in 2018 came too soon, and it is a great regret of mine that he never got see this book or the accompanying exhibition, which was canceled as the result of a global pandemic. The extraordinary events of 2020 have provided a pertinent reminder that time is artificial and uneven. Despite all our confidence about our ability to anticipate the future based upon what we have learned from the past, we have no mastery over what is to yet come. We can only mark the present.

Edward Town
NEW HAVEN, CONNECTICUT
LADY DAY (MARCH 25) 2020

Literature

Paul Glennie and Nigel Thrift, *Shaping the Day: A History of Timekeeping in England and Wales 1300–1800* (Oxford: Oxford University Press, 2009)

Tara Hamling and Catherine Richardson, *A Day at Home in Early Modern England: Material Culture and Domestic Life, 1500–1700* (New Haven, CT: Yale University Press, 2017)

David Olusoga, *Black and British: A Forgotten History* (London: Macmillan, 2016)

Stuart Sherman, *Telling Time: Clocks, Diaries, and English Diurnal Form, 1660–1785* (Chicago: University of Chicago Press, 1996)

James Walvin, *Slavery in Small Things: Slavery and Modern Cultural Habits* (Chichester, UK: John Wiley and Sons, 2017)

Keith Wrightson, ed., *A Social History of England, 1500–1750* (Cambridge: Cambridge University Press, 2017)

Maps & Key to Places

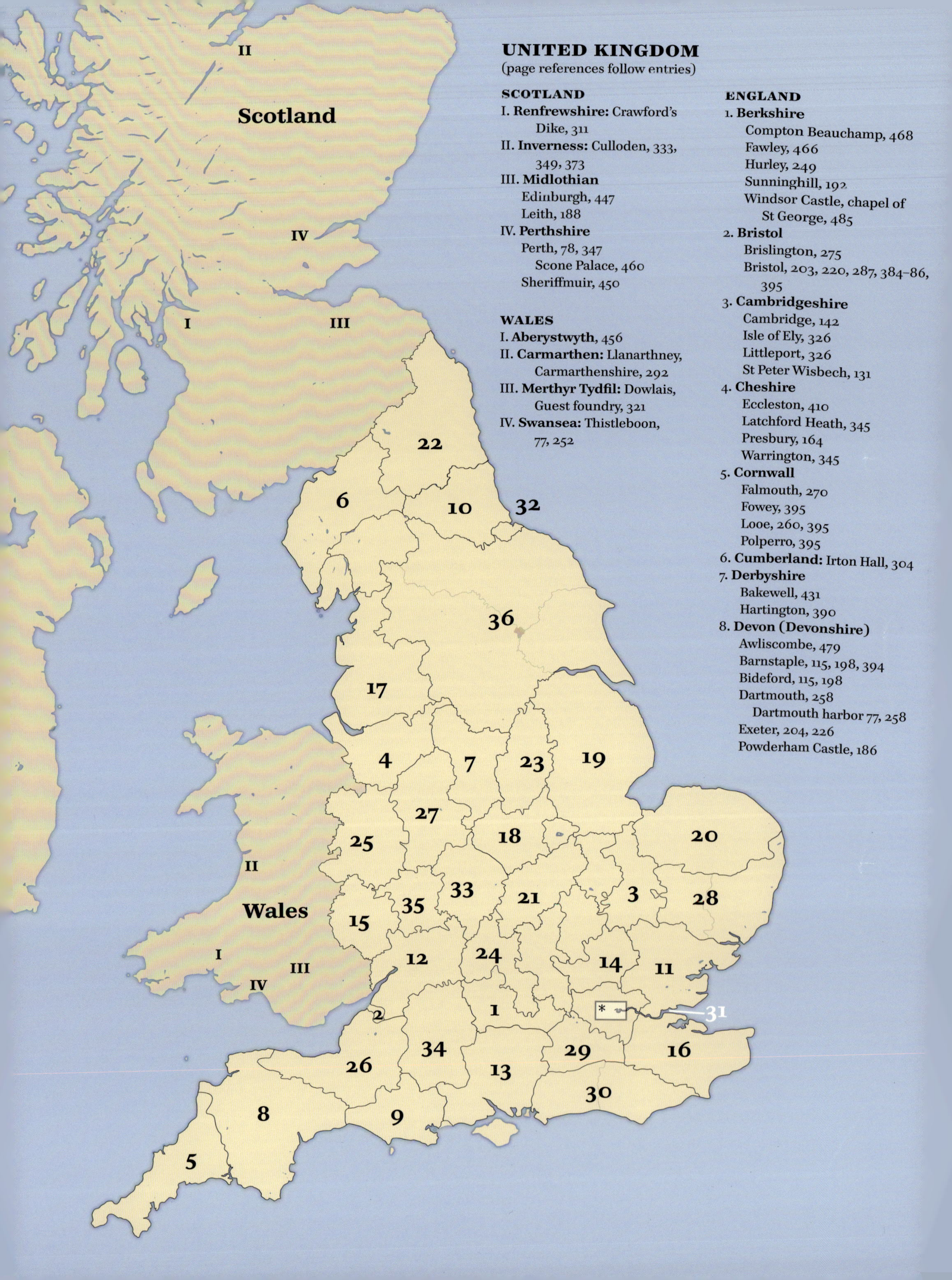

UNITED KINGDOM
(page references follow entries)

SCOTLAND
I. **Renfrewshire:** Crawford's Dike, 311
II. **Inverness:** Culloden, 333, 349, 373
III. **Midlothian**
Edinburgh, 447
Leith, 188
IV. **Perthshire**
Perth, 78, 347
Scone Palace, 460
Sheriffmuir, 450

WALES
I. **Aberystwyth**, 456
II. **Carmarthen:** Llanarthney, Carmarthenshire, 292
III. **Merthyr Tydfil:** Dowlais, Guest foundry, 321
IV. **Swansea:** Thistleboon, 77, 252

ENGLAND
1. **Berkshire**
Compton Beauchamp, 468
Fawley, 466
Hurley, 249
Sunninghill, 192
Windsor Castle, chapel of St George, 485
2. **Bristol**
Brislington, 275
Bristol, 203, 220, 287, 384–86, 395
3. **Cambridgeshire**
Cambridge, 142
Isle of Ely, 326
Littleport, 326
St Peter Wisbech, 131
4. **Cheshire**
Eccleston, 410
Latchford Heath, 345
Presbury, 164
Warrington, 345
5. **Cornwall**
Falmouth, 270
Fowey, 395
Looe, 260, 395
Polperro, 395
6. **Cumberland:** Irton Hall, 304
7. **Derbyshire**
Bakewell, 431
Hartington, 390
8. **Devon (Devonshire)**
Awliscombe, 479
Barnstaple, 115, 198, 394
Bideford, 115, 198
Dartmouth, 258
Dartmouth harbor 77, 258
Exeter, 204, 226
Powderham Castle, 186

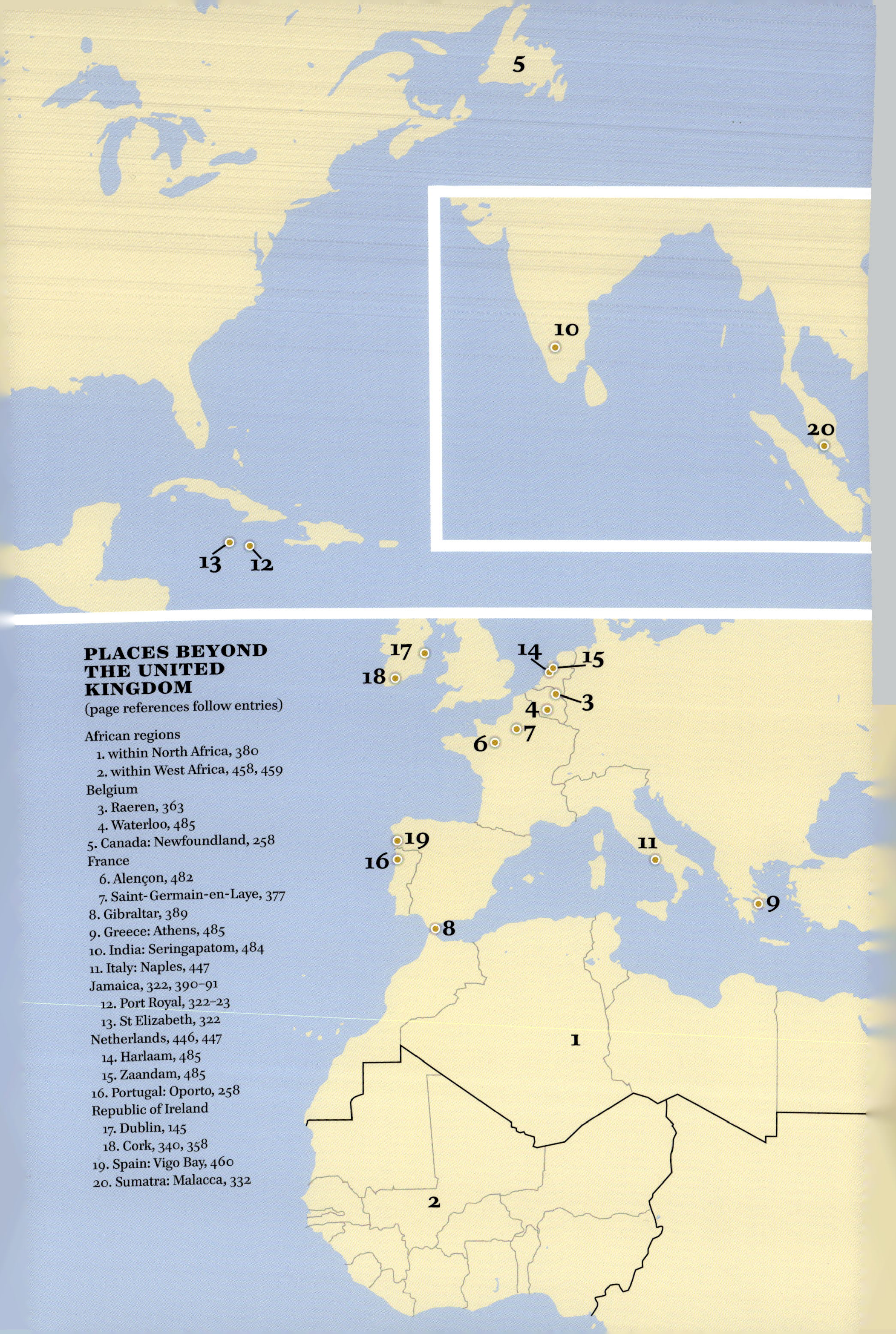

PLACES BEYOND THE UNITED KINGDOM

(page references follow entries)

African regions
- 1. within North Africa, 380
- 2. within West Africa, 458, 459

Belgium
- 3. Raeren, 363
- 4. Waterloo, 485

5. Canada: Newfoundland, 258

France
- 6. Alençon, 482
- 7. Saint-Germain-en-Laye, 377

8. Gibraltar, 389
9. Greece: Athens, 485
10. India: Seringapatom, 484
11. Italy: Naples, 447

Jamaica, 322, 390–91
- 12. Port Royal, 322–23
- 13. St Elizabeth, 322

Netherlands, 446, 447
- 14. Harlaam, 485
- 15. Zaandam, 485

16. Portugal: Oporto, 258

Republic of Ireland
- 17. Dublin, 145
- 18. Cork, 340, 358

19. Spain: Vigo Bay, 460
20. Sumatra: Malacca, 332

LONDON

(page references follow entries)

1. Bloomsbury: King Street, 279

City of London

- 2. Aldersgate Street, 215
- 3. Aldgate, 353, 392
- 4. Bishopsgate, 109, 443
  - 5. parish of St Helen, 181
- 6. Cheapside, 442, 445
- 7. Cornhill: St Michael Cornhill, 443
- 8. Farringdon Without: Chancery Lane, 344, 444
- 9. Foster Lane, 182–83, 359
- 10. Knightrider Street, 293
- 11. parish of St Alban Wood Street, 381
- 12. parish of All Hallows London Wall, 116, 160–61, 215
- 13. parish of St Augustine, 306
- 14. parish of St Benet Paul's Wharf, 177
- 15. parish of St Botolph-without-Bishopsgate, 408
- 16. parish of St John Zachary, 182
- 17. parish of St Giles-without-Cripplegate, 326
  - 18. Whitecross Street, 442
- 19. parish of St Margaret Lothbury, 299
- 20. parish of St Mary Woolchurch, 117
- 21. parish of St Mary Woolnoth: Exchange Alley, 126
- 22. parish of St Peter Paul's Wharf, 117
- 23. St Bartholomew's Hospital, 406
- 24. St Paul's Cathedral, 485

25. Clerkenwell: parish of St James, 424

26. Fleet Prison, 132, 300, 400

27.* Hammersmith, 307

- 28.* church of St Paul, 329

Holborn

- 29. Clement's Inn, 306
- 30. parish of St Andrew, 201
- 31. parish of St Sepulchre-without-Newgate, 398, 424

32.* Hounslow: Isleworth, 183, 185

33. Islington, 126

- 34. church of St Luke, Old Street, 147

35. Lambeth, 113, 265, 283

- 36. church of St Mary Lambeth, 162
- 37. Lambeth High Street, 185, 221

38.* Newham [New Canton]: Bow Porcelain Manufactory, 405

39. Newington, 113

40. Piccadilly, 444

- 41. parish of St James, 300

42.* Richmond upon Thames

- 43.* Hampton Court Palace, 417
- 44.* Hampton House (Garrick's Villa), 424
- 45.* Mortlake, 185

Southwark

- 46. Montague Close, 296
  - 47.* parish of St Saviour, 180
- 48. Pickleherring pottery, 275, 296
- 49. Rotherhithe pottery, 275, 296

50. Stepney, 132, 144

51. Tower Hill, 445

52. Tower of London, 139, 177

- 53. Royal Mint, 452, 455, 457

54.* Uxbridge, 389

55. Vauxhall pottery, 185, 470

Westminster

- 56. Bow Street Magistrates Court, 446
- 57. Devereux Court, 294
- 58. Exchequer, 72, 76, 419, 428, 430
- 59. parish of St Margaret, 478
- 60. Round Court, the Strand, 442
- 61. St Martin-in-the-Fields, 181, 347, 400, 407, 417, 442, 444
- 62. Whitehall, 474–75
  - 63. New Palace Yard, Queens Bridge, 401

64. Westminster Abbey, 451

65. Whitechapel

- 66. parish of St Mary Whitechapel, 132
- 67. Whitechapel Foundry, 201

*on London Region Map

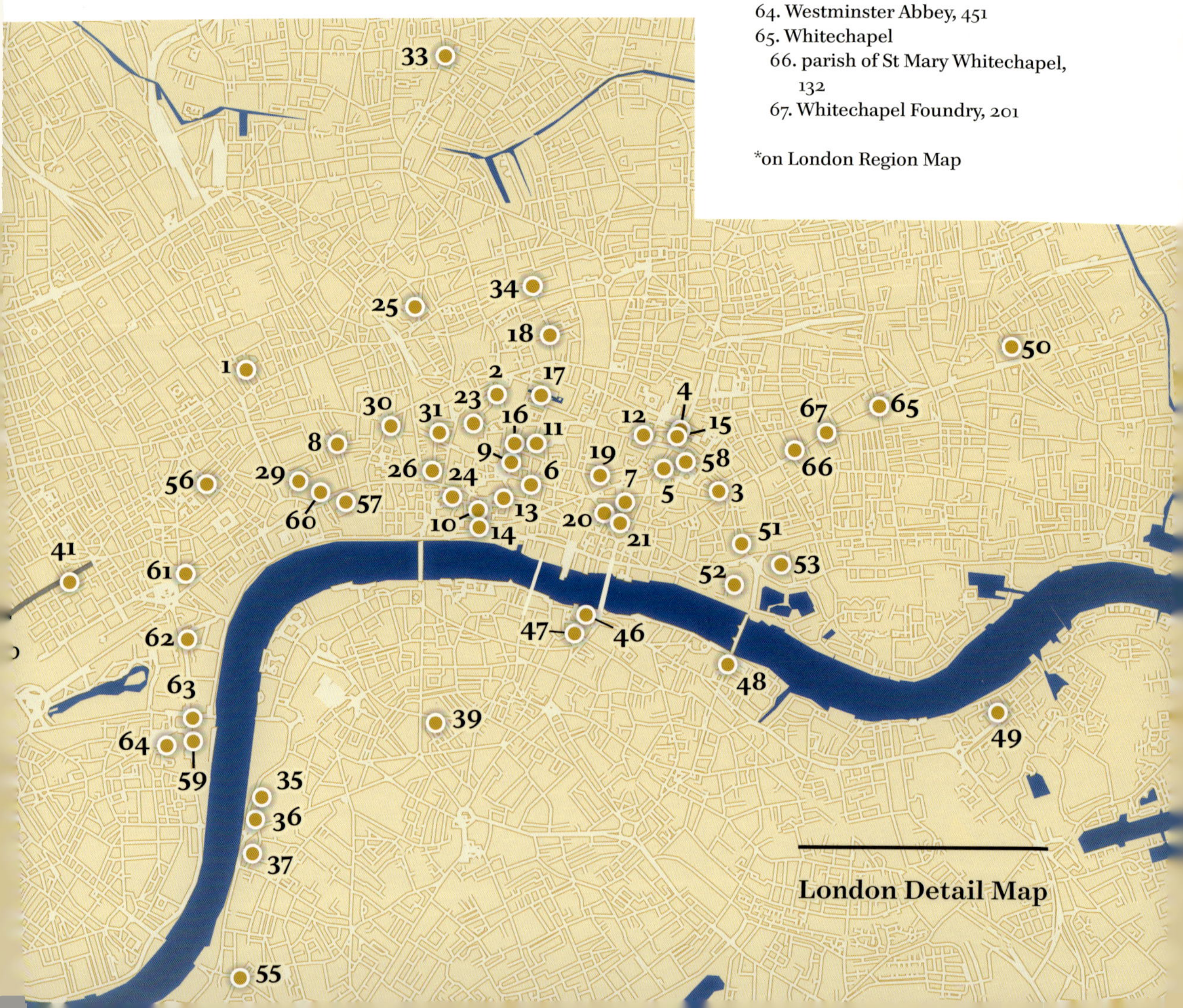

* London Region Map

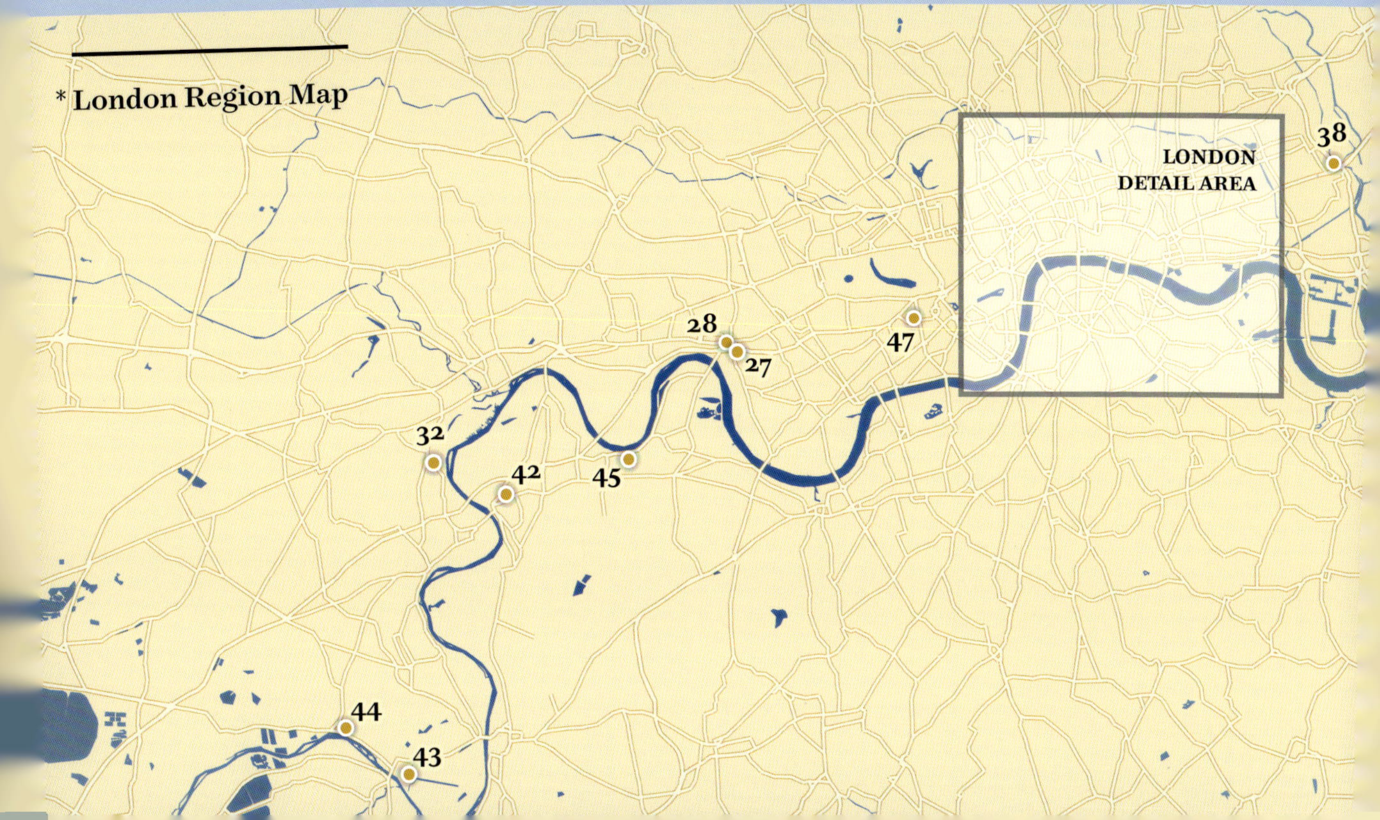

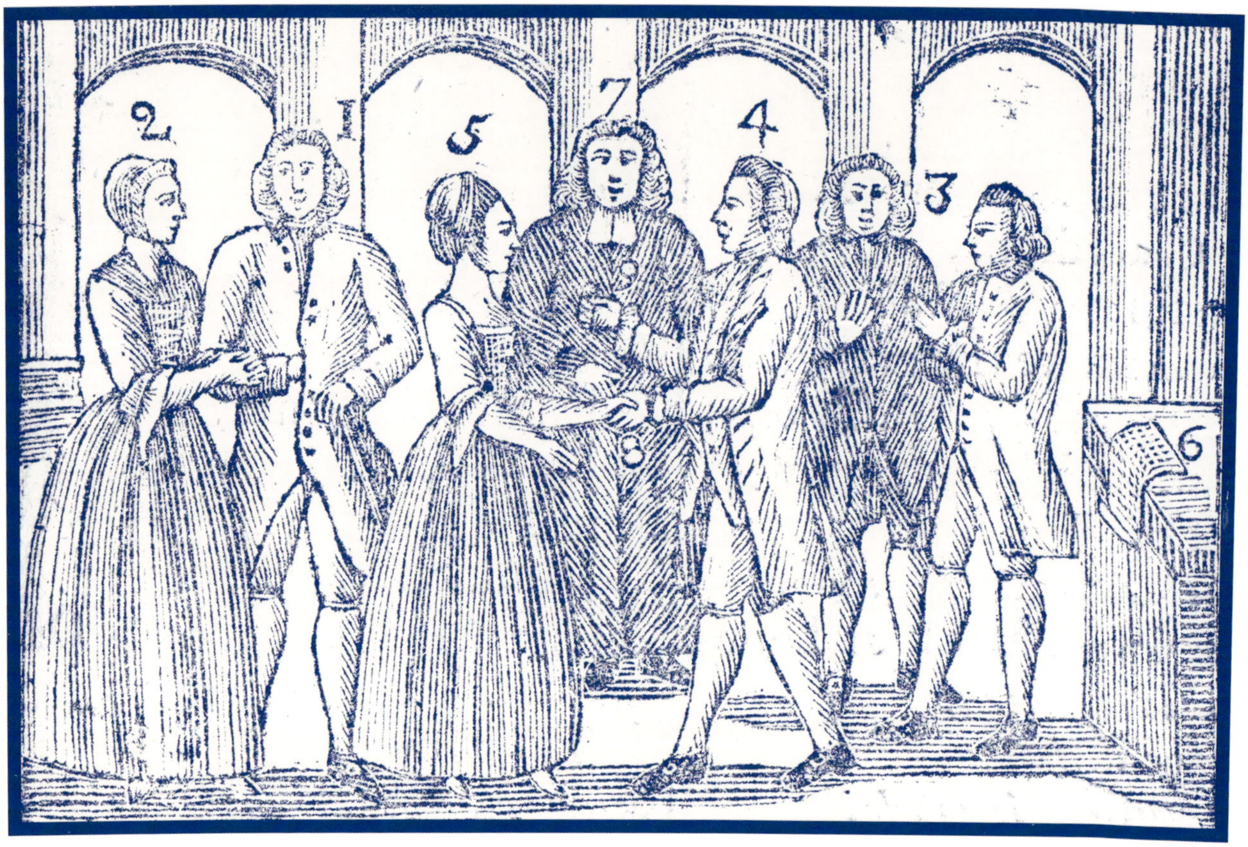
2
1
5
7
4
3
6

INTRODUCTION

Past Times

Temporalities in Early Modern England

KEITH WRIGHTSON

RICHARD GOUGH OF MYDDLE, IN THE COUNTY OF SHROPSHIRE, KNEW his place in time. The year was "Anno Dm 1700," and he was sixty-six years old. Both pieces of information were included in the frontispiece he drew for his manuscript book *Antiquities and Memoyres of the Parish of Myddle*.

Gough's book reveals a complex apprehension of time. As a man of some education, he was aware of the classical past and the enduring wisdom of ancient authors. Having had some legal training, he appreciated the importance of precedent. As a Christian, he knew how the world had begun and how God's providential purposes were unfolding in time. Gough read works of antiquarian scholarship (notably those of William Camden) and chronicles of English history. He took an interest in ancient deeds and knew the descent of kings. He was a parish officer who consulted, copied, and helped to create the records stored in Myddle's parish chest. As an inhabitant, he was intimately acquainted with the topography of his parish, past and present, and with its customs and traditions. As a neighbor, he knew Myddle's people and the posthumous reputations of their forebears. As an elderly man, he had a long memory.

Each element of Gough's identity can be said to have contributed to his sense of time. Time was present in the places and people around him, in the things he saw and touched, in his own aging body. He observed it. More, he gathered it together, above all in the second and most original and telling part of his book: "Observations concerning the Seates in Myddle and the familyes to which they belong," written in 1701. Beginning with a sketched pew plan of his parish church, he used it as a template for what he knew of the families, including his own, that occupied (or had occupied) those spaces and the landholdings to which they were attached. He described the living and the dead, the virtuous and the notorious, those whose lines continued and those "extinct in this parish." And all this sprang from and was anchored in material things: wainscoted pews. His first sentence began "A Piew is. . . ."[1]

Gough's now famous text makes him exceptional. He created a unique repository of information for his family and community and later for historians. But he was what microhistorians call a "normal exception," in that the unusual mark he left on the historical record provides a point of entry to more widespread features of the culture of his day. And because his subjects were antiquity, memory, and descent, that access is particularly evident in what his work reveals about early modern temporality, about a former temporal experience, the culture of time, and the practice of memory.

Three characteristics of that culture stand out both in Gough's writings and in other evidence of how early modern people thought about and engaged with time. The first is what has been called *temporal plurality*: the diversity of temporal frameworks that people used alongside one another.[2] The second is *approximation*: the fact that people's sense of time was commonly imprecise. For the most part, they did not need to aspire to temporal exactitude or foster the skills that would enable them to achieve it. The third is *association*: their habit of locating events in time by pegging memories in relation to other events, places, and things. None of these characteristics was peculiar to the early modern period. They had long existed, and in part they still persist, albeit subject to changes that began in this period. But between the sixteenth and eighteenth centuries, they were present in forms and combinations that permit us to speak of a distinctive early modern culture of time, one that is strangely alien in some respects, yet immediately familiar in others.

Anyone picking up a printed almanac, the ubiquitous pocket calendar of this period, would quickly have been reminded that there were many ways of reckoning time.[3] That almanacs contained many kinds of time was evident on every page of their monthly calendar sections. These provided guidance on the numbered days of the month and, alongside them, the equivalent days of the week; festivals and saints' days of the ecclesiastical year; "law days," bracketing the four terms of the legal calendar; the phases of the moon; planetary motions (important for the practice of astrological medicine); and the predicted times of sunrise. Each monthly calendar combined in tabular form a variety of temporal conventions, structuring time for a number of particular purposes. Additional tables offered detailed information on legal terms, movable church feasts, the dates of the fairs held in England's principal towns, tide tables for major ports, and astrological "prognostications" for each season, mostly concerning predicted weather events and the prevalence of diseases. If users cared to read on, they might find supplementary disquisitions on the differences between the Julian calendar (still used in England) and the reformed Gregorian calendar (used in much of continental Europe after 1582), the calendrical systems of the ancient world, and Christian versus "Turkishe" or "Arabique" systems of dating.[4]

Almanacs thus represented—visually, numerically, and verbally—the reality of what has been termed "a world of multiple temporal coexistence." Isaac Newton may have promulgated his concept of "absolute, true and mathematical time," but, as he recognized, for most people the reckoning of time was a relative and variable thing. They lived with "an ungainly mixture of times—unfolding at different speeds in different places—which intersect and interact in all manner of ways."[5] Within that context of temporal diversity, almanacs focused primarily on the *cyclical* dimensions of time—astronomical, seasonal, ritual, institutional—and brought them into relation with one another. With or without their aid, a "layered and structured" calendrical consciousness "permeated people's lives," serving in different ways to structure time and direct and coordinate their activities.[6]

People located events in time with reference to these cycles and the tasks and obligations associated with them. Called upon to date an incident, court witnesses in the sixteenth and early seventeenth centuries most commonly did so by referencing an ecclesiastical feast or the seasons of the agricultural year. They might remember that it was "about shrovetide," "between St Ellenmas and Christmas," "thannunciation day before," "the Satturdaie next after midsummer," "at the beginning of hay harvest," or "about in chirrie [cherry] time last."[7] Within the household, memories were pegged with references to the habitual tasks of the daily and weekly round. A woman might recall "breaking dowe for bread," being busy "upon a washing daye," or "having a child upon her knee crying." A man might have been taking "his morning draught," "setting up his shop stall," "washing his sheep," or coming home wet and "warming himself at his ffier." A child might remember she was "coming home from milking kyne [cows] with divers other maids," or "going down the staires to go to school"; a servant might have been "carreing a pale of water" and a couple "newlye gone to bed."[8]

Such a cyclical perception of time was perhaps the dominant reality—reflected in the popular visual representations of the four seasons, the times of the day, and the seven ages of man—but that did not preclude simultaneous awareness of *linear* time.[9] Witnesses unable to offer the exact date of a past event commonly reckoned back from their present in terms of units of time (days, weeks, months, or years) "past," "since," "gone," and "agoe," indicating a sense of regular linear progression and knowledge of the units by which it was measured.[10] Almanacs were compiled for specific years Anno Domini (AD), a linear method of ordering time long known but little used before the fourteenth century.[11] From around 1600, they commonly included a chronology or "computation" of the "yeares expired" or "time compleat" since the occurrence of particularly notable historical events. These chronologies began with the creation of the world (estimates varied) and proceeded to braid together biblical events with significant moments in classical history (starting with the fall of Troy). Then came important reference points in national history, mostly since the Norman Conquest, multiplying as the progression reached more recent times: "the beginning of blessed Queen Elizabeths raigne," for example, or "the damnable Powder Treason." Compilations aimed at regional audiences included happenings of local note, such as "the great Plague at Newhaven." Two Durham astrologers ended theirs with "the last great snow" in the northern counties. It had been six years ago in 1621, seven years ago in 1642: a "great snow" was a movable feast in a cold climate. Awareness of the course of national history was further enhanced by tables of regnal years, listing all the monarchs since William the Conqueror, their accession days, the lengths of their reigns, and how many years had passed "since their reign began."[12] Those wishing to extend their knowledge of such things could turn to published chronicles of the kingdom, like Holinshed's, or the popular historical dramas based upon them.

Both linear and cyclical perceptions of time jostled together in Gough's writings. His reading of antiquarian and historical works gave him a broad awareness of English history. He alluded to nineteen English rulers between William the Conqueror and William III, "our graciouse King," and had almost certainly consulted a regnal table when he observed that the Lords Strange had held the manor of Myddle for "about 420 yeares and dureing the Reigne of eighteen kings." He mentioned important national historical events from the Reformation onward, though predominantly those that had occurred during his own long life or those of people he had talked to: the Civil Wars, the Regicide and Interregnum, the Restoration of Charles II, and the Revolution of 1688. His historical perspective could be called parochial in the sense that these references were almost invariably in the context of events affecting Myddle or moments in the careers of Myddle's inhabitants. But it was a parochialism framed by a linear vision of time. The same was likely true of those who shared their memories with him, even if they lacked his depth of historical awareness.[13]

Gough's perception of Myddle's families was also linear to a degree. His accounts of them were driven by a preoccupation with descent. Individuals were descended of "good," "poor," or "obscure" parentage. To designate a family as "an ancient family in Myddle" was a compliment, and he used dynastic terminology to describe them: "William the First," "Richard the Third," and "Richard the Fift" were not kings but heads of the Gittins clan. Yet, although they were framed by linear sequences of descent, Gough's family histories rarely contained dates. For the most part, their chronology (which frequently must have extended back a century or more) can only be inferred. What concerned Gough more, and gave internal structure to his "Observations," was the repetitive cycle of the life course, which emerged in the words he used repeatedly to describe lives. Children were "borne," "brought upp," "sett apprentice." When "growne," young men were "sett upp," "succeeded," or "came to be tenant." They "marryed" a young woman with a specified number of "pounds to her portion." Together they "lived loveingly" (or not) and "had ishue." Some "lived pretty well," in "good estate." Others "got farre in debt" and "broke." Eventually they "dyed," "left behind" children, "bequeathed" their estates, and were "buryed" or "interred," usually in Myddle. These experiences were timeless, and so too was the moral universe that shaped Gough's pithy judgments of those he had known or (most often) knew only from their posthumous reputations among the neighbors: "a reasonable good husband"; "a sober man and painefull labourer"; "a good husband by fits"; "a provident and prudent gentlewoman"; "a sickly woman, but... a religious person and a good housekeeper"; one "more commendable for her beauty than her chastity." Such human characteristics were as perennial as the life events of those described, repeated with variants one generation after another, like the church feasts and seasons that marked the turning of the year in Myddle and elsewhere.[14] For Gough, as for his contemporaries, then, there was "a diversity of temporalities operating in any present."[15]

If time was heterogeneous in nature, it was also characteristically imprecise: "more proximate than exact."[16] To be sure, time could be exactly measured: compilers of almanacs could be precise to the second in their calculations. The discipline of chronology excited Renaissance scholars who fretted over the construction of "a single coherent time line" that could reconcile astronomy with ancient texts, achieve a "reformation of historical chronology," and produce "a cartography of time."[17] Dated documents were created and records kept "to the end the things therein comprised might both more certainlie be kept in memorie and moore easilie... be proved."[18] Genealogies were constructed and monuments erected for members of the social elite, often incorporating celebratory inscriptions, to "materialize" memory and "block the work of forgetting."[19] As we have seen, people knew, and often used, the conventional units of time measurement. Clock time was also "a familiar general idea," marked by church clocks chiming the hour, and was a resource for the ordering of at least some social activities.[20]

Nevertheless, the need for chronological precision was limited. The exactitude to which the compilers of almanacs aspired was more a demonstration of their mathematical credentials than a requirement of their readers. Most people's sense of historical time did not involve exactly measured chronological progression. It was more diffuse than that: localized, episodic, and chronologically vague as memory dissolved into legend.[21] The keeping of chronological records was rare outside legal and administrative contexts. Most lives were not commemorated beyond a few entries in parish registers. That most people were only loosely aware of their own dates of birth is evident from the fact that when asked to state their age they commonly fell back on rounded decadal numbers (twenty, thirty, forty, and so on). Burials were registered, but most graves were unmarked.

As Daniel Woolf puts it, "Diurnal precision in most (though not all) contexts, mattered very little," and calendrical accuracy was relatively rare for most of the population.[22] Witnesses asked to date events in court were rarely able to do so precisely. Their statements usually began with the word "about" and ended "or thereabouts." William Symons was unusual in being able to state that the will of Edward James had been written on October 20, 1601. But he was a clergyman, he had written the will, and he was shown the dated document to confirm that fact. More typical was the testimony of John Jeffrey, an illiterate laborer who had been "threshing in the barne" when called in to witness James's will. He could recall only that it was made "about xvi or xvii yeares since... not long after Michaelmas."[23] Most people, as we have seen, could venture an adjacent church feast or a seasonal referent, or reckon back roughly "the subjective measure of distance from the speaker's present."[24] But they were rarely exact, and sometimes their uncertainty was acute. It was "either six weekes before michaelmas or six weekes after." The previous tenants were "all longsince dead, but how many years in particular it is since their... deaths this deponent cannot set forth."[25] Richard Gough was little different. He referred to specific numerical years only occasionally and provided day, month, and year dates only fifteen times, usually when he had clearly had occasion to consult documents held in the parish chest. For the most part, he oriented himself in time less specifically, using such expressions as "in the time of... Lord Strange," in "the Parliament time," in "the Protectors time," or (most often) "in the time of the wars." For the rest, he distinguished "our Fathers time" (meaning the early seventeenth century) from "my time" (from the 1630s). Earlier occurrences were mostly "long since."[26]

As for clock time, people may have been familiar with its conventions, but it was rarely referenced by court witnesses, and then mostly in towns, where the chimes of public clocks were features of the aural environment. That clock time was heard rather than seen is implicit in people's tendency to approximate to preceding and succeeding hours: "about six or seaven of the clocke," "between the hours of eight or twelve." And even then the hours specified had meaning only for the place concerned, since clock time was not yet subject to regional or national standardization.[27] For the most part, references to the time of day were still more approximate: "before it was daye light," "about the sonne rise," "in the forenoone," "about the midd time of the day," "in the afternoon," "towards the evening."[28]

Time, then, was measurable, but awareness of time was generally approximate and involved uneven valuation. Some days weighed more than others. Tasks in hand mattered more than the hours devoted to them. Time descended in irregular units of varying duration: reigns, generations, tenancies granted for a number of lives rather than a term of years. "My time" (defined by an Essex husbandman as "since he was a boy and of remembraunce") shaded into "our Fathers time" and then faded into "time beyond the memory of man."[29] The relative importance and meaningfulness of time varied with context—that is why it was recalled principally by *association*.

Locating past events in time by association with the high points of the ecclesiastical year, seasonal festivals, and agricultural cycles was an established feature of late medieval time consciousness, and one, as we have seen, that persisted into the sixteenth and seventeenth centuries. Witnesses were also prone to place occurrences by associating them with other significant events. Before the introduction of parish registers in 1538, those attesting the age of heirs in "proof of age" proceedings frequently oriented themselves temporally by citing associated events in their own life course: a birth, death, or marriage; being apprenticed or dubbed a knight.[30] This practice also continued. Deponents in Elizabethan and seventeenth-century court cases often cited such events: "a year or more after the death of George Thompson," "soon after that Margaret Garnet was delivered," "the tyme as he tooke his jorney to sea." One London woman, married and aged thirty-four (or thereabouts), organized her memory with reference to the successive masters she had served when a single woman.[31] Others alluded to significant local events: "at such time as Lammas sessions was kept at Nottingham," "in the sicknesse yeare" (the plague of 1665), "about the later end of the last great frost."[32] A few associated personal memories with events of national significance: "the commocion tyme in thes north partes . . . xxx yeres ago" (the Pilgimage of Grace rebellion of 1536), "the year of the insurrection or rebellion in the north that last was" (the Northern Rising of 1569, recalled in 1620), "a little after Easter before the death of King Charles II" (meaning 1684, remembered nine years later), "the Troubles in Ireland" (1689–90, associated with an alleged clandestine marriage).[33] For many, the association of past events came more naturally than determining "an objective point in a sequence of the years."[34]

People did not just recall such events when called upon to testify. They did it constantly. They told stories and passed them on to the young, tales that were memorable because they contained lessons, judgments, and the moral freight of conventional master plots.[35] Through frequent interaction, "daily recounting," and longstanding "mutual familiarities," they came to "remember in common," thereby creating "social memory," a common past central to group identities.[36] Richard Gough's book grew out of such informal memories. It is punctuated not by references to the sources he consulted but by phrases such as "in my memory," "which I saw," "I have heard," "he tould me," "I had from," "by report," and "we have a tradition."[37] Such narratives allowed "events themselves to create the order of time," experienced as a shared memory of episodes not clearly arranged in sequence or readily datable—"a vague background awareness of the past" that was itself a "way of living in time."[38] These stories also had material contexts: they were preserved by associations with places and with things.

First and foremost, there was the landscape. Landscapes "lay at the core of early modern popular memory." They presented "topographies of remembrance," "rich in commemorative association."[39] "Personal experiences and memories became entwined, indeed inseparable from" spaces and places.[40] Landscape was "a repository of memories, historical associations, and legendary tales that invested it with meaning for local people."[41] They knew that, as Tim Ingold puts it, "to perceive the landscape is...to carry out an act of remembrance," to engage with "an environment that is itself pregnant with the past."[42] England's post-Reformation landscape was "littered with mutilated remains" of the sacralized landscape of a religious past that had been repudiated but not forgotten, ruins that in time came to generate a "discourse of nostalgic lament."[43] Persisting agrarian custom was "grafted onto the environment" of particular manors. So, indeed, was agrarian change, as in the depopulated village of Whatborough, Leicestershire, where, in the 1580s, local people could point to the "signes or likelihoods" that "howses have there stood" and to the "ridge and furrow" of former plowland in what had become a vast sheep pasture.[44] When Gough walked his parish in his mind, Platt Bridge and Gallow-tree Stile triggered tales of medieval border raiding by the Welsh, preserved "by tradition." The ruined castle recalled the career of the outlaw Wild Humphry Kinaston, of whom "the people tell almost as many romantick storyes as of the great outlawe Robin Whood." Haremeare pool and the Mosse brought to mind a gruesome homicide a century before. Bristle Bridge and the nearby Goblin Hole cave stirred memories of former soldier William "Scoggan" Preece and "the storyes that were told of him."[45]

In short, "the land told tales."[46] And so did buildings and their furnishings. Parish churches symbolized a community of the living and the dead. The living gathered there; the dead lay around and beneath them. Testators routinely instructed their executors to bury them in their parish churchyard. Some were much more specific, requesting burial "under the pew where I did sit," "near unto the burials of my ancestoures," "as neare unto my late wife as maye be," "so neare unto my late husband as conveniently may be," "nere to my brother Nicholas," "neere unto my mother and sisters."[47] Graves might be unmarked, but family members knew where they were, just as Gough could visualize the generations that had occupied a pew.

The town hall symbolized the autonomous urban community and its history, often built or rebuilt to commemorate its legal incorporation. So did some of its furnishings, like the elaborate mayor's settle in Beverley, inscribed with the date 1604 and the initials of the mayor who had commissioned it; the town "hutch" of Yarmouth, which contained the corporation's records; and the ranks of portraits of former mayors of Norwich and elsewhere.[48] The livery halls of city companies were also "theatres of memory," where the exemplary careers and charitable acts of former members were commemorated. Objects presented to the company for display or use in rituals both proclaimed the standing of their donors and preserved their memories. David Gittings left the Vintner's Company of London a ceramic pot with silver mounts, inscribed "Thank David Gittings for ys [this]. Think and Thank 1563." One goldsmith gave a silver cup and cover to his company with the message, "This guifte I leave amongst my friends...That when I dye this guifte of myne Amongst my friends may live." Both men knew that their gifts would have "active social and ritual lives" long after their deaths.[49] Narratives were contained in such objects.

So too with houses and what contemporaries called "householde stuffe." If landscape "created a spatial pattern within which collective memory and social relations could operate," the same was surely true of domestic space.[50] Domestic architecture and its constituent spaces (hall, kitchen, parlor, chambers, and lofts) channeled the household's activities and also structured relationships among its members.[51] Objects located and grouped in particular spaces revealed the "normative ways of conducting domestic life": provisioning, processing, cooking, eating and drinking, washing, sleeping, engaging in hospitality.[52] The materiality of the household could connect memories to things—things encountered, seen, used. Sometimes the meaning of things was explicit. The heraldic displays of gentry households celebrated descent, dynastic marriages, and family estates. The union of married couples was symbolically displayed in initials and marriage dates inscribed on oak furniture, silver plate, ceramics, or the fabric of the house itself (commonly above the door). But messages were also embodied in everyday objects that "communicate[d] their significance implicitly."[53] Some spoke of particular phases of the life course: a cradle, a child's chair, a servant's trundle bed. Others simply carried an emotional association. And the symbolic significance of objects could be extended beyond the human life span. Goods were passed on, handed down within households and between generations. Those regarded as heirlooms were acknowledged "vessels of family memory," sometimes presented at marriage or when leaving home, sometimes bequeathed at death.[54] Other things had affective significance of a less formal nature but sufficient to ensure that testators specified that a certain object—a looking glass, a book, an embroidered cushion, a bed and its "furniture"—should go to a particular child or relative. Some items carried remembrance beyond the household to the friends and neighbors among whom testators had lived, in the form of bequests of small goods or items of clothing, "for a token." The best gown Margery Learner of Terling, Essex, left to her friend Agnes Biggin in 1556 was passed on to Agnes's daughter ten years later as "the kyrtell that I had of Mother Lerner." She had not forgotten Margery.[55]

Family members, kinfolk, and friends were the vital connections between objects and memory. Things contained associations and held meanings. Their continued presence created communities among those who saw, held, wore, and used them, groups who remembered their significance and communicated that memory, probably when talking around a table or before the hearth. Family memory "[held] the dead in the minds of the living and transmit[ted] responsibility for remembrance forward to children and grandchildren."[56] It was the most particular manifestation of the larger phenomenon of practicing memory by association through which early modern people "availed themselves of an enormous range of objects and artefacts and features in the landscape, which could attain long-term significance as mnemonic markers as long as there were people to tell the tales associated with them."[57]

It is possible, then, to describe an early modern culture of time. The temporal environment was characterized by a plurality of coexisting calendrical frameworks and temporal orientations, by approximation in the estimation of time, and by a tendency to place events and conjure memories by association. At the same time, however, it must be recognized that each of these characteristics existed to different degrees and that they varied in their relative importance and were combined in different ways by different people. Time, it has been said, is "embedded in practices."[58] "Systems of time reckoning reflect the social activities of the group" and, in particular, the need to coordinate its practices.[59] This being so, it is only to be expected that people's experience of and relationships with time would have varied greatly. There would have been a range of coexisting "communities of practice," differentiated by their place in the social structure, by education, by location, by occupation, by gender, and by age.[60] People would have apprehended and experienced time in different ways in accordance with the needs, demands, and rhythms of their lives. They would have engaged in different temporal practices, exhibiting different levels of temporal skill. Jacques Le Goff famously distinguished "the church's time" from "the merchant's time" in medieval Europe.[61] We might add that there was also the time of the lawyer, the administrator, the scholar, the farmer, the artisan, the sailor, the day laborer, the housewife, the servant, and the time of children and youths, of villagers and townspeople. In sum, time consciousness was segmented, hierarchical, and gendered—although also subject to overlapping features, since people might belong to more than one of these communities.[62] This characteristic was an additional dimension of temporal plurality, a mosaic of differing temporalities variously situated within the broad parameters of the early modern culture of time. And its complexity was enhanced by the fact that while the central characteristics of that culture certainly endured into the eighteenth century, they were also subject to change. That peculiar blend of continuity and transformation was perhaps what made the period distinctive.

Three forms of change deserve particular emphasis. First, there were shifts in the relative importance of different modes of time reckoning and temporal awareness. Calendars remained diverse, but in the aftermath of the English Reformation the ecclesiastical calendar was drastically pruned, loosening the hold of the church on the annual cycle.[63] Slowly—very slowly—the use of ecclesiastical referents in the dating of events diminished. Whereas in the later sixteenth century it was still dominant, by the early eighteenth century it was rare outside of a religious context, except for references to such major feasts as Christmas and Easter or to those (like Lady Day or Michaelmas) that had a secular meaning as conventional dates for the payment of rents or taxes. Conversely, dating by day of the month and year, Anno Domini, previously a highly restricted practice, rose in frequency. This practice was helped along by the growing use of almanacs, which long retained information on ecclesiastical feasts but clearly privileged days of the month as their primary time referent. Day/month/year dating was increasingly common by the early seventeenth century, especially in urban contexts; by the turn of the eighteenth century, it was dominant.[64]

The stronger sense of the linear progression of time implied by these changes was enhanced by greater historical awareness in the English population. The later sixteenth century witnessed "a conspicuous renewal in historical studies," stimulated in part by the cataclysm of the Reformation and a nationalistic preoccupation with the uniqueness and antiquity of English institutions.[65] Antiquarians undertook "the conscious pursuit and study of old things," emulating William Camden's mission to "restore antiquity to Britaine, and Britaine to his antiquity."[66] Chorographers began "discovering English history in its places."[67] Local and national histories multiplied. Many of these were disseminated in print, and the outcome was the development of an "historical culture," "the progressive saturation of public and private discourse by historical references," and "an increasingly homogenized and chronologically rigorous (even if ideologically fragmented and contentious) national sense of the past."[68] The core audience was learned gentlemen but extended to include interested men like Richard Gough, and even those who took little personal interest in history cannot have escaped exposure to a broadly shared litany of historical reference points. Almanacs listed them. Church services and bell ringing commemorated dynastic anniversaries (the accession of Elizabeth, the Regicide, the Restoration of Charles II) and celebrated Protestant deliverances (in 1588, 1605, and 1688). Some historical events were also the subjects of widely distributed popular prints.[69] This was everyday history—familiar dates to which meanings were attached but which also carried messages about Englishness. Social memory was becoming "historically tutored."[70]

The second significant change in temporal culture was enhanced precision in the recording and apprehension of time. Anthony Grafton discerns a "newfound mania for precision" among the chronologers of Renaissance Europe, and it was a trend not confined to scholars.[71] One indication is the keeping of more records: a "surge in public and private documentation" deriving from "an impulse to create a permanent record of the present for future use."[72] On the one hand, the increasingly elaborate administrative and judicial infrastructure of the English state, which reached into every parish, involved a growing number of chief inhabitants (men like Gough) in governmental duties requiring forms of recording, listing, and reporting. In these "local paper regimes," information was recorded in ways that often demanded "exactitude" and "precision," both numerical and chronological.[73] On the other hand, forms of documentation became "deeply... integrated into the daily routines of ordinary people."[74] Intensified commercial activity meant increased use of dated "instruments": bills, bonds, conveyances, acquittances, and the like. It involved greater regularity and systemization in the keeping of accounts.[75] Staying afloat in a more dynamic economic environment, in which time was money, also required commercial farmers and tradespeople to keep notes and "memorials" of appointments; money lent and owed; and payments made, received, or pending. Almanacs catered to that need by providing blank pages opposite each monthly calendar to "keepe notes of any actes, deedes, or thinges that passeth from time to time (Worthy of memory, to be registered)."[76] They were specifically designed for the recording of past or the anticipation of future events and their precise placement in time. And that, to judge by surviving annotated copies, is exactly how their owners used them.

Most almanacs were probably discarded at the year's end, but others were retained—sometimes because they included the exact dates of significant family events, sometimes because their owners' numerous entries had created personal diurnals. Almanacs had become records rather than tools, and as such, they had an affinity with the spiritual diaries and personal journals also emerging during this period.[77] All of these documents recorded the events of specific days. But chronological precision was also advancing *within* the day itself in the form of a growing emphasis upon clock time. The "horological revolution" of the seventeenth century—which involved the adoption of the pendulum and the anchor escapement in standing clocks and the balance spring in pocket watches—made timepieces more accurate and eventually more affordable and available. In the later seventeenth and eighteenth centuries, clock ownership became more widespread. Environments, especially urban environments, became more "dense with private time." Clock time was "increasingly used as an organizational tool," and social and commercial activity became "increasingly saturated" with clock time information.[78] By the mid-eighteenth century, then, present time was more precisely specified, and past time was more precisely recorded. Future time became more precisely imaginable. Approximation persisted but it was ceasing to be the cultural norm, gradually acquiring the taint of inadequacy.

The third form of significant change was in the prevalence of mnemonic markers, those things to which memory was attached or with which memory could be associated. Sometimes this involved their loss. The transformative impact of economic change on landscapes could challenge, erode, and even obliterate memories of what had been.[79] Conversely, the period also witnessed a burgeoning culture of commemoration. From the late sixteenth century, portraits were more frequently commissioned, first among the social elite but gradually extending in social range. A portrait often marked or celebrated significant moments in the life course of the sitter and commonly included, alongside the sitter's likeness, the date, the sitter's age, and texts and symbols indicating his or her social standing, virtues, and achievements.[80] Whatever the occasion of the commissioning, portraits were meant to be seen by a privileged audience and to be kept. They were "authorized versions of the self," demonstrating "how people wanted to be remembered."[81]

Elaborate church monuments to members of the rural and urban elites—which included effigies of the deceased and members of their families, inscriptions detailing lifespans and accomplishments, and (from the later seventeenth century) pithily flattering epitaphs—served a similar purpose. Sometimes, they were commissioned before death.[82] Less declaratory, but perhaps more socially significant, were the headstones marking the graves of people of the "middle sort," which proliferated in parish churchyards from the later seventeenth century and were commonplace by the eighteenth. Their inscriptions were simple—name, age, date of death, or sometimes simply initials and year of death—but they marked places of memory and mourning in a new way, as did the more perishable wooden headboards and planted flowers of the poor.[83]

Like the erection of permanent grave markers, the practice of inscribing dates and initials on houses and household furnishings was a distinctive development of this period and one that diffused socially in a similar manner.[84] But even when things were not inscribed to commemorate particular occasions, the increasingly elaborate material culture of the middle-rank household provided more objects to which memory could be attached. From the later seventeenth century, more households acquired "new" things—earthenware, pictures, looking glasses, knives and forks, tea and coffee sets, china dishes, clocks—which carried a symbolic charge and played "expressive roles" in what Lorna Weatherill calls the "frontstage activities" of the household. These things created new mnemonic springs alongside the "old" things through which memory continued to flow by virtue of their enduring "backstage" presence.[85] In the household, as elsewhere, the media of memory were accumulating.

Explaining fully these changes in the early modern culture of time would involve every dimension of the history of this period. They were born out of the shifting "matrix of everyday life" in a complex society undergoing protracted and entangled processes of economic, social-structural, political, and cultural change, and were diffused by "modest increments in practice."[86] Like all innovations in social practice, these changes involved material objects: documents, almanacs, clocks, portraits, gravestones, domestic possessions. They involved the acquisition of new "competencies": growing levels of literacy and numeracy and improved temporal skills. They also involved modifications in the meaning of time.[87] Time was becoming more secular in expression and more exact, regular, and linear. It was more closely recorded, more crowded with historical reference points, more *known*. It was becoming busier, and more urgent in its passage. But this displacement was not simply that of one culture of time by another—time remained heterogeneous. Rather, it was a shift in the mixture of forms within the early modern culture of time. Some were finding greater salience and authority and new kinds of expression. They were favored by those who exercised political and economic power and cultural dominance, those who sought to exert control over time and social memory. Others were diminishing in value and becoming associated with members of socially subordinate groups who lacked the assets, ability, or aspiration to emulate the practices of their "betters." Yet they persisted.

Richard Gough straddled both worlds. He could navigate the emergent temporalities and temporal practices of his time. He knew what could be learned from the traces of the past that were preserved in records, buildings, and artifacts. But he was also instinctively in tune with those whose sense of time was less disciplined and who left few traces behind except their children and the memories retained by their neighbors. In an act of astonishing originality, he wrote their stories down and passed them on.

Notes

1. Richard Gough, *The History of Myddle*, ed. David Hey (Harmondsworth, UK: Penguin, 1981). For a digital reprint of the complete first edition of 1875, see Richard Gough, *Antiquities and Memoirs of the Parish of Myddle, County of Salop* (Cambridge: Cambridge University Press, 2015). All subsequent citations refer to the 1981 edition.

2. The term is taken from Andy Wood, "Popular Senses of Time and Place in Tudor and Stuart England," *Insights* 6, no. 3 (2013): 4–5. The same point, variously expressed, is made by many other students of temporality in this period.

3. For almanacs, see Bernard Capp, *English Almanacs, 1500–1800: Astrology and the Popular Press* (Ithaca, NY: Cornell University Press, 1979); Adam Smyth, *Autobiography in Early Modern England* (Cambridge: Cambridge University Press, 2010), chap. 1; and Lauren Kassell, "Almanacs and Prognostications," in *The Oxford History of Popular Print Culture: Cheap Print in Britain and Ireland to 1660*, ed. Joad Raymond, vol. 1 (Oxford: Oxford University Press, 2011).

4. This account of the contents of almanacs is based on seventeenth-century almanacs included in Early English Books Online or held by the Beinecke Rare Book & Manuscript Library, Yale University, in particular, John Vaux, *A New Almanack, and Prognostication for This Present Year of Our Lord God, 1621* (London: Company of Stationers, 1621); Richard Allestree, *A New Almanacke and Prognostication for the Yeere of Our Lord God 1634, and from the Creation 5596, Being the Second from the Bissextill or Leape Yeere* (London: Company of Stationers, 1634); George Naworth, *Naworth, 1642: A New Almanacke and Prognos-*

tication for the Yeere of Our Lord and Saviour Iesus Christ, 1642, Being the Second from the Bissextile or Leape-yeere, and from the Creation of the World, 5591 (London: Company of Stationers, 1641); and George Wharton, *Calendarium Ecclesiasticum: Or, a New Almanack after the Old Fashion… to Which Is Added, Gesta Britannorum: Or, a Briefe Chronologie for 56 Yeares Last Past, viz. from the Yeare 1600 (in Which the Late K. Charls Was Born) until the Present 1657* (London: John Grismond, 1657).

5. Paul Glennie and Nigel Thrift, *Shaping the Day: A History of Timekeeping in England and Wales 1300–1800* (Oxford: Oxford University Press, 2009), 66. For the reference to Isaac Newton, see Donald J. Wilcox, *The Measure of Times Past: Pre-Newtonian Chronologies and the Rhetoric of Relative Time* (Chicago: University of Chicago Press, 1987), 16.

6. David Cressy, *Bonfires and Bells: National Memory and the Protestant Calendar in Elizabethan and Stuart England* (Berkeley, CA: University of California Press, 1989), 13–14. For an account of the conventional cycles of the year, see 16–33.

7. London Metropolitan Archives (hereafter LMA) DL/C/225, folios 31v, 33v, 85v, 147r; Durham University Library Archives and Special Collections (hereafter DUL) DDR/EJ/CCD/1/2, folio 183v; and Keith Wrightson, "Popular Senses of Past Time: Dating Events in the North Country, 1615–1631," in *Popular Culture and Political Agency in Early Modern England and Ireland*, ed. Michael J. Braddick and Phil Withington (Woodbridge, UK: Boydell Press, 2017), 95–96. In London, some referenced the legal calendar, as in "about the beginning of Easter Term," or "in Trinitie Term," LMA DL/C/225, folios 15r, 111v.

8. LMA DL/C/214, 49, 75; LMA DL/C/225, folios 90r, 210r; DUL DDR/EJ/CD/1/2, folio 305v; and Wrightson, "Popular Senses of Past Time," 98–99.

9. Malcolm Jones, *The Print in Early Modern England: An Historical Oversight* (New Haven, CT: Yale University Press, 2010), 19–20, 24–25, 27–34.

10. Wrightson, "Popular Senses of Past Time," 97.

11. Guido Ruggiero, *The Renaissance in Italy: A Social and Cultural History of the Rinascimento* (Cambridge: Cambridge University Press, 2015), 7. Compare Robert Bartlett, *The Hanged Man: A Story of Miracle, Memory, and Colonialism in the Middle Ages* (Princeton, NJ: Princeton University Press, 2004), 55. For the earlier history of year dating, see Deborah Mauskopf Deliyannis, "Year-Dates in the Early Middle Ages," in *Time in the Medieval World*, ed. Chris Humphrey and W. M. Ormrod (York, UK: York Medieval Press, 2001).

12. Capp, *English Almanacs 1500–1800*, 215–23. The quotations are taken from Richard Allestree, *Allestree 1628: A New Almanack and Prognostication, for the Ye[a?]re of Our Lord God 1628 and from the Creation 5590, Being Bissextile or Leap-yeere: Calculated and Properly Referred to the Longitude; Sublimity of the Pole Articke of 52; 53 Deg.* (London: Printed for the Company of Stationers, 1628), n.p; Vaux, *A New Almanack*, n.p.; and Naworth, *Naworth, 1642: A New Almanacke*, n.p.

13. Gough, *History of Myddle*, passim, quoting 49, 190.

14. Gough, *History of Myddle*, passim. For the Gittins dynasty, see 190; for the posthumous reputations quoted, see 112, 126, 145, 193, 206. There are many more examples.

15. The phrase quoted is used (in a different context) in William H. Sewell Jr., *Logics of History: Social Theory and Social Transformation* (Chicago: University of Chicago Press, 2005), 9.

16. H. Porter Abbot, *The Cambridge Introduction to Narrative*, 2nd ed. (Cambridge: Cambridge University Press, 2008), 4.

17. Anthony Grafton, "Dating History: The Renaissance and the Reformation of Chronology," *Daedalus* 132, no. 2 (Spring 2003): 77, 79; Daniel R. Woolf, "From Hystories to the Historical: Five Transitions in Thinking about the Past, 1500–1700," *Huntington Library Quarterly* 68, nos. 1–2 (March 2005): 41.

18. William West, "Of Instruments," in *Symbolaeographia… Or the Notarie or Scrivener* (London: Richard Tothill, 1590), book 1, section 1.

19. For genealogy and monuments, see Daniel Woolf, *The Social Circulation of the Past: English Historical Culture, 1500–1730* (Oxford: Oxford University Press, 2003), chaps. 3 and 4; Ralph Houlbrooke, *Death, Religion, and the Family in England, 1480–1750* (Oxford: Oxford University Press, 1998), chap. 11; and Nigel Llewellyn, *Funeral Monuments in Post-Reformation England* (Cambridge: Cambridge University Press, 2000). The phrases quoted are from Pierre Nora, "Between Memory and History: Les Lieux de Mémoire," *Representations* 26 (Spring 1989): 14, 19.

20. Glennie and Thrift, *Shaping the Day*, chap. 4, quoting 126.

21. Woolf, "From Hystories to the Historical," 41; Adam Fox, "Remembering the Past in Early Modern England: Oral and Written Tradition," *Transactions of the Royal Historical Society*, 6th ser., 9 (1999): 241.

22. Daniel Woolf, "Afterword: Shadows of the Past in Early Modern England," *Huntington Library Quarterly* 76, no. 4 (2013): 645–46.

23. LMA DL/C/225, folios 167v–168r.

24. Bartlett, *Hanged Man*, 55.

25. LMA DL/C/24, 15; The National Archives, Kew (hereafter TNA), DURH 7/68, *Mickleton v. Mickleton* (1701).

26. Gough, *History of Myddle*, passim. For some specific examples of these usages, see 43, 53–54, 58, 71, 114, 145, 152, 160, 167, 193, 220.

27. LMA DL/C/225, folios 67r–68r, 103, 163v. For the later standardization of clock time, see Eviatar Zerubavel, "The Standardization of Time: A Sociohistorical Perspective," *American Journal of Sociology* 88, no. 1 (July 1982): 5–19.

28. LMA DL/C/214, 15, 52, 59, 140; LMA DL/C/225, folio 25r; and TNA STAC 5/S2/13.

29. LMA DL/C/213, 85. The final phrase was that conventionally used to designate immemorial custom.

30. John Bedell, "Memory and Proof of Age in England 1272–1327," *Past & Present* 162, no. 1 (February 1999): 17–18; Joel T. Rosenthal, *Telling Tales: Sources and Narration in Late Medieval England* (University Park: Pennsylvania State University Press, 2004), 15–57; and M. McGlynn, "Memory, Orality, and Life Records: Proofs of Age in Tudor England," *Sixteenth-Century Journal* 40, no. 3 (Fall 2009): 687–91.

31. Wrightson, "Popular Senses of Past Time," 96; LMA DL/C/225, folio 48r; and LMA DL/C/241, folio 55r.

32. LMA DL/C/214, 53–54; LMA DL/C/241, folios 30v, 53r.

33. Wrightson, "Popular Senses of Past Time," 97, 99; LMA DL/C/244, folios 14r, 33r, 36r, 37v.

34. Bartlett, *Hanged Man*, 57.

35. Abbot, *Cambridge Introduction to Narrative*, 46, 67.

36. Paul Connerton, *How Societies Remember* (Cambridge: Cambridge University Press 1989), 16–17; James Fentress and Chris Wickham, *Social Memory* (Oxford, UK: Basil Blackwell; Cambridge, MA: Three Cambridge Center, 1992), ix–x; Eviatar Zerubavel, *Time Maps: Collective Memory and the Social Shape of the Past* (Chicago: University of Chicago Press, 2003), 63; and David Carr, *Time, Narrative, and History* (Bloomington: Indiana University Press, 1986), chap. 6.

37. Gough, *History of Myddle*, passim. For specific examples, see 43, 54, 57, 58, 72, 77, 84, 104, 212.

38. Abbot, *Cambridge Introduction to Narrative*, 4; Carr, *Time, Narrative, and History*, 3, 177.

39. Andy Wood, *The Memory of the People: Custom and Popular Senses of the Past in Early Modern England* (Cambridge: Cambridge University Press, 2013), 188, 223.

40. Nicola Whyte, "An Archaeology of Natural Places: Trees in the Early Modern Landscape," *Huntington Library Quarterly* 76, no. 4 (2013): 502.

41. Adam Fox, "Vernacular Culture and Popular Customs in Early Modern England: Evidence from Thomas Machell's Westmorland," *Cultural and Social History* 9, no. 3 (2012): 337.

42. Tim Ingold, quoted in Wood, *Memory of the People*, 189.

43. Alexandra Walsham, *The Reformation of the Landscape: Religion, Identity and Memory in Early Modern Britain and Ireland* (Oxford: Oxford University Press, 2011), 232, 234. See also Nicola Whyte, *Inhabiting the Landscape: Place, Custom and Memory, 1500–1800* (Oxford, UK: Oxbow Books, 2009), chap. 2.

44. Wood, *Memory of the People*, 188, 223–24.

45. Gough, *History of Myddle*, 32, 54, 56–57, 59–60.

46. Wood, *Memory of the People*, 224.

47. David Levine and Keith Wrightson, *The Making of an Industrial Society: Whickham, 1560–1765* (Oxford, UK: Clarendon Press, 1991), 292; Keith Wrightson, *Ralph Tailor's Summer: A Scrivener, His City, and the Plague* (New Haven, CT: Yale University Press, 2011), 95–96.

48. Robert Tittler, *Architecture and Power: The Town Hall and the English Urban Community c. 1500–1640* (Oxford: Oxford University Press, 1991), chap. 6, and for the Beverley settle, see 115 and Plate 9; Andy Wood, "Tales from the 'Yarmouth Hutch': Civic Identities and Hidden Histories in an Urban Archive," *Past & Present* 230, supplement 11 (November 2016); and Robert Tittler, *The Face of the City: Civic Portraiture and Civic Identity in Early Modern England* (Manchester: Manchester University Press, 2007), esp. 47–52.

49. I. W. Archer, "The Arts and Acts of Memorialization in Early Modern London," in *Imagining Early Modern London: Perceptions and Portrayals of the City from Stow to Strype, 1598–1720*, ed. J. F. Merritt (Cambridge: Cambridge University Press, 2001), 90; Jasmine Kilburn-Toppin, "Material Memories of the Guildsmen: Crafting Identities in Early Modern London," in *Memory before Modernity: Practices of Memory in Early Modern Europe*, ed. Erika Kuijpers and others (Leiden, The Netherlands: Brill, 2013), 165, 167, 171.

50. Wood, *Memory of the People*, 235.

51. Anthony Buxton, *Domestic Culture in Early Modern England* (Woodbridge, UK: Boydell and Brewer, 2015), chap. 6; Matthew Johnson, "Living Space: The Interpretation of English Vernacular Houses," in *The Uses of Space in Early Modern History*, ed. Paul Stock (New York: Palgrave Macmillan, 2015), 29.

52. Buxton, *Domestic Culture*, chaps. 3–6, quoting 206.

53. Adrian Green, "Consumption and Material Culture," in *A Social History of England, 1500–1750*, ed. Keith Wrightson (Cambridge: Cambridge University Press, 2017), 242, 251. For the ways in which dated domestic objects illuminate women's experience of time, see Sophie Cope, "Women in the Sea of Time," in *Gendered Temporalities in the Early Modern World*, ed. Merry E. Weisner-Hanks (Amsterdam: Amsterdam University Press, 2018).

54. Green, "Consumption and Material Culture," 243, 245.

55. Wrightson, *Ralph Tailor's Summer*, 91–94; and Keith Wrightson and David Levine, *Poverty and Piety in an English Village: Terling, 1525–1700*, 2nd ed. (Oxford, UK: Clarendon Press, 1995), 38.

56. Katharine Hodgkin, "Women, Memory and Family History in Seventeenth-Century England," in Kuijpers and others, *Memory before Modernity*, 312.

57. Judith Pollmann and Erika Kuijpers, "Introduction: On the Early Modernity of Modern Memory," in Kuijpers and others, *Memory before Modernity*, 22.

58. Glennie and Thrift, *Shaping the Day*, 68.

59. Pitirim A. Sorokin and Robert K. Merton, "Social Time: A Methodological and Functional Analysis," *American Journal of Sociology* 42, no. 5 (March 1937): 620.

60. Glennie and Thrift, *Shaping the Day*, 5. A "community of practice" is described by Etienne Wenger as one "created over time by the pursuit of a shared enterprise" of any kind, defined above all by "mutual engagement" in particular practices. Wenger, *Communities of Practice: Learning, Meaning, and Identity* (Cambridge: Cambridge University Press, 1998), 45, 73.

61. Jacques Le Goff, "Au moyen âge: Temps de l'église et temps du marchand," *Annales: Économies, sociétés, civilisations* 15, no. 3 (1960).

62. That people belong to multiple communities of practice is emphasized in Wenger, *Communities of Practice*, 6, and Elizabeth Shove, Mika Pantzar, and Matt Watson, *The Dynamics of Social Practice: Everyday Life and How It Changes* (London: Sage Publications, 2012), 68.

63. Cressy, *Bonfires and Bells*, 4–9.

64. These shifts are fully explored in my forthcoming essay, "Reckoning Time: 'Dating Statements' and the Rise of 'Almanac Time' in England, c. 1565–c. 1720."

65. R. W. Southern, "Presidential Address: Aspects of the European Tradition of Historical Writing: 4. The Sense of the Past," *Transactions of the Royal Historical Society*, 5th ser., 23 (1973): 246.

66. Woolf, *Social Circulation of the Past*, 147; William Camden, "The Author to the Reader," in *Britain, Or a Chorographical Description of the Most Flourishing Kingdomes, England, Scotland, and Ireland*, trans. Philemon Holland (London: Georgii Bishop and Ioannis Norton, 1610).

67. Rebecca Brackmann, *The Elizabethan Invention of Anglo-Saxon England* (Woodbridge, UK: D. S. Brewer, 2012), 147.

68. Woolf, *Social Circulation of the Past*, 9, 12–13, 274.

69. Capp, *English Almanacs 1500–1800*, 217–22; Cressy, *Bonfires and Bells*, xi, chaps. 4–11, passim; and Jones, *The Print in Early Modern England*, 60–63, 125–132.

70. Connerton, *How Societies Remember*, 16.

71. Grafton, "Dating History," 77.

72. Alexandra Walsham, "The Social History of the Archive: Record-Keeping in Early Modern Europe," *Past & Present* 230, supplement 11 (November 2016): 12, 45.

73. Paul Griffiths, "Local Arithmetic: Information Cultures in Early Modern England," in *Remaking English Society: Social Relations and Social Change in Early Modern England*, ed. Steve Hindle, Alexandra Shepard, and John Walter (Woodbridge, UK: Boydell and Brewer, 2013), 115, 116, 118, 122–25. For the processes of "state formation" and their impact on local office holders, see Steve Hindle, *The State and Social Change in Early Modern England, c. 1550–1640* (Basingstoke, UK: Macmillan, 2000), and M. J. Braddick, *State Formation in Early Modern England c. 1550–1700* (Cambridge: Cambridge University Press, 2000).

74. Walsham, "Social History of the Archive," 12.

75. Smyth, *Autobiography*, 3, 60; for contemporary guides to account keeping, see 61–72.

76. Thomas Purfoote, quoted in Smyth, *Autobiography*, 19.

77. See the discussion of some outstanding examples in Smyth, *Autobiography*, 23ff. For the remarkable case of the tradesman Joseph Bufton, who amassed twenty-two densely annotated almanacs between 1678 and 1716, see H. R. French, *The Middle Sort of People in Provincial England, 1600–1750* (Oxford: Oxford University Press, 2007), 243–46, and Brodie Waddell, "Writing History from Below: Chronicling and Record-Keeping in Early Modern England," *History Workshop Journal* 85 (Spring 2018).

78. Glennie and Thrift, *Shaping the Day*, 35, 108–15, 122, 124–25, 127, 131.

79. Whyte, *Inhabiting the Landscape*, 162–64; Wood, *Memory of the People*, 236–46, 348–51; and Ian Waites, *Common Land in English Painting, 1700–1850* (Woodbridge, UK: Boydell Press, 2012), chap. 5.

80. Tarnya Cooper, *A Guide to Tudor and Jacobean Portraits* (London: National Portrait Gallery, 2008), 8–9, 13; Tarnya Cooper, *Citizen Portrait: Portrait Painting and the Urban Elite in Tudor and Jacobean England and Wales* (New Haven, CT: Yale University Press, 2012), 3, 14, 90; Tittler, *Face of the City*, 134–39; and Robert Tittler, *Portraits, Painters, and Publics in Provincial England, 1540–1640* (Oxford: Oxford University Press, 2012), 142ff.

81. Cooper, *Citizen Portrait*, 66; Cooper, *Guide*, 4.

82. Houlbrooke, *Death, Religion, and the Family*, 344–57.

83. Houlbrooke, *Death, Religion, and the Family*, 360–66; Green, "Consumption and Material Culture," 255.

84. Green, "Consumption and Material Culture," 252.

85. Lorna Weatherill, *Consumer Behaviour and Material Culture in Britain, 1660–1760* (London: Routledge, 1988), esp. chaps. 2–4, 7–8, quoting 165. See also Mark Overton and others, *Production and Consumption in English Households, 1600–1750* (Abingdon, UK: Routledge, 2004).

86. Glennie and Thrift, *Shaping the Day*, 19, 78.

87. Shove, Pantzar, and Watson, *Dynamics of Social Practice*, 14–16, 54.

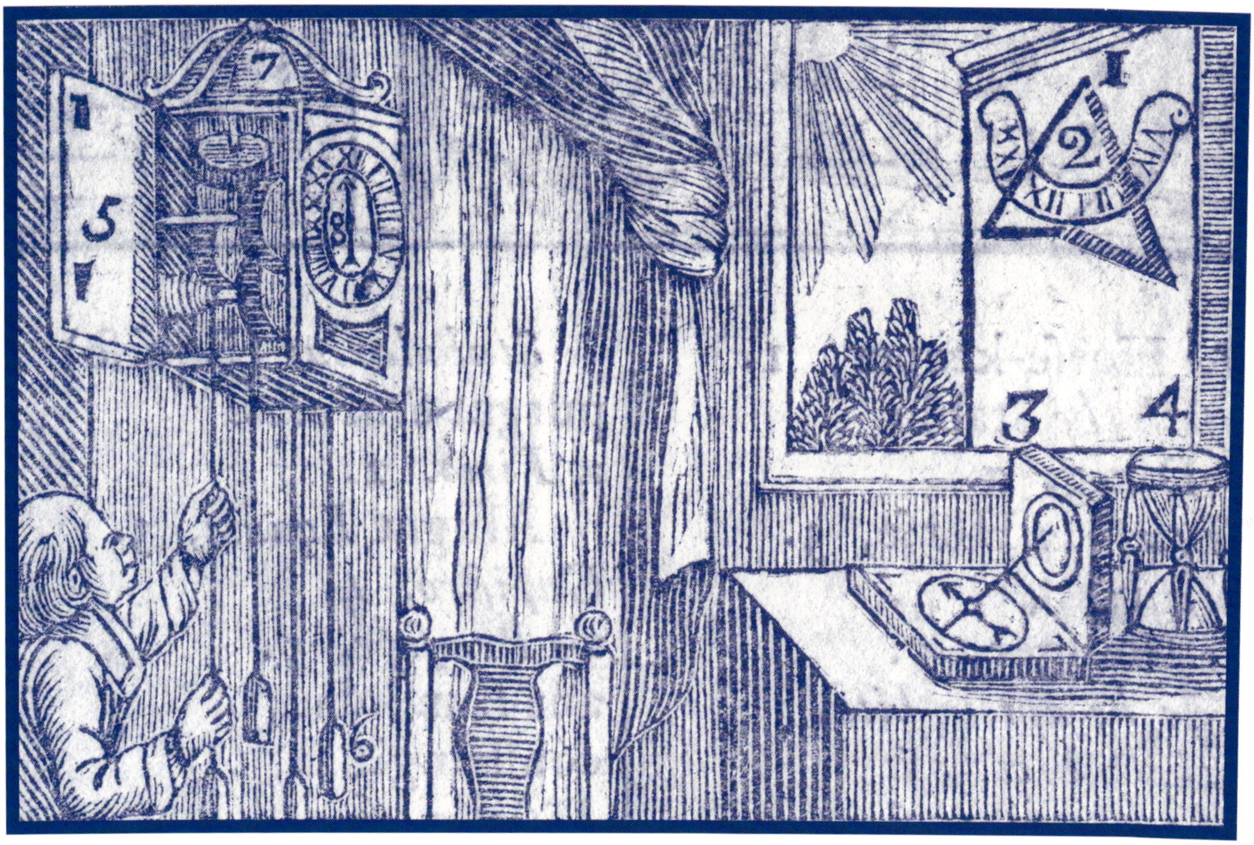

1
2
3
4
5
6
7

Running Like Clockwork?

Objects and "Early Modern" Time

GLENN ADAMSON

EARLY MODERN IS A TERM WITH A TROUBLED PAST. IT CAME INTO general use in the 1970s, just as progressive definitions of modernity were under attack. At issue in this debate was the presumption that history had traveled in an upward curve for the past five hundred years, as indexed by increases in industrial production, urbanization, life expectancy, scientific knowledge, globalism, secularization, and individualism—in short, that the standard of living had gotten steadily better over the past half millennium or so, and could be counted on to keep doing so. Even today this optimistic view has its adherents.[1] But, at least in the precincts of academic history, it is a distinctly minority position. "Theories of modernization enjoyed a relatively short heyday," as Hamish Scott writes. "It was soon evident that they were linear in structure, teleological in nature, and Western-centric in their assumptions, while actual developments in the Third World obstinately refused to conform to the theoretical paradigm." That the term *early modern* is rarely used outside of German and Anglo-American scholarship is itself a worrying indication of its underlying Eurocentrism.[2] In Britain, a consensus along these lines has long been developing among historians: progressive accounts, like those established by writers affiliated with the Whig party in the eighteenth and nineteenth centuries, have been vigorously rejected as part of a self-serving elite ideology.[3]

The critique of modernization theory came from many directions, most importantly from feminist and postcolonial studies, which by the late 1970s had come to be associated with "postmodernism," marking a departure from grand narratives of ceaseless improvement.[4] And while one might initially think that "early modern" is the natural pendant of postmodernism—two chronological bookends—in fact, their usage is nearly opposite in practice. To speak of postmodernism implies a critique of modernity as uniform, often denying its validity as a concept. Indeed, some postmodernists denied the very legitimacy of objective historical explanation as such.[5] The term *early modern*, by contrast, is implicitly homogenizing and positivist. It portrays the period from 1500 to 1800 as if in a rearview mirror, framing it as a protracted preparatory phase for the great upheavals to come.

So much for early modern, one might think, but at the same time, it is undeniable that these centuries did witness massive change, along the vectors mentioned above and many others, the ramifications of which we still feel today. In recognition of this, the tendency in contemporary scholarship has been to preserve admittedly problematic terminology, while resisting its teleological and universalizing implications. Such a move would avoid treating modernity "as a period in human history, rather than a culture," as Chandra Mukerji puts it in *Modernity Reimagined*, attending to "the power of the cultural imaginaries that drive it, the improvised nature of modern life."[6]

This notion of modernity as a distinct culture *within* the early modern, limited in its reach, contested at every turn, has been central to our approach in *Marking Time*. Read most anything about the history of temporal experience and you will encounter the same story: In medieval Europe, time was experienced cyclically, according to the rhythms of agriculture and the ecclesiastical calendar. Temporality was both approximate and fragmentary, such that most people had only an imprecise sense of the hour or day, uncalibrated with others elsewhere. With modernity came linear time, more finely divided (reckoned in seconds rather than seasons) and gradually more exact and universal. It also brought an orientation away from past tradition and toward future risk and opportunity, as exemplified by commodities markets. Joyce Appleby summarizes it like this: "With time now parceled out in dependable processes, sequential patterns, and irreversible transformations, all the rivulets of human activity could be seen flowing into the great river of improvement."[7] The advent of rail transport in the nineteenth century is often claimed as a watershed moment, as it necessitated coordination of schedules from town to town, but long before that, time was already being disciplined. In the early factory, as E. P. Thompson writes in his influential 1967 essay, "Time, Work-Discipline, and Industrial Capitalism," temporality as such was remade as something quantifiable rather than qualitative: "Time is now currency: it is not passed but spent."[8]

The standard history of modernization places emphasis on time as a topic of increasingly self-conscious interest, driven by a sensation of ever-increasing speed. This emphasis was manifested in such phenomena as daily newspapers (the first was *The Daily Courant*, published by Elizabeth Mallet of St. Martin, Ludgate, in 1702). "Such are these base ephemera, so born / To die before the next revolving morn," as George Crabbe put it in his satirical poem *The News-Paper* (1785).[9] Wherever the forces of modernization went—whether to Britain's own rural districts or to far-flung colonies—the increased rapidity of production, communications, and transport repeated itself. Even Fredric Jameson, one of the key theorists of postmodernism, subscribed to this monolithic view: "When the premodern vanishes, when the peasantry shrinks to a picturesque remnant, when suburbs replace the villages and modernity reigns triumphant and homogeneous over all space, then the very sense of an alternate temporality disappears as well."[10]

But there is another way to look at this story, one that tells it, indeed, in more than one way. We can recognize the emergence of a "stronger sense of the linear progression of time"—as Keith Wrightson puts it in his essay (page 19)—and find evidence of the very diverse ways in which this sense was experienced and embodied. The merchant and the blacksmith, the mother and the wet nurse, the slaver and the enslaved were defined partly according to their different experiences of time. The factors of identity and place were particularly important in this respect. The feminist historian Rita Felski, for example, emphasizes the "uneven temporality" of modernity, "the sense of being caught between contradictory modes of thought, behavior, and historical understanding," a description that she sees as applicable "perhaps above all to women," who have historically been expected to embrace the repetitive habits of domesticity while also traversing historical change.[11] Paul Gilroy, meanwhile, describes in *The Black Atlantic* the nonsynchronous, ritualized temporalities of the black diaspora as a means of building "communities of sentiment and interpretation." This observation is usefully juxtaposed to Justin Brown's essay (page 43), which considers "global time" in the context of imperialism. If devices like ship's logs and ankle irons were ways of imposing both physical and temporal control on the enslaved, as Brown argues, then those people could respond in kind, understanding their own time in a protective and private way.[12]

One thing is certain: living in a house filled with dated objects was not like inhabiting a calendar. A typical household likely had many artifacts marking different moments in time, and the dates they bore might refer to the moment of a thing's making, its acquisition, or (just as likely) an event disconnected from the object's specific history. In her essay (page 57), Gavi Levy Haskell assesses particular dated objects in the Bryan Collection—a set of bed hangings, a cradle, a spit jack—and notes that different members of the same household, of different social stations and at different stages of life, would have had disparate relationships to these temporal markers. Their associations would have been guided by gender and social status as well as by the course of life itself, sometimes grounded in memory and at other times only received through narrative.

Even the aforementioned advent of the daily newspaper, which might initially seem such a clear manifestation of shared temporality, proves on closer examination to be anything but. As Tony Claydon writes in "Daily News and the Construction of Time in Late Stuart England," the eighteenth-century press gave "a highly fractured presentation of time." A single issue might report on the previous day's events (if they had also occurred in the newspaper's city of origin), as well as include items that were days, weeks, or even months old, depending on geography and the vagaries of sourcing.[13] Editorial habits exacerbated this temporal fragmentation. Some narratives might carry through from one issue to the next, contributing to a sense of a connective flow, but readers were just as likely to encounter a jumble of disparate contents. In these early papers, as William Warner writes in his essay "Truth and Trust and the Eighteenth-Century Newspaper," "letters, articles, advertisements, and grain prices are assembled in an additive, disconnected fashion and organized with such a weak principle of subordination that their rhetorical effect is paratactic."[14]

We might take these newspapers as an analogy for the world of dated objects, which were similarly layered and multiperspectival. For the historian, it is precisely these qualities that make these objects a sensitive register of temporal experience. Time was definitely changing in the years from 1500 to 1800, but not moving at a uniform speed or in a uniform direction. Dated artifacts not only recorded this variance but also affected their makers' and users' sense of time. Rather than seeing objects as passive, we can understand them as having an active role in shaping daily temporal experience. It is particularly important to understand these artifacts as operating within a relational field—much as a handful of pebbles tossed into still water creates a highly complex pattern of overlapping ripples.

Just such an understanding did circulate in the eighteenth century, as in Joseph Addison's "Adventures of a Shilling," published in the *Tatler* in 1710. Framed explicitly as an allegory for "busie men of the age, who only valued themselves for being in motion," the brief tale is an early example of a so-called it-narrative, in which an inanimate object takes on the role of narrator. Addison relates a dream in which the coin in his pocket speaks to him: "Methoughts the Shilling that lay upon the table reared itself upon its edge, and turning the face towards me, opened its mouth, and in a soft silver sound gave me the following account of his Life and Adventures." It tracks its own hand-to-hand passage, from apothecary to butcher to brewer: "I sometimes fetched in a shoulder of mutton, sometimes a play-book." At one point in the story, the shilling is bent, and a young woman offers it to her sweetheart, but, the coin says, he "pawned me for a dram of brandy, and drinking me out next day, I was beaten flat with an hammer, and again set a running." Addison's witty parable positions currency as the motor force of economic exchange, as if humans were abiding by its desires: "We shillings love nothing so much as travelling." And although he doesn't mention it, we can be fairly sure that such a coin would have borne the date of its minting, an anchor in its life of constant motion.[15]

Ideally, we would approach each of the objects in *Marking Time* as imaginatively as Addison did, attempting to understand their roles in shaping temporal experience. A logical place to start is with those objects that are themselves timekeepers. As Angela McShane writes in her essay (page 69), measurement of all things was a pressing issue in the seventeenth and eighteenth centuries; a bewildering miscellany of so-called standards presented real challenges in the marketplace. As with weight, volume, and length, the period saw concerted efforts to improve the counting of time. This practice was not just a learned pursuit but a widely shared experience, as the precipitous rise in the ownership of domestic longcase clocks and pocket watches attests, the latter becoming commonplace possessions even among working-class men. On the face of it, one might think this shift represented a big step toward a new understanding of time as precise, consistent, and pervasive.[16] But, in fact, technical limitations meant such devices could not suture their owners into a single shared temporal fabric. A clock or watch did assert "independence from a day's, a season's, nature's dictates," but it did so primarily symbolically, not instrumentally. If timepieces were first and foremost "refinements" that reinforced "modern" self-conscious individualism, they did so without disrupting temporal fragmentation.[17]

Did other dated objects also function as temporal "accessories" in this way? That is, could the inscription of a year on a timber chest or a Delftware plate have been just as much a fashion statement as the decoration elsewhere on the object's surface? This question propels us in the direction of the "consumer revolution" that many historians have identified as taking place in the eighteenth century. Taking modernization theory to an extreme, one could hypothesize that dated artifacts manifest an early form of planned obsolescence, objects initially declaring their own novelty and then quickly fading out of style, like a chrome-bedecked 1950s automobile or a quickly superseded electronic device.[18] Their physical assertions of newness would have been all the more potent in the context of an accumulated stratigraphy of earlier dated objects. Such a relational array embodied the conscious separation of present from past that is meant to be central to modernity. There is a problem with this neat picture, though, which immediately becomes clear when we recall that the date on an artifact might well signify the date of a birth, attainment of professional status, marriage, or other major life event constitutive of ongoing identity. Some dated objects spoke clearly of political conviction and loyalty; others were religious in character, expressing faith in the eternal power of God. Whatever fashionability might have been attached to such dated objects—and it was certainly always a factor, even in ecclesiastical contexts—the sense of novelty existed in tension with various assertions of permanence.

In approaching this seeming contradiction, it is helpful to think of dated objects as *using* time, both in their crafting and their ongoing care. The proper employment of temporality was a much-discussed topic in the early modern period, which prompted careful contemplation by leading scientific minds. Consider Benjamin Franklin's half-joking proposal that all of humanity should rise at the crack of dawn year-round and hence save the expense of burning candles in the evening.[19] Or consider Edmond Halley's "Life Table," published in the *Philosophical Transactions* of the Royal Society in 1693. In this paper, Halley (who accurately predicted the periodic reappearance of the now eponymous comet) tried to establish a mathematical basis for life insurance rates. Basing his work on detailed demographic data from Breslau (today Wrocław), Poland—then an ideal case study because of its unusually stable population—he calculated "at what number of *Years*, it is an even Lay that a Person of any *Age* shall die." He found, for example, that an average thirty-year-old man could expect to live to little more than fifty-seven. On this basis, Halley argued that "the *Price* of *Insurance* upon *Lives* ought to be regulated, and the difference is discovered between the *price* of ensuring the *Life* of a *Man* of 20 and 50, for Example: it being 100 to 1 that a Man of 20 dies not in a year, and but 38 to 1 for a Man of 50 Years of Age."[20]

In this impressive yet unnerving computation, Halley succeeded in cutting right through the emotive ideals of the "ages of life," reducing them to sheer probability. Time really could be equated to value. In the face of this rigor, as Bill Schwarz comments, "subjective time, the time of the everyday and of the self, memory included, could appear *only* as dysfunctional, working to interrupt the clear geometrical abstractions of the time of history."[21] But, of course, one could (and even today, probably most people would) look at things exactly the other way round: although Halley's statistical "Life Table" might help secure profit margins, it captures nothing of life as it is actually lived, or death as it is actually witnessed. Even Franklin, who fetishized economy in all things, conceded that "people are apt to be obstinately attached to old customs, and that it will be difficult to induce them to rise before noon." He facetiously suggested posting armed guards in the chandler shops, to enforce candle quotas.[22] More seriously, as Amy Witherbee points out, early financial instruments like that described by Halley were notoriously uneven in their reliability, often resulting in financial bubbles and other wild miscalculations: "When in the late seventeenth and early eighteenth centuries, Britain deployed mathematics for state financing plans, long-term mortgages, insurance, and even gaming theories, mathematicians' functional conception of time came up against a human temporal consciousness that it was never designed to address."[23]

In objects, one can see these competing ideals converge. Every object in *Marking Time* was a commodity, an investment of labor and material that could be tallied up—although, as Edward S. Cooke Jr. explains in his consideration of artisan temporality (page 83), tallying was not always a straightforward process. When an embroidered cap was made for an unborn child, or a tomb monument carved for a person still living, those objects, and the expenditure they represented, were more than commodities. They were objects set against an uncertain future. There was no equation, here, that men like Franklin or Halley could have solved. Nathan Flis tracks this reality in his affecting study of seventeenth- and eighteenth-century commemorative practices (page 91). In this fundamental area of human concern, there was no one dominant direction of change but instead a diversity of practices, ranging from the quietly private to the grandly ostentatious, the secular to the transcendent. Early modern people lived in hope, each in their own way, and they marked objects in time partly as a way of articulating that hope, sometimes against the odds.

This is what it meant to use time well, or in period parlance, to "redeem" time.[24] In the popular conception, time was beyond any earthly price, for it was God-given and could only be repaid through devotion. "Redeeming denotes the excellent worth and preciousness of Time; a Commodity to be valued above the Gold of Ophir," explained John Fox in 1670. "To redeem a thing, is to buy it for a price; the price we redeem it with, is labour, travel, faithful and serious diligence, and greater activity and vigour in the prosecutions of our duty."[25] The same year that Fox wrote these words, a three-penny coin was struck in silver—and the next year, another. At some point thereafter, an artisan cunningly wrought a glass goblet, such that the little hoard of two was captured forever in its hollow stem (page 276). Among the hundreds of artifacts included in *Marking Time*, this one seems especially emblematic of the project: a fragile time capsule whose survival, three hundred fifty years on, is as remarkable as its creation was. The currency contained within the glass positions the goblet in time, but only inexactly; we know that it was made no earlier than 1671 but cannot be certain how much later. And although these coins were never terribly valuable—only sixpence all told—the gesture of imprisoning them feels momentous. Remember the words that Addison put in his shilling's imagined mouth: it loved "nothing more than traveling." Well, these two coins are out of circulation. They have been removed from the flow of exchange, which is as much a temporal structure as a spatial and social one. In a metaphorical sense, they have been removed from time itself. (Think of this: the air inside the glass also dates to the seventeenth century.) And yet, as with so many of the objects included in *Marking Time*, the object reflects a conscious desire to communicate to the future—a future that has now become our own.

Notes

1. A prominent recent example of this view is found in Steven Pinker, *Enlightenment Now: The Case for Reason, Science, Humanism, and Progress* (New York: Penguin, 2018).

2. Hamish Scott, "Introduction: 'Early Modern' Europe and the Idea of Early Modernity," in *The Oxford Handbook of Early Modern European History, 1350–1750*, ed. Hamish Scott, vol. 1, *Peoples and Place* (Oxford: Oxford University Press, 2015), 7.

3. This critique was first influentially made by Herbert Butterfield in *The Whig Interpretation of History* (London: G. Bell and Sons, 1931). For a review of subsequent literature, see Adrian Wilson and T. G. Ashplant, "Whig History and Present-Centred History," *Historical Journal* 31, no. 1 (March 1988): 1–16; for an attempt to recuperate "Whiggish" progressivism, see Annabel Patterson, *Nobody's Perfect: A New Whig Interpretation of History* (New Haven, CT: Yale University Press, 2003).

4. The term *postmodernism* was first popularized in architectural circles by Charles Jencks and quickly gained broader usage. See Glenn Adamson and Jane Pavitt, eds., *Postmodernism: Style and Subversion, 1970 to 1990* (London: V&A Publishing, 2011).

5. Hayden White, *Metahistory: The Historical Imagination in Nineteenth-Century Europe* (Baltimore: Johns Hopkins University Press, 1973).

6. Chandra Mukerji, *Modernity Reimagined: An Analytic Guide* (London: Routledge, 2016), xxiii.

7. Joyce Appleby, "The Intellectual Underpinnings of American Democracy," *Daedalus* 136, no. 3 (Summer 2007): 18.

8. E. P. Thompson, "Time, Work-Discipline, and Industrial Capitalism," *Past & Present* 38, no. 1 (December 1967): 56–97.

9. George Crabbe, *The News-Paper: A Poem* (London: J. Dodsley, 1785).

10. Fredric Jameson, "The End of Temporality," *Critical Inquiry* 29, no. 4 (Summer 2003): 699.

11. Rita Felski, *Doing Time: Feminist Theory and Postmodern Culture* (New York: New York University Press, 2000), 100.

12. Paul Gilroy, *The Black Atlantic: Modernity and Double Consciousness* (London: Verso, 1993), 197. Gilroy (202) quotes Ralph Ellison's *Invisible Man*: "Invisibility, let me explain, gives one a slightly different sense of time, you're never quite on the beat. Sometimes you're ahead and sometimes behind. Instead of the swift and imperceptible flowing of time, you are aware of its nodes, those points where time stands still or from which it leaps ahead. And you slip into the breaks and look around." Ellison, *Invisible Man* (Harmondsworth, UK: Penguin, 1976), 11.

13. Tony Claydon, "Daily News and the Construction of Time in Late Stuart England, 1695–1714," *Journal of British Studies* 52, no. 1 (January 2013): 55–78.

14. William B. Warner, "Truth and Trust and the Eighteenth-Century Newspaper," in *Travelling Chronicles: News and Newspapers from the Early Modern Period to the Eighteenth Century*, ed. Siv Gøril Brandtzæg, Paul Goring, and Christine Watson (Leiden, The Netherlands: Brill, 2018). See also Daniel Woolf, "News, History, and the Construction of the Present in Early Modern England," in *The Politics of Information in Early Modern Europe*, ed. Brendan Dooley and Sabrina A. Baron (London: Routledge, 2002), 80–118.

15. Joseph Addison, "Adventures of a Shilling," *Tatler* (November 11, 1710), in *The British Essayists*, vol. 5 (London: Printed for T. and J. Allman, 1823), 166.

16. John Styles, "Time Piece: Working Men and Watches," *History Today* 58, no. 1 (January 2008): 44–50. Thorstein Veblen was more right than he knew when he observed in 1914 that a modern mechanical timepiece "marks the passage of time with a degree of exactness that would have seemed divertingly superogatory to the common man of the high tide of handicraft." Veblen, *The Instinct of Workmanship* (New York: Macmillan, 1914), 311.

17. Martin Bruegel, "Time That Can Be Relied Upon: The Evolution of Time Consciousness in the Mid-Hudson Valley, 1790–1860," *Journal of Social History* 28, no. 3 (Spring 1995): 551. See also Richard Bushman, *The Refinement of America: Persons, Houses, Cities* (New York: Vintage, 1993).

18. For a history of the concept of planned obsolescence, see Giles Slade, *Made to Break: Technology and Obsolescence in America* (Cambridge, MA: Harvard University Press, 2006).

19. Franklin to the *Journal of Paris*, letter to the editor, 1784, reprinted in *Lapham's Quarterly* 7, no. 4 (Fall 2014): 108–10. As the editors of *Lapham's* point out, Franklin anticipated daylight savings time, which was finally introduced in the United States in 1918.

20. Edmond Halley, "An Estimate of the Degrees of Mortality of Mankind," *Philosophical Transactions* 17 (1693): 596–610.

21. Bill Schwarz, "Memory, Temporality, Modernity: Les Lieux de Mémoire," in *Memory: Histories, Theories, Debates*, ed. Bill Schwarz and Susannah Radstone (New York: Fordham University Press, 2010), 43.

22. Franklin to the *Journal of Paris*.

23. Amy Witherbee, "The Temporality of the Public in *The Tatler* and *The Spectator*," *The Eighteenth Century* 51, nos. 1–2 (Spring/Summer 2010): 173.

24. This idea derives from the book of Ephesians: "See then that ye walk circumspectly, not as fools, but as wise, Redeeming the time, because the days are evil. Wherefore be ye not unwise, but understanding what the will of the Lord is." Eph. 5:15–17 (King James Version).

25. John Fox, *Time and the End of Time* (London: Printed for George Calvert and Samuel Sprint, 1670), 6. Ophir is a legendarily wealthy city, mentioned in the book of Genesis, from which King Solomon obtained his gold.

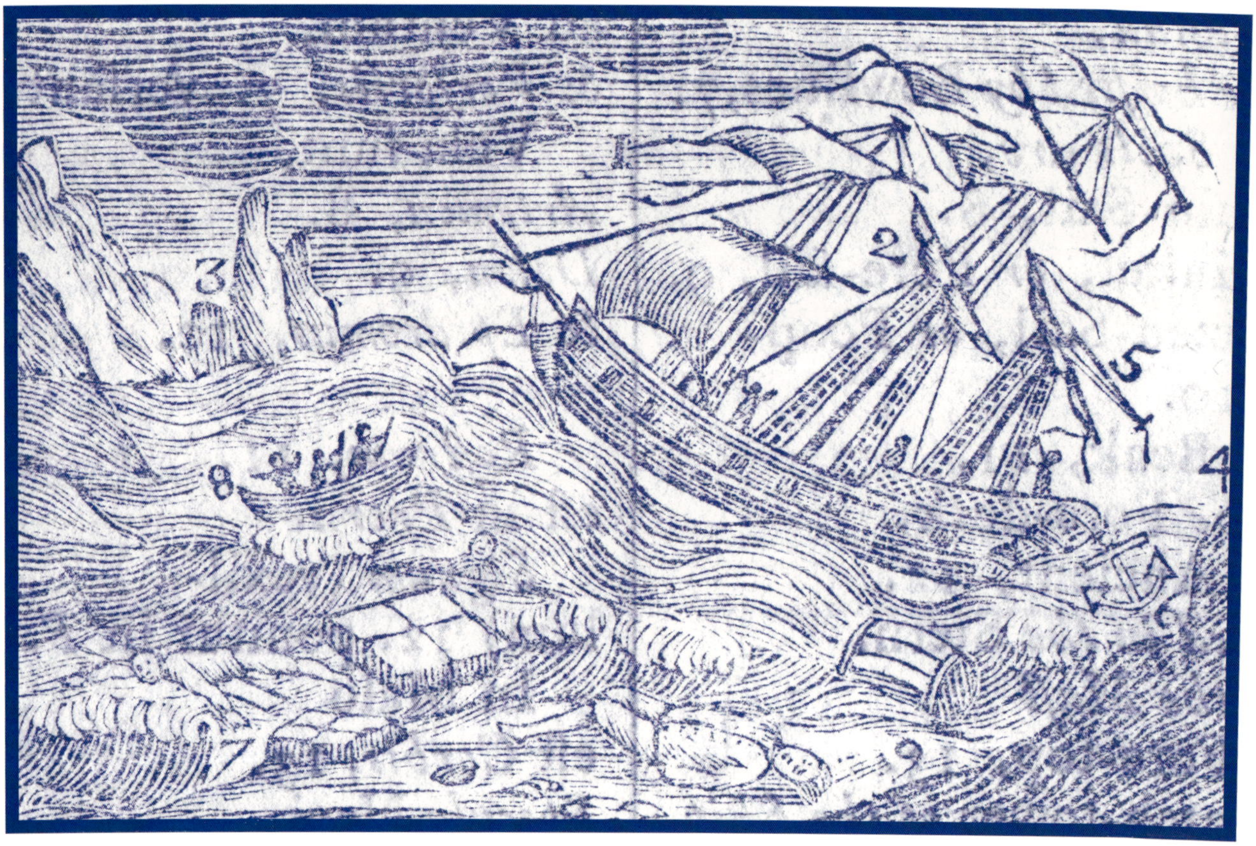

3
2
5
4
8
9

Marking Transatlantic Time

Accounting for Slavery in Britain and the Caribbean

JUSTIN M. BROWN

MARY PHILLIPS SURROUNDED HERSELF WITH TYPICAL GENTEEL comforts. Her will reveals the sundry valuables that she bequeathed to family and friends, including such things as "my goldwatch and a diamond ring that was my Mothers and a Diamond ring of my owne," which she left to her brother, John Hothersall, and "my Silver Chafeing Dish," given to her uncle, John Farmer. To this list of her dear possessions, she added a set of more substantial movable items, including "my best bed and Furniture and the Furniture of the Chamber," which she gave to her aunt, Margaret Holder. By all appearances, Phillips maintained an agreeable lifestyle. In addition to the precious and mundane things she owned, her will tells us that her interpersonal connections formed an important part of her life.[1]

Phillips was not a metropolitan woman; rather, her home lay many miles from any city in Britain, across the Atlantic on the small island of Barbados. As a member of the island's leading plantocracy—her uncle was governor of Barbados between 1701 and 1703—Phillips inhabited a truly transatlantic world, with family ties that extended as far as England and New York. Her will thus bears unusual testimony to how material things shaped transatlantic life in the early modern world and the ways in which they allowed people like her to construct and articulate their social identities on the periphery of empire.[2] But more than having historical import, her account gives an affecting testimony to the symbolic dimensions of the value that things acquired during a period marked by global exchange.

Because documents and objects often bear the literal mark of time, they help us to reconstruct material histories, not only by telling us who owned what but also by revealing how objects moved across space and time. As people traveled ever-greater distances, so did the objects they cherished. Dated documents like wills, diaries, and account books likewise reveal the long-distance movement of both things and people in the British Atlantic during this period. It took several years, for example, and travel on the part of the executor and two witnesses, before Phillips's will—written shortly before her death in 1698—was processed by the probate court in Britain in January 1705. Thomas Foulerton, a member of one of the Inns of Court in London, was her only surviving executor. He and the witnesses to the will—William Clealand of Lime Street, London, and Thomas Garth of Navestock, Essex—were living in Barbados when Phillips died there. Because of her wealth, the will was to be proved at the premier probate court in England: the Prerogative Court of Canterbury. It took

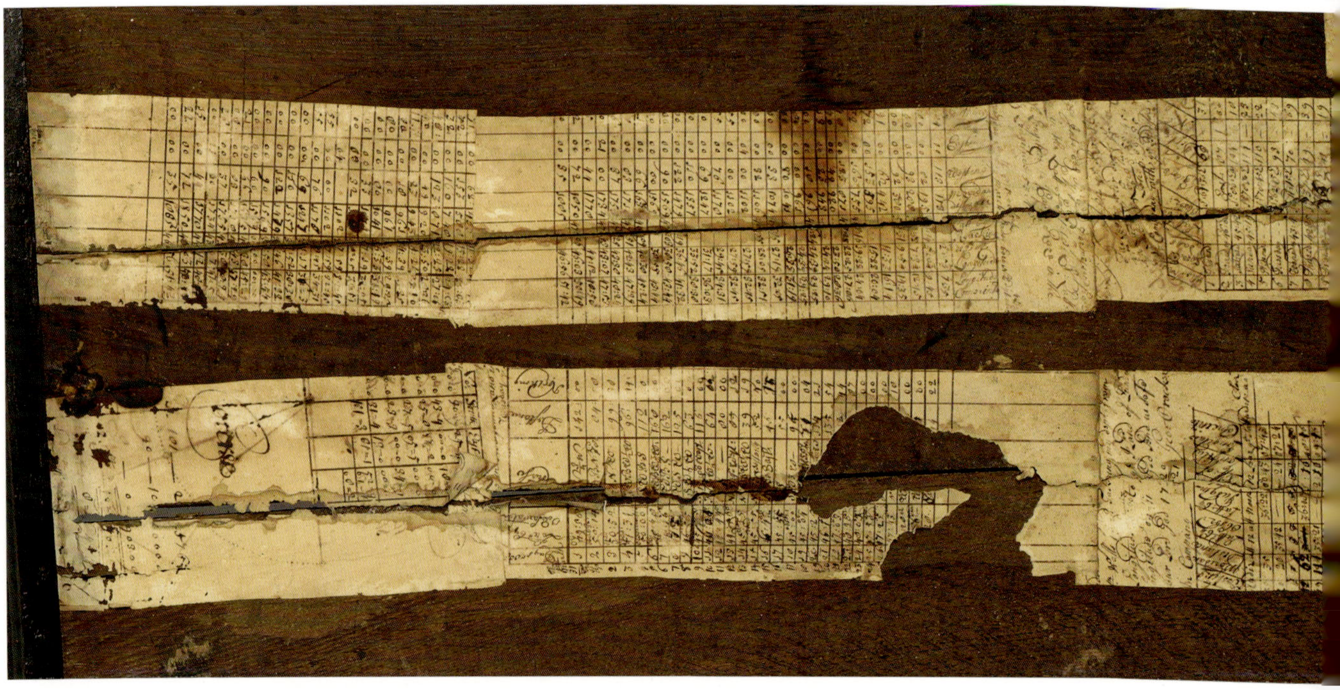

some time for the three men to travel back across to England to verify the will, thus evincing the inherent "time delay" in transatlantic transactions.

Significantly, wills and other documents frequently reveal more brutal contours of transatlantic history by calling attention to the fact that human beings, in addition to material objects, were taken as possessions. Phillips herself was a slaveholder.[3] Among her many possessions, she included a "Negroe Girle" named Priss, who, along with "my Gold Locketts and the Chest of Drawers," she bequeathed to her aunt for the full duration of Priss's lifetime.[4]

This essay explores the impact of time on the things and people that moved across the Atlantic Ocean in the early modern period. For those like Mary Phillips, a lifetime's worth of cherished possessions extended their intimate, although wide-reaching, worlds. For those who profited from the institution of slavery, time was simply money. And for the many people, like Priss, whose lives were lived in unrelenting bondage, time was a stolen commodity. The following discussion considers how early modern people confronted their global present—recognizing the great toll of the slave trade, a harsh reality that defined important aspects of possession and ownership—and the role that time played in reconciling the global with the local. Finally, this essay suggests that the things that people kept, regardless of on which side of the Atlantic they stood, often told complicated stories.

Moving Things

The titular hero of Daniel Defoe's *Robinson Crusoe*, first published in 1719, spends a great deal of time throughout the book collecting items from a shipwreck on a remote Caribbean island. Eventually he amasses a remarkable collection of objects, ranging from chests loaded with provisions to pistols and powder horns, lists of which he diligently commits to paper.

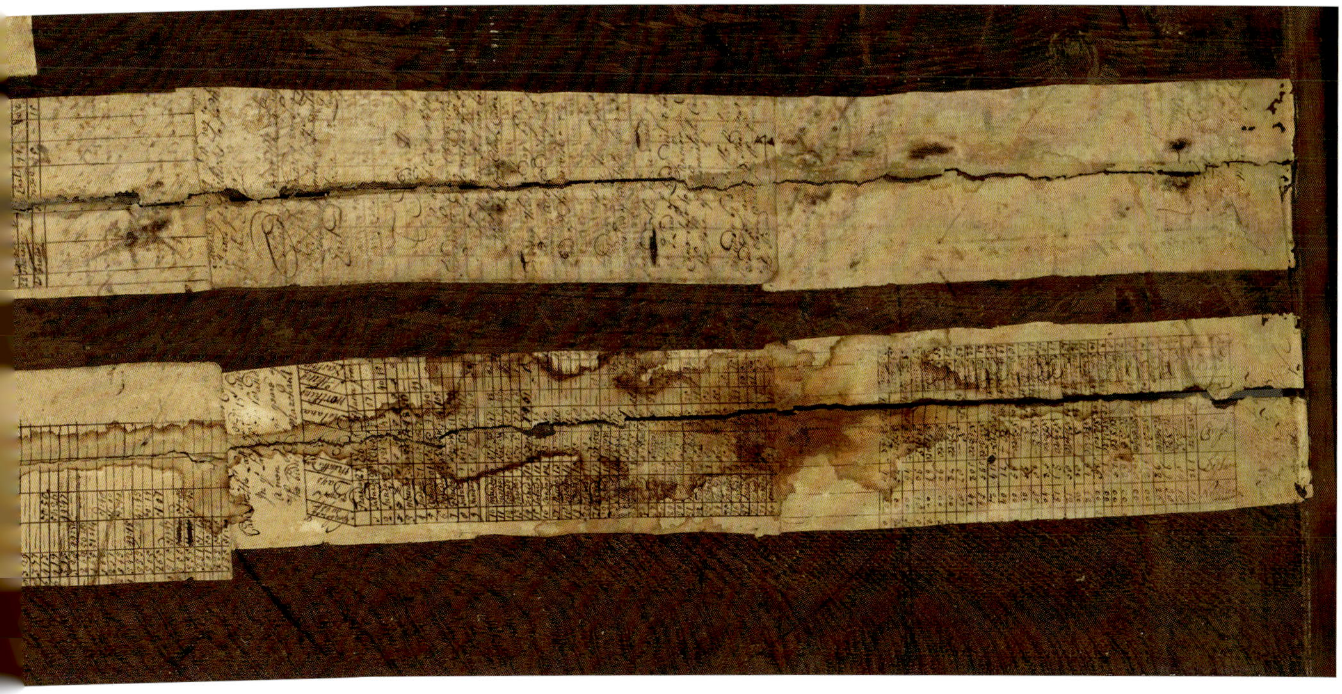

FIG. 1 Chest, 1609, oak, iron, 30½ × 60 × 23½ in. (77.5 × 152.4 × 59.7 cm). The Bryan Collection, Lake Bluff, IL

Like Crusoe, early modern travelers kept records of the things they moved from one place to another. Although record keeping was by no means a new development, the example of Crusoe demonstrates that by the eighteenth century it served as an important administrative technology in transatlantic commerce.

Both personal and executive archives, such as diaries and account books, became increasingly important in a world defined by the perpetual movement of things and people.[5] Early modern merchants, sailors, and transatlantic travelers indexed their personal cargos and tracked their long-distance voyages. Such methods ensured not only the value of the things they carried but also ensured their timely delivery. A chest bearing the date 1609 would have been familiar to someone, like Crusoe, traveling across the ocean (FIG. 1 and page 198). A ship's log dating from the 1720s offers a clue to the chest's itinerant history: adhered to the interior of the lid is a diurnal record of a voyage, enumerating distances traveled at sea and observations of latitude. Significantly, the log also suggests the chest's possible use during transatlantic shipments. Two ships are named on the log: the *John of Barnstaple* and the *Bidford*, each linked to an important port for the eighteenth-century American tobacco trade in North Devon. The only named person on these scraps of paper is George Croker, plausibly identified as a mariner of Fremington, North Devon.

Although we can only speculate as to what types of things the chest might have held, objects like it were essential cargo on board ships bound for a transoceanic voyage. In addition to the ship's log, an inventory of goods might have also accompanied the chest. On his journey to Jamaica in 1750, for example, the English planter Thomas Thistlewood (1721–1786) listed in a handwritten inventory the things he stored in his shipping chests, including an impressive collection of books, hats, linens, and even a couple of periwigs ill suited for the tropical climate. Thistlewood, like many entrepreneurial individuals of his time (including the fictional Crusoe),

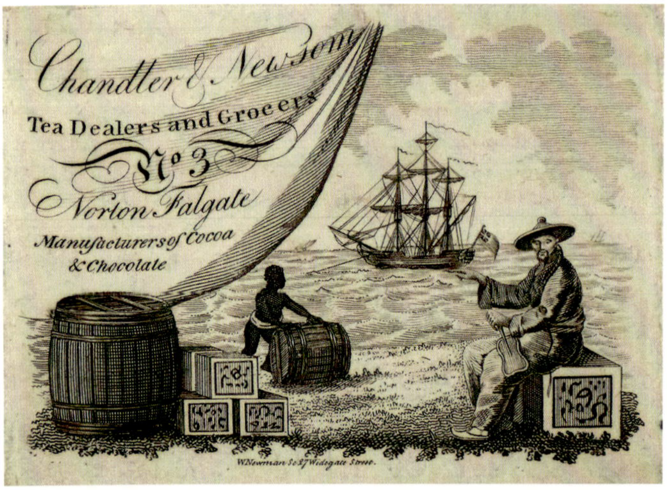

FIG. 2 Trade card for Chandler & Newsom, Tea Dealers and Grocers, London, eighteenth century, engraving. Bodleian Library, John Johnson Collection, University of Oxford

set out from Britain seeking his fortune in the Caribbean sugar economy. His extensive records—comprising inventories, diaries, and account books—shaped his very meticulous world. And, as we will see, they offer some of the most grisly depictions of eighteenth-century plantation slavery.

Even as Britons established more connections across the Atlantic, the West Indies remained by and large a world apart in the popular imagination. On one hand, stories like *Robinson Crusoe* spurred the adventure-driven imagination of British readers, even if the things Crusoe listed would have been strikingly ordinary to them: planks of wood, pieces of iron, loaves of bread, and wheels of cheese, to name a few. On the other hand, West Indian goods like sugar—or, as one critic dubbed it, "the inseparable companion of Tea"—continued to evoke images of the foreign and exotic even as they became everyday staples.[6] Advertisements from the period offer a telling picture: the trade card of the London tea dealers Chandler & Newsom depicts a black slave bringing a hogshead of sugar ashore, and a figure outfitted in ostensibly East Asian costume sits on a box of tea marked with pseudocalligraphic script (FIG. 2). Such images frame distinct foreign types as exotic companions within a single commercial iconography, making plain the associations that Britons made between imported commodities and ideas of foreignness.[7] In other words, sugar produced by enslaved black people came from a far-off world.

At the same time, the seemingly irreconcilable distance between Britain and the sugar colonies—the centers of sugar production in the West Indies—colored the geographic imagination. Illustrated maps often foregrounded the natural riches of the colonies through allegorical imagery. In a map of Jamaica from the 1740s, black slaves labor in a sugarcane field in the lower-right cartouche (FIG. 3). Turtles crawl from beneath the draped canvas bearing the map's title, and a playful mermaid swims near the shore. Yet, while images like these aimed to pique the imagination, they also sought to inform viewers. Indeed, the shells of Caribbean hawksbill turtles were valued for their "strength and beauty" and became fashionable decorative accessories in the seventeenth century, thus contributing to a novel material culture comprising, among other things, combs.[8] A late seventeenth-century tortoiseshell comb case is inscribed with the name of the city of Port Royal, Jamaica, where it was likely crafted, and records the year of its manufacture, 1688 (page 323).[9] This information is not included

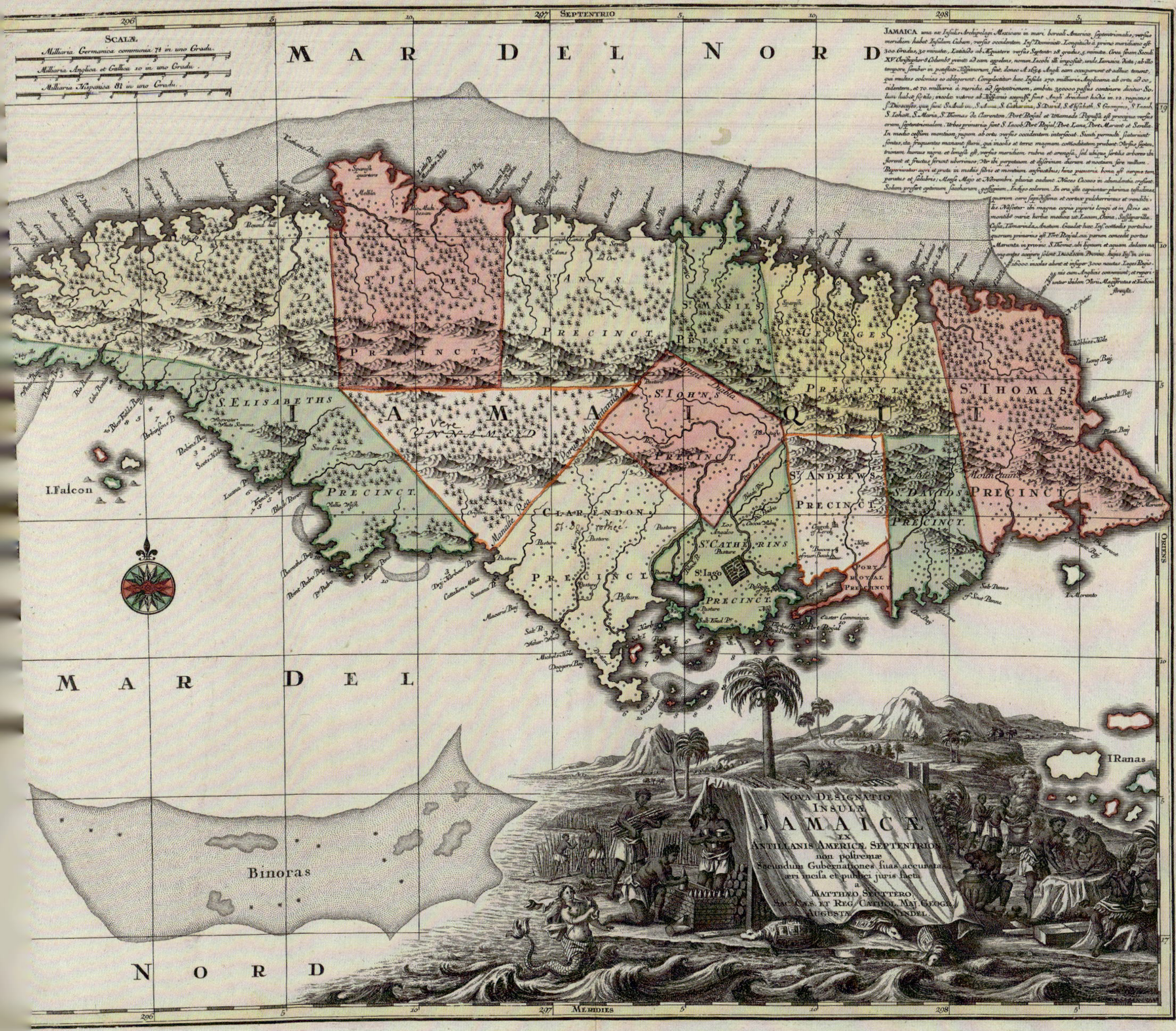

FIG. 3 Matthaeus Seutter, *Nova Designatio Insulae Jamaicae ex Antillanis Americae Septentrion*, ca. 1745, hand-colored map, 19 × 21¾ in. (48 × 55 cm); sheet: 21¾ × 25⅛ in. (55 × 64 cm). Beinecke Rare Book & Manuscript Library, Yale University

arbitrarily. Rather, the indication of both place and time serves the double function of affirming the comb's exotic appeal and its novelty.

The map of Jamaica also informs viewers of sugar's key role in driving the British colonial enterprise in the West Indies. More than any other agricultural commodity, slave-produced sugar shaped new habits and social attitudes in early modern Britain.[10] Refined sugar was widely used as a sweetener and in confectioneries, while rum made from molasses—a by-product of sugar production—was consumed in ever-greater volumes throughout the eighteenth century.[11] The material culture that developed around these imported commodities appealed to the refining tastes of elite and nonelite consumers alike. For example, sugar bowls and punch bowls made of finely decorated earthenware reveal the growing role of sugar and rum in domestic rituals (pages 277, 292, and 387). Nevertheless, these handsome objects and their sweet contents belie the ugly methods of their production, resulting in the very real traumas of plantation slavery.

Human Property

Items like inventories, maps, and even combs can sometimes reveal to us where things come from but they do not always communicate their price. The transatlantic slave trade had both direct and indirect effects on the material world of early modern Britain. To satisfy the demands of British consumers across all levels of society, providers of goods depended on the extraction of increasingly more labor from enslaved Africans who had been forcibly brought by ship to the sugar plantations of the West Indies. This veritable human cargo gives a haunting testament to how disparate experiences of time defined the transatlantic trade: for the enslaved, time was a stolen commodity; for those who profited from their enslavement, time was money.

For West Indian merchants, the traffic in human lives was the foundation of their success. Writing from St. Christopher's (St. Kitts) in 1741, the Bristol merchant Henry Bright informed his business partner, Richard Meyler, of an outgoing shipment that included "thirteen hogsheads of sugar . . . twelve tierces [casks] & three hogsheads of sugar & three barrels of rum." The same letter contained a request for equipment, including "10 grindstones & 10 iron furnaces of the size mentioned with 2 of 25 gallons 2 chests dry bacon 5 hogsheads boiling lime," which Meyler was expected to send with the next shipment from Bristol.[12] Indeed, such transactions were quite ordinary: colonial merchants supplied sugar and other goods for British markets, and their metropolitan connections sent them supplies and provisions. Extraordinary, however, is the magnitude of these individuals' involvement in the shipment of human beings: between 1749 and 1766, Bright's vessels delivered more than five thousand enslaved Africans to the sugar, tobacco, cotton, and indigo plantations of the West Indies and North America.[13]

Merchants like Bright maintained important connections in cities like Bristol, London, and Liverpool, all of which were major gateways to the Atlantic in the seventeenth and eighteenth centuries.[14] In fact, a great deal of slave-trading activity took place in these cities. Many absentee planters—plantation owners who resided in Britain rather than on their plantations—conducted the bulk of their business from the metropole, often operating through merchants and agents involved in the so-called Guinea trade, that is, the African slave trade. The transatlantic connections between West Indian planters and merchants were therefore essential to the vitality of the colonial enterprise.[15] Ships departing from Britain and bound for transatlantic voyages stopped along the coast of West Africa, bringing American- and European-produced goods like tobacco and textiles to exchange for slaves.[16]

In what is commonly called the Triangular Trade, the traffic of enslaved Africans was by all means the foundation of the British Atlantic world. In the seventeenth and eighteenth centuries, British slave traders bought and sold enslaved Africans at every corner of the Atlantic rim. Merchants, ship captains, and sailors helped mobilize a vast movement of enslaved people across the Atlantic Ocean and, not infrequently, either kept slaves for themselves or delivered them to wealthy British families. "Negro sales," held at

FIG. 4 *Elihu Yale; William Cavendish, the Second Duke of Devonshire; Lord James Cavendish; Mr. Tunstal; and an Enslaved Servant*, ca. 1708, oil on canvas, 79 ¼ × 92 ¾ in. (201.3 × 235.6 cm). Yale Center for British Art, Gift of Andrew Cavendish, eleventh Duke of Devonshire

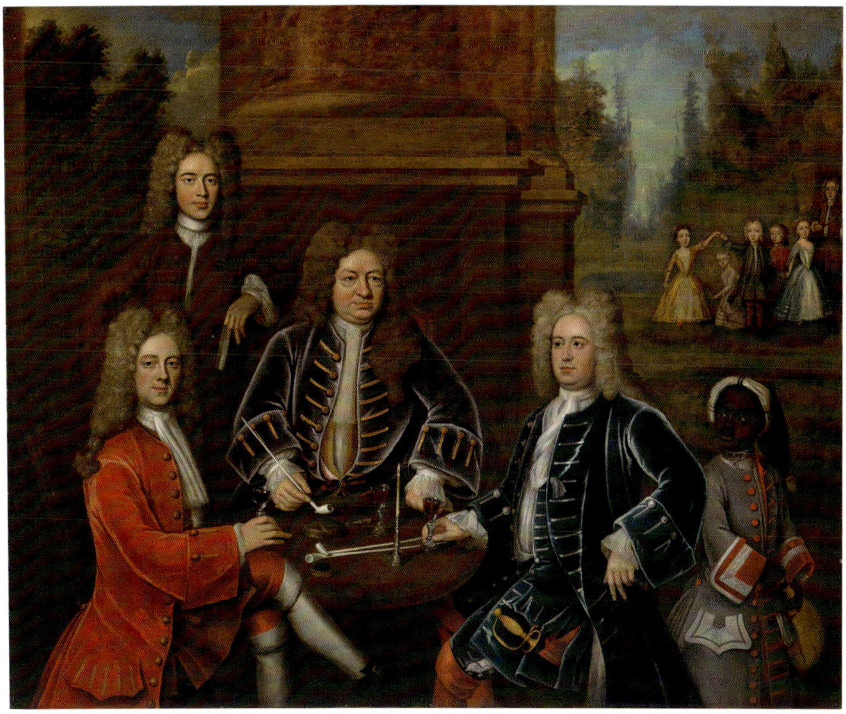

markets, pubs, and coffeehouses, offered private citizens direct access to the slave trade in the metropole.[17]

Slavery in Britain was multifaceted. On one hand, white Britons sought enslaved Africans to fulfill practical roles as domestic servants and skilled laborers.[18] On the other hand, the wealthy employed them as a means of articulating social rank. The pervasiveness of black servants in genteel portraits and conversation pieces of the period attests to this currency.[19] An eighteenth-century group portrait that includes Elihu Yale shows a young black man wearing a silver collar and carrying a wineskin (FIG. 4). The white men appear to be drinking port or Madeira wine, probably sweetened with Caribbean sugar, and are smoking tobacco from clay pipes. A tobacco box lies open on the surface of the table. The presence of an African-descended person and exotic stimulants suggests the wealth and social prestige that Yale had amassed as a slave trader and merchant for the successful East India Company.[20]

Similarly, a view of Bristol from 1734 shows a genteel family with an enslaved servant in the center foreground, advertising the wealth of the city's global economy in the early eighteenth century (FIG. 5). One could imagine that this family compelled the enslaved person to wear some type of constraining device to identify him as property, much like the collar worn by the enslaved boy in Yale's portrait. This device might have resembled an inscribed ankle iron that possibly was manufactured in Bristol (FIG. 7 and page 384). Unlike chained shackles, which would have too closely resembled the brutal constraints of plantation slavery, this manacle was likely meant to communicate ownership rather than restrict movement; it reads simply, "Deverall Corn street Bristoll, 1733."

John Deverell (b. ca. 1696/8; d. 1760) was a surgeon of Bristol. He was the only resident of Corn Street by that name between 1720 and 1740, aside from his first wife, Sarah Raintorp (d. 1729), who was the daughter of the

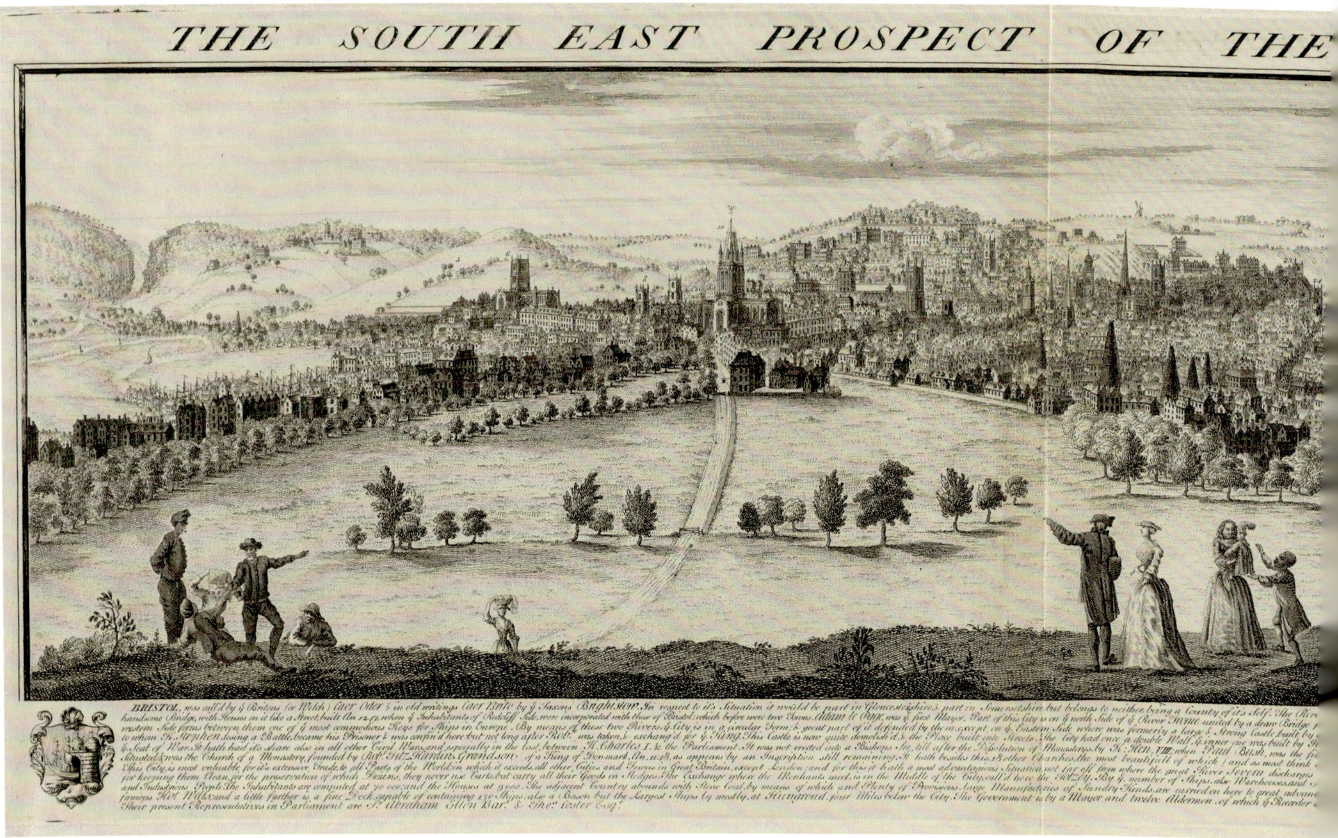

FIG. 5 Samuel Buck, *The South East Prospect of the City of Bristol*, 1734, etching and engraving on laid paper, $12\frac{1}{4} \times 31\frac{1}{2}$ in. (31 × 79.8 cm); sheet: $17\frac{5}{8} \times 33\frac{3}{8}$ in. (44.8 × 84.9 cm). The Lewis Walpole Library, Yale University Topographical Prints Collection

mariner Hugh Raintorp (d. 1718), and Deverell's second wife, Rachel Gwatkin (d. 1779). John had served his apprenticeship as a surgeon under his father, Jeremiah, and between 1724 and 1739 took on five of his own apprentices. There is the possibility that either John or—more likely, given the record of Deverell's continuous presence in Bristol during this period, one of his apprentices—served as a surgeon aboard ships bound for West Africa. Apprentice surgeons were often sent on such trips as part of their training.

A grim perk of this dangerous and miserable job was the right to trade in two slaves on the voyage, one male and one female, for payment over and above the surgeon's regular pay, or "head money," which was based on the number of enslaved people who survived the Middle Passage. Thus the iron would indicate which of the slaves was deemed his property. The other alternative—and perhaps the more feasible one, given the survival of a similar steel manacle bearing the inscription "S. Bosanquet Layton 1746" (FIG. 8)—is that the iron was worn by an enslaved servant in Deverell's household. "S. Bosanquet" can be identified as Samuel Bosanquet (1700–1765) of Leyton, Essex, a near-contemporary of Deverell and a member of a family embedded in the slave trade, who does not seem to have acted as a captain, surgeon, or in any other seafaring capacity with perquisite to trade on a voyage.[21]

Indeed, enslaved people—including children—were not an uncommon sight in eighteenth-century Britain. For the most part, the presence of bonded people in Britain conveyed the development of new social norms, such as the articulation of wealth and status. But in the West Indies, where chattel slavery was the way of life, human bondage was not defined by the

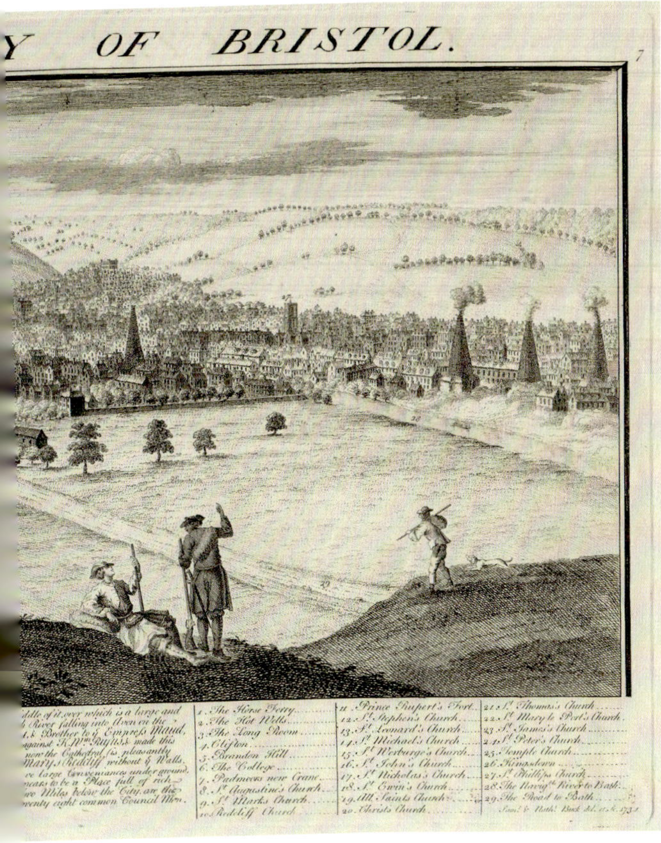

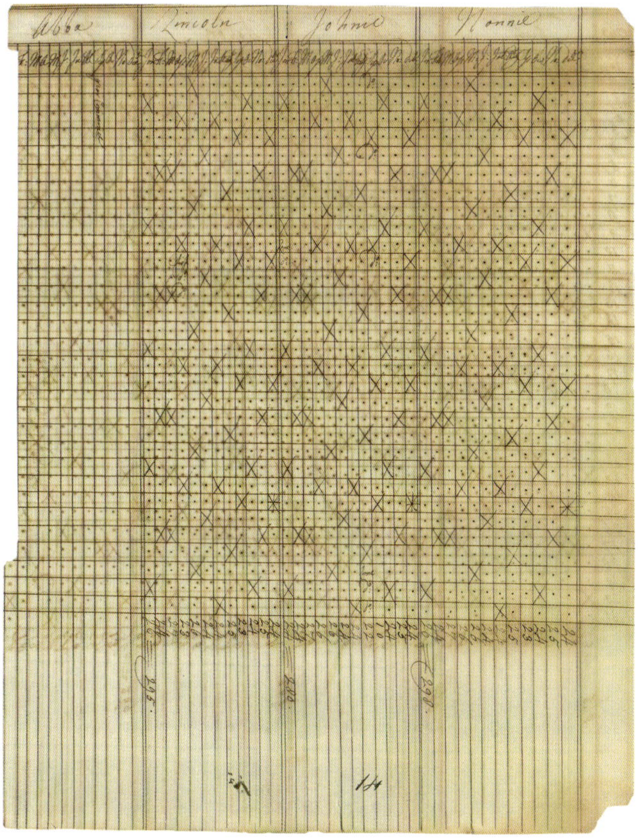

FIG. 6 Thomas Thistlewood, *List of Negroes belonging to Mr. Thomas Thistlewood, formerly of Tupholme Lincolnshire, with a Tabular account of their weekly labour, between 1758 & 1766*, n.p. Beinecke Rare Book & Manuscript Library, Yale University

sentimental preference for fashionable silver collars over iron chains and shackles. Instead, the conditions of slavery were explicitly premised on coercive labor extraction, capital accumulation, and a racial caste system that normalized the wholesale dehumanization of African and African-descended people. The royalist exile Richard Ligon explained in 1657 how Barbadian planters "choose [slaves] out of the Ship...as they do Horses in a Market; the strongest, youthfullest, and most beautiful, yield the greater price."[22] In eighteenth-century Jamaica, Thistlewood was methodical in his selection, writing: "I would choose men-boys and girls, none exceeding 16 or 18 years old...those which are in a moderate condition hold their flesh better and are commonly hardier."[23] The monetary and aesthetic value assigned to enslaved people was therefore contingent on their systematized objectification and the arbitrary evaluation of their physical traits. Put more plainly, slaveholders highly valued those who were young, strong, unmarred, and industrious.[24]

Physical characteristics ostensibly communicated to planters and slaveholders which slaves were most productive—that is to say, which could perform the most work in the least amount of time. To ensure productivity, planters developed sophisticated systems of record- and timekeeping, for example, ledgers that documented slaves' work times and work days (FIG. 6). Planters understood time to be an important asset: "Like money, it could be saved, exchanged, or spent."[25]

Still, levels of productivity varied throughout the year. For example, Thistlewood's records show that July was on average a less productive month than most (it was the beginning of cane-growing season, which was generally a slower period for the plantation).[26] Meanwhile, disease and

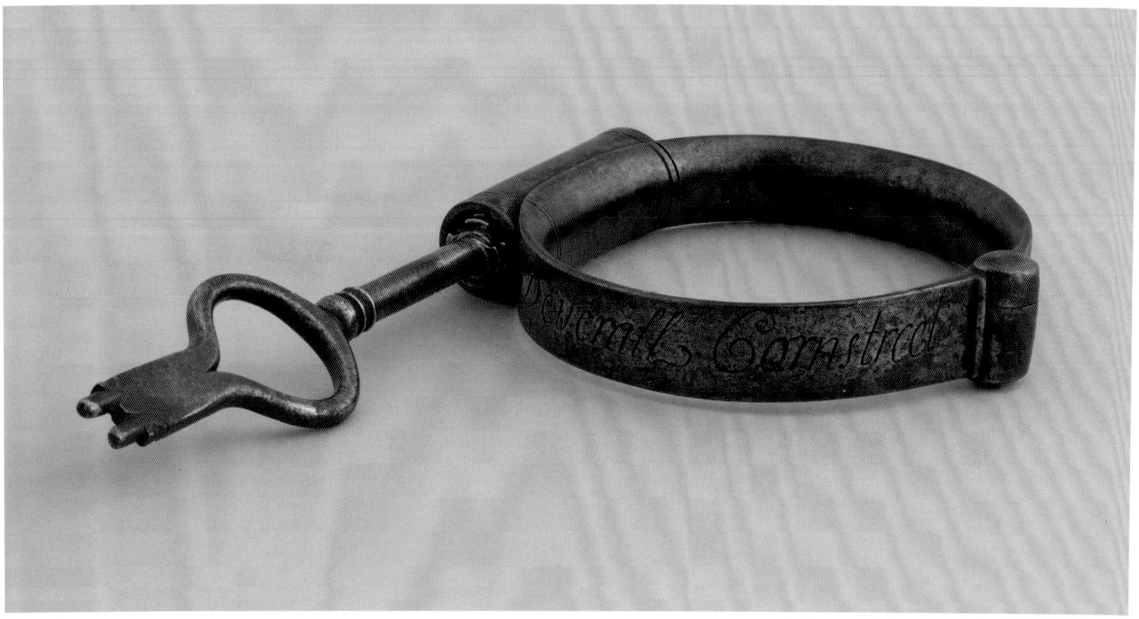

FIG. 7 Ankle iron and key, 1733, steel, $1\frac{1}{4} \times 4\frac{1}{4} \times 3\frac{1}{2}$ in. ($3.1 \times 10.8 \times 8.8$ cm). The Bryan Collection, Lake Bluff, IL

illness had a huge effect on productivity, especially in Jamaica where sickness rates tended to be higher than on other Caribbean islands in the eighteenth century.[27] Hans Sloane, an Irish physician who spent several years in Jamaica, reported cases of yaws, cholera, tertian, chincough (whooping cough), violent fevers, pleurisy, and dropsy, among other illnesses. He also purported to observe the adverse effects of the tropical climate on newcomers, both Africans and Europeans.[28]

In addition to these physical afflictions, the psychological toll of plantation life was considerable. Escape was an important cause of fluctuating productivity rates. Enslaved people frequently ran away from plantations to escape the harsh conditions and cruel treatment. As a consequence, slaveholders and overseers developed draconian methods of enforcing punishment and retribution. Thistlewood, for instance, related how he branded the bodies of those who tried to run away and bound them with devices like collars and yokes to restrict their mobility.[29] Other accounts record the brutal measures slaveholders took to impose order on plantations. In late seventeenth-century Barbados, an English observer related how a mistress "burned her *Negro* for running away."[30]

In Britain, by comparison, newspaper advertisements announced runaways with astounding frequency, attesting to the cruel form of slavery exercised there. These advertisements suggest that constraining devices like collars could aid in identifying runaways. In one instance, an enslaved boy ran away from Colonel Percy Kirke (d. 1691), who, in March 1685, took out an advertisement in the *London Gazette* that read, "He has a Silver Collar about his Neck, upon which is the Colonel's Coat of Arms and Cipher."[31] On another occasion—September 27, 1701—the London newspaper *The Post Man* reported: "A Tall Slender Negro, on *Sunday* last near to a Clock [*sic*] in the morning, absented himself from his Master, viz. John Gandy, at Popler near the Pond."[32] Though the time of day appears in this narrative only incidentally, it indexes yet another example of the pervasive time-governed system that imposed order on people viewed as mere possessions.

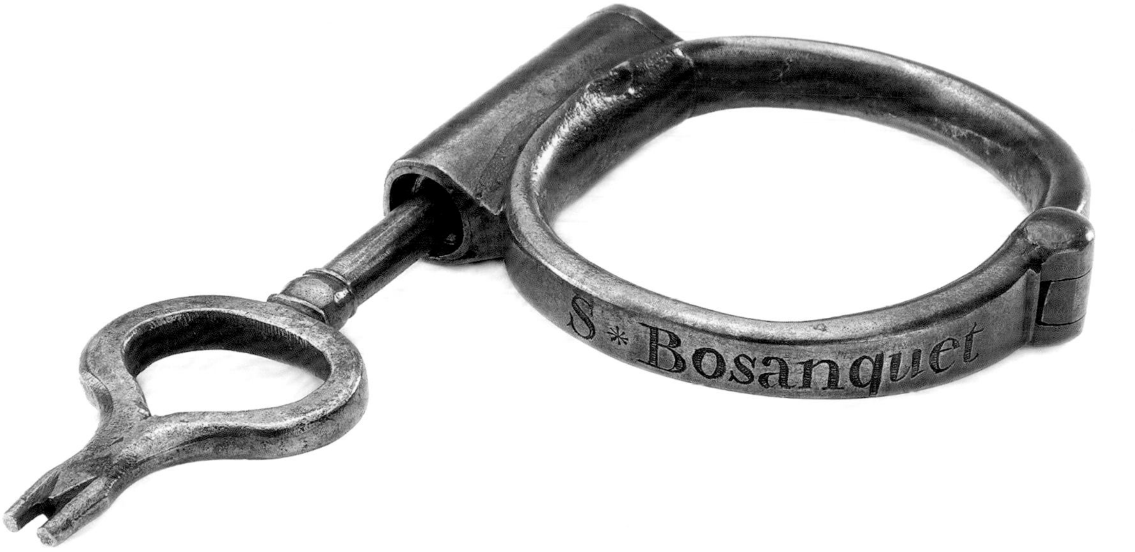

FIG. 8 Ankle iron and key, 1746, steel, $\frac{3}{4} \times 5\frac{7}{8} \times 3\frac{3}{8}$ in. (1.8 × 15 × 8.7 cm). National Maritime Museum, Greenwich, London, Michael Graham-Stewart Slavery Collection

A New World of Goods

The inscribed steel manacles give poignant evidence of how both place and time shaped Britain's experience of the Atlantic world. Like the silver collar in Colonel Kirke's advertisement, they served to prevent escape and ensure accountability. But all these objects communicate the vicious asymmetries of power that fueled the British Atlantic enterprise. Cargo lists, ledgers, and newspaper advertisements provide archival examples of dated materials that imposed discipline and order on enslaved people. Time, understood along these lines, was a systemic means of regulating a rapidly expanding society.

In the eighteenth century, time was both order and currency. The ceremony for the laying of the foundation stone of the Exchange of Bristol on the morning of March 10, 1740, carried this message clearly. It was a momentous occasion. The mayor, Henry Combe, had invited the whole city to attend, with special place given to the Corporation of the City of Bristol and the Society of Merchants. The weather seemed to provide a good omen of things to come: the skies were clear and the air was mild. John Wood, the building's architect, instructed the mayor to throw "some of the newest Coins under the First Stone," a traditional means of confirming the date of the foundation. The inscription on the foundation stone, too, was a temporal marker, recording the reign of George II, the mayoralty of Henry Combe, and the year, 1740.[33]

This new monument stood at the center of a huge network of global exchange that included tobacco from Virginia, sugar from the West Indies, and enslaved people from West and Central Africa. The construction of the Exchange, its inscribed foundation stone, and the coins laid by the mayor all reified a common system of time—indeed, a global system of time. The interior courtyard was decorated with stucco festoons depicting "three quarters of the world," namely Asia, Africa, and America. Such allegories of the continents had long been the subject of sculptures, printed frontispieces, and maps.[34] But, positioning them on this building—in a space where things and people, free and unfree, passed on a regular basis—gave time a new relation to value, an imposition of order that could be felt no matter where one stood in the world.

With sugar, slaves, and other commodities, Britain engaged in a cruel exchange of things and people, giving rise to unprecedented wealth and global power. The things that early modern Britons bought and sold bear testimony to this newfound agency. In essence, time helped to make sense of it all, measuring local events against a rapidly expanding world—but time was also deeply complex. As in the case of Mary Phillips of Barbados, objects transgressed the limits of time and space: her belongings were surely not all locally produced; rather, they illustrate a broad network of production and exchange. For people like Thomas Thistlewood, by comparison, time operated like clockwork, structured around strict systems of order, value, and discipline. In sharp contrast, the countless lives that were lived in enslavement reveal to us that, for others, time was all too often taken away. Whether silver collars or steel shackles, the objects that burdened enslaved people's bodies dramatically defined how they experienced time—indeed, as an unyielding force.

Notes

1. The National Archives, Kew (hereafter TNA), PROB 11/480/233, Will of Mary Phillips, Widow of Saint Thomas and Island of Barbados, February 3, 1705.

2. See Jennifer Van Horn, *The Power of Objects in Eighteenth-Century British America* (Chapel Hill: University of North Carolina Press, 2017).

3. Phillips's will mentions that she held a share of a plantation called Springhead in St. James Parish, Barbados. Slaves worked on this plantation throughout the eighteenth century. See "Spring Head (or Springhead) [Barbados | St James]," Legacies of British Slave-ownership Database, University College London, http://wwwdepts-live.ucl.ac.uk/lbs/estate/view/828.

4. TNA PROB 11/480/233.

5. On the notion of movement and circulation in early modern economic thought, see Lars Magnusson, *Mercantilism: The Shaping of an Economic Language* (New York: Routledge, 1994).

6. Sidney W. Mintz, *Sweetness and Power: The Place of Sugar in Modern History* (New York: Penguin Books, 1986), 114; and Erika Rappaport, *The Thirst for Empire: How Tea Shaped the Modern World* (Princeton, NJ: Princeton University Press, 2017), 23–56.

7. Beverly Lemire, *Global Trade and the Transformation of Consumer Cultures: The Material World Remade, c. 1500–1820* (New York: Cambridge University Press, 2018), 239–40. The trajectory of tobacco provides a valuable lesson; see Kristen G. Brookes, "Inhaling the Alien: Race and Tobacco in Early Modern England," in *Global Traffic: Discourses and Practices of Trade in English Literature and Culture from 1550 to 1700*, ed. Barbara Sebek and Stephen Deng (New York: Palgrave MacMillan, 2008), 157–78.

8. Mark Catesby, *The Natural History of Carolina, Florida, and the Bahama Islands* (London: Benjamin White, 1771), 2:39. On the use of tortoiseshell in furniture objects, see Clive D. Edwards, *Eighteenth-Century Furniture* (Manchester: Manchester University Press, 1996), 97–99.

9. Port Royal remained the principal hub of British Jamaica until a disastrous earthquake left the city in ruins in 1692; see Donald F. Johnson, "Combing the Roots of Colonialism: Jamaican Tortoiseshell Combs and Social Status, 1655–1692," *Winterthur Portfolio* 43, no. 4 (2009): 313–34.

10. See Mintz, *Sweetness and Power*, 131–39.

11. Maxine Berg, *Luxury and Pleasure in Eighteenth-Century Britain* (New York: Oxford University Press, 2005), 224–45.

12. Henry Bright, St. Kitts, to Richard Meyler II, Bristol, April 19, 1741, in *The Bright-Meyler Papers: A Bristol-West India Connection, 1732–1837*, ed. Kenneth Morgan, Records of Social and Economic History: New Series 40 (Oxford, UK: British Academy, Oxford Scholarly Editions Online, 2016), 18, https://www.oxfordscholarlyeditions.com/view/10.1093/actrade/9780197264058.book.1/actrade-9780197264058-div1-25.

13. Morgan, *The Bright-Meyler Papers*, 113.

14. Bristol provides a classic example. For a historical overview of Bristol's involvement in the eighteenth-century Atlantic economy, see Kenneth Morgan, *Bristol and the Atlantic Trade in the Eighteenth Century* (New York: Cambridge University Press, 1993).

15. S. G. Checkland, "Finance for the West Indies, 1780–1815," *Economic History Review* 10, no. 3 (1958): 461–69.

16. Morgan, *Bristol and the Atlantic Trade*, 129. See also Stanley B. Alpern, "What Africans Got for Their Slaves: A Master List of European Trade Goods," *History in Africa* 22 (1995): 5–43.

17. Peter Fryer, *Staying Power: The History of Black People in Britain* (London: Pluto Press, 1984), 59–64. Urban coffeehouses became hubs for affairs related to the West Indies and Africa in the late seventeenth and early eighteenth centuries.

18. Fryer, *Staying Power*, 72–77.

19. See David Dabydeen, *Hogarth's Blacks: Images of Blacks in Eighteenth Century English Art* (Manchester: Manchester University Press, 1987); David Bindman, "Subjectivity and Slavery in Portraiture: From Courtly to Commercial Societies," in *Slave Portraiture in the Atlantic World*, ed. Agnes Lugo-Ortiz and Angela Rosenthal (New York: Cambridge University Press, 2013), 71–88; and Catherine Molineux, *Faces of Perfect Ebony: Encountering Atlantic Slavery in Britain* (Cambridge, MA: Harvard University Press, 2012).

20. See Richard B. Allen, *European Slave Trading in the Indian Ocean, 1500–1850*, Indian Ocean Studies Series 6 (Athens: Ohio University Press, 2014).

21. Douglas Hamilton and Robert J. Blyth, eds., *Representing Slavery: Art, Artefacts and Archives in the Collections of the National Maritime Museum* (London: Lund Humphries, 2007), 59.

22. Richard Ligon, *A True and Exact History of the Island of Barbados*, ed. Karen Ordahl Kupperman (Indianapolis, IN: Hackett Publishing, 2011), 97.

23. Douglas Hall, *In Miserable Slavery: Thomas Thistlewood in Jamaica, 1750–86* (Kingston, Jamaica: University of the West Indies Press, 1999), 119.

24. For an account of how the bodies of enslaved people were evaluated or "read" in nineteenth-century America, see Walter Johnson, *Soul by Soul: Life inside the Antebellum Slave Market* (Cambridge, MA: Harvard University Press, 1999), 135–61; and Daina Ramey Berry, *The Price for Their Pound of Flesh: The Value of the Enslaved, from Womb to the Grave, in the Building of a Nation* (Boston, MA: Beacon Press, 2017).

25. Justin Roberts, *Slavery and the Enlightenment in the British Atlantic, 1750–1807* (New York: Cambridge University Press, 2013), 27.

26. Richard B. Sheridan, *Sugar and Slavery: An Economic History of the British West Indies, 1623–1775* (Kingston, Jamaica: Canoe Press, University of the West Indies, 1974), 111–12; and Richard S. Dunn, *Sugar and Slaves: The Rise of the Planter Class in the English West Indies, 1624–1713* (Chapel Hill: University of North Carolina Press, 1972), 200.

27. Roberts, *Slavery and the Enlightenment*, 67.

28. Hans Sloane, *A Voyage to the Islands Madera, Barbadoes, Nieves, St. Christophers, and Jamaica* (London: B. M., 1707), 1:xc–cliv.

29. Hall, *In Miserable Slavery*, 286.

30. Thomas Tryon, *Friendly Advice to the Gentleman-Planters of the East and West Indies: In Three Parts* (London: Andrew Sowle, 1684), 111.

31. Fryer, *Staying Power*, 22.

32. "Advertisements," *Post Man and the Historical Account*, September 27–September 30, 1701; and Simon P. Newman, "Freedom-Seeking Slaves in England and Scotland, 1700–1780," *English Historical Review* (October 3, 2019), https://doi.org/10.1093/ehr/cez292.

33. John Wood, *A Description of the Exchange of Bristol, Wherein the Ceremony of Laying the First Stone of That Structure* (Bath, UK: J. Leake, 1743), 11–13.

34. See, for example, Valerie Traub, "Mapping the Global Body," in *Early Modern Visual Culture: Representation, Race, and Empire in Renaissance England*, ed. Peter Erickson and Clark Hulse (Philadelphia: University of Pennsylvania Press, 2000), 44–97.

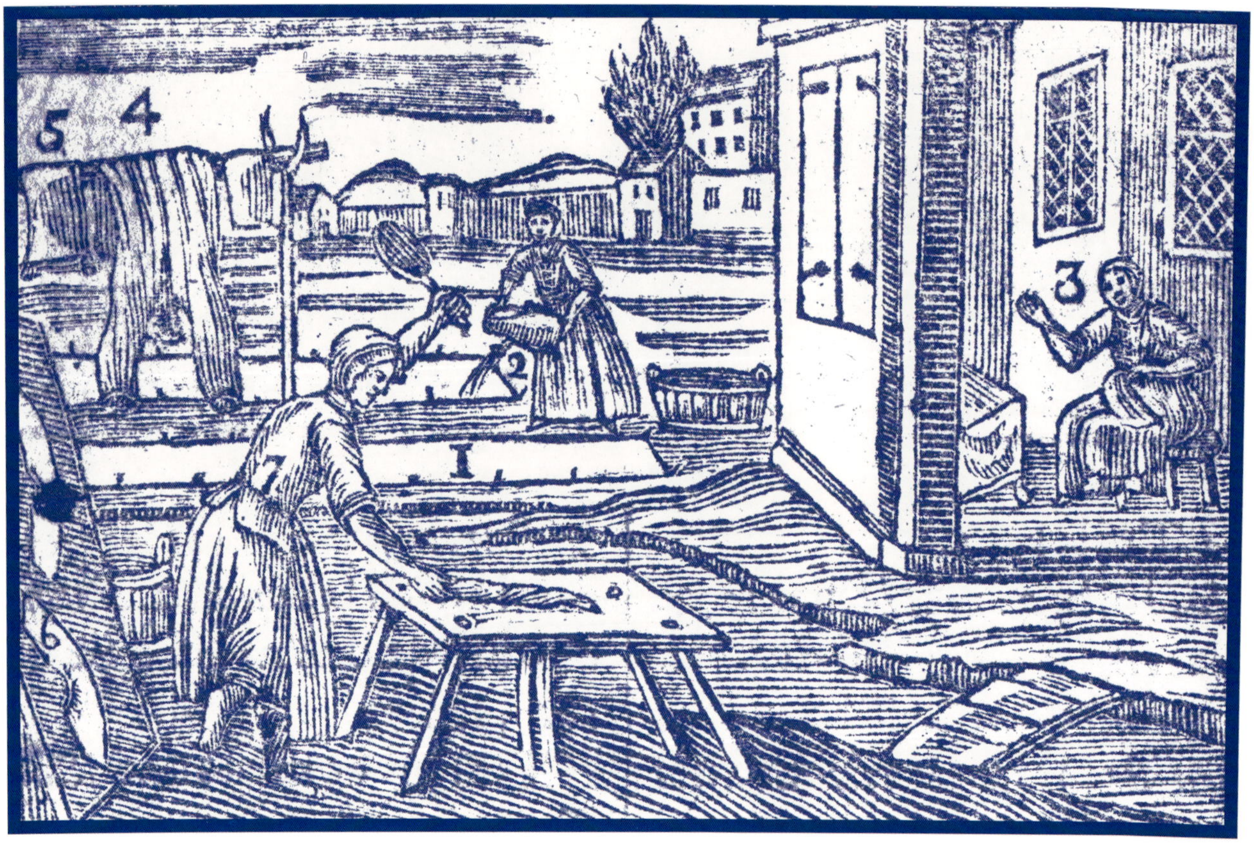
5
4
3
2
1
7
6

Narrative Time

The Intersecting Lives of Early Modern Objects

GAVI LEVY HASKELL

THE ORGANIZING PRINCIPLE OF THE CATALOGUE IN THIS BOOK—THE stages of an early modern life—naturally presents each object through a particular perspective: generally, that of the person who owned it. A cradle is viewed through the implied eyes of the child and, later, through the nostalgic eyes of that child in adulthood. Bed hangings are seen through the eyes of a wife and husband, the decorations perhaps made by the wife herself or a female progenitor; bed hangings also required frequent washing and airing by household servants. By its very nature, this act of focalization suggests the many possible alternative views: the cradle occupied the time not just of the child but also of the child's mother, father, and wet nurse. As the word *focalization* begins to suggest, this essay will approach these alternative views through the lens of narrative, arriving at an understanding of dated objects as temporally and narratively dense, reflecting the intersecting lifetimes of the many people who made, used, and interacted with them.[1]

This density results in part from the structure of the early modern household, which might comprise not only immediate family members but also a large cast of servants, apprentices, and workers. This range of roles captures a broad swath of individuals at many points throughout their lives: for example, as R. C. Richardson notes in *Household Servants in Early Modern England*, a majority of people in this period, especially women, would at some point work as domestic servants.[2] Furthermore, as Justin Brown suggests in his essay (page 43), these objects intersected not only with the lives of people within a household—and within Britain—but also throughout an increasingly global network of trade and enslaved labor, especially in the Caribbean. This reality was far from invisible even in its time: British families, especially the wealthy, sometimes purchased enslaved African people to work as household servants.

The notion of narrative time offers an opportunity to explore some of these intersecting lives and to imagine the richness of the early modern world as reflected in its material culture. Narrative often implies dialogue, but such accounts from the socially peripheral, if physically and economically central, individuals who are most of interest here are hard to come by. Witness depositions offer one of the richest sources. Naturally they are imperfect records, even of those whose stories they purport to tell. Court clerks no doubt filtered the information they transcribed, and witnesses themselves shaped their answers, the facts they emphasized, and even their speaking styles for the contexts in which they spoke.[3] In *Road to Divorce*, Lawrence Stone furthermore points out that "bribery, either to give false testimony or not to appear in court at all, was inevitably a common occurrence in a situation in which the reward could easily amount to a year's wages." Witnesses were rarely tried, and even less often convicted, for providing false evidence.[4] Depositions, however, still provide useful information for understanding early modern lives, offering narratives in the voices of people who might otherwise remain invisible to history.

The record itself is shaped by time. Many depositions cite specific hours "of the Clock." In one electronic corpus of English-language depositions from 1560 to 1760, there are 161 instances of time-related language.[5] Roughly speaking, those deposed cite hours of the day unevenly: the most frequent mentions of time begin at the hour of 1 a.m. and gradually increase to 9 p.m., after which they quickly taper off, with sharp rises and plunges at particular hours.[6] This pattern may point either to a conventional preference for certain times over others when estimating time of day or to an accurate daily rhythm of witnessing. Similarly, the fact that reports of the same event do often cite the same hours may suggest either accurate recall or the narrator's awareness of the temporal story expected by the court (or paid for by a disputant).[7] One deposition from a twenty-four-year-old laborer in 1654 includes a fascinatingly precise division of measured time: "lesse then halfe a quarter of an houre."[8] While the legal system relied on and no doubt to a certain extent induced it, such specificity nonetheless reveals a broad awareness of measurements of time among husbandmen, gentlemen, and widows.[9]

Depositions, while suggesting the prevalence of time keeping, also hint at some of the narratives enfolded in the objects presented in this book. Servants' interventions in the lives of objects—their use, preparation, and repair—can in a sense loop some of these narratives back on themselves, increasing the density of the narrative. The stories of objects include not only the stories of those who owned them but also those of the owners' servants and many others. After all, in the early modern world, many people worked as servants only for a limited period and then went on to become tradespeople, wives of tradesmen, or other members of the middling sort.[10]

Some objects in this book readily reveal the presence of servants: an oak cradle (page 112), crewelwork bed hangings (page 152), and a mechanical spit jack (page 214). The first two objects relate to laundering, one of the continual tasks of the early modern household and, at the time, a complex and arduous undertaking. In 1770, Anne Barker, who "for many Years discharged the Office of Housekeeper in the most respectable Families," wrote a guide entitled *The Complete Servant Maid: Or Young Woman's Best Companion*.[11] Between the sections on lady's maids, chambermaids, and laundry maids, nearly half of the volume comprises instructions for washing textiles.[12] Of course, this advice may have been aspirational as much as practical, and some of Barker's recommendations (especially concerning behavior) seem to fall into this category. However, the steps she enumerates and the cautions she offers hint at the realities of the tasks. Washing linen required first mending it, "otherwise if it is washed, the rents will be much worse than before."[13] Although clothes were washed in tubs, water was often collected from a stream or communal pump. Barker warns her reader not to wash "with any sort of water, unless it has stood two or three days, for when newly taken in, it is always thick and muddy; if it is from a stream where there is a muddy bottom, it will be better to let it stand four days."[14] This part of the process made laundering a task that required travel and heavy lifting in addition to the work of scouring itself. The need for repetitive laundering often required routine travel to public or semipublic places that had access to water. As such, the work not only occupied the time of the women who performed it, it also provided them opportunities to interact with neighbors and friends, adding to the wealth of stories captured in the narrative time of the textiles they washed.

As previously suggested, the task of washing laundry fell to different people in different households—and sometimes to more than one person. Small, modestly wealthy households might only have one servant (usually female) who would act as housemaid, laundry maid, and nursemaid all in one.[15] Larger households might have a separate laundry maid, but even in this case she would not necessarily do all the washing. The washing of silks and laces and spot treatments on finer clothes might fall under the auspices of a lady's maid or a chambermaid. Even with the work so divided, laundering required substantial time and energy, and servants would sometimes hire day laborers to assist them—evidenced by the fact that Barker warns against it, writing, "never take a woman to assist you without leave from your mistress."[16]

Although bed hangings, unlike linens, would likely be washed only rarely, they and the cradle suggest two of the main sources of early modern laundry: beds and babies. Infants, in particular, would have produced considerable quantities of dirty laundry. Parents, nurses, or maids—who often took on childcare duties in addition to their other tasks—wrapped infants tightly in cloth.[17] Such swaddling is evident at every level of the socioeconomic scale in early modern Britain, including in the painting popularly known as *The Cholmondeley Ladies* (FIG. 9), in which two lavishly dressed women in elaborate ruffs sit in bed holding tightly swaddled babies. This wrapping—like a long linen bandage—had to be changed frequently. The surgeon Robert Barret suggested in 1699 that "if the Nurse

FIG. 9 *The Cholmondeley Ladies*, ca. 1600–10, oil on panel, 34 7/8 × 67 7/8 in. (88.6 × 172.3 cm). Tate, London

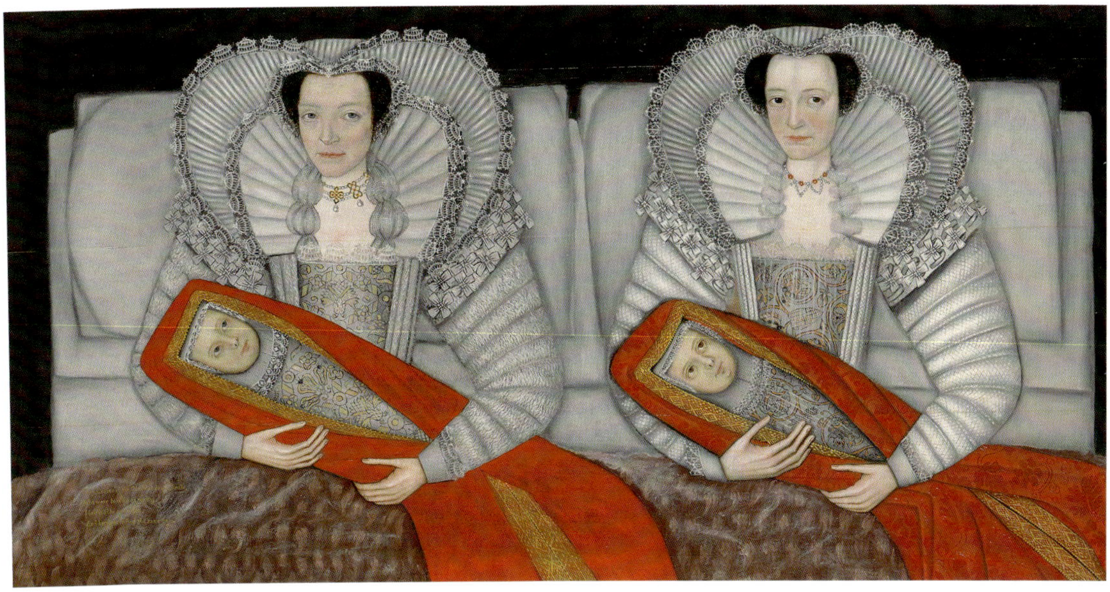

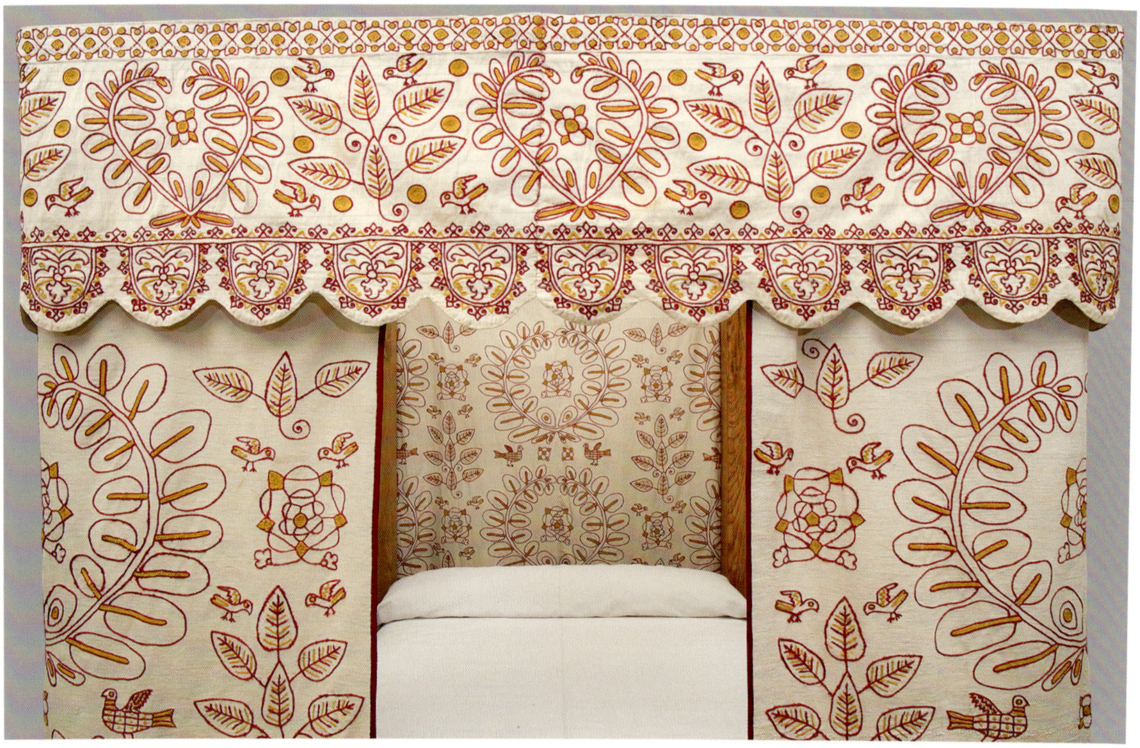

FIG. 10 Crewelwork bed hangings, 1670, wool thread on cotton; head cloth: 67 × 56 in. (170.2 × 142.2 cm); pelmets: $15\frac{3}{4}$ × 103 in. (40 × 261.6 cm); side panels: 65 × $37\frac{1}{4}$ in. (165.1 × 94.6 cm). The Bryan Collection, Lake Bluff, IL

be cleanly, and mind her Business, she will shift the Child three or four times a Day, or oftener."[18] Even given the relatively small amount of cloth required—such that households might easily have kept more than one set—swaddling clothes would have produced the need for a great amount of arduous and probably unpleasant laundering.

In addition to suggesting a cast of characters—chambermaids, laundry maids, wet nurses, midwives, and so on—the objects themselves also hold the histories of their own making, another layer in the overlapping narratives they contain. It required considerable knowledge to craft a full set of bed hangings. Among the accomplishments quoted in the unauthorized version of Hannah Woolley's work *The Gentlewomans Companion* (1673) is her skill in making "feathers of Crewel for the corner of Beds," which she claims to have acquired while working as a governess for a "Noble Lady in this Kingdom" (possibly Anne, Lady Maynard of Little Easton, Essex, d. 1647, or Anne's daughter Anne, Lady Wroth, 1632–1677).[19] There were professional embroiderers—both male and female. Embroidery as elaborate as that on the bed hangings shown (FIG. 10), however, is much more likely to have been the work of a gentlewoman's daughter, who would have had time to work them for her own use. She might have learned the skill from her family or from the increasing number of pattern books available at the time. Many crewelwork designs of the period also borrow from Indian textiles brought to Britain (page 195), suggesting the interweaving of a far-distant person's knowledge and the time required for the labor into the narrative of the bed hangings.

The spit jack shown on page 214 also represents the acquisition of complex knowledge by its maker, Lewes Predham (active 1697; d. 1749), a Devon clockmaker.[20] Predham would have learned the trade as an apprentice to another clockmaker, who would have learned his trade the same way. In turn, his son Noah worked as a clockmaker in Devon. As with the embroidered bed hangings, this making of this device relies on cumulative knowledge and invention. The long history of clockmakers—who devised, honed, and passed on the technologies that drive the spit jack, from the flywheel that stores its energy to the gears that direct it—converges in its mechanism, drawing together a network of narrative time that passed from master to apprentice.

Recipes, or "receipts," represent another apparent instance of knowledge embedded in household objects.[21] Both cooking and healing—skills for the hearth and for the sickbed—required women to learn to combine ingredients, often seasonally dependent, to make food and medicine. This knowledge might have been acquired in various ways. Books of recipes did exist—with cooking and medicine sometimes covered in the same volume—and recipes were also included in more general books of advice.[22] Wealthier women might keep their own manuscripts of recipes, gleaned from various sources, and even pass on the resulting collections to others.[23] Even for women who could read, however, these volumes likely only supplemented knowledge that they gained in other ways, such as from friends or through household service. This knowledge thus indicates another set of stories about the lives of the women who possessed it—including the wife of the house and the servants, each with her own recipes at her disposal.

The weight of overlapping narratives, of layered human time, manifested itself in early modern anxieties. These anxieties often collected at sites of particular narrative density—spaces also of physical and social vulnerability—beds and cradles among them. Both types of objects bore witness to some of the most momentous, and dangerous, occasions in an early modern life—birth and death, joyfulness and sickness—involving the interaction of a potentially disparate array of people. In these vulnerable moments, trust would have been a scarce virtue. Servants often worked in a household for only a year, sometimes less, and during that time might well bring strangers into the home or themselves commit crimes.[24] They could, too, act as witnesses either for or against the other occupants of the house, exposing the household to casual surveillance. This pragmatic anxiety could translate into a more metaphysical one. One purported case of witchcraft in 1618 involved a woman who was dismissed from her job as a laundry maid for stealing a glove. The woman had passed the glove on to her mother, who, aping the act of laundering, reportedly rubbed it over the back of her familiar—an animal believed to have supernatural powers and to assist with magic. The mother then plunged the glove in boiling water, poked holes in it with a needle, and buried it, thereby, allegedly, making its owner ill.[25]

Many similar claims of witchcraft focused on objects like cradles and beds, related to those vulnerable moments that also would require a large cast of hired assistants. The nebulous danger of introducing potentially unpredictable strangers into the intimate workings of the household transformed into a more tangible threat of supernatural violence. A book printed in 1579 records an allegation against a suspected witch executed in Chelmsford. After a woman refused to give the accused woman some yeast, her

> yonge child in the Cradle was taken vehemently sicke, in a meruelous strange maner, wherevppon the mother of the childe tooke it vp in her armes to comforte it, whiche beyng doen, the Cradle rocked of

> it self, sixe or seuen tymes, in presence of one of the Earle of Surreis gentilmen, who seyng it stabbed his dagger three or fower tymes into the Cradle ere it staied: Merily iestyng and saiyng, that he would kill the Deuill, if he would bee rocked there.[26]

Thus, concerns about the often-unobserved people who interact with objects, like cradles, in vulnerable points of the home—providing the narrative weight that makes these objects of interest—appear through the more direct pretext of a supernatural threat.

The narrative time enfolded in the objects described here has a geographic reach. All are touched in one way or another by stories that extend beyond Britain: the designs of Indian textiles inspired the bed hangings; the spit jack would have enabled the preparation of spices and teas secured through trade and nascent imperial efforts. Furthermore, the British prosperity that purchased all these belongings resulted from international trade and enslaved labor. Some objects, however, have much more immediate relationships to the labor and materials that Britain drew from the world. Three staves in the Bryan Collection—two of lignum vitae and one of mahogany—served as signs of power (pages 390–92). Each incorporates wood likely imported from Jamaica or Belize. In the eighteenth century, the British used enslaved African labor to find and collect hardwoods in the Caribbean. A skilled searcher, known as a huntsman, would find a tree, and another person would cut it down. Others would trim the logs and send them down the river.[27] The process not only produced valuable wood for export, but it also cleared land that the British would later use to cultivate sugar and other agricultural products.

Each of the enslaved people who worked on the timber for these staves—as well as the merchants, traders, carvers, and others who handled the wood on its journey—adds to the objects' narrative density. In contrast to many of these people, who lived constrained lives, the staves arrived in England and were elevated to symbols of power. Each of the owners—two mayors, "J. L." of Huntingdon and Edward Markland of Leeds, and Daniel Munton, an inspector of the pavement—recorded his name on his staff's brass fitting. Each of these men held political or guild authority and each carried a truncheon, a weapon signifying power backed by violence.

The View from the Nineteenth Century

Many popular notions about early modern Britain have their roots in the nineteenth century. Some stem from a tacit assumption that the British past is monolithic, that the conditions of domestic life during the Victorian period also apply to the early modern period. The center of the home—which was the bed in the early modern period—relocated to the hearth in the Victorian era. This shift was captured, in all its gendered ideology, in Coventry Patmore's popular poem *The Angel in the House* (1854–62).[28] Similarly, the patterns of early modern domestic service, mentioned previously, must be distinguished from the professionalized model of the nineteenth and twentieth centuries.[29] Television series like *Upstairs, Downstairs* and *Downton Abbey* more often represent the latter style of service, suggesting to many that it was a timeless historical norm.

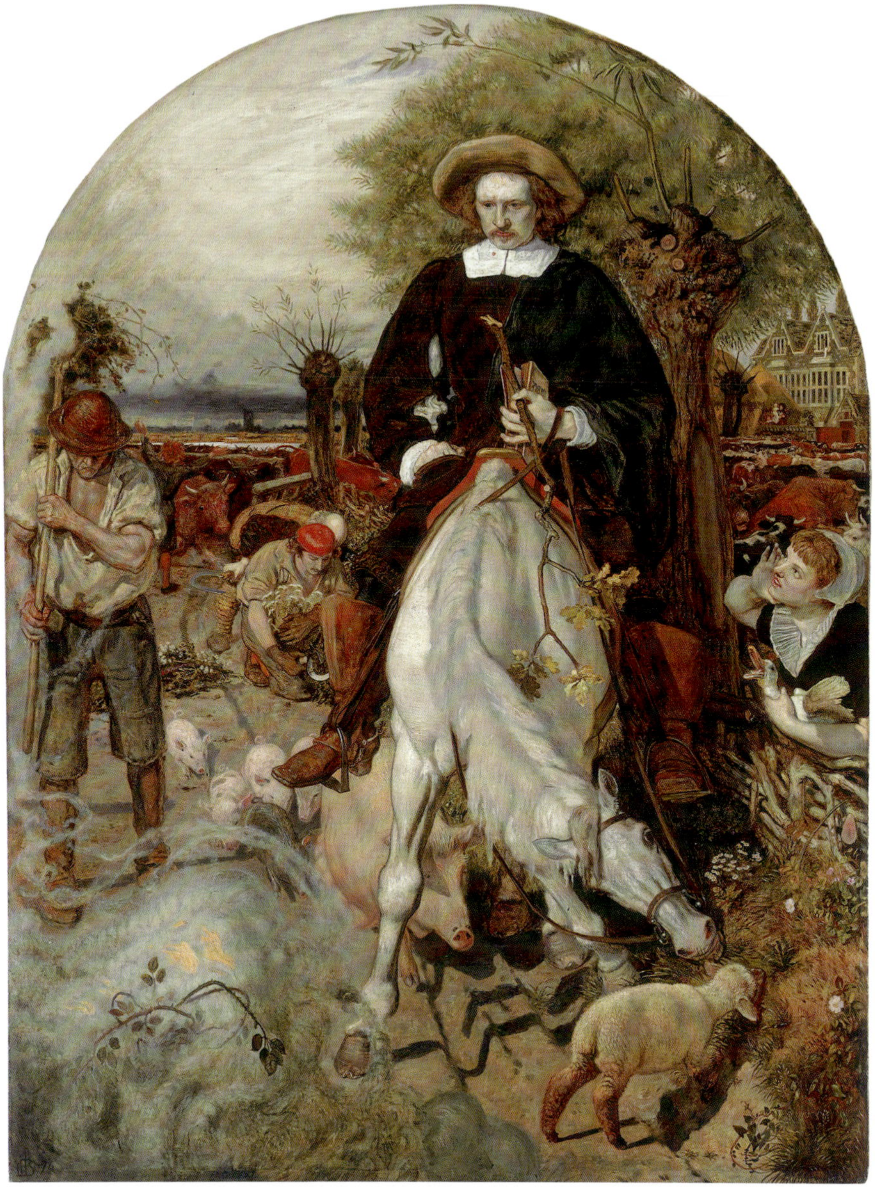

FIG. 11 Ford Madox Brown, *Cromwell on His Farm*, 1874, oil on canvas, $56\frac{1}{4} \times 41$ in. (143 × 104.3 cm). National Museums Liverpool, Lady Lever Art Gallery

Some notions of early modern life were forged in the Victorian period itself. The historical novelist Sir Walter Scott may have popularized a romantic image of the medieval British past—amplified by Thomas Carlyle, John Ruskin, and others—but Scott, inspired by the early modern books he edited for Victorian publication, also depicted early modern Britain.[30] His romantic view of the period percolated into Victorian visual culture, which often replicated his emphasis on the English Civil War and the Elizabethan reign. Painter Ford Madox Brown took up Carlyle's interest in Oliver Cromwell in paintings like *Cromwell on His Farm* (FIG. 11), which monumentalizes Cromwell's life prior to the start of his political career. Alongside other members of the St. John's Wood Clique, William Frederick Yeames painted dramatic reenactments of historic and imaginary moments in sixteenth- and seventeenth-century British history. His painting *Amy Robsart* (FIG. 12) was probably inspired by Scott's novel

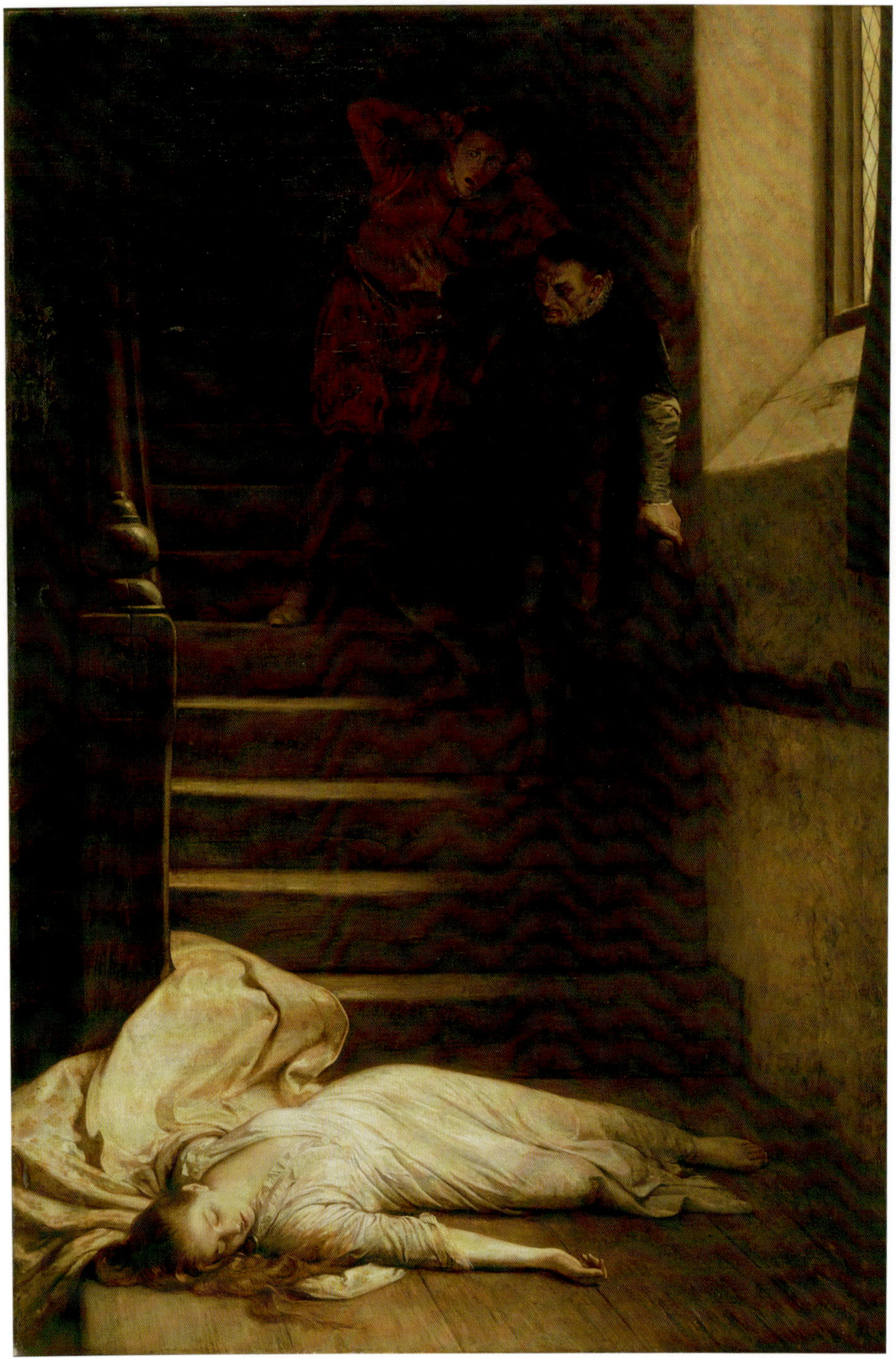

FIG. 12 William Frederick Yeames, *Amy Robsart*, 1877, oil on canvas, 110 7/8 × 74 1/4 in. (281.5 × 188.5 cm). Tate, London

Kenilworth; another work, *And When Did You Last See Your Father?* (FIG. 13), depicts Parliamentarian soldiers interrogating a Royalist child as his mother and sisters nervously look on.

This Victorian interest in early modern life no doubt aided in the survival of many of the objects presented in this volume. Without great care (or benign neglect and good luck), potentially ephemeral belongings—some of them cloth or wood and therefore prone to pests and the usual environmental threats—would have perished. Adding to the layers of human

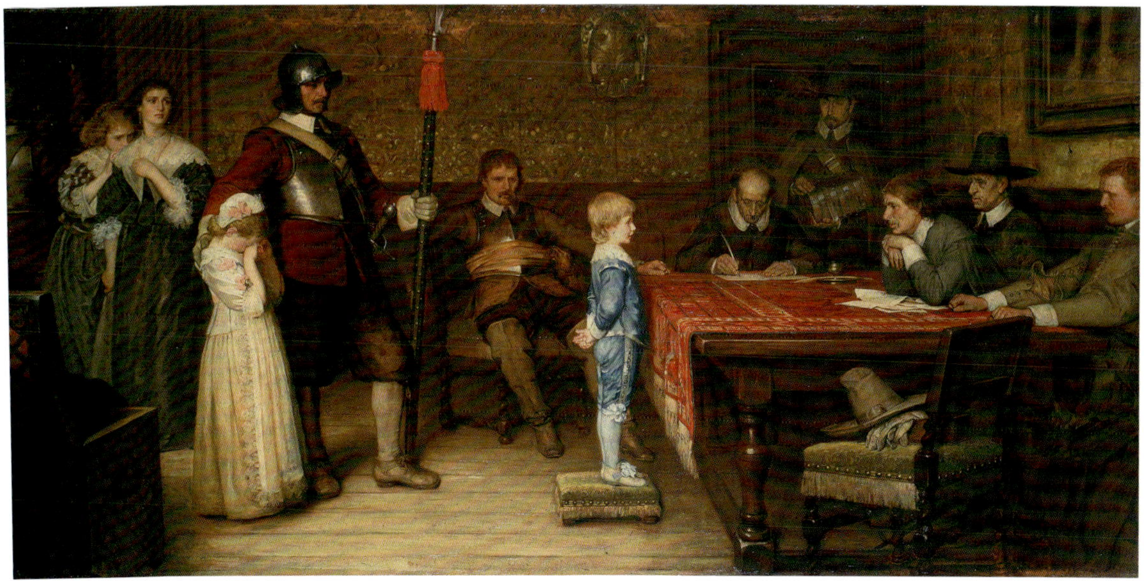

FIG. 13 William Frederick Yeames, *And When Did You Last See Your Father?*, 1878, oil on canvas, $51\frac{1}{2} \times 99$ in. (131 × 251.5 cm). National Museums Liverpool, Walker Art Gallery

narrative imbued in each object, then, are the narratives of the descendants, collectors, and others who shepherded it into the present day, after its useful life was done.

Notes

1. Narrative is often defined in terms of temporality. In 1979, a symposium at the University of Chicago described it as the "illusion of sequence." See W. J. T. Mitchell, "Editor's Note: On Narrative," *Critical Inquiry* 7, no. 1 (1980): 1–4; and Paul Ricoeur, "Narrative Time," *Critical Inquiry* 7, no. 1 (1980): 169–90.

2. R. C. Richardson, *Household Servants in Early Modern England* (Manchester: Manchester University Press, 2010).

3. Tom Johnson, "The Preconstruction of Witness Testimony: Law and Social Discourse in England before the Reformation," *Law and History Review* 32, no. 1 (February 2014): 127–47. In addition to the errors made by clerks, Johnson has pointed out that early modern witnesses often shaped their statements to suit what they believed the courts wanted to hear. The possibility that court clerks unintentionally misreported depositions resonates with recent concerns about court reporting and African American Vernacular English; see Taylor Jones and others, "Testifying while Black: An Experimental Study of Court Reporter Accuracy in Transcription of African American English," *Language* 95, no. 2 (June 2019), 1–37.

4. Lawrence Stone, *Road to Divorce: England 1530–1987* (Oxford, UK: Clarendon Press, 1990), 30.

5. Merja Kytö, Peter J. Grund, and Terry Walker, "An Electronic Text Edition of Depositions 1560–1760," in *Testifying to Language and Life in Early Modern England* (Amsterdam: John Benjamins Publishing, 2011), CD-ROM. The phrasing, it should be noted, included "o'Clock," "a Clock," and "of the Clock," among others.

6. Kytö, Grund, and Walker, "An Electronic Text Edition of Depositions."

7. Mentions of days of the week also reveal variation. The tally is Monday: 80; Tuesday: 30; Wednesday: 40; Thursday: 68; Friday: 58; Saturday: 60; Sunday: 53. Kytö, Grund, and Walker, "An Electronic Text Edition of Depositions."

8. Kytö, Grund, and Walker, "An Electronic Text Edition of Depositions," F_3EC_Colchester_009. Although several depositions refer to fractions of minutes, none refers to seconds. Some clocks, despite their imprecision, marked seconds—pendulum clocks, which could keep time by this unit, were invented about 1657 by Christiaan Huygens—but the marking of time in seconds remained largely inexact and unavailable to laypeople through the early modern period.

9. For more on the meaning and recording of clock time in the early modern period, drawn from a broader range of evidence, see Paul Glennie and Nigel Thrift, "Clock Times in Everyday Lives," in *A History of Timekeeping in England and Wales, 1300–1800* (Oxford: Oxford University Press, 2009), 181–243.

10. For exceptions to this "life-cycle model" of service, see Charmian Mansell, "The Variety of Women's Experiences as Servants in England (1548–1649): Evidence from Church Court Depositions," *Continuity and Change* 33, no. 3 (December 2018): 315–38.

11. Anne Barker, *The Complete Servant Maid: Or Young Woman's Best Companion* (London: J. Cooke, 1770), title page. Notably the guide is aimed at "young women" in general—as stated, many spent time as servants in their youth and went on to other careers or marriage, as the author's name on the title page ("Mrs. Anne Barker") is perhaps meant to suggest.

12. Barker, *The Complete Servant Maid*, 13–14, 18–29, and 40–43. Eighteen of forty-three pages are devoted to the topic of washing textiles.

13. Barker, *The Complete Servant Maid*, 40.

14. Barker, *The Complete Servant Maid*, 41. For a more contemporaneous account of tubs, with the two objects described, see also Hannah Woolley, *The Gentlewomans Companion; Or, a Guide to the Female Sex* (London: Dorman Newman, 1673), 215–16.

15. Peter Earle, *The Making of the English Middle Class: Business, Society, and Family Life in London, 1660–1730* (Berkeley: University of California Press, 1989), 219. Of the households Earle tallies in two parishes of the London Marriage Duties Assessments, more than half have only one servant. Of these households, 96 percent have a female servant, while only 4 percent have a male servant.

16. Barker, *The Complete Servant Maid*, 40. By the end of the early modern period, if not before, women seem to have begun hiring day laborers—in lieu of having a servant's assistance—as the main way of accomplishing the task of laundering. Dorothy Wordsworth writes to Jane Pollard, "We were a whole month without servant, but now we have got one of the nicest girls I ever saw; she suits us exactly, and I have all my domestic concerns so arranged that everything goes on with the utmost regularity. We wash once a month. I hire a woman, to whom I give ninepence for one day, to wash, on the next we have got the clothes dried and on the third have finished ironing." Wordsworth to Pollard, November 30, 1795, in *Letters of Dorothy Wordsworth: A Selection*, ed. Alan G. Hill (Oxford, UK: Clarendon Press, 1985), 200.

17. Jane Sharp, *The Midwives Book: Or the Whole Art of Midwifery Discovered* (London: Simon Miller, 1671), 372. The process called "swaddling" would begin almost as soon as the infant was born. "Take notice of all the parts of it, and see all be right; and the Midwife must handle it very tenderly and wash the body with warm wine, then when it is dry roul it up with soft cloths, and lay it into the Cradle: but in the swadling of it be sure that all parts be bound up in their due place and order gently, without any crookedness, or rugged foldings; for infants are tender twigs, and as you use them, so they will grow straight or crooked."

18. Robert Barret, *A Companion for Midwives, Child-Bearing Women, and Nurses Directing Them How to Perform Their Respective Offices* (London: Thomas Ax, 1699), 80. Barret's advice should perhaps be taken with a grain of salt, as he was a surgeon rather than a midwife.

19. Woolley, *The Gentlewomans Companion*, 11–13.

20. The inscription on the object reads, "JD/Lewes Predham Sandford fecit: 1697." Elsewhere the name appears as "Lewis Pridham" or "Lewis Prydham."

21. For an etymology and historical usages, see *OED Online*, s.v. "receipt, *n*.," IV 12 and 14, https://www.oed.com/view/Entry/159401?rskey=FWvq1s&result=1#eid; s.v. "recipe, *n*.," 1 and 2, https://www.oed.com/view/Entry/159522?rskey=u9BELU&result=1#eid.

22. See, for example, Gervase Markham, *The English Huswife* (London: Roger Jackson, 1615).

23. For more on this practice of manuscript writing, see Sara Pennell, "Perfecting Practice? Women, Manuscript Recipes and Knowledge in Early Modern England," in *Early Modern Women's Manuscript Writing: Selected Papers from the Trinity/Trent Colloquium*, ed. Victoria E. Burke and Jonathan Gibson (London: Routledge, 2004), 237–58; Elaine Leong, "Collecting Knowledge for the Family: Recipes, Gender and Practical Knowledge in the Early Modern English Household," *Centaurus* 55, no. 2 (May 2013): 81–103; and Elaine Leong, "'Herbals She Peruseth': Reading Medicine in Early Modern England," *Renaissance Studies* 28, no. 4 (September 2014): 556–78.

24. Amy Froide, *Never Married: Singlewomen in Early Modern England* (Oxford: Oxford University Press, 2005), 90. Froide cites "a very high turnover rate, with servants rarely working for an employer for more than a year, and some working for just a few months in one household before they moved on." For more on crimes committed by servants, see J. M. Beattie, *Policing and Punishment in London, 1660–1750: Urban Crime and the Limits of Terror* (Oxford: Oxford University Press, 2001), esp. 35–39.

25. *The Wonderful Discoverie of the Witchcraft of Margaret and Phillip Flower, daughters of Joan Flower neere Bever Castle: Executed at Lincolne, March 11, 1618* (London: 1619).

26. *A Detection of Damnable Driftes, Practized by Three Witches Arraigned at Chelmifforde in Essex, at the Laste Assises there Holden, Whiche Were Executed in Aprill* (London: Edward White, 1579), 15; a somewhat similar event in Chelmsford was recorded in 1652. Even more violently, in 1606, a witch supposedly induced "the cradle wherein [a] child lay… [to be] trowne ouer shattered all to péeces, the child vpon the face whelmed vnder it, & killed"; see *The Witch of Wapping, Or an Exact and Perfect Relation, of the Life and Devilish Practises of Joan Peterson* (London: Th. Spring, 1652), 5–6. See also *The Most Cruell and Bloudy Murther Committed by an Inkeepers Wife, Called Annis Dell,*

and Her Sonne George Dell, Foure Yeeres Since (London: William Firebrand and John Wright, 1606), 23–24.

27. Jennifer L. Anderson, *Mahogany: The Costs of Luxury in Early America* (Cambridge, MA: Harvard University Press, 2012), esp. 156–83.

28. Coventry Patmore, *The Angel in the House* (London: Macmillan, 1863).

29. See Peter Laslett, *Family Life and Illicit Love in Earlier Generations* (Cambridge: Cambridge University Press, 1977); Ann Kussmaul, *Servants in Husbandry in Early Modern England* (Cambridge: Cambridge University Press, 1981).

30. Walter Scott, *Peveril of the Peak*, 4 vols. (Edinburgh: Archibald Constable, 1822); Walter Scott, *Kenilworth: A Romance*, 3 vols. (Edinburgh: Archibald Constable and John Ballantyne, 1821); and Walter Scott, *Woodstock, or The Cavalier: A Tale of the Year Sixteen Hundred and Fifty-One*, 3 vols. (Edinburgh: Archibald Constable, 1826). For works edited by Sir Walter Scott, see Samuel Rowlands, *The Letting of Humours Blood in the Head Vaine, &c.* (Edinburgh: William Blackwood, 1814); Robert Cary and Sir Robert Naunton, *Memoirs of Robert Cary, Earl of Monmouth, Written by Himself, and Fragmenta Regalia; Being a History of Queen Elizabeth's Favourites* (Edinburgh: Archibald Constable, 1808); and *A Collection of Scarce and Valuable Tracts, on the Most Interesting and Entertaining Subjects: But Chiefly Such as Relate to the History and Constitution of These Kingdoms, Selected from an Infinite Number in Print and Manuscript, in the Royal, Cotton, Sion, and Other Public, As Well As Private Libraries; Particularly That of the Late Lord Somers*, 13 vols. (London: T. Cadell and W. Davies, 1809–15).

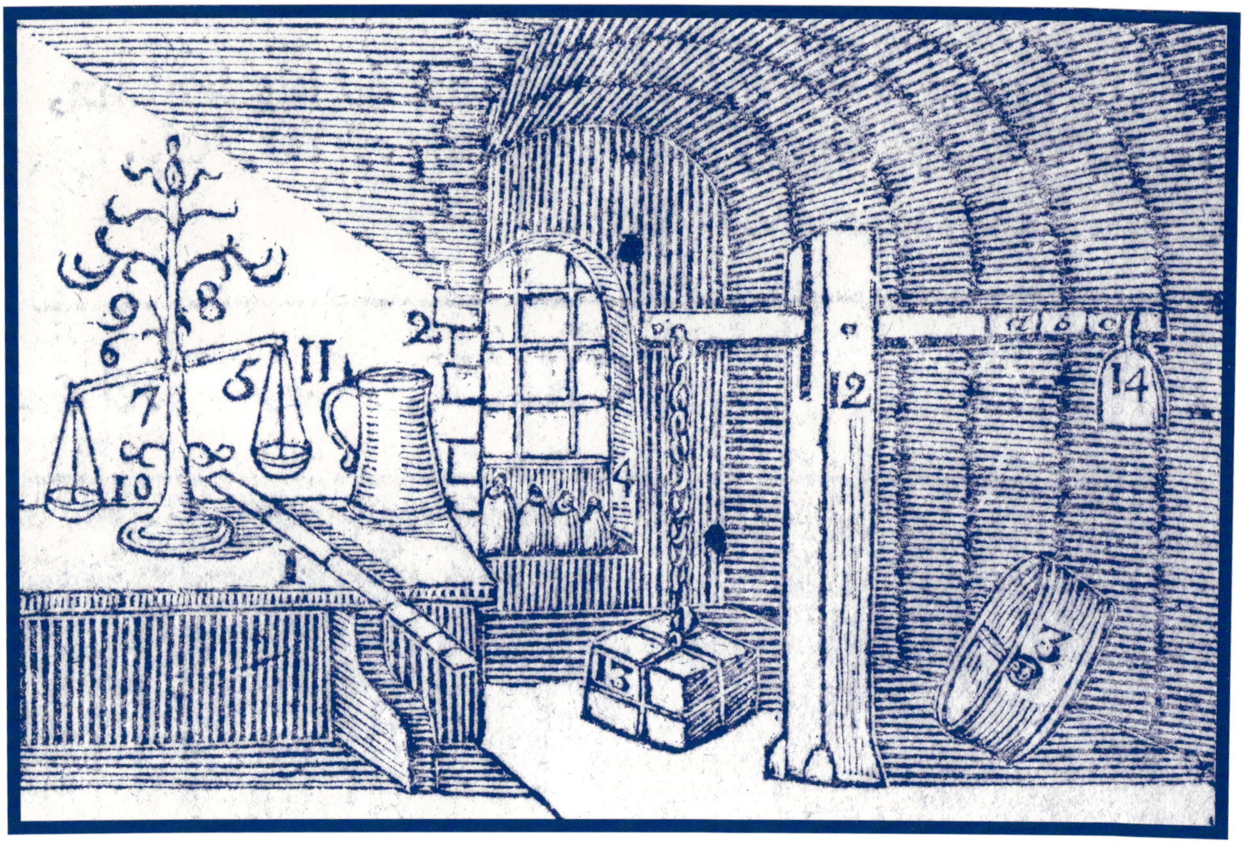
1
2
3
4
5
6
7
8
9
10
11
12
13
14
a
b
c

Time, Measurement, and Authority

Regulating Early Modern Society with Dated Objects

ANGELA McSHANE

THE EARLY MODERN WORLD WAS CONSTANTLY PREOCCUPIED WITH and yet endlessly frustrated by problems of measurement. An enthralling topic for salons, scientific societies, engineers, and inventors, it was necessarily also a central theme for national, local, secular, and religious governing authorities who wanted to enforce and uphold standards judiciously. Measurement was an issue of everyday concern, too—for kitchen cooks, dispensing apothecaries, skilled craftspeople, and anyone who bought and sold merchandise in markets, fairs, and shops. International merchants and government excisemen struggled to work out what duty was owed or what profit could be made on a cargo load of goods. Money was also subject to questions of accurate measurement, as the value of silver and gold coins related directly to their weight.

In the seventeenth century, there was both an intellectual and technological revolution in the measurement of time itself. In 1650, James Ussher developed the Before Christ (BC) / After Christ (AD) dating system. Influenced by the early modern mania for chronologies, Ussher sought to date the point of creation accurately—as 4004 BC, according to his calculations. This date was immediately seized upon by almanac makers, who typically began their books with an outline history of the world.[1] The wider effect of Ussher's dating system, however, was to bring all of time (regardless of when it supposedly began or ended) into a single, universally applicable continuum.[2] This concept of "absolute" time, which Isaac Newton in 1687 described as flowing "equably without regard to anything external," was further boosted by technological advancements in the accurate measurement of clock time, especially by Christiaan Huygens's invention, in about 1657, of the pendulum clock and by the introduction of the minute hand to the pocket watch.[3] These changes contributed to an ever-growing awareness among individuals of their own place in time, which we see, not least, in the practice of dating objects.

This essay considers ways in which early modern people used dated objects to apply continually shifting concepts of measure to their everyday lives: in trading goods and commodities, in calculating time and weather, and in measuring distance and size. It shows how the application of dates to objects of measurement functioned quite differently than did the application of commemorative dates pertaining to "life cycle," the former operating instead to link time, measurement, and authority. In the first section, we investigate dated standard measures, which were intended to perform as governmental assurances of accuracy and legitimacy at both local and national levels but, in fact, revealed a confused world of contested authority. In the second section, we see how renowned makers of watches and scientific instruments developed a range of alternative dating systems that included symbols and serial numbers, which acted both to guarantee authenticity and also to place their work within the continuum of time. In contrast, dates on works that were constantly in need of updating, such as maps and globes, acted to assure buyers of the maker's authoritative knowledge of and engagement with the latest scientific developments. Meanwhile, for those who used objects of measurement in their everyday life and work, personalized naming and dating were less concerned with temporal authority; instead they reflected, as did so many of the objects discussed throughout this book, a desire to fix important moments in their own lives within a Christian time continuum, between Creation and the Second Coming.

Weights, Measures, and Contested Authority

In 1816, in a book entitled *Metrology; Or, An Exposition of Weights and Measures*, schoolmaster and renowned meteorologist Patrick Kelly wrote:

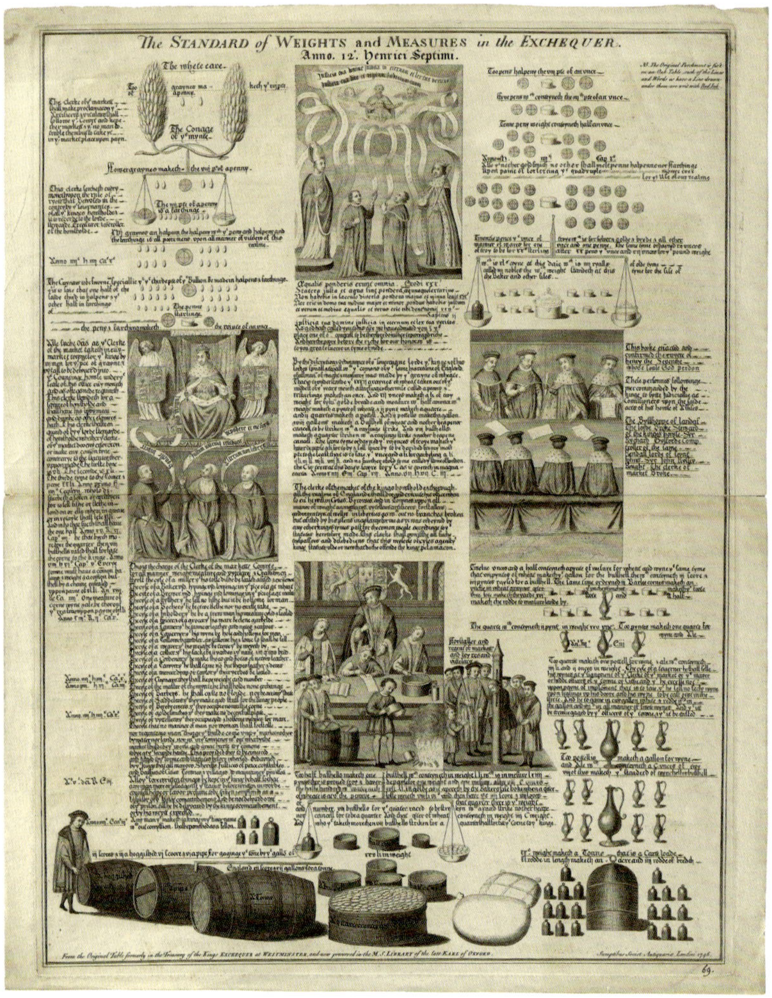

FIG. 14 Probably Isaac Basire, after a manuscript, standard weights and measures in use during the twelfth year of the reign of Henry VII, ca. 1497, engraving on paper. Wellcome Collection, London

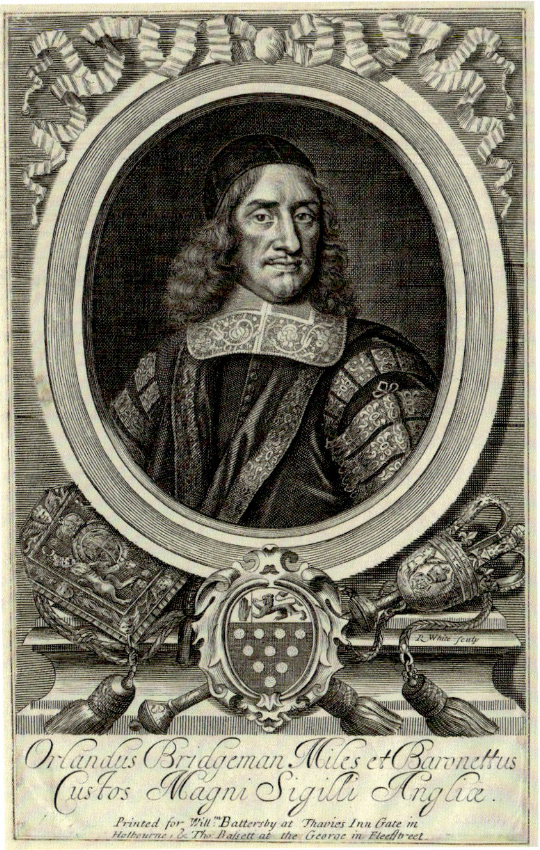

FIG. 15 Robert White, after William Faithorne, *Sir Orlando Bridgeman, 1st Bt*, late seventeenth century, line engraving on paper, 9¾ × 6 in. (24.7 × 15.1 cm). National Portrait Gallery, London, Given by the daughter of compiler William Fleming MD, Mary Elizabeth Stopford (née Fleming), 1931

> the importance of [weights and measures] mingles more or less with every concern, and condition of life. . . . They have been very properly defined [as] the foundation of justice, the safeguard of property, and the rule of right.[4]

From medieval times, English governments had established an apparently foolproof system for setting and enforcing standards of dry and liquid measure across the kingdom. This system was first documented in 1215 in King John's *Magna Carta*:

> There shall be one measure of wine throughout our whole realm, and one measure of ale, and one measure of corn—namely, the London quart—and one width of dyed and russet and hauberk cloths—namely, two ells bellow the selvage. And with weights, moreover, it shall be as with measures.

Official standard measures relating to these quantities were kept by the Crown. Exact copies were to be "lodged in the principal cities and towns throughout the kingdom, from which other standards were to be made, sealed and marked, by persons appointed for the purpose."[5] What could be simpler? And yet, six hundred years later, Kelly could fill a large book cataloguing the enormously diverse array of English and Scottish measures in use and the many modifications they had undergone over time. Until the "imperial" system was introduced in 1824, there was endless variation and confusion across the board, even among those whose job it was to enforce the standards.[6]

Certainly, confusion is what Chief Justice Orlando Bridgeman, Lord Chief Baron of the Exchequer, discovered in 1662 when asked to review the case of Gabriel Beckman from Maidstone in Kent. Beckman was a local distiller, and his entire set of pewter measures—a gallon (eight pints), a pottle (four pints), a quart (two pints), and a one-pint measure—had been destroyed by order of the local "court of the market." The jurymen and clerk of the market court found that, according to their officially tested and dated eight-gallon bushel standard, Beckman's vessels were the wrong size. Beckman, however, vehemently protested that his pewter pots had been precisely the right size for legally measuring wine and distilled spirits. Justice Bridgeman went to considerable trouble to find out the truth of the matter. He consulted with government officials, met with brewers, and talked with the artisans who made measuring vessels to understand the practical application of volume, capacity, and weight.[7]

Measurement, Bridgeman discovered, was a complicated business. Liquid and dry goods were measured in gallons (eight pounds or pints) and bushels (eight gallons) and were calculated by avoirdupois (literally, "to have weight").[8] In theory, then, eight pounds of wheat and eight pints of water should weigh exactly the same. But, for a number of "common sense" reasons, gallon and bushel measures differed in size depending on the type of goods being measured. For example, measures of ale were adjusted to take into account that the liquid was frothy and contained "filth" (or lees). Similarly, measures for corn took into account "the hollowness in dry things."[9] In contrast, wine and spirits did not require any allowances to be made, whether for air, froth, or lees. Consequently, in practice, a bushel measure for corn was smaller in volume than an ale bushel, but both held considerably more than a bushel measure for wine. The common-sense approach affected measurement of length also. For example, woolen cloth lengths were measured in ells, which were slightly larger than yards, to take into account the shrinkage that occurs when woolen textiles are washed.

Justice Bridgeman next investigated what standards really were held in the governmental vaults. His officials directed him to three places: the wine gallon standard was in the Tower, the ale gallon standard was in Winchester, and the corn gallon standard was in the Exchequer. Yet, when Bridgeman visited the Exchequer, he found a mixed bag of objects there, including broken vessels and several old brass standards dated 1601. These standards comprised a bushel and a gallon measure for corn and a quart measure for ale. At the conclusion of his investigations, Bridgeman found that an injustice had indeed been done to Beckman: the Maidstone court had tested Beckman's wine and spirit measures using a standard bushel measure meant for corn.

The ale quart measure marked "16 ER 01" (page 428) replicates precisely the royal standard that Bridgeman saw in the treasury, made in response to Elizabeth I's 1601 reaffirmation of the customary Winchester standards. It is notable that the capacities of the officially marked ale measures in the Bryan Collection, which date from 1601 to 1707, remained uniform (page 430). Made of durable bronze or brass, so that their volume could not be altered by knocks and dents, and decorated with the royal insignia and legislative date, these objects unambiguously denoted legal authority but gave no indication of which kinds of goods they were intended to measure. This lack of specificity made it difficult for local officials to be certain as to the purposes of the standards they owned. As

Bridgeman had discovered in 1662, unless great care were taken, each new measure might simply add to the jumble of locally held standards that were already not entirely clear to officials.

Whereas weighty goods were calculated in pounds, lighter things, like small quantities of precious metals, were calculated in "troy" weight (one troy pound was just 0.82 avoirdupois).[10] This variation mattered to everyone in early modern Britain, because it was not just commodities but also the silver and gold coins used as payment for them that needed to be valued by weight. "Clipping," or counterfeiting, coins was treasonous and punishable by death, but, despite this, the practice was rife, and customers and traders alike feared being cheated, especially in port towns and cities, where shopkeepers and other traders dealt with strangers they might never see again.

One remedy for this problem was the portable and foldable coin balance (pages 453–54). With each balance was an officially sanctioned set of weights that could be used to identify counterfeit currency. As the printed instructions pointed out, if customers distrusted the trader's weights, the scales

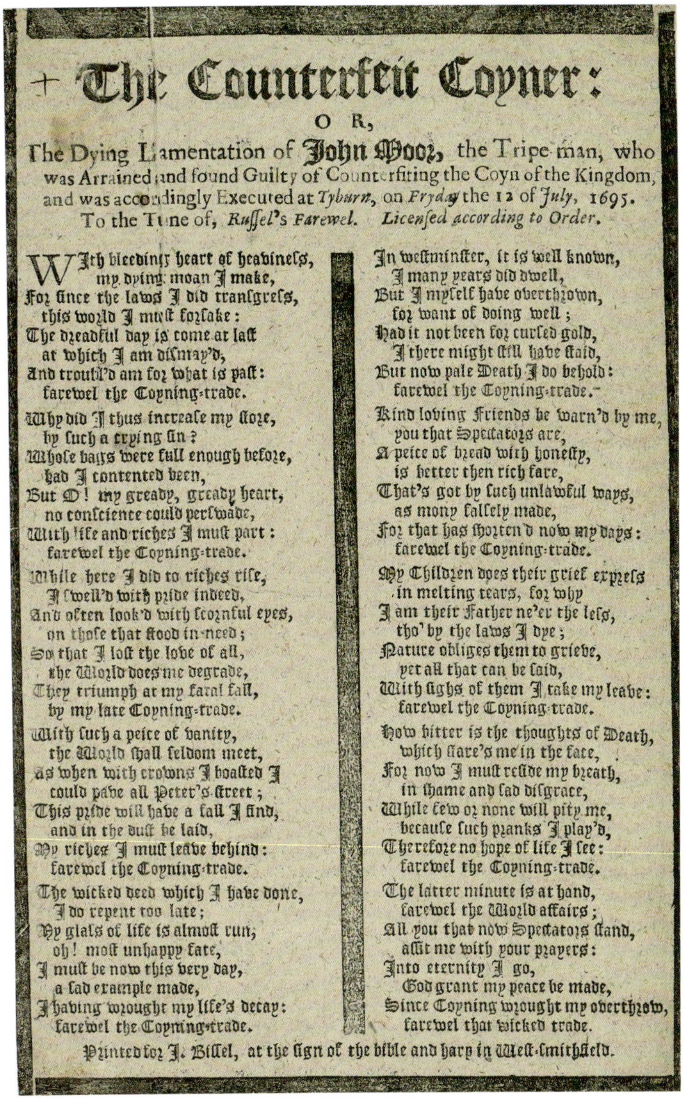

The Counterfeit Coyner:
OR,
The Dying Lamentation of John Moor, the Tripe-man, who was Arrained and found Guilty of Counterfiting the Coyn of the Kingdom, and was accordingly Executed at *Tyburn*, on *Fryday* the 12 of *July*, 1695.
To the Tune of, *Russel's Farewel*. *Licensed according to Order.*

With bleeding heart of heaviness,
my dying moan I make,
For since the laws I did transgress,
this world I must forsake:
The dreadful day is come at last
at which I am dismay'd,
And troubl'd am for what is past:
farewel the Coyning-trade.

Why did I thus increase my store,
by such a crying sin?
Whose bags were full enough before,
had I contented been,
But O! my gready, gready heart,
no conscience could perswade,
With life and riches I must part:
farewel the Coyning-trade.

While here I did to riches rise,
I swell'd with pride indeed,
And often look'd with scornful eyes,
on those that stood in need;
So that I lost the love of all,
the World does me degrade,
They triumph at my fatal fall,
by my late Coyning-trade.

With such a peice of vanity,
the World shall seldom meet,
As when with crowns I boasted I
could pave all Peter's-street;
This pride will have a fall I find,
and in the dust be laid,
My riches I must leave behind:
farewel the Coyning-trade.

The wicked deed which I have done,
I do repent too late;
My glass of life is almost run,
oh! most unhappy fate,
I must be now this very day,
a sad example made,
I having wrought my life's decay:
farewel the Coyning-trade.

In westminster, it is well known,
I many years did dwell,
But I myself have overthrown,
for want of doing well;
Had it not been for cursed gold,
I there might still have staid,
But now pale Death I do behold:
farewel the Coyning-trade.

Kind loving Friends be warn'd by me,
you that Spectators are,
A peice of bread with honesty,
is better then rich fare,
That's got by such unlawful ways,
as mony falsely made,
For that has shorten'd now my days:
farewel the Coyning-trade.

My Children does their grief express
in melting tears, for why
I am their Father ne'er the less,
tho' by the laws I dye;
Nature obliges them to grieve,
yet all that can be said,
With sighs of them I take my leave:
farewel the Coyning-trade.

How bitter is the thoughts of Death,
which stare's me in the face,
For now I must reside my breath,
in shame and sad disgrace,
While few or none will pity me,
because such pranks I play'd,
Therefore no hope of life I see:
farewel the Coyning-trade.

The latter minute is at hand,
farewel the World affairs;
All you that now Spectators stand,
assist me with your prayers:
Into eternity I go,
God grant my peace be made,
Since Coyning wrought my overthrow,
farewel that wicked trade.

Printed for J. Bissel, at the sign of the bible and harp in West-smithfield.

FIG. 16 *The Counterfeit Coyner: OR, The Dying Lamentation of John Moor, the Tripe man, who was Arrained and found Guilty of Counterfiting the Coyn of the Kingdom, and was accordingly Executed at Tyburn, on Fryday the 12 of July, 1695*, broadside printed for I. Bissell, 93⅜ × 57⅞ in. (237 × 147 cm). Pepys Library, Magdalene College, Cambridge

could be “tried with sealed weights at any time for the satisfaction of such as refuse to take money by them.”[11] Eighteenth-century Liverpool was a fast-growing port with many strangers who were passing through and dealing with local shops and businesses. Local shoemaker John Kinsey protected himself by acquiring a coin balance (page 453). It was dated 1774, the same year in which he was recorded as being at 12 Church Street, Liverpool, and perhaps first setting up business there.[12]

The many dated measures that survive today were commissioned either by local authorities or by traders themselves as part of the constant struggle to establish what the proper volume or capacity should be and to issue an accurate and replicable standard to enforce it. As such, measures provide unique insight into the day-to-day business and governance of early modern England. Whenever new legislation enforcing standard measures was passed by central government—usually because of changes to taxation or other reforms—local governors, who were responsible for “courts of the market,” manorial courts, or “courts leet,” commissioned their own copies. These measures were inscribed with the date of the new legislation and often too with the name of the manorial lord or local office holder that had provided them. As in Maidstone, these new measures would then be used to test the vessels, weights, and other objects relied upon by local traders in shops, markets, and fairs and by victuallers in alehouses, taverns, and inns.

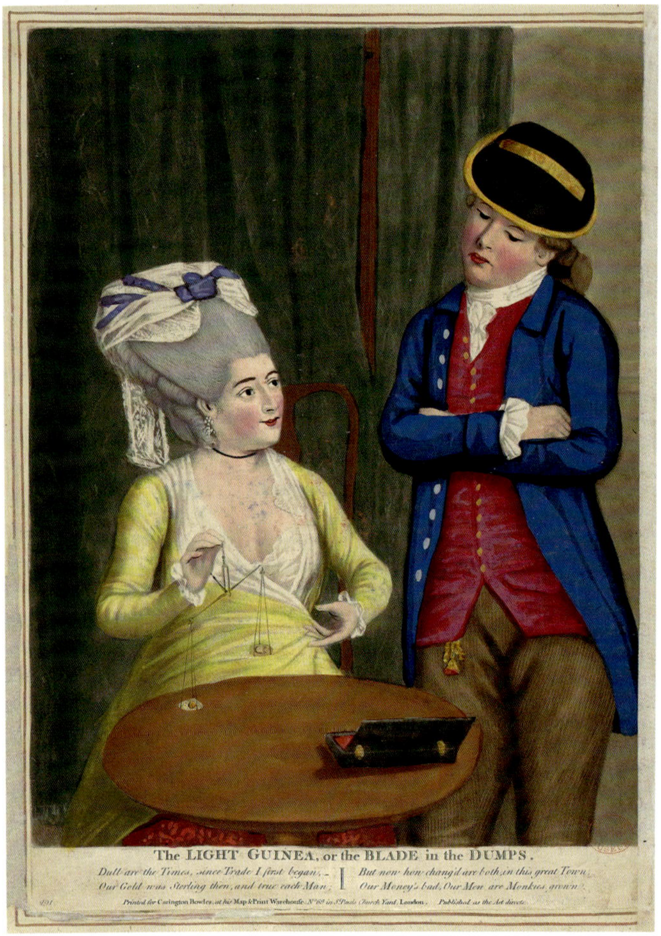

FIG. 17 *The Light Guinea, or the Blade in the Dumps*, 1773–74, hand-colored mezzotint, 13⅛ × 9⅞ in. (35.3 × 25.2 cm). British Museum, London

For example, the bronze Winchester ale measures (page 437) that were used to police Wivenhoe's numerous, busy little alehouses and larger inns were inscribed with "1766," the date of the enforcing legislation, and the name "Nicholas Corsellis," then the lord of Wivenhoe's manor.[13] Similarly, a 1709 corn bushel (page 435) is inscribed "THOMAS : GOUGE : ESQR." Gouge was the lord of the manor for Dartford, Kent, where there was a busy market for meat, fish, corn, and other provisions every week in the High Street.[14] When wielded by the authorities, these official objects struck fear into the heart of any dealer or retailer who tried to sell short measures to their customers.

Some measures named an office holder rather than a manorial lord, a sure sign of the growing strength of local civic powers (page 436). For example, Spilsby in Lincolnshire was a prosperous and independent little town. The local people built a combined town hall, courthouse, and covered marketplace, which still stands today. The power to nominate officers to Spilsby's market court rested with the lord lieutenant of the county, the Duke of Ancaster. Over time, however, leading figures in the town appropriated the duke's powers. In particular, the local Brackenbury family acquired control over the office of Clerk of the Peace, giving the family authority over the market courts. From 1787 until 1826, the lawyer Joseph Brackenbury held the post. Consequently, when legislation required the renewal of the town's standard bushel in 1793, Brackenbury's name was inscribed upon it as a mark of his hereditary office.[15]

Providing the locality with accurate and authoritative measures was immensely important (as Gabriel Beckman discovered), but because manorial lords and officials often inherited their status, as happened in Spilsby, disputes sometimes arose over authority. Just such a dispute led to the making, in 1670, of a uniquely decorated, newly standardized measure for dry goods in response to new legislation. The statute enforcing the use of the "Winchester bushel" was particularly stringent, with an enormous fine of forty shillings (two pounds) for noncompliance.[16] For the community of New Malton in Yorkshire, the new legislation presented considerable difficulties. The lordship of their manor had been in dispute since 1652, when Lord Eure died, leaving his estate to two sisters: Mary Eure, who married William Palmes of Lindley in 1663, and Margaret Eure, who married Thomas Danby of Farnley in 1659. Margaret Eure was a formidable character and a notoriously aggressive litigant. Over many years, she disputed every inch of her legacy: in 1664, coal mines in Evenwood, Durham, and, in 1668, property in Old and New Malton, Yorkshire. In 1675, in resolution of the disputes, the court ordered the family mansion to be pulled down stone by stone and divided between the sisters. By the time the "New Malton bushel" measure was created in 1670, Margaret was a widow and in sole charge of her own affairs; her sister Mary's interests were looked after by her husband, William Palmes, who was also the local member of parliament.[17]

For New Malton, the long-running dispute between the sisters had made local governance almost impossible. In 1668, local magistrates had complained that

> the common well in the town of New Malton is in great decay for want of repairing, and, by reason of the contest and difference betwixt the Lords of the Manor of Malton, no Court-leets of late hath been kept there.[18]

In desperation, the local tradespeople, many of them brewers and corn factors, decided to provide their own authenticated bushel measure and inscribe it with the names of both claimants: "WILLIAM PALMES ESQUR : & MADAM MARGRET DANBY NEW MALTON." But they added a telling message: the Latin motto "EADEM MENSVRA QVA METIMINI CONTRA METIETVR VOBIS" is a biblical quotation from Matthew 7:2—"the same measure that ye mete withal, it shall be measured to you again." This quotation was perfectly applicable to the circumstances of the bushel's use, but it also held an underlying rebuke, which would have been obvious to every educated observer, warning that New Malton's warring "lords" would eventually get their just deserts.

New official measures were not always welcome, especially in places where customs differed.[19] When, in 1707, the Act of Union joined England, Wales, and Ireland with Scotland to form the United Kingdom, Article XVIII of the treaty stipulated that "from and after the Union, the same Weights and Measures shall be used throughout the United Kingdom" and were to be based upon the "Standards kept in the Exchequer at Westminster."[20] In response, new ale quart measures were issued with decorations that combined the English rose and the Scottish thistle (page 430)—but the new measures were not very popular in Scotland, where, for centuries, measures had been at least one-third larger than those in England.[21]

Systematic measurement of all the dated objects in the Bryan Collection that relate to weights, measures, drink, and tobacco allowed us to assess the success (and failure) of standardization. The process also allowed us to track shifts in drinking tastes and practices.[22] For example, surprising conclusions can be drawn about the likely function of large leather jugs and bottles, which are usually presumed to be for ale, beer, or water. Based on their capacity, however, two such objects in the Bryan Collection seem to have been designed for different kinds of drink and, perhaps, different kinds of drinking situations, too. The first of these (page 280), shaped as a costrel, has just the right capacity for a gallon of wine. It is stamped with the year 1659, a crucial year politically, as the last in a series of republican governments set up since the outbreak of Britain's civil wars in 1642 was in the final throes of collapse. During the interregnum period, the ritual of excessive wine drinking was a particularly Royalist trait, and we might speculate that this vessel was perfectly designed for making loyal toasts to the exiled King in hopes that he might return. The second object (page 281), a large jug, dated 1662 and decorated with crown and fleur-de-lis designs, as well as ownership initials ("TP"), is exactly the right size for a gallon of ale or beer. It would have been ideal for an officially sponsored celebration—perhaps held in a parish or on a manorial estate—in honor of the restored King Charles II's wedding to the Portuguese princess Catherine of Braganza in that year.[23]

Glass bottles impressed with seals offer unique insights into the spread of standardization and the drinking fashions in private and public houses.[24] Four bottles in the Bryan Collection, with seals dated after 1700, were ideally suited to ale or beer measures (three quarts and one half pint).[25] We know from court records that bottled beer was increasingly popular in taverns by the eighteenth century, but what is particularly notable is that the seals indicate that all the bottles were owned by private individuals, who did not need to worry about official measures—seeming to suggest that standardization was having an impact on domestic consumption practices. This adherence to official standards also applied to wine or spirit bottles in quart, pint, three half-pint, and single half-pint measures. Court records suggest that it was not uncommon for a person to order a pint of wine in a tavern for breakfast, drink half, and then take the other half pint away in a bottle.[26] Just such a wine-pint-size "pocket bottle" was acquired by James Gouldsbury in 1724 (page 279).

The seal, which often incorporates a name and a date, can also help to identify the geographical and social range of the people who used the bottle. For example, one wine quart bottle, dated 1674 (page 252), was owned by Thomas Bydder, a yeoman farmer in Thistleboon, near Oystermouth, Swansea (d. 1681). A wine pint bottle dated 1723 (page 258), found covered with barnacles in Dartmouth harbor, was owned by Robert Newman (1676–1739), a merchant-mariner of that town. He specialized in port wine and was a well-known citizen, elected mayor in 1725, 1728, 1730, and 1731. Other bottles come from as far afield as Cornwall (page 260); one magnum bottle was owned by a market gardener in Herefordshire (page 271). Two of the dated bottles in the Bryan Collection (pages 267 and 269) were made for James Oakes (1741–1829) of Bury St Edmunds, Suffolk, a banker and the owner of a yarn-making business. Many bottles sealed with Oakes's name and dated between 1770 and 1795 are known—some were surely from his large wine cellar but some intended for commercial purposes. In 1784, Oakes became the landlord of the Green Dragon Public House, located next to his yarn-combing warehouses, ideally positioned to catch the business of his thirsty employees.[27]

Scientific Authority and Authenticity

Among the many things that early modern people hoped to better understand, and measure, was time itself. For most of human history, the most accurate way of telling time was with sundials, which served to order and, to some extent, regulate everyday life.

During the Christian era, the Catholic Church was at the forefront of scientific inquiry into the measurement of time, as it sought to regulate the liturgical calendar and enforce canonical routines of daily worship.[28] Working closely with astronomers, mathematicians, and engineers, architects incorporated sundials into early modern churches and cathedrals, enabling the science of accurate time telling to advance apace. Protestants were initially suspicious of these Catholic-sponsored developments, delaying acceptance of the more accurate Gregorian calendar for hundreds of years.[29] Nevertheless, the first book on the subject of sundials appeared in English in 1593. By the early 1600s, in Scotland, makers were drawing upon the latest European advances in mathematical knowledge to specialize in highly sophisticated sundials with multiple faces set around a standing stone.[30] The proliferation of clocks and watches in early modern households was matched by an ever-increasing number of sundials, which were used to regulate the mechanical timepiece.[31] In 1698, the French polymath Philippe de la Hire explained the importance of sundials in his horological treatise: "Here would be the place to report all the uses of sundials, but I believe it suffices to say that there is no state in life in which exact knowledge of the time is unwarranted, be it for regulating matters of religion and the Divine Office, or for apportioning work and rest fairly."[32] Sundials were often dated, denoting the time of their making, gifting, or installation in a private garden or a public space.[33] In contrast, few surviving clocks or watches appear to be dated in this way. Instead, clock- and watchmakers devised alternative ways of marking time.[34] Notable exceptions are the clocks of Richard Savage, who habitually put his name, the date, and the recipients' names on the dials (page 420).

Although watches needed to be checked regularly against a sundial, these timepieces were nevertheless widely sought after, not least because they could give an indication of the time even when there was no sun to cast shadows on the face of a dial. By the 1700s, a very large proportion of the population, including masters, servants, and even paupers, owned watches.[35] Often beautifully worked, pocket watches were decorated inside and out, encouraging an appreciation of the internal mechanism as much as the external face. Rather than marking their works as having a fixed place in time, high-end watchmakers adopted systems of authentication that positioned their works within a continuum, linked to the maker's life span and beyond. A fine pocket watch in the Bryan Collection offers an excellent case in point (page 347). The inner workings are inscribed with the watchmaker's name, "And^{r} Dunlop," which confines the moment of its making to a date within the span of the celebrated maker's life (Dunlop died in 1735).[36] A serial number, "1628," marked on the workings and the inner and the outer surfaces of the case, places the watch in a sequence within Dunlop's oeuvre.

As intended, Dunlop's timepieces long outlasted his own life. The ongoing functioning of the Bryan watch was captured by two "watch papers" (essentially, a sort of trade card) hidden inside the case. Dated 1740 and 1744, the papers record the regular maintenance of the watch by McNabb's, a specialist firm on George Street, Perth. As was common practice, the owner kept the papers inside the watch to provide additional protection for the inner workings. Only the watchcase is dated to a single year, using a system of dating that required specialized knowledge to interpret. It is stamped inside with the London gold assay hallmark for the year 1723 and the goldsmith's initials, "IW." The identifying marks of Dunlop's name and serial number not only offered guarantees of authenticity and reliability, but they would also have been useful if the watch ever went missing, as we see in a case reported in the *Caledonian Mercury* on April 26, 1744:

> Lost on Saturday, April the 21st, betwixt the Gray's Close and Miln's Square, betwixt 8 and 9 o'clock at night, a large gold watch with Andrew Dunlop's name, with two gold cases and a green silk string with a small seal and an inscription on it, Love for Love.[37]

A further revolution in early modern measurement, which overturned long-held Aristotelian ideas, came about when scientists, having discovered the nature of air and how to create a vacuum, used this new knowledge to measure atmospheric pressure. A major breakthrough in domesticating this knowledge came in 1643 when Italian scientist Evangelista Torricelli invented the mercury barometer. It was not until 1666, however, that the Royal Society's *Philosophical Transactions* announced that "A Barometer or Baroscope" (derived from the Greek word *baros*, meaning "weight"), was "first made publick by that Noble Searcher of Nature, Mr. Boyle."[38] Like the watch, the barometer required a daily routine of checking due to the "diurnal variation of pressure," which naturally rises and falls throughout the day. Over the course of the eighteenth century, the ownership of barometers spread quickly from elite households to middling-sort homes, and some makers—such as Charles Orme (page 440)—became so well known that their works were authenticated with signatures and dates. Large "almanac" barometers (many made by Francis Watkins) combine a fine barometer and, in the center, a finely engraved plate of tables, some movable. The device, which was called a Perpetual Regulation of Time, gave calendrical, lunar, tidal, and zodiacal information. These spectacular items of interactive furniture brought together several of the most up-to-date measurement systems, which a family could regularly monitor and update—profoundly suggestive of how early modern people were increasingly imposing intellectual control over the elements.[39]

The desire to display knowledge, as well as to acquire it, led to growing competition in the creation of and market for scientific instruments. Dates added value to objects of science: the more recent the date, the more it could be relied upon to incorporate the most advanced knowledge. Globes, for example, mapped the fullest extent of man's knowledge of the heavens and earth at the time they were made. Once owned only by the rich, globes became increasingly accessible, thanks to the development of print-transfer technology. From the mid-1700s, copyright laws required printed and engraved items to be dated. The notice "Published as the Act directs by Harrison & Co N° 18 Paternoster Row, Augt 1st 1783," for example, secured the rights of the original mapmakers for fourteen years. Printed preambles and publication dates drew attention to the most-current knowledge that maps contained and also indicated when they were out of date. For instance, the "NEW Terrestrial GLOBE by Nath Hill 1754" (page 344), was made before Cook and Vancouver's discoveries (made between 1768 and 1795) were known. Consequently, the coast of Alaska is missing, and Greenland is joined to eastern North America.[40] The wide dissemination of spatial and geographical knowledge, laying out the full measure of the earth and planets, is best indicated in the Bryan Collection by a pair of globes (page 439). They were created by William Bardin to accompany the *Geographical Magazine*, written by the Reverend William Mavor and published in forty parts beginning in 1782.[41] By 1785, geographical knowledge had become so widespread that thirteen-year-old Martha Chittenden of Sittingbourne, Kent, was able to depict her country to scale using needle and thread (page 137).

Although dates and dating systems on objects of measurement performed a wide range of official and unofficial functions, personal "life cycle" dates can still be found on a wide variety of measuring tools—such as the magnifying lens owned by the forty-six-year-old laborer Thomas Gill of Carnaby, Yorkshire (page 346); the shot gauge inscribed "Nathaniel : Foster. 1693" (page 426);and the tape measure inscribed with the name of a Somerset schoolmaster, "George Mullins Box 1748" (page 427). These objects enabled their owners to engage directly with scientific revolutions in measurement and precision, while the dates with which they were inscribed declared each individual's unique position in the brave new world of early modern time.

Notes

Research for this essay was generously supported by John H. Bryan II through the American Friends of the Victoria and Albert Museum (V&A). My thanks to Danai Synadinou, MA, and Dr. Roxanne Ravenhill, both of the V&A/Royal College of Art. My thanks also to the team at Crab Tree Farm, especially to curator Dru Muskovin, for their invaluable assistance.

1. Bernard Capp, *English Almanacs, 1500–1800: Astrology and the Popular Press* (Ithaca, NY: Cornell University Press, 1979).

2. David S. Landes, *Revolution in Time: Clocks and the Making of the Modern World* (Cambridge, MA: Harvard University Press, 1983).

3. Newton, quoted in Lynn Hunt, *Measuring Time, Making History* (Budapest: Central European University Press, 2008), 26; and Eric Bruton, *The History of Clocks and Watches* (Kent, UK: Grange Books, 2002). Further, as Edward Town notes in the descriptive entry for the wall clock on page 420, in addition to the introduction of the pendulum, improvements in the design of the escapements dramatically reduced the slippage of seconds over the course of a day, making clocks more reliably accurate.

4. P. Kelly, *Metrology; Or, An Exposition of Weights and Measures* (London: J. Whiting for the Author, 1816), vii.

5. "Magna Carta (1215 A.D.)," in *Select Historical Documents of the Middle Ages*, trans. and ed. Ernest F. Henderson (Honolulu: University Press of the Pacific, 2004), 141.

6. Julian Hoppit, "Reforming Britain's Weights and Measures, 1660–1824," in *English Historical Review* 108, no. 426 (1993): 82–104.

7. "*Beckman v. Maplesden*. (a) Tr. 10 Car. 2. Rot. 1746," in *Reports of Judgments Delivered by Sir Orlando Bridgman when Chief Justice of the Common Pleas from Mich. 1660 to Trin. 1667. Edited from the Hargrave Manuscripts*, ed. S. Bannister (London: Joseph Butterworth and Son, 1823), 60–81.

8. Jim Andrew, *Old Weights and Measures*, rev. ed. (Birmingham, UK: Birmingham Museum of Science and Industry, 2008), 2.

9. "*Beckman v. Maplesden*," 77.

10. Andrew, *Old Weights and Measures*, 3–4.

11. Printed instructions for the coin scale, page 454.

12. See the coin scale on page 453. *Liverpool's Fifth Directory: A Reprint of the Names and Address from Gore's Directory for 1774* notes, "John Kinsey, shoemaker, 12 Church Street" (Liverpool, UK: Henry Young and Sons, 1930), 32.

13. See the ale quart measure on page 437. Pat Marsden, "Wivenhoe House: A Tale of Two Lost 18th Century Mansions and the Sea-Captains Who Built Them," *Essex Society for Archaeology and History Transactions* 4, no. 1 (2010): 284–302.

14. See the corn bushel measure on page 435. John Burke, *A Genealogical and Heraldic History of the Commoners of Great Britain and Ireland*, vol. 1 (London: Henry Colburn, 1836), 87; and Edward Hasted, "Parishes: Dartford," in *The History and Topographical Survey of the County of Kent*, vol. 2 (Canterbury: W. Bristow, 1797), 286–328.

15. Lincolnshire Archives' Committee, *Archivists' Report* 8 (1956–57): 17.

16. Danby Pickering, *The Statutes at Large, from the Twelfth Year of King Charles II. to the Last Year of King James II. Inclusive* (Cambridge: Joseph Bentham, 1763), 8:287–88.

17. The National Archives, Kew, c 6/171/80, c 5/469/38, c 5/52/19, and c 7/92/71; William Page, ed. "Parishes: New Malton," in *A History of the County of York North Riding* (London: Victoria County History, 1914), 1:532; Joseph Brogden Baker, *The History of Scarborough* (London: Longmans, Green and Co., 1882), 200; and M. W. Beresford, G. C. F. Forster, and J. W. Kirby, eds., *Leeds in the Seventeenth and Early Eighteenth Centuries* (Leeds: Thoresby Society, 2008), lxiv.

18. J. C. Atkinson, ed., *The North Riding Record Society for the Publication of Original Documents Relating to the North Riding of the County of York, Quarter Sessions Records* (London: Printed for the North Riding Record Society, 1888), 130.

19. Hoppit, "Reforming Britain's Weights and Measures," 90.

20. Daniel Defoe, *The History of the Union between England and Scotland* (London: Printed for John Stockindale, 1786), 356.

21. R. E. Zupco, "The Weights and Measures of Scotland before the Union," *Scottish Historical Review* 56 (1977): 119–45.

22. My gratitude to Dr. Jenny Saunt, Dr. Jasmine Kilburn-Toppin, and Dru Muskovin for their enthusiastic help and support in measuring all the Crab Tree Farm intoxicants vessels.

23. Angela McShane, "Material Cultures and 'Political Drinking' in Seventeenth-Century England," in *Cultures of Intoxication*, ed. P. Withington and A. McShane, *Past & Present* 222, supplement 9 (Oxford: Oxford University Press, 2014), 247–76.

24. Sealed bottles have now been studied in considerable depth: see David Burton, *Antique Sealed Bottles 1640–1900: And the Families That Owned Them*, 3 vols. (London: Antique Collectors' Club: 2014).

25. For examples of quarts, see pages 255 (1705), 263 (1745), and 268 (1769). For an example of a half pint, see page 260 (1725).

26. Trial of George Furnier, December 1690, in Old Bailey Proceedings Online, t16901210-19, www.oldbaileyonline.org, version 8.0.

27. Jane Fiske, ed., *The Oakes Diaries: Business, Politics and the Family in Bury St Edmunds, 1778–1827*, vol. 32 (Woodbridge, UK: Boydell Press for the Suffolk Records Society, 1990), 47.

28. Landes, *Revolution in Time*, 60–65.

29. Sara J. Schechner, "The Material Culture of Astronomy in Daily Life: Sundials, Science, and Social Change," *Journal for the History of Astronomy* 32 (2001): 193–96.

30. Christopher St. J. H. Daniel, *Sundials* (Princes Risborough, UK: Shire Publications, 1986), 11–22.

31. Schechner, "The Material Culture of Astronomy," 209.

32. Schechner, "The Material Culture of Astronomy," 210.

33. Schechner, "The Material Culture of Astronomy," 189–222. A notable English example, dating from the eighteenth century, can be seen on the front wall of the Church of King Charles I the Martyr in Royal Tunbridge Wells, http://kcmtw.org/heritage/architecture/.

34. Richard Savage, clockmaker of Much Wenlock in Shropshire, turned out at least five dated clocks for a number of his country clients—in several cases as wedding gifts. See Brian Loome, "The Clocks of Richard Savage," *Clocks* 20, no. 12 (November 1997): 28–31.

35. Laurence Fontaine, "The Circulation of Luxury Goods in Eighteenth-Century Paris: Social Redistribution and an Alternative Currency," in *Luxury in the Eighteenth Century: Debates, Desires and Delectable Goods*, ed. M. Berg and E. Eger (Basingstoke, UK: Palgrave Macmillan, 2003), 89–102.

36. John Smith, *Old Scottish Clockmakers from 1453 to 1850*, 2nd ed. (Edinburgh: Oliver and Boyd, 1921), 245–46.

37. Smith, *Old Scottish Clockmakers*, 165; see also 191–92.

38. *OED Online*, s.v. "barometer, *n.*," https://www.oed.com/view/Entry/15654.

39. Nicholas Goodison, *English Barometers 1680–1860: A History of Domestic Barometers and Their Makers and Retailers* (Woodbridge, UK: Antique Collectors' Club, 1985), 269, 273.

40. Gerard L'E Turner, *Scientific Instruments, 1500–1900: An Introduction* (Berkeley: University of California Press, 1998), 27–28.

41. Obituary for Rev. William Mavor LLD, *The Gentleman's Magazine*, vol. 9 (London: William Pickering, 1838), 434–35; and John R. Millburn and Tor E. Rössaak, "The Bardin Family, Globe-Makers in London, and Their Associate, Gabriel Wright," *Der Globusfreund* 40/41 (November 1992): 25, 36.

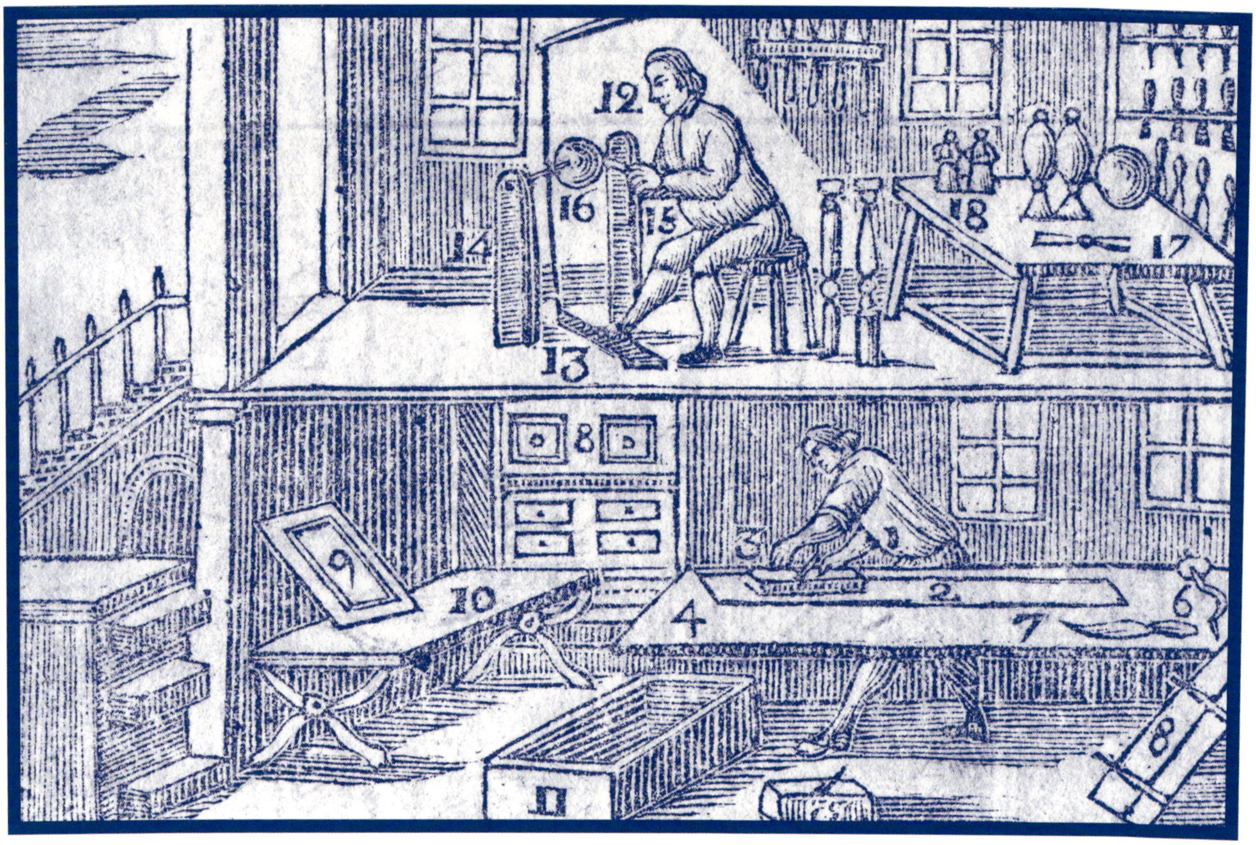
12
16
15
14
13
18
17
8
9
10
3
2
4
7
6
8
11

Artisanal Time

Cumulative, Partially Invisible, Nonlinear, and Episodic

EDWARD S. COOKE JR.

AN ENGLISH COPPER-ALLOY SUNDIAL FROM THE LATE SEVENTEENTH century explicitly refers to natural time, tracked by the sun's shadow cast by the gnomon, and to calibrated time, measured in units of hours in a surrounding circular ring (page 423). As such, it embodies the contrasting notions of "agricultural task-orientation" and "industrial clock-orientation" articulated so clearly by E. P. Thompson, the renowned British labor historian. In his article "Time, Work-Discipline, and Industrial Capitalism," Thompson points out that farmers organized their work according to the availability of daylight, the seasons, and the weather, whereas professionals—merchants, lawyers, clergy, and, eventually, factory owners—divided their time into hourly segments regardless of the natural environment.[1] The sundial thus speaks to a historical moment when the two systems of measurement coexisted.

Yet the gnomon and the numbered dial have something in common, too: both were cast in a brazier's foundry. The engraved inscription above the point of the gnomon also indexes a third daily rhythm, one more difficult to penetrate: artisanal time. Francis Cleaton, the sundial's maker, seemingly asserted his active role, cutting his name, a mark of authorship, and the date into the metal, within a break between the numerical registers of time and cardinal direction: "Fran. Cleaton in Salop Fecit 1696." Nevertheless, while the engraved inscription marks the year in which the object was completed, it reveals the complicated nature of artisanal time in the early modern period. In spite of Cleaton's claim as the maker (*fecit*), he was not the brazier who cast and assembled the pieces. Cleaton—a surveyor and mathematician who provided the necessary calculations and angles for the proper functioning of the sundial in its appointed location—acted more as a designer or installer of the pieces made by another craftsman. This fact raises questions about what constitutes making, who gets credit for it, and how much time is involved.[2]

The inscription, in addition to drawing attention to the designer rather than the brazier, also emphasizes the year of completion rather than the time required to make the sundial. How long did it take to make the master casting pattern? Were the patterns for the gnomon and dial developed years before and simply reused for this commission? How much time was required to prepare the mold and cast the parts? How much labor was contributed by other members of Cleaton's foundry, and how long did it take to clean up and assemble the elements? Much of that time is hidden by the year date. The example of a copper-alloy mortar made by Edward Neale offers a constructive comparison (page 365). The son of a brazier, Neale possessed generational knowledge about alloys, which was an important advantage in an era without scientific understanding of metallurgy.[3] His tacit knowledge allowed him to know materials by sight, sound, and through his other senses, which enabled him to make highly esteemed bells for churches near his Gloucestershire foundry. When casting bells, Neale was in the habit of setting letters and numbers into his molds in order to identify the church and the year of the bell's construction. When making the mortar, he simply applied these typesetting skills to mark a domestic object that was made through a similar molding process. Although the mortar is marked with the year 1666, it does not account for the longer period of time entailed in skill acquisition and pattern making. Both of these dated objects—the sundial and the mortar—demonstrate how measuring the time devoted to making things can be an elusive endeavor for historians.

While we may wax eloquent and even romantic about the development and application of craft skills—detailing the importance of practice, the value of dexterity, and the necessity of flow—understanding artisanal time has proven to be difficult. We struggle to assign to a finished product a measurable value that accounts for the total amount of time invested in it.[4] In the twenty-first century, we focus upon hourly wages and the total number of observable physical hours invested in the object. This calculation of time determines market price, which in turn affects consumer decisions. Indeed, one of the most common questions the general public asks about objects, both historic and contemporary, is, "How long did it take to make that?" However, this conception of "exchange value," to use Karl Marx's terminology, is a fairly recent development. Historically, the hours invested in an object were less important than the materials incorporated and the object's "use value."[5]

It is hard to tally the cumulative time that makers took to create an object, particularly if we include the time required to develop the necessary tacit knowledge or the time spent working intermittently on a large project. Whereas the work of a brazier is a good example of the former, an excellent example of the latter is the needlework screen wrought by Elizabeth Wentworth (page 192). Like other genteel women of the period, Wentworth acquired needleworking skills not as a form of employment but as a symbol of her accomplishment, leisure class, and knowledge of beauty. These women made needlework pictures and screens to decorate their homes, devoting odd hours to these tasteful examples of their prowess.[6] They worked as true amateurs, without regard to the hours spent on the needlework and without the need for wages. They worked to their own satisfaction instead of for an employer and at their own pace rather than to earn an hourly or piece rate. According to the upper spandrels of the two end screens, Wentworth began work on the object in 1738 and finished it four years later. In the top corner of the first panel, she embroidered with pride her commencing of work: "EliZath fecit; and "Wentworth 1738." In the top corner of the fourth panel, she announced "fine Sept 1742." Although the elapsed time is thus recorded, the time devoted to the design of the panels, the days invested in the processing and weaving of the cloth ground and in the spinning and dyeing of the wool—and the cumulative years tied up in the growth, harvesting, shipping, and working of the mahogany frame—are all uncounted.

The needlework screen thus draws attention to the difficulty of calculating the developmental time devoted to conceptualizing a new design and solving technical problems and the time required for nonlinear and episodic methods of fabrication. The time necessary for developing the idea behind an object is often obscured by the passage of time and the lack of a paper trail. Even practiced workmanship that seamlessly links design and technical issues does not necessarily result in a higher object value. For example, superior craftsmanship in plain utilitarian objects like ceramic or wooden bowls may often be taken for granted because it can be misconstrued as merely an approximation of manufactured work (FIGS. 18 and 19). When contemporary makers seek to charge clients for design time in addition to fabrication time, they are often unsuccessful and simply have to absorb the expense. Clients tend to judge an object primarily based on how many hours of physical work are invested.

Finally, craft work has always relied on nonlinear and episodic methods of fabrication that are often linked to efficiency and rhythm. Most artisans do not work on a single piece at a time but rather fit in craft work according to other demands on their time. They may make parts in batches, when certain tools or setups are in use, or respond to different seasonal needs or the availability of other artisans' labor. The making of a turned chair, for example, while ostensibly a straightforward task drawing on simple technology, often stretched out over months: cutting and riving out parts in the winter and then setting them aside to dry, turning stock legs and stretchers in the spring while the lathe is set up, assembling the whole, and perhaps customizing the splat and crest rail, as orders came in or as the demands of agricultural work lessened (FIG. 20). A craftsman did not make one chair at a time but rather made stockpiles of parts, eventually completing a group of chairs together. Given the complexity of this interconnected process, it is nearly impossible to specify the exact amount of hours devoted to one particular chair.[7]

These dynamics are reflected in the widely ranging ways in which artisans were compensated for their work. In the seventeenth century, the architect Christopher Wren explained that there were three means by which craftsmen could be paid: by the day, by a set price, or by measure.[8] The first was the most prevalent way to pay unskilled or semiskilled laborers or farm laborers. The second was typical when paying for large, complicated

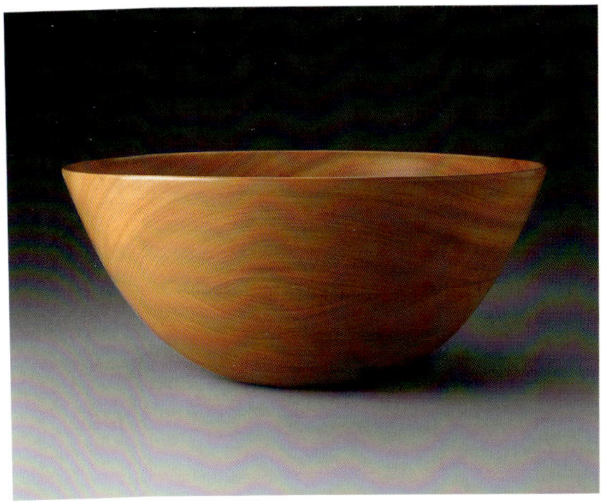

FIG. 18 James Prestini, *Bowl*, 1922–53, turned laminated birch, $4\frac{3}{4} \times 10\frac{1}{2}$ in. (12.1 × 26.7 cm). Museum of Fine Arts, Boston, Gift of James Prestini

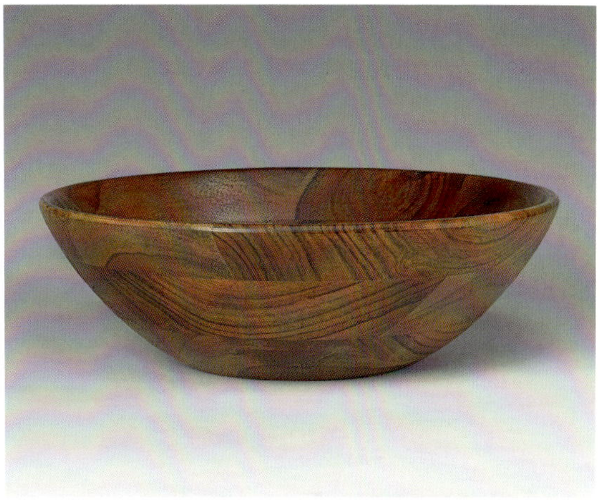

FIG. 19 *Chateau Wood Bowl*, sold by Pottery Barn, ca. 2019, turned laminated acacia, $5\frac{1}{4} \times 14$ in. (13.3 × 35.6 cm). Yale Center for British Art

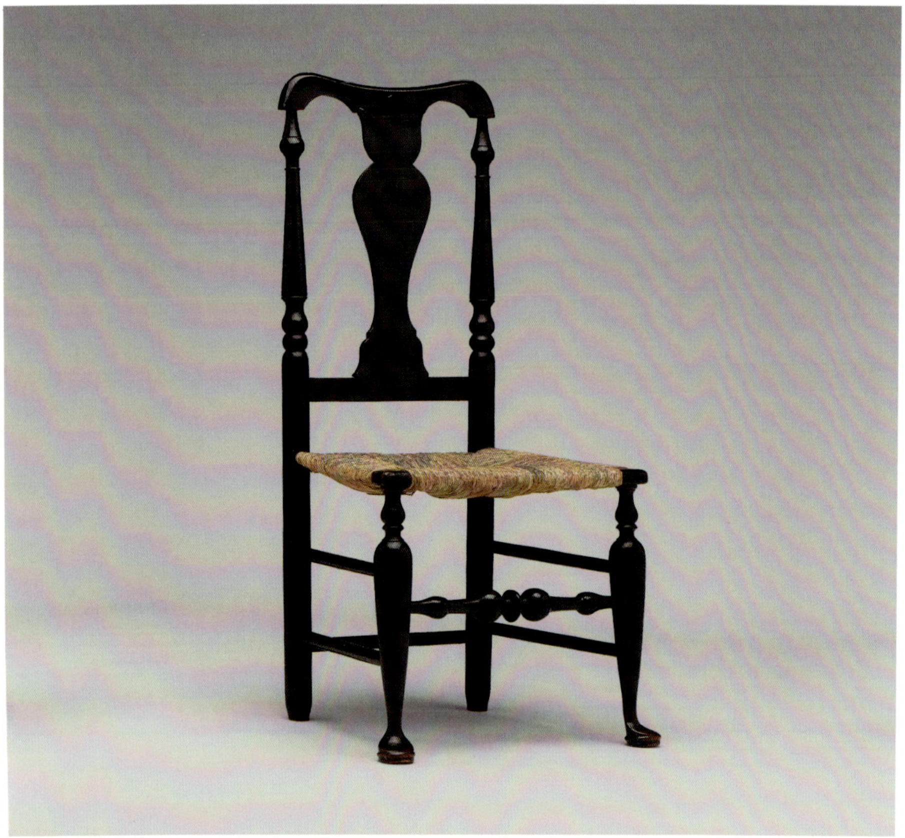

FIG. 20 Side chair, Hudson Valley, New York, 1790–1830, soft maple, yellow poplar, hickory, birch, 39 7/8 × 15 1/8 × 14 5/8 in. (101.4 × 38.3 × 37.1 cm); seat: 17 × 18 7/8 in. (43.2 × 47.9 cm). Yale University Art Gallery, Mabel Brady Garvan Collection

projects, like a building, which required the work of many different tradespeople.[9] The head contractor would give the client a set price and then apportion work to various contractors, also at set prices. Furnituremakers also seemed to be paid set prices, acknowledgment that making furniture often required the work of several specialists, including carvers, turners, and upholsterers. The third means of payment, by measure, was the most common pay system for other types of craftsmen: spinners involved in the "putting out," or "cottage," system of production for textile firms were paid by the weight of thread they spun; pewterers and metalsmiths by the weight of their metalwares; weavers by the yardage, regardless of the type of fiber; and bricklayers or plasterers by the running foot.[10]

At the heart of the early modern economy in Britain were credit, reputation, and trust. As the historian Craig Muldrew argues in his book *The Economy of Obligation*, participants in a local, small-scale economy tended to communicate through "structural chains of obligation."[11] As circles of exchange expanded slightly, craftsmen, farmers, and shopkeepers increasingly turned to account books that had a double-entry system of debits and credits. Throughout the seventeenth and eighteenth centuries, account books in Britain and America used the term *cash* in the double-entry system, but *cash* did not necessarily mean the actual exchange of coinage or other physical currency. Rather it was more a general unit of reckoning—a temporary placeholder that allowed for settlement at a later date, often through the exchange of services or products.[12]

Before the advent of the factory time clock in the nineteenth century, people who were not craftsmen found it difficult to understand artisanal time and the networks in which it took place. In the late seventeenth and eighteenth centuries, many gentlemen engaged with an *arts et métiers* ("arts and trades") genre of literature, which compiled information about various crafts. To understand the tools, steps, and flow of work—what apprentice agreements referred to as "the art, trade, and mystery" of a particular craft—these texts broke down a process into its constituent elements and then abstracted them, denying their very tangible, interconnected nature. Even pictorial depictions of the inside of a shop presented stages of work in a frozen, isolated, and sanitized manner. In this way, a gentleman might convince himself that he could fully comprehend and analyze the mechanical arts. One example is the taxonomical laying out of the tools of a specific craft in an organized, flat, graphic composition. These diagrams were linked to natural-history illustrations popular during the Enlightenment: a means of gaining, displaying, and contributing to knowledge.[13] In the minds of these intellectuals, visualizing a tool was enough to understand the trade. Of course, this way of thinking ignored the real and subtle linkages among tools, the rationale behind certain preferred techniques and tools, the way work was divided within a shop, and the amount of time the work required. Much like watching how-to videos today, it would have been impossible to have learned a craft simply by reading Joseph Moxon's *Mechanick Exercises, or The Doctrine of Handy-Works*, a late seventeenth-century serial publication by one of England's prominent natural philosophers, a member of the Royal Society, or Denis Diderot's *Encylopédie*.[14]

With the emergence of industrial capitalism and wage labor in the nineteenth century, the project of measuring artisanal time became increasingly important. As the marketplace placed a premium upon minimizing cost, maximizing profit, and ensuring consistency, value was explicitly tied to the number of hours spent making.[15] This drive to efficiency ultimately led to the 1911 publication of Frederick Winslow Taylor's *Principles of Scientific Management*, which applied principles of time and motion studies to "improve" the work of factories. Educated professionals would observe the everyday work of skilled and semiskilled workers, develop a list of the components of each task, and then produce a chart with all of the elements itemized, with a recommended time allotted to each. The premise was that there was a single best way to do every job, and a worker who had trained in an apprenticeship system would not instinctively know it. Efficiency and command structure were valued over tacit knowledge and judgment at the bench.[16]

In this new paradigm, craft knowledge—accumulated through training with a master, experience with materials, and years of problem solving—was considered old-fashioned and inefficient. Progressives instead believed that analysis was the best way of gaining expertise, which could then be disseminated in pieces to workers. Habit, which comprised an important part of an artisan's skills and developed cumulatively over generations, was suddenly cast as a form of laziness or complacency. The efforts of the growing union movement to preserve workers' rights ironically focused on hourly wages and benefits as the primary issues and did little to question the systematic division of skills into hourly segments. The unions did not focus on the preservation of tacit knowledge or the protection of artisanal autonomy. Artisanal time and control of work instead became the calling cards of small-shop craftsmen, Arts and Crafts practitioners, hobbyists, and studio craftsmen.[17] In the 1880s, the potter Hugh Cornwall Robertson bankrupted his Chelsea Keramic Art Works business in Massachusetts by

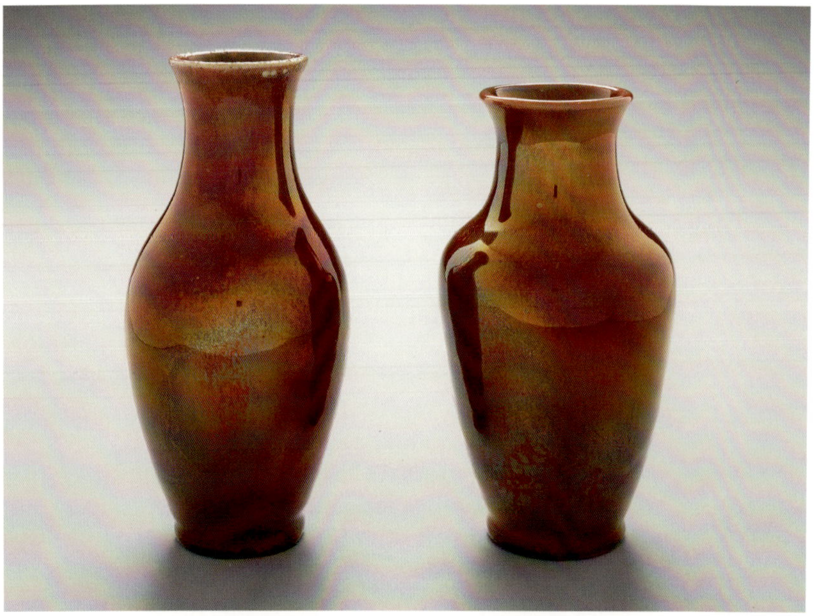

FIG. 21 Hugh Cornwall Robertson, Chelsea Keramic Art Works, *Twin Stars of Chelsea*, 1884–89, wheel-thrown stoneware with sang-de-boeuf glaze; left: 7 × 3¼ in. (17.8 × 8.3 cm); right: 6¼ × 3¼ in. (15.9 × 8.3 cm). Museum of Fine Arts, Boston, Gift of Miss Eleanor M. Hearn

seeking to perfect certain historic glazes, especially the Chinese-inspired oxblood-red glaze (FIG. 21). He could not price his simple vessels in a way that accounted for all of his trials and failures and thus he accumulated a private archive of his experiments.[18] Such continued adherence to artisanal time, and the public's difficulty in understanding any basis for cost beyond hours, confined makers like Robertson to niche markets where price is not the sole determining factor for a purchase. Rather it is the aura of craftsmanship that motivates the consumption of this work.

Even scholarship analyzing historic craft work is beholden to the more recent notions of industrial capitalism, breaking down artisanal time into the categories of materials, labor, and profit. Many scholars simply fall back on the familiar formula of one-third materials, one-third labor, and one-third profit when calculating the time spent making an object.[19] But estimating hours should not be the goal for scholars of historical craft work. It is more instructive to understand the sources of materials, the organization of work, the networks among suppliers and makers, and the rhythms of life and work. Artisanal time must be measured according to the context of the era. For the early modern period, measuring means considering chains of obligation, applying qualitative considerations rather than quantitative ones both to the study of social networks that linked makers and clients and to the extrapolation of technical chains of relations from surviving objects. A craftsman might inscribe the date or year of completion for an object, but that inscription simply marks the end date of his role; often the date is more germane to the client or owner. Less obvious are the various stages in the object's development and fabrication. While manuscripts might help scholars recover some of these steps and networks, that research should be undertaken comparatively. We also need to look at artifactual evidence, which contains traces of these networks of production and suggests the complex nature of artisanal time.

Notes

1. E. P. Thompson, "Time, Work-Discipline, and Industrial Capitalism," *Past & Present* 38, no. 1 (December 1967): 56–97.

2. My thanks to Edward Town for his research on Francis Cleaton. See Town's entry for the sundial on page 423.

3. For biographical information on Neale, see Edward Town's entry for the mortar on page 365.

4. A classic example of the wonderment of craft is Richard Sennett's *The Craftsman* (New Haven, CT: Yale University Press, 2008). The attraction of modern craft has been the uncoupling of efficient time from craft and the valorization of slow time. See also Stephen Knott, "Give It Time/ Watching Making," in *Watching Making* (Moss, Norway: Punkt Ø Galleri F 15 and Momentum, 2018), 10–18, exhibition catalogue.

5. On use value and exchange value, see Karl Marx, "The Commodity," *Capital: A Critique of Political Economy* (1867; repr., New York: Penguin Books, 1976), 1:125–77.

6. On needlework as women's accomplishments, see Pamela Parmal, *Women's Work: Embroidery in Colonial Boston* (Boston: Museum of Fine Arts, 2012); and Laurel Thatcher Ulrich, *The Age of Homespun: Objects and Stories in the Creation of an American Myth* (New York: Knopf, 2001).

7. On the rhythms of turned chair work, see Benno M. Forman, *American Seating Furniture, 1630–1730: An Interpretive Catalogue* (New York: W. W. Norton, 1988), esp. 49–54; and Bernard D. Cotton, *The English Regional Chair* (Woodbridge, UK: Antique Collectors' Club, 1990), 13–31.

8. David Woodward, *Men at Work: Labourers and Building Craftsmen in the Towns of Northern England, 1450–1750* (Cambridge: Cambridge University Press, 1995), 35.

9. Woodward, *Men at Work*, 2.

10. For different measuring systems, see Craig Muldrew, "'Th'ancient Distaff' and 'Whirling Spindle': Measuring the Contribution of Spinning to Household Earnings and the National Economy in England, 1550–1770," *Economic History Review* 65, no. 2 (May 2012): 498–526. In her discussion of contemporary brass craftsmen in Madhya Pradesh, India, Katherine Hacker discusses how retailers pay artisans by the pound and then turn around and sell the products by a set price. "Traveling Objects: Brass Images, Artisans, and Audiences," *RES: Anthropology and Aesthetics* 37 (Spring 2000): 159.

11. Craig Muldrew, *The Economy of Obligation: The Culture of Credit and Social Relations in Early Modern England* (New York: Palgrave, 1998), esp. 152–59.

12. On double-entry accounting, see Christopher Densmore, "Understanding and Using Early Nineteenth Century Account Books," *Midwestern Archivist* 5, no. 1 (1980): 5–19. Helpful explorations of the historical meaning of cash within a social economy are Michael Merrill, "Cash Is Good to Eat: Self-Sufficiency and Exchange in the Rural Economy of the United States," *Radical History Review* 17 (Winter 1977): 42–71; and Craig Muldrew, "'Hard Food for Midas': Cash and Its Social Value in Early Modern England," *Past & Present* 170, no. 1 (February 2001): 78–120.

13. For a recent study of the *arts et métiers* genre, see Paola Bertucci, *Artisanal Enlightenment: Science and the Mechanical Arts in Old Regime France* (New Haven, CT: Yale University Press, 2017).

14. Joseph Moxon, *Mechanick Exercises, or The Doctrine of Handy-Works* (1678; repr., New York: Praeger Publishers, 1970); and Denis Diderot, *Encyclopédie, ou dictionnaire raisonné des sciences, des arts et des métiers, par une société de gens de lettres* (Encyclopedia, or an Analytical Dictionary of the Sciences, Arts and Trades, by a Society of Men of Letters) (Paris: Briasson/David/Le Breton/Durand, 1751–1772). On the seductive power of YouTube videos in regard to craft skills, see Michael Kardas and Ed O'Brien, "Easier Seen Than Done: Merely Watching Others Perform Can Foster an Illusion of Skill Acquisition," *Psychological Science* 29, no. 4 (April 2018): 521–36.

15. Alfred Chandler, *The Visible Hand: The Managerial Revolution in American Business* (Cambridge: Belknap Press, 1977).

16. Frederick Winslow Taylor, *The Principles of Scientific Management* (New York: Harper and Brothers, 1911).

17. David Montgomery, *Workers' Control in America: Studies in the History of Work, Technology, and Labor Struggles* (New York: Cambridge University Press, 1979); Eileen Boris, "Dreams of Brotherhood and Beauty: The Social Ideas of the Arts and Crafts Movement," in *"The Art That Is Life": The Arts & Crafts Movement in America 1875–1920* ed. Wendy Kaplan (Boston: Museum of Fine Arts, 1987), 208–22; Edward Cooke Jr., "Arts and Crafts Furniture: Process or Product?," in *The Ideal Home: 1900–1920*, ed. Janet Kardon (New York: American Craft Museum, 1993), 64–76; Stephen Knott, *Amateur Craft: History and Theory* (London: Bloomsbury Academic, 2015); and Vicki Halper and Diane Douglas, eds., *Choosing Craft: The Artist's Viewpoint* (Chapel Hill: University of North Carolina Press, 2009).

18. On Hugh Cornwall Roberston's obsession with red glazes, see Edwin Atlee Barber, *The Pottery and Porcelain of the United States* (New York: G. P. Putnam's Sons, 1901), 260–67; and Ellen and Bert Denker's catalogue entry on "The Twin Stars of Chelsea" in *"The Art That Is Life": The Arts & Crafts Movement in America, 1875–1920*, ed. Wendy Kaplan (Boston: Museum of Fine Arts, 1987), 148.

19. This proportioning was proposed by Charles F. Montgomery in *American Furniture: The Federal Period* (New York: Viking, 1966), 23, 26. It has continued to inform discussions of pricing, as in Charles Hummel, *With Hammer in Hand: The Dominy Craftsmen of East Hampton, New York* (Charlottesville: University Press of Virginia, 1968), 235.

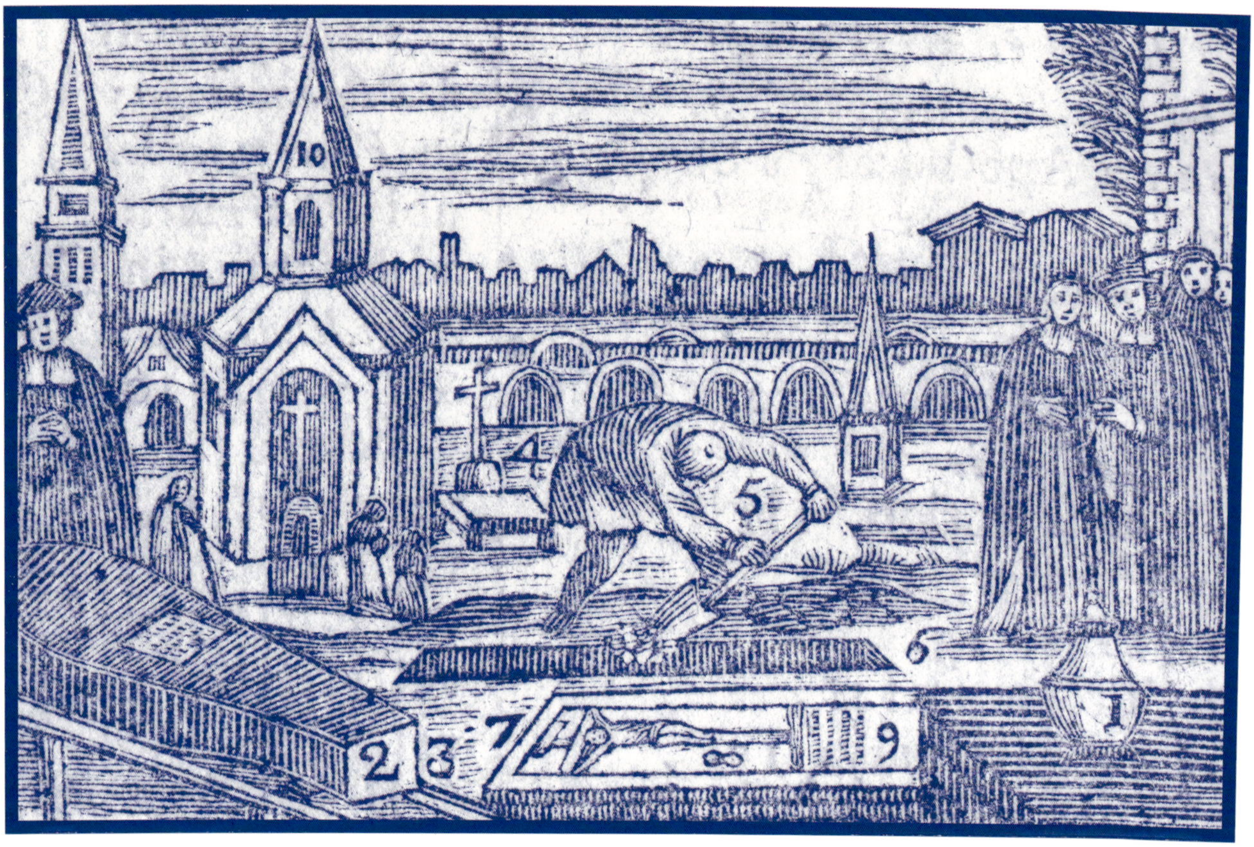
10
4
5
6
1
2
3
7
9

Stopping Time

The Material Culture of Death and Commemoration

NATHAN FLIS

> But thought's the slave of life, and life, time's fool,
> And time, that takes survey of all the world,
> Must have a stop.
>
> —*Henry IV, Part 1*, 5.4.80–82

MORE THAN A CENTURY AFTER HER COPY OF THE BISHOP'S BIBLE WAS printed, Mary Coles inscribed the leaf facing the title page of the Psalter (page 481). The neatly penned lines read:

> Mary: Coles her book · god give her
> grace theairin to kook and when the
> bell for her douth toule lord iesus
> christ recive her soule the rose is
> read the gras is grene the days are
> past which i have seene november the 28
> 1692

The literary devices are not sophisticated, and the moralizing tone recalls an earlier inscription in the same book, penned by a previous owner.[1] Nevertheless, Coles's inscription is a complicated act, which permits us to enter the history of the object at a certain point in time and to understand the ways in which early modern Britons thought about their own mortality.

By the late seventeenth century, the memento mori—a reminder of the inevitability of death—was a well-worn trope. Coles prophesies her own death through quotations of popular poetry. Her line "when the / bell for her douth toule" echoes John Donne's "Ask not for whom the bell tolls" (Meditation 17), and "the rose is read" recalls Edmund Spenser's "In a fresh fountaine, farre from all mens vew . . . / She bath'd with roses red, and violets blew" (*The Faerie Queene*, 3.6.6). Only a month before her forty-seventh birthday, Coles writes as if she has already gone to the grave: "the gras is grene the days are past which i have seene." Sixteen years a widow, conscious of the passage of time, she addresses an imaginary reader, the inheritor of her legacy: the Bible in which she has left her mark.

Encountering an inscription like this one is a tangible reminder of our own mortality. The effect is like the experience of reading a gravestone: the words and the person's name connect us to another human life, while the date separates us from a distant time. As we consider the past life of the object—like Hamlet holding the skull of Yorick—we are made aware of our own position in the present, relative to the past. And so we may turn to contemplate that unknown moment in the future when time will also stop for us.

In modern Western culture, death is an avoided subject, and the business of dying is usually a clinical, secular, and discreet event.[2] Death was understood very differently in early modern Britain, and both its anticipation and commemoration found expression in a rich visual and material culture. As Nigel Llewellyn has demonstrated, rituals, monuments, and artifacts of the period reflect a continuum of life and death, seen as "a single cultural process, not as opposing values in a crude binary model of the human condition."[3] Later ritual "tended to stress the abnormality and deep difficulty of death." In the Victorian period, death seems to have emerged as the taboo subject it is now. In the early modern period, however, ritual placed "death *in* life in order to soften its blow."[4] This essay surveys the visual, material, and literary culture around death and commemoration in Britain from about 1600 to 1800, placing a selection of dated objects in the context of historical events and developments.[5]

Subtle developments in word usage over time reveal a gradual change in the way commemoration was thought about; it is probably no coincidence that these shifts in usage have parallels in evolving concepts of time, as discussed by others in this volume. As we might expect, *memory* is at the root of *commemoration*, a word that first appeared in English in the late fourteenth century. Specifically, *commemoration* derives from the Latin *commemoratio* ("reminding" or "mention") and is also related to the verb *commemorare* ("to call to mind").[6] It was originally used in the context of a church service or festival, so the earliest thinking identified it with an event.

Surprisingly, the English verb "to commemorate" only emerged in the 1590s, again, in the context of a "call to remembrance," and in the 1630s—in a notable shift—it gained the meaning of "perpetuating the memory" of something or someone, especially by a "solemn act."[7] Its meaning was further circumscribed by 1766: "to serve as a memento of." I suggest that this narrowing usage of the word was connected with the gradual emergence of new and diverse material expressions of commemoration, including a burgeoning market for "mementos" in the late seventeenth century.

The word *memento* had a similar development.[8] It was originally associated (in the late fourteenth century) with Psalm 131, a chant that begins with the word *memento*, "remember."[9] By the 1580s it meant "a hint or suggestion to awaken memory, a reminder, an object serving as a warning," especially, perhaps, in relation to memento mori ("remember death"). The sense of the word as "keepsake" is first recorded in 1768. This shift in language was complemented by the proliferation of memento mori imagery and symbols: skeletons and death's heads, spades (for digging graves), arrows and scythes, and sandglasses (which connect impending death with the passage, and stopping, of time). Most of these images are found on a horn tobacco box produced in the heyday of memento mori decoration (page 472). This object is dated 1674 and bears the motto: "SICVT FVMVS SIC VITA," which is an abbreviated quotation from the Latin Vulgate version of Psalm 101:4, which reads, "For my days are vanished like smoke: and my bones are grown dry like fuel for the fire." Although inspired by a solemn *vanitas* concept, the expression here seems playful, perhaps reflecting a culture that had gone so far in its acceptance of death as to delight in the macabre.

One of the most complex efforts of commemoration from early modern England survives as a cluster of artifacts that pertain to Thomas Brathwaite of Ambleside Hall in Cumbria. He died in December 1607, aged thirty-one. Although Brathwaite's primary heir was his brother Gawen, he bequeathed ten pounds to his friend and kinsman George Preston (d. 1640) of Holker Hall, Cark-in-Cartmel, who used the inheritance to honor Brathwaite's memory.[10] An illustration now in the collection of the Yale Center for British Art shows a thin and sickly Brathwaite writing his will, assisted by Preston (FIG. 22). Brathwaite's canopied bed is covered in a fur that resembles a mound of green grass, suggesting that Brathwaite is already half in the grave. The family motto, *Vita vt herba* (which is an anagram of their surname) incidentally translates as "Life is like grass" (or, loosely, "So Life, so grass"). The motto appears in a separate illumination of the Brathwaite coat of arms, which was likely meant to be viewed together with the illustration, bound facing each other in what was then the latest version of the Bible in English (the Geneva edition, printed in London by Robert Barker in 1607). At the front of the book is a specially commissioned printed leaf that reads, "This Booke was giuen by Legacie from THOMAS BRATHWAITE of Ambleside Gent. Who died in the yeare of our Lord 1607, vnto GEORGE PRESTON of Houlker Gent." When Preston died, he bequeathed the Bible to the Brathwaite family.

At Abbott Hall, there is a second version of the composition in oil on canvas (FIG. 23).[11] A third version, in the form of a wood-carved panel of about the same size, was originally installed in the Brown Room at Holker Hall and is now also at Abbott Hall.[12] The image of Brathwaite writing his will in Preston's presence commemorated not only the friendship between the two men but a bond of kinship between families, as the will itself forged a legal bond. The repetition of the image in these different forms amplified its meaning. In the family Bible, it acquired heraldic significance through its juxtaposition with the coat of arms; as painting and panel in the country house, the image would readily be seen by visitors and understood as an important part of the Preston legacy.

One of the most dramatic commemorative gestures of the period is the funerary monument of the poet and clergyman John Donne (1572–1631). Knowing he was going to die, Donne arranged for his likeness, with eyes shut, to be drawn as a corpse wrapped in a shroud, standing upon an urn. This portrait or sketch served as the basis for a sculpture made by Nicholas Stone the Elder—one of the few monuments to survive the destruction of Old St Paul's Cathedral in 1666 (FIG. 24). As Izaak Walton related in his 1640 life of Donne,

> several charcoal fires being first made in his large study, he brought with him into that place his winding-sheet in his hand, and having put off all his clothes, had this sheet put on him, and so tied with knots at his head and feet, and his hands so placed as dead bodies are usually fitted, to be shrouded and put into their coffin, or grave. Upon [the] urn he thus stood, with his eyes shut, and with so much of the sheet turned aside as might show his lean, pale, and deathlike face, which was purposely turned towards the east, from whence he expected the second coming of his and our Saviour Jesus.[13]

An engraving by Martin Droeshout depicts the head and bust of the monument (FIG. 25).[14] This print served as the frontispiece to the posthumously published *Deaths Duell* (1632), which contains the final sermon Donne preached, delivered before Charles I at Whitehall on February 25,

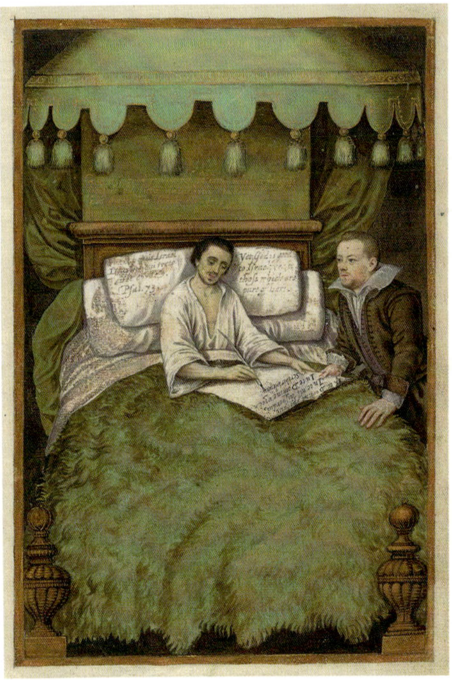

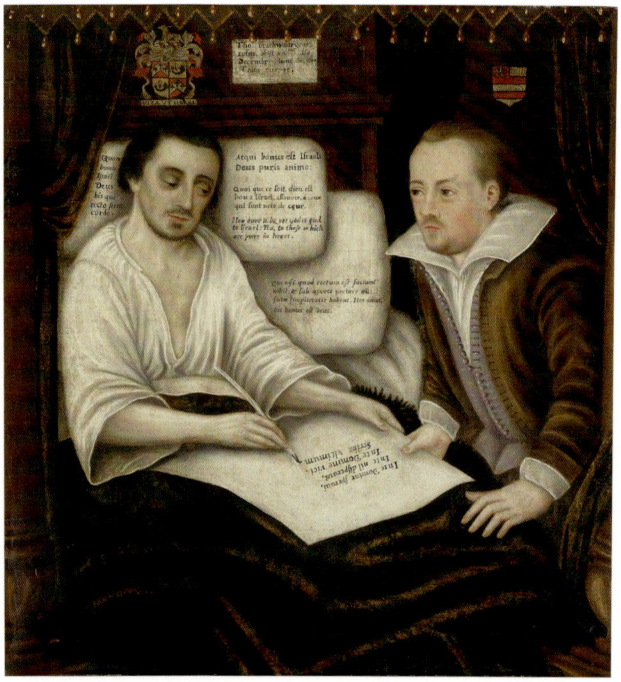

1631, just a month before the poet's passing.[15] The sermon is Donne's meditation on the circumstances of his own death, and the long title of the published sermon explains that the king's household afterward called it "the doctors owne funerall sermon." Droeshout's print is just one of a few contemporary images made after the sepulcher at St Paul's; their multiplication reveals how public fascination was stirred by Donne's prescient, self-orchestrated commemoration.[16]

In the year that Donne was buried, the antiquary and poet John Weever (1575/76–1632) published the first examination of British funeral monuments and inscriptions, revealing a growing historical consciousness of commemoration practices.[17] The book's frontispiece shows Weever, aged fifty-five, with his hand resting on a skull; the accompanying verses trace the landmarks in his own life: "Lanchashire gave him breath, /And Cambridge education. / His studies are of Death. / Of Heaven his meditation" (FIG. 26). Weever died a year later, and in little more than a decade, commemorative practices would be radically altered. The iconoclasm of the English Civil War (1642–51) marked a break with the past, and for a time burial sites and funerary practices became less ostentatious, in line with Puritanical belief and custom. However, in the wake of Charles I's execution at the hands of the Parliamentarians on January 30, 1649 (FIG. 27), there emerged among Royalists a wealth of commemorative materials marking the king's death: prints and illustrated books, medals, jewels, and embroidery. Examples in the Bryan Collection include a pendant in gold and jet that bears an effigy of the king (page 474) and two portraits in less conventional media: one in silk, silver gilt, and glass (page 475); the other in ink on vellum, said to depict the king as he sat for his trial on January 23 (page 476). All bear the date 1648 (Old Style).

That year, which divides the century in half just as its events divided the nation, became a touchstone for many lives and biographies. The precarious politics of the late seventeenth and early eighteenth centuries—

FIG. 22 *Thomas Braythwait Making His Will*, after 1607, watercolor on paper with gold illumination, 9 ½ × 6 ¼ in. (24 × 15.8 cm); sheet: 13 ⅛ × 8 ¾ in. (33.3 × 22.1 cm). Yale Center for British Art, Paul Mellon Fund

FIG. 23 *Thomas Braithwaite of Ambleside Making His Will*, after 1607, oil on canvas, 26 × 23 in. (66 × 58.5 cm). Abbott Hall Art Gallery and Museum, Kendal, Cumbria

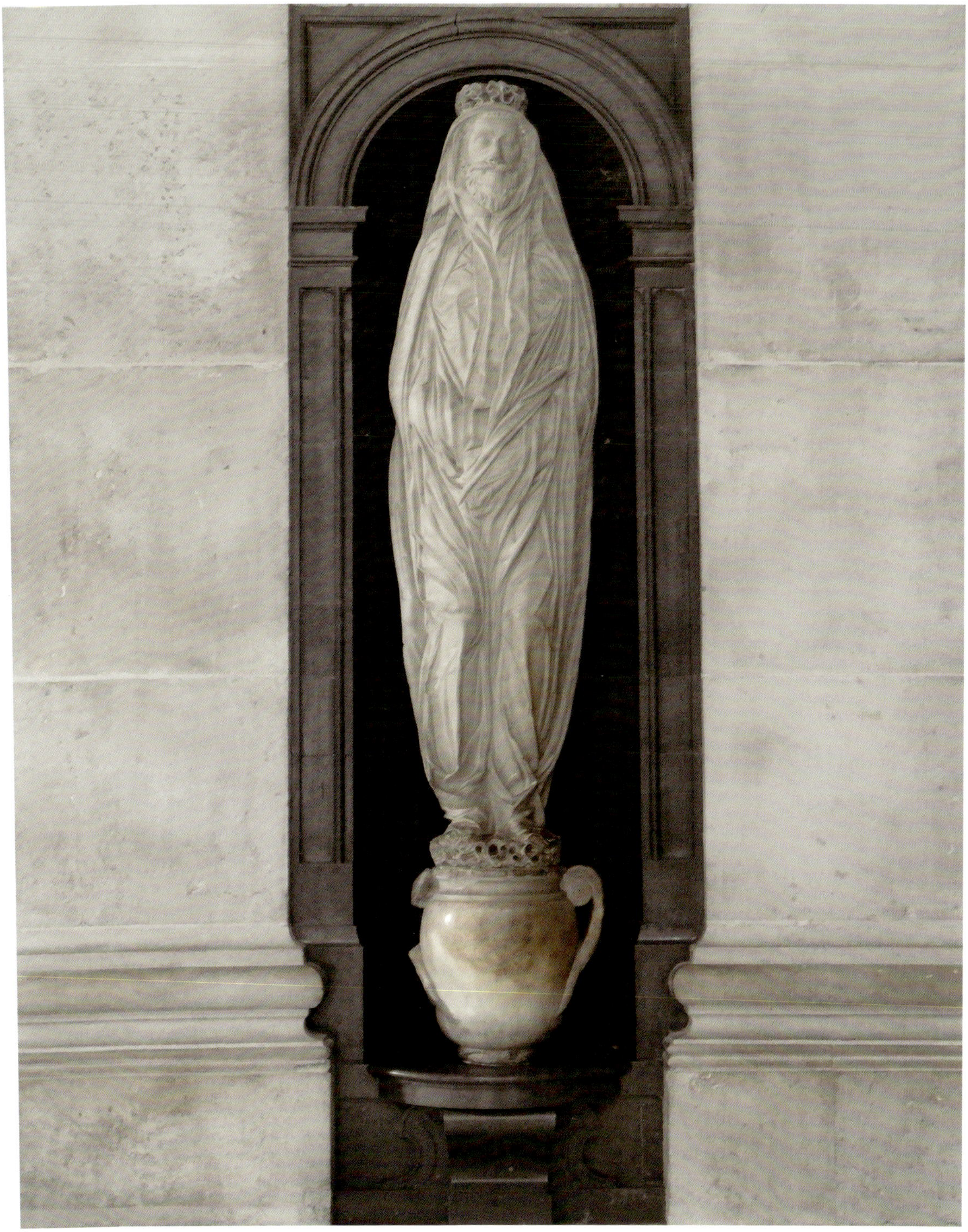

FIG. 24 Nicholas Stone the Elder, effigy, mural monument to John Donne, 1631, carved marble, height: 118 $\frac{1}{8}$ in. (305 cm). St Paul's Cathedral, London

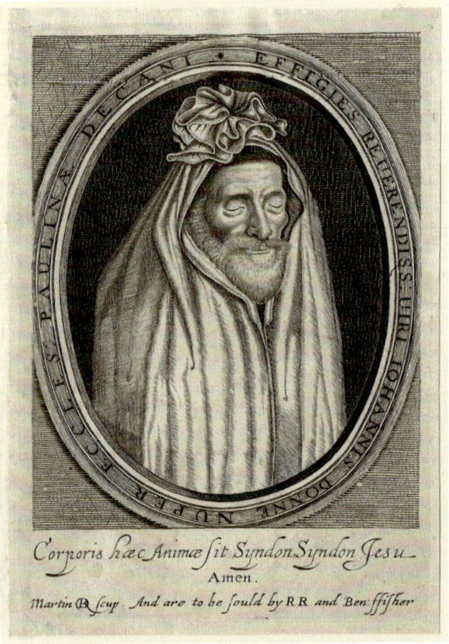

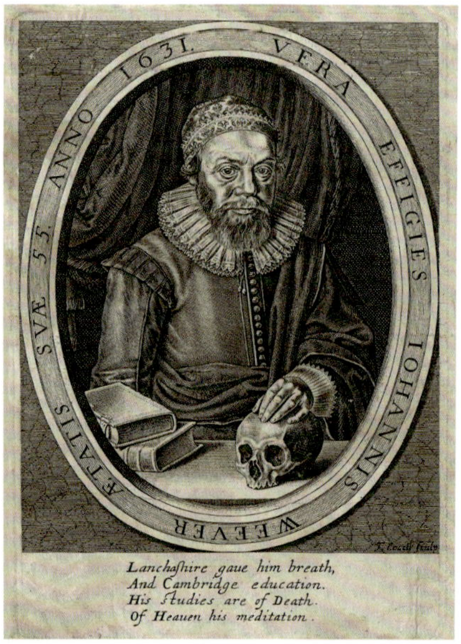

FIG. 25 Martin Droeshout, *John Donne*, 1633, engraving on paper, 6⅜ × 4¼ in. (16.2 × 10.7 cm). National Portrait Gallery, London, Given by the daughter of compiler William Fleming MD, Mary Elizabeth Stopford (née Fleming), 1931

FIG. 26 Thomas Cecill, *John Weever*, 1631, engraving on paper, 6¾ × 4⅝ in. (17.2 × 11.8 cm). National Portrait Gallery, London, Given by the daughter of compiler William Fleming MD, Mary Elizabeth Stopford (née Fleming), 1931

especially the Catholic scare of the 1670s and 1680s (including the Popish Plot of 1678 and resulting Exclusion Crisis), the Monmouth Rebellion in 1685, and the deposition of James II (the so-called Glorious Revolution) in 1688—recalled, for many who had lived through them, the troubles of the 1640s. The image of Charles I as martyr was frequently re-inscribed as an emblem: not just as a memento mori but as an expression of longing among Jacobites for the return of the Stuart dynasty. The death of the king haunted political memory for generations. A needlepoint portrait of the king (after a painting by Edward Bower), inscribed "ANNA SKINNER IN THE 69 YEAR OF HER AGE 1716," is a case in point (page 478), although Skinner's choice of subject may have had as much to do with personal history as with political allegiance.

After the restoration of the monarchy in 1660, commemorative culture became increasingly diverse, at least judging from what survives. The market brimmed with new accessories that the middling sort could afford, including a variety of memento mori jewels, vessels, medals, and tokens. The pair of "mourning slides" dated January 28, 1684, are a prime example; each slide is made of gold and glass and features a tiny enamel skeleton that carries a death's arrow and sandglass (page 467). These skeletons are set against precious fretworks woven from the deceased's hair. A similar slide, fashioned into a ring in the twentieth century, is in the Yale Center for British Art's collection (FIG. 28). The initials of the dead person, "F E," are formed from twists of gold wire, and between the letters is an ivory skeleton; the underside is engraved with the date of the death (*obit*), "obt. 28 Nov.[16]89." A pair of rings fashioned from gold and rock crystal, dating much later (1753), are likewise containers, or lockets, for preserving the hair of the dead (page 468); the materials are reminiscent of a medieval reliquary. The rings and the slides (worn on a band of fabric) were typically given to pallbearers, family members, and friends as mementos of their loved ones. These objects commemorate the dead at the same time as they anticipate the deaths of those left behind.

Literature associated with death and dying reflected this proliferation of commemorative accessories. The most popular books of the period have titles named for objects: for example, John Flavel's *A Token for Mourners* (1674) and John Dunton's *The Mourning-Ring, In Memory of Your Departed Friend* (1692).[18] Flavel (ca. 1627–1691), a Presbyterian clergyman, was concerned about the material and social "excesses" of mourning practices, a perspective he shared with the Nonconformist minister John Owen (1616–1683), author of *Immoderate Mourning for the Dead, Prov'd Unreasonable and Unchristian* (1680).[19] Both works were aimed at easing parents' grief after the death of children, and both emphasized spiritual counsel and faith, rather than elaborate material ritual, as a means of grieving.

On the other hand, John Dunton (1659–1733), a highly creative publisher and bookseller, famed for his travels around New England (1685–86), told his readers that he called his book *The Mourning-Ring* so "that it might be given at Funerals, instead of Gloves, Bisquets, Wine." He added, "Those that think it proper, may Print in a Sheet of Paper, the most Material Passages, in the Life of their Dead Friend, and bind it up with those [copies of the] Mourning-Ring, they give away at his Funeral."[20] Dunton further explains that there is "room left in the Title [page] of this Book, for inserting his Name and Place of Burial" and recommends that copies be "Bound in Black, with a Cypher of Mortality."

The Mourning-Ring is actually a composite publication containing shorter tracts that Dunton published before his time in New England, including *The House of Weeping, or, Mans Last Progress to His Long Home* (1682).[21] A substantial section of the new book was devoted to the "Death[s] and last Sayings

FIG. 27 *Death Warrant of Charles I*, dated 1648, ink and wax seals on parchment membrane, $8\frac{1}{4} \times 17\frac{1}{4}$ in. (21 × 43.8 cm). Parliamentary Archives, London

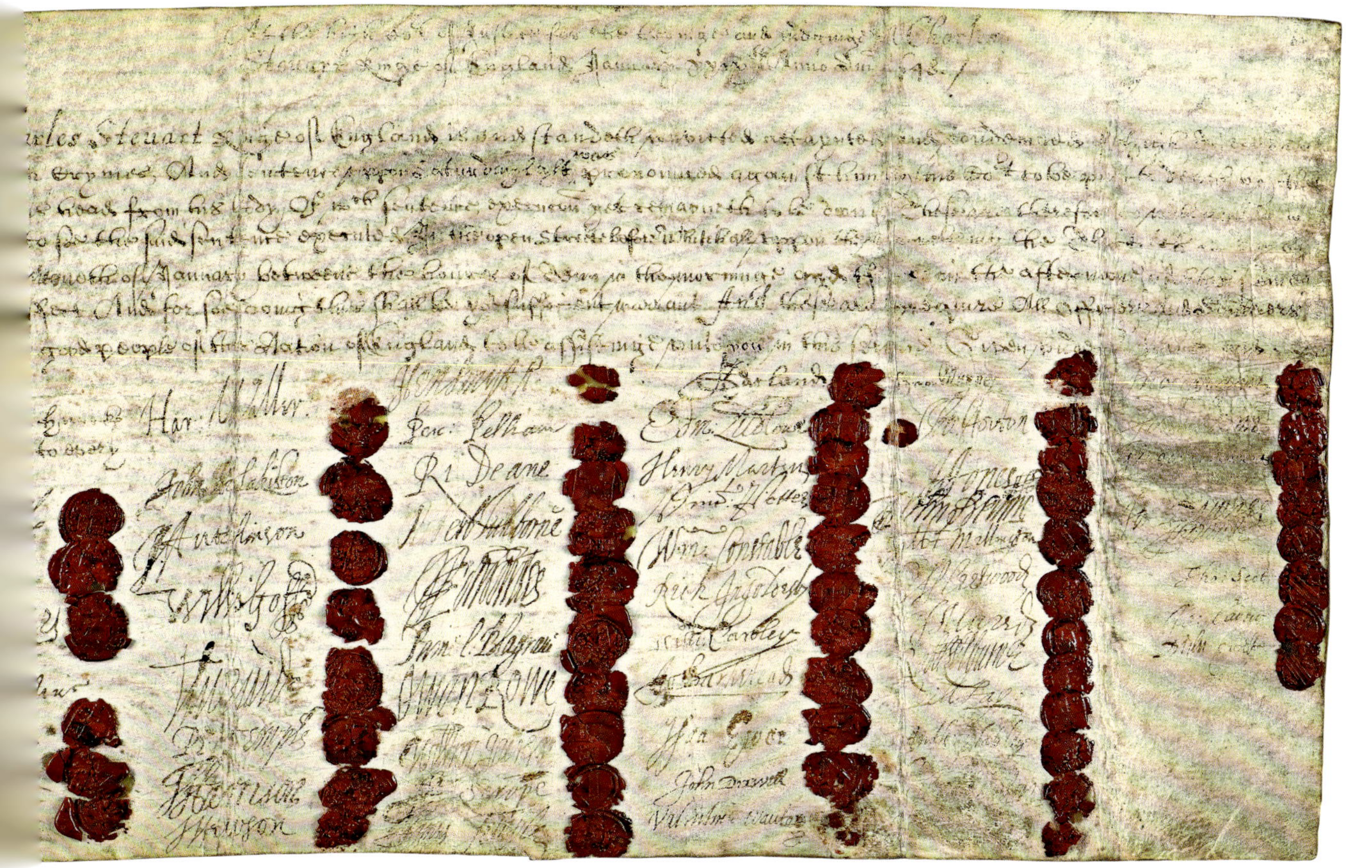

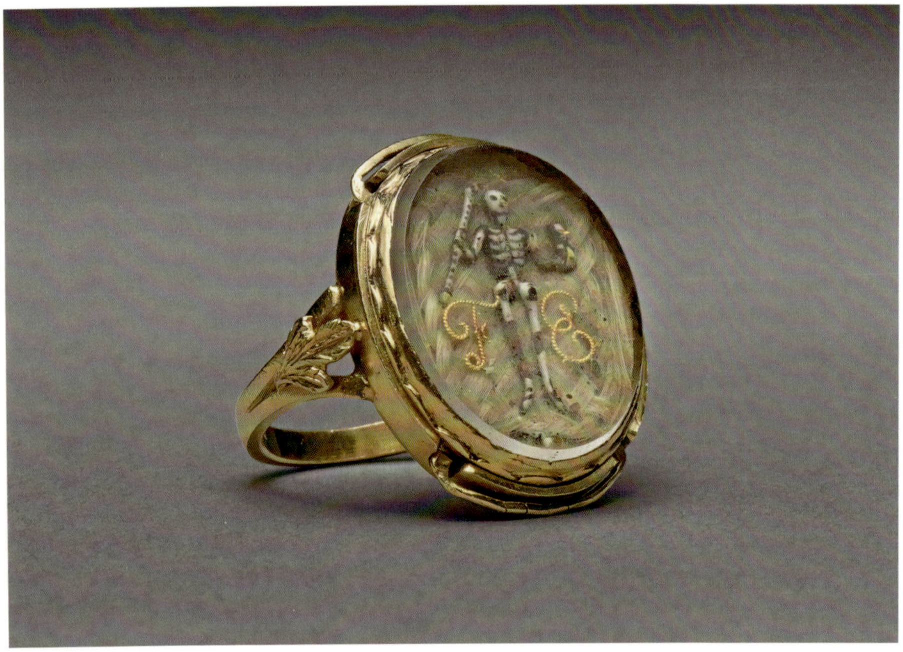

FIG. 28 Memento mori ring, 1689, gold, glass, ivory, gold wire, and hair; face: $\frac{7}{8} \times \frac{5}{8}$ in. (2.22 × 1.59 cm); band: $\frac{3}{4}$ in. (1.9 cm) in depth. Yale Center for British Art, Gift of Susan M. Yecies, Yale B.A. 1971

of Eminent Persons," beginning with Jesus Christ, running through several ancient and modern figures, and ending with Charles I, Charles II, and, finally, "Old Mr. Eliot, of New England," aged eighty-six. This man was John Eliot (ca. 1604–1690), the "apostle of the Indians," who had translated the Bible into Algonquian and with whom Dunton spent time in Roxbury, Massachusetts. Eliot's *Mamusse Wunneetupanatamwe Up-Biblum God* (1663) was the first Bible to be printed in British North America.

The "lively" frontispieces Dunton commissioned for *The House of Weeping* and *The Mourning-Ring* constitute important visual evidence of funerals in the British Isles during the period (FIGS. 29 and 30).[22] Each of the images is divided into sections showing different stages of the funeral rite, with depictions of black grieving garments (including head coverings for both men and women) and the decoration of the hearse with heraldic insignia.

Dunton's frontispieces also exhibit the mode for binding up the body of the dead in wool. Parliament had recently passed the Burial in Woollen Acts (1666, 1678, 1680), which required burial in British wool—strictly excluding the use of, or mixture of wool with, any foreign textile—under pain of a five-pound fine. Those who could not afford a woolen shroud were to be buried naked. An affidavit dated April 1, 1680, certifying the burial in wool of Mary Harvey of Gamlingay, Cambridgeshire, survives at the Yale Center for British Art (FIG. 31). The border is decorated with the imagery of death, including coffins, scythes, and arrows, and the depiction of a female body in a winding sheet.

Dunton provides a connection to Elizabeth (née Singer) Rowe (1674–1737), one of the most important British writers on the subject of death in the eighteenth century. In 1691, at the age of seventeen, Singer anonymously contributed verses to Dunton's periodical, the *Athenian Mercury*. She only later revealed her identity to Dunton, who published her first independent collection, *Poems on Several Occasions: Written by Philomela* (1696).[23] In 1710, Singer married fellow writer Thomas Rowe, who was thirteen years younger, but he tragically died five years later of consumption, aged twenty-eight. His death inspired one of her most successful poems, "On the Death of Mr Thomas Rowe" (1717).

Elizabeth Rowe's most enduring work, however, is *Friendship in Death; in Twenty Letters from the Dead to the Living* (1728). During the eighteenth century, the book was available in more editions than Daniel Defoe's *Robinson Crusoe*, and the number grew to nearly ninety before 1840.[24] One of the most poignant letters is addressed to the "Countess of ****" from her only son and heir, the two-year-old "Narcissus." The infant admonishes his mother for her immoderate grief, explaining that any legacy he would have received on earth pales in comparison to the life he now leads in heaven. The letter ends, "Pardon me, Madam, 'tis you now seem the Infant, and I repay you that superior regard and tenderness which you lately bestowed on me."[25] Most of the letters refer to the bliss of eternal life, suggesting that death is not to be feared but welcomed. Rowe herself died on February 20, 1737, aged sixty-two, having made "meticulous preparations" for her own funeral. She left behind several letters to friends. One of these was to the minister Isaac Watts, whom she appointed to edit her religious meditations.[26]

Rowe's letters reflect a shift in focus from the death and burial of the body to the immortality of the soul. She provides an important link with another literary giant of the period, the clergyman and poet Edward Young (1683–1765), who was the dedicatee of Rowe's *Friendship in Death*. Young's

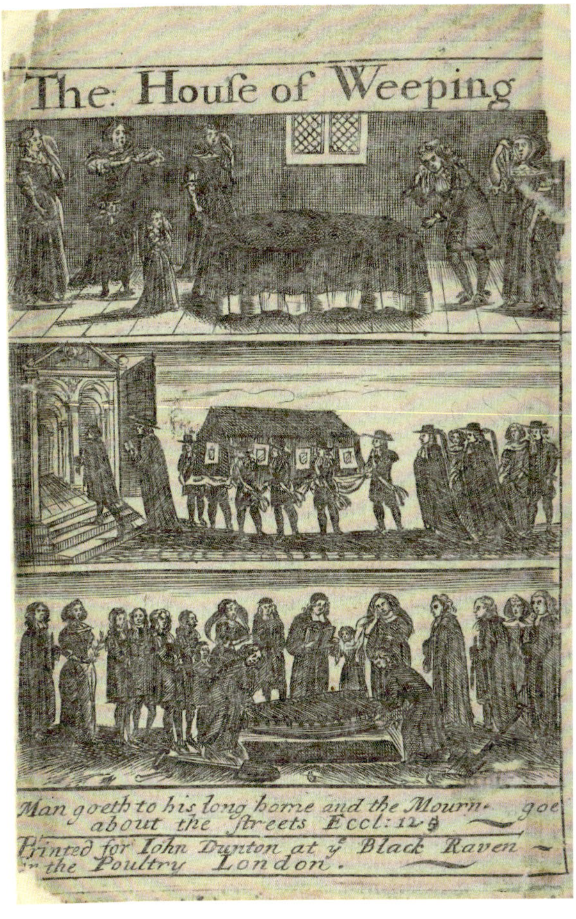

FIG. 29 Frontispiece, *The House of Weeping, or, Mans Last Progress to His Long Home* (London: Printed for John Dunton, 1682), etching on paper. British Library

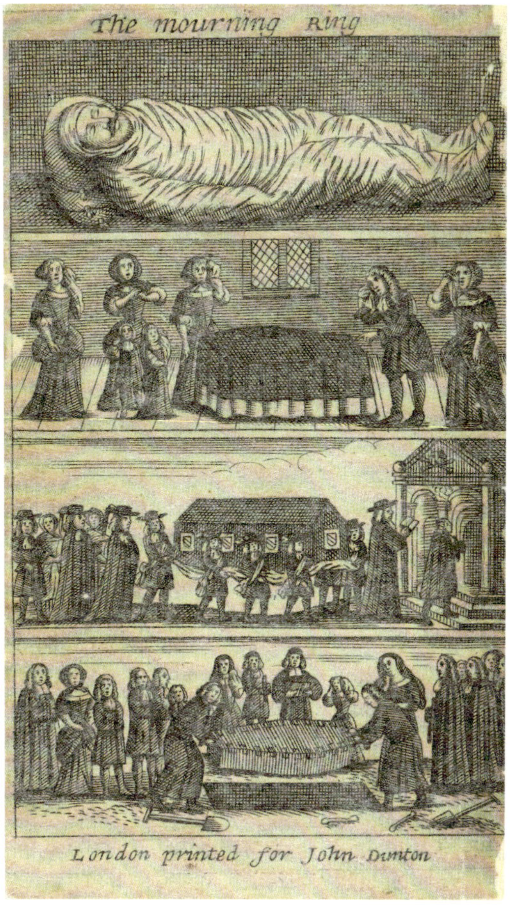

FIG. 30 Frontispiece, *A Mourning-Ring, in Memory of Your Departed Friend* (London: Printed for John Dunton, 1692), etching on paper. British Library

most famous work, *The Complaint; or Night Thoughts, on Life, Death and Immortality* (1742)—another bestseller of the period—is said to have been inspired by the death, in 1740, of his wife, Elizabeth (née Lee), the daughter of the first Earl of Lichfield. It takes the form of a long poem, reminiscent of John Milton's *Paradise Lost*, on the power and mystery of death: death as sublime. Admired by the Romantics—by Edmund Burke, in particular—for its imagery of melancholy, moonlight, and terrors, Young's book, like Rowe's letters, shifts the focus from the decay of the material body to the intangible afterlife of the spirit. Young goes as far as to comment on "needless monuments," asking, "What are our noblest ornaments, but deaths / Turn'd flaterers of life, in paint, or marble, / The well-stain'd canvas, or the featur'd stone?" Instead, for inspiration, he looks to nature and to the higher power that created it: "Teach me by this stupendous scaffolding, / Creation's golden steps, to climb to THEE."[27]

FIG. 31 Affidavit certifying the burial in wool of Mary Harvey of Gamlingay, Cambridgeshire, 1680, engraving, ink and wax seals on paper, 12 ¼ × 8 ⅝ in. (31 × 22 cm). Yale Center for British Art, Paul Mellon Fund

It is tempting to see the cultural perception of death in the eighteenth century as gravitating toward a new Romantic emphasis on the immortality of the spirit—which might lead one to expect a decrease in the material culture around death and commemoration—but, in fact, these phenomena continued to find complex material expression. When Young died at Welwyn, Hertfordshire, on April 12, 1765, his parishioners could not help but adorn his resting place. "The bell did not toll at his funeral, nor was any person allowed to be in mourning," according to a later remembrance. Nonetheless, the "altar-piece [under which Young is buried] is reckoned one of the most curious in the kingdom, adorned with an elegant piece of needlework by the late lady Betty Young."[28]

In fact, material expressions of death and its commemoration seem to have been as diverse as ever by the end of the eighteenth century. In 1782 alone, we find two dated objects that serve very different commemorative functions. One, an oval-shaped silver box, formerly contained (according to its engraved inscription) "some of the remains of a most kind, virtuous & affectionately loving Wife, M^rs: Sarah Kingsford in whom centered every female & desirable excellency... died much lamented by all that knew her Ap. 29th 1782, & left one daughter, Sarah Maria born April 16. 1782" (page 469). The other is a horn cup commemorating the execution of David Tyrie, the last man to be hanged, drawn, and quartered in England. Tyrie had been convicted of betraying military secrets to the French (page 471). The silver box is an object of devotion, a vessel to contain the remains of the dead (perhaps a lock of hair, now missing). The cup is a souvenir, putting us in mind of the public house—though, doubtless, a sober reminder of the horrors resulting from disloyalty to the state. The cup's message is a far cry from the high-minded sentiments of Rowe's letters or Young's poem, revealing of the contradictions inherent in eighteenth-century society's thinking about death.

There is an everlasting human need to commemorate a life (or life lost), be it by fashioning or inscribing an object or by marking a grave. Strangely, there is a sense of security bestowed by acknowledging the moment of death and positioning it in space and time. Dated objects from the early modern period indicate a diverse and complicated expression across British society, even as literary culture gradually shifted its emphasis from the decay of the physical body to metaphysical considerations of the transcendence of the soul. In the popular culture of our increasingly secular society, there seems to have been a return to a focus on the body—as

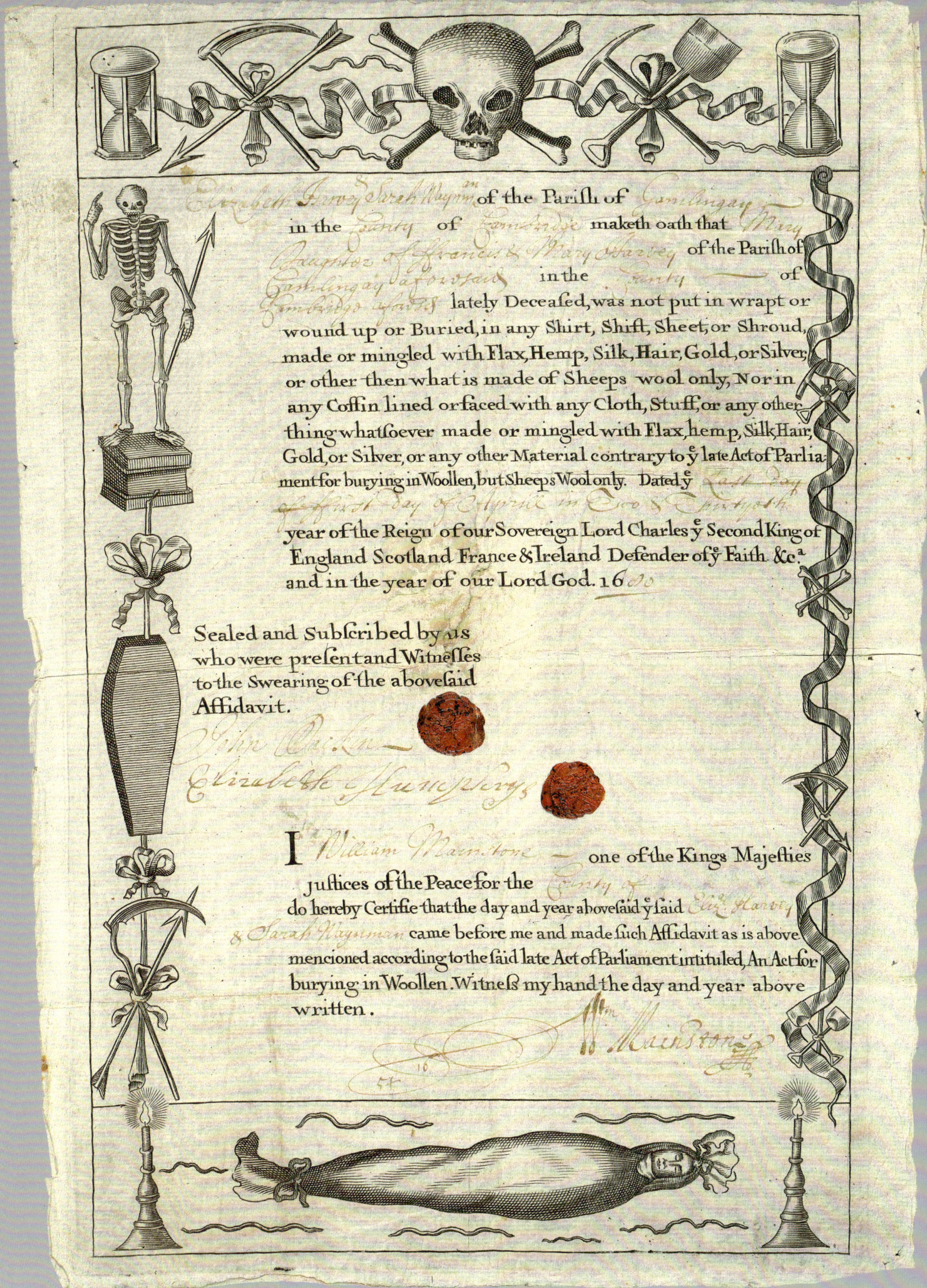

Elizabeth Harvey & Sarah Waynman of the Parish of Gamlingay in the County of Cambridge maketh oath that Mary Daughter of Francis & Mary Harvey of the Parish of Gamlingay aforesaid in the County of Cambridge aforesd lately Deceased, was not put in wrapt or wound up or Buried, in any Shirt, Shift, Sheet, or Shroud, made or mingled with Flax, Hemp, Silk, Hair, Gold, or Silver, or other then what is made of Sheeps wool only, Nor in any Coffin lined or faced with any Cloth, Stuff, or any other thing whatsoever made or mingled with Flax, hemp, Silk, Hair, Gold, or Silver, or any other Material contrary to ye late Act of Parliament for burying in Woollen, but Sheeps Wool only. Dated ye ~~Last day~~ of ye first day of April in Two & Thirtyeth year of the Reign of our Sovereign Lord Charles ye Second King of England Scotland France & Ireland Defender of ye Faith &ca. and in the year of our Lord God. 1680

Sealed and Subscribed by us who were present and Witnesses to the Swearing of the abovesaid Affidavit.

John Parker

Elizabeth Humphreys

I Wm William Mainstone one of the Kings Majesties Justices of the Peace for the County of do hereby Certifie that the day and year abovesaid ye said Eliz. Harvey & Sarah Waynman came before me and made such Affidavit as is above mencioned according to the said late Act of Parliament intituled, An Act for burying in Woollen. Witness my hand the day and year above written.

Wm Mainstone

10

54

in film, for example—but these artistic expressions only occupy a circumscribed part of contemporary culture.

Even despite the continuing threats of global pandemics, reminders of death are no longer as widely visible as they once were in Western culture. In more economically developed countries, where people now live much longer—and some with a sense of invincibility—there is no daily reckoning with fate, which was unavoidably the way of life for our ancestors.

Notes

Epigraph: These words are spoken by Hotspur after he is mortally wounded by Henry at the Battle of Shrewsbury. Arthur Henry Bullen, ed., *The Works of William Shakespeare Gathered into One Volume* (New York: The Shakespeare Head Press Edition Published by the Oxford University Press, 1938), 446.

1. On June 23, 1666, Elizabeth Hawgood wrote in the margin of Psalm 104, page 68, "Som learne to love thy leson it is most plesant / having once learned it you never will forget it / againe."

2. Julie-Marie Strange, "Grave Expectations: Have We Lost the Ability to Handle Death in the Modern World?," *Times Literary Supplement*, June 8, 2018, 3.

3. Nigel Llewellyn, *The Art of Death: Visual Culture in the English Death Ritual: c. 1500–c. 1800* (London: Reaktion Books, 1997), 10.

4. Llewellyn, *The Art of Death*, 136 (italics added for emphasis).

5. It is worth mentioning that the diversification of the kinds of commemorative objects discussed here resonates with the commodification of gravestones in New England during the same period; see Edward S. Cooke Jr., "Slate," in *Inventing Boston: Design, Production, and Consumption* (New Haven, CT: Yale University Press, for the Paul Mellon Centre for Studies in British Art, 2019), chap. 2. Looking beyond the common historical opinion that gravestones simply reflected a "Puritan view of death," Cooke traces the development of gravestone imagery and places our understanding of that imagery—and the shapes of the gravestones themselves—into the wider context of visually similar household objects. For example, he notes the similarity between the shapes of gravestones and the shapes of contemporary cane chairs. He also makes a telling comparison between the decorative borders of gravestones and those on the title pages of early modern books; in both, the decoration serves as a frame and draws the reader to the information contained within it.

6. *Online Etymology Dictionary*, s.v. "commemoration, *n.*," https://www.etymonline.com/word/commemoration.

7. *Online Etymology Dictionary*, s.v. "commemorate, *v.*," https://www.etymonline.com/search?q=commemorate.

8. *Online Etymology Dictionary*, s.v. "memento, *n.*," https://www.etymonline.com/search?q=memento.

9. Latin Vulgate version of the Bible.

10. Robert Harding, "The Art of Dying: The Brathwaite Bible Miniature," *British Art Journal* 1, no. 1 (Autumn 1999): 16–19. Brathwaite's will was proved June 20, 1608; see York Register V30, ff. 61–62, Borthwick Institute of Historical Research, University of York.

11. See Llewellyn, *The Art of Death*, 38, fig. 27; and Robert Tittler, *Portraits, Painters, and Publics in Provincial England, 1540–1640* (Oxford: Oxford University Press, 2013), 23–24, fig. 2.3.

12. Unknown carver, *Thomas Brathwaite and George Preston*, before 1640, wood panel, Abbott Hall Art Gallery and Museum, Kendal, Cumbria.

13. Izaak Walton, *The Life of John Donne, Dr. in Divinity, and Late Dean of Saint Pauls Church London*, 2nd ed. (London: Printed for R. Marriot, 1658), 112–13.

14. The translation of the inscription around the edge of the portrait is "Portrait of the most reverend John Donne late Dean of St Paul's Church," and the inscription in the print's caption reads, "This is the shroud of my body, may Jesus be the shroud of my soul."

15. John Donne, *Deaths Duell, or, A Consolation to the Soule, against the dying Life, and living Death of the Body. Deliuered in a Sermon at White Hall, before the Kings Maiesty, in the beginning of Lent, 1630, By … Iohn Donne … Being his last Sermon, and called by his Maiesties household 'The Doctors Owne Funerall Sermon'* (London: Richard Redmer and Beniamin Fisher, 1632).

16. On Nicholas Stone's funerary monument to Donne, see most recently Philip Cottrell, "John Donne, Undone, Redone: The John Donne Monument Reconsidered," in *Death, Burial, and the Afterlife: Dublin Death Studies*, ed. Philip Cottrell and Wolfgang Marx (Dublin: Carysfort Press, 2019), 33–64. Other contemporary images of Donne's sepulcher include the decorated title page made by William Marshall for a posthumous edition of *Devotions* (London: August Mathewes, 1634) and Wenceslaus Hollar's illustration in William Dugdale, *The History of St Pauls Cathedral in London, From its Foundation untill these Times …* (London: Printed by Thomas Warren, 1658), 62.

17. John Weever, *Ancient Funerall Monuments within the United Monarchie of Great Britaine, Ireland, and the Islands Adiacent* (London: Thomas Harper, 1631).

18. John Flavel, *A Token for Mourners, or, the Advice of Christ to a Distressed Mother Bewailing the Death of Her Dear and Only Son*... (London: Robert Boulter, 1674); John Dunton, *The Mourning-Ring, In Memory of your Departed Friend*... (London: Printed for John Dunton, 1692).

19. John Owen, *Immoderate Mourning for the Dead, Prov'd Unreasonable and Unchristian*... (London: John Williams, 1680).

20. Dunton, "To the Reader," in *The Mourning-Ring*, n.p.

21. Dunton, *The House of Weeping, or, Mans Last Progress to His Long Home*... (London: Printed for John Dunton, 1682).

22. On the title page to *The House of Weeping*, Dunton describes that work's frontispiece as a "Lively Emblem of a Funeral Solemnity."

23. Jonathan Pritchard, "Rowe [née Singer], Elizabeth (1674–1737), poet and devotional writer." *Oxford Dictionary of National Biography*, https://doi.org/10.1093/ref:odnb/24198.

24. Paula R. Backscheider, *Elizabeth Singer Rowe and the Development of the English Novel* (Baltimore: Johns Hopkins University Press, 2013), 2.

25. Elizabeth Rowe, "Letter 3," in *Friendship in Death; in Twenty Letters from the Dead to the Living*... (London: Henry Lintot, 1750), 12.

26. Pritchard, "Rowe [née Singer], Elizabeth."

27. Edward Young, "Night 9," in *The Complaint; or, Night Thoughts, on Life, Death, and Immortality* (London: Thomas Tegg, 1812), 227, 242.

28. George Wright, "Memoirs of the Late Dr. Edward Young," introduction to Edward Young, *The Complaint*, vii.

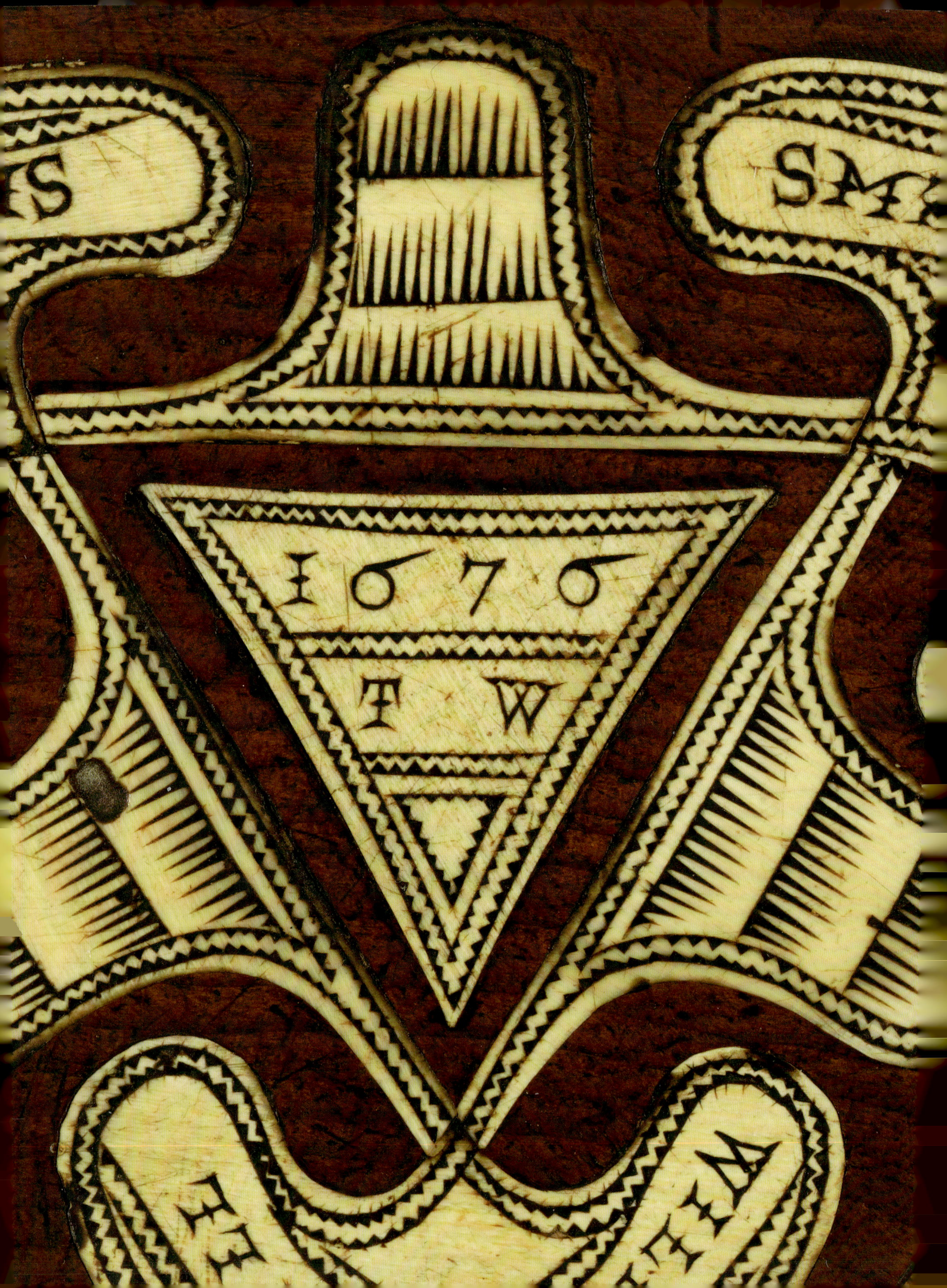
1676
T W

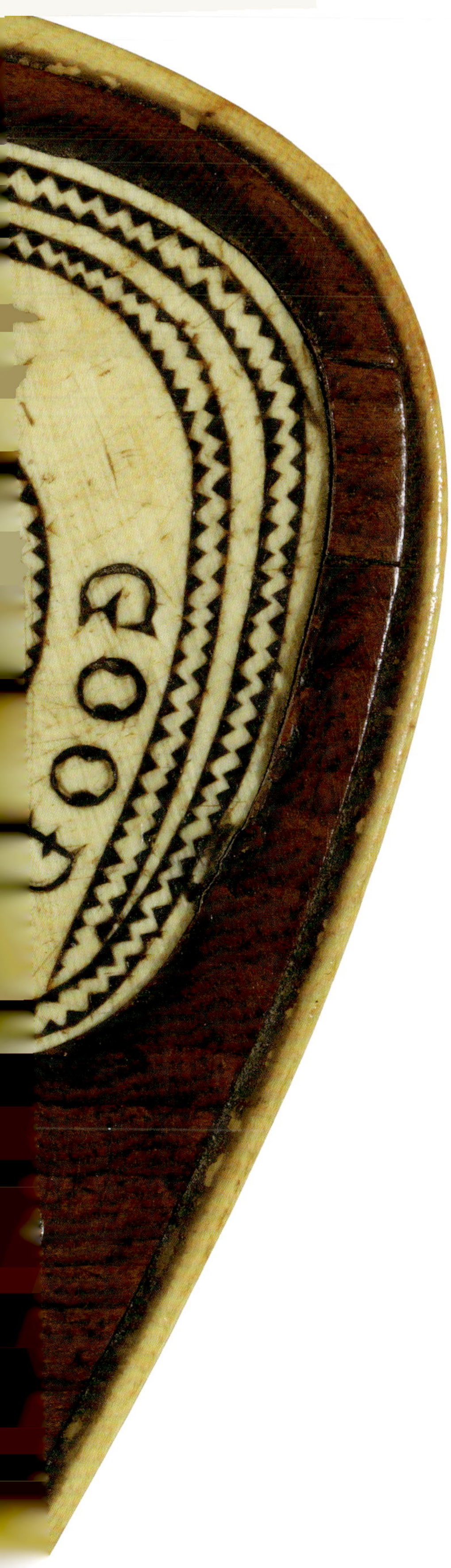

Catalogue of Objects

Introduction to the Catalogue

RATHER THAN BEING ORGANIZED SIMPLY BY DATE, MATERIAL, OR TYPE, the objects in this catalogue are ordered in sections that mirror successive stages of life in early modern Britain, beginning with birth and ending with death. Some of the stages in between are more obvious than others, for there are also sections relating to domestic, professional, social, and religious life. Although none of these was considered a specific moment in the early modern conceptualization of a life cycle, they were all among the pervasive aspects of life that warrant discussion.

Dates

The inscribed date, as the fundamental premise for each object's inclusion within this publication, is prominent at the start of each catalogue entry. Although often the case, dates do not always reflect when the object was made, as an object may carry a date, or dates, that significantly post- or predate its creation. The catalogue entry makes note of any discrepancy between the date of production and the date inscribed. When more than one date is inscribed on an object, the dates are listed in chronological order, irrespective of their position on the object.

When three or more objects of the same type appear in the same section, they are presented together, within close chronological order, to demonstrate the changes in form, use, and design that developed over time.

Dates that appear within the entries remain in Old Style, with the year beginning on March 25 rather than January 1. This treatment is in recognition of the fact that it was not until the year following the so-called Calendar Act of 1751 that Britain adopted the Gregorian Calendar. Accordingly, until 1752, dates between January 1 and March 24 are expressed as double years (for example, "February 1, 1683/4"). However, the parenthetical life dates for makers, found at the head of each entry, have been modernized to New Style, as have those dates included within the provenance descriptions. Some fiscal reckonings and office appointments ran across two years (for example, from Michaelmas of one year to Michaelmas the following year). In these cases, the years are presented as inclusive: "1683–84."

Object Names and Dimensions

Within each entry, the name of the object follows the date or dates. A concerted effort has been made to name things as simply as possible, using period-appropriate language in step with Victor Chinnery's *Names for Things: A Description of Household Stuff, Furniture and Interiors, 1500–1700* (West Yorkshire: Oblong Creative, 2016). At all times, attempts have been made to avoid language used in the antiques trade, such as "Charles II cabinet." Where an object's intended purpose is ambiguous, this uncertainty is acknowledged by providing alternative names. Materials are likewise described in the simplest language possible.

Dimensions are in order of height, width, and depth, provided both in inches and centimeters. As part of the research for this project, vessels were measured for their "dry capacity." Based on that measurement, suggestions have been made as to what the vessel may have been intended to contain.

Inscriptions and Marks

Inscriptions have been recorded as accurately as possible and are presented in what has been deemed to be the most logical order, even when there is no clear hierarchy as to their placement on the object. Line breaks are indicated with pipes (|), and semicolons indicate the end of one inscription and the beginning of another in a different place on the object. Inscriptions are often visible in the featured photograph, but when not, their placement is described in the entry. When a triad of initials is recorded, like the letters "T|W|A" cast onto the water cistern shown on page 109, they are read from left to right, with the central letter both above and between the first and the third letters. Where inscriptions have been damaged or are simply unintelligible, square brackets and an underscore indicate the missing or illegible section. Nearly all these inscriptions were made freehand, so distinctions are made between words deemed to be superscript and subscript. Likewise, transcriptions here faithfully record upper- and lowercase letter forms but not italicization. The eccentricities of sixteenth-, seventeenth-, and eighteenth-century spellings have been maintained, although antiquated letter forms, such as the long *S*, have been rendered in their modern equivalents.

Hallmarks, touchmarks, and other dating, making, and ownership marks are described in the prose entry.

Entry Text

Although every effort has been made to achieve consistency, in a catalogue that includes such a diverse group of objects it is difficult to follow a single cataloguing convention that serves each object equitably. Nor can any entry be claimed the definitive account of the object's material properties or history. Rather, personal stories and biographical information, particularly those of makers, have been privileged over rich descriptions of the object's decoration or construction. The latter sort of information can often be found elsewhere, and where useful sources exist, they are recorded in the bibliographic information that follows the entry under the heading "Literature."

Parish registers are an important source for the information presented in the catalogue. The decision was made not to cite each archival reference for these registers because these sources are easily accessible online (albeit through subscription services) and are sufficiently specific in place and time as to be readily discoverable. References to probate wills proved at the Prerogative Court of Canterbury and now kept at the National Archives, Kew, are cited as: "(TNA PROB 11/.../...)" and references to the State Papers as "(SP.../.../...)."

Where an individual is known to have been a member of a corporate body, such as a livery company, the name of the company is capitalized (for example, "William Hole, Goldsmith").

Provenance and Literature

The ownership history of objects of everyday use is often hard to recover. Where an object has a traceable provenance back to its commissioner, that information is provided. Where first owners cannot be identified with total confidence, however, they are not recorded in the provenance information.

The section "Literature" lists those published works specifically cited in the description of the individual object but does not include the various general texts that informed the editors' understanding of the object and its context.

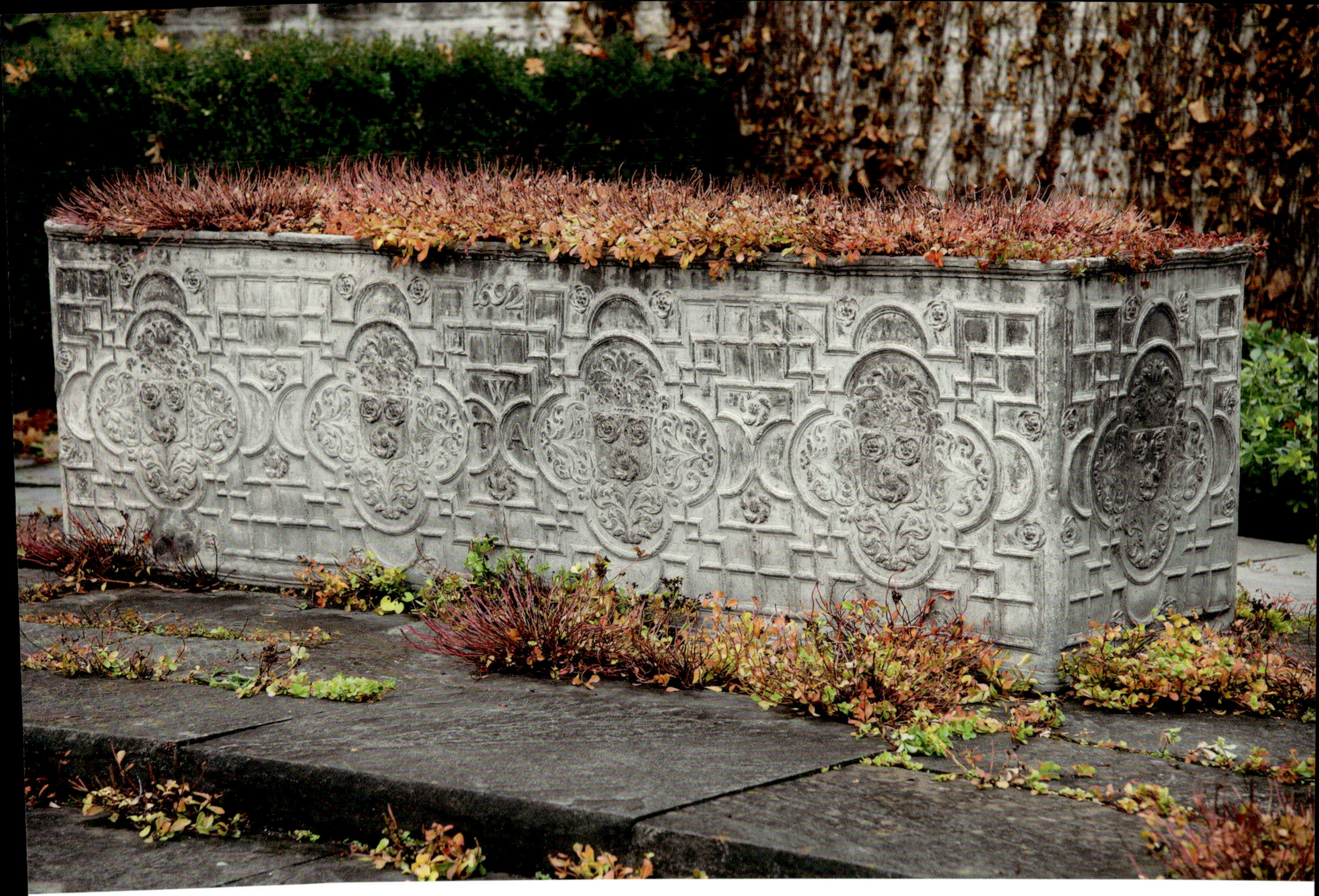

1692

Often located in courtyards—and sometimes gilded, burnished, and brilliantly colored—lead cisterns were used to collect water for drinking, washing, cooking, and other domestic activities. Rainwater caught in lead gutters traveled through downpipes by way of rainwater heads (or hoppers), which, in elite houses, were often decorated with the owners' initials and a date. Mindful of the dangers of lead poisoning, plumbers were careful to tin the inside of cisterns intended for drinking water. The cistern was raised above the ground so a bucket could be placed under the draw-off tap. This cistern formerly resided in the alleyway of 8 Bishopsgate, London, which, in the early nineteenth century, became the headquarters of Barings Bank. Given that the cistern predates the founding of Barings by seventy years, it presumably had belonged to a previous resident of the Bishopsgate property. It was sold after the collapse of the business that had financed the purchase of Maine, the American war effort in 1812, and the Louisiana Purchase.

PROVENANCE

Baring's Bank, London; Purchased November 11, 1996, from Crowther of Syon Lodge by John H. Bryan of Lake Bluff, IL (1936–2018); gifted November 23, 2012, to John H. Bryan III of Nashville, TN

LITERATURE

Lawrence Weaver, *English Leadwork: Its Art & History* (London: B. T. Batsford, 1909), 72–85

Water cistern

Lead

29 ½ × 95 × 27 ½ in. (74.9 × 241.3 × 69.9 cm)

Inscription: 1692; T|W|A

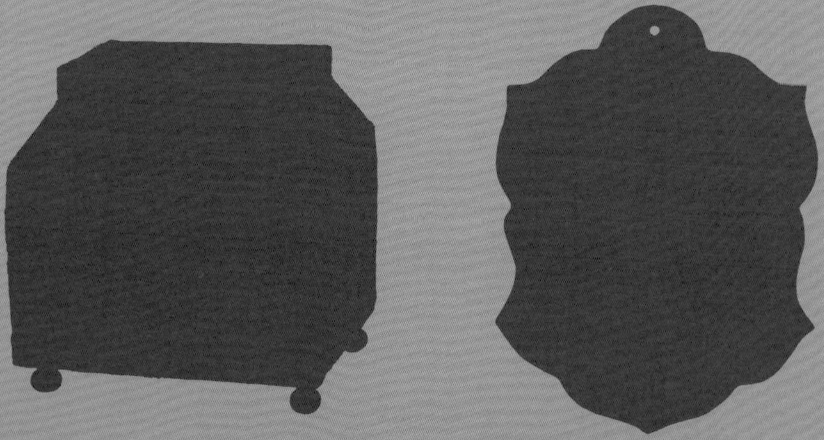

I. CHILDHOOD & YOUTH

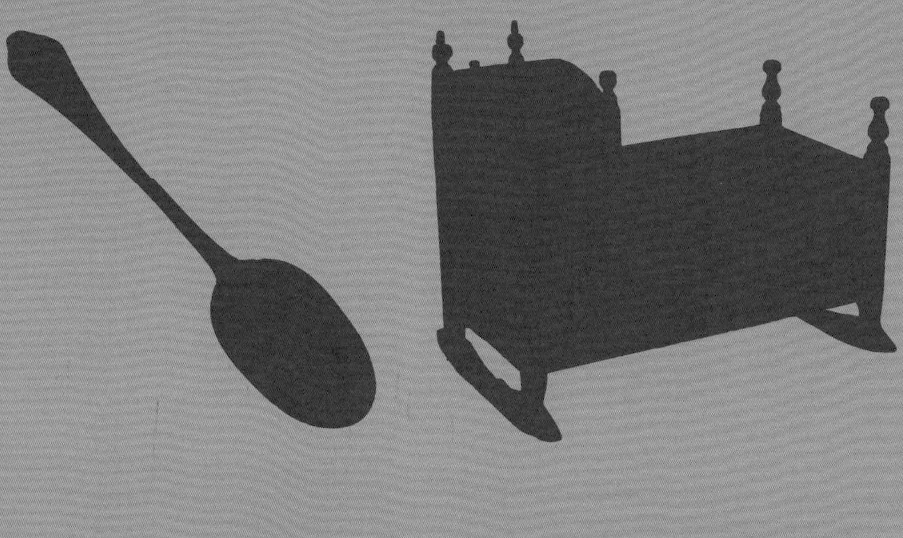

JOY AND TREPIDATION WENT HAND IN HAND WITH PREGNANCY IN early modern Britain. More than one-third of children died before the age of six, and fewer than half reached the age of sixteen. Soon after giving birth, a mother could expect to be visited by a number of "gossips"—a word derived from *God-sibs* or *God siblings*—local women of standing who attended on the mother and welcomed the child into the world. Gossipings, as the visits were known, provided women the opportunity to give gifts to the mother and child.

The high mortality rate created deep-seated anxiety about a child's dying before baptism. Licensed midwives were permitted to baptize the child if it appeared the infant was not going to live long enough for a church ceremony—and, in extremis, even a punch bowl could serve as a font. Although many nonconformists shunned the rituals of the Church of England, most children born during this period were baptized in the church and a record duly made in the parish register. A church baptism allowed the parents to choose godparents, which behooved those chosen to furnish the child with gifts—a porringer or baptismal spoon, which were often personalized through an inscription (pages 114, 116, and 117).

Education could be had either at school or at home. Schools for boys and girls, founded by charitable gifts, grew in number throughout the period. For example, in the mid-seventeenth century, a number of girls' schools sprang up around London, where young women of means could learn to write, play music, and hone their skills with the needle.

Needlework was considered to be a powerful deterrent against idle behavior, and aptitude with the needle was read as an index of a young woman's virtuosity. As a girl's proficiency grew, so the number and complexity of the stitches in her repertoire would increase. A demonstration of these skills could be writ large in the production of a sampler, which commemorated the occasion upon which this mastery had been achieved, as does the sampler made by the ten-year-old Sarah Quare in 1700 (page 126). Likewise, teenage girls could work panels of embroidery to be incorporated into cabinets in which they could store letters, writing implements, keys, jewelry, and other small keepsakes and possessions (page 124).

Boys, particularly those at grammar schools, could expect a more thoroughgoing education. Young men of the aristocracy, gentry, and clergy matriculated at university and, if destined for a career in government or law, might spend time at one of the various Inns of Court.

Most boys and some girls were sent for trade or craft apprenticeships, often at about the age of fourteen, the age at which childhood, such as it was, was considered to have ended. A formal apprenticeship, which could last anywhere from seven to twelve years, required a contract between master and apprentice, ratified through a bond and sealed through payment to the master, the sum proportional to the status of the latter. At the completion of an apprenticeship, however, it was customary for either the apprentice or master to present a spoon to the livery company as a material record of the apprentice's admission. At this stage, most apprentices then became "journeymen," working by contract for more-established company members. Yet some individuals had the means to strike out on their own and would issue plates and other objects inscribed with the year, announcing their arrival on the professional stage (page 147).

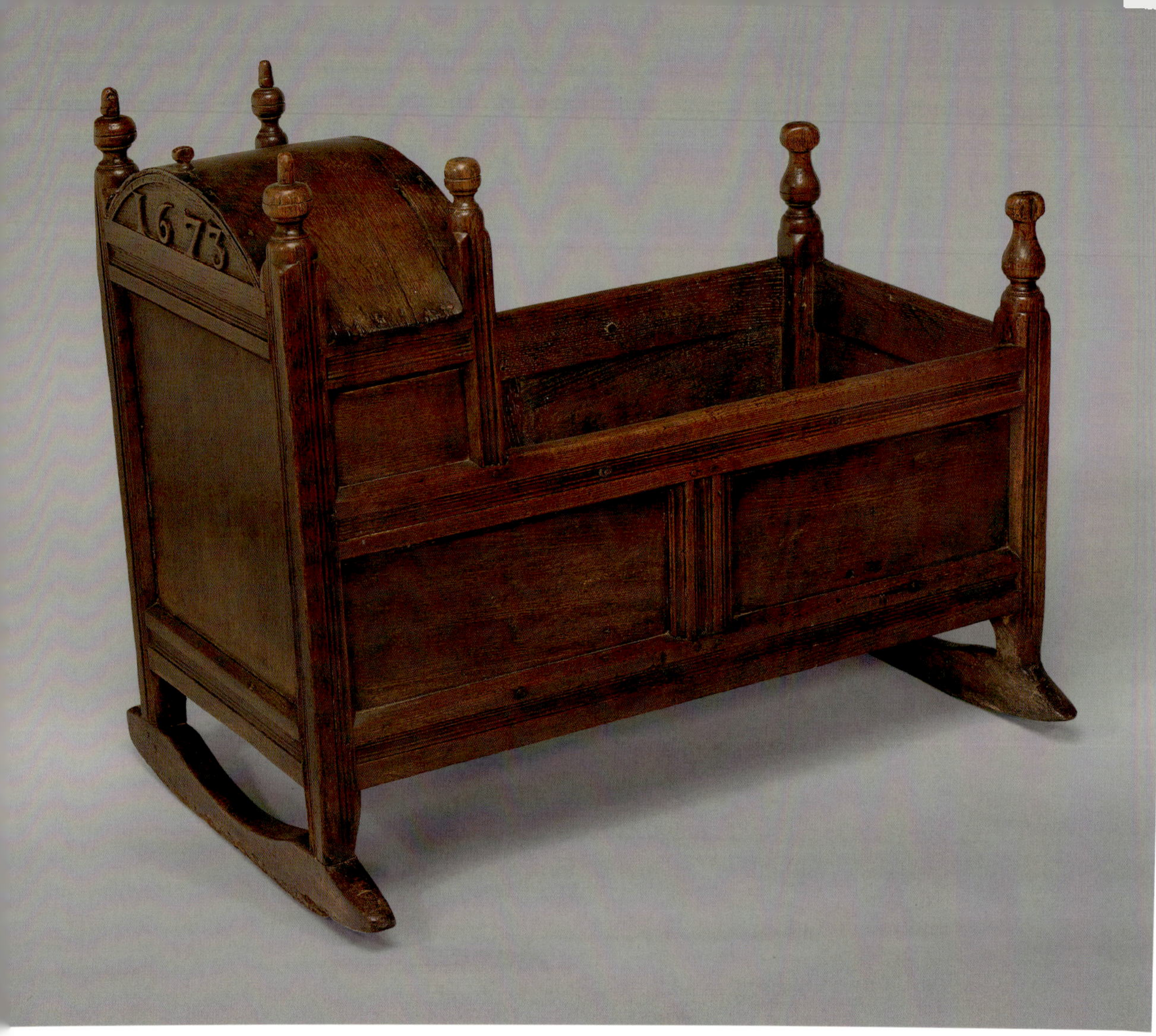

1673

Cradle
Oak
$32\frac{3}{4} \times 24\frac{7}{8} \times 37\frac{5}{8}$ in.
(83.2 × 63.2 × 95.6 cm)

Inscription: 1673

Cradles in this period were either suspended on a frame or built on rockers, enabling mothers and nurses to soothe the infant inside with gentle motion. The rockers on this example may at one point have had an iron lining to protect them from direct contact with the floor, although they nonetheless show a great deal of wear. The date inscribed on the back of the hood, "1673," likely records the date of the occupant's birth—or rather, the birth of the first occupant; this cradle has been repaired numerous times, which suggests it held more than one generation of infants.

PROVENANCE
Bonhams, London, September 20, 2017 (lot 491); bt. John H. Bryan III of Nashville, TN, through Tobias Jellinek Antiques, London

LITERATURE
Victor Chinnery, *Oak Furniture: The British Tradition, A History of Early Furniture in the British Isles and New England* (Woodbridge, UK: Antique Collectors' Club, 1979), 399

1747

Alexander Tillet, the son of Wayman and John Tillet, was baptized on January 24, 1746/7, at St Mary, Newington. South of the River Thames and not far from the London residence of the Archbishop of Canterbury at Lambeth Palace, Newington was where Alexander Tillet would spend his life.

Lambeth was the center of eighteenth-century London's tin-glazed-earthenware industry. The pottery made there sought to imitate Dutch Delftware and is now known as Lambeth Delft. In 1762, at age ten, Tillet was apprenticed to Adam Cain, a Citizen and Cooper of London. Kent's London directories for 1813, 1814, and 1820 list Tillet as Cooper (barrel maker) and saltpeter refiner of Vine Street, Lambeth. Saltpeter (potassium nitrate) was used as an explosive and a fertilizer, and refining the chemical was a profitable business. Tillet's will shows him to be a wealthy property developer in Lambeth, having major dealings with the archbishop. Having lived well into his eighties, Tillet died on September 21, 1833, the year of the Abolition Act, and so he would have known Britain only as a slave economy.

PROVENANCE

H. Truscott; Sotheby's, London, April 28, 1964 (lot 69); Thomas George Burn, Rous Lench Court, Worcestershire; Sotheby's, London, July 1, 1986 (lot 44); bt. John H. Bryan of Lake Bluff, IL (1936–2018) from Alistair Sampson Antiques Ltd., London, December 29, 1986

LITERATURE

Sally Kevill-Davies, *Yesterday's Children: The Antiques and History of Childcare* (Woodbridge, UK: Antique Collectors' Club, 1991), 284

Louis L. Lipski and Michael Archer, *Dated English Delftware: Tin-Glazed Earthenware, 1600–1800* (London: Sotheby Publications, 1984), 407, no. 1697

Plaque

Delftware (Lambeth Delft)

$7\frac{7}{8} \times 5\frac{3}{4} \times \frac{1}{4}$ in. (20 × 14.6 × .64 cm)

Inscription: Alexander| Tillet;Born Ian^{y} y^{e} 22^{d} |1747

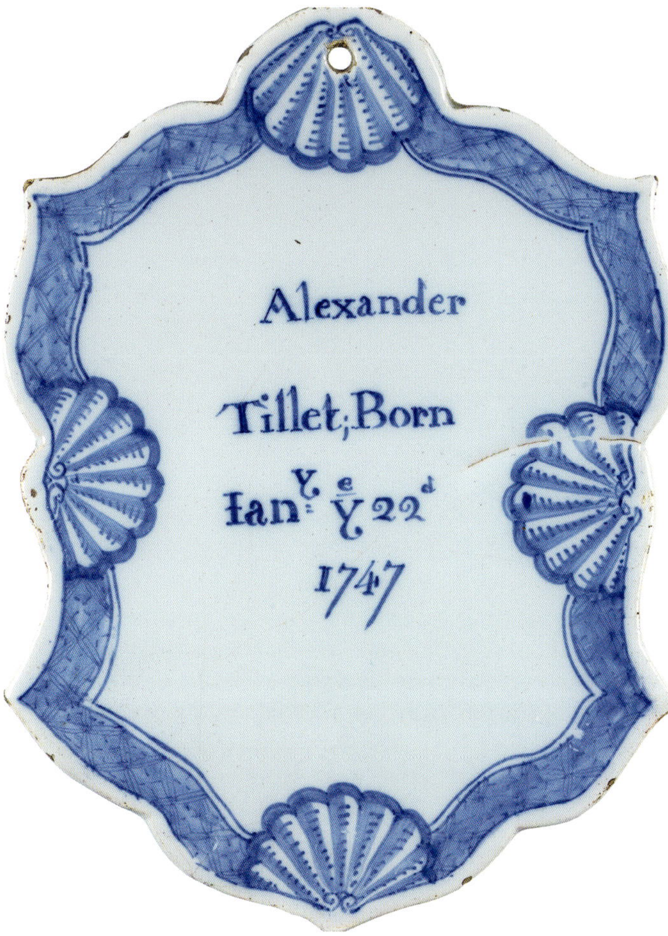

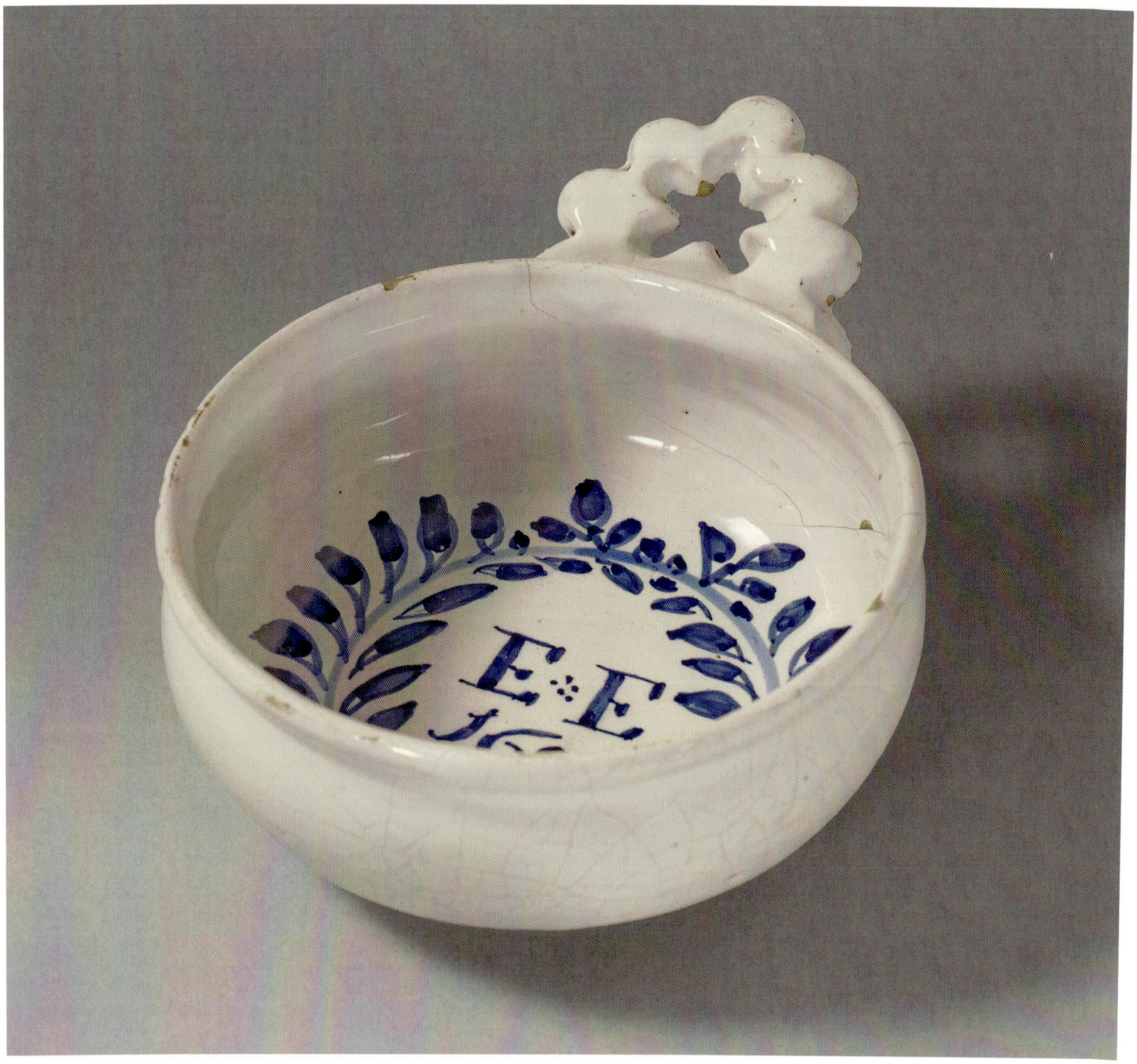

1696

Porringer
Delftware
$2\frac{7}{8} \times 6\frac{7}{8} \times 5\frac{1}{4}$ in.
(7.3 × 17.5 × 13.3 cm)

Inscription: E E | 1696

The porringer was ubiquitous in early modern England. This type of bowl, with a single small handle, could hold soup, stews, broth, pottage, sweets, puddings ("poudingers"), and porridge. As such, it was ideal for feeding the sick, the small, and the elderly.

The same form, in pewter, was used to catch blood; throughout this period, bloodletting was believed to be a cure-all for most illnesses and ailments. Pewter porringers were often incised with small, graduated rings that could measure the amount of blood that had been drawn.

PROVENANCE

Christie's, London, July 12, 1982 (lot 102); bt. John H. Bryan of Lake Bluff, IL (1936–2018) through Jellinek & Sampson Antiques, London

LITERATURE

Walter R. Brown, *The Stuart Legacy: English Art 1603–1714* (Birmingham, AL: Birmingham Museum of Art, 1991), 136

1698

Ceramic cradles were associated with courtship, betrothal, and marriage. They were frequently given as gifts to celebrate a birth, being more affordable than silver spoons. Examples from Staffordshire date from the 1670s, yet this object is the only North Devon slipware cradle known to survive. The decorative motifs are typical of the pottery design of this region. It has been suggested that this cradle must be by the same anonymous maker of the harvest jug now in the Winterthur Museum, Winterthur, Delaware (1964.0025), which carries the same date and a nearly identical depiction of the woman in profile. A similar figure appears on a North Devon jug with incised decoration in the Fitzwilliam Museum, Cambridge, inscribed "CL|1703/4" (GI.C.58-1928).

Large quantities of this type of pottery were manufactured in the port towns of Bideford and Barnstaple and traveled from there to other ports in England and to colonies across the Atlantic.

PROVENANCE

Collection of Joseph P. Gromacki, Kelton House Farm, WI

LITERATURE

Alison Grant, *North Devon Pottery: The Seventeenth Century* (Exeter: University of Exeter Press, 1983)

Robert Hunter, "17th-Century North Devon Slipware and the 'Keep Me' Factor," in *A Glorious Empire: Archaeology and the Tudor-Stuart Atlantic World, Essays in Honour of Ivor Noël Hume*, ed. Eric Klingelhofer (Oxford, UK: Oxbow Books, 2013), 169–72

Model cradle

Slipware

$5\frac{3}{4} \times 8\frac{3}{8} \times 4\frac{5}{8}$ in.
(14.6 × 21.3 × 11.7 cm)

Inscription: 98; SD

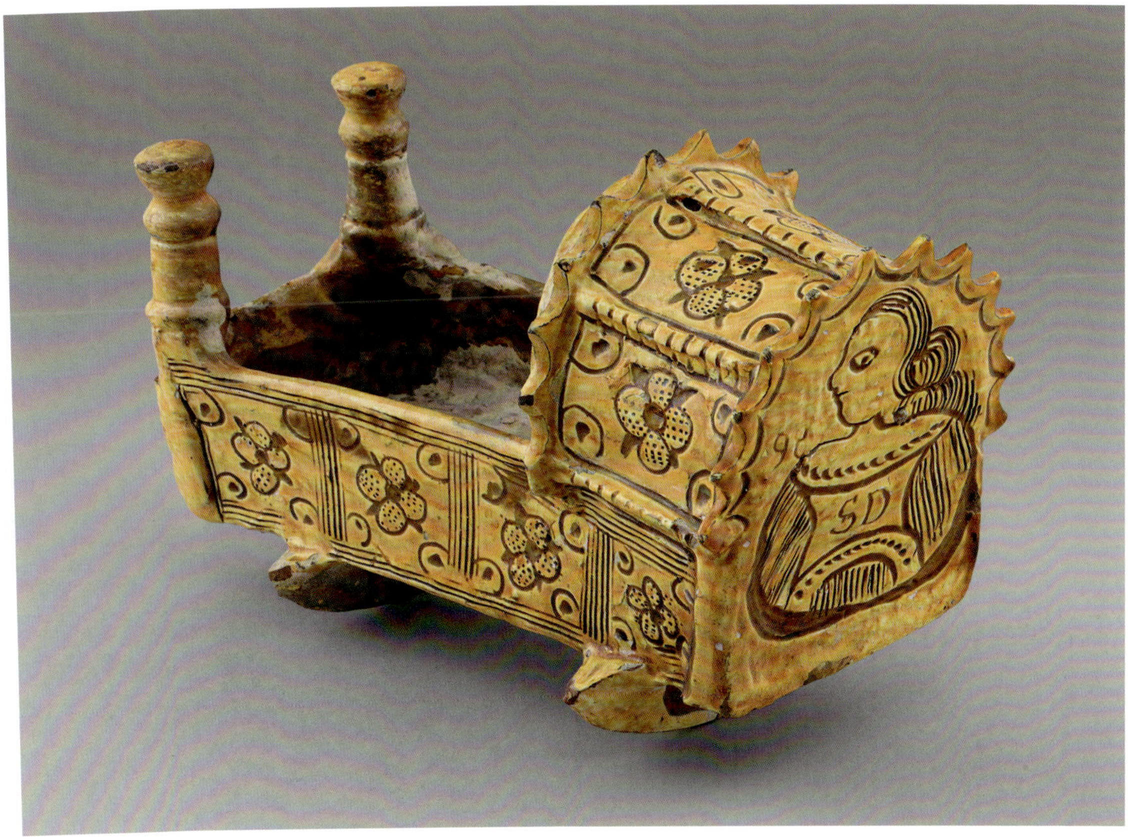

1706

Spoon
Silver gilt
$7\frac{7}{8} \times 1\frac{3}{4} \times \frac{1}{2}$ in. (20 × 4.5 × 1.27 cm)
William Juson, (active 1705; d. 1740)

Inscription: The Gift of|Mr William Harcum|Uncle; 20s 5p½; Marey Harcum born y^e 20 June at one 'Clock|1706; Elizabeth Harcum born y^e 23 June Between two & three of y^e Clock|In y^e Afternoon,

The inscription on this spoon tells a tragic story, one that was all too familiar in the early modern period. It records the time and date of the births of Mary and Elizabeth Harcum, twin daughters of Anne (née Noblett) and her husband, John. The couple had married in July the year before and were living in the parish of All Hallows London Wall. Mary was born at one o'clock on June 20, 1706, but her twin did not arrive until June 23. Although the baptisms of both children appear in the register—on June 21 and 23, respectively—Mary, Elizabeth, and their mother were buried together only days later. As was so often the case, the birth of these children proved fatal for the mother, and John Harcum was suddenly bereft of both his wife and his children. He may have kept this silver-gilt spoon, an object synonymous with celebrations of birth, gifted by his brother William, in memory of the tragedy.

The maker, William Juson, began his apprenticeship in 1694 under William Scarlett, a member of the Embroiders' Company of London, but he practiced as a silversmith, outside of the auspices of the Goldsmiths' Company. In 1704, Juson married Elizabeth West at St Giles-without-Cripplegate and struck out on his own, registering his own hallmark and taking on apprentices. Spoons and forks bear dates from 1705 to 1711. By 1720, Juson had begun a new profession as a corn chandler (a retailer of grain stuffs). He died in 1740.

PROVENANCE
Tennants Auctioneers, Leyburn, April 7, 2018 (lot 2326); bt. John H. Bryan of Lake Bluff, IL (1936–2018) through Peter Cameron, London

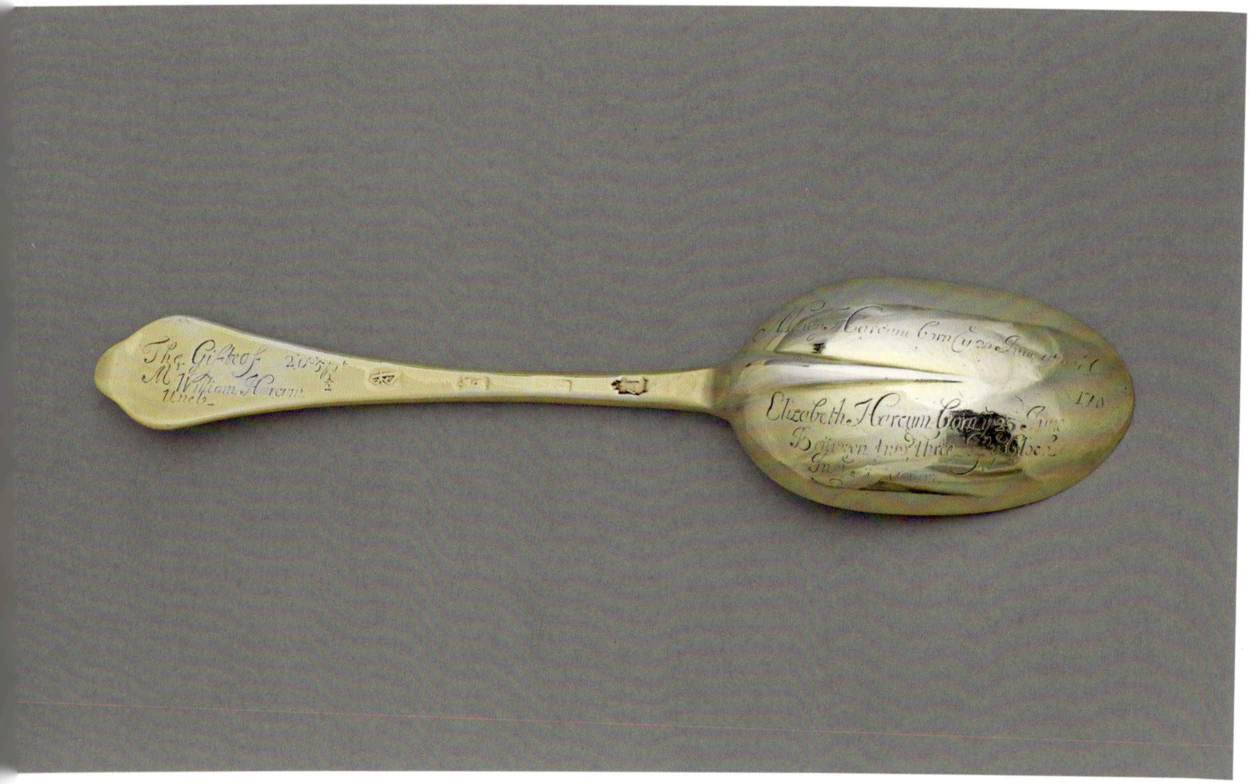

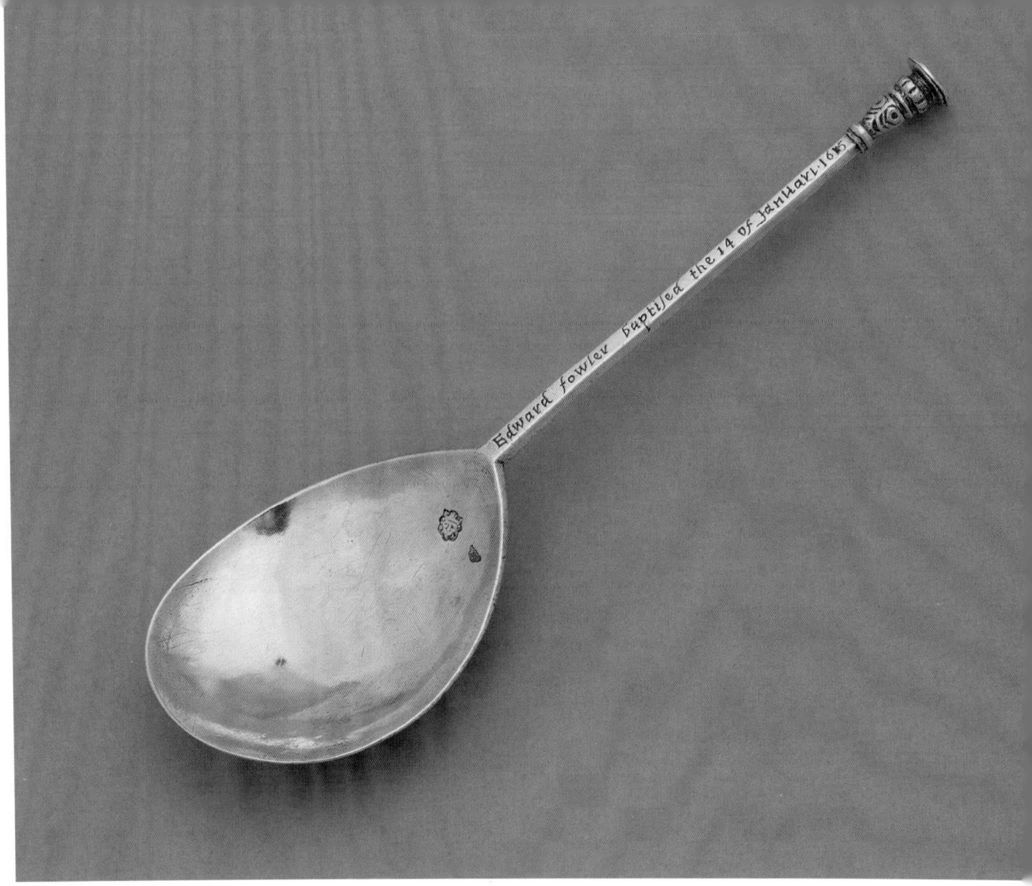

1615

Spoon

Silver

6 × 1⅞ × ¾ in. (15.2 × 4.8 × 1.9 cm)

John Saunders (active 1612–1630)

Inscription: Edward fowler baptised the 14 of Januari 1615; E&F

The ubiquity of the name Edward Fowler makes it impossible to trace this child in the records, but we know from this object that he was born with, as the expression goes, a silver spoon in his mouth.

The mark on this seal-top spoon reads "I. S." in monogram, the mark of John Saunders, son of John Saunders, born in Castle Cary, Somerset. On July 26, 1605, Saunders was apprenticed to Daniel Carey (or Carie or Cary) (active 1604–1641). Saunders was later turned over to William Hole but gained his freedom in 1612, when he began to take on his own apprentices. The first, on April 15, 1614, was Edward Hole, son of William and a yeoman of Castle Cary, Somerset. It seems likely that Saunder's master was the William Hole, Goldsmith, who was the versatile engraver responsible for the maps in *Camden's Britannia* (1607) and the cover of *Coryats Crudities* (1611). The affiliations give rise to the possibility that there was a relationship between the two families in Somerset.

On March 9, 1613/14, Saunders married Emma Marberye (or Marbury) at the parish of St Peter Paul's Wharf. Between 1614 and 1629, at least seven children were baptized at Saunders's home parish of St Mary Woolchurch. In 1630, his apprentice Richard Vaux was made free after serving his term of indenture, and records of Saunders's activities went silent. Spoons bearing his mark include one sold at Christie's, London, November 25–26, 2014 (lot 449) (seal top, 1615); one with M. Ford Creech Antiques and Fine Arts (the apostle St. James, 1616); and one with Antique Silver Spoons (the apostle St. James).

PROVENANCE

bt. John H. Bryan of Lake Bluff, IL (1936–2018) from Tobias Jellinek Antiques, London, May 12, 1996

LITERATURE

Timothy Arthur Kent, *London Silver Spoonmakers* (London: The Silver Society, 1981), 27

1787

Pincushion

Silk with metal beads, silk thread, spangles

$5\frac{5}{8} \times 3\frac{3}{4} \times 2\frac{3}{8}$ in.
(14.3 × 9.5 × 6 cm)

Inscription:
C|M|H|1787;
Wellcome / Little / Stranger

In England, a pincushion was a typical gift to celebrate a child's birth, but there was a commonly held belief that if given to the mother before labor began it would—much like a voodoo charm—exacerbate the pain of childbirth. As the saying went, "more pins, more pain."

PROVENANCE

bt. John H. Bryan of Lake Bluff, IL (1936–2018) from Jellinek & Sampson Antiques Ltd., London, October 23, 1980; gifted to Carolyn Bryan Harrell, 1980

LITERATURE

Sally Kevill-Davies, *Yesterday's Children: The Antiques and History of Childcare* (Woodbridge, UK: Antique Collectors' Club, 1991), 277–79

Sue Laurence, *The Hand That Rocked the Cradle* (Norwich: Unicorn Press, 2018), 84

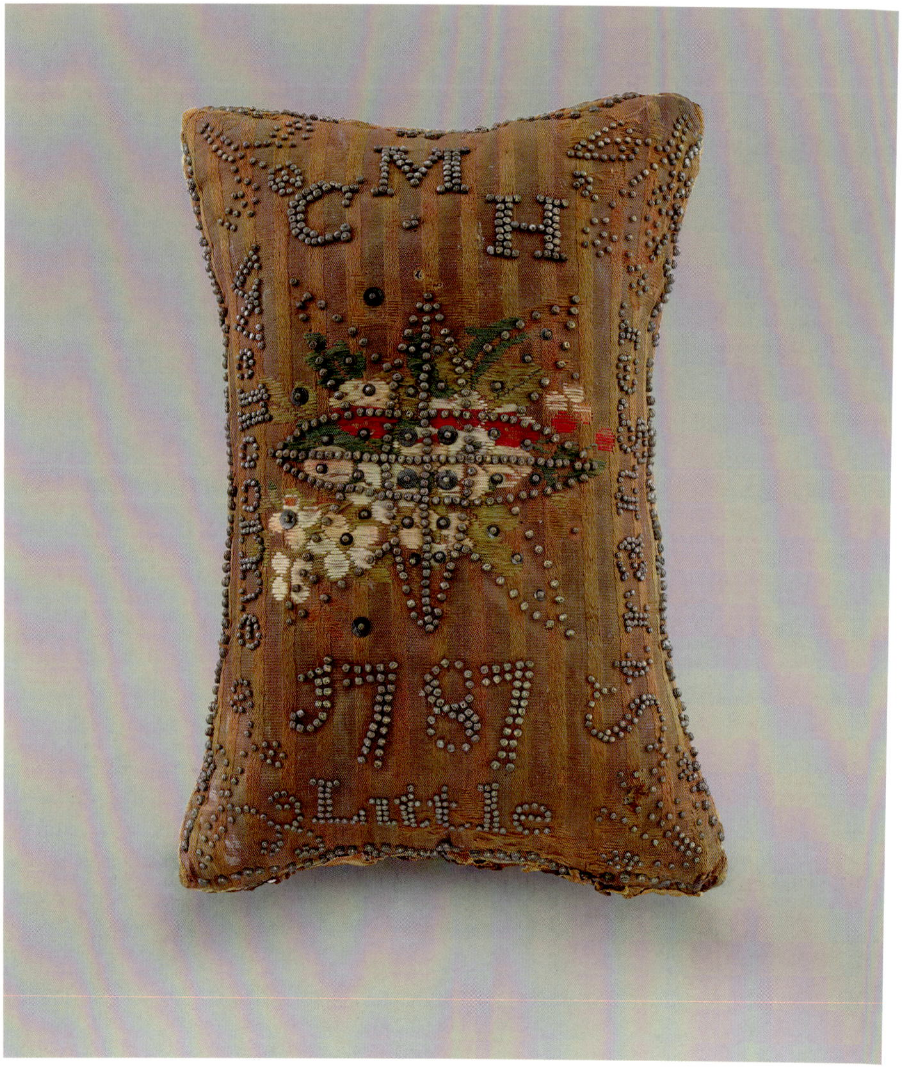

1728

This embroidery dated 1728 depicts the Pharoah's daughter (who wears the crown) and her attendant discovering Moses in the bulrushes. Although the stylized figures have a naive quality, the work demonstrates great skill. The complex diaper-stitch patterns and swirling forms are related to those of the bizarre silks of the eighteenth century, which drew their inspiration from Eastern textiles. The silk stitching is on linen ground, with long-and-short, stem, double running, raised-cross, and overcast stitches and double backstitches. The gilded gesso frame is early and presumed original.

PROVENANCE

The Marcelli-Shaw Collection; Alistair Sampson Antiques Ltd., London; bt. John H. Bryan of Lake Bluff, IL (1936–2018), September 5, 1989

Embroidered picture

Linen with silk thread

6 7/8 × 7 5/8 in. (17.5 × 19.4 cm)

Inscription:
MOSES; EW 1728

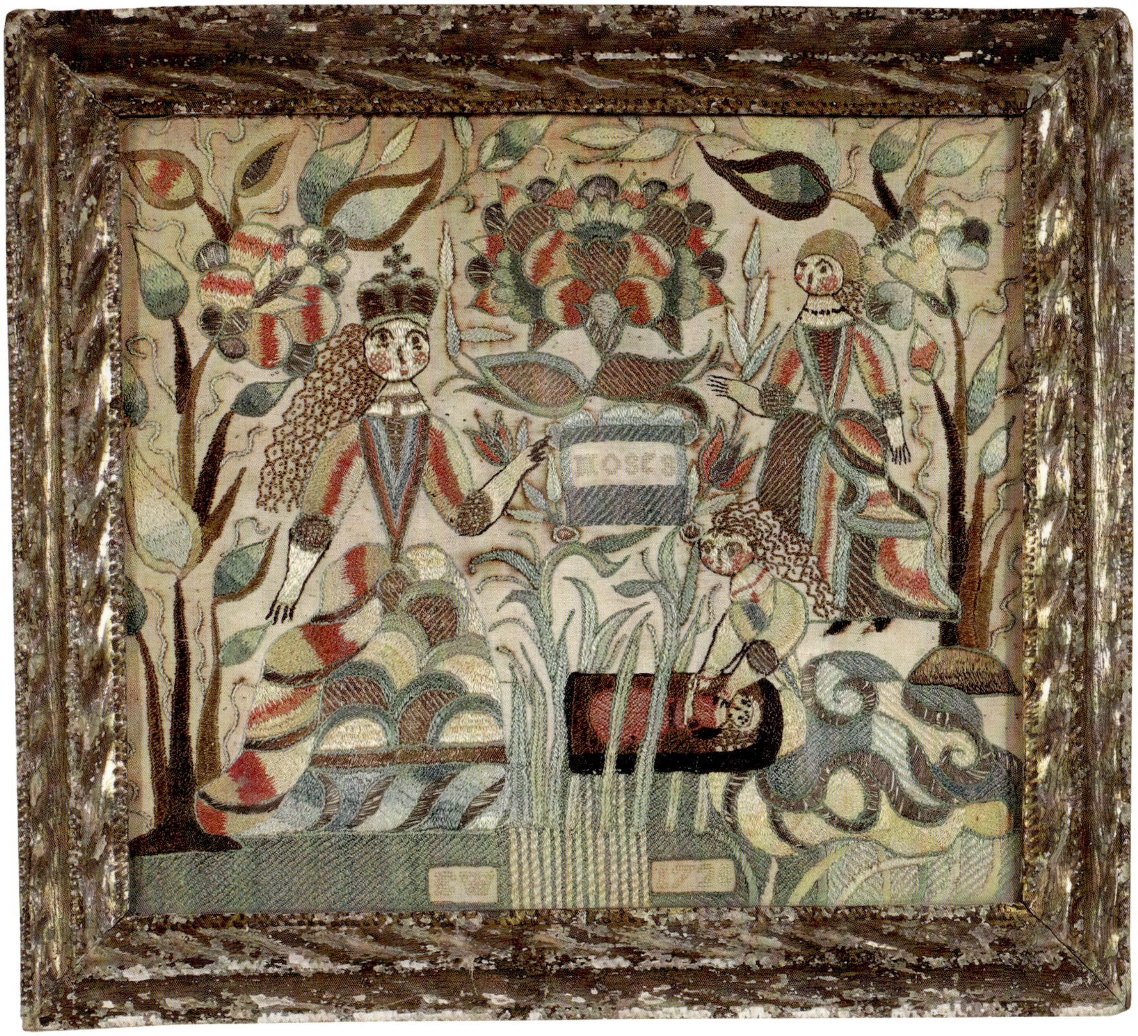

1730

Plate

Delftware (Liverpool Delft)

9 1/8 × 1 in. (23.2 × 2.5 cm)

Inscription: Elizabeth Pearso[n]|Born 19th:July|1730

A series of plates by the same maker commemorate the births of Elizabeth (1730), Ann (1732), and Sarah Pearson (1734). These plates in the Bryan Collection are three known examples of what might have been a larger set. Presumably the three women were sisters, but their baptisms cannot be traced in the parish registers of England.

That the plates have identical dimensions suggests they were made at the same time. There were clearly multiples because there is a plate in another collection that is identical to the one commemorating Sarah's birth in 1734.

PROVENANCE

Louis L. Lipski, London; Peter Clark; Alistair Sampson Antiques Ltd., London; bt. John H. Bryan of Lake Bluff, IL (1936–2018), June 20, 1995

LITERATURE

Louis L. Lipski and Michael Archer, *Dated English Delftware: Tin-Glazed Earthenware, 1600–1800* (London: Sotheby Publications, 1984), 86–87, no. 362

1732

Plate

Delftware (Liverpool Delft)

9 1/8 × 1 in. (23.2 × 2.5 cm)

Inscription: Ann Pearson|Born 17th May|1732

PROVENANCE

Bearnes, Hampton, and Littlewood, Exeter, July 14, 2015 (lot 432); Private Collection, UK; bt. John H. Bryan III of Nashville, TN, through Martin P. Levy, London

1734

Plate

Delftware (Liverpool Delft)

9 1/8 × 1 in. (23.2 × 2.5 cm)

Inscription: Sarah Pearson|Born 17th:Agust|1734

Another plate with an identical inscription was in the collection of C. H. Spiers and sold at Sotheby's, London, November 10, 1977 (lot 85), sold subsequently with Benjamin Ginsburg and Cora Ginsburg Collection, and sold again at Christie's, New York, October 14, 1983 (lot 194).

PROVENANCE

Sotheby's, London, March 26, 1968 (lot 64); Thomas George Burn, Rous Lench Court, Worcestershire; Christie's, London, May 29–30, 1990 (lot 19); bt. John H. Bryan of Lake Bluff, IL (1936–2018) through Alistair Sampson Antiques, London

LITERATURE

Louis L. Lipski and Michael Archer, *Dated English Delftware: Tin-Glazed Earthenware, 1600–1800* (London: Sotheby Publications, 1984), 93, no. 395A

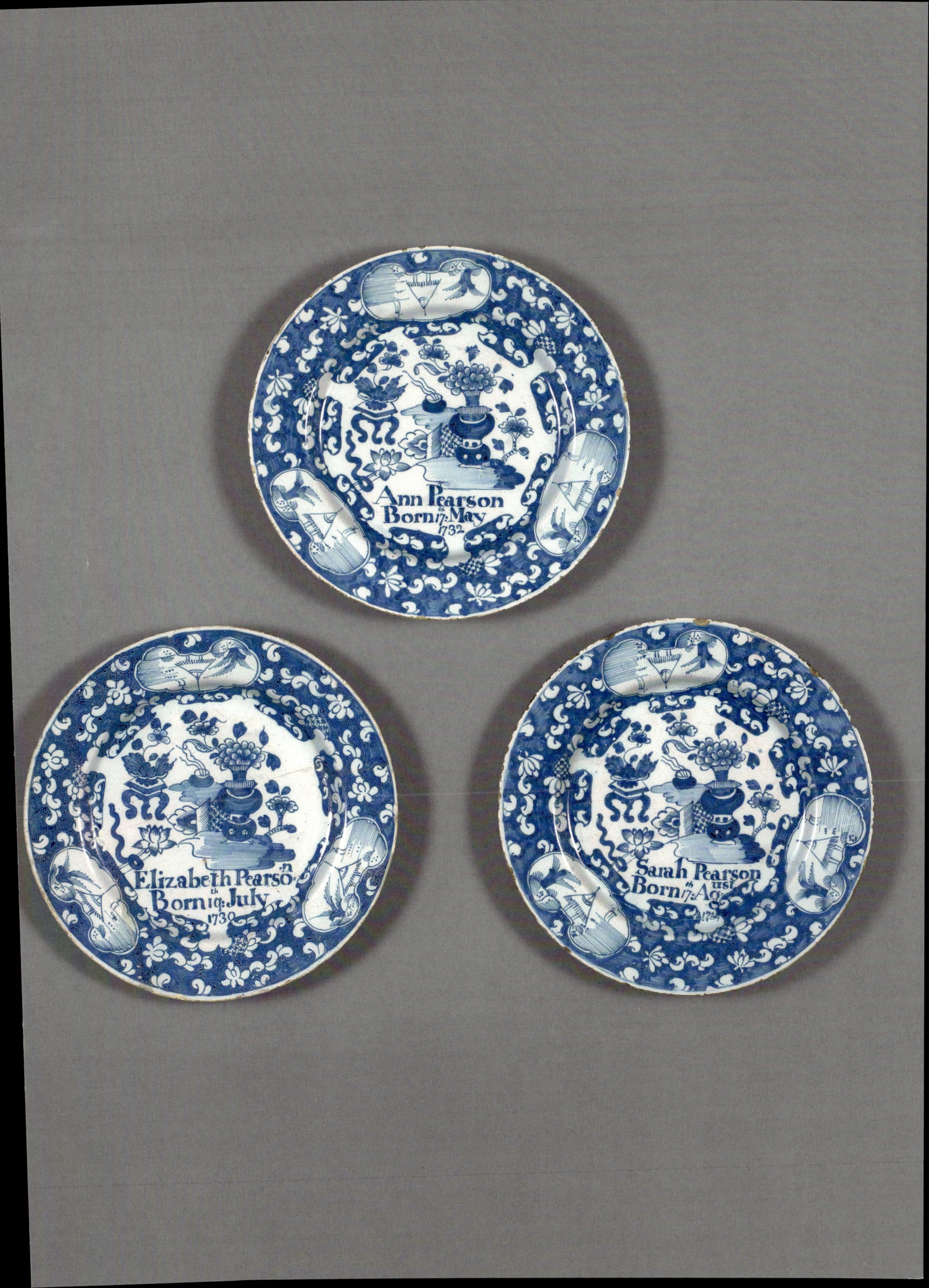

Ann Pearson
Born 17th May
1732
Elizabeth Pearson
Born 19th July
1730
Sarah Pearson
Born 17th Agust

1754

Plate
Delftware
9 × $1\frac{1}{4}$ in. (22.9 × 3.2 cm)

Inscription: SALLY, TAYLOR, 1754,

"Sally Taylor" was a very common name, which makes it impossible to determine whose birth this plate commemorates. It may well have been part of a larger set, but no other plates of this type are known to have survived.

PROVENANCE
Sotheby's, London, May 21, 1948 (lot 18); T. M. Ragg; Sotheby's, London, March 23, 1954 (lot 31); Jellinek & Sampson Antiques, London; bt. John H. Bryan of Lake Bluff, IL (1936–2018), July 14, 1981

LITERATURE
Louis L. Lipski and Michael Archer, *Dated English Delftware: Tin-Glazed Earthenware, 1600–1800* (London: Sotheby Publications, 1984), 132, no. 585

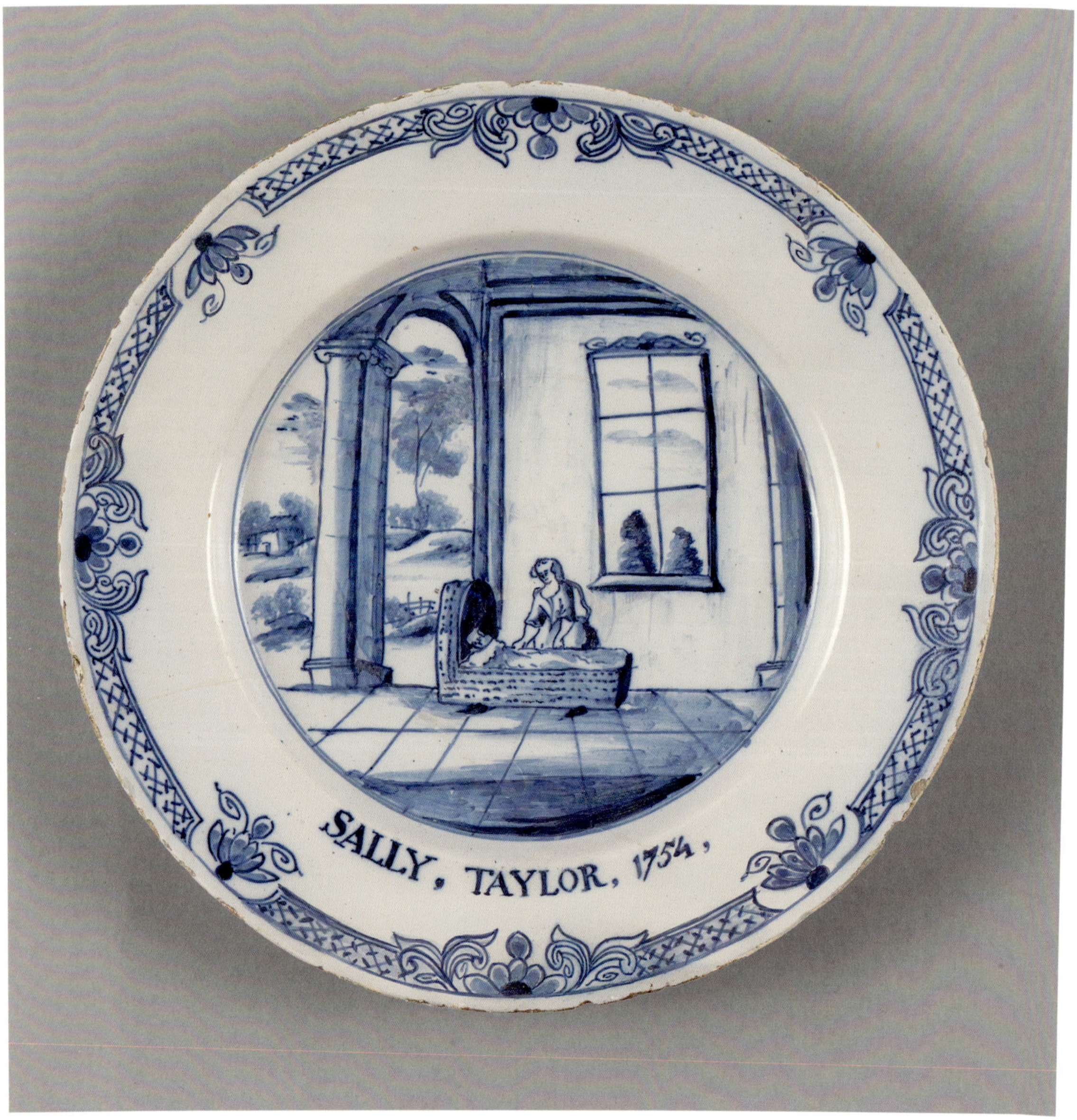

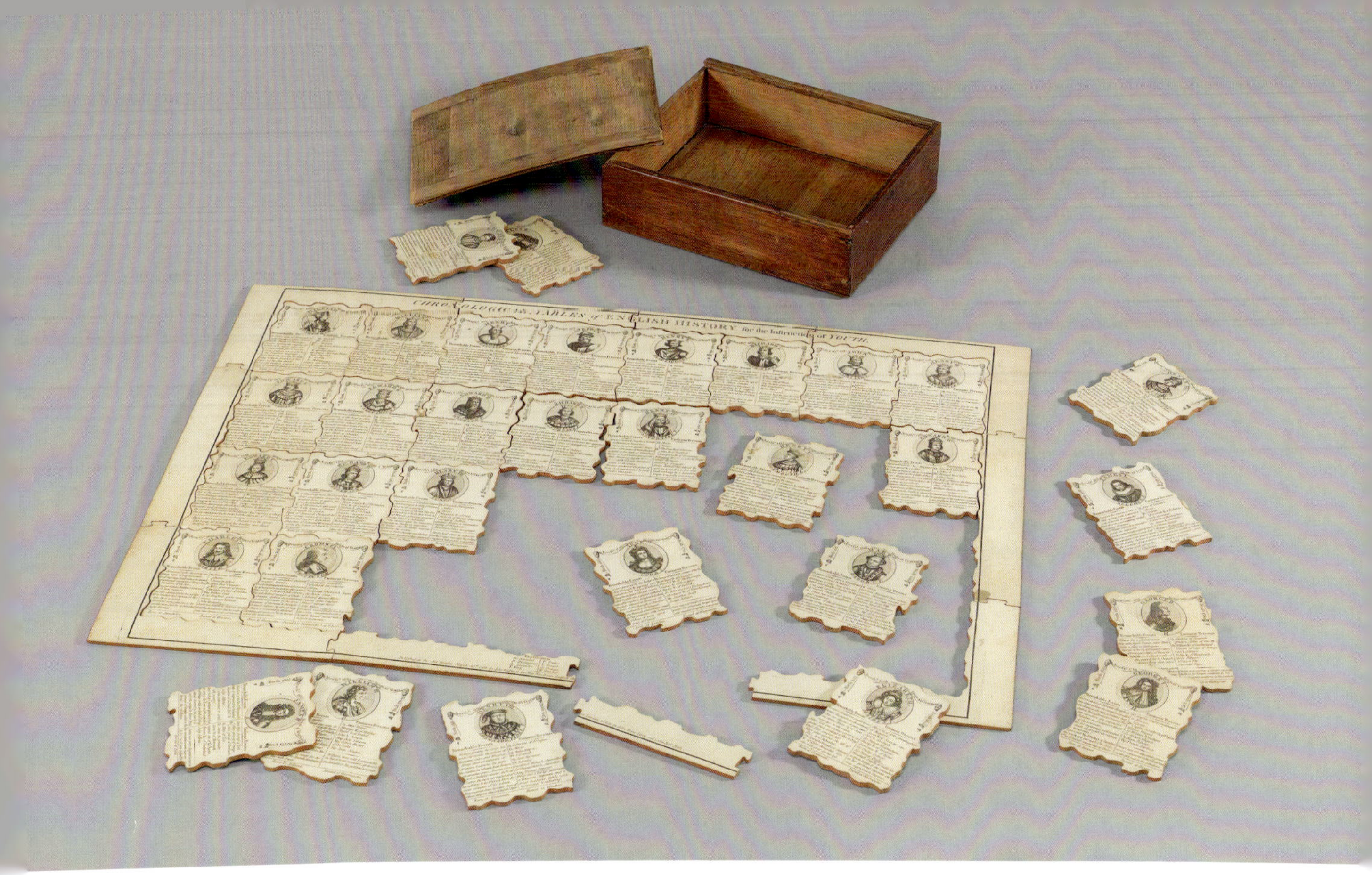

1788

Puzzle

Engraving on wood

16¼ × 24 × ⅛ in. (41.3 × 61 × .32 cm); box 23⅞ × 16 × ⅜ in. (60.6 × 40.6 × .95 cm)

John Wallis (d. 1818)

Inscription: Published as the Act Directs March 25th. 1788 by John Wallis, at his Map Warehouse Ludgate Street, London; & Sold by John Binns, Leeds; and Lewis Bull, Bath; Sold by|CHARLES VINE|N°. 9 Bond Street|BATH

Developments in historical education encouraged a patriotic perspective on the British monarchy by linking the twists and turns of the past into one long, glorious, national story. Influential written histories of Britain began in the sixteenth century. Their "drum and trumpet" narratives became so well known that by the 1630s popular songs were sold on the street to help children remember all the kings and queens of England, with a line or two indicating whether they were "good" or "bad." In the eighteenth century, schoolboys recited facts and wrote essays while schoolgirls worked with their needles to learn their history lessons. These simple outline histories of great men and women were also presented in jigsaw puzzles like this one, published in 1788 by John Wallis, a specialist educational toy maker. It proffered "Chronological Tables of English History for the Instruction of Youth . . . each piece depicting an English monarch with relevant Remarkable Events, and Eminent Persons."

PROVENANCE

bt. John H. Bryan of Lake Bluff, IL (1936–2018) from Lindsay Antiques, London, July 2, 1985

LITERATURE

Sally Kevill-Davies, *Yesterday's Children: The Antiques and History of Childcare* (Woodbridge, UK: Antique Collectors' Club, 1991), 246

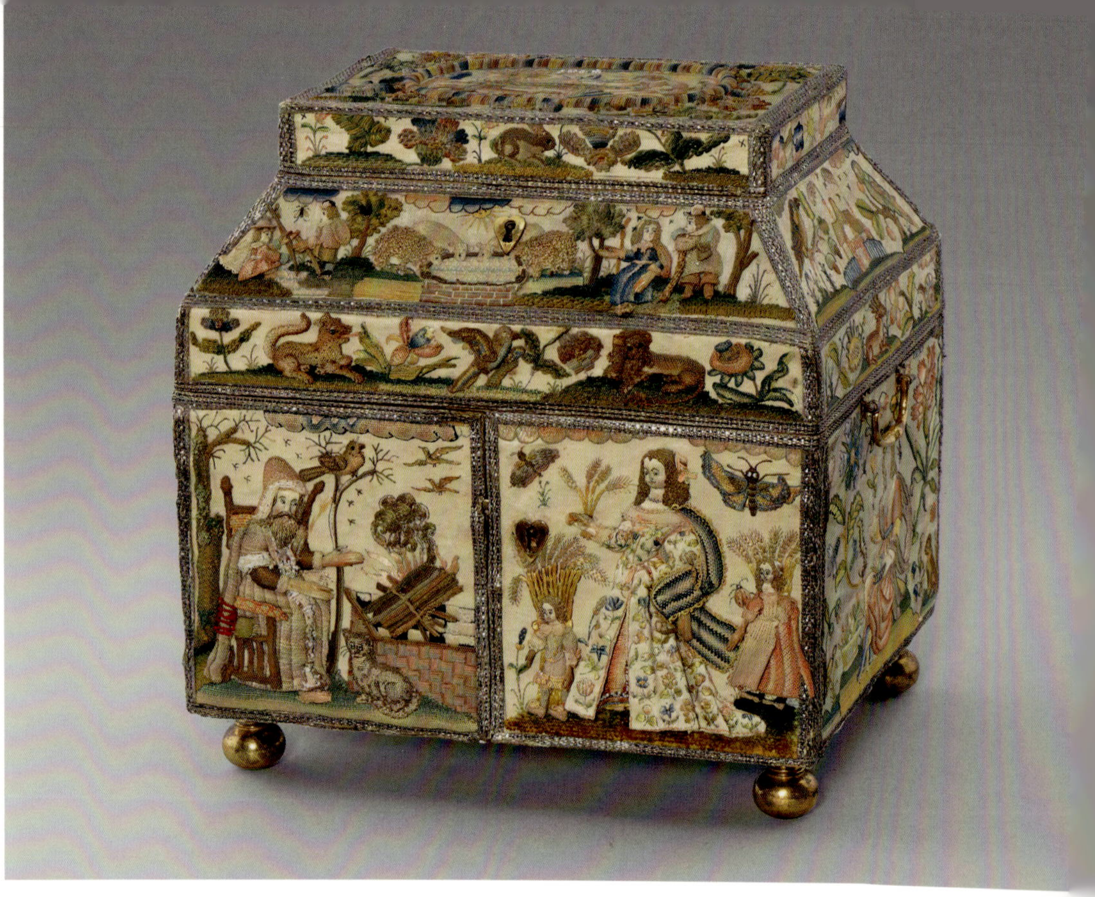

1660

Cabinet

Silk with silk and metal threads, seed pearls, glass, beads, metal thread trim, wood, wood frame, silk lining, print

11¾ × 10¾ × 8½ in. (29.8 × 27.3 × 21.6 cm)

Inscription: 1660

The date of 1660, set in seed pearls on the top lid of this cabinet, marked the completion of a project to hone embroidery skills, undertaken by a girl of eleven or twelve as part of her education at home or at school. Similarly, a cabinet (ca. 1645–55) in the Los Angeles County Museum of Art is stitched with the coat of arms of the Perwich family, which ran a school for girls in Hackney. That embroidery is attributed to Susanna (1636–1661), who, along with her parents, John and Mary Perwich, taught at the school. She was lauded as a musical prodigy, and her short life and virtuosity with the needle were commemorated in *The Virgin's Pattern: In the Exemplary Life, and Lamented Death of Mrs. Susanna Perwich* (London: Simon Dover, 1661).

Cabinets like this one survive in significant numbers from the second half of the seventeenth century. They were assembled from sets, with patterns provided by professional embroiderers, who may also have provided other parts; the use of woven tape and stamps suggests that bookbinders may have had a hand in the final construction. The patterns, some of which survive in an unfinished state, were drawn in ink on satin ground, and once worked, were cut out and set onto the cabinet. These patterns, often depicting Old Testament and mythological scenes, derived at least in part from print sources, many from northern Europe. The lid of this cabinet depicts Phaethon attempting to drive the chariot of his father, the Greek sun god Helios, across the sky to herald the passing day—a fitting backdrop to an inscribed date. The front panels of the doors bear stumpworked allegorical figures representing summer and winter, and their hands, presumably pulls, are made of wood. Abraham and Isaac are shown on the inside of one door, and Cain and Abel on the other. Unlike the scenes on some cabinets of the period, with a cohesive iconographic scheme, these form an eclectic mix of patriarchal and whimsical imagery.

The inside lid is set with five mirrors and a section of a hand-colored print (reversed) of Aegidius Sadeler II's engraving after Jacopo Tintoretto's *Massacre of the Innocents*. This arrangement of print and mirrors emulated some of the sophisticated perspectival tricks achieved on large collectors' cabinets of ebony and marquetry. Another cabinet of very similar dimensions at the Metropolitan Museum of Art, New York (39.13.3a–k), with a near-identical arrangement of mirror-lined upper compartment and hand-colored landscape print, suggests that the same craftsperson may have provided both the design and materials.

The drawers would have contained letters, writing implements, keepsakes, cosmetics, and other small possessions precious to the young owner. There are two pewter pots for ink and pounce (powder) and two glass jars with pewter tops—in the same location as on the Metropolitan's cabinet, which also shares the same salmon-pink lining and removable compartments. The cabinet's original outer case of pine and gilded, embossed leather is lined with marbled paper and fastened with a gilded metal clasp that measures 13 inches high and 12 ½ inches wide (33 × 31.8 centimeters).

PROVENANCE

Dreweatt Neate Holloways, September 16, 1998 (lot 206); bt. John H. Bryan of Lake Bluff, IL (1936–2018) through Rupert Gentle

LITERATURE

Andrew Morrall and Melinda Watt, eds., *English Embroidery from the Metropolitan Museum of Art, 1580–1700: 'Twixt Art and Nature* (New Haven, CT: Yale University Press, 2008), 245

Isabella Rosner, "'A Cunning Skill Did Lurk': Susanna Perwich and the Mysteries of a Seventeenth-Century Needlework Cabinet," *Textile History* 49, no. 2 (2018): 140–63

Lanto Synge, *Art of Embroidery: History of Style and Technique* (Woodbridge, UK: Antique Collectors' Club, 2001), 134–35

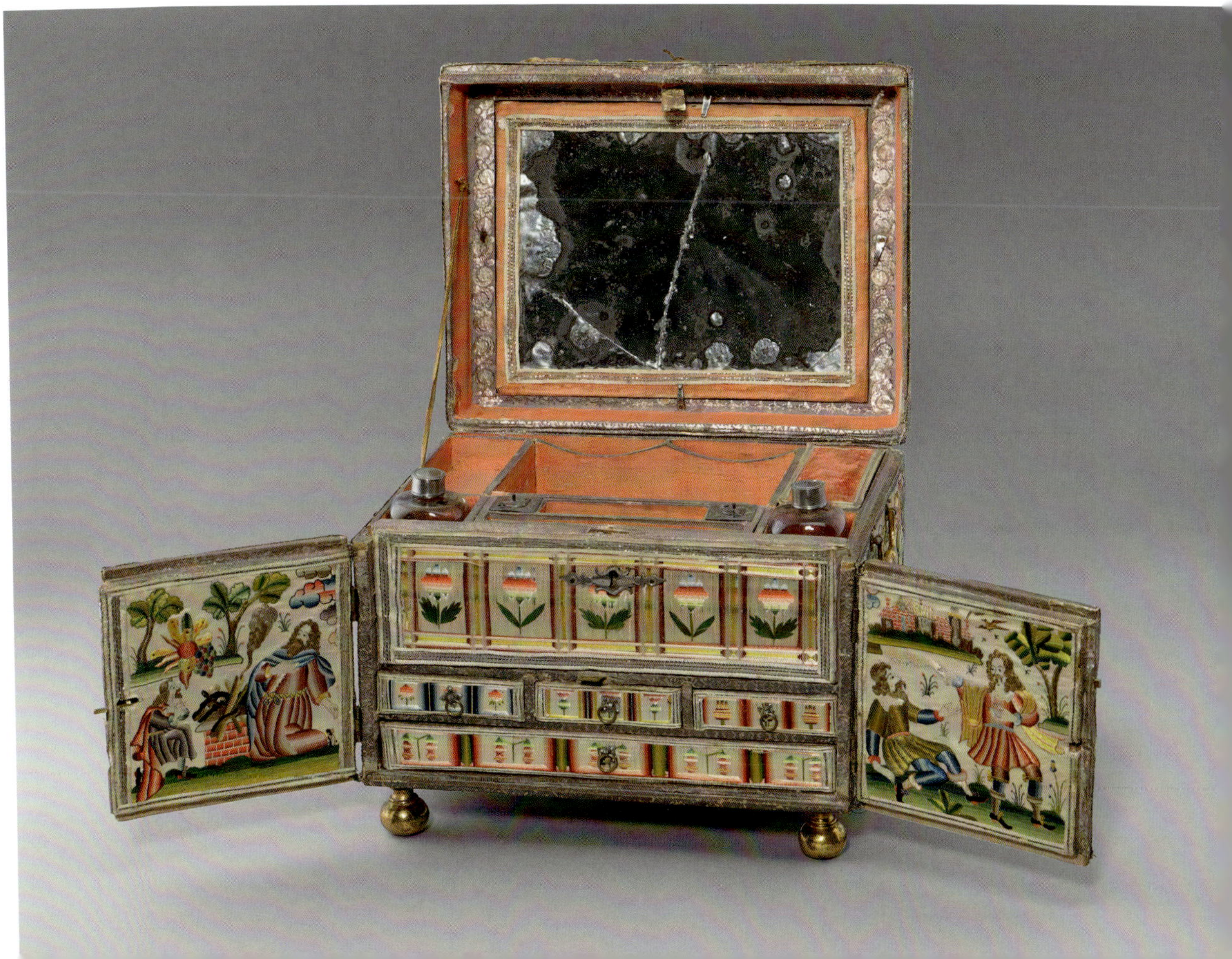

1700

Sampler

Linen with silk thread

17 × 8 in. (43.2 × 20.3 cm); framed dimensions: 19 ½ × 10 ½ × ⅞ in. (49.5 × 26.7 × 2.22 cm)

Sarah Quare (1690–1717)

Inscription: AaBCDEFGHIKLM NOPQRSTVWXYZ| Blessed is the pure in heart for they shall see god Blessed are the peace|makers for they shall Be called the|children of God 123456789 MP SQ B; Keep On Thy Tongue A Bridle That|It Talk Not Idle Sarah Quare 1700

Sarah Quare and her twin sister, Elizabeth, delivered by a midwife named Margaret Harris, were born on March 24, 1689/90, to Mary and Daniel Quare (1648 or 1649; d. 1724). The twins' father was one of the most famous clock- and instrument makers of his day, who supplied timepieces to the Crown and Court. At the time his daughters were born, the family was living in Exchange Alley in the parish of St Mary Wolnoth, London. Quare was the first watchmaker to use jewel end stones, which improved the action and durability of escapements; he also introduced dust caps to keep movements clean.

In this sampler, there are many areas of detached-buttonhole stitching, which was very labor-intensive; this technique (far more finely executed) is found earlier in the century on professionally embroidered elite garments. There are also many details executed with satin stitch, which, although not complex in structure, creates a luxurious surface appearance and shows off the abundant use of silk. These types of stitches contrast with the simple cross-stitch that predominates in many later samplers. In general terms, the words, alphabet, and other characters in this example indicate a shift in the focus of samplers from a means of recording and perfecting stitches to a more general demonstration of education, discipline, and virtue.

Sarah would have been about ten years old when she made this sampler. This sampler is similar to one made by the Quaker Mary Milner in 1694, now in the collection of the Cooper Hewitt, Smithsonian Design Museum, New York (1981-28-49). The two samplers share the same pictorial band at the bottom, suggesting that these girls either attended the same school—possibly the first official Quaker school in Shaklewell—or at least had the same teacher.

Learning the stiches, the letter and number forms, and the religious maxims (like the quote here from Matthew 5:8–9) would all have been considered an important part of a girl's education. As a young woman growing up in a Quaker household, Sarah would have been aware that women in her community enjoyed greater freedom than most—in particular, the freedom to preach in public, a right forcefully defended by Margaret Fell in *Women's Speaking Justified* (1666). This pamphlet was written and published while Fell was in prison, having been convicted on a charge of praemunire (the offense of recognizing jurisdiction other than the monarchy) for holding Quaker meetings. Such views ran counter to the maxim "Keep on thy tongue a bridle, that it talk not idle," which Sarah embroidered onto the sampler. The "bridle" alluded to here is the "scold's bridle," a particularly unpleasant iron-work contraption, placed over the head and into the mouth to hold the tongue in place and thus prevent speech. It was the punishment of choice deployed on women Quaker preachers in Britain. Olaudah Equiano (ca. 1745–1797), abolitionist and onetime slave of a Philadelphia Quaker, noted that the bridle was also used on slaves in Virginia.

On March 15, 1711/2, Sarah, just shy of her twenty-second birthday, married Jacob Wyan (1688–1730), Draper of the Wheatsheaf and Anchor, Gracechurch Street, London. Wyan was a small wholescale mercer whose stock-in-trade was the overseas exchange of silks and fabrics. Sarah died on July 17, 1717. By report of the nurse, she had suffered a fever. She was buried at the Quaker burial ground at Bunhill Fields, Islington.

PROVENANCE

Marcelli-Shaw Collection; bt. John H. Bryan of Lake Bluff, IL (1936–2018) from Alistair Sampson Antiques, London, July 17, 1986

LITERATURE

Walter R. Brown, *The Stuart Legacy: English Art 1603–1714* (Birmingham, AL: Birmingham Museum of Art, 1991), 82

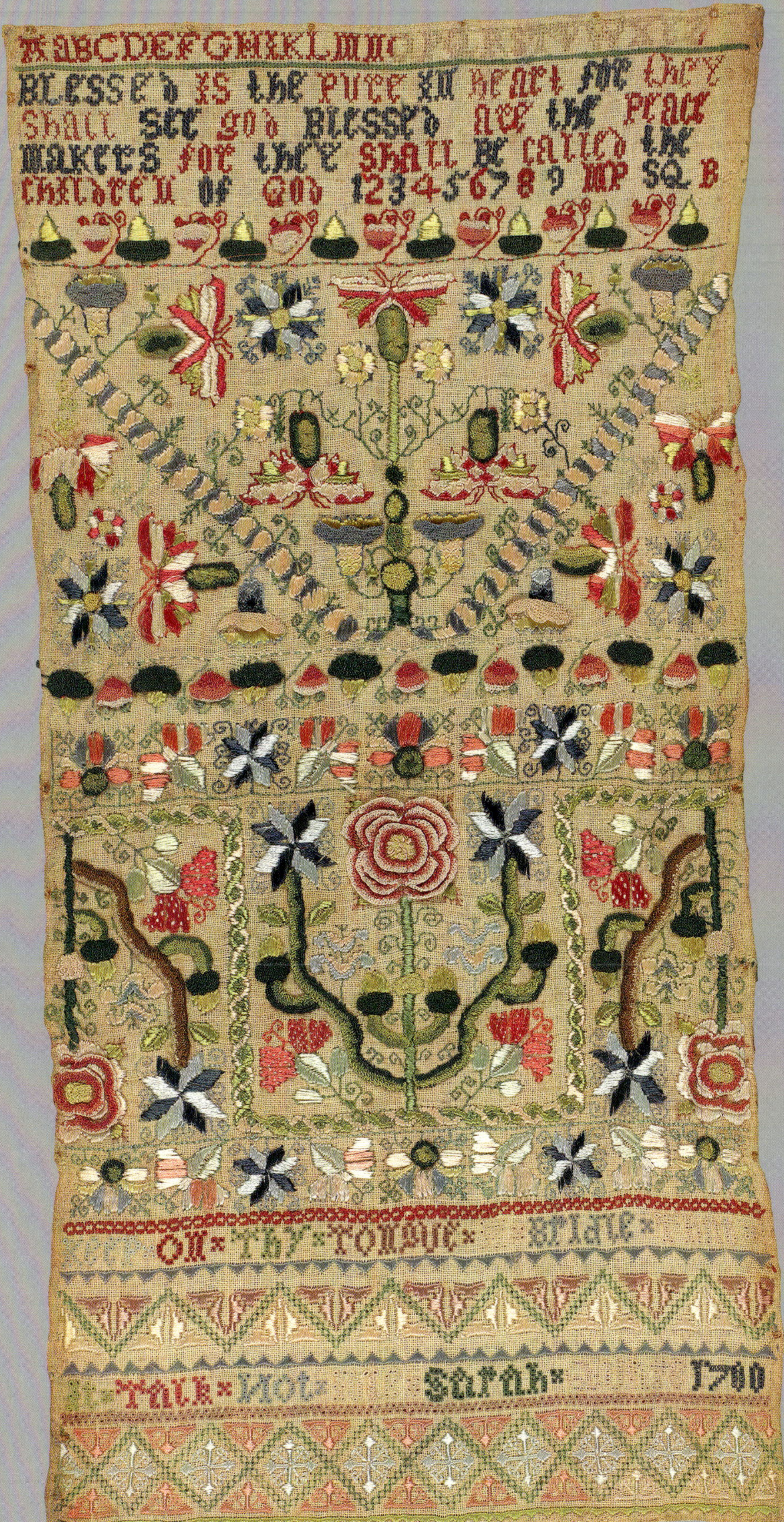
BLESSED IS THE PURE IN HEART FOR THEY
SHALL SEE GOD BLESSED ARE THE PEACE
MAKERS FOR THEY SHALL BE CALLED THE
CHILDREN OF GOD 123456789 MP SQ B
ON THY TONGUE BRIDLE
TALK NOT SARAH 1700

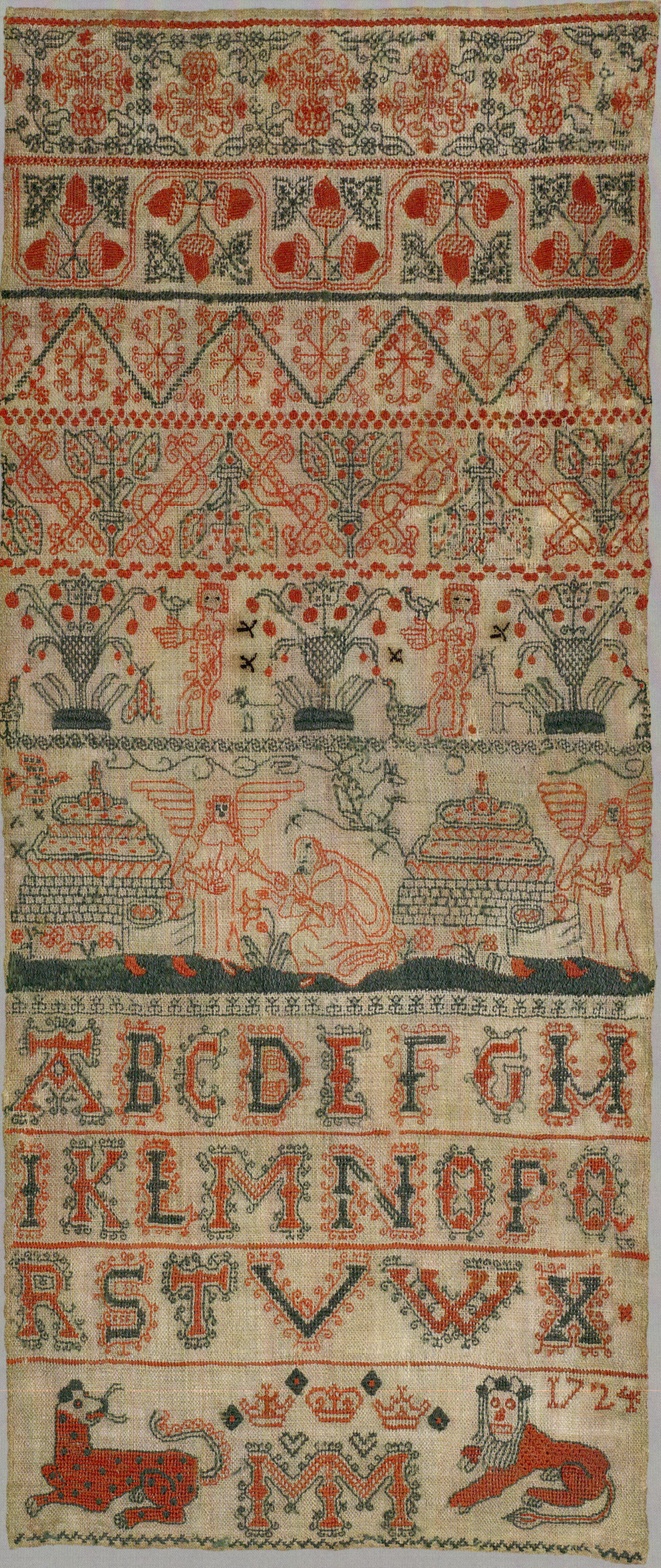
ABCDEFGH
IKLMNOPQ
RSTVWX
1724
MM

1724

Sampler

Linen with wool thread

$25\frac{1}{2} \times 10\frac{1}{2}$ in. (64.8 × 26.7 cm); framed dimensions: $31\frac{1}{4} \times 16\frac{1}{4} \times 1\frac{1}{2}$ in. (76.8 × 41.3 × 3.8 cm)

Inscription: ABCDEFGH| IKLMNOPQ| RSTVWX; MM [in monogram]| 1724

The scene depicted in the central band of repeating decoration is difficult to identify—it might be the angel that appeared to Abraham to prevent him from sacrificing his son, Isaac; the pyre that Abraham had prepared is visible behind them. Neither the bird nor the squirrel appears in the biblical account, but both are welcome interlopers in this scene. The curly-haired characters holding small birds have, on account of the size of their hands, been called "boxers," something of a misnomer as they probably derive from sixteenth-century Venetian pattern-book designs that had been misunderstood and bastardized. It has been suggested that this sampler may have been made in Scotland, as the red-and-green color scheme is typical of Scottish work of this period.

Some confusion as to where this sampler was made is caused by the unusual ligature of the initials "MM," presumably those of the young woman who made the piece. The three coronets above the initials, which loosely approximate those of the aristocracy, were frequently stitched above family initials, although they do not relate to the rank of the maker or her family. As such, stitching initials and coronets would have been deemed a useful skill for a young woman to learn, for if she entered service in a large house, there would be a formidable stack of linen to monogram.

PROVENANCE

Cora Ginsburg LLC, New York; bt. John H. Bryan of Lake Bluff, IL (1936–2018) from Mallett & Son Antiques Ltd., June 14, 2001

LITERATURE

Clare Browne and Jennifer Wearden, *Samplers from the Victoria and Albert Museum* (London: V&A Publications, 1999)

Carol Humphrey, *Sampled Lives: Samplers from the Fitzwilliam Museum* (Cambridge: Fitzwilliam Museum, 2017)

Naomi E. A. Tarrant, *"Remember Now Thy Creator": Scottish Girls' Samplers 1700–1872* (Edinburgh: Society of Antiquaries of Scotland, 2014)

Helen Wyld, *Embroidered Stories: Scottish Samplers* (Edinburgh: National Museum of Scotland, 2018)

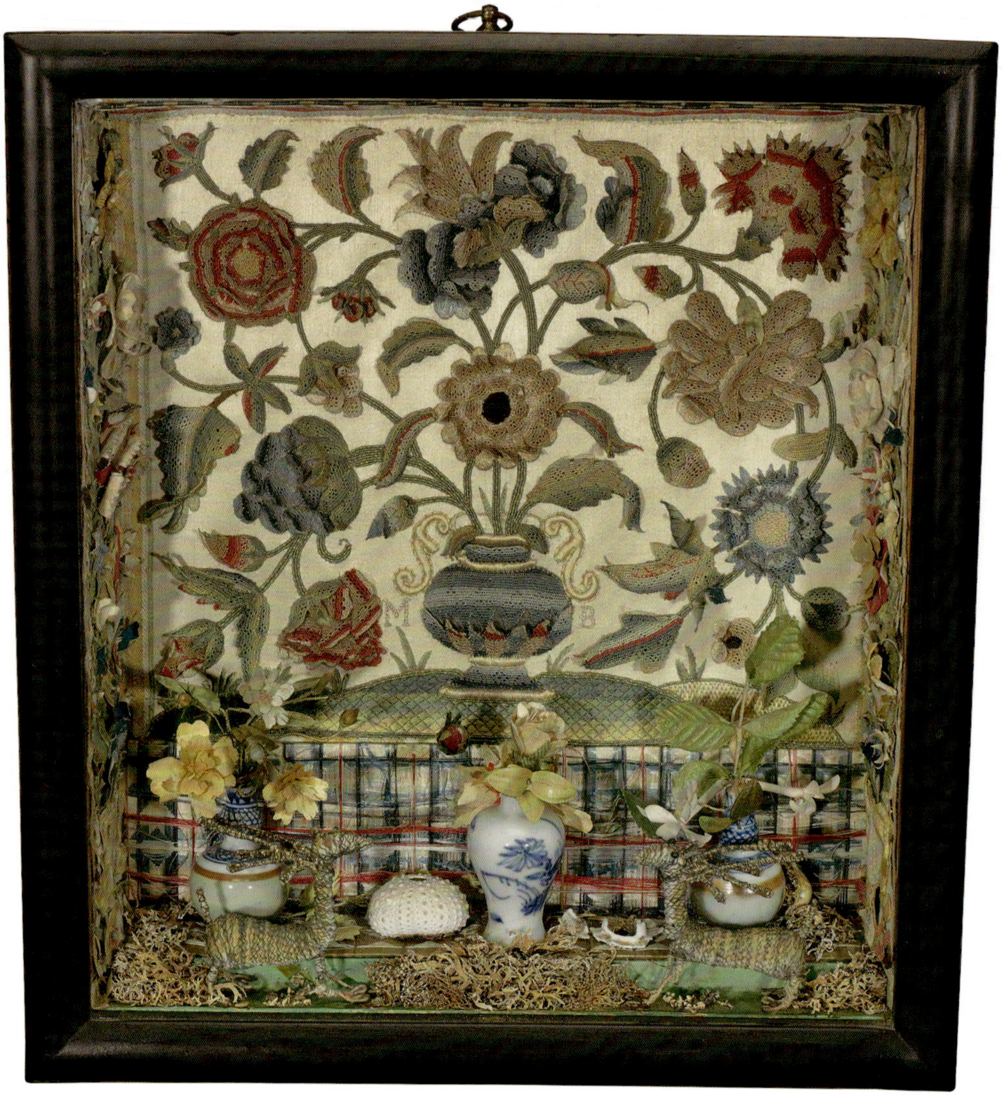

1712

Shadowbox

Silk with silk thread, porcelain, metal thread, paper, lichens, seashells

$13\frac{3}{4} \times 12\frac{1}{4} \times 3\frac{1}{4}$ in. (34.9 × 31.1 × 9.5 cm)

Inscription: 1712| M B

This collaged picture box is set within an ebonized and satin-lined case that bears traces of original gold chinoiserie decoration. The back panel is embroidered and appliqued; at center is a shaded blue vase that bears the date 1712, on either side of which are the initials "M B," presumably those of the maker of this ambitious diorama. The bottom and side panels of the shadowbox are decorated with wired silk blossoms; wire stags; lichens and seashells; and three miniature, blue-and-white Ming vases. Other eighteenth-century shadowboxes of this type incorporate a larger number of shells and lichens, creating grotto-like interiors.

PROVENANCE

New York, Sotheby's, London, March 5, 1998 (lot 411); bt. John H. Bryan of Lake Bluff, IL (1936–2018) through Cora Ginsburg LLC, New York

1728

Sampler
Linen with silk thread
16 ½ × 9 ¼ in. (41.9 × 23.5 cm); framed dimensions: 22 × 14 ½ in. (55.9 × 36.8 cm)
Elizabeth Coker (b. 1716)

Inscription: ABCDEFGHIK JLMNOPQRSTVWYZ&|ABCDEF GHIKJLMNOPQRSTVWYZ|abc defghiljklmnopqrstuwxyz&|AB CDEFGHIKJLMNOP|QRST VWYZ&|ABCDEFGHIKJLM NOPQRST|VWYZ|1234567891011 12131415161718192021222324; Notre Pere qui es aux Cieux Ton Nom soit sancti|fie Ton Regne vienne Ta volonte soit faite en la|Terre comme au Ciel Donne nous au jourd hui|notre Pain quotidian Et nous pardonne nos Offenses com|me nous pardonnons a ceux qui nous ont offense Et|ne nous Indui point en Tentation mais delivre nous|du malin Car a toi est le Regne la Puissance & la| Gloire aux Siecles des Siecles Ainsi soit il; Elizbeath Coker Aged 11 years 1728

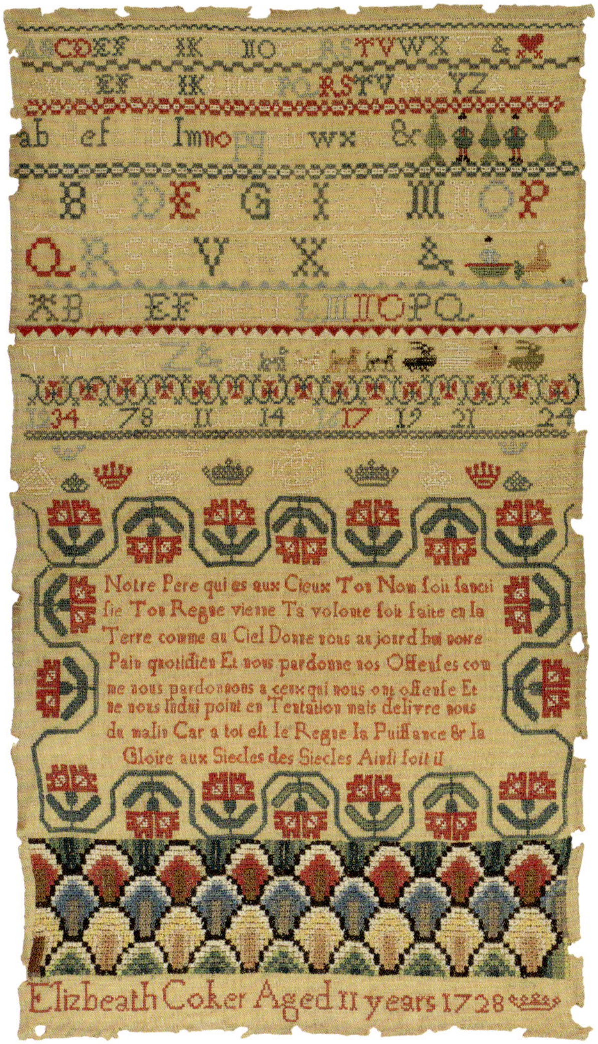

An essay in repetition, this sampler cycles through the alphabet in upper and lower case—an alphabet that, at this time, was only twenty-four letters long. The embroiderer then moves through the first twenty-four numbers before beginning a rendition of the Lord's Prayer in French. Girls who received education were taught how to read at an early age, from about four years old or younger. Embroidery was taught at school, and the tedium of needlework was frequently the cause of complaints among young women.

French was considered to be the most polite and refined European language, and for centuries it had been prerequisite for any participation in genteel company. As such, it was taught throughout the eighteenth century with increasing frequency by private tutors and at schools for both boys and girls. Whereas boys were expected to perfect their spoken French while abroad in France, girls were limited to learning *en Angleterre*. As it happens, samplers derive their English name from the French *essamplaire*, meaning any form of work that served as a model to be copied or imitated.

An Elizabeth Coker was baptized on December 16, 1716, at St Peter Wisbech in Cambridgeshire. Her parents, Anne and John Coker, had married at the same church in 1711. Nothing more of their lives is known.

PROVENANCE

bt. John H. Bryan III of Nashville, TN, from Witney Antiques, Witney, Oxfordshire, June 18, 1999

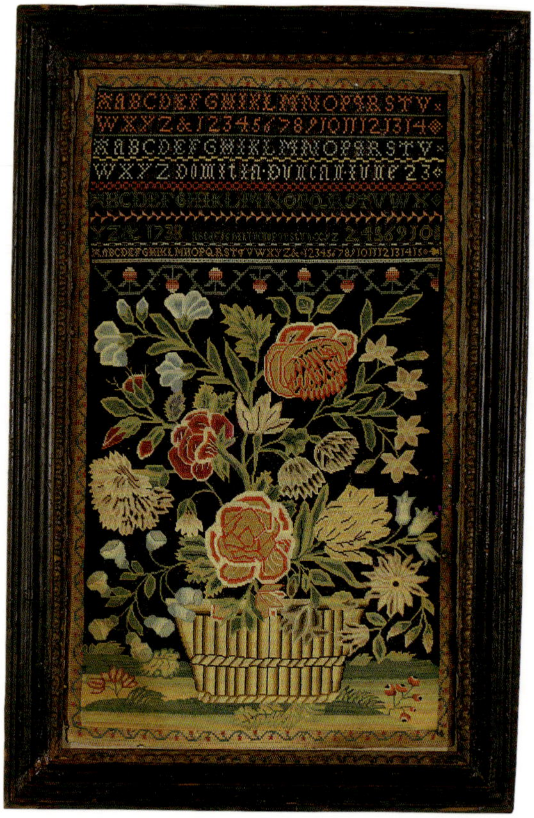

1738

Sampler

Linen or cotton with silk and wool threads

$15\frac{1}{2} \times 9$ in. (39.4 × 22.9 cm); framed dimensions: $18\frac{1}{4} \times 11\frac{3}{4} \times 1$ in. (46.4 × 29.8 × 2.5 cm)

Domitia Duncan (1725–1792)

Inscription:
AABCDEFGHIKLM
NOPQRSTV|
WXYZ&12345
67891011121314|
AABCDEFGHIKLM
NOPQRSTV|WXYZ
Domitia·Duncan·
Iune 23|ABCDEFGHI
KLMNOPQRSTVWX|
YZ& 1738 abcdef
ghiklmnopqrst
vwxyz2456910|
AABCDEFGHIKLM
NOPQRST[?]
VWXYZ&.12345
6789101112131415

The lower two-thirds of this sampler can be clearly linked to domestic needlework of the early eighteenth century—which carried floral designs or biblical scenes stitched in "petit point" (or tent stitch) on a dark brown ground—but there is much about this sampler that is unusual.

For one, it was made by a young woman with an unusual name: Domitia. She was thirteen years old when she embroidered the date—June 23, 1738—into the sampler. Black wool yarn was used to fill the background, which was usually left as bare linen. In the sixth band of the decoration, the embroiderer included only the numbers "2456910," omitting the figures already worked on that line within the date.

Domitia was baptized on June 1, 1725, at St Dunstan and All Saints, Stepney, London. Her parents, Andrew and Eleanor Duncan, both from the parish of St Mary Whitechapel, had conducted a clandestine marriage on April 8, 1716—a marriage without banns or a license from their parish church. Although theirs was likely not, marriages like these were often conducted in the vicinity of Fleet Prison, which was deemed outside of parish jurisdiction. These ceremonies were easier, cheaper, and more straightforward. By the time the Duncans married, there was a small army of clergy, many of dubious credentials, operating in the area of the Fleet.

Andrew Duncan was a surgeon. His will was proved in Dublin in 1740, which makes it likely that the maker of this sampler was the "Domitia Dunkin, spinster" of Dublin whose will was proved in 1792—in turn, this fact raises the possibility that the sampler may have been made in Ireland.

PROVENANCE

bt. John H. Bryan of Lake Bluff, IL (1936–2018) from Cora Ginsburg LLC, New York, September 30, 1999

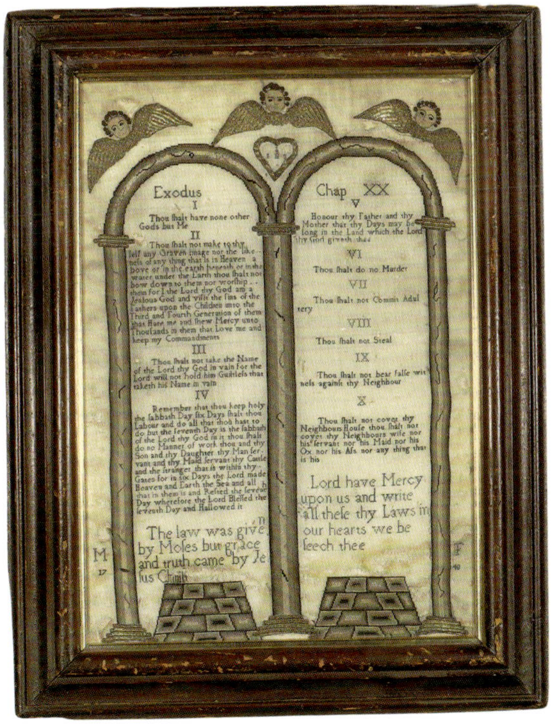

1748

Sampler

Linen with silk and metal thread

$17 \times 11\frac{3}{4}$ in. (43.2 × 29.8 cm); framed dimensions: 22 × $16\frac{1}{2}$ in. (55.9 × 41.9 cm)

Inscription: M F|17 48; I H S; Exodus|I|Thou shalt have none other|Gods but Me|II|Thou shalt not make to thy|slef any graven Image nor the like..|ness of any thing that is in Heavn a|bove or in the earth beneath or in the|water under the Earth thou shalt not|bow down to them nor worship..|them for I the Lord thy God am a|Jealous God and visit the sins of the|Fathers upon the Children into the|Third and Fourth Generation of them|that Hate me and shew Mercy unto|Thousands in them that Love me and|keep my Commandements|III|Thou shalt not take the Name|of the Lord thy God in vain for the|Lord will not hold him Guiltless that|taketh his Name in vain|IV|Remember that thou keep holy|the sabbath Day six Days shalt thou|Labour and do all thou hast to|do but the seventh Day is the sabbath|of the Lord thy God in it thou shalt do no Manner of work thou and thy|Son and thy Daughter thy Man ser|vant and thy Maid servant thy Cattle|and they stranger that is within thy Gates for in six Days the Lord made|Heaven and Earth the Sea and all that in them is and Rested the sevent^{h}|Day wherefore the Lord Blessed the|seventh Day and Hallowed it; The law was give^{n}|by Moses but grace|and truth came by Je|sus Christ; Chap XX; Honour thy Father and thy|Mother that thy Days may be|long in the Land which the Lord|thy God giveth thee|VI|Thou shalt do no Murder|VII|Thou shalt not Commit Adul|tery|VIII|Thou shalt not Steal|IX Thou shalt not bear false wit|ness aginst thy neigbour|X|Thou shalt not covet thy|Neighbour House thou shalt not|covet thy Neightbours wife nor|his servant nor his Maid nor his Ox nor his Ass nor any thing that|is his; Lord have Mercy|upon us and write|all these Laws in|our hearts we be|seech thee

In the early modern period in Britain, exposure to the Ten Commandments would have come with grim regularity and from an early age. The Ten Commandments were printed in the Catechism, which was used to teach children to read.

This sampler mirrors the form of wooden boards (decalogues) that contained the Ten Commandments, the Apostles' Creed, and the Lord's Prayer. These were placed at the east end of the chancel, where the Catholic altar had once stood, and their presence there reinforced the authority of the state over the sacramental power of the church. During the same period, there were other samplers of a similar design imitating the stone tablet form, including one worked by Sarah Broadway, age ten, dated 1763, in the collection of the Lady Lever Art Gallery, Liverpool (LL5329).

PROVENANCE

Wilkinson's Auctioneers, Doncaster, December 3, 2017 (lot 311); bt. John H. Bryan of Lake Bluff, IL (1936–2018)

1776

Sampler

Cotton with silk thread

8 ½ × 9 in. (21.6 × 22.9 cm); framed dimensions: 12 ¾ × 13 × 1 ¼ in. (32.4 × 33 × 3.2 cm)

Inscription:
MR. 1776

From the 1750s, some embroiderers developed a type of sampler to demonstrate their darning technique, a skill essential to efficient housekeeping and thrift. This example, embroidered in silk on a cotton ground, is similar in style and stitching to a sampler in the Victoria and Albert Museum, London (T.36-1945). In the center is a tied spray of flowers, surrounded by some evenly spaced petal forms. The corner squares and rectangles served as darning exercises, imitating a variety of woven cloth patterns. It is unlikely that the embroiderer of this sampler was exhibiting her darning skills for professional purposes; instead she likely made this as a decorative piece, to display her level of proficiency with needle and thread.

PROVENANCE

John H. Bryan of Lake Bluff, IL (1936–2018)

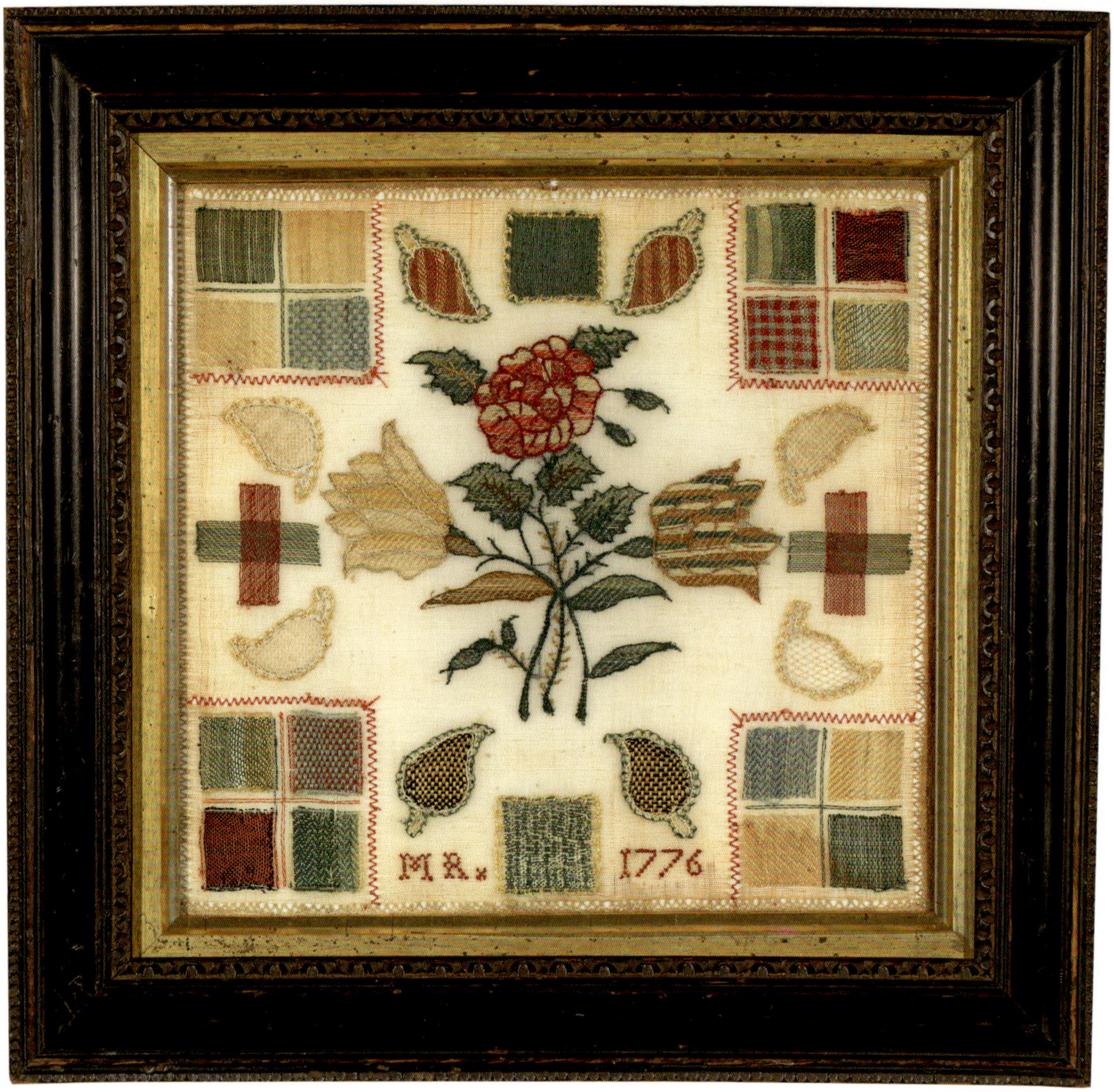

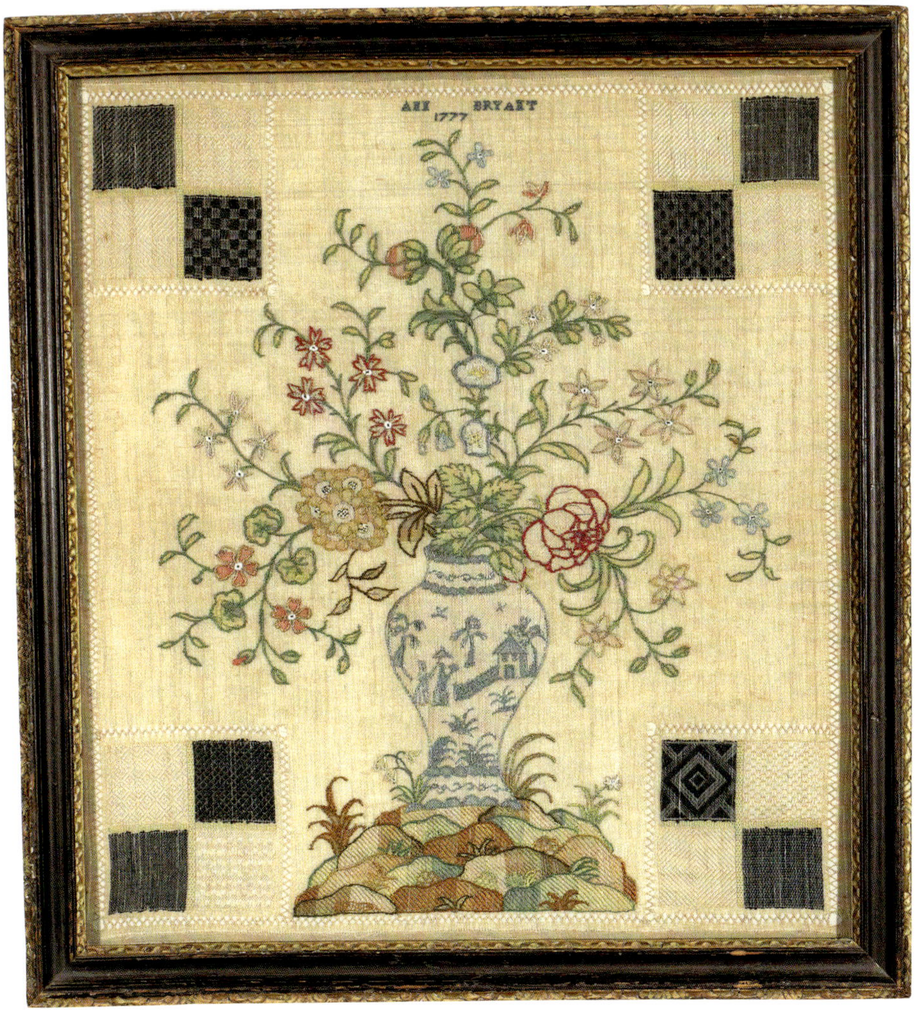

1777

Still in its original black, gilded, molded-wood frame, this sampler depicts a Chinese blue-and-white vase filled with floral sprays. The central design was worked with polychrome silks in darning and stem stitches, the name and date in cross-stitch, and the borders and corner panels in drawn-thread work, which was oversewn in a zigzag pattern. The corner squares contain four different darning patterns. They were worked alternately in black and cream silks in a variety of plain, twill, and diaper-stitch patterns. Many of the patterns are designed to imitate the weave of linen damask, used to mend household items. The central image makes this embroidery more similar to a needlework picture than to a typical sampler. It carries the type of decoration that would have been seen on a domestic fire screen; the distinctive mound beneath the vase recalls the stylized depictions of landscape in late seventeenth- and early eighteenth-century crewelwork embroideries.

PROVENANCE

bt. John H. Bryan of Lake Bluff, IL (1936–2018) from Cora Ginsburg Inc., New York, February 15, 1989

Sampler

Wool with silk thread

15 × 13 in. (38.1 × 33 cm); framed dimensions: $16\frac{3}{4}$ × $14\frac{7}{8}$ × 1 in. (42.5 × 37.8 × 2.5 cm)

Ann Bryant

Inscription:
ANN|1777|BRYANT

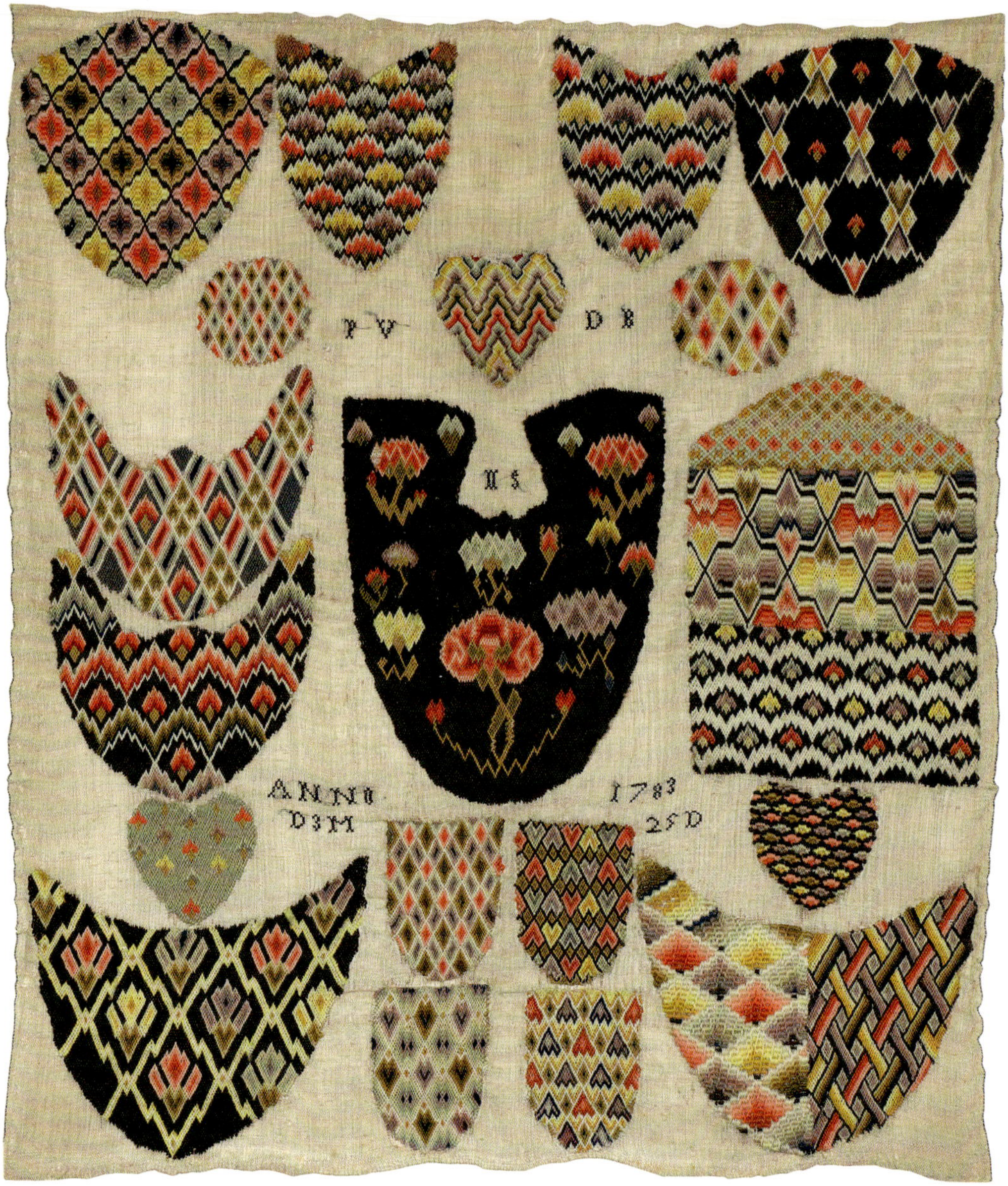

1783

Sampler

Linen with wool thread

31 × 25½ in. (78.7 × 64.8 cm)

Inscription: PV DB|OS|ANNO 1783|D3M 25D

The date on this sampler is rendered in the Quaker manner, which gives the year, the month ("3M" for March), and the day ("25"). The Quaker preference for numbers rather than names for the months derives from the desire to avoid using pagan language. This sampler may have been made in one of the various Quaker schools. It contains a group of panels for pockets, a folding pocketbook, drawstring purses, pincushions, and a stomacher, all worked in brilliant colors with geometric patterns in flame stitch.

PROVENANCE

bt. John H. Bryan of Lake Bluff, IL (1936–2018) from Alistair Sampson Antiques Ltd., London, August 14, 1996

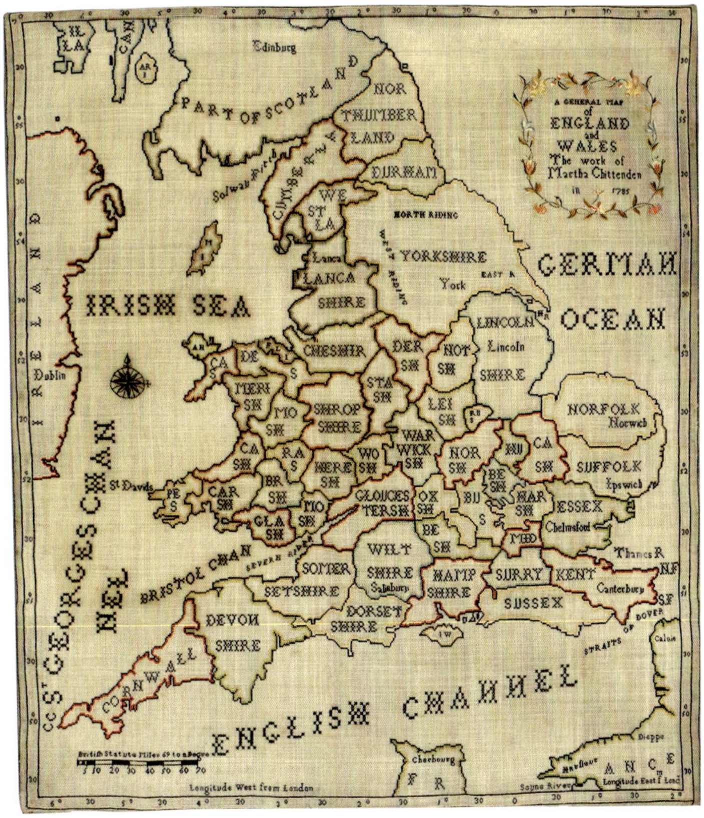

1785

The popularity of map samplers during the 1780s reflected the increased importance of education, which coincided with the belief that it was profitable for boys, particularly those earmarked for the new professions serving the expansion of trade and empire, to learn geography as a subject, even at the expense of Latin. There was, at the same time, increased education for young women, who also had more contact with maps.

Such was the vogue for this new type of sampler that, by the following decade, firms like Laurie and Whittle in London began to produce sampler patterns of the maps of England, Scotland, Ireland, and Europe, although most samplers would have been copied from existing maps, with varying degrees of accuracy. This example is fastidious in the exactitude of its degrees of latitude and scale, which, as Martha Chittenden's biography might suggest, probably speaks more to her discipline and skill than to the pattern's precision.

Martha was baptized on December 9, 1772, at Milton Regis, Sittingbourne, in Kent. She was the daughter of John and Thamar Chittenden. She never married and appears to have lived her entire life in Sittingbourne. She was buried there at the church of Holy Trinity, Milton Regis, at age ninety on February 24, 1863. We know nothing more about her aside from that, in 1851, at the time of the census, when she was age seventy-nine, she visited Mary Lake, a widow two years older than she. Lake was described as a land proprietor living in Bapchild, about a mile and a half east of Sittingbourne.

PROVENANCE

Dorothy Rothschild; bt. John H. Bryan of Lake Bluff, IL (1936–2018) from Leslie Hindman Auctioneers, Chicago, February 20–22, 2005 (lot 1402)

LITERATURE

Donald King, *Samplers: Victoria and Albert Museum* (London: Her Majesty's Stationery Office, 1960)

Sampler

Linen with silk thread

24¾ × 20½ in. (62.9 × 52.1 cm)

Martha Chittenden (1772–1863)

Inscription: A GENERAL MAP | of | ENGLAND | and | WALES | The work of Martha Chittenden | in 1785

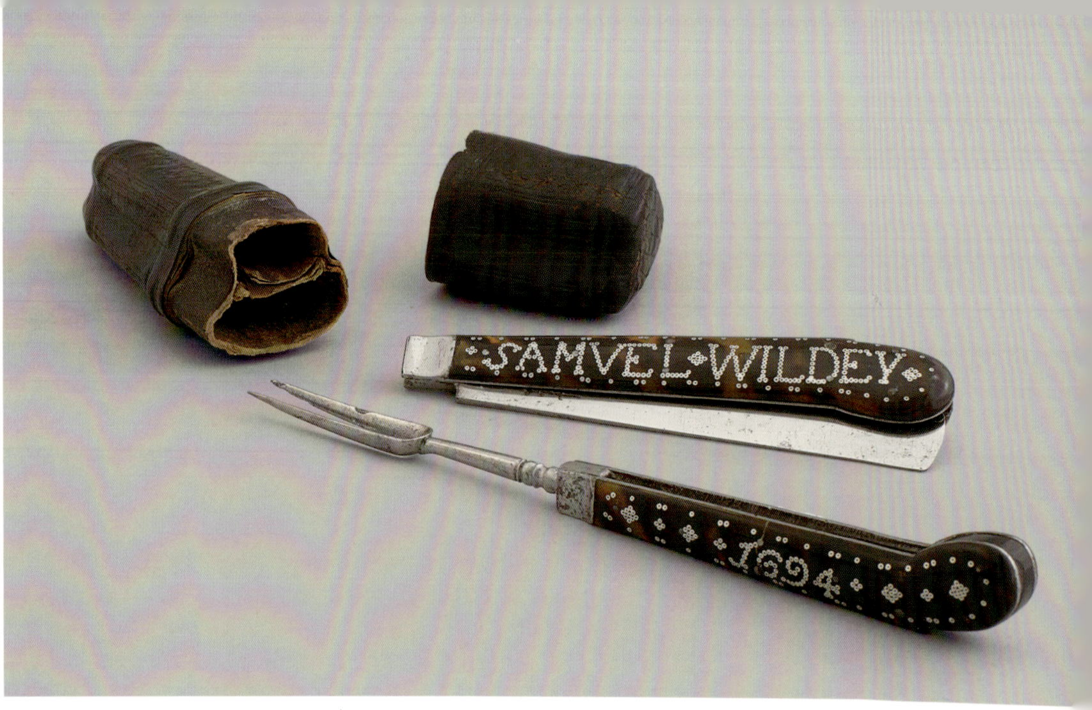

1694

Cutlery set

Steel, tortoiseshell, silver, leather

Knife (open): 9 3/8 × 1/2 in. (23.8 × 1.27 cm); fork: 7 1/2 × 1/2 in. (19.1 × 1.27 cm)

Richard (d. 1660) or Elizabeth Kettlebuter (d. 1685)

Inscription, on both handles: SAMVEL WILDEY; 1694

Prior to the mid-eighteenth century, cutlery was a personal, portable possession that traveled with the owner; it also indicated degrees of wealth, status, and taste. Therefore, great effort, expense, and design went into the production and purchase of cutlery.

This set includes a knife and fork made of steel with tortoiseshell handles and silver piqué (inlay) work, which was at the height of fashion in 1694. These types of objects were typically French products, probably made locally by the French Huguenots who lived in East London, having escaped religious persecution after the Revocation of the Edict of Nantes in 1685. The tortoiseshell came from the tropical hawksbill turtle, which seventeenth-century Europeans exploited in the Caribbean. The shells of these turtles were extremely valuable, and objects made from the material required a specialized method of production.

This set was owned by Samuel Wildey (d. 1752), who was baptized on March 23, 1679. He was the son of woodmonger Richard (1641–1704) and Dorothy Butterfield (d. 1699). Wildey's family members, many of whom were sea captains and mariners, were wealthy and armigerous. Through his wife, Lydia, Wildey was related to the painter John Wotton (1682–1764), who married her sister, Rebecca. Wildey would have received this set of cutlery at about age fifteen, by which point he would have begun his own apprenticeship. In 1703, he acted as steward of the Stepney Feast, or "Cockneys' Feast," an annual gathering to raise money for apprenticeships to the sea service for impoverished children of East London. At such an event, all the great and good of the area would meet to talk, dine, and watch each other's manners—and no doubt Wildey's cutlery would have caught the eye.

Wildey made his money as a coal merchant, and his will of 1752 attests to his wealth. He left strict instructions that his body should be "inclosed in a plain English Oak Coffin and to be carried in a hearse with four horses and two Mourning Coaches following with four horses to each Coach and the Corps to be removed out of my house at Eleven o'Clock at night." His servant was to be the pallbearer. He was buried between his wife (on his left) and his son Richard.

The blade displays the "half-moon and heart" mark designated to Richard Kettlebuter by the Cutlers' Company of London on November 29, 1636. From

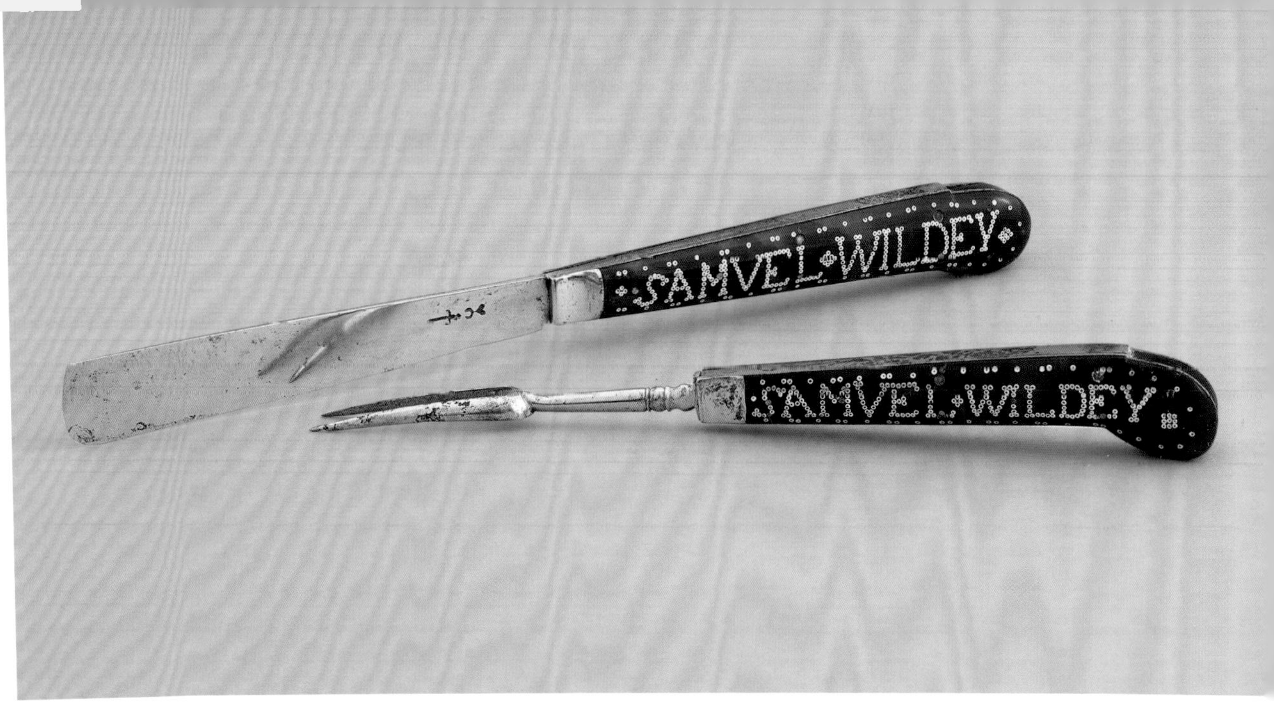

May 4, 1664, onward, the mark was used by Kettlebuter's widow, Elizabeth, one of five women cutlers operating in London at the time. Richard was the son of a gunmaker who, to judge from his name, Harmon, was of German descent; the family name was probably originally Ketelbueters. Richard had begun an apprenticeship under gunmaker Richard Briganshawe, but in August 14, 1627, the apprenticeship was turned over to Arnold Cornelius of the Cutlers' Company. By 1658, Kettlebuter was resident in St Peter ad Vincula in the Tower ("St Peter in Chains" within the walls of the Tower of London), which indicates that Wildey's cutlery was made and sold locally.

As his will of 1660 records, Kettlebuter had considerable property holdings and also held shares in ships returning from Virginia, one of which was captained by his son-in-law John Fox. Elizabeth, his "most loving and deare wife," was to receive all his "rings jewells household goods and my stocke of Iron Steele instruments or what ever else belongethe to my Trade All which I leave to her disposal as she shall see cause." Elizabeth died in 1685 and was buried in the Chancel of St Katharine by the Tower on October 27 that year.

Their son, John (d. 1723), continued in the trade and served as Master of the Cutlers' in 1692. His mark of the "half-moon and P" was registered with the Cutlers' Company on May 1, 1688. Given that Wildey's set bears the mark used by Kettlebuter's parents and not the one used by John, the set was presumably kept in stock and sold a decade after Elizabeth's death. The other possibility is that it was bought at about the time of Wildey's birth and given its piqué decoration later.

PROVENANCE

Donahoe Collection; bt. John H. Bryan of Lake Bluff, IL (1936–2018) from Alistair Sampson Antiques Ltd., London, June 15, 1990

LITERATURE

Peter Brears, *Food & Cooking in Seventeenth Century Britain* (London: English Heritage, 1995)

Peter Brown, ed., *British Cutlery: An Illustrated History of Design, Evolution and Use* (London: Philip Wilson, 2001)

Simon Moore, *Cutlery for the Table: A History of British Table and Pocket Cutlery* (Sheffield, UK: Hallamshire, 1999)

Tessa Murdoch, "Ivory Piqué: The Huguenot Tradition 1680–1760," *Huguenot Society Journal* 33 (forthcoming)

1689 | 1766 | 1797 | 1810

Friendship book

Silk, with silk and metal thread, glass beads, metal-thread trim, wood, paper, silver

$5\frac{1}{2} \times 3\frac{7}{8} \times 1$ in. (14 × 9.8 × 2.5 cm)

Inscription: Anne Warren Mary Smith|In y^r absens this fredon we take|for if you was here oure hads ye wolld breke|Catherine Willughby|1797; The owner of these very|witty Conceits is an asse|I vow a mighty good Joke|ha ha ha h-a|June y^e 16th|1689; Jane Cotton is a person|of singular worth &|goodnesse this I must|a vow to bee most furiously|true|June ye 16th; Francis Cholmeley|October 12|anno domini|1797|Thomas; Simon Scroope of Danby|1766; More than my self|I Love Katherine Willoughby|Elizabeth Hartopp; Catherine Willughby|Ann Pontrell; Abstant tis true|but still I find|A Kind remebrance|of you in my Mind|which bids me [——]|Like [——] bed a dur|Jan 8 1797; 13 pieces of gold|2 pare of gold buttens 15|piece of silver a|gold scall a gold booke|my father pickter in|gold a pickter that belongs to Cathy Willoughby|of gold 2 gold|chaine with 3 lokits at them; 4 dimand rings|a murning ring|4 dosen of groats; Anne Mellish|fortin w^hy y frounest|thou so on me [——] ever wass mad un fortanat|by thee|June 30 1810

The inscriptions on the pages of this friendship book were written with a silver stylus, which made its mark on pasteboard coated with Chinese white. The messages and notes contained within the book range considerably in their dates and content.

There is the childish and whimsical phrase, "The owner of these witty conceits is an ass, I vow a mighty good joke, ha ha ha"; and the crudely spelled note endorsed by Anne Warren, Mary Smith, and Catherine Willoughby, that "In y^r absens this fredon we take for if you was here oure hads ye wolld breke," which conjures a scene of three giggling girls given a moment's reprieve from a draconian schoolmaster or governess.

There is also the tender note, "Jane Cotton is a person of singular worth." The banal comes in the form of a list of gold items and pictures belonging to a member of the Willoughby family.

Who many of these people were remains a puzzle; the only name that can be made out with any certainty is that of Simon Scroope of Danby (1758–1838), the son of Simon Scroope, Esquire, (d. 1788), and his wife, Lady Ann Clementina. The appearance of other names of Yorkshire families—Willoughby (of Birdsall) and Francis Cholmeley (of Brandsby Hall, d. 1854)—point strongly to this object's being owned by one of these families for a significant period of its history. Both the Scroope and the Cholmeley families were Catholic; the latter had married into the Willoughby family in the previous century.

The embroidered scenes on either side of the book are of the late seventeenth century and derive from patterns: both the fishing lady and the piping man appear as a pair on the top of the lid of an embroidered cabinet in the Royal Scottish Museum, Edinburgh (A.1961.502). The lady appears on another embroidered cabinet formerly in the Talbot Hughes Collection.

PROVENANCE

bt. John H. Bryan of Lake Bluff, IL (1936–2018) from Alistair Sampson Ltd., London, April 28, 1992

LITERATURE

G. Saville Seligman and Talbot Hughes, *Domestic Needlework: Its Origins and Customs throughout the Centuries* (London: Country Life, 1926), plate 92A

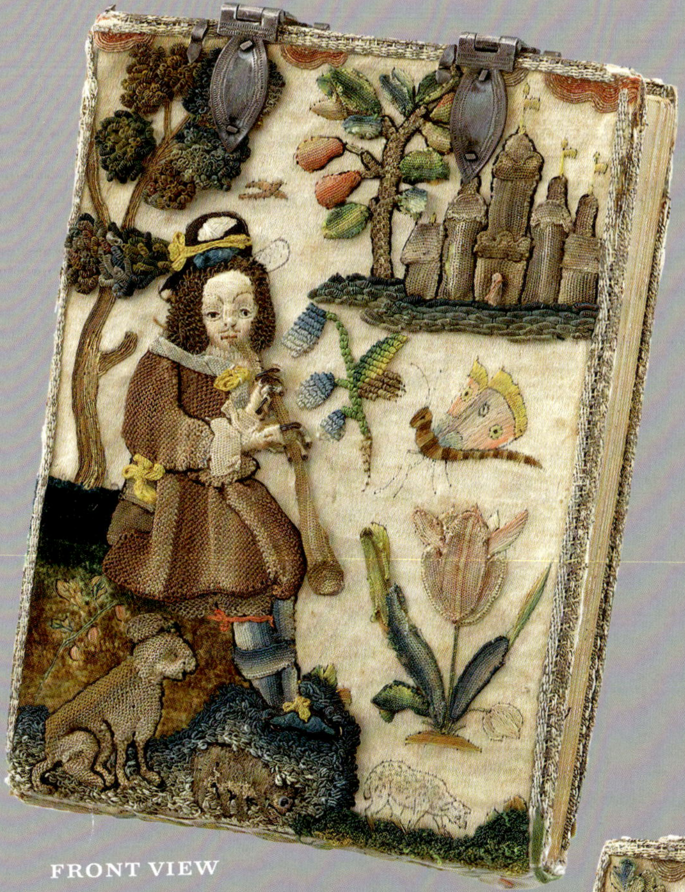

FRONT VIEW

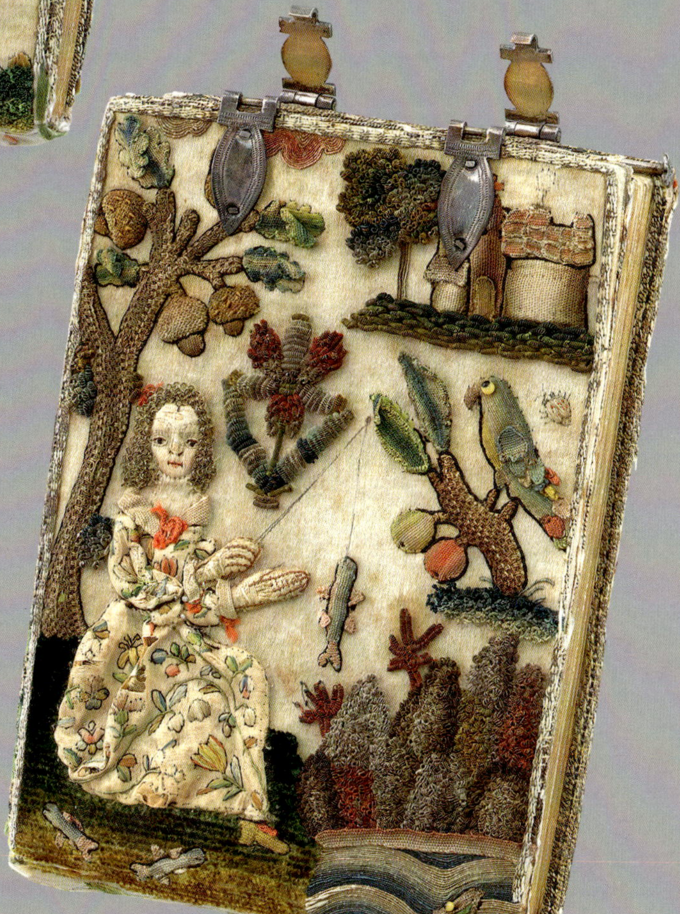

BACK VIEW

1734

Portrait

Oil on canvas

21¼ × 16½ in. (55.2 × 41.9 cm); framed: 25½ × 20⅞ in. (64.8 × 53 cm)

William Hogarth (1697–1764)

Inscription, on verso: This Picture of Mr. Western|was Finish'd by Mr. Hogarth, Oct| 16 1734 Æt Sua 20 years 10 months| and given by him to me Wm Cole this 18 day Nov: 1734; [on a damaged label]: This Canvas is [not] to|Be removed from|the picture

As the painted note on the back of this picture records, this portrait of Thomas Western (1714–1766) was made by William Hogarth. When Hogarth completed it on October 16, 1734, Western was in his twentieth year and a couple of months shy of his birthday, when he would reach his majority and enter a new phase of life. His father, Thomas Western (d. 1733), had been a member of Parliament for Sudbury in Suffolk but had died when his son was still in his minority, so his widowed wife, Mary Shirley, held responsibility for the estates at Rivenhall, Essex, at the time this picture was made. The family was among the emerging middle class of the early eighteenth century, having made its fortune as foundry masters of government ordnance and timber suppliers to the navy in Sussex.

The note was made by Western's friend, Rev. William Cole (1714–1782). He received this painting as a gift on November 18, 1734—an event he dutifully noted alongside Western's coat of arms. As a child, Cole had transcribed inscriptions on monuments and copied coats of arms he saw in church windows. Such inclinations made him eminently qualified for his subsequent admission as a Fellow of the Society of Antiquaries of London.

The portrait depicts Western in his study at Clare Hall, Cambridge, wearing gold-laced and tasseled academic garb. He holds a letter with a broken seal, which, as Elizabeth Einberg suggests, may have been the family request to leave Cambridge and return home. Given the timings Cole recorded on the back of the painting, the portrait was likely Western's parting gift to Cole as Western departed for Rivenhall to marry and assume his responsibilities as the head of a gentry family. Cole was devoted to Western, believing him the most handsome man in college, and wrote fondly of their time together at Rivenhall and at Cambridge, where the two students shared rooms. On Western's birthday in 1735, Cole learned of the death of his father; his inheritance put him in position to become a Fellow Commoner. He then took over sole use of Western's chambers at Clare but, inconsolable over the absence of his friend, he transferred to King's College, where he remained until 1753.

For reasons that are unclear, the men's intense friendship was ruptured abruptly in about 1736, which seems to have resulted in Cole's portrait being redacted and replaced with Western's wife's portrait as the family conversation piece. Initially commissioned in 1735, that portrait is now in the National Gallery of Ireland. Cole downplayed the nature of their falling-out as no more than "some trifle," but Einberg has raised the possibility that, following his departure from Cambridge, Western may have wished to distance himself from bachelor friends and friendships that may have hitherto also been homoerotic.

Cole held on to the painting for many years and considered it "one of the most resembling portraits I ever met with." It hung in his gallery until his death, whereupon it was bequeathed to Western's grandson and remained in that family until the early twentieth century.

PROVENANCE

Painted for Thomas Western of Rivenhall Place, Essex (1714–1766) and presented on November 18, 1734, to William Cole (1714–1782), by whom bequeathed in 1782 to the sitter's grandson, Charles Callis Western, first Baron Western (1767–1844); thence to his cousin, Sir Thomas Burch Western, first Baronet (1795–1873), and by descent to his son Sir Thomas Western, second Baronet (1821–1877); presumably to his sister Margaret Mary, who married Robert Capel Cure (1823–1878); by

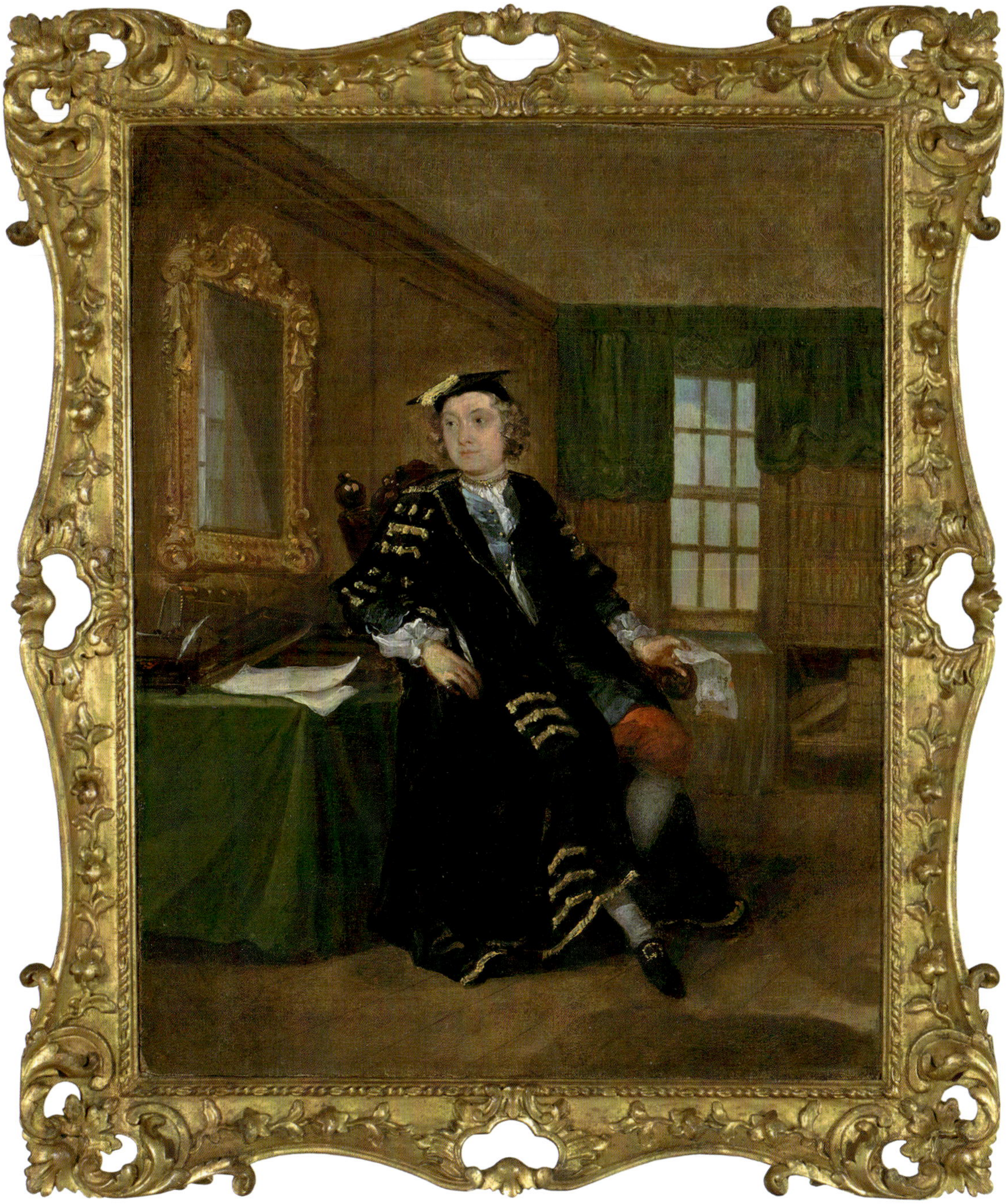

descent to Sir Thomas Charles Callis Western, third Baronet (1850–1917) of Felix Hall, Essex; his sale, Christie's, London, June 1913, 13 (lot 92); with Knoedler, New York.; Alfred Morell; John Hemming Fry (1860–1946), by whom gifted in 1946 to the Canton Museum of Art, OH; Christie's, London, December 2015, 9 (lot 150); Rafael Valls Ltd. and Derek Johns Ltd., London; bt. John H. Bryan III of Nashville, TN, November 18, 2016, through Gurr Johns, London

LITERATURE

Elizabeth Einberg, *William Hogarth: A Complete Catalogue of the Paintings* (New Haven, CT: Yale University Press, 2016), 142–43, 156

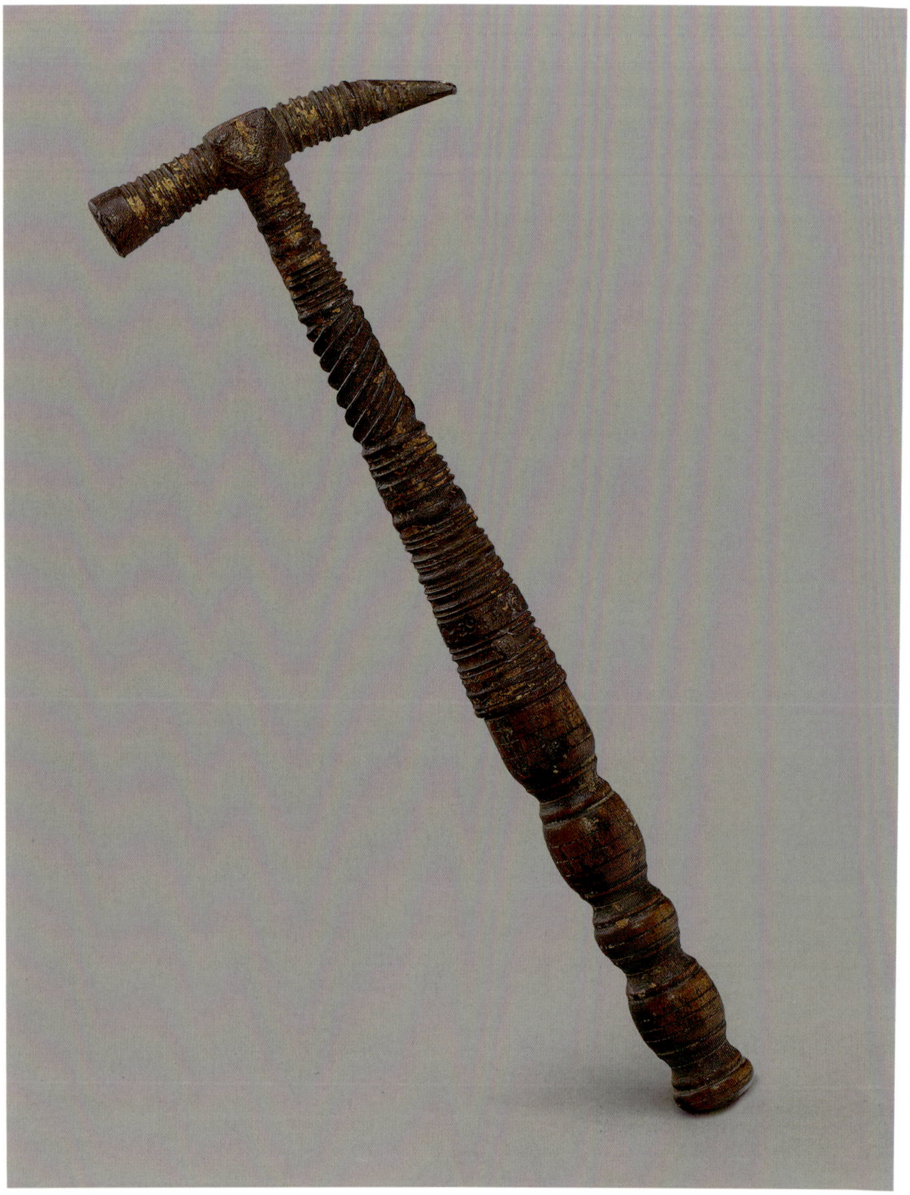

1729

Hammer

Wrought iron, wood

12 × 4 × 1 in. (30.5 × 10.2 × 2.5 cm)

Inscription:
THO·FLETCHER·1729

Another hammer—with the same wrought-iron shaft, claw, and head and a similar turned wood handle—is found at the Victoria and Albert Museum, London (W.94-1922). That hammer bears the inscription "Leander Green" and the date "1694." Leander Green was a glazier of Mile End who, on June 24, 1697, married Elizabeth Meues at St Dunstan, Stepney, just east of London. Green made his will on February 28, 1701/2, when he was "Outward bound to Sea," bequeathing all his possessions to his wife. The lack of any record of his burial suggests that he died overseas. It is possible that Green's hammer was made for him around the time he completed his apprenticeship. Presumably this hammer, dated 1729, was made in the London area for someone in a similar profession to that of Leander Green.

PROVENANCE

bt. John H. Bryan of Lake Bluff, IL (1936–2018) from Alistair Sampson Antiques Ltd., London, June 7, 1996

1765

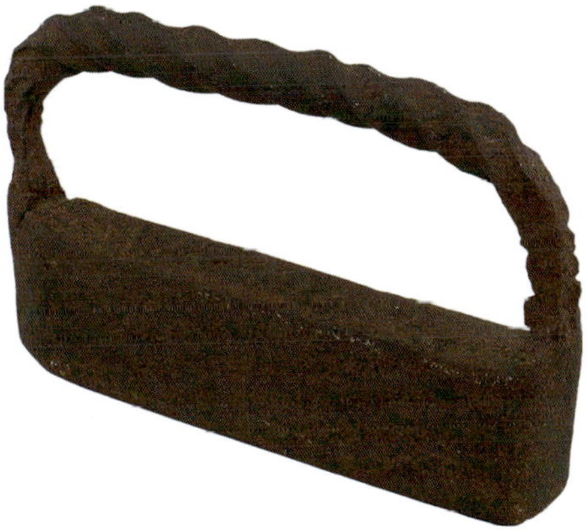

Tailor's smoothing iron
Iron
$5\frac{1}{2} \times 1\frac{1}{2} \times 9$ in. (14.6 × 3.8 × 22.9 cm)

Inscription: W R 1765

A tailor's goose or weasel was heated and used as a smoothing iron for fabrics. The last line of the nursery rhyme, "Half a pound of tuppenny rice / Half a pound of treacle / That's the way the money goes / Pop goes the weasel," refers to the weasels used by the tailors in and around City Road, London.

PROVENANCE
bt. John H. Bryan of Lake Bluff, IL (1936–2018) from Christopher Bangs Ltd., Suffolk, May 16, 2017

1777

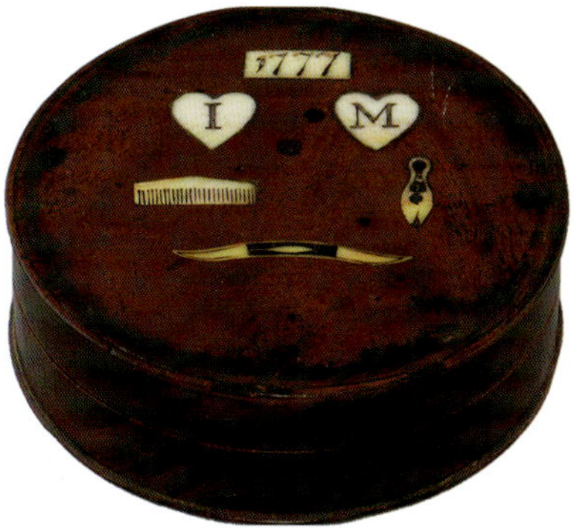

Tobacco box
Mahogany, bone
$1 \times 2\frac{7}{8}$ in. (2.5 × 7.3 cm)

Inscription: 1777|I M

Weavers' tools are depicted in bone on the lid of this tobacco box: wool shears, a wool comb, and a shuttle.

PROVENANCE
bt. John H. Bryan of Lake Bluff, IL (1936–2018) from Alistair Sampson Antiques, London, August 6, 1987

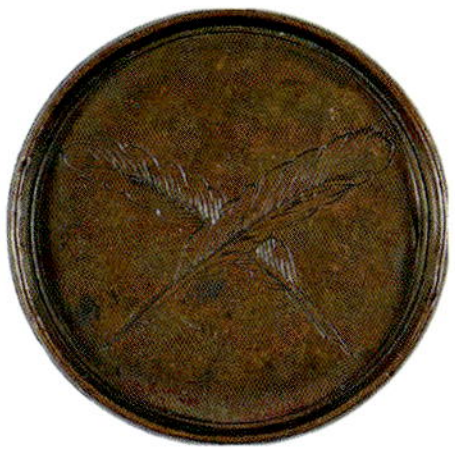

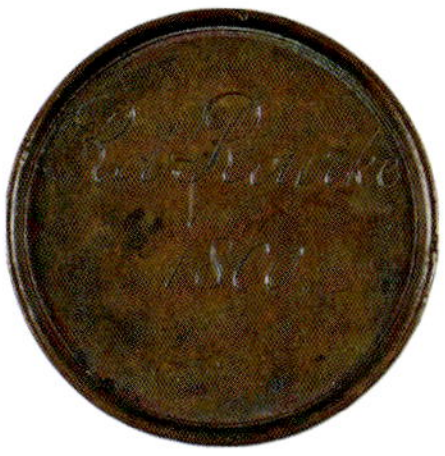

1800

Medal
Brass
$1\frac{1}{4}$ in. (3.2 cm) diameter

Inscription: Robt Rourke|1800.

The crossed quills on the reverse suggest that this medal was a prize awarded for penmanship in 1800 to young Robert Rourke, who might be the son of Bryan Rourke and Honora Dowling of North Anne Street, baptized at St Michan, Dublin, on April 18, 1791.

PROVENANCE
bt. John H. Bryan of Lake Bluff, IL (1936–2018) from Jim Marshall, Pen & Pencil Gallery, The Wirral, Merseyside, May 30, 2017

1695

Plate
Delftware
9 × 1 in. (22.9 × 2.5 cm)

Inscription: H|M|I; 1695

The figures painted on this plate are in the process of surveying. This new profession was associated with the mapping of great estates and illustrates the skillful handling of a cross-staff, a geometric square, and a theodolite.

PROVENANCE

D. M. & P. Manheim, New York; Private Collection, London, acquired from Jellinek & Sampson Antiques, London, February 15, 1979; Christie's, London, Sainsbury Sale, June 18, 2008 (lot 139); bt. John H. Bryan of Lake Bluff, IL (1936–2018) from Sampson & Horne Antiques, London, June 20, 2008

LITERATURE

Louis L. Lipski and Michael Archer, *Dated English Delftware: Tin-Glazed Earthenware, 1600–1800* (London: Sotheby Publications, 1984), 61, no. 204

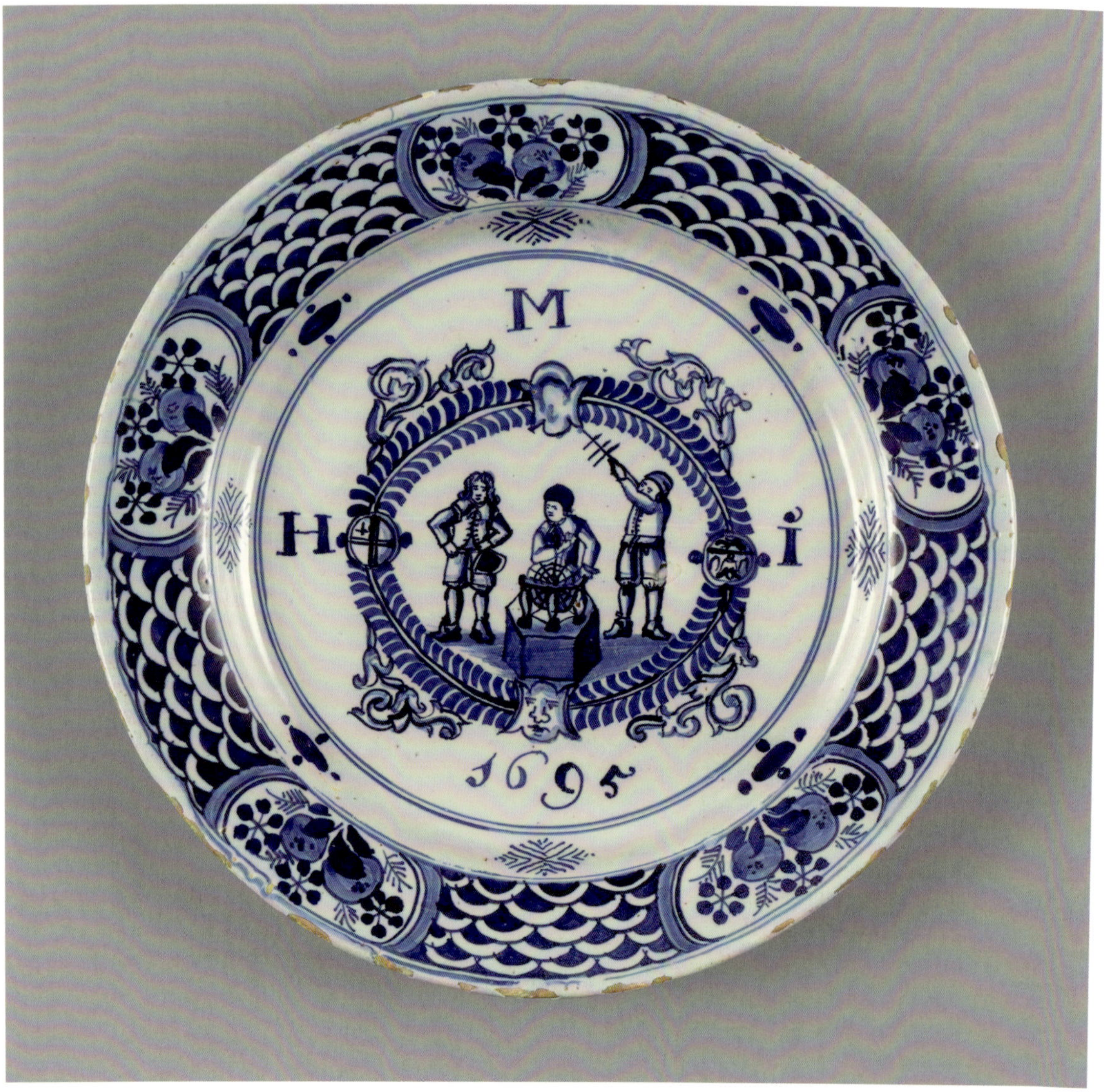

1753

Plate
Delftware
8 ¾ × 1 in. (22.2 × 2.5 cm)

Inscription: S|H|E; 1753

There are two more plates from the same set that bear the same initials and decoration: one was formerly in the collection of Thomas George Burn at Rous Lench Court, Worcestershire, and the other is in the Bristol Collection. Frank Britton has identified the initials on the plate in the Bristol Collection as those of Samuel Hall, Butcher, who married Elizabeth Audsley on October 14, 1750, at St Luke, Old Street, London. He was the son of Nathaniel Hall (d. 1761), a Butcher of Newgate Market, who himself had married the daughter of a butcher. Samuel became a freeman by patrimony on April 5, 1753, and this plate was made to celebrate that occasion. He became a livery man in 1762 and took on six apprentices between that year and 1780. Samuel Hall died in 1789.

PROVENANCE

bt. John H. Bryan of Lake Bluff, IL (1936–2018) from Jonathan Horne, London, February 17, 1983

LITERATURE

Frank Britton, *English Delftware in the Bristol Collection* (London: Sotheby Publications, 1982), 154

Louis L. Lipski and Michael Archer, *Dated English Delftware: Tin-Glazed Earthenware, 1600–1800* (London: Sotheby Publications, 1984), 128, no. 564

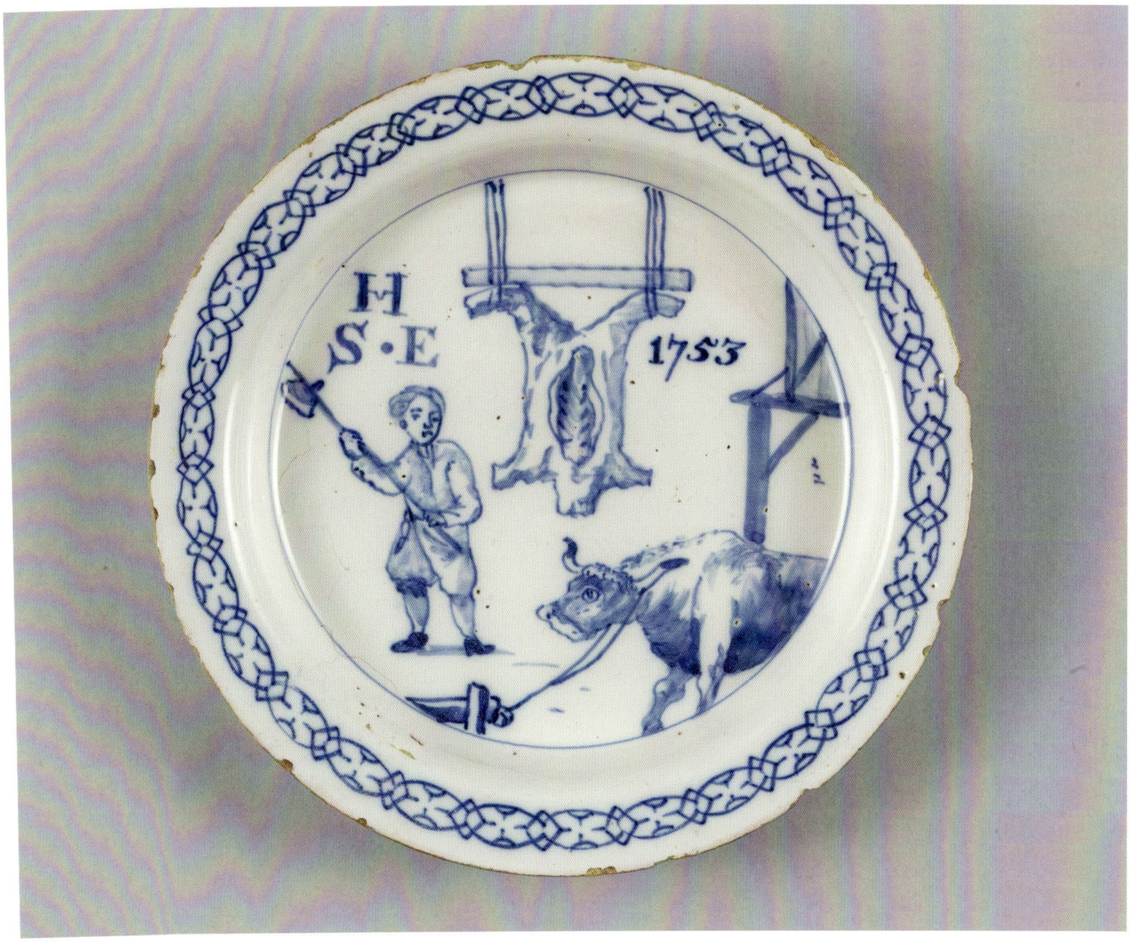

1790

Bloodletting knife

Brass, steel

Closed: ¼ × ¾ × 3½ in. (.64 × 1.9 × 8.9 cm)

Inscription:
T Wilson 1790

This late eighteenth-century folding knife has two hinged steel blades enclosed in the brass case. This style of knife is described as a surgeon's knife or bloodletting instrument, perhaps due to the shape of the long blade, which often would have a sharp, protruding cutting edge.

Bloodletting was practiced in England into the nineteenth century. It was conducted in a variety of ways, including by cutting, by cupping, and by application of leeches. Many saw it (although not without criticism) as a cure for a host of complaints, including fevers and seizures. This kind of tool would have been a central part of any physician's practice, perhaps explaining why it was thus named and why it was kept in a finely engraved case.

PROVENANCE

bt. John H. Bryan of Lake Bluff, IL (1936–2018) from Lindsay Antiques, London, November 11, 1991

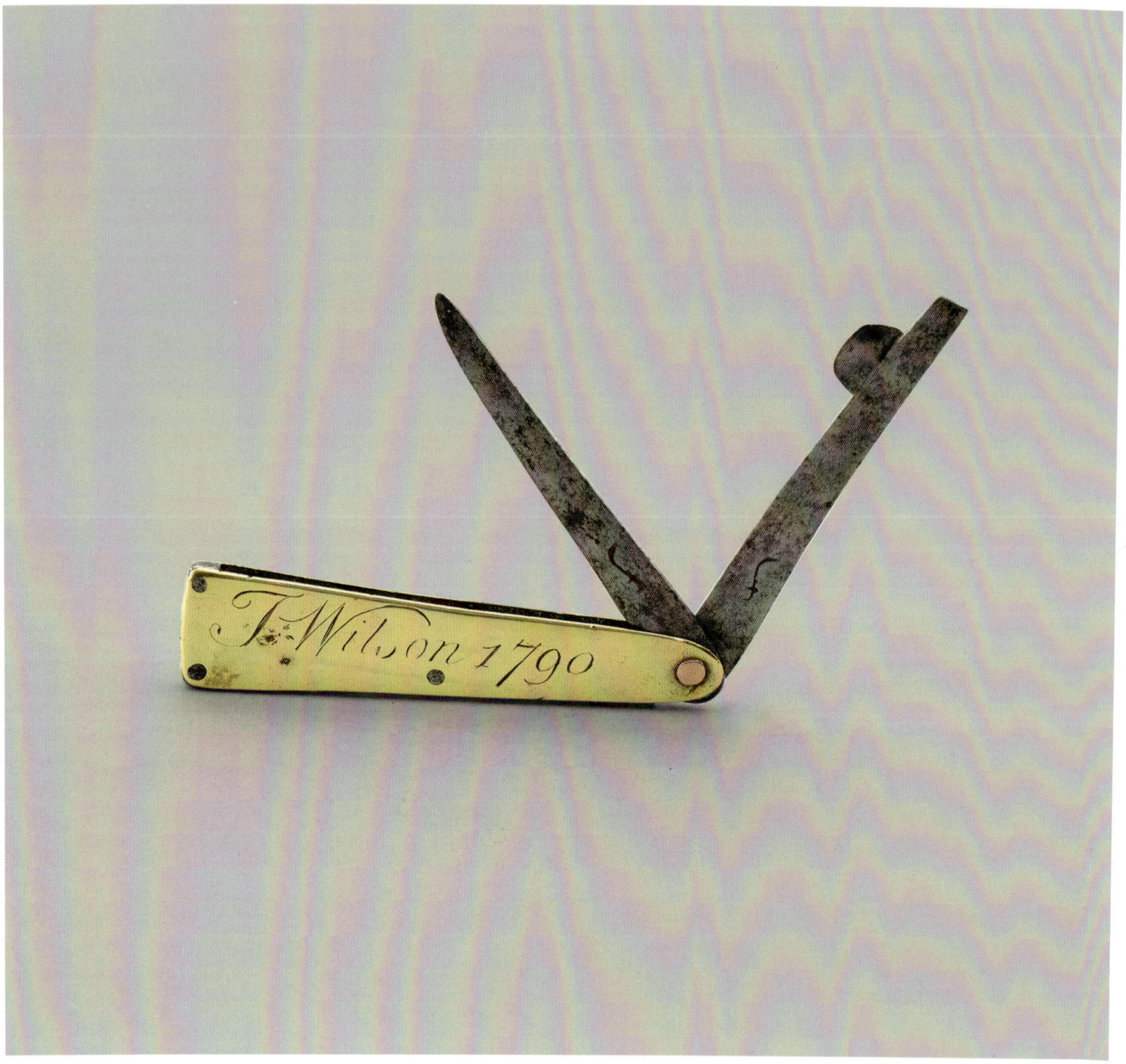

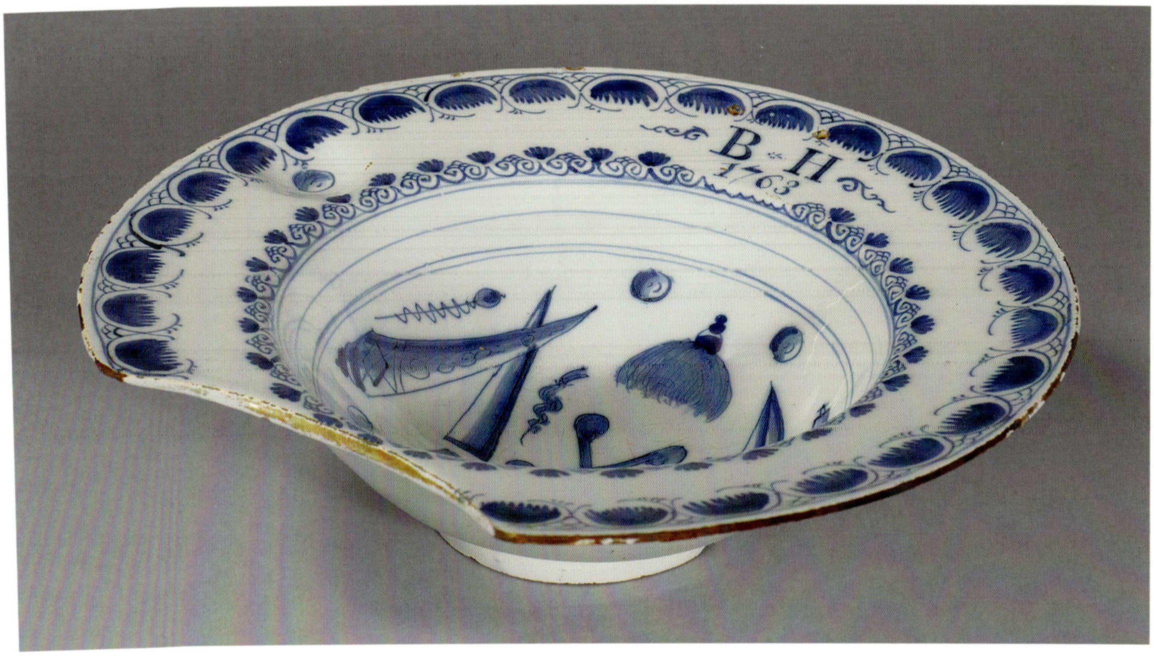

1763

Bowl

Delftware

$10\frac{1}{4}$ × 3 in. (26 × 7.6 cm)

Inscription:
B H|1763

This bowl is one of a small group of barber's bowls that share similar border decorations found on pottery shards at the Temple Back site in Bristol. The deep well is painted in blue with the tools of the barber's trade—a cut-throat razor, scissors, comb, and brush—and a pair of crossed bones. The rim is inscribed within stylized lappet and foliate borders. It also has a small depression to hold shaving soap. The hemispherical cutout was made to accommodate the customer's neck. Two holes were made before firing so the bowl could be hung when not in use.

The initials "B H" raise the possibility that the bowl belonged to the Bristol Barber-Surgeon Barnet Hughes (1733–1783), who began his apprenticeship under Thomas Ellis in 1751; the date plausibly relates to his freedom following his term of servitude.

These bowls might also have been used by professional barber-surgeons in the performance of basic medical procedures, such as collecting blood when bloodletting. Another barber's bowl, of an earlier date and by a different maker, in the collection of the Victoria and Albert Museum, London (297-1869), was clearly for professional use. Its inscription and those on similar bowls make reference to payments due from regular customers who paid bills quarterly: "SIR: YOVRE : QVARTER IS : VP :" and "Quarter Day Pray Gentlemen Pay 1716" (Christie's, London, June 10–11, 2010 [lot 1119]).

PROVENANCE

Mrs. Caldwell; Captain D. Cooper; Sotheby's, London, November 11, 1969 (lot 93); Thomas George Burn, Rous Lench Court, Worcestershire; Sotheby's, London, July 1, 1986 (lot 58); bt. John H. Bryan of Lake Bluff, IL (1936–2018) through Lindsay Antiques, London

LITERATURE

Louis L. Lipski and Michael Archer, *Dated English Delftware: Tin-Glazed Earthenware, 1600–1800* (London: Sotheby Publications, 1984), 296, no. 1224

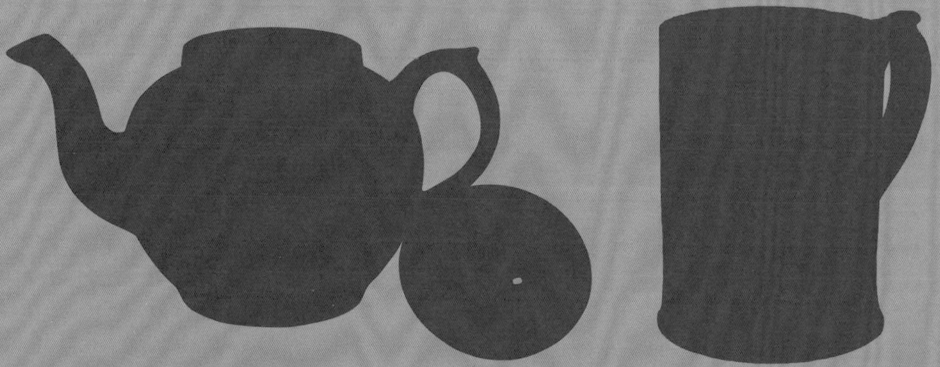

II. COURTSHIP & MARRIAGE

ALTHOUGH SOME PEOPLE MARRIED IN THEIR TEENS, MOST MARRIED after the age of twenty-five, with men tending to marry a couple of years later than women, at twenty-seven or twenty-eight. Marriage conveyed status and, particularly for women, afforded participation in various social circles and activities hitherto unavailable. As divorce was not permitted and separation looked down upon, the choice of one's partner had to be made carefully. Choice was, nevertheless, predicated on social status, and it was extremely uncommon for serious courtship to occur between two people of demonstrably different means. Although consent and free choice were supposed to be essential in the decision to marry, in a highly patriarchal society, parental intervention was often heavy-handed, and parents' approval was required. Many couples would eventually lose a spouse to disease or sudden death, but marriage was otherwise intended to be a permanent state—one that governed a shared economic and social future—making the decision of whom to marry the most important one of most people's lives.

The church provided no prescriptive guide to courtship, although there was a series of unwritten rules to which both parties were expected to adhere; the higher the status of the couple, the more formalized the series of approaches and enquiries. Yet, up and down the social ladder, the period of courtship occasioned the exchange of gifts, which were given as tokens of affection—by both men and women. Willing acceptance of such a token—for example, the carved stay busks shown on pages 169–71—was considered tantamount to an acceptance of the proposal, although such gifts might then be returned, signaling the end of the courtship.

Betrothal, if and when it came, might be celebrated by bestowing another set of gifts and by breaking a coin into two symbolic halves. Or the engagement might be formally declared with the assistance of a prayer book and priest. Once this formal step was taken and the banns of marriage were read in church, many couples wasted no time in beginning their sexual partnership: roughly one-quarter of baptisms recorded in parish registers took place within the first eight months of the marriage ceremony.

Although marriage is not a sacrament in the Church of England, conventional marriages were solemnized in the church ceremony through the exchange of vows and the man's placement of a ring on the fourth finger of the woman's left hand. The ring, because of its form, symbolized the continuous and unbroken nature of marriage, and the object often bore an inscription that spoke to contentedness, partnership, and reciprocity. Not everyone married, however: in the seventeenth century, for example, one-quarter of women remained single. A significant number of marriages were conducted in clandestine fashion, many officiated by disreputable clergy at the precinct of Fleet Prison, which was deemed outside of the jurisdiction of the London parishes.

Once a couple was united in marriage, the focal point in their household was the marital bed. As the site of conception, birth, and death, the bed was the setting for the most significant moments of the early modern life cycle. In affluent homes, the bed was the largest item of furniture in the house, replete with frame, stacked mattresses, layers of textiles, and curtains to draw around it when the occupants retired. Dates, often embroidered into the bed hangings, as on those shown on page 153, underscored the significance of the bed and also provided a temporal marker upon which future events could be pegged.

1670

Bed hangings

Cotton with wool thread

Coverlet: 64½ × 74½ in. (163.8 × 189.2 cm); headcloth: 67 × 56 in. (170.2 × 142.2 cm); pelmets: 15¾ × 103 in. (40 × 261.6 cm); side panels: 65 × 37¼ in. (165.1 × 94.6 cm)

Inscription, on each piece: IG 1670 EG

Crewelwork, made with wool yarn in often-elaborate, freehand embroidered designs, became popular in England in the last quarter of the seventeenth century. These bed hangings are among the earliest dated examples of crewelwork, which include a workbag at the Metropolitan Museum, New York (2006.263), made one year earlier, in 1669.

Women would decorate textiles for their own use, and a project of this scale would have required substantial time and effort. The initials on these bed hangings—"IG" and "EG"—and the date, "1670," repeat on each piece. These items may commemorate the marriage of their embroiderer, charmingly referenced with the matching pairs of birds in the design, heart-shaped motifs, and the central rose. These design elements shift in scale and pattern to accommodate the varying shapes of the pelmets (valances) and panels, incorporating the imagery of flowers, foliage, and birds.

PROVENANCE

Kerry Taylor Auctions, London, November 20, 2017 (lot 144); bt. John H. Bryan of Lake Bluff, IL (1936–2018) through Cora Ginsburg LLC, New York

LITERATURE

Joan Edwards, *Crewel Embroidery in England* (New York: Morrow, 1975)

Sylvia Houghteling, "Origins in Entanglement: Connections between English Crewel Embroidery and Indian Chintz," in *Cloth That Changed the World: The Coloured Cottons of India*, ed. Sarah Fee (New Haven, CT: Yale University Press, 2019), 183–91

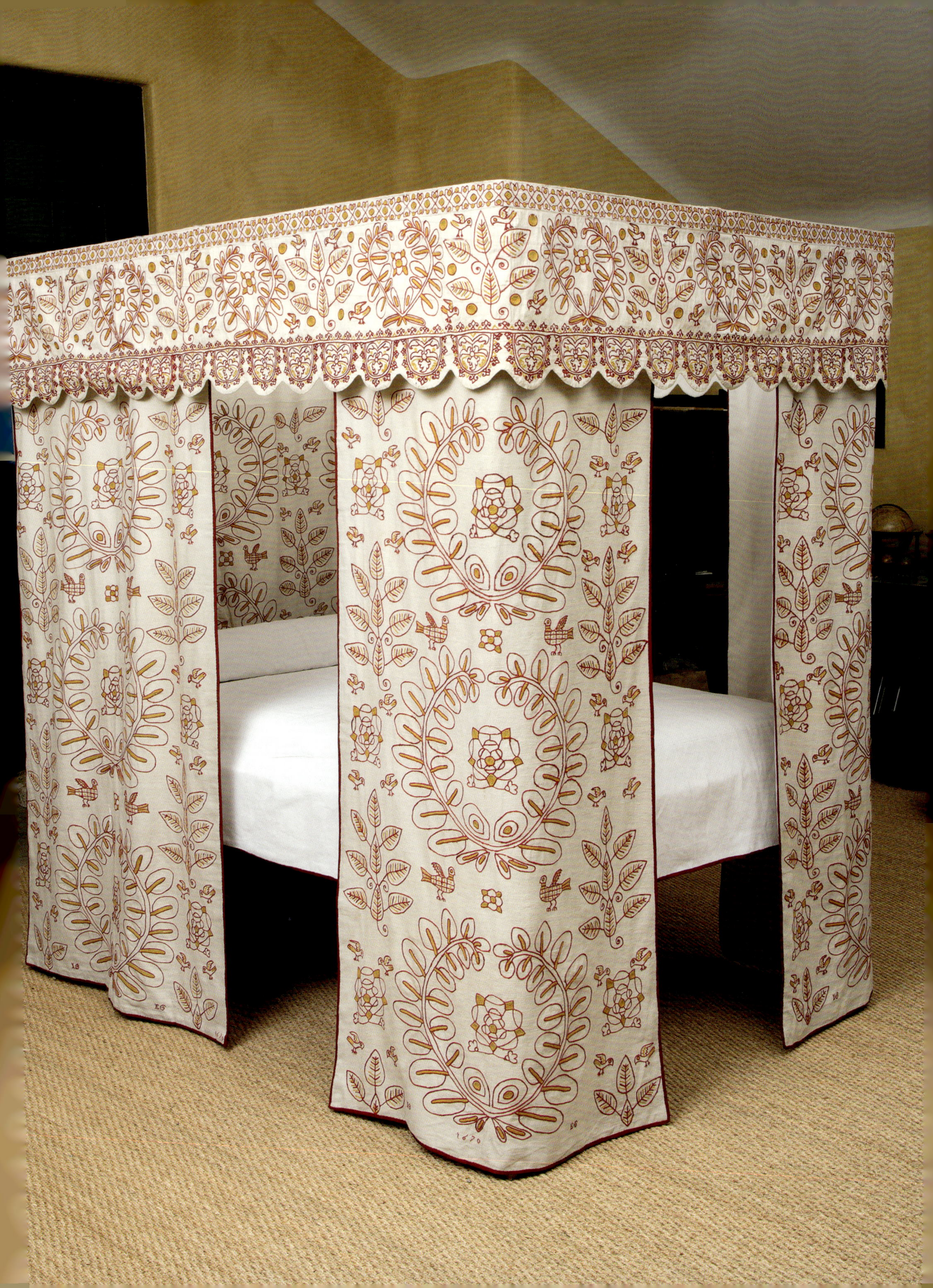

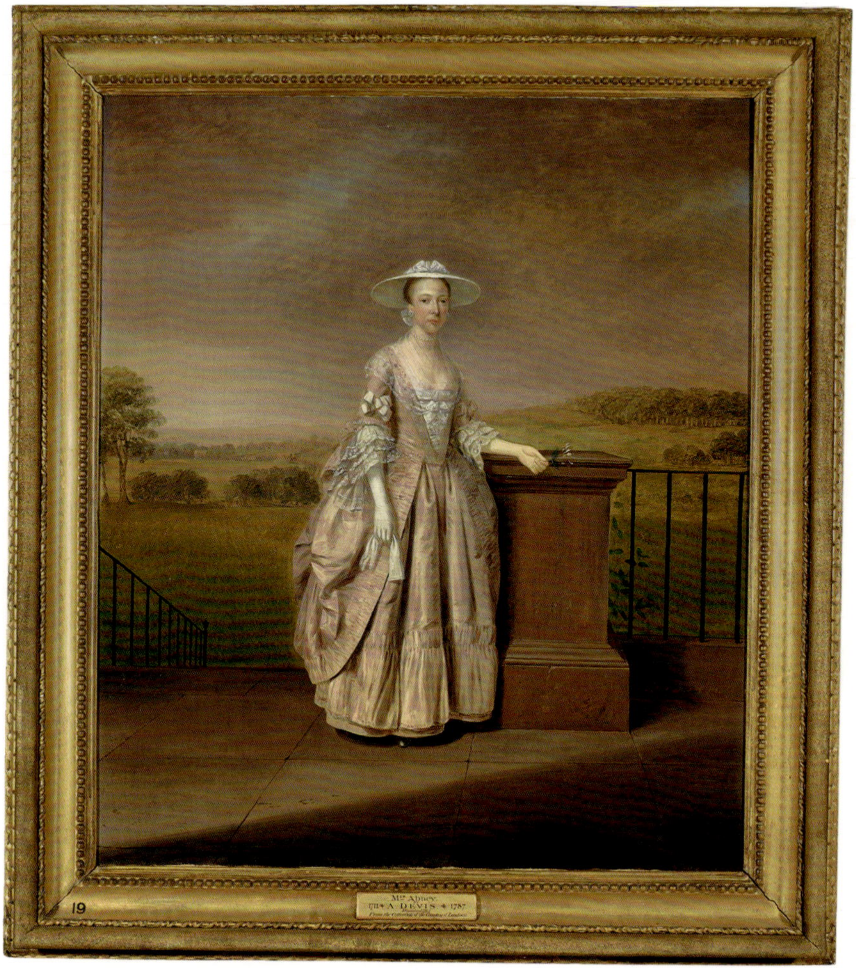

1764

Portrait

Oil on canvas

$30 \times 24\frac{7}{8} \times \frac{7}{8}$ in. ($76.2 \times 63.2 \times 2.22$ cm)

Arthur Devis (1712–1787)

Inscription: Arthur Devis. fe 1764.; Parnell the Wife of Thomas Abney of Willesley Hall in the County of Derby Esqr 1764.;
G S 7 GD

Arthur Devis (1712–1787) was born in Preston in Lancashire. He trained with the sporting and topographical artist Peter Tillemans. Devis's early works were bird's-eye views and landscapes. He later moved to portraits and became the most prolific exponent of a new form: the conversation piece. He made more than three hundred paintings of members of the landed gentry and professional classes, many of whom were Tories.

Lacking training in depicting the human body from life, Devis worked with a wooden mannequin clothed with costumes. He compensated for his limitations by situating figures within an interior or, as is the case here, in a landscape.

These pendant portraits were made for Thomas Abney (d. 1791) of Willesley Hall, near Ashby-de-la-Zouch on the Leicestershire/Derbyshire border, and his wife, Parnell Villiers (d. 1798), daughter of George Villiers of Hanbury in Staffordshire. The Abneys had been at Willesley since the early fifteenth century, serving as mayors of London, members of Parliament, and judges in the Court of Common Pleas. Thomas's father, Sir Thomas Abney (1691–1750), had inherited the family estates—in Derbyshire, Leicestershire, and Staffordshire—because his older brother, deemed mentally insecure, was excluded. Thomas, in turn, inherited from his father. On May 11, 1754, as a member of Middle Temple, he married Parnell at St Clement Danes in Westminster. In his will, he named his "affectionate wife" as executor and asked to be

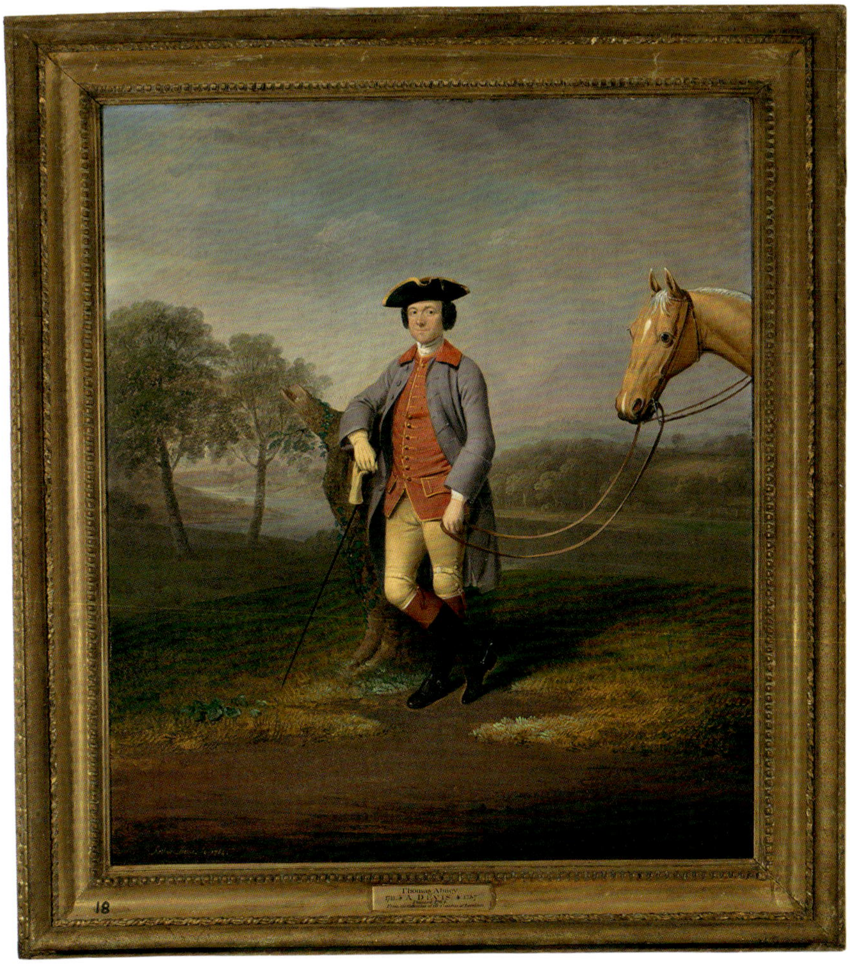

1764

buried "as privately as may be consistent with decency and that all my Men Tenants may have hatbands" (TNA PROB 11/1210/139). He had no male heir; his estate passed to his son-in-law, Charles Hastings, the "natural" (illegitimate) child of Francis Hastings, tenth Earl of Hastings.

PROVENANCE

Painted for Thomas Abney of Willesley Hall, Leicestershire (d. 1791); his son-in-law, Sir Charles Hastings, first Baronet of Willesley Hall (1752–1823); thence by descent to Edith Rawdon-Hastings, tenth Countess of Loundon (1833–1874); Mrs Ernest Innes; Christie's, London, December 13, 1935 (double lot 100); bt. Martin; Walter H. Samuel, second Viscount Bearsted of Upton House, Warwickshire (1882–1948), and thence by descent; with Gurr Johns, London; bt. John H. Bryan of Lake Bluff, IL (1936–2018) May 21, 2014

LITERATURE

Sydney Herbert Pavière, "Biographical Notes on the Devis Family of Painters," *Volume of the Walpole Society* 25 (1936): 115–66, 125

Portrait

Oil on canvas

30 × 24 7/8 × 7/8 in. (76.2 × 63.2 × 2.22 cm)

Arthur Devis (1712–1787)

Inscription: Arthur Devis. fe 1764.; in pen on verso: Thomas Abney of Willesley Hall in the County of Derby Esq[r] 1764.; Tho[s]. Abeny Esq[r] Father of Parnall Abney|Lady Hastings; stamped on verso: G S 7 GD

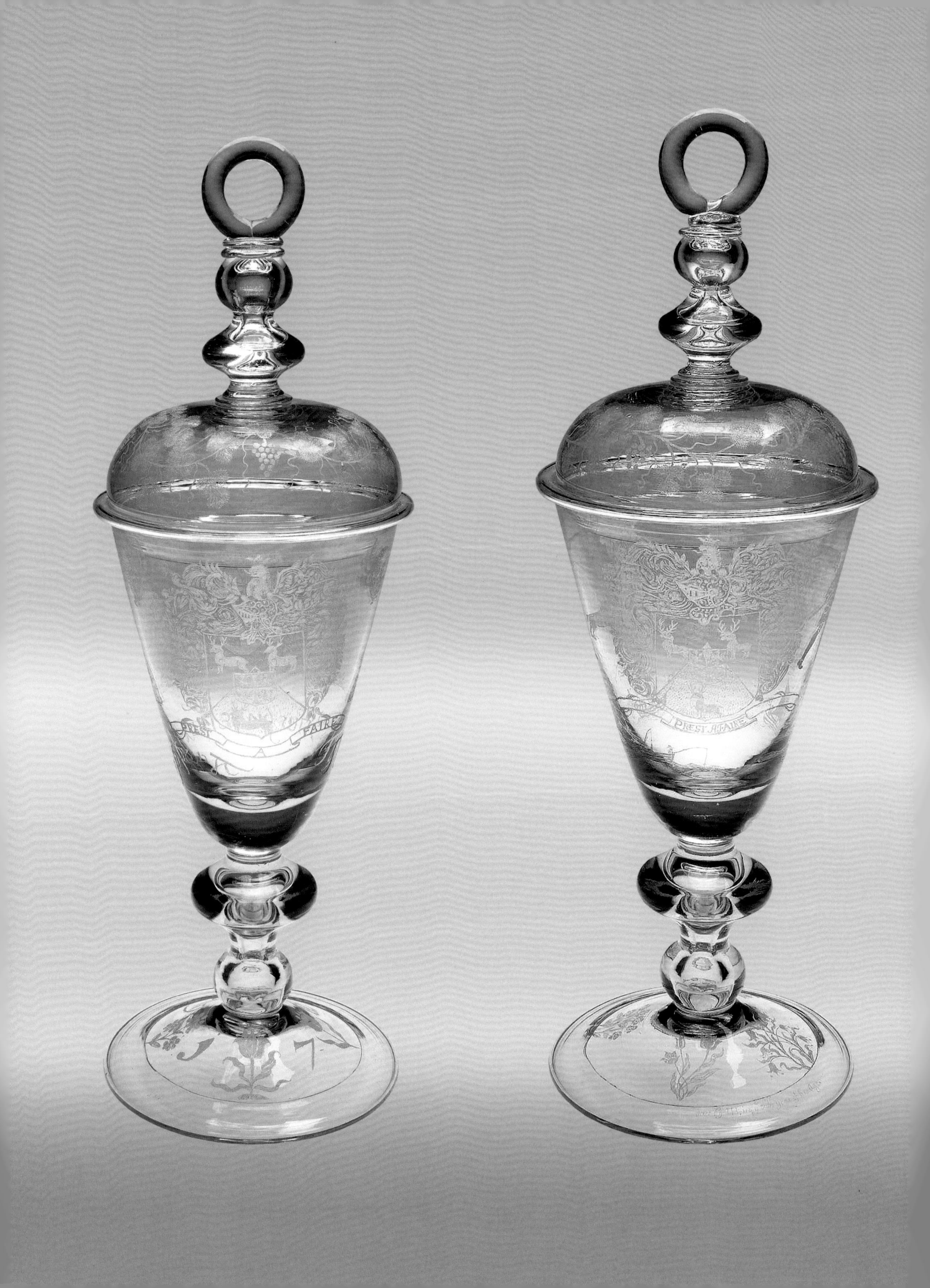
PREST A FAIRE
PREST A FAIRE

1709

Wedding gifts often included vessels for eating or drinking. Some might be expensive gifts like this elegant and lavishly decorated pair of diamond-engraved marriage goblets. Dated 1709, the glasses were probably decorated in celebration of the twentieth anniversary of the wedding of Percival Hart (1666–1738) of Lullingstone Castle, Kent, and his wife, Sarah, daughter of Edward Dixon of Helden, Tonbridge. The engraver dated the larger glass and inscribed it with a badly garbled attempt in Cyrillic of the Russian word for "health" (a drinking toast); on the smaller glass he inscribed his initials, "IH," and his age, seventy-eight.

Another glass vessel with decoration attributed to the same engraver is a large bowl with a crenellated rim, presumed to be a monteith (punch bowl), bearing inscriptions in French, Greek, Dutch, Russian, and Italian, none of which demonstrates the inscriber's mastery of the language or of Greek or Cyrillic letter forms.

Lullingstone is not a castle in any meaningful sense, but instead an early sixteenth-century country house with two outer courtyards (of which only one remains), which was remodeled in the late Stuart period. Queen Anne visited Hart at Lullingstone, who, after her death, harbored sentimental sympathies for the Jacobite cause.

PROVENANCE

Made for Percival Hart of Lullingstone Castle, Kent (1666–1738) and his wife, Sarah Dixon (d. 1720); presumably by descent and sold following the death of Sir William Hart-Dyke (1837–1931) at auction at Lullingstone Castle, Kent, December 9–10, 1931; bt. John H. Bryan of Lake Bluff, IL (1936–2018) from Sheppard and Cooper Ltd., London, October 22, 1992; Art Institute of Chicago, IL, gift of Mr. and Mrs. John H. Bryan, 2012 (2012.867.1)

LITERATURE

Dwight P. Lanmon, *The Golden Age of English Glass: 1650–1775* (Woodbridge, UK: Antiques Collectors' Club, 2011), 124–28

Goblet

Glass

18 ½ × 6 ⅛ in. (47 × 15.6 cm)

Inscription: PREST A FAIRE; 1709 [with garbled Cyrillic text]

Goblet

Glass

19 ⅛ × 6 in. (48.6 × 15.2 cm)

Inscription: dwn by IH in ye 78th: year of his Age|IH

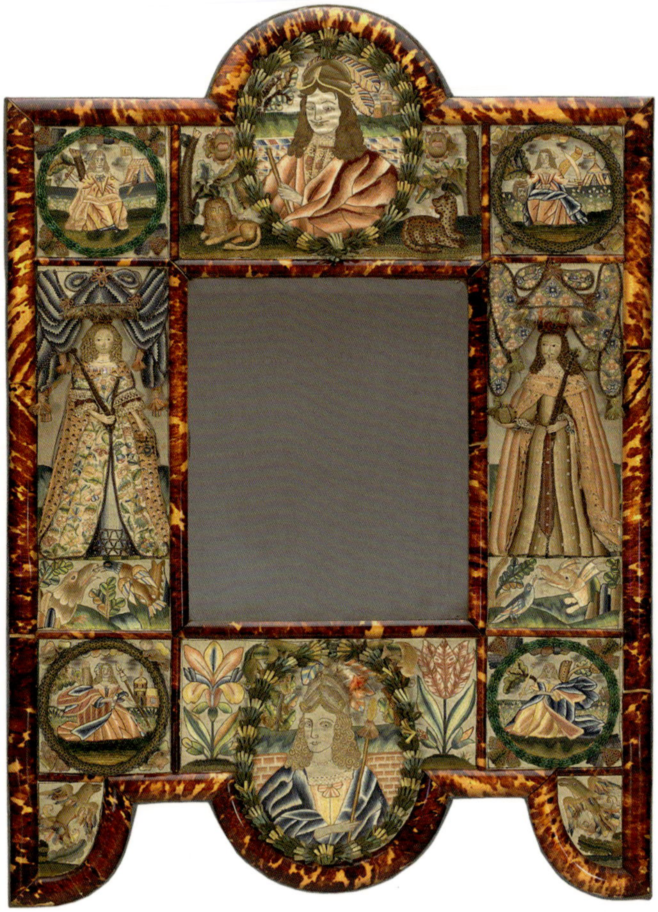

1670

Mirror

Silk, with silk, linen, and metal thread; paper; feathers; glass beads; seed pearls; glass; imitation celluloid tortoiseshell

33 × 24 × 1 in. (83.8 × 61 × 2.5 cm)

Inscription: 1670; IH

During the 1660s, improvements in manufacturing made mirror production cheaper and, with that, mirrors became more affordable. The monarch, on either side of the glass and at top and bottom, was an established trope in frame design of this period. Although the depictions of the monarchs here are generic, the figures embroidered in the roundels in each corner are less common. In the top-left roundel, which contains the date, is Jael, who, as told in Judges 4:17–21, drove a stake into the head of Sisera, general of the enemies of the Israelites, while he slept in the tent she had offered him as refuge. An equally gruesome death befell Holofernes, whose head is held by the other Old Testament heroine, Judith, in the adjacent roundel. Both women saved the Israelite nation by enticing men with their beauty—fitting subjects for the decoration of a looking glass. The figures in the remaining roundels are less easy to identify but might be Esther (with her scepter) and Deborah (the prophetess who predicted the victory of Barak over Sisera), completing the quartet of women who fulfilled God's will as intermediaries and who serve as exemplars to the woman looking at herself in the mirror.

PROVENANCE

William Randolph Hearst (1863–1951); Cora Ginsburg Inc., New York; bt. John H. Bryan of Lake Bluff, IL (1936–2018), March 13, 1992

LITERATURE

Walter R. Brown, *The Stuart Legacy: English Art 1603–1714* (Birmingham, AL: Birmingham Museum of Art, 1991), 87

Andrew Morrall and Melinda Watt, eds., *English Embroidery from the Metropolitan Museum of Art, 1580–1700: 'Twixt Art and Nature* (New Haven, CT: Yale University Press, 2008), 219–20

1682

Made to celebrate either the year of union or the ongoing success of a marriage, this plate was probably produced in London. It is closely comparable to another octagonal plate of 1685, illustrated in Lipski and Archer.

PROVENANCE

Lady Henderson Collection; Phillips, London, February 3, 1993 (lot 29); bt. John H. Bryan of Lake Bluff, IL (1936–2018) through Alistair Sampson, London

LITERATURE

Louis L. Lipski and Michael Archer, *Dated English Delftware: Tin-Glazed Earthenware, 1600–1800* (London: Sotheby Publications, 1984), 23, no. 159

Plate

Delftware

$7\frac{7}{8} \times 7\frac{3}{4} \times 1$ in. (20 × 19.7 × 2.5 cm)

Inscription: I|B|I; 1682

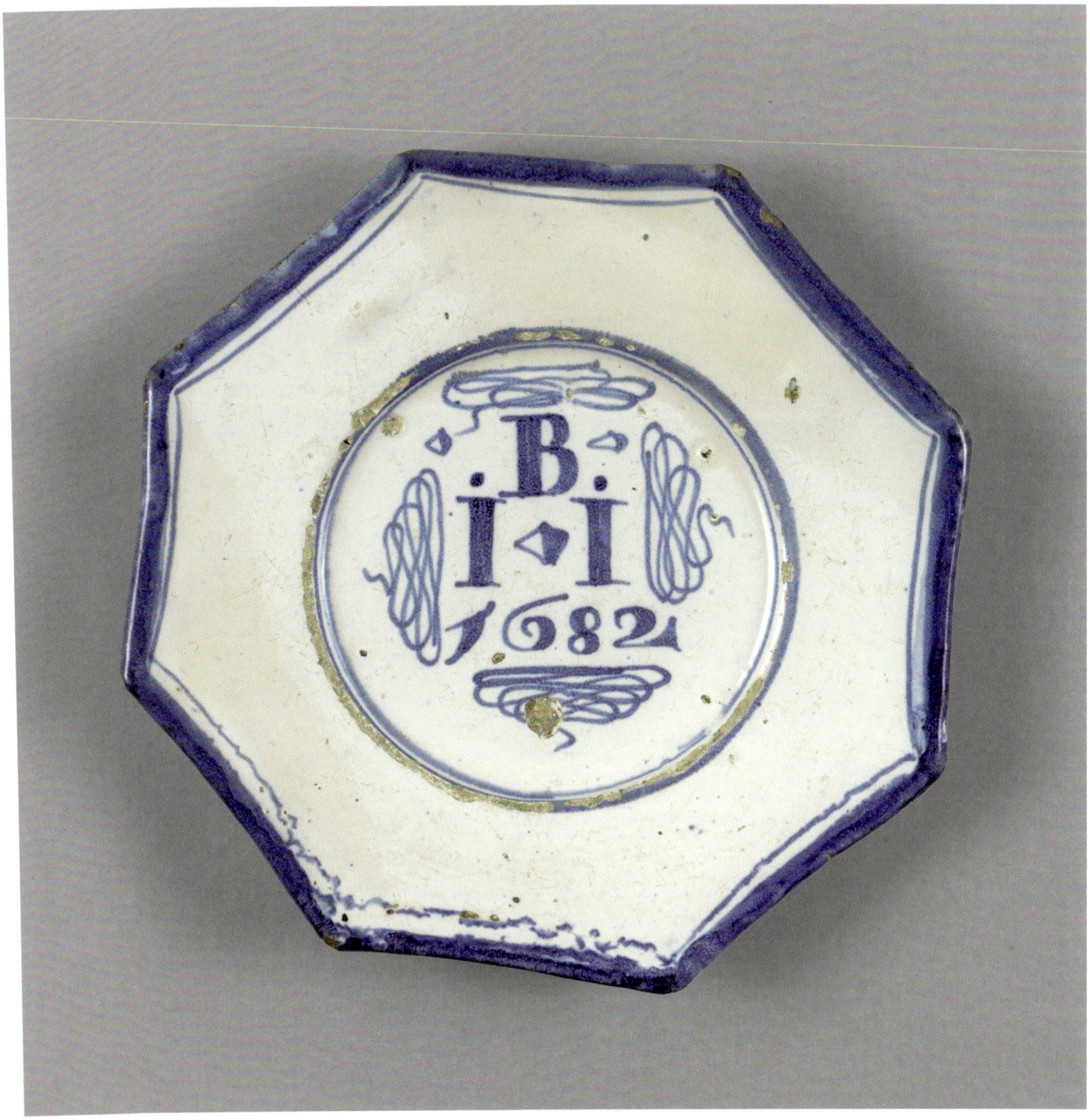

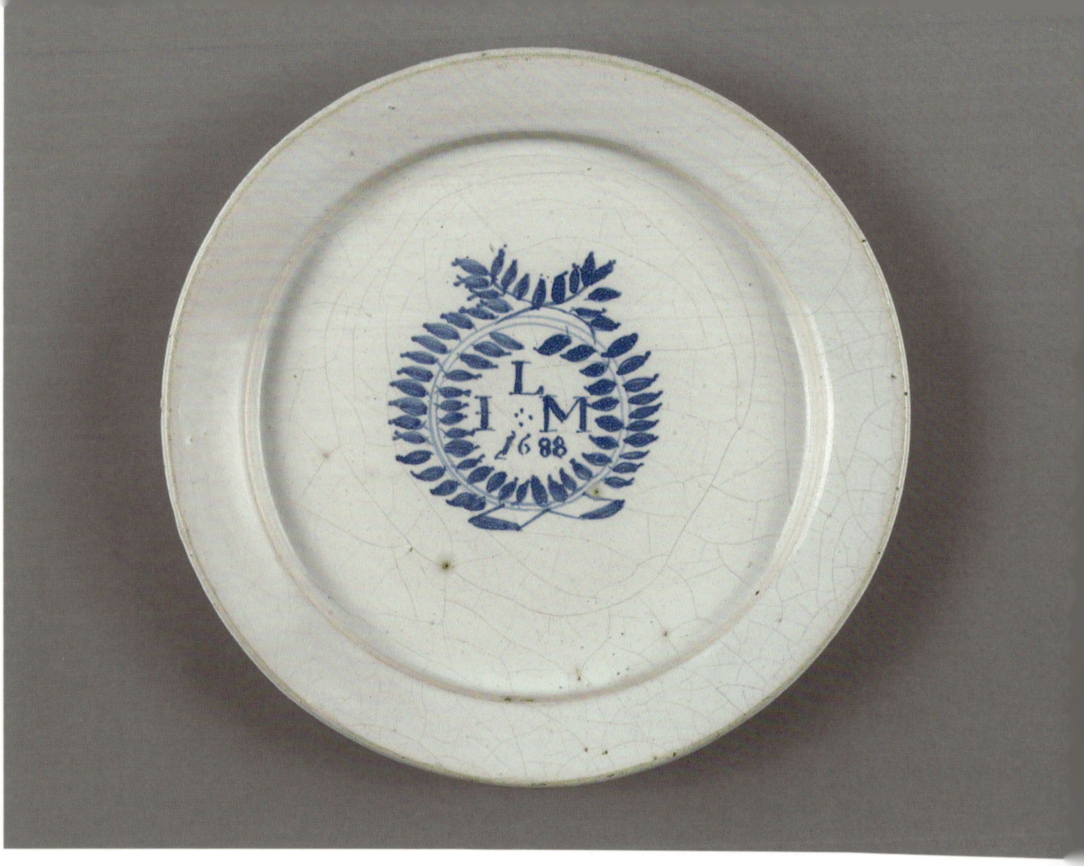

1688

Plate

Delftware

8 ½ × 1 in. (21.6 × 2.5 cm)

Inscription: I|L|M; 1688

The decorative motive of crossed palms was popular from the 1670s and endured in the early decades of the following century. Another plate from the same set was sold at Sotheby's, London, on November 20, 1979 (lot 16).

PROVENANCE

Christie's, London, November 2, 1981 (lot 205); bt. John H. Bryan of Lake Bluff, IL (1936–2018) through Jellinek & Sampson Antiques, London

LITERATURE

Louis L. Lipski and Michael Archer, *Dated English Delftware: Tin-Gazed Earthenware, 1600–1800* (London: Sotheby Publications, 1984), 56, no. 173A

1696

Plate

Delftware

8 ¾ × 1 ¼ in. (22.2 × 3.2 cm)

Inscription: I|B|M; 1696

John and Mary were two of the most common names in seventeenth-century England, so perhaps it is unsurprising that the initials "I B M" appear on a number of dated objects, Delftware and otherwise, from the last two decades of the century. (At this time, the name John was represented with the letter *I*.) These dated and initialed objects survive in sufficient volume, however, to suggest that at least some of them may have belonged to the same owners.

What is more, in the Museum of London (no. 7134), there is a house sign of painted stone for the Ape and Apple Inn, showing a crouching ape devouring an apple, which carries the inscription "I B M; 1670." It had been built into a wall at No. 14 on the west side of Philip Lane, near London Wall, sometimes referred to as the Old Ape; a Wood Street firm gave it to what was then the Guildhall Museum.

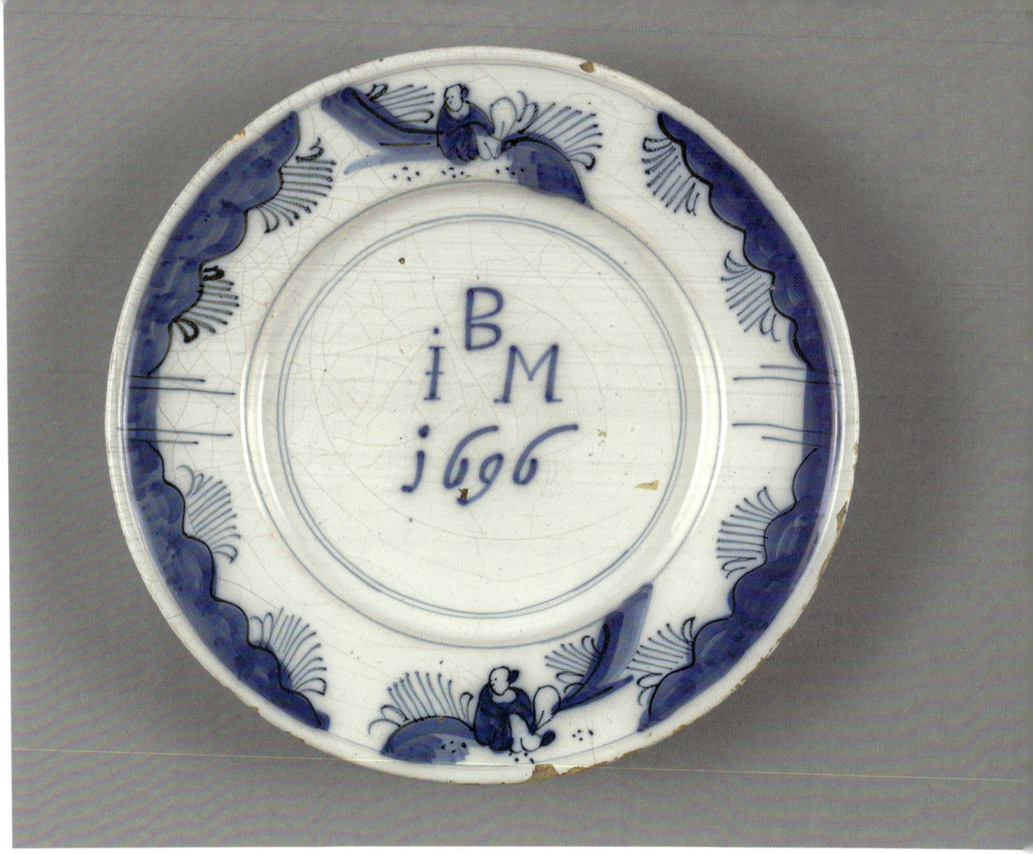

Other objects bearing identical initials and similar dates include a porringer and cover by John Duck of 1680, sold at Chiswick, October 31, 2017 (lot 499); a silver beaker, also by Duck, of 1688, inscribed "I B M 1708," sold at Bonhams, London, November 4, 2015 (lot 108); a Delftware bust of a woman wearing a necklace and a loose robe, with an inscription dated 1688, at the Fitzwilliam Museum, Cambridge (C.1444-1928); a pewter tankard of the second half of the seventeenth century, also at the Fitzwilliam Museum (NAV.164-1933), "crudely engraved with the initials IBM"; and a silver tumbler cup with marks for "Leeds 1680" and the mark "RW" (probably for Robert Williamson), sold at Christie's, London, May 30–31, 2012 (lot 317).

The two plates listed in Lipski and Archer, painted blue within blue-black outlines, decorated with two herons, plants, flowers, and rocks, and inscribed "I B M" are now in the collection of the Museum of Fine Arts, Houston, Texas (nos. B.59.86.1, B.59.86.2). These two were part of a set of at least four. Others can be found at Colonial Williamsburg, Williamsburg, VA (no. 1957–83), and one appeared at Sotheby's, London, October 5, 1955 (lot 10). Could the proprietor of the Ape and Apple have owned some of these plates? Only further research could prove this one way or another.

The design here is similar to two plates painted in blue (one with dark manganese outline), with a "Chinaman" among stylized rocks and grasses, inscribed "R G 1684" and "R C 1684," respectively.

PROVENANCE

F. H. Garner of Wymeswold, Leicestershire (1893–1964); John Philip Kassebaum of Mount Pleasant, SC (1932–2016); Sotheby's, London, October 1, 1991 (lot 23); bt. John H. Bryan of Lake Bluff, IL (1936–2018) through Lindsay Antiques, London

LITERATURE

Louis L. Lipski and Michael Archer, *Dated English Delftware: Tin-Gazed Earthenware, 1600–1800* (London: Sotheby Publications, 1984), 52–53, nos. 157, 158; 62, nos. 209, 902A

1704

Plate

Delftware

8 ⅞ × 1 in. (22.5 × 2.5 cm)

Inscription: E|T|E; 1704

This plate is part of a set containing at least two other plates, one depicting a couple walking in a landscape and the other with the same illustration of a man with a dog. Although it is impossible to prove, these plates might commemorate the marriage of Edward Thomas and Elizabeth Barton on November 16, 1704, at St Mary Lambeth.

PROVENANCE

bt. John H. Bryan of Lake Bluff, IL (1936–2018) from Tristram Jellinek Antiques, London, November 20, 1982

LITERATURE

Louis L. Lipski and Michael Archer, *Dated English Delftware: Tin-Glazed Earthenware, 1600–1800* (London: Sotheby Publications, 1984), 68, nos. 239, 240

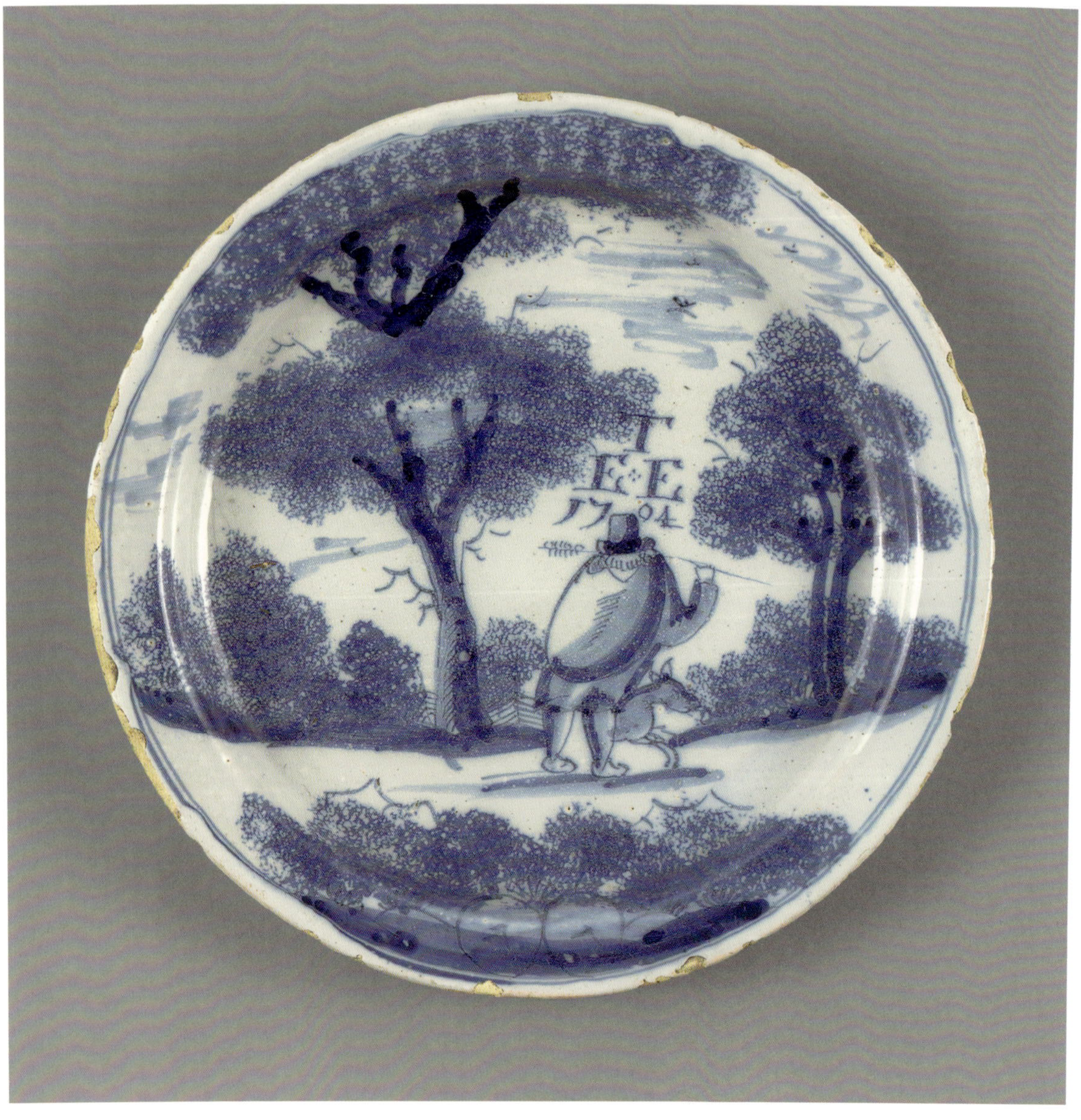

1712

This plate, one of a pair, features four cruciform panels with ogival ends. Each panel contains flower sprays painted in green, red, and blue, set against a diaper-pattern background.

PROVENANCE

Sotheby's, London, February 13, 1979 (lot 17); bt. John H. Bryan of Lake Bluff, IL (1936–2018) from Jellinek & Sampson Antiques, London, February 13, 1979

LITERATURE

Louis L. Lipski and Michael Archer, *Dated English Delftware: Tin-Glazed Earthenware, 1600–1800* (London: Sotheby Publications, 1984), 72

Plate

Delftware, polychrome

8 7/8 × 1 in. (22.5 × 2.5 cm)

Inscription: T|T|A; 1712

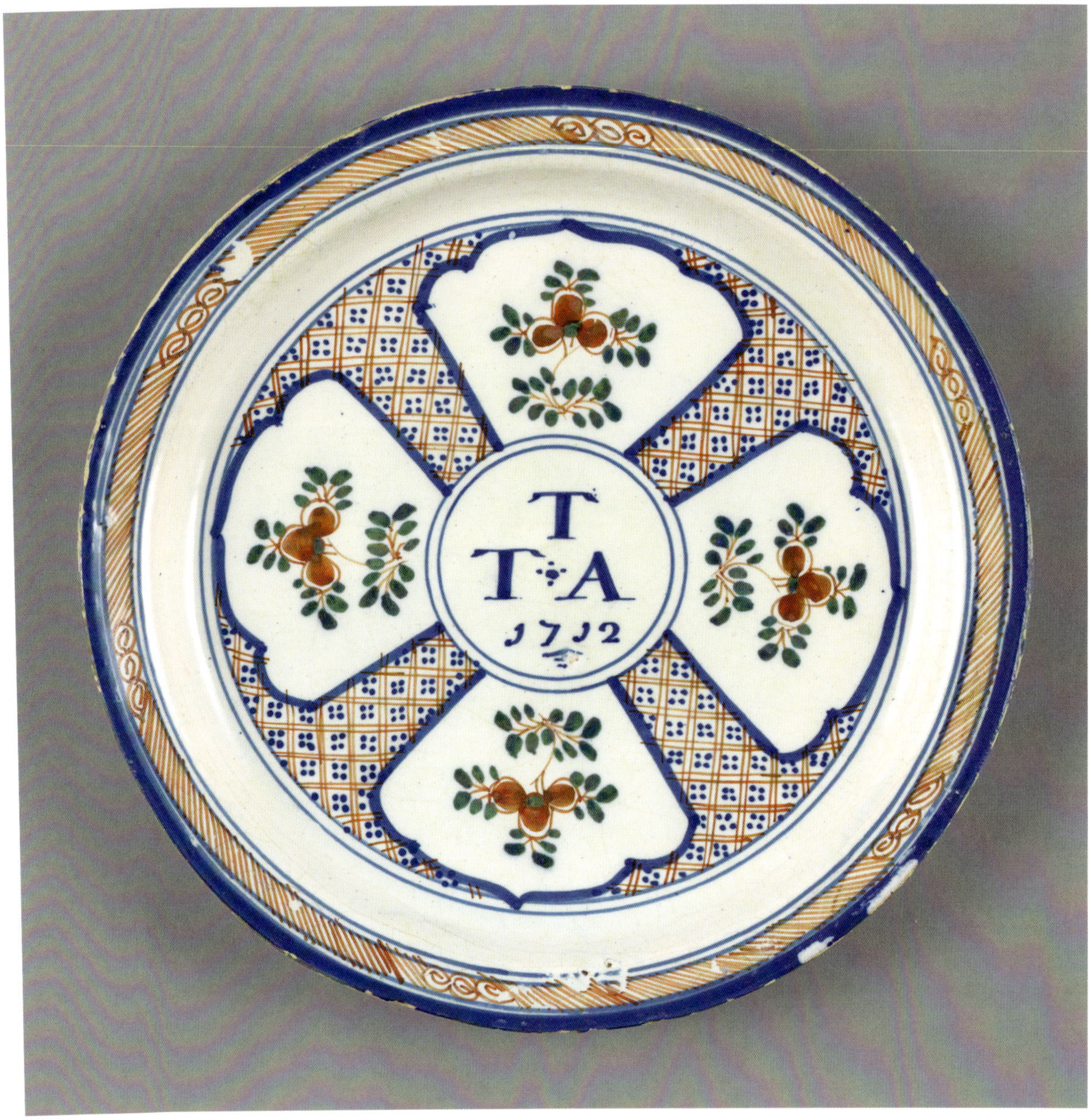

1764

Plate

Delftware, manganese decoration

$9\frac{1}{4} \times 1\frac{1}{8}$ in. (23.5 × 2.9 cm)

Inscription: Gilbert & jane Dane; 1764

Gilbert Dane, a chapman of St Peter Alstonefield, Staffordshire, married Jane Mason of Presbury, Cheshire, on October 27, 1761. A chapman was an itinerant seller, and perhaps Dane chanced upon Jane while hawking his wares across the Peak District and later commissioned this plate to celebrate their marriage. To judge from the fact that they could only sign the registry with their marks, neither of them was able to write.

PROVENANCE

B. Harland; Sotheby's, London, February 11, 1931 (lot 30); Puttick and Simpson, March 13, 1933 (lot 70); Mrs. Donald F. Carpenter, PA; bt. John H. Bryan of Lake Bluff, IL (1936–2018) from Mark & Marjorie Allen Antiques, Laconia, NH, March 4, 1997

LITERATURE

Louis L. Lipski and Michael Archer, *Dated English Delftware: Tin-Glazed Earthenware, 1600–1800* (London: Sotheby Publications, 1984), 142, no. 635

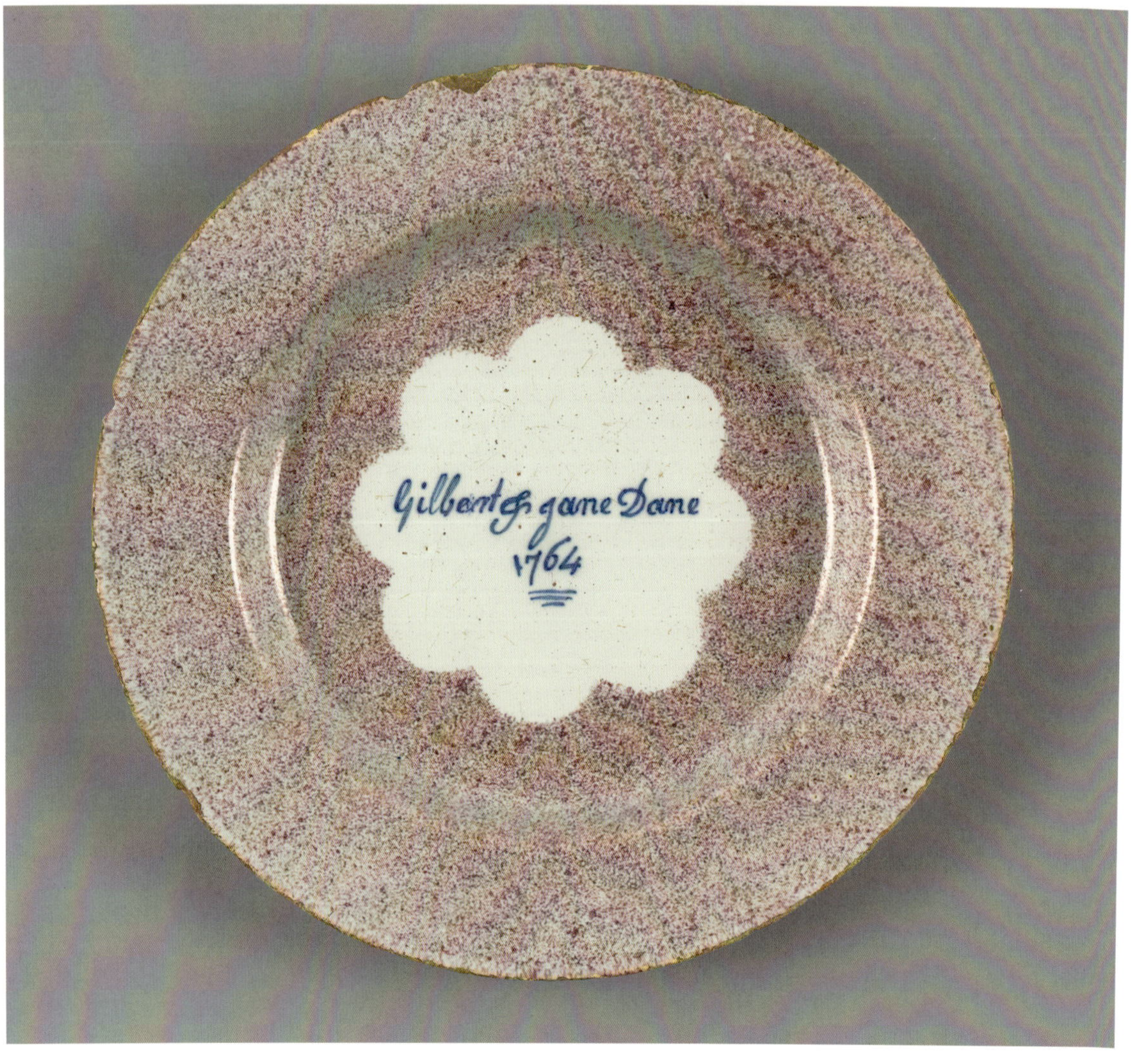

1768

In Chinese culture, cracked ice is the symbol for spring, and this decorative motif is found on the porcelain that Delftware manufacturers in England sought to emulate.

PROVENANCE
bt. John H. Bryan of Lake Bluff, IL (1936–2018) from Lindsay Antiques, London, March 1, 1988

Plate
Delftware
$9\frac{1}{4} \times 1\frac{1}{8}$ in. (23.5 × 2.9 cm)

Inscription: I|G|L; 1768

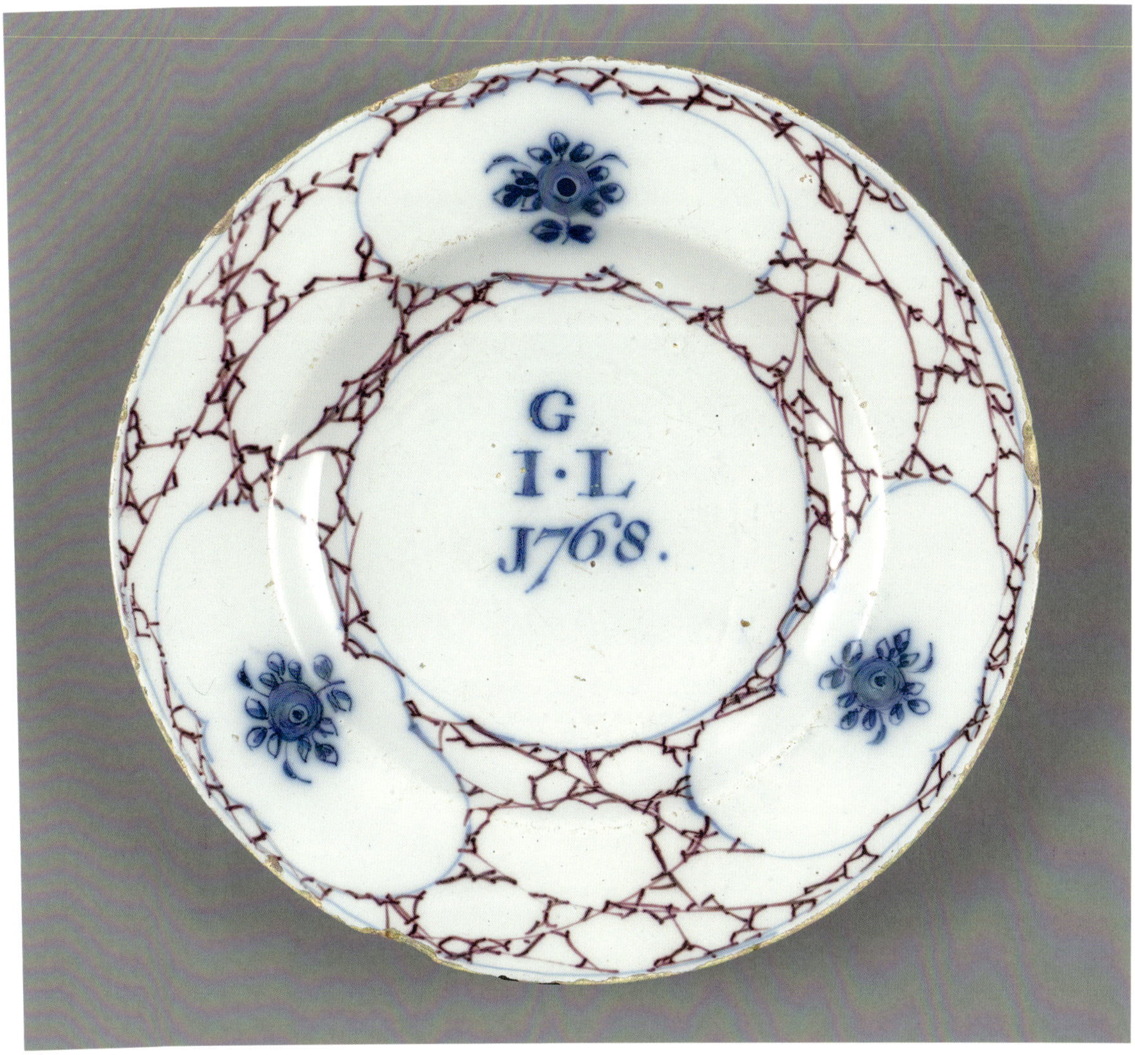

1739

Model shoe

Delftware (Liverpool Delft)

4½ × 2½ × 6½ in. (11.4 × 6.4 × 16.5 cm)

Inscription: ·E S·|1739

Whether buried in a wall or placed on a mantel, shoes had for centuries provided apotropaic properties and ensured good luck. The intimate size and the limited capacity of these Delftware models to serve as either paperweights or receptacles was in keeping with their whimsical, sentimental nature. Surviving examples show that they were made in pairs (Colonial Williamsburg, Williamsburg, VA, 1960-359,1), and it is believed that they were given as gifts, sometimes for marriage (those with three painted initials and a date) and perhaps also sometimes as tokens of affection (those with only two initials). The date of this shoe places it toward the end of the fad for Delftware ornaments of this type, which had begun about 1654 and endured to as late as 1768.

PROVENANCE

Christie's, London, March 18, 1985 (lot 15); bt. John H. Bryan of Lake Bluff, IL (1936–2018)

LITERATURE

Leslie B. Grigsby, *The Longridge Collection of English Slipware and Delftware* (London: Jonathan Horne Publications, 2000), 1:398

1760

Model shoe

Delftware (Bristol or Liverpool Delft)

3½ × 2⅛ × 5⅝ in. (8.9 × 5.4 × 14.3 cm)

Inscription: E·K 1760

The flower sprays on either side of this shoe are painted in what is known as the Fazakerley palette, named for what was then a small town in Lancashire, north of Liverpool. Fazakerley was also the name of a family resident there. The colors associated with Fazackerley were print-enamel purples, greens, yellows, and blues.

PROVENANCE

Sotheby's, London, November 15, 1994 (lot 110); bt. John H. Bryan of Lake Bluff, IL (1936–2018) through Alastair Sampson Antiques Ltd., London

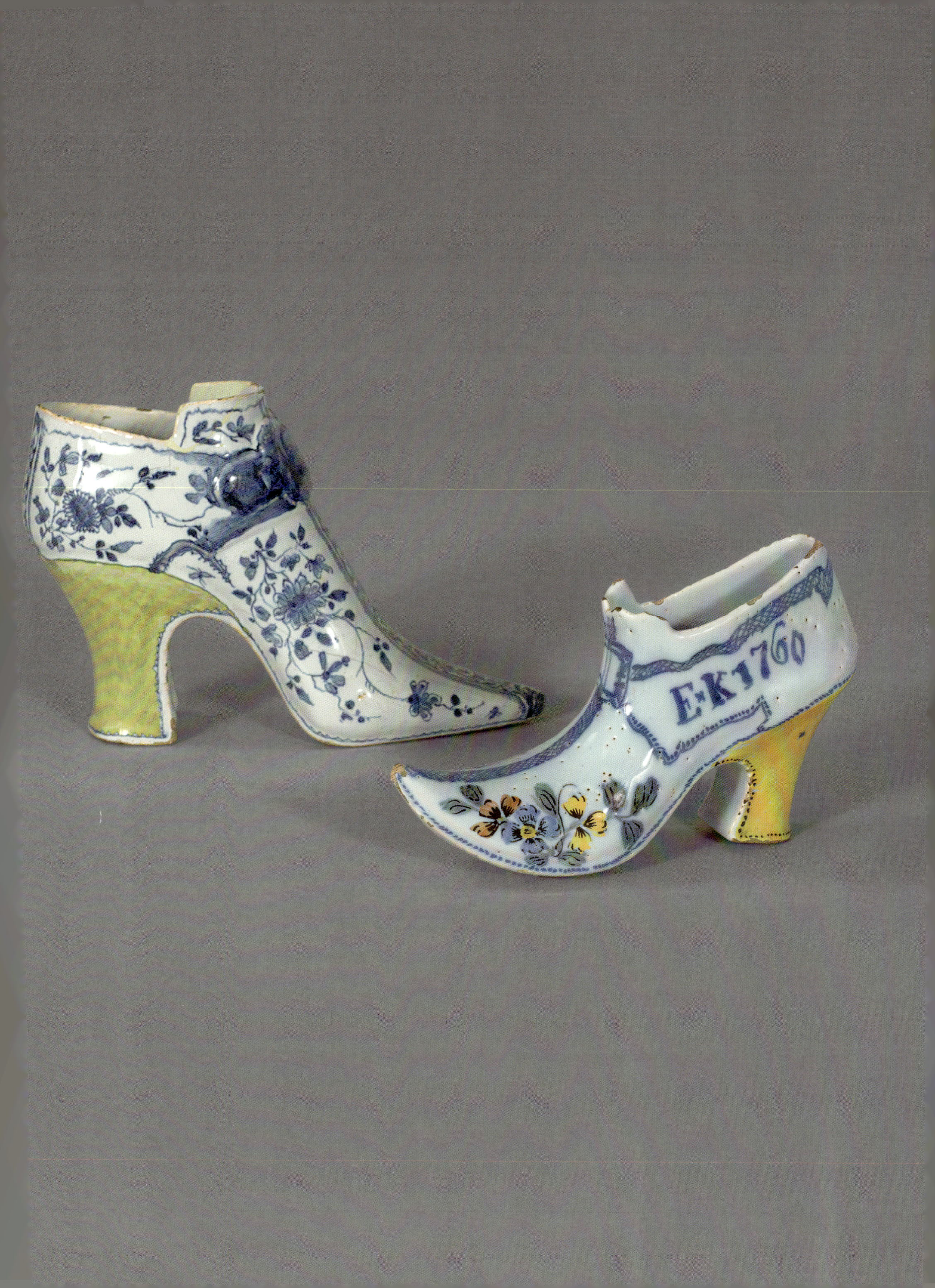
E·K 1760

1792

Bird feeder

Delftware (Lambeth Delft)

$3\frac{3}{4} \times 2 \times 2\frac{3}{8}$ in. (9.5 × 5.1 × 6 cm)

Inscription:
J Giles|June:7:1792

This feeder, filled with seeds, would have been placed inside a birdcage. On the back is a painting, in blue, of a bird on a branch of a tree. The keeping of birds, particularly exotic species, was fashionable among the elite throughout this period.

PROVENANCE

Oliver Van Oss (1903–1992); Louis L. Lipski (1914–1979); bt. John H. Bryan of Lake Bluff, IL (1936–2018) from Jonathan Horne, London, June 15, 1983

LITERATURE

Louis L. Lipski and Michael Archer, *Dated English Delftware: Tin-Glazed Earthenware, 1600–1800* (London: Sotheby Publications, 1984), 423, no. 1776

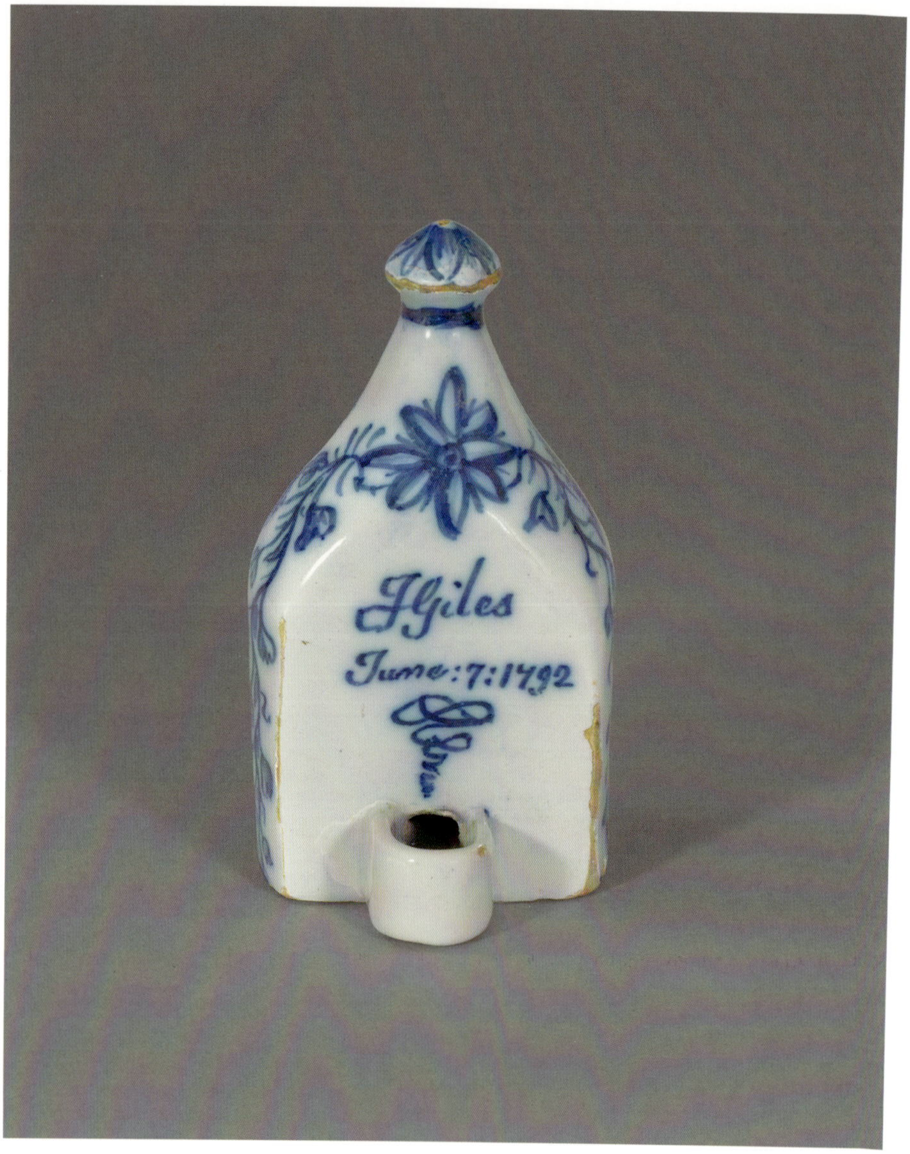

1715

Stay busk

Wood

$12\frac{7}{8} \times 1\frac{3}{8} \times \frac{1}{4}$ in. (32.7 × 3.5 × .64 cm)

Inscription: G R; M H; R E; MDCCXV; GTMWC; No Joy in Life; Like a good wife

Prior to the nineteenth century, a corset was called a stay. Stay busks like this one were inserted into the front of the stay to stiffen the garment's structure and flatten the appearance of the wearer's stomach. Their intimate position next to the skin and heart made stay busks popular as love tokens, and often they were carved with hearts, short rhymes, or sets of initials.

As was typical, this stay incorporated the initials of two unmarried people: "M H" and "R E." Less typical are the other decorative features, which convey the impression of an earnest young man who wished to establish himself and his intended bride as loyal to both church and state. The year of the gift, 1715, is inscribed in Roman numerals along with a rhyme drawn from the apocryphal book of Ecclesiasticus: "No joy in life like a good wife." This well-known text contrasted the behaviors of the good and bad wife, making clear the giver's expectations of the recipient. The function of the gift itself and the slender figure of the young "M H," whose portrait appears at the top of the stay busk, unambiguously projected the ideal physical attributes of that good wife. Also carved into the wood, and doubtless meaningful for the people concerned, is a frustratingly mysterious string of letters: "GTMWC."

PROVENANCE

Tristram Jellinek of London (1933–1995); Sotheby's, London, May 10, 1996 (lot 689); bt. John H. Bryan of Lake Bluff, IL (1936–2018) through Tobias Jellinek Antiques, London

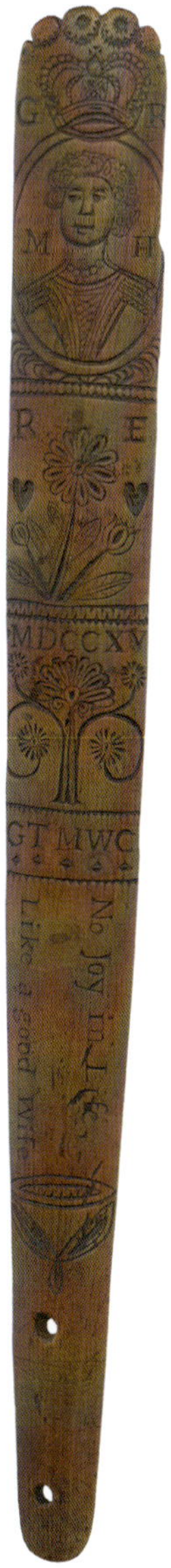

1724

Stay busk

Wood

13 × 1⅞ × ¾ in. (33 × 4.8 × 1.9 cm)

Inscription:
E H|1724; I B

Made from sycamore, limewood, pear wood, or yew, stay busks were chip-carved with simple decorations, which included hearts and other intertwined motifs, alluding to the notion of "tying the knot." The stay busks in the Bryan Collection are relatively early examples of what remained a popular, customary gift into the first quarter of the nineteenth century.

PROVENANCE

bt. John H. Bryan of Lake Bluff, IL (1936–2018) from Polly de Courcy-Ireland, Hampshire, March 18, 1999

LITERATURE

Owen Evan-Thomas, *Domestic Utensils of Wood: From the XVIth to XIXth Century* (London: Owen Even-Thomas Ltd., 1932), 136–41

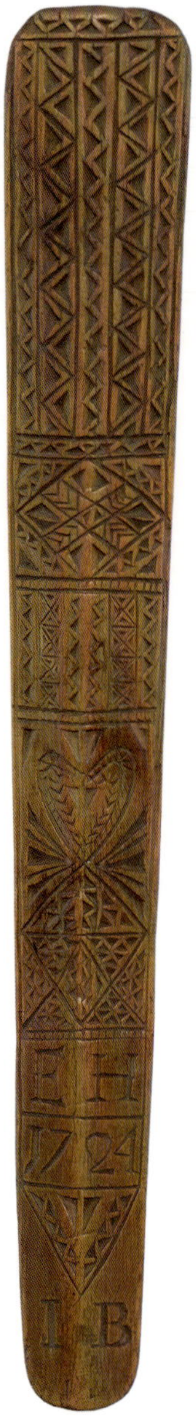

1776

Stay busk

Mahogany, brass

$13\frac{1}{4} \times 1\frac{5}{8} \times \frac{3}{8}$ in.
(33.7 × 4.1 × .95 cm)

Inscription: MARY| BACKHOUSE|Æ ·22·|1776

There were too many women by the name Mary born across the country in or about 1754 to identify the owner of this stay busk with any certainty. But it is tempting to believe that she was the woman who, on May 15, 1777, married John Grubb at St Marylebone, London. If so, Mary Backhouse would have received this stay busk from Grubb as a token of affection—shaped at top in the form of a heart, with interwoven hearts pierced with two arrows carved at center. The church, then outside London, was in William Hogarth's day notorious for the vicar's willingness to perform clandestine marriages; it was immortalized by the artist as the setting for the fifth painting in the series *A Rake's Progress*, in which the Rake marries an ancient bride in a desperate attempt to revive his fortunes.

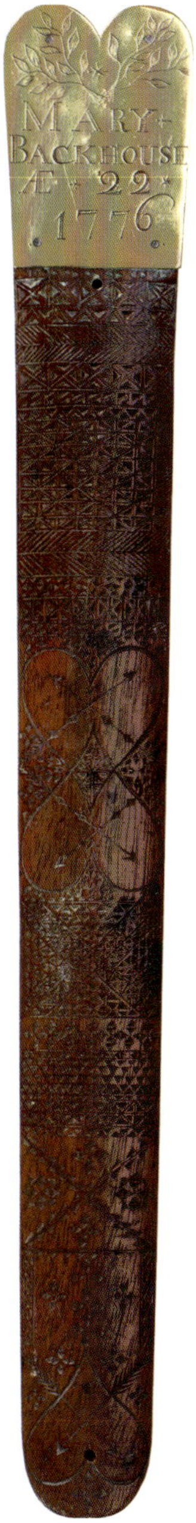

We know from the records of the Old Bailey that, in 1744, Mary was living in the nearby house of Nicholas Viel, the son of a Frenchman and a brewer's clerk of Castle Street, St Giles-in-the-Fields. On October 12 of that year, Viel's house was burgled by John Sampson, who liberated a number of items of linen but was promptly accosted in the street by John Viel (presumably a brother of Nicholas). John had a tip-off from a boy named Hamilton who had seen Sampson enter the house. At court, Mary Backhouse gave testament: "I was present when the prisoner was taken; there was a cap and a pair of sleeves belonging to me in his apron." The prisoner called several witnesses to attest to his good character but was found guilty nonetheless.

This stay busk is similar to another of about 1770, illustrated in Gentle and Field, which is inscribed in brass with the initials "P. S." and the warning "PRAY LOSE NO TIME FOR IF YOU DOW AGE WILL COME ON THEN YOU WILL RUE."

PROVENANCE

bt. John H. Bryan of Lake Bluff, IL (1936–2018) from Alastair Sampson Antiques, London, February 28, 1984

LITERATURE

Rupert Gentle and Rachel Field, *Domestic Metalwork: 1640–1820*, revised and enlarged by Belinda Gentle (Woodbridge, UK: Antiques Collectors' Club, 1994) 410

1738

Tea caddy
Delftware
$4\frac{3}{4} \times 3\frac{3}{4} \times 2\frac{1}{4}$ in.
(12.1 × 9.5 × 5.7 cm)

Inscription: I|H|M; 1738

The success of tea as the national drink of choice began with the opening of the English East India Company's regular Canton trade to China in 1717. The faraway origins of this new commodity are referenced in the decoration of this caddy. Such was its cost that tea was kept under lock and key in caddies such as these. They became increasingly common in affluent homes during the second and third decades of the seventeenth century. Typically, they were fitted with Delftware lids and were sometimes inscribed with the name of the type of tea contained within.

PROVENANCE
bt. John H. Bryan of Lake Bluff, IL (1936–2018) from Jonathan Horne, London, March 15, 1988

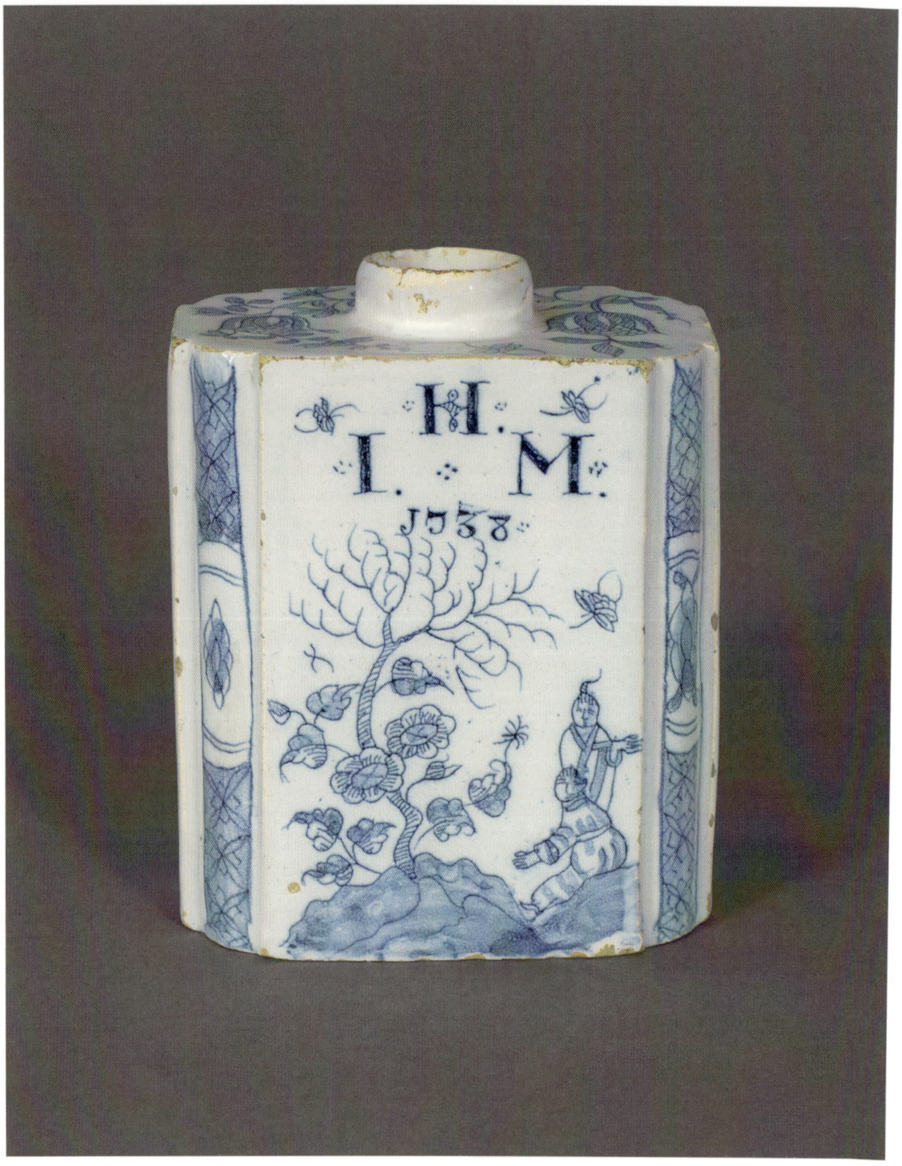

1763

Teapot
Soft paste porcelain
$5\frac{1}{2} \times 4\frac{3}{8} \times 7\frac{3}{8}$ in.
(14 × 11.1 × 18.7 cm)

Inscription: R|B|E; 17 63; R|B|E

This teapot is painted deep blue with floral vignettes on a white background on the body and lid, in direct imitation of Chinese Kangxi period "bleu soufflé" glaze. It was possibly produced by the Lowestoft Porcelain Factory in Suffolk for Robert Browne, a partner and manager of the factory. His initials appear on the base and on the inside of the lid with those of his wife, Elizabeth. An inkpot in Norwich Castle Museum, dated 1762, likewise bears an inscription of the initials "R B" and, in the *Catalogue of Lowestoft China in the Possession of Frederick Arthur Crisp* (1907), it is described as made for the same Robert Browne.

PROVENANCE

Bonham's, London, November 15, 2017 (lot 263); bt. John H. Bryan of Lake Bluff, IL (1936–2018) from English Porcelain, Roderick Jellicoe, London, November 27, 2017

LITERATURE

Frederick Arthur Crisp, *Catalogue of Lowestoft China in the Possession of Frederick Arthur Crisp* (London: Privately printed, 1907)

Geoffrey Godden, *English China* (London: Barrie and Jenkins, 1985)

Bernard Watney, *English Blue and White Porcelain of the Eighteenth Century* (London: Faber and Faber, 1963)

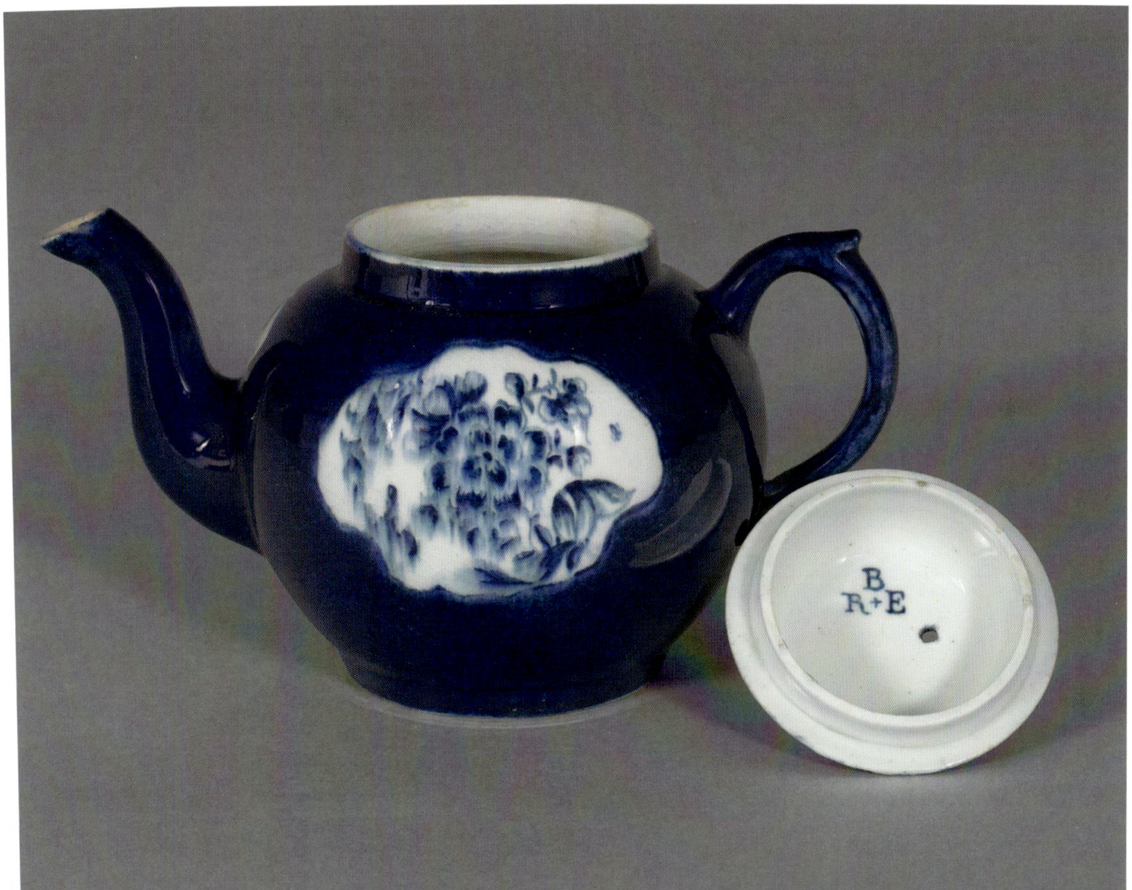

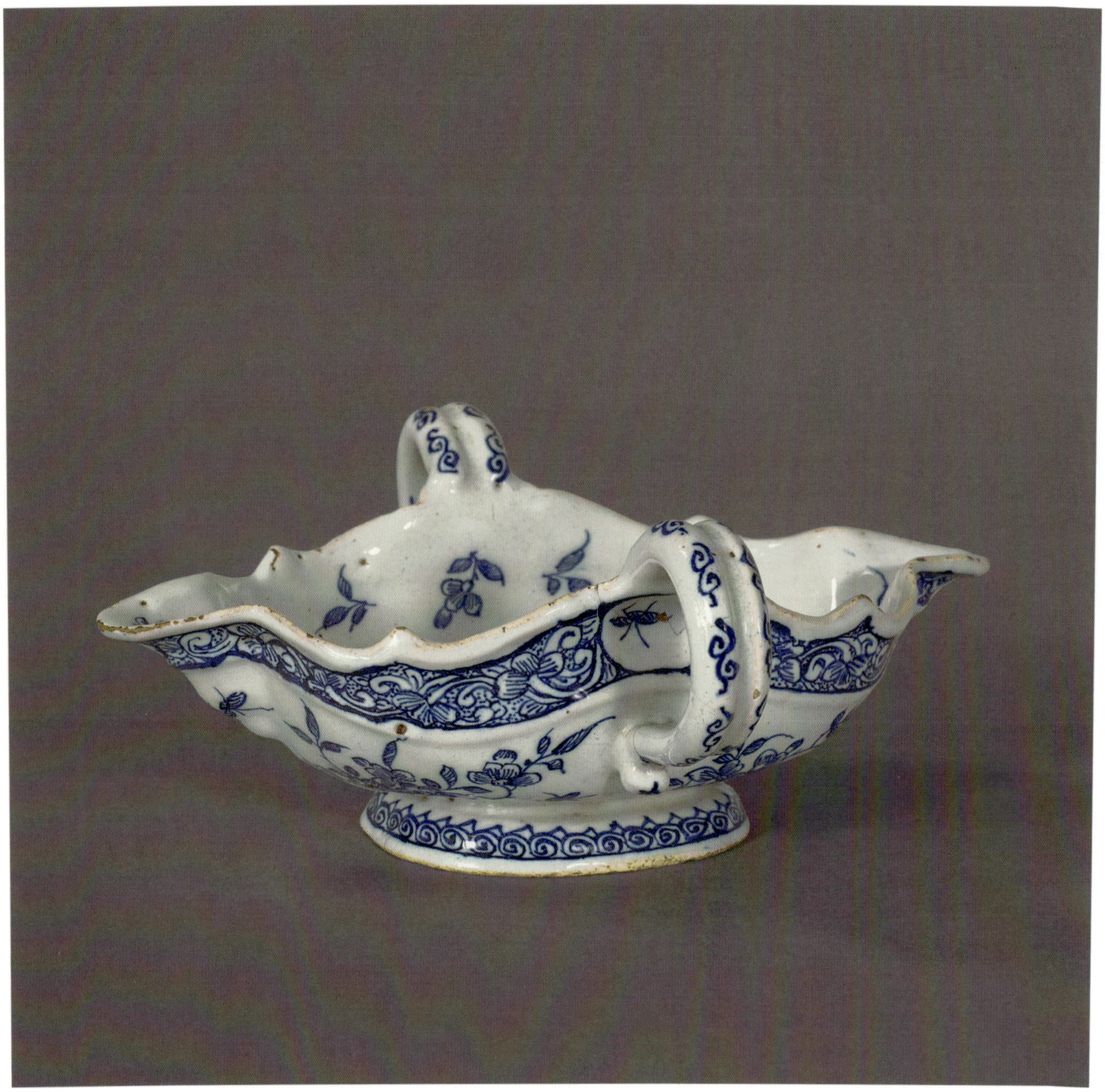

1736

Sauceboat

Delftware

$3 \times 7\frac{1}{4} \times 6\frac{1}{2}$ in.
(7.6 × 18.4 × 16.5 cm)

Inscription: G|P|E; 1736

This object may the earliest dated Delftware sauceboat. It shares a form that, established in silver, became one of the most common shapes for this type of tableware. The shape, handles, and spout suggest that it was designed to contain a light, liquid sauce, possibly taken from the pot in which the meal had been cooked and thickened with flour in the French manner. Mustard, horseradish, and other condiments also had their place on the table, while various other types of sauces made from herbs mixed with white wine vinegar were also commonplace.

PROVENANCE

bt. Nicholas Panes, Sanderstead, from Garry Atkins, November 24, 2011; bt. John H. Bryan of Lake Bluff, IL (1936–2018) from Nicholas Panes, Sanderstead, February 22, 2017

1668

Posset and syllabub were drinks made from milk or cream that had been curdled with alcohol, often spiced or sweetened to taste. They were believed to provide a remedy for colds and other common ailments. The liquid could be poured out of the spout of the pot, and the thicker mixture, forming a creamy crust of curd, was spooned out of the top. Randle Holme thought these pots equivalent to wassail cups, which suggests they were also used to contain spiced ciders, ales, and wines.

Surviving posset pots range in date from 1631 to 1766, making this pot a relatively early example.

PROVENANCE

Sotheby's, London, November 20, 1979 (lot 38); Dolz collection, Alexandria, VA; Nick Narganes; bt. John H. Bryan of Lake Bluff, IL (1936–2018) from Lindsay Antiques, London, June 10, 1988

LITERATURE

Michael Archer, *Delftware in the Fitzwilliam Museum* (London: Philip Wilson Publishers, 2013), 187

Louis L. Lipski and Michael Archer, *Dated English Delftware: Tin-Glazed Earthenware, 1600–1800* (London: Sotheby Publications, 1984), 202, no. 897

Posset pot

Delftware

8 × 9½ × 7 in. (20.3 × 24.1 × 17.8 cm)

Inscription: I|A|A; 1668

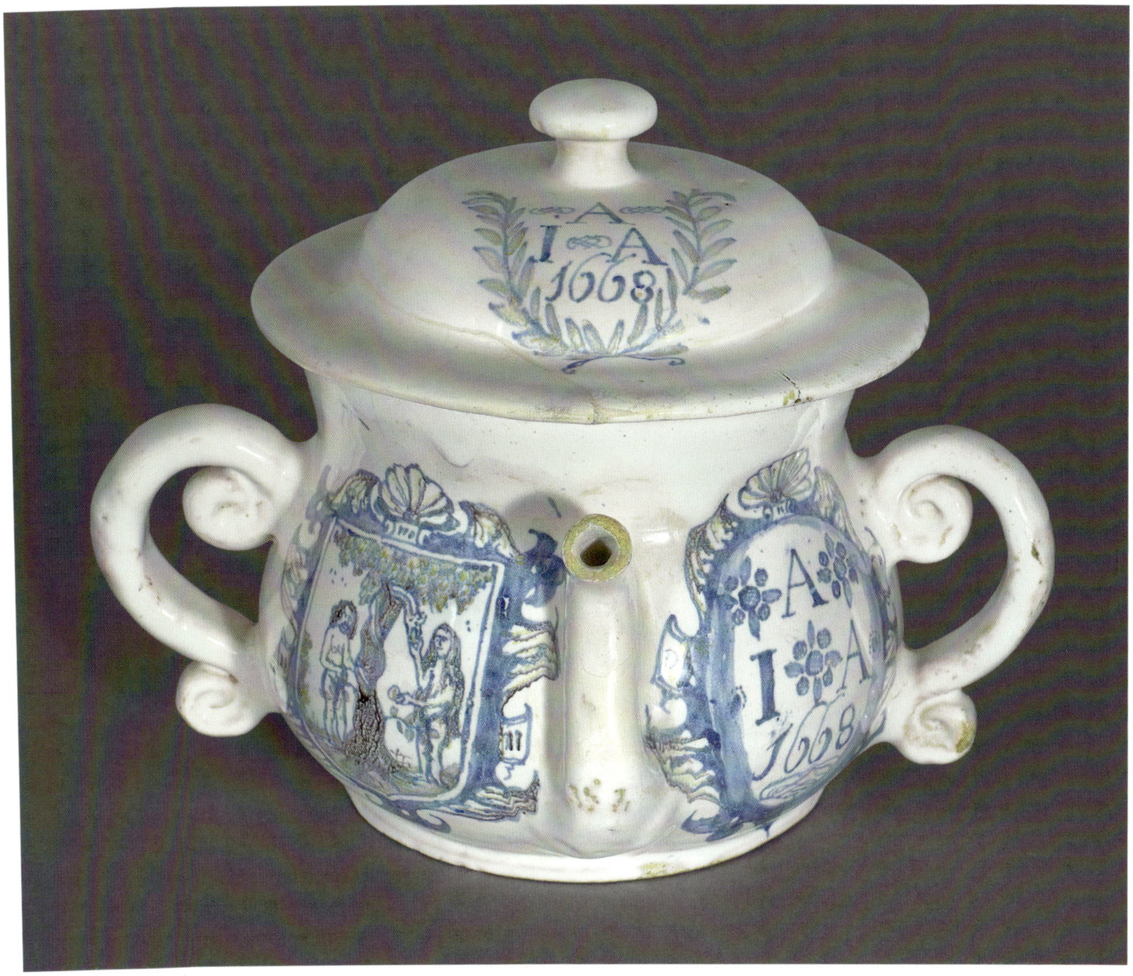

1688

Cup
Delftware
$5\frac{3}{8} \times 6\frac{1}{4} \times 10\frac{3}{8}$ in.
(13.7 × 15.9 × 26.4 cm)

Inscription: A|G|A; 1688

This type of large cup, intended for consuming alcohol, was a perfect object for a marriage, which the initials and date on this object may celebrate. This cup was owned by the antiquarian Sir John Evans and then by his granddaughter Susan Minet, daughter of William Minet and his wife, Alice Evans. The Minets were landowners of Huguenot descent who helped establish and then subsequently contributed to the Huguenot Society of London, forming the collections now in the care of University College London.

PROVENANCE

Sir John Evans of Britwell, Hertfordshire (1823–1908); Christie's, London, February 14, 1911 (lot 47); Susan Minet (1844–1976) in 1914; Sotheby's London, July 17, 1950 (lot 51); Louis L. Lipski (1914–1979); Sotheby's, London, March 10, 1981 (lot 57); bt. John H. Bryan of Lake Bluff, IL (1936–2018) through Jellinek & Sampson Antiques, London

LITERATURE

Louis L. Lipski and Michael Archer, *Dated English Delftware: Tin-Glazed Earthenware, 1600–1800* (London: Sotheby Publications, 1984), 208; 209, no. 922

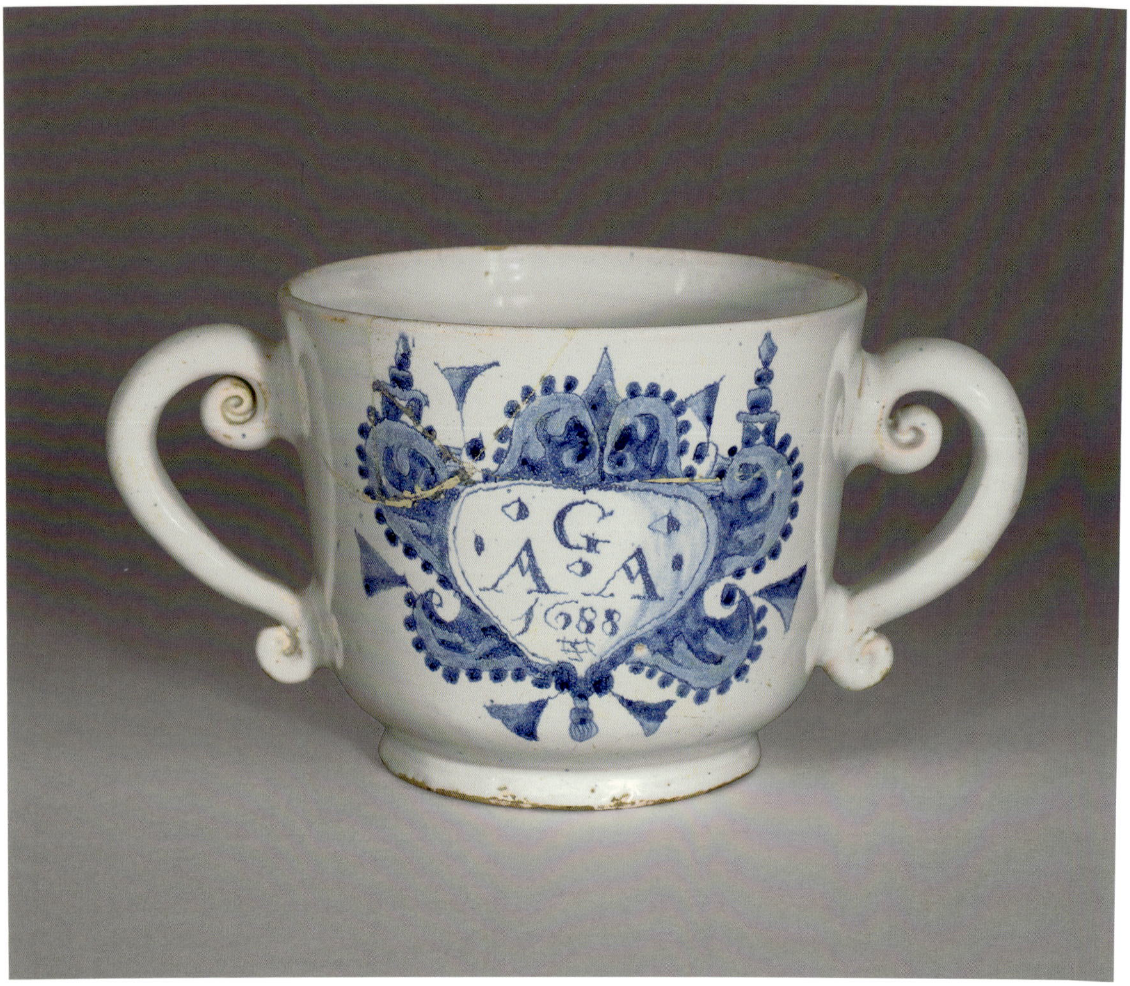

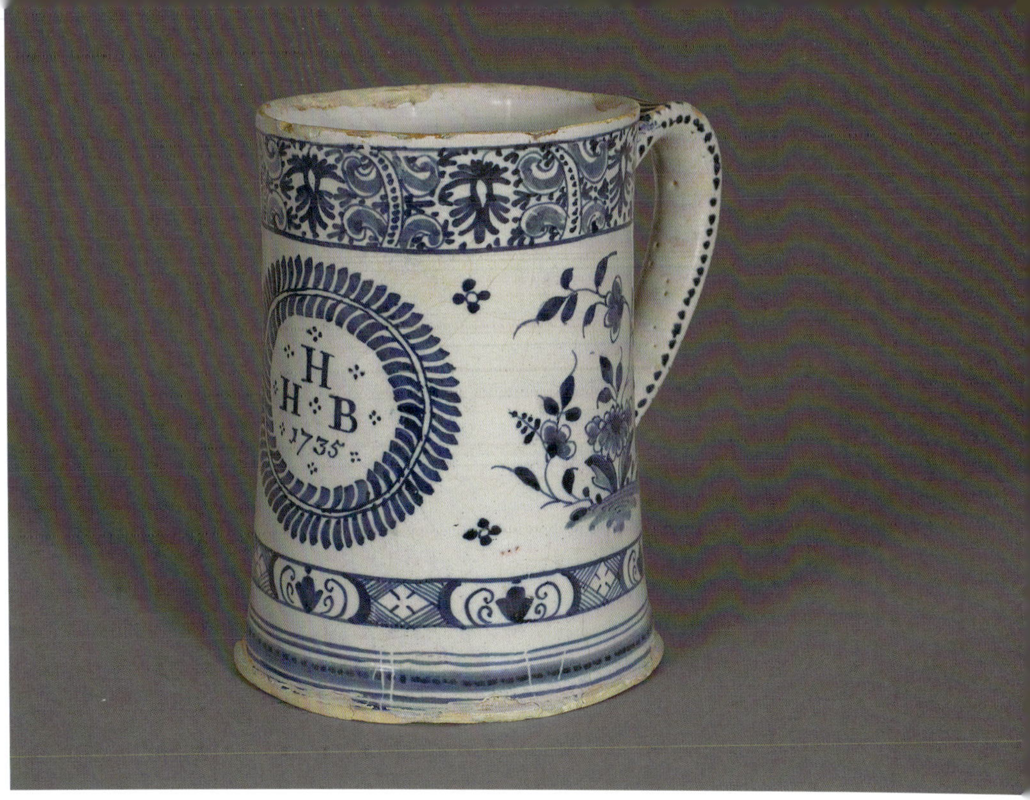

1735

Mug
Delftware
$7\frac{5}{8} \times 5 \times 7\frac{3}{8}$ in.
(19.4 × 12.7 × 18.7 cm)

Inscription: H|H|B; 1735

This large mug seems to have been made for someone with a suitably sized appetite. The initials and date suggest it could commemorate the marriage of Henry Hoadly and Bridget Blake in London on August 19, 1735, at St Benet Paul's Wharf. Bridget's will of 1776 (TNA PROB 11/1026/73), made when she was a widow living at the Tower of London, stipulates the bequest to her niece of all her china "with her uncles and my arms thereon." She also left her niece "her uncles Effigy in a glass case" and left the Reverend John Hoadly a "large picture of my late husband," possibly made by her relative Sarah Hoadly (1676–1743), wife of Bishop Benjamin Hoadly and a pupil of the portraitist Mary Beale.

A publication of Bishop Hoadly's works includes a history of a family heavily embedded in Jamaica as planters. It recalls how Henry, born October 1699, was bred to the sea and, "from the lowest Station of the Ship, arrived to be some Years Captain of the *Royal Guardian*" in the East India Company, referring to his apprenticeship to mariner Richard Else of Gosport, Hampshire. The short biography goes on to say that "his surprising Bulk was thought to shorten his Days, though by his great Strength he went with Alacrity through all the Offices of the Ship. He died before he was completely 42 years old, of the amazing Weight of 33 Stone, March 25, 1742" (John Hoadly, *The Works of Benjamin Hoadly* [London: 1773], 1:lxv).

PROVENANCE

A. E. Clark; Sotheby's, London June 18, 1919 (lot 8); Cecil Baring, third Baron Revelstoke (1864–1934); Puttick and Simpson, London, November 20, 1934 (lot 509); Puttick and Simpson, London, July 12, 1935 (lot 128); F. H. Garner; Sotheby's, London; Puttick and Simpson, London, October 6, 1964 (lot 71); The Moor Wood Collection; Sotheby's, London, May 15, 1979 (lot 16); bt. John H. Bryan of Lake Bluff, IL (1936–2018) from Jellinek & Sampson, London, May 16, 1979

LITERATURE

Louis L. Lipski and Michael Archer, *Dated English Delftware: Tin-Glazed Earthenware, 1600–1800* (London: Sotheby Publications, 1984), 183, no. 822

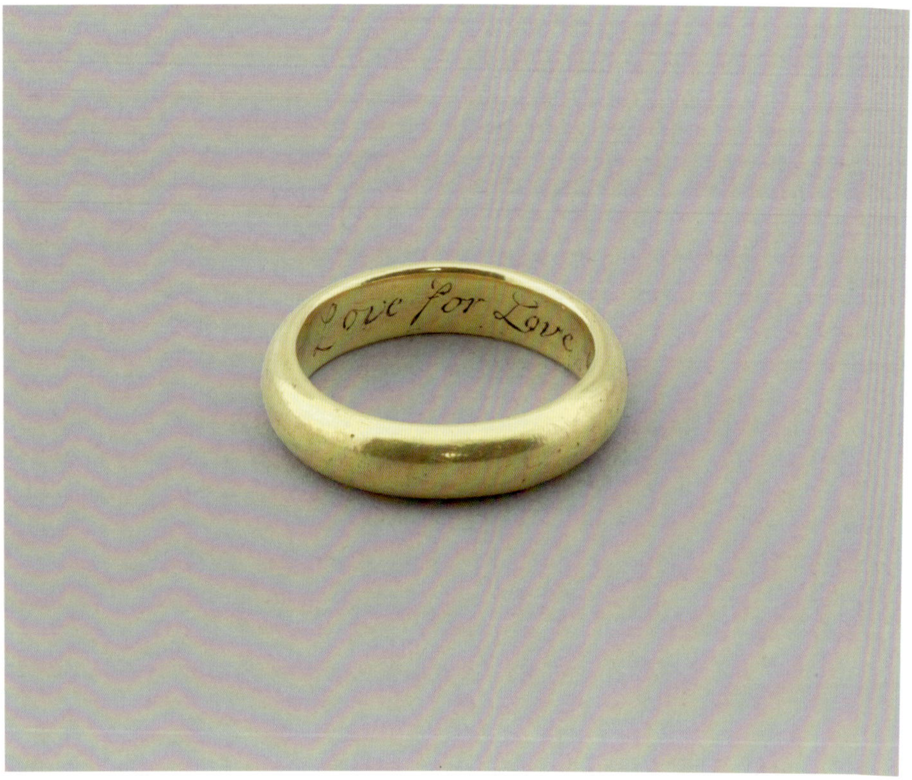

1719

Posy ring
Gold
$\frac{1}{8} \times \frac{7}{8}$ in. (.32 × 2.22 cm)

Inscription: Love for Love 1st Septr 1719

The posy motto "Love for Love" and the iconography of two clasped hands suggest that this gold band was a wedding ring. In post-Reformation liturgy, after the husband made his vow, the couple parted hands; the bride took the groom's hand again before making her own vow. This reciprocal gesture is echoed in the prose inscription on this ring and those on other rings from the same period: "I chuse thee till life refuse me" (British Museum, London, 1961, 1202.26); "God for me appoint thee" (British Museum AF.1248); "In thy Sight is my delight" (British Museum AF.1302); and "Lord lincke our harts in lasting love" (British Museum 1961, 1202.230).

Although mutual consent to marry was required by law, love and personal choice were not the main factors in determining a match in early modern Britain. Affinities in social standing, religious allegiance, material worth, honor, and the influence of friends and family were the driving forces when making marital choices. The presence of a wedding ring on the finger would garner for the wearer respect and recognition, affording women the right to attend the lying-in, gossiping, and churching rituals from which single women were excluded.

PROVENANCE

bt. John H. Bryan of Lake Bluff, IL (1936–2018) from Les Enluminures, Chicago, October 13, 2016

LITERATURE

David Cressy, *Birth, Marriage, and Death: Ritual, Religion, and the Life-Cycle in Tudor and Stuart England* (Oxford: Oxford University Press, 1997), 287–88

Joan Evans, *English Posies and Posy Rings: A Catalogue with an Introduction* (London: Oxford University Press, Humphrey Milford, 1931), 72

Christine Peters, "Gender Sacrament and Ritual: The Making and Meaning of Marriage in Late Medieval and Early Modern England," *Past & Present* 169, no. 1 (November 2000): 63–96

1751

Candle boxes were hung in various rooms of the house to hold supplies of candles and tapers. Wax candles would rot if they were left in the open air for too long, so correct storage was paramount for their safekeeping. Boxes like these were hung high on the wall to secure them from rats and mice—tallow, an edible fat, was very attractive to rodents. Boxes were also hung in a cool place as tallow has a low melting point.

PROVENANCE
bt. John H. Bryan of Lake Bluff, IL (1936–2018) from Alistair Sampson Antiques, London, June 22, 1998

Candle box
Oak, fruitwood
$8\frac{1}{4} \times 12\frac{5}{8} \times 5\frac{1}{2}$ in.
(21 × 32.1 × 14 cm)

Inscription: ANNO P|S|H 1751

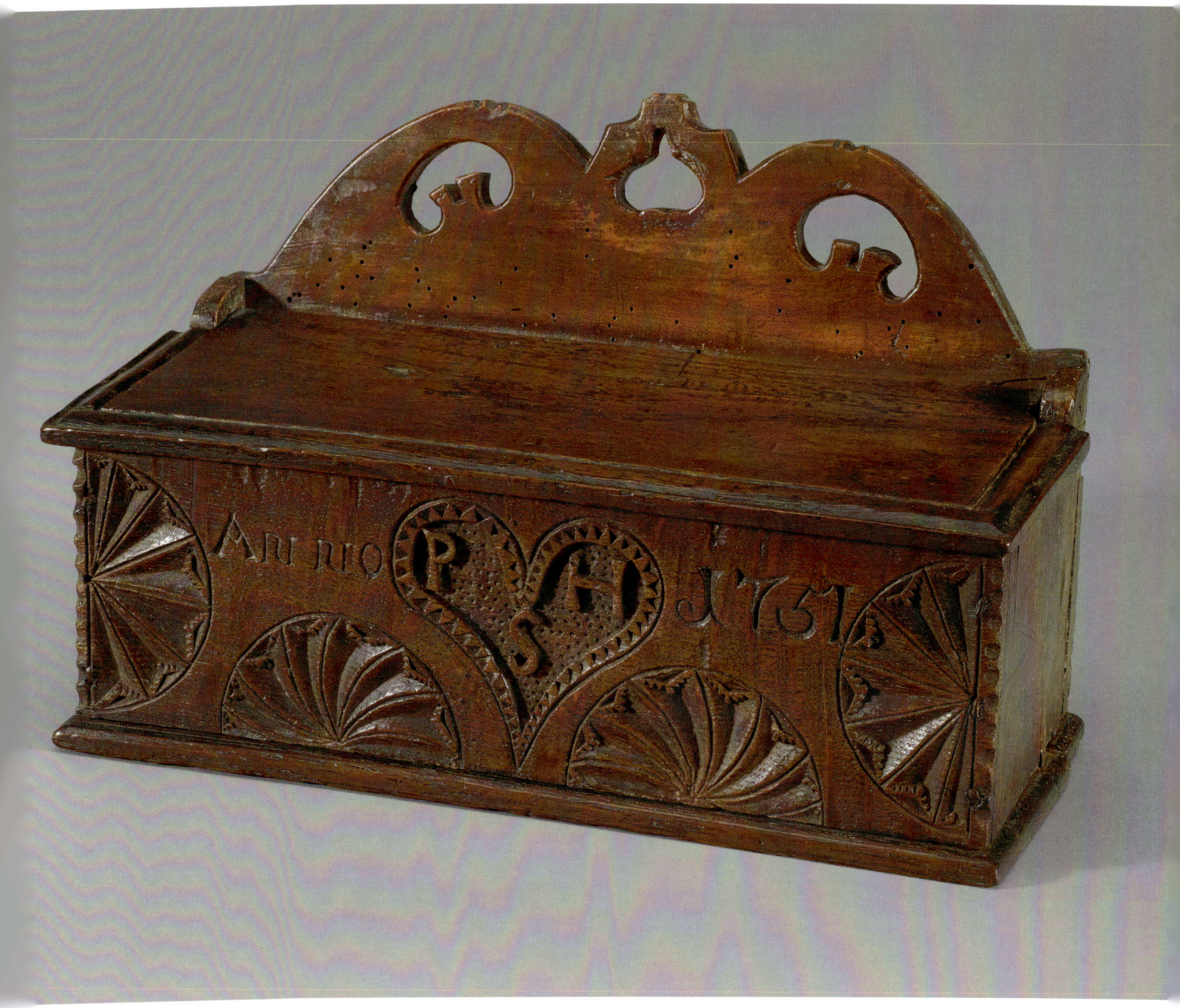

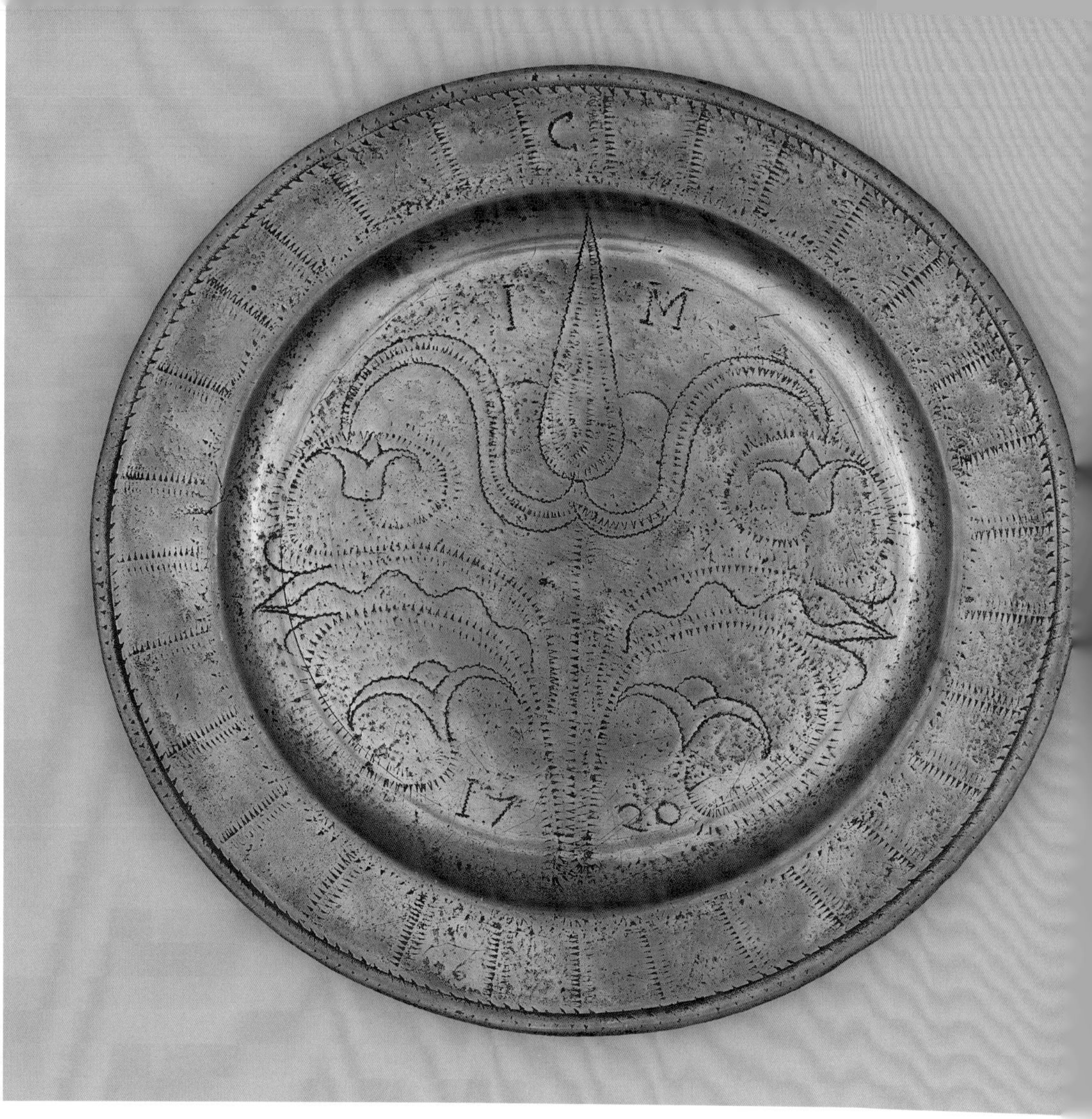

1720

Plate
Pewter
8 ½ × ½ in. (21.6 × 1.27 cm)
James Hitchman (1679–1735)

Inscription: I|C|M; 1720

The decoration and inscription on this pewter plate is wrigglework, made with a method called "wriggle engraving." The maker, James Hitchman, is known for his plates engraved with peacocks and flowers, but he also made shaving bowls, as seen in the example at the Victoria and Albert Museum, London (M.99-1945).

Baptized at Shipton-under-Wychwood, Oxfordshire, on January 28, 1678/9, Hitchman was the namesake of James Hitchman of Ascott, Oxfordshire, husbandman (d. 1706). He was apprenticed to Thomas Buttry on August 9, 1694, and from 1706 took on his own apprentices, many of whom were from Oxfordshire. Hitchman died in the parish of St Saviour Southwark in 1735. The bequests in his will record the "Moulds, Blocks, Anvills, Hammers, Turning Wheels, Hooks, Pitts, and all the other Tools of my Trade of a Peweterer whatsoever" that were to be delivered to his son Robert when he reached the age of twenty-four (TNA PROB 11/672/70).

PROVENANCE

bt. John H. Bryan of Lake Bluff, IL (1936–2018) at Bonhams, London, February 24, 2016 (lot 14) through David Moulson, Alcester, Warwickshire

LITERATURE

Charles Welch, *History of the Worshipful Company of Pewterers of the City of London*, vol. 1 (London: Blades, East and Blades, 1902)

1707 | 1755

Mug
Pewter
$6\frac{1}{4} \times 4 \times 6\frac{3}{4}$ in.
(15.9 × 10.2 × 17.1 cm)
Henry Hammerton I
(b. ca. 1685;
d. 1750)

Inscription: D|H|H; 1755.

This mug held a quart of ale and bears the hallmark of Henry Hammerton. His initials appear on either side of a crowned tun and the date "1707," which are integral to the pewterer's touchmark.

The Hammerton dynasty of pewterers was founded at the beginning of the eighteenth century when Henry, son of William Hammerton of Cookham, Berkshire, was apprenticed to Lawrence Child on February 26, 1699/700. Hammerton later lived and was buried in the parish of St Helen Bishopsgate, London.

Next came Henry Hammerton II (b. 1719), who was granted his father's touchmark in 1748. Henry's son, Richard Hammerton, Pewterer, of St Martin-in-the-Fields, was apprenticed on April 4, 1737, and then was turned over to Richard Midwinter of the Founders' Company on May 2, 1737. His touchmark was recorded in 1751.

PROVENANCE

J. L. Grant Collection (no. 33); Bonhams, London, September 20, 2017 (lot 192); bt. John H. Bryan of Lake Bluff, IL (1936–2018) through David Moulson, Alcester, Warwickshire

LITERATURE

Howard Herschel Cotterell, *Old Pewter, Its Makers and Marks in England, Scotland and Ireland: An Account of the Old Pewterer & His Craft* (repr., Rutland, VT: Charles Tuttle, 1963), 225

Henri J. L. J. Massé, *The Pewter Collector* (London: H. Jenkins, 1921), 175

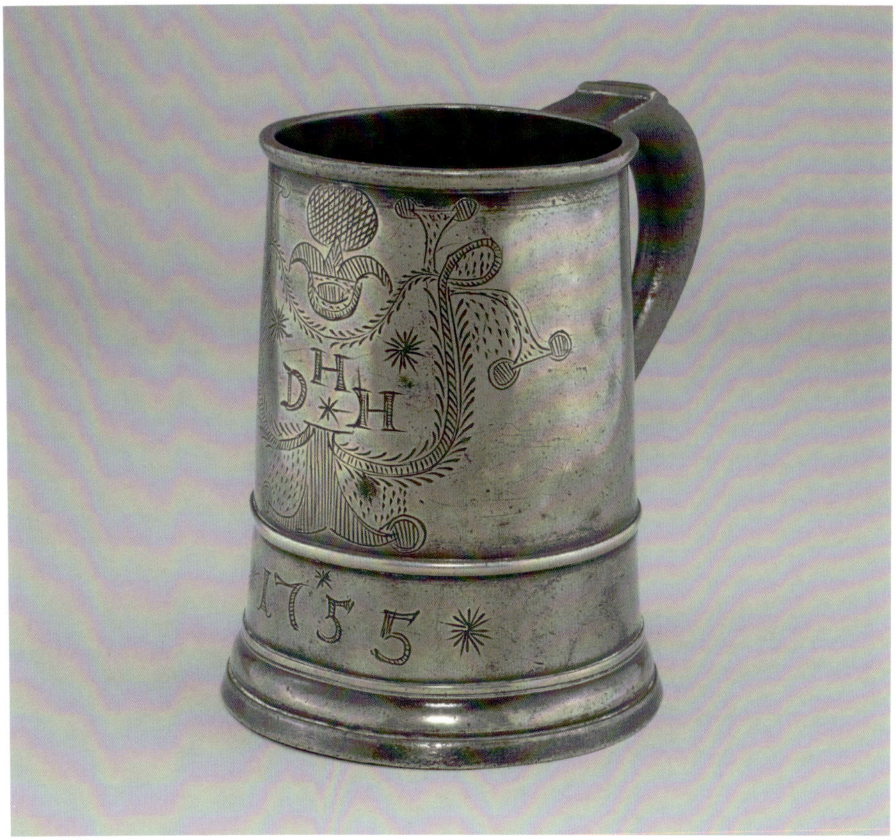

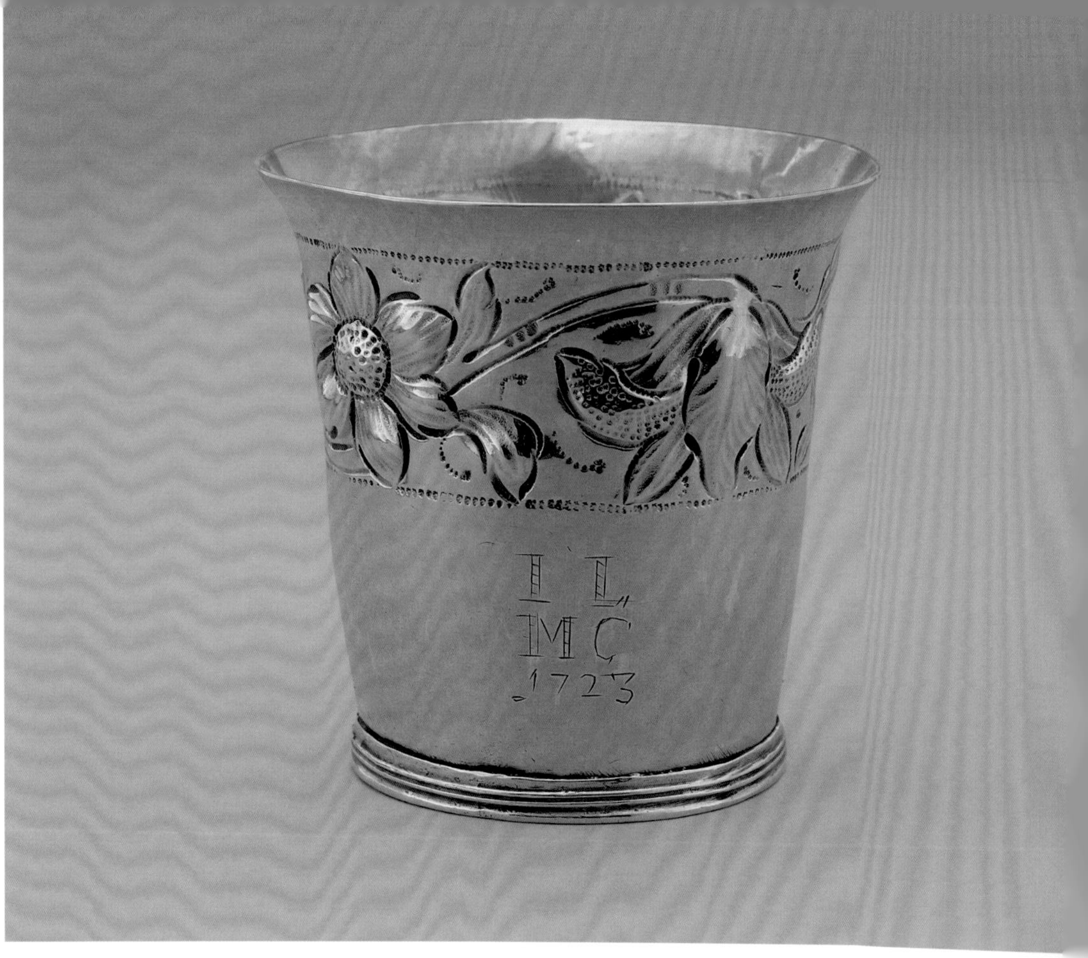

1723

Beaker
Silver
3 × 3⅛ in. (7.6 × 7.9 cm)
John Duck (1653–1745)

Inscription:
I|M|L|C; 1723

The strange inscription on this silver beaker remains a puzzle. Initially it seemed to have read "I M L," although the placement of the letters seems a little odd in comparison to other three-character inscriptions. The clumsily scrawled "C" and the year "1723" seem to have come later; the date is some twenty-five years after the beaker was made. Hallmarked for the year 1688, it bears the maker's mark—a goose in a dotted circle—identified as that of the aptly named John Duck (1653–1745).

Duck himself is intriguing. The first part of his life seems conventional: He was the son of Edward Duck, a barber of Marlborough in Wiltshire. On May 5, 1669, at about age fifteen, he was apprenticed in London to a fellow West Countryman, Roger Stevens (active 1616–1673). Stevens was a member of the Goldsmiths' Company and practiced as a silversmith; he was probably the artisan by that name that Samuel Pepys consulted for appraisals and, in November 1660, treated to a pint of wine at a London tavern.

Shortly before Stevens died in 1673, he made a nuncupative will, stipulating that his widow, Catherine, should take over the running of his workshop, located at the sign of the Angel and Trumpet at the east end of Foster Lane in the parish of St John Zachery, a stone's throw from Goldsmiths' Hall. The neat signatures of his daughters, Hannah and Elizabeth, who witnessed the spoken will, attest to the women's literacy. The inventory of goods made after Stevens's death records eighteen anvils, two organ bellows, brass and iron, pans, beams, nails and weights, and other hammers and pots. Also in the shop were wearing apparel, books, and weapons.

After serving out his apprenticeship, Duck was made free in 1677. On September 6, 1678, he followed what was an established rite of passage by marrying the daughter of his former master, her mother granting permission in the father's stead. The couple married at Isleworth, then still a village to the west of London, where the Stevens family may have had a country abode. For the following decade, much of Duck's time must have been spent running the Angel and Trumpet workshop. Thomas Dare (d. 1685), the former Goldsmith and radical exiled to Amsterdam, wrote to Duck at that address in August 1683—the year of John Locke's flight to that city and Shaftesbury's death there shortly after the failure of the Rye House Plot. To judge from the letter (which appears to have been intercepted or possibly betrayed to authorities), Duck and Dare were on very familiar terms and had several mutual acquaintances, including Sir John Trenchard (1649–1695), the nonconformist politician, member of the Green Ribbon Club, and later secretary of state (1693). This affiliation would suggest that, like Dare and Locke, Duck was embedded in West Country Whiggism and may have been briefly implicated in the Rye House Plot. Dare later served as secretary to the ill-fated Duke of Monmouth; Monmouth met with the Earl of Argyll at Dare's house to discuss their invasion of Britain. Soon after arriving in England with the invasion force, Dare was shot in the head by a co-conspirator due to an argument over a horse.

To judge from the number of extant silver vessels bearing Duck's mark, the Foster Lane workshop was a thriving business throughout the 1680s, producing not only beakers but also tankards, mugs, porringers, chalices, and other church plates. The date "1691" inscribed on a silver flagon (Bonhams, London, November 3, 2010 [lot 157]) is overstruck, Duck having taken advantage of the fact that the letter date for 1691 was exactly the same as for 1571, so he reused, refashioned, and rehallmarked an Elizabethan flagon to pass it off as his own work.

There is a degree of irony here. Duck later embarked upon a career as assistant to Nathaniel Bowles, the assayer of the Goldsmiths' Company. After eighteen months, Duck succeeded him as the new assayer, an appointment he may have owed to his old acquaintance Trenchard, now secretary of state and privy councilor. During Duck's tenure, sterling silver moved to the new standard of Britannia silver, which was softer and had a higher silver content—part of an incentive by William III of recoinage begun by act of Parliament in 1697. As a result, sterling silver soon grew unpopular.

Duck lived to the ripe age of ninety-two. The last notice of his activities is on July 6, 1715, when he put an advertisement in the *Daily Courant* announcing the sale of "a large barge (as good as new)" that could be viewed at his Foster Lane address. The barge, with plate and other goods, had been mortgaged to him by the Goldsmiths' Company in lieu of a salary, which was 300 pounds in arrears. In 1716, after twenty-three years, citing ill health, Duck was allowed to retire from the Assay Office. Bowles then returned to his former role, having provided Duck with a healthy 70-pound bribe to help secure his reelection. The bribe was discovered, and Duck duly fined.

PROVENANCE

bt. John H. Bryan of Lake Bluff, IL (1936–2018) from Peter Cameron, London, December 30, 2016

LITERATURE

John Culme, "The 'Goose in a Dotted Circle': A Mystery of the Seventeenth Century Investigated," *Silver Society Journal*, no. 14 (2002): 97–105

J. S. Forbes, *Hallmark: A History of the London Assay Office* (London: The Goldsmiths' Company and Unicorn Press, 1998)

Mark Goldie, "John Locke's Circle and James II," *Historical Journal* 35, no. 3 (1992): 557–86

Beth Carver Wees, *English, Irish, & Scottish Silver at the Sterling and Francine Clark Art Institute* (New York: Hudson Hills Press, 1997)

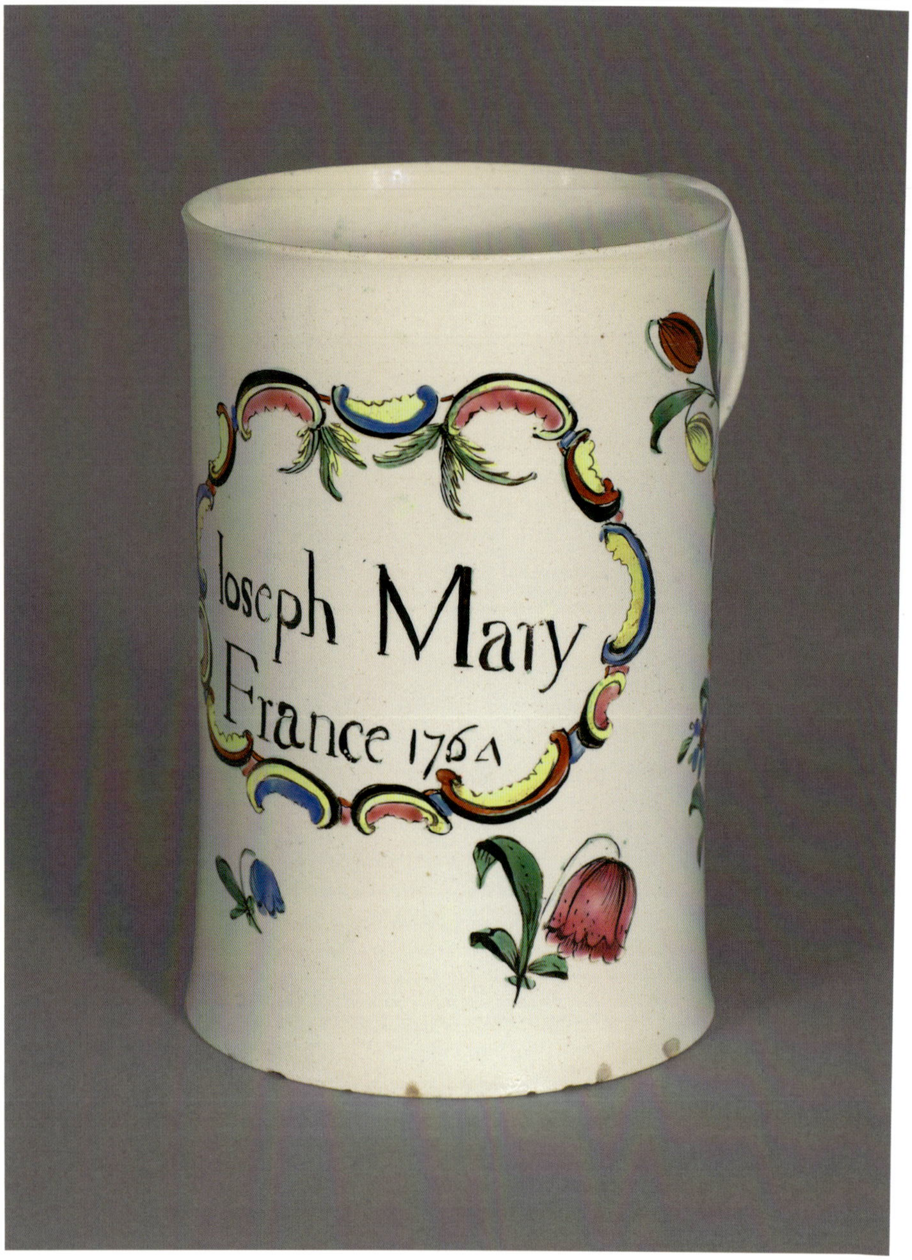

1764

Mug

Salt-glazed stoneware

$6\frac{3}{8} \times 4 \times 5\frac{1}{2}$ in. (16.2 × 10.2 × 14 cm)

Inscription: Ioseph Mary|France 1764

On January 12, 1764, Joseph France of Whiston, a widower, age forty, married Mary Stacey, widow, at Rotherham, Yorkshire. Although the signature lacked confidence, Joseph was able to sign his name in the registry. Mary did not sign.

PROVENANCE

bt. John H. Bryan of Lake Bluff, IL (1936–2018) from Andrew Dando Antiques, Wiltshire, May 25, 2017

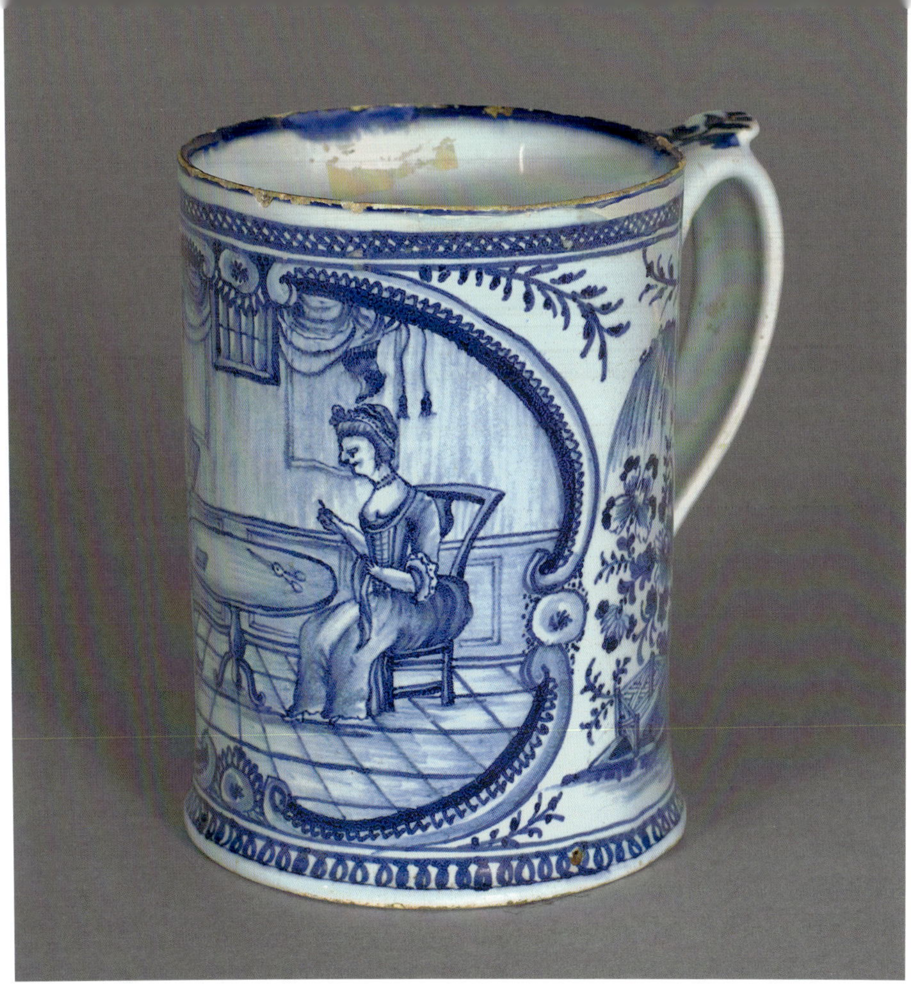

1783

Mug
Delftware, glass
$5\frac{1}{8} \times 3\frac{3}{4} \times 5$ in. (13 × 9.5 × 12.7 cm)

Inscription: D|P|L; 1783

This mug has a similar shape and floral decoration as a group that bears dates ranging from 1777 to 1793. Many of those mugs have glass bottoms, and some are gilded. The painted vignette within the cartouche depicts a quiet domestic scene in which a woman concentrates on her needlework while her male companion is preoccupied with his book. A mug depicting a similar interior scene in a very comparable manner is in the collection of the Fitzwilliam Museum, Cambridge (C.1559-1928), and it is likely that both scenes derive from print sources.

Pottery shards with this form of floral and foliate decoration have been excavated in London, which suggests that they were made there. Michael Archer has suggested that, because of their very late date, they must have come either from Lambeth High Street, Vauxhall, Mortlake, or from Isleworth, as these were the London potteries still in operation at that time. It may be no more than a coincidence, but the terminating date for this group of mugs is the same year as John Brayne's death in 1793 and, with that, the closure of his pottery at Lambeth High Street.

PROVENANCE

Louis L. Lipski (1914–1979); Sotheby's, London, March 10, 1981 (lot 211); bt. John H. Bryan of Lake Bluff, IL (1936–2018) through Jellinek & Sampson Antiques, London

LITERATURE

Michael Archer, *Delftware in the Fitzwilliam Museum* (London: Philip Wilson Publishers, 2013), 184

Louis L. Lipski and Michael Archer, *Dated English Delftware: Tin-Glazed Earthenware, 1600–1800* (London: Sotheby Publications, 1984), 191, no. 856

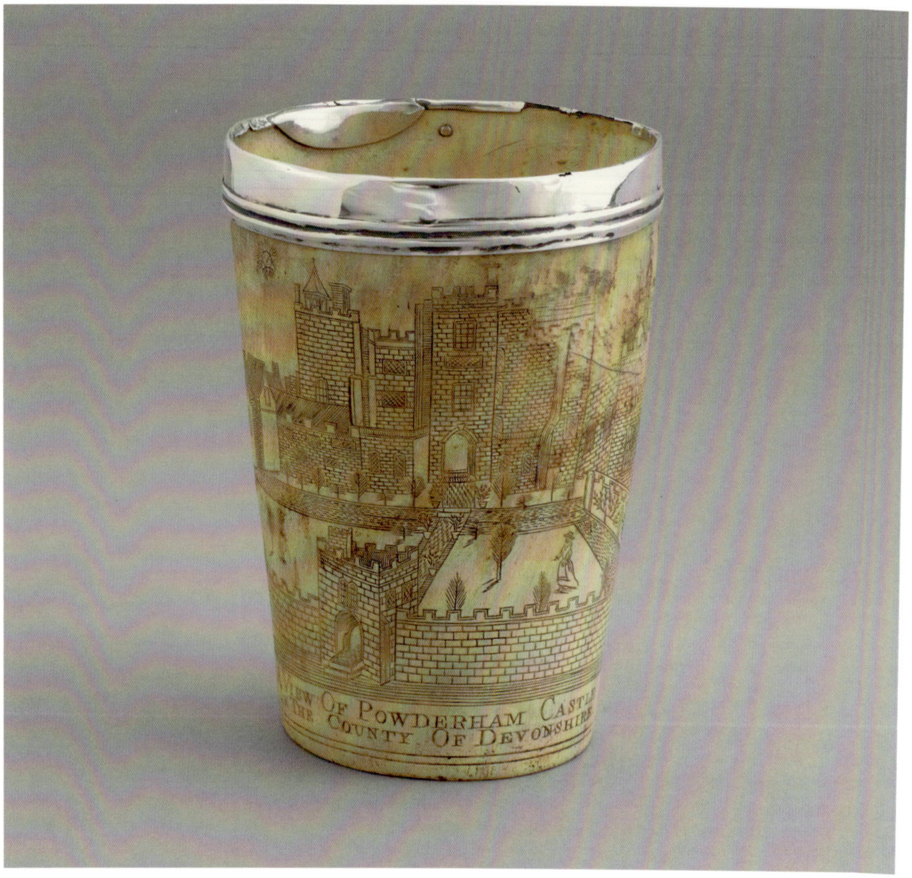

1788

Cup

Horn, wood, silver

4 7/8 × 3 3/8 in. (12.4 × 8.6 cm)

Nathaniel Spilman (1755–1836)

Inscription: A VIEW OF POWDERHAM CASTLE|IN THE COUNTY OF DEVONSHIRE.; JOHN &|MARY ANN|PAGE,| LUDHAM| NORFOLK| JUNE. 7|1788.

In the eighteenth century, England was the center of the global horn trade: every year raw horns flooded into major cities, mostly from Africa, by the hundreds of thousands. To meet rising demand on the Continent and in America, the horns were molded into shoehorns, snuff mulls, lanthorns (lanterns that used horn as reflective material), and other objects. Simple, widely produced horn cups like these—for everyday drinking—were often engraved with scenes of hunting or of the owner at work at his trade or profession.

The inscribed recipients of this cup—John Page and Mary Ann Page, née Barber—were married in Great Yarmouth, Norfolk, on December 11, 1768, so, when this cup was made, they were about half a year shy of their twentieth wedding anniversary. The image of Powderham Castle was probably based on Samuel and Nathaniel Buck's engraving *The East View of Powderham-Castle in the County of Devon* (ca. 1734). It is unclear what significance, if any, Powderham Castle had for John and Mary Ann Page in 1788—but presumably the maker, Nathaniel Spilman (1755–1836), had it to hand to use as a source.

PROVENANCE

Woolley & Wallis, Salisbury, October 4, 2017 (lot 29); bt. John H. Bryan of Lake Bluff, IL (1936–2018) through Tobias Jellinek Antiques, London

LITERATURE

Trevor Fawcett, *Music in Eighteenth Century Norwich and Norfolk* (Norwich, UK: Center of East Anglian Studies, University of East Anglia, Norwich, 1979), 32

Paula Hardwick, *Discovering Horn* (Guildford, UK: The Lutterworth Press, 1981)

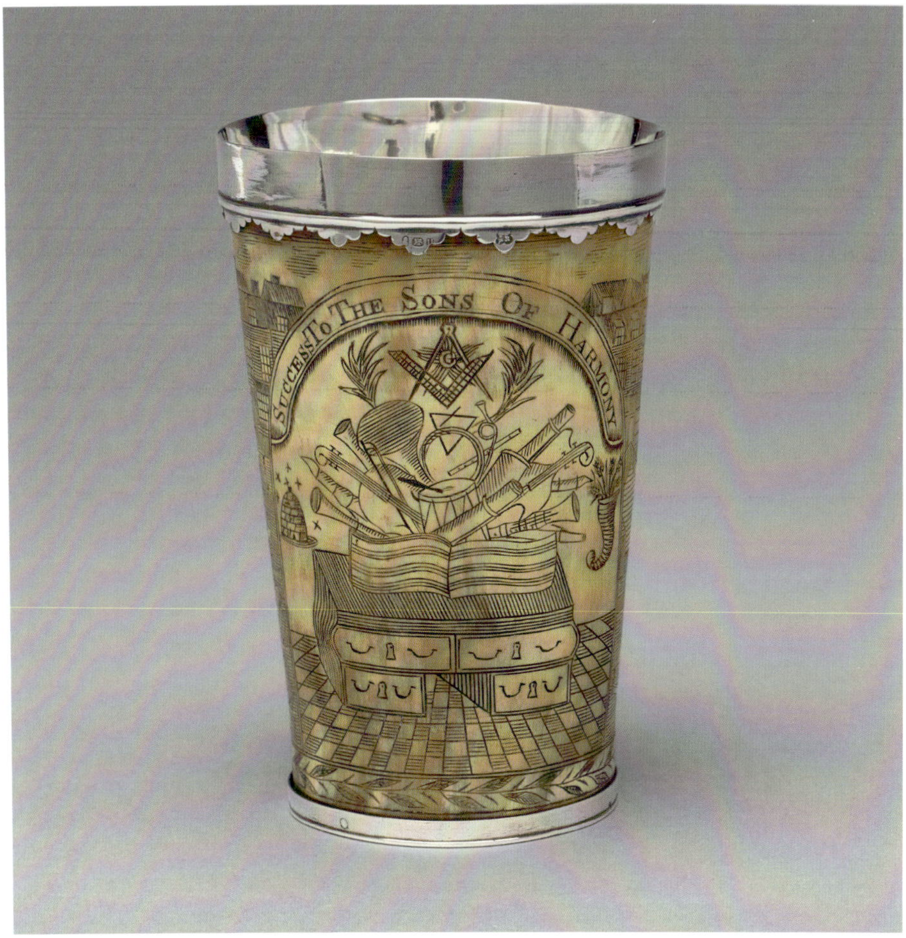

Cup

Horn

5 × 3 in. (12.7 × 7.6 cm)

Nathaniel Spilman (1755–1836)

Inscription: SUCCESS TO THE SONS OF HARMONY; DOWNES

The engraving on this undated cup is so similar to that on the other signed and dated horn cups in the Bryan Collection that all three can confidently be attributed to Nathaniel Spilman. At least ten engraved horn cups with silver rims, ranging from ca. 1785 to 1815, can be ascribed to Spilman. Some include charming depictions of comings and goings along the quayside of Great Yarmouth, Spilman's home and place of work.

One of these, commemorating Nelson's victory at Trafalgar in 1805, is signed "N.SPILMAN SCUP YARMO" and is in the collection of the National Maritime Museum at the Royal Museums Greenwich (OBJ0435). Other cups by Spilman are found in the Brighton Museum & Art Gallery and in private collections, including a signed cup dated August 11, 1788, made for John Cubitt, a hairdresser who worked at number 25, London Lane, in Norwich. Another cup, also depicting the crow's nest on the spire of the Newcastle Exchange and attributed to Spilman, was sold Bonhams, Oxford, January 22, 2015 (lot 396).

This cup is inscribed "SUCCESS TO THE SONS OF HARMONY." The Sons of Harmony were a band of musicians associated with the annual river procession of Freemasons downstream from Norwich, which culminated with a venison feast. In 1791, the Sons of Harmony accompanied the Freemasons as part of a flotilla of eighteen boats.

PROVENANCE

bt. John H. Bryan of Lake Bluff, IL (1936–2018) from Alistair Sampson Antiques Ltd., London, February 24, 1997

1666 | 1783 | 1791

Cup
Horn
$4\frac{3}{8} \times 2\frac{3}{4}$ in.
(11.1 × 7 cm)
Nathaniel Spilman
(1755–1836)

Inscription: ALEX[R] NEILSON | LEITH FEB[RY] 14th 1791; PRÆSTO PROPATRIA; FESTINA LENTE; GOD SPEED THE PLOUGH; A VIEW OF THE CROWE NEST BUILT IN MARCH 1783 ON THE TOP OF NEWCASTLE EXCHANGE SPIRE; LONDON; LONDON | MONUMENT | BUILT 1666; YARMOUTH; YARMOUTH TOWN HALL; A DUTCH SHOE

Nathaniel Spilman was the son of a Presbyterian mariner of Great Yarmouth, Norfolk, himself the son of a sailor of the same port town. On September 16, 1782, Nathaniel married Mary Hobbs at All Saints, Newmarket, Suffolk. Their son, also named Nathaniel, was baptized on August 31, 1792. He became an artist and may have trained under a relative called Samuel Spilman, a painter in Great Yarmouth. At the time of his death in October 1835, Nathaniel the younger was a member of two of the Masonic lodges in Great Yarmouth. Nathaniel the elder died the following year, aged eighty-two, and was buried February 8, 1836.

Spilman also made a horn cup for the Society of Universal Goodwill, originally named the Scots Society, founded in Norwich in 1784 by Dr. John Murray. Family ties to the Presbyterian Church might explain Spilman's work for Scottish societies within England and patrons like Alexander Neilson, the owner of this cup, who was from Leith, Scotland. Neilson was a solicitor-at-law and, in 1793, a subscriber to the first volume of *The Theory and Practice of Finding the Longitude at Sea or Land*. His name and coat of arms are on the cup, which is dated February 14, 1791, as are engraved views of London, Great Yarmouth, and Newcastle, again all seemingly derived from print sources.

PROVENANCE
Collection of Stuart Kirby of Long Valley, NJ (1952–2019); bt. John Bryan II of Nashville, TN, from Adam Ambros, New York, 2019

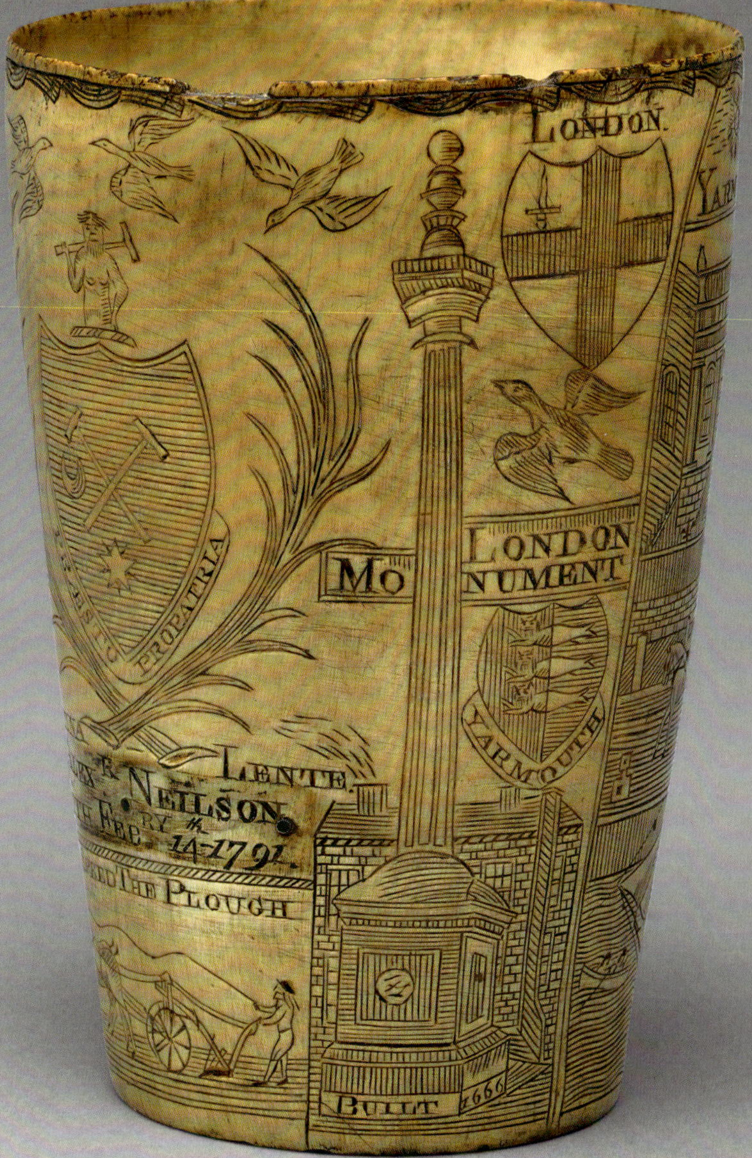

LONDON.
LONDON
MO
NUMENT
YARMOUTH
LENTE.
NEILSON
THE PLOUGH
BUILT

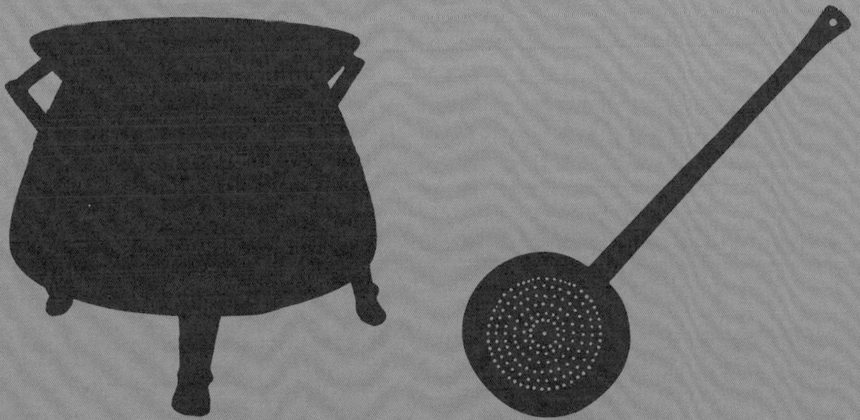

III. HEARTH & HOUSEHOLD

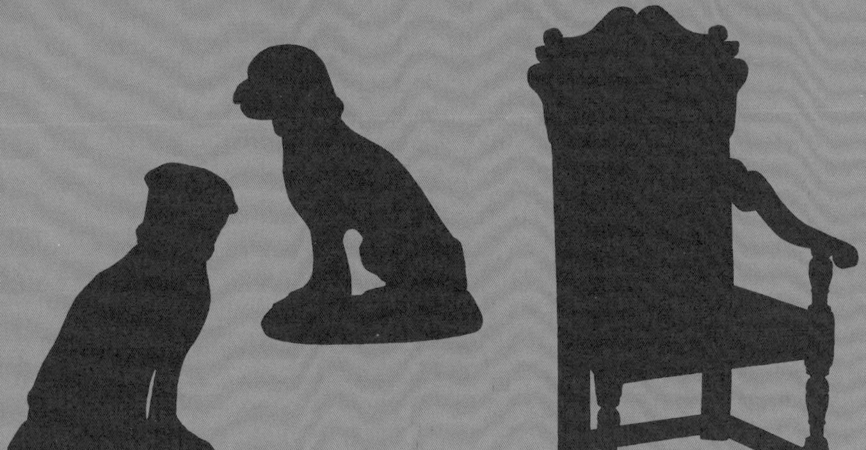

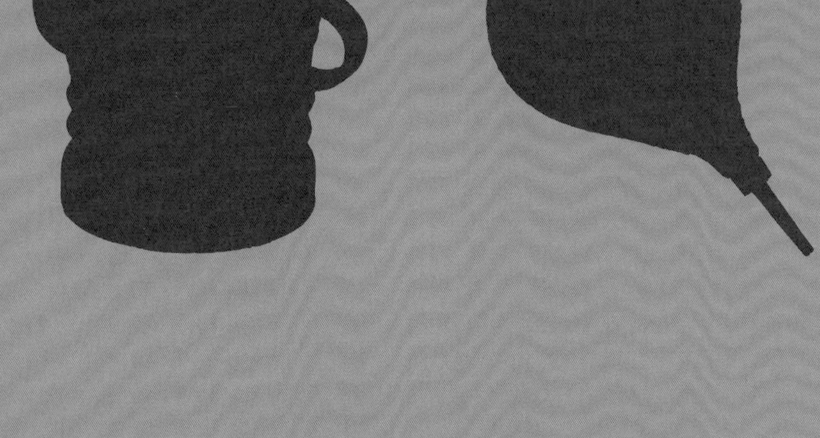

ALTHOUGH THE BED WAS THE PRIMARY FOCAL POINT OF THE HOME, the other essential locus was the hearth—and, within a home with multiple hearths, the kitchen hearth was the most important. The kitchen was where warmth, light, and nourishment were found, and in a medium-size to large home, this room was a hub for the various activities associated with feeding the household: boiling, roasting, broiling, and, if there was an oven built into the hearth, baking. At a time in which about 40 percent of homes had servants, maids who lived as part of the household would have done much of the work of food preparation.

The kitchen hearth contained a spit, upon which to turn meat to roast it evenly; a spit jack, a device that automated the work (page 214), would provide assistance. There were racks to support the spits, kettles, skillets, gridirons, and the chafing dishes that kept food warm. Alongside the fire furniture—firebacks, andirons, and tongs—there would also be numerous utensils, such as ladles, forks, skimmers, and dredgers. All might be marked with initials and dates to help in the process of making inventories, which, in a large household, was a frequent activity; it was also conducted in a home upon the death of the householder when an itemized probate inventory was required by law. For example, in 1704, when Richard Wildey left three silver spoons to his wife in his will, he noted that those spoons were marked "R W C"—their shared initials. Marking objects also deterred sticky-fingered visitors from liberating items from the house and helped speed their return if ever temptation got the better of self-restraint.

The household, the most important social and economic entity of the early modern world, was the unit around which society was ordered. Within each household was its patriarchal head, who worked with a companionate spouse to manage religious observance, financial management, and orderly behavior in the home and beyond. Such partnerships were frequently celebrated in the triad of initials added, often with dates, to household objects and sometimes integrated into the fabric of the building itself, declaring to visitors, with demonstrations of wealth and taste, the success of those unions. Providing hospitality—a fundamental tenet of early modern society—gave a family the opportunity to make a fulsome display of its prosperity through the objects shared with and used by guests. The hosting of guests was particularly prevalent during times of seasonal or liturgical significance; the occasions of Christmas and Twelfth Night, for example, called for celebration with wassailing sets of the type included in this section of the catalogue (page 194).

Married couples often made decisions together about household furnishings and decorative schemes, although the husband, who held veto over all household expenditures, often made the decisions about larger transactions—for items like plate. Female heads of household led the consumption of material culture within the home, whether through purchases or through management of production. All the women in the household—whether family members or servants—gave significant amounts of their time to the creation, repair, and maintenance of various types of textiles. Thus, knitting needles and sheaths (pages 228–30) would have been in use throughout the day and into the night.

1738|1742

Screen

Canvas with wool thread, mahogany

59¾ × 98 × ¾ in. (151.8 × 248.9 × 1.9 cm)

Elizabeth Wentworth

Inscription: EliZath|fecit; Wentworth|1738; fine Sept 1742

From what little is known about the very few surviving needlework screens of this period, they were made by women of genteel birth. Only those born into privilege could afford the expensive materials and had the luxury of time. This screen, begun in 1738, was finished in 1742.

A comparable screen dated 1727, at Wallington Hall, Northumberland (National Trust, NT 582790), was embroidered by Julia, Lady Calverley (1706–1768). She was the daughter of Julia Conyers and William Blackett, first Baronet. Likewise, in 2018, a screen with the dealer Keshishian, London, had come out the aristocratic collection of Lord Harlech of Glyn Cywarch, Gwyneed.

The Wentworth name is synonymous with the ancient and august family of Wentworth Castle and Wentworth Woodhouse in Yorkshire, the latter being the largest house in private ownership in Britain. Yet no member of this family is a likely candidate for maker of this screen. There were other women from cadet branches of the same family, however, who may be candidates, including one that could credibly claim descent from the time of the Norman Conquest. One possibility is Elizabeth, daughter of Robert Lord of London, who, in 1720, married Thomas Wentworth, MP (ca. 1693–1747) of Sunninghill, Berkshire. He was the younger son of William Wentworth, third Baronet of Bretton, Yorkshire. In addition to representing Whitchurch, Hampshire, in the Parliament of 1743 to 1747, Wentworth was a lieutenant general in the army. He was the officer and eventual commanding officer of the Twenty-Fourth Regiment of Foot, serving in the War of Jenkins' Ear (the first war begun by Britain solely to benefit trade and in response to populist agitation). He also helped quash the final Jacobite rising of 1745.

Although it is impossible to prove this Elizabeth was the woman who made this screen, the War of Jenkins' Ear, which started a year after Elizabeth Wentworth began her needlework, was still raging when she put in the final stitches. By then, the conflict had been subsumed into the larger War of the Austrian Succession, and Thomas Wentworth had returned

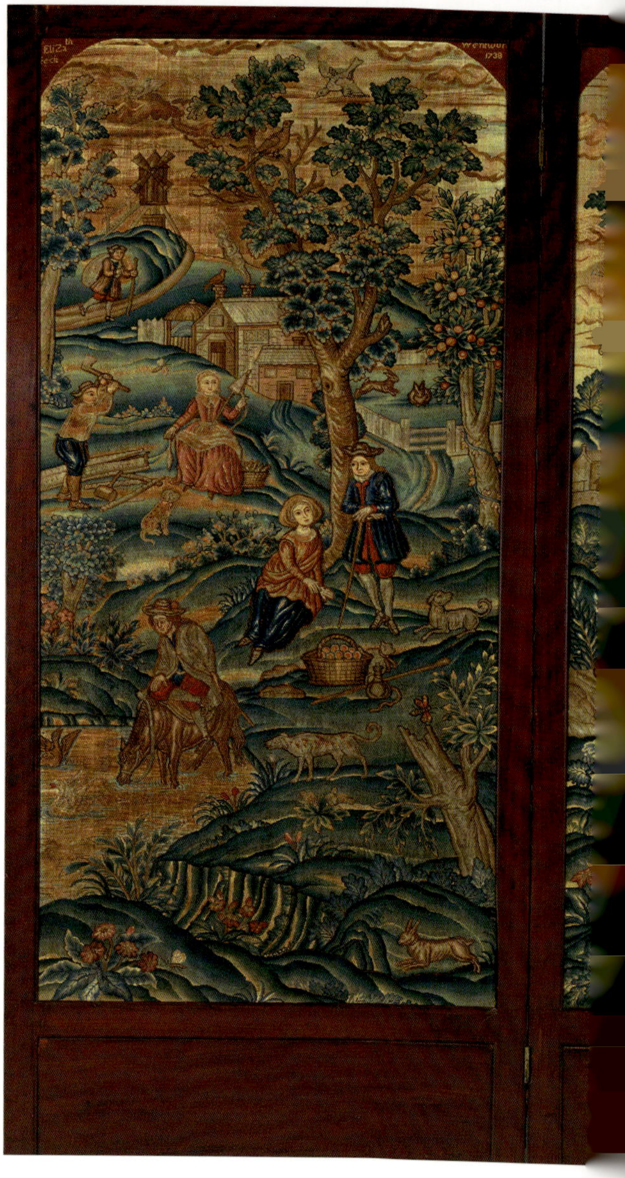

from dismal campaigns in New Grenada against colonial Spain and begun his Parliamentary career. A plausible scenario is that Elizabeth began this embroidery shortly after her husband became an officer in the Twenty-Fourth Foot Regiment (1737), mindful that the likelihood of war with Spain was imminent and, with that, the prolonged absence of her husband.

PROVENANCE

Antony Pitt Cholmondeley of South Lopham, Norfolk (1908–1986); his sale, Sotheby's, London, February 28, 1969 (lot 121); Dreweatt Neate, Donnington Priory, Berkshire, October 11, 2000 (lot 129); bt. John H. Bryan of Lake Bluff, IL (1936–2018) from S. Franses, London, May 16, 2001

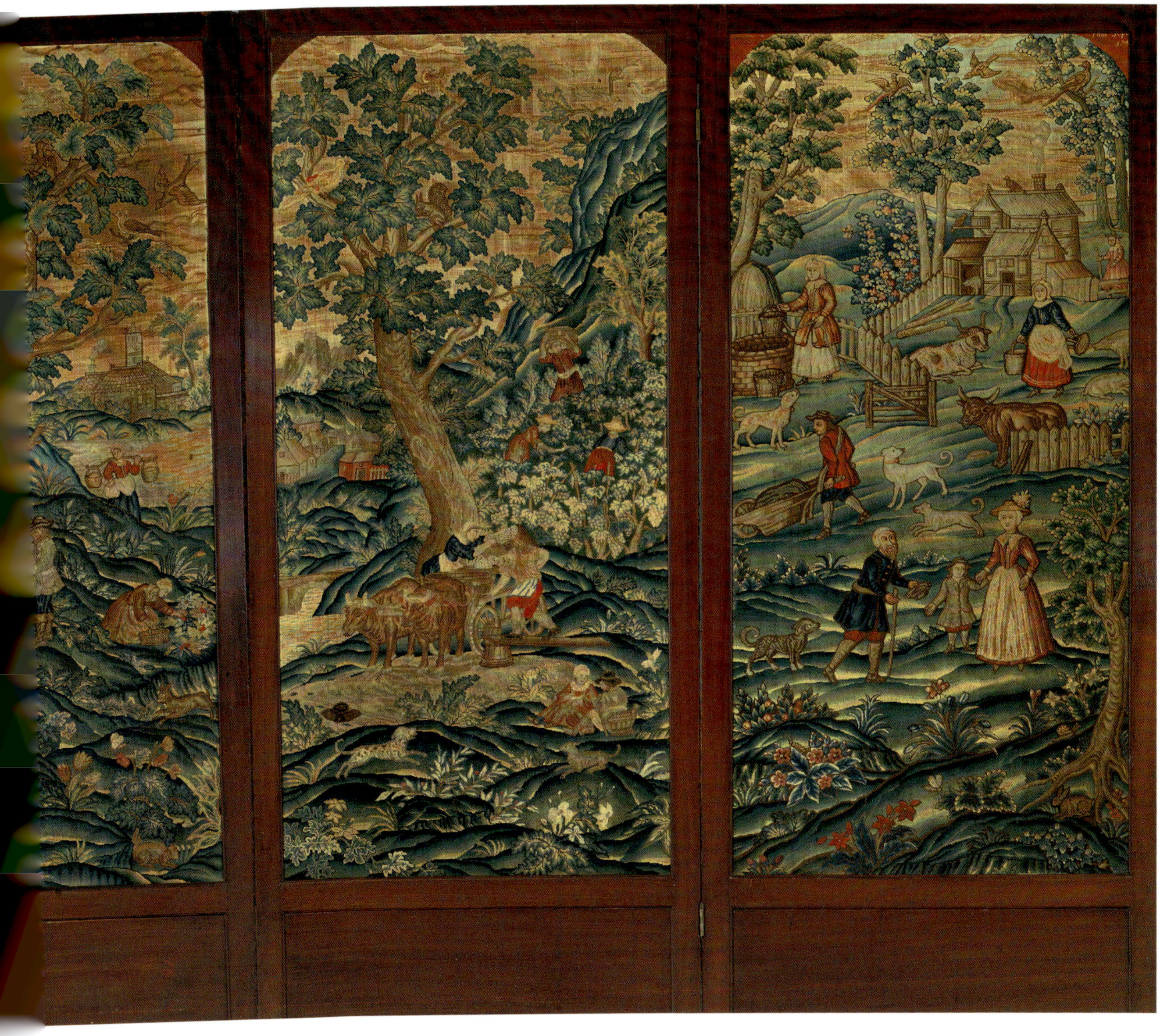

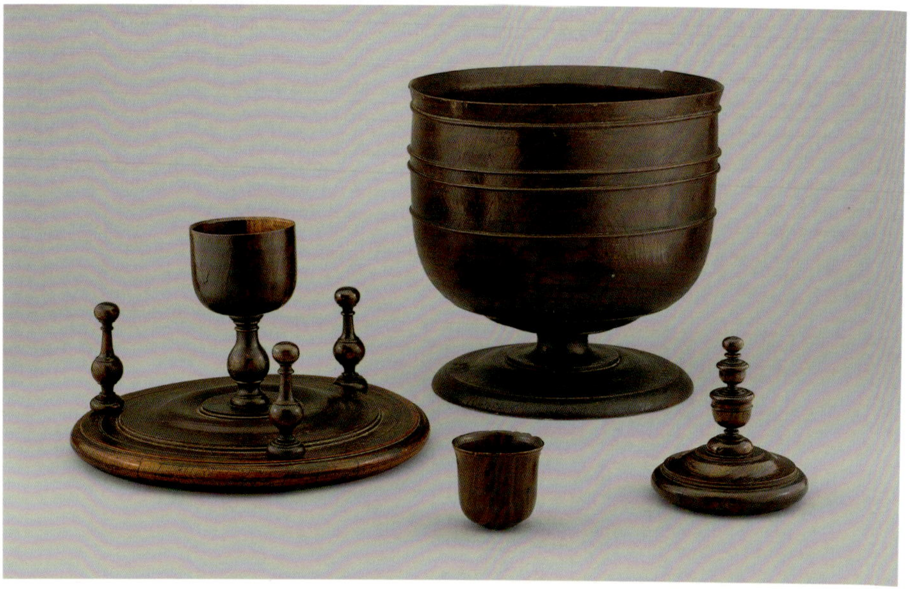

1685

Wassail bowl-and-cup set

Lignum vitae

Assembled: 20 3/8 × 10 5/8 in. (51.8 × 27 cm)

Inscription: C T of Boton fecit 1685; W|S|A

The word *wassail* comes from the Anglo-Saxon *waes hael*, meaning "be of good health." According to the first written description, by Peter de Langtoft in the 1320s, the leader of a gathering took the bowl (essentially, a massive cup) and cried, "Wassail," to which the response was, "Drinkhail." The gatherers drank and then passed the bowl to the next person with a kiss, a ritual repeated by all in attendance. The ceremony of wassailing was popular during Christmas and New Year's celebrations, particularly Twelfth Night. The wassail bowl would be decorated with ribbons and rosemary, then carried around the streets by maidens singing carols. Payments to "wassailers" are recorded in the household accounts of the gentry throughout the late Stuart and Hanoverian periods, and the custom continues in some parts of Yorkshire today. One traditional recipe contained spiced and sweetened ale, into which apple pulp was added, but mulled wine and cider were also common.

A wassail bowl-and-cup set like this one might also be used to perform rites of purification and blessing. Various writers of the seventeenth century describe men who traveled to the fields and orchards "wassailing" the fruit trees to ensure a good crop in the coming year—and sometimes toasting the roots for good measure.

Wassail bowls are almost exclusively made from the guaiacum tree, indigenous to Brazil, which is commonly known as the "tree of life," or lignum vitae, and thought to possess medicinal qualities. The small urn on top of this lid contains a small, wooden drinking cup. The maker, "C T of Boton," who inscribed his name into the underside of the lid, has not been identified. Here, Boton is understood as the Norfolk town in the Hundred of South Erpingham. The parish registers for marriages between 1682 and 1723 are missing, so it is impossible to know whether this set was made to celebrate a wedding as the three initials, "W S A," would suggest.

PROVENANCE

bt. John H. Bryan of Lake Bluff, IL (1936–2018) from Sampson and Horne Antiques, London, June 24, 2008

LITERATURE

Ronald Hutton, *The Stations of the Sun: A History of the Ritual Year in Britain* (Oxford: Oxford University Press, 2001), 13, 46

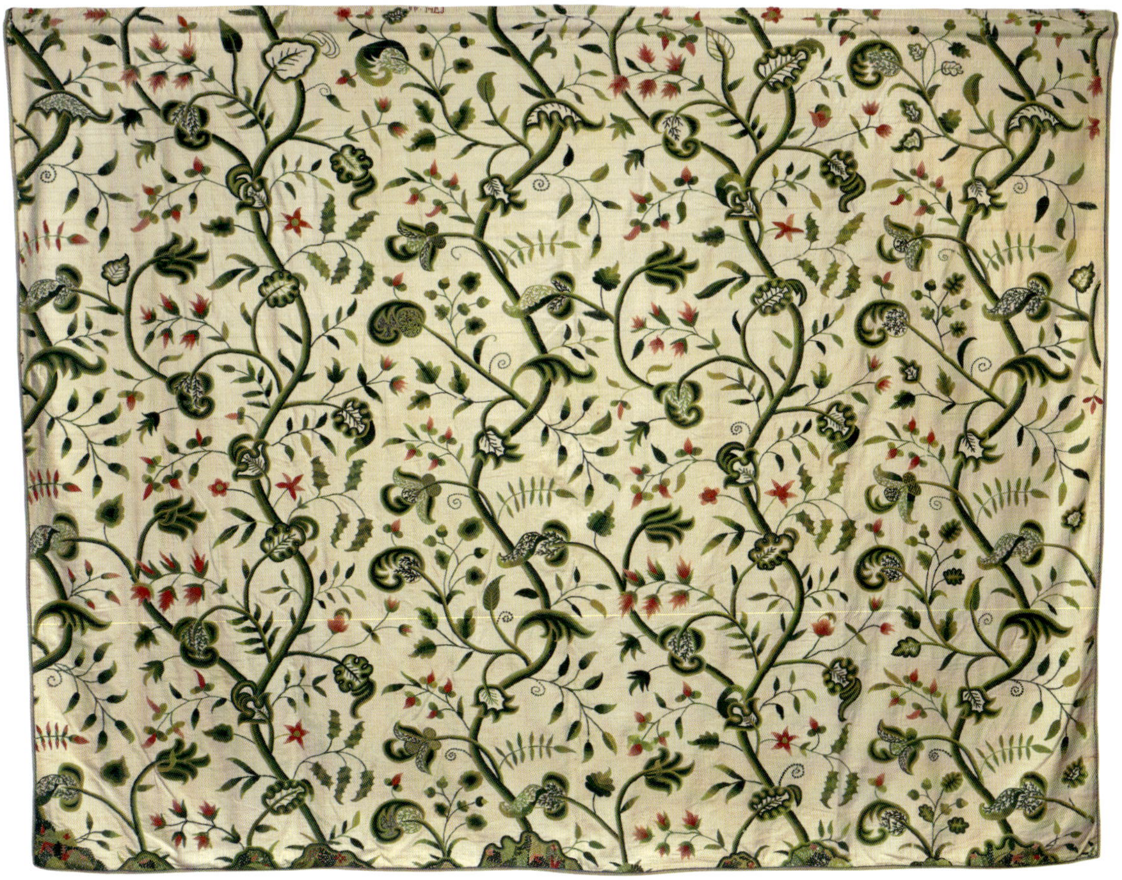

1723

Wall hanging

Cotton with wool thread

74 × 91¼ in. (188 × 231.8 cm)

Inscription: SV 1723

Like much of eighteenth-century crewelwork, this wall hanging drew inspiration from imported Indian textiles, similar to those likely made in the Coromandel Coast in the last decades of the seventeenth century. Examples formerly at Ashburnham House, Sussex, are now found at the Victoria and Albert Museum, London (IS. 156-1953); the Calico Museum of Textiles, Ahmedabad, India (324); the Museum of Fine Art, Boston (53, 2001); and the Cooper Hewitt, Smithsonian Design Museum, New York (1953.123.1). These textiles were made for the export market, which closed in 1721 when Britain outlawed all Indian imports. This wall hanging is derivative of the Indian precedent but employs a more regular, repeating, and sparse pattern. Similar compositions appear frequently on English crewelwork of the period.

PROVENANCE

Loaned by Cora Ginsburg, New York, to John H. Bryan of Lake Bluff, IL (1936–2018)

LITERATURE

Sylvia Houghteling, "Origins in Entanglement: Connections between English Crewel Embroidery and Indian Chintz," in *Cloth That Changed the World: The Coloured Cottons of India*, ed. Sarah Fee (New Haven, CT: Yale University Press, 2019), 183–89

Amelia Peck, *Interwoven Globe: The Worldwide Textile Trade, 1500–1800* (New Haven, CT: Yale University Press, 2013), 230–32

1621

Armchair
Oak, painted
$38\frac{5}{8} \times 20\frac{3}{8} \times 20\frac{1}{2}$ in.
(98.1 × 51.8 × 52.1 cm)

Inscription: 16 21

The coat of arms painted on this armchair belongs to Sir Richard Wilbraham Baronet (1579–1643), and his life dates are painted on the back panel. The paint has dulled to the point where the inscription of the date is only visible by X-ray. The chair itself is thought to date to the first two decades of the seventeenth century, probably decorated on the occasion of Wilbraham's ascension to the baronetcy on May 5, 1621. A squabble between the Wilbraham and Playters families—stemming from the similarity of their respective coats of arms (Wilbraham: argent three bends wavy azure; Playters: bendy wavy of six pieces argent and azure)—reached the Court of Chivalry in 1635 and resulted in a small change to the coat of arms on this chair, in accordance with the ruling by the earl marshall, head of the court. Presumably, the Playters had tired of seeing so many Wilbraham shields that too closely resembled those of their family and so they sought redress.

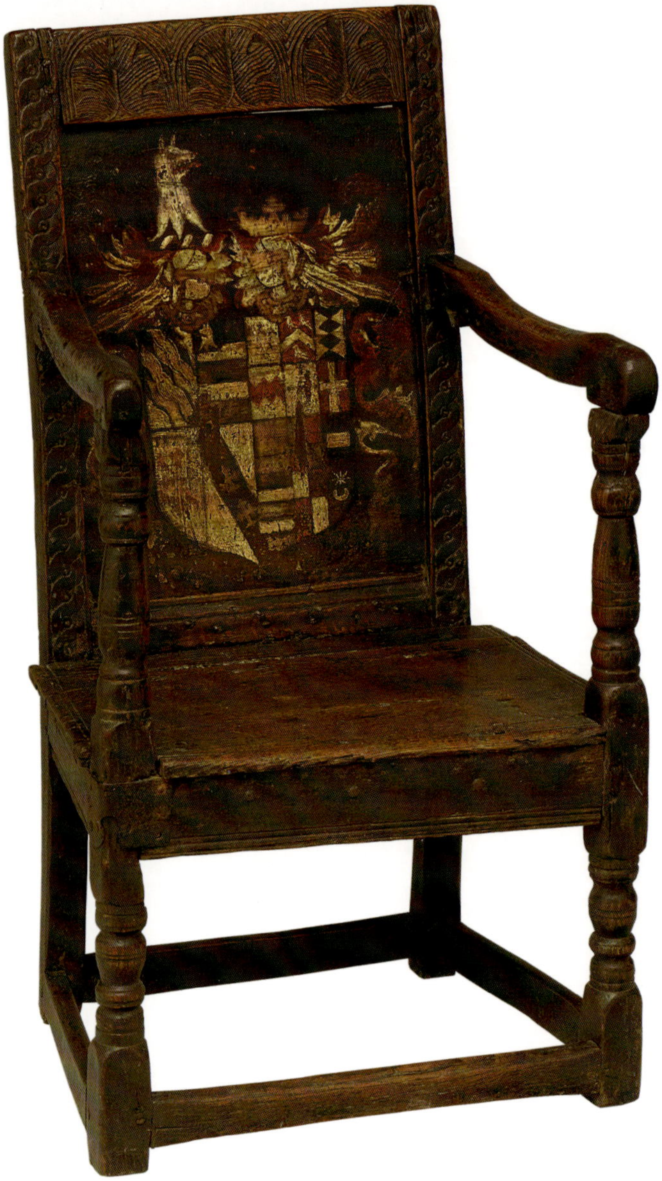

PROVENANCE
Dreweatts 1759, Donnington Priory, Berkshire, July 15, 2010 (lot 564); bt. John H. Bryan of Lake Bluff, IL (1936–2018) from Tobias Jellinek Antiques, London, October 14, 2010; bt. John H. Bryan III of Nashville, TN, September 5, 2013

LITERATURE
Adam Bowett, "An Early Seventeenth-Century Armchair from Cheshire," *Regional Furniture*, no. 28 (2014): 89–96

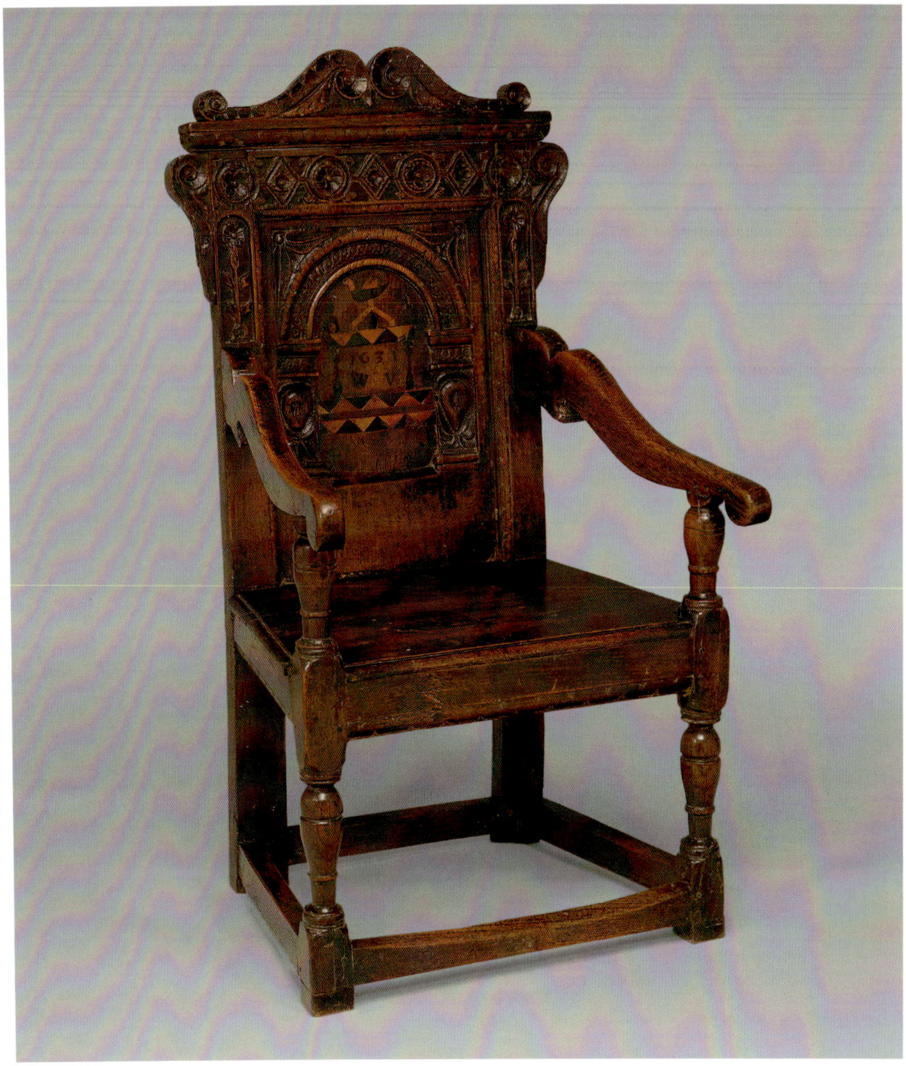

1631

Armchair

Oak, bog oak, and other woods

$45\frac{1}{4} \times 23\frac{7}{8} \times 24$ in. (114.9 × 60.6 × 61 cm)

Inscription: 1631; W V

The design and construction of this inlaid-panel-backed chair has led to the suggestion that it was made in one of the neighboring counties of Yorkshire or Derbyshire. The back panel is carved with a guilloche-filled arch; the pillars are carved with a highly stylized male face and a conforming female face, each haloed by a ruff. The simple checkered inlay at center depicts a building facade with columns, towers, and a gabled roof with a bird perched atop and the dates and initials "1631 W V" within. Portions of the inlay, originally made with bog-wood, have been replaced, as have the front seat board, the proper left ears, both back feet, and the back and side stretchers.

PROVENANCE

Alistair Sampson Antiques, London; Graham and Susan James; Bonhams, Oxford, May 13, 2015 (lot 108); bt. John H. Bryan III of Nashville, TN, through Tobias Jellinek Antiques, London

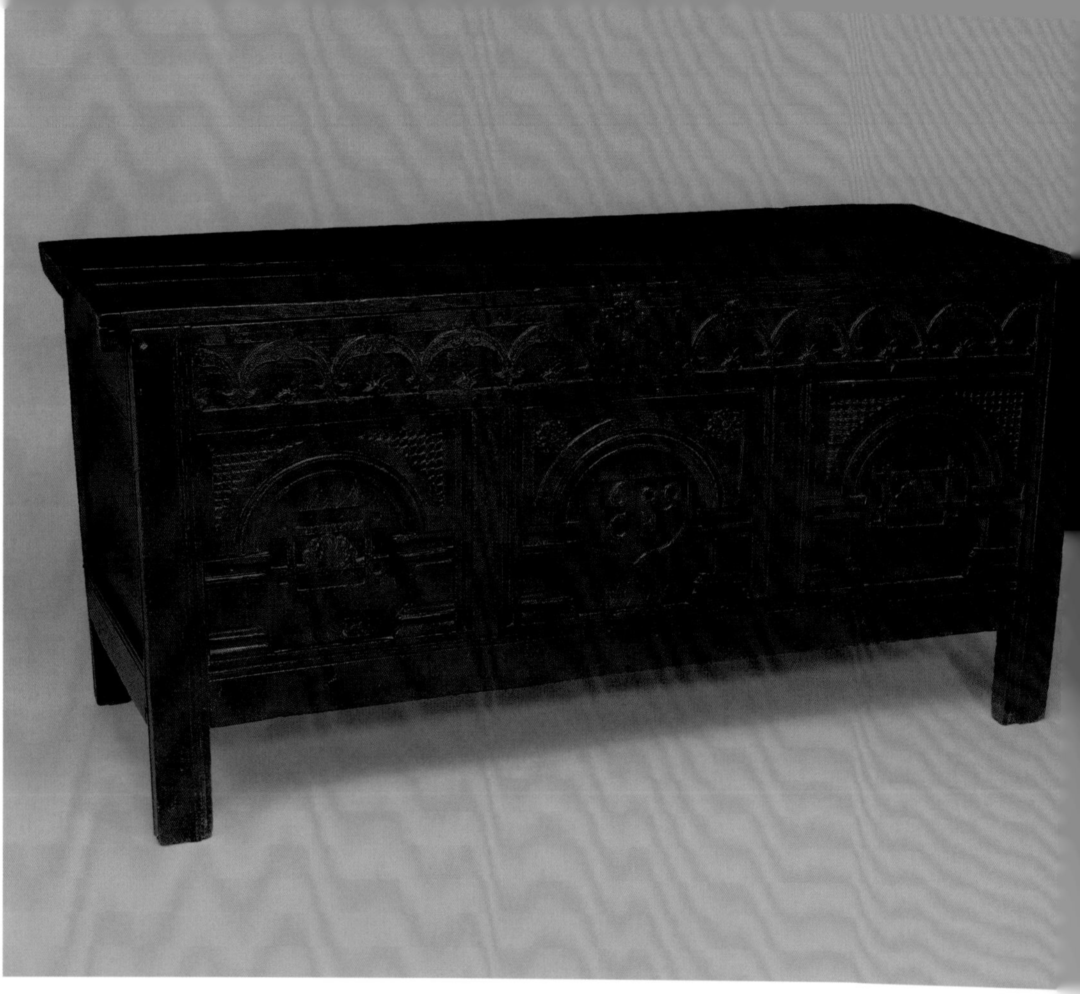

1609

Chest

Oak, iron

30½ × 60 × 23½ in.
(77.5 × 152.4 ×
59.7 cm)

Inscription: 1609|
C R

To judge from its construction and style of decoration, this joined oak chest, or coffer, was made in the West Country. It has been suggested that its plain, uncarved arcade columns and hatched arch section point to Taunton, Somerset, as the place of manufacture.

The joins of the boards of the triple-plank, end-cleated hinged top are patched with various strips of a ship's log, one of which is inscribed, "God Send us Safe (___) our Lord God 1720/21" (pages 44–45). These journals made a diurnal record of the voyage: the days of the week, the course, the distances, latitude by observation, latitude by judgment, differences of longitude, and voyage duration. The quest to accurately determine longitude while at sea was one of the greatest scientific endeavors of the eighteenth century, which many Western powers viewed as essential in establishing maritime supremacy.

"George Croker" is the only name that can be made out in these journal strips. He was probably the mariner of Fremington who was party to leases and indentures in 1742, 1752, and 1761. The names of the ships—the *John of Barnstaple* and the *Bideford*—indicate the chest was aboard a ship at one of these North Devon ports. Given that the neighboring port towns of Barnstaple and Bideford specialized in tobacco importation, it is likely that tobacco was the lading on the ship that carried this chest.

PROVENANCE

Bonhams, London, September 20, 2017 (lot 6); bt. John H. Bryan III of Nashville, TN, through Tobias Jellinek Antiques, London

LITERATURE

Gabriel Olive, "West Country Chests, Coffers and Boxes," *Regional Furniture* 4 (1990): 49–70

1602

Bell-metal mortars like this one are made of a composite metal similar to that used for church bells. It is a bronze alloy composed primarily of copper and tin in the proportion approximately of seventy-seven parts of copper to twenty-three parts of tin. Both bells and mortars were exported to be melted down and cast to make cannons, so the inscribed date went some way toward ensuring the provenance of the material.

Francis Howles, the bronze founder who made this mortar, is yet to be identified beyond name.

PROVENANCE

bt. John H. Bryan of Lake Bluff, IL (1936–2018) from Rupert Gentle, Pewsey, Wiltshire, May 12, 1998

LITERATURE

Michael Finlay, *English Decorated Bronze Mortars and Their Makers* (London: Plains Books, 2010), 63, no. 83

Mortar

Bronze

8 × 10 in. (20.3 × 25.4 cm)

Francis Howle

Inscription:
FRAVNCIS HOWLE MADE. MEE 1602

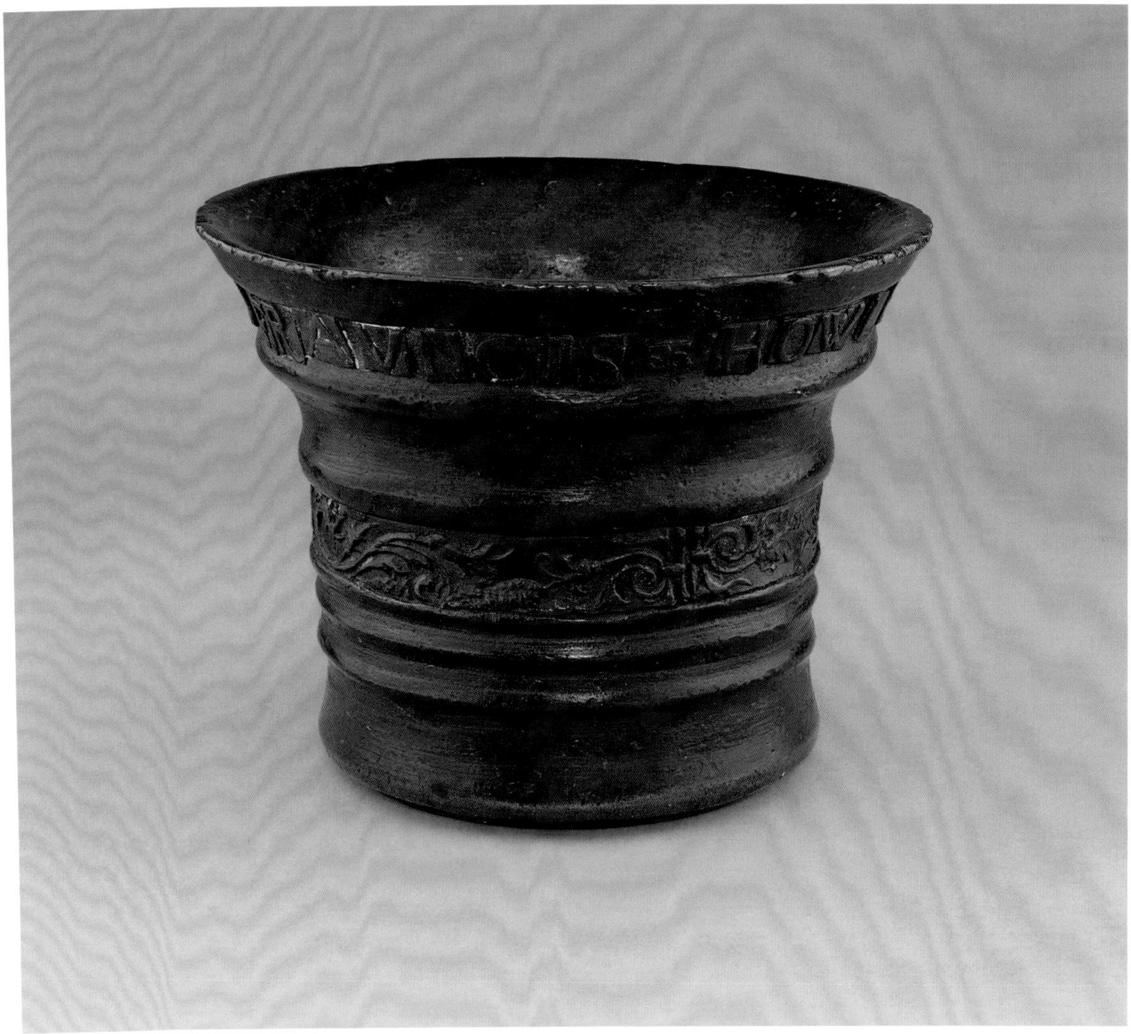

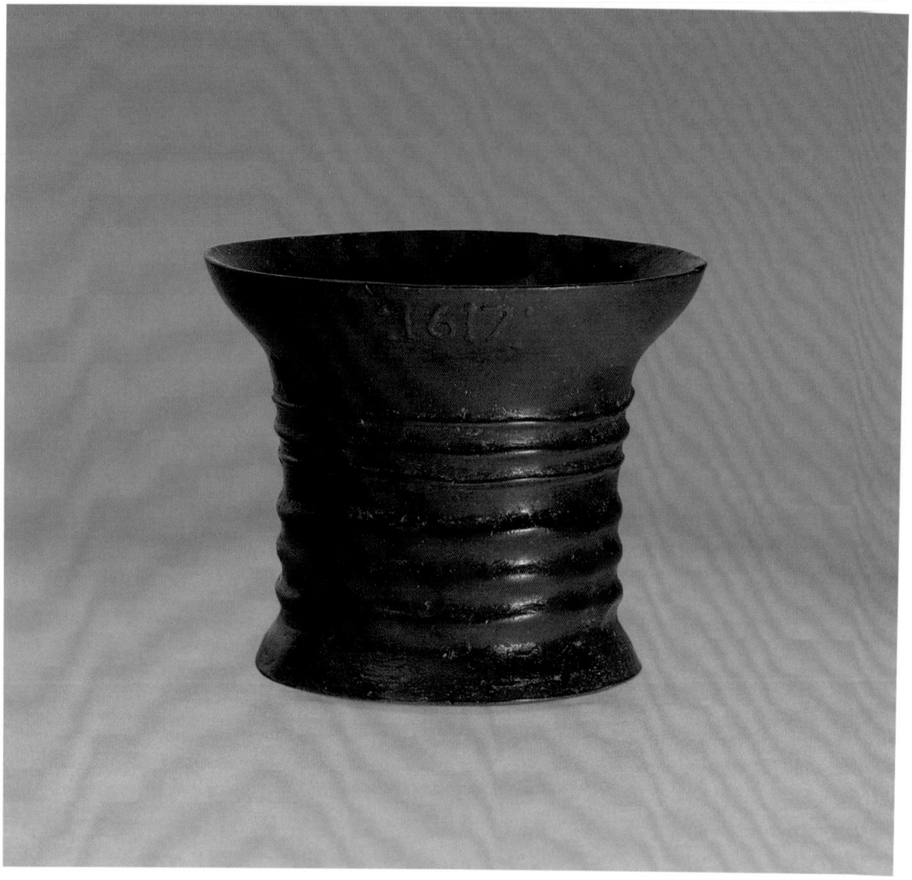

1617

Mortar

Bronze

$5\frac{1}{8} \times 4\frac{3}{8}$ in.
(13 × 11.1 cm)

Inscription: 1617

It has been suggested that, based on the style of the numerals, this mortar can be attributed to George Oldfield II of the Nottingham bell foundry on Long Row. This dynasty of founders included George Oldfield I (active 1537–1558), Thomas Oldfield, Henry Oldfield (1535–1615 or 1619), and George Oldfield II (1601–1680).

Various church bells of the East Midlands and beyond bear the mark, between the initials "H" and "O," of the cross on Calvary with a crescent star above (the star belonging to the town seal of Nottingham). George, son of Henry, used the mark of his father, replacing the "H" with a "G." George's mark appears on bells in Leicestershire from 1620 to 1673.

PROVENANCE

Woolley and Wallis, Salisbury, April 12, 2017 (lot 60); bt. John H. Bryan of Lake Bluff, IL (1936–2018)

LITERATURE

Michael Finlay, *English Decorated Bronze Mortars and Their Makers* (London: Plains Books, 2010)

Arthur G. Hemming, "Dated English Bell-Metal Mortars," *Connoisseur* 38 (1929): 158–66

1632

The owner of this formidably heavy mortar was Allen Talbott, a Grocer of St. Andrew Holborn, who, on December 4, 1628, married Mary Sawyer of St Vedast Foster Lane, the daughter of a wealthy Goldsmith. The couple, who lived near the Bell in Holborn, were armigerous, entering their pedigree to the Heralds' Visitation of London in 1633. The Grocers' Company of London traded in large quantities of spices, herbs, and minerals, all of which could be finely ground within this mortar. The maker, John Clifton, was the master of the Whitechapel Foundry from 1632 to 1640.

PROVENANCE

bt. John H. Bryan of Lake Bluff, IL (1936–2018) from Christopher Bangs, London, October 12, 1993

LITERATURE

Michael Finlay, *English Decorated Bronze Mortars and Their Makers* (London: Plains Books, 2010), 66, no. 87

Mortar

Bronze

$12\frac{1}{2} \times 15\frac{1}{2}$ in. (31.8 × 40 cm)

John Clifton (active 1632–1640)

Inscription: IC: MADE ME FOR ALLEN TALBOTT 1632

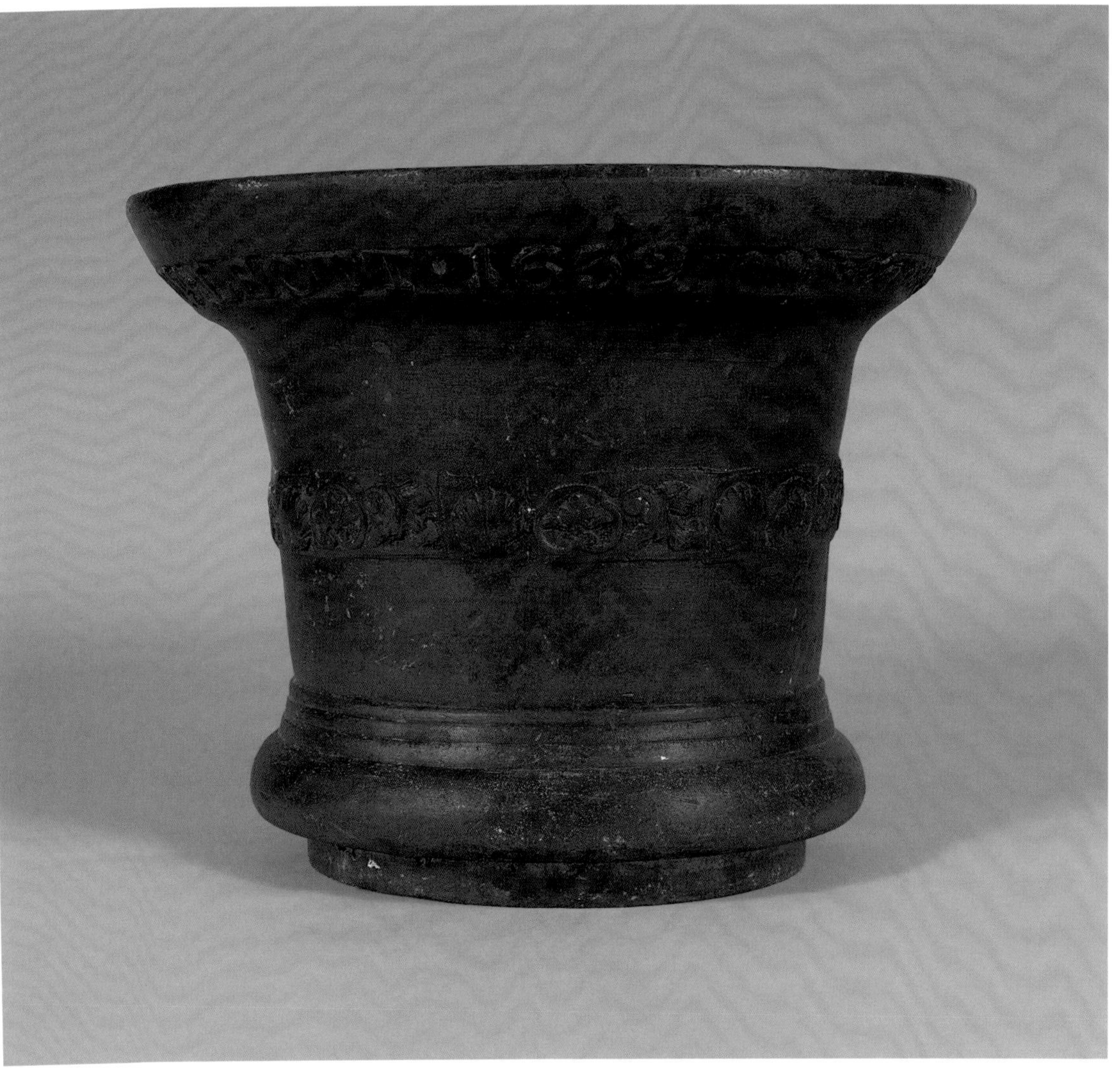

1637

Mortar
Bronze
$6\frac{3}{4} \times 4\frac{3}{8}$ in.
(17.1 × 11.1 cm)

Inscription: F|P|A; 1637

It has been suggested that there are Doncaster mortars that resemble this one, having similar letter sets and similar handles—the work of William Oldfield (active 1599–1652) and his son William (ca. 1607–1666/7). However, this foliate motif has been found on a mortar attributed to George Oldfield II and also on a bell dated 1649 at East Bridgeford, Nottinghamshire.

PROVENANCE
Bonhams, London, September 20, 2017 (lot 365); bt. John H. Bryan of Lake Bluff, IL (1936–2018)

LITERATURE
Michael Finlay, *English Decorated Bronze Mortars and Their Makers* (London: Plains Books, 2010), 96

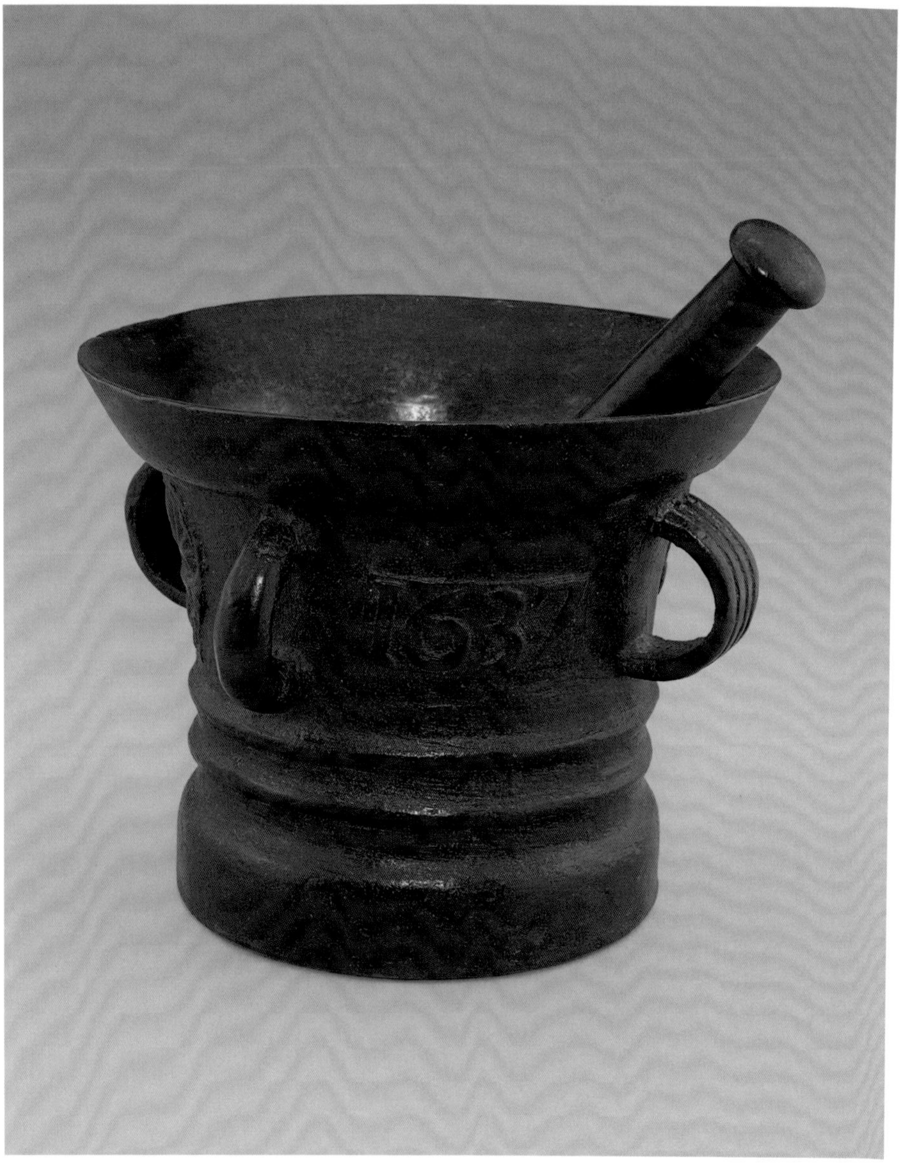

1667

"Hill" probably refers to the owner rather than to the maker. A Quaker named William Hill, who was buried in Bristol 1679, is a candidate, although the ubiquity of the name makes it impossible to be sure.

PROVENANCE

Michael Wakelin, Petworth, Sussex; bt. John H. Bryan of Lake Bluff, IL (1936–2018) from Lindsay Antiques, London, November 11, 1991

LITERATURE

Rupert Gentle and Rachel Field, *Domestic Metalwork: 1640–1820*, revised and enlarged by Belinda Gentle (Woodbridge, UK: Antiques Collectors' Club, 1994), 236

Mortar

Bronze

$5\frac{1}{2} \times 7\frac{1}{4}$ in.
(14 × 18.4 cm)

Inscription:
William Hill 1667
Bristoll

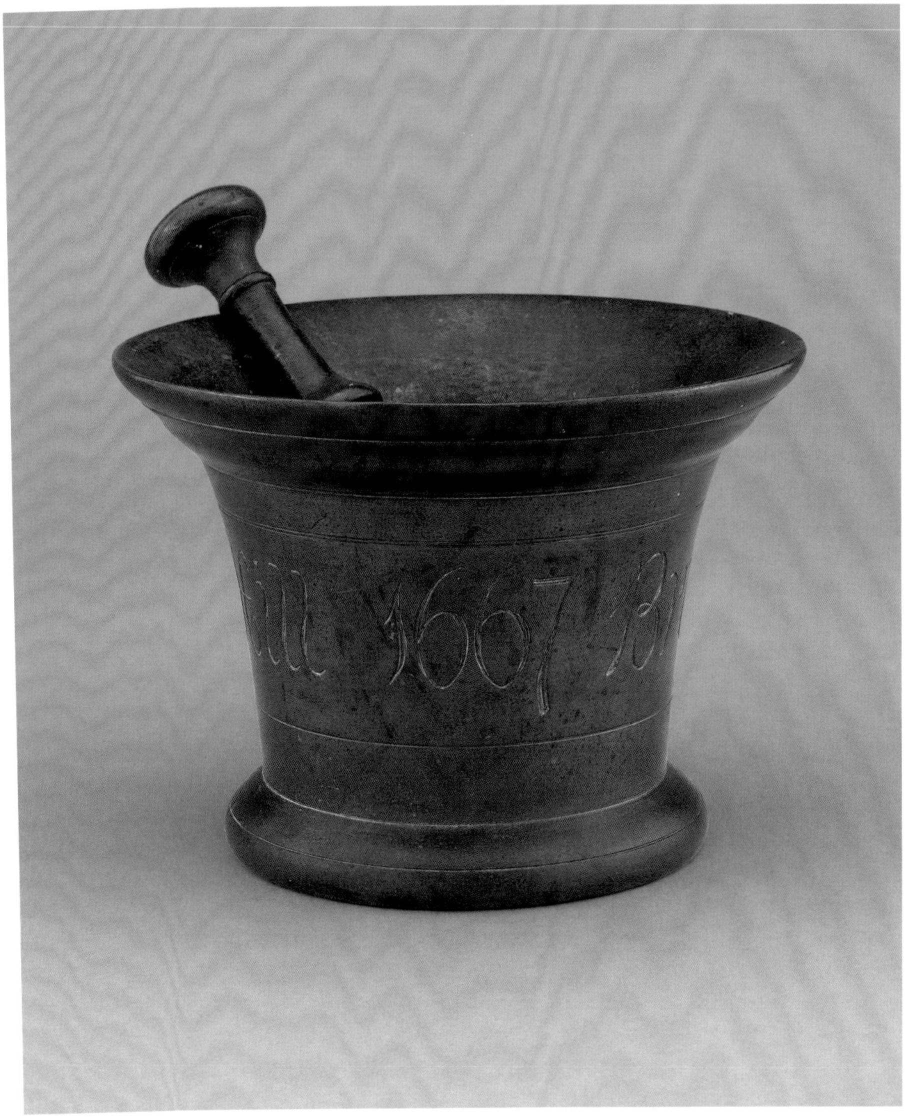

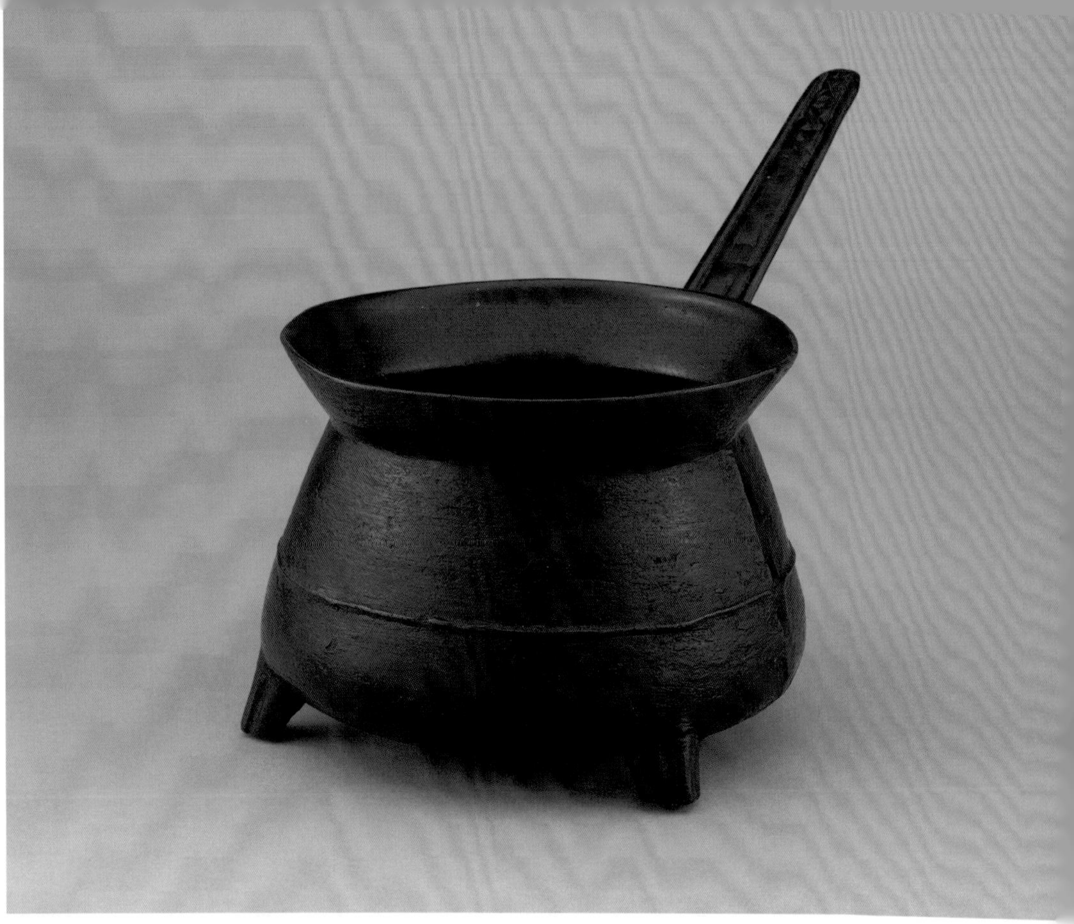

1592

Posnet
Bronze
$7\frac{1}{2} \times 7 \times 12\frac{1}{2}$ in.
(19.1 × 17.8 × 31.8 cm)

Inscription: 1592

The scratched mark of a pentagram on this posnet has led to the tentative attribution to the Birdall foundry on Cowick Street, Exeter. This foundry was run by three generations of the Birdall family between ca. 1525 and 1624. The parish registers of St Thomas, Exeter, record the burials in 1609 of Thomas Birdall, "ye brasier" (April 18), John Bridal, "ye elder" (November 6), and Margaret Birdall, wife of Thomas (November 1).

PROVENANCE
bt. John H. Bryan of Lake Bluff, IL (1936–2018) from Roderick Butler, Honiton, Devonshire, June 14, 2017

LITERATURE
Roderick Butler and Christopher Green, *English Bronze Cooking Vessels & Their Founders: 1350–1830* (Honiton, UK: Roderick and Valentine Butler, 2003)

1592

Skillet
Bronze
6 ¼ × 20 ¼ × 12 in.
(15.9 × 51.4 × 30.5 cm)

Inscription: HENRI APSLI ANNO DMI 1592

Typically, a skillet (a long-handled saucepan or cooking pot) had three integral legs to help it rest on the hearth. This example is attributed to Edmund Giles (active in Lewes, Sussex, 1592; d. 1614), who was also a bell founder. His bells bear dates that range from 1595 to 1614. Giles was buried at St Michael, Lewes, on February 27, 1613/14.

It is possible that the owner, whose name is cast across the skillet, was Henry Aspley of Ore, Sussex (1561–1605), member of Parliament for Steyning (1589) and Hastings (1593). In 1592, Aspley was involved in negotiating a local commons enclosure, an event that this pot perhaps commemorates—but the date more likely simply marks the acquisition of an object fine enough to display by the parlor or bedroom hearth, where it could be used to warm drink or broth without necessitating a trip to the kitchen.

PROVENANCE

Clive Sherwood (1929–2001), Sotheby's, Olympia, May 22, 2002 (lot 230); bt. John H. Bryan of Lake Bluff, IL (1936–2018) through Tobias Jellinek Antiques, London

LITERATURE

Roderick Butler and Christopher Green, *English Bronze Cooking Vessels & Their Founders: 1350–1830* (Honiton, UK: Roderick and Valentine Butler, 2003), 69

Amherst Daniel-Tyssen, "The Church Bells of Sussex," *Sussex Archaeological Society* 16 (1864): 138–232, 158–59

Christopher Green, "Crocks and Kettles: Documentary Records of Domestic Metalware," *Journal of the Antique Metalware Society* 23 (2015): 50–57

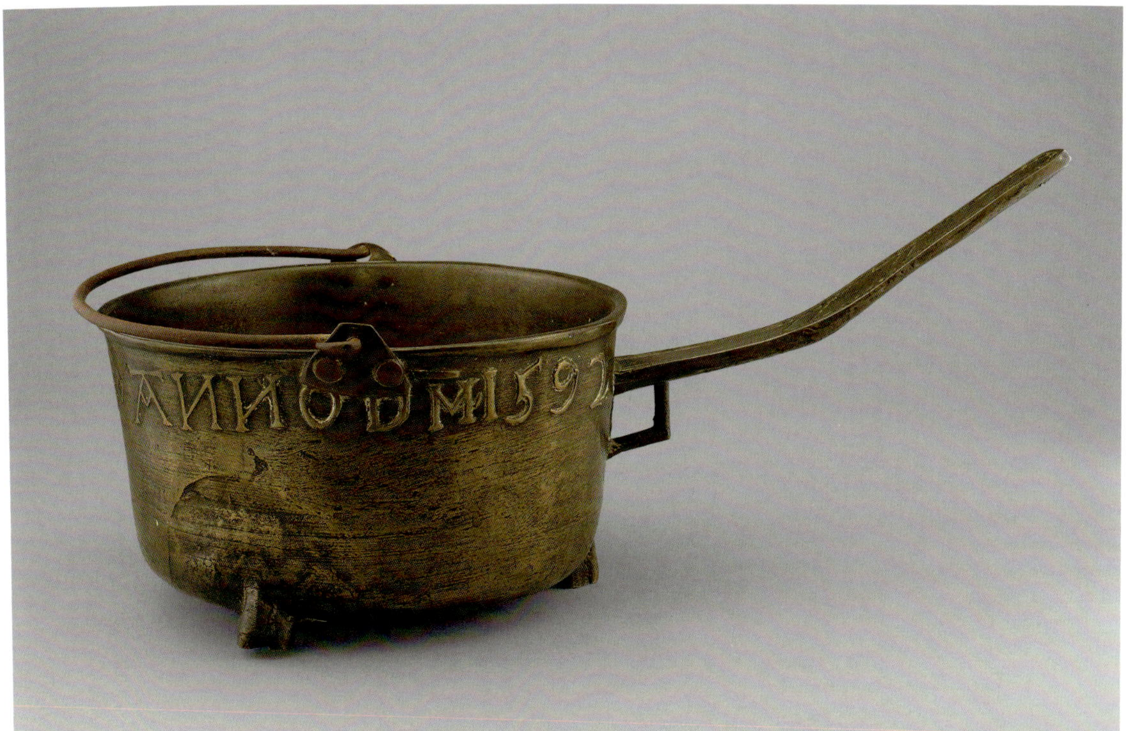

1658

Cauldron
Bronze
14¼ × 18 in. (36.2 × 45.7 cm)
Thomas Sturton I (active 1630–1658)

Inscription: MG 1658

Held aloft above the flames of the hearth by means of hanging lugs (or "ears"), this cauldron would have been used for boiling or stewing. It bears the mark of Thomas Sturton and his foundry at South Petherton, Somerset. Inscribed and dated vessels seem to have been a specialty of this foundry, which also made posnets, skillets, and bells. A very similar cauldron from the same foundry, dated 1657, bears an ownership triad of three initials. Various examples of Sturton's work can be found in the collection of the Museum of Somerset, Taunton.

PROVENANCE
bt. John H. Bryan of Lake Bluff, IL (1936–2018) from David Pickup Antiques, Cheltenham, Gloucestershire, June 15, 2017

LITERATURE
Roderick Butler and Christopher Green, *English Bronze Cooking Vessels & Their Founders: 1350–1830* (Honiton, UK: Roderick and Valentine Butler, 2003), 102, no. 85

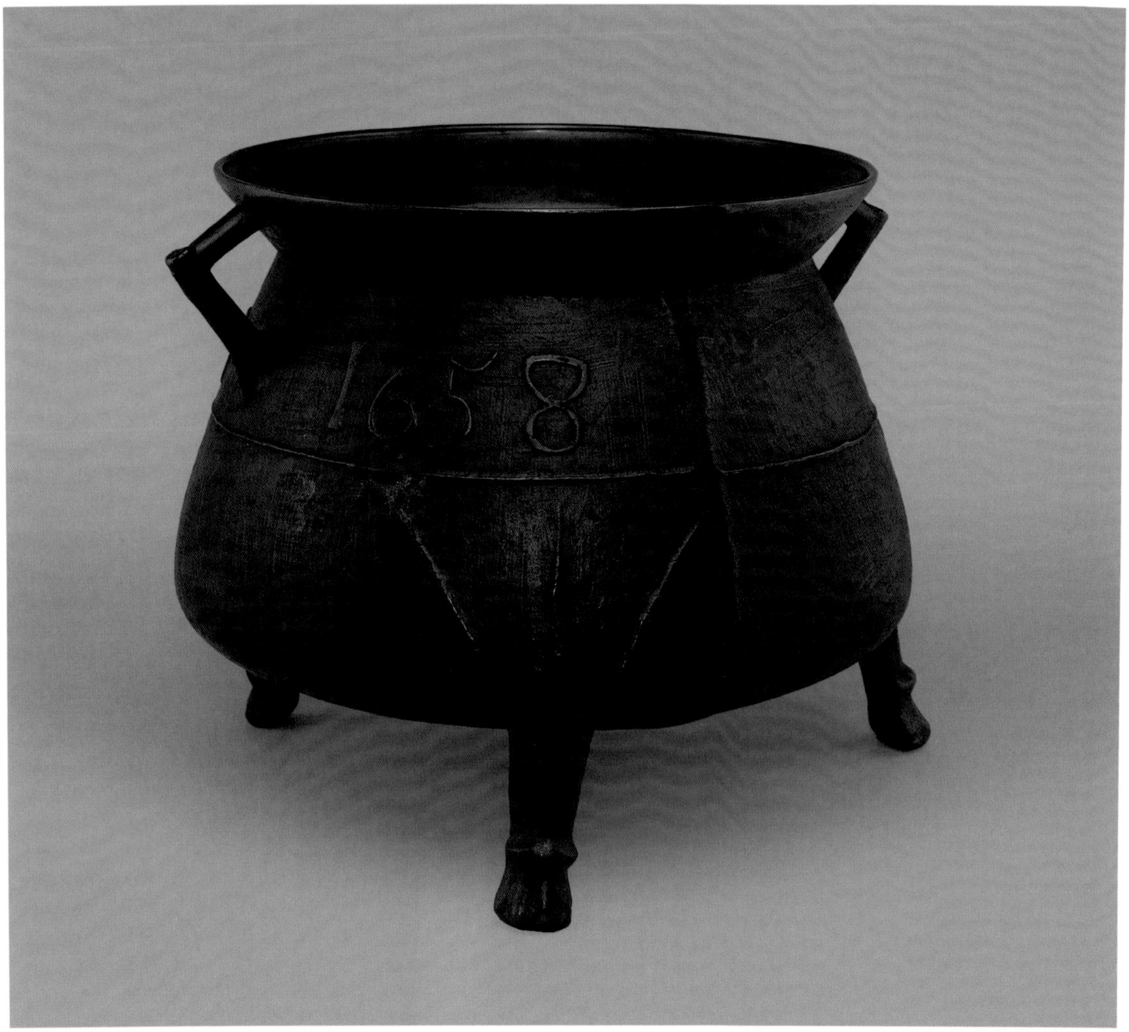

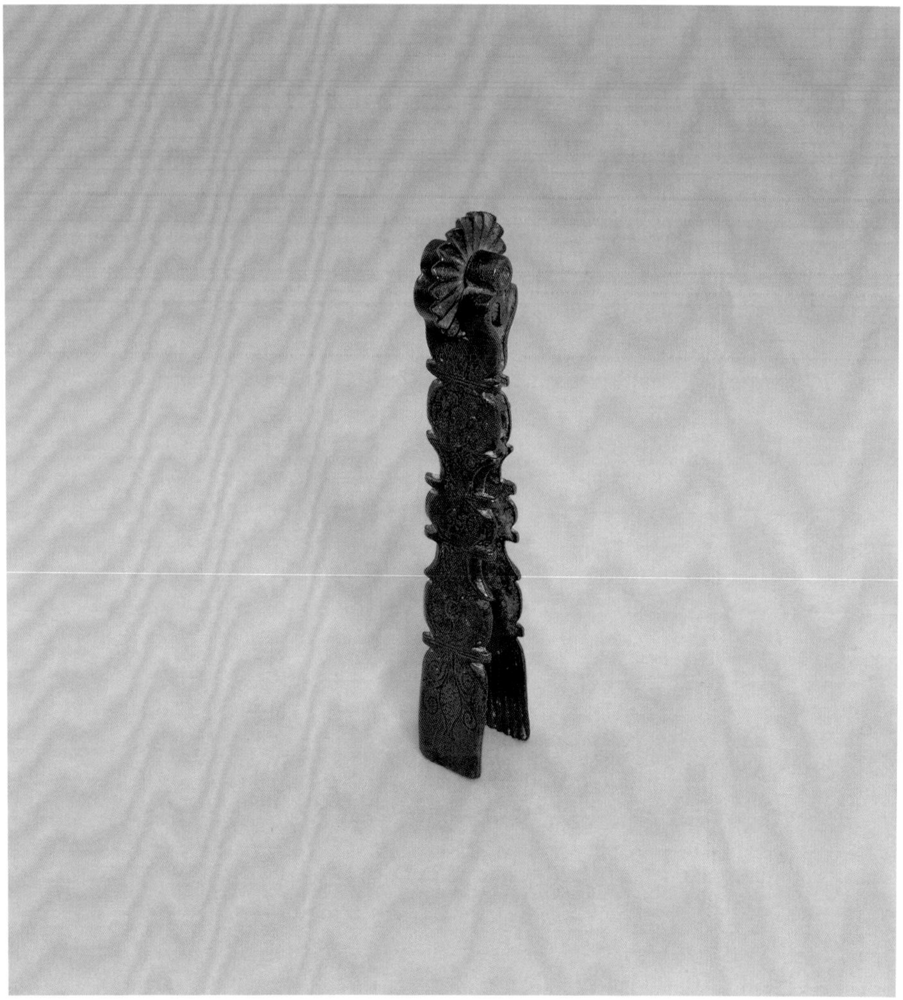

1656

Pastry jigger

Brass

4 × ¾ × ½ in. (10.2 × 1.9 × 1.27 cm)

Inscription: E. S.; 1656

In Britain, pies were—and remain—a popular form for both sweet and savory dishes. From mincemeat (spiced beef tongue, suet, and spices) to cucumbers, most foodstuffs could be encased in a full covering of pastry, then crimped on the edges with a jigger to seal the top. The pastry casing also helped prolong the shelf life of the pie and made it eminently portable. Pies were often given as gifts. They were also carried in ceremonial fashion on special occasions like weddings, when pies made with prodigious amounts of currants and pastry dough were featured in the marriage processions.

This jigger, recovered from the floorboards of an old home, has two sprung blades with faceted and shaped edges. At one end is a pair of tongs and at the other a serrated wheel. The style of punchwork decoration is very reminiscent of the work of Thomas Madin, a Sheffield maker whose work bears dates from 1652 to 1658. A nearly identical example, inscribed "1652," was sold at Bonhams, London, on March 15, 2017 (lot 207).

PROVENANCE

Warrington Collection; bt. John H. Bryan of Lake Bluff, IL (1936–2018) from Christopher Bangs, London, October 10, 1995

LITERATURE

Michael Finlay, *Pastry Jiggers and Pastry Prints* (London: Plains Books, 2014)

1776

Shaker or muffineer

Brass

4 × 2¾ × 2⅞ in. (10.2 × 7 × 7.3 cm)

John Dunkerton I (1723–1793) or John Dunkerton II (1753–1825)

Inscription: JN^{o} DUNKERTON; PILTON|1776

John Dunkerton belonged to a dynasty of blacksmiths who lived and worked in Pilton from the late seventeenth to early nineteenth century. At the head of this family was James Dunkerton (1698–1756), whose son John (1723–1793) was both a blacksmith and an innkeeper. This combination of roles was by no means uncommon: coaching inns needed a farrier on-site, and Pilton was on a well-traveled connecting road between Shepton Mallet and Glastonbury. In the next generation, the family enterprises were divided between John's two boys—James ran the inn, and John continued at the forge. This shaker, or muffineer, may have been used at the inn.

It is impossible to know whether this shaker was made by John Dunkerton or by his son John. The younger was born in 1753 and in 1773 married a woman named Sarah Higgins, three years his junior. Both died in 1825.

PROVENANCE

bt. John H. Bryan of Lake Bluff, IL (1936–2018) from Alistair Sampson Antiques, London, February 24, 1989

LITERATURE

Marcia Evans, *The Place of the Rural Blacksmith in Parish Life, 1500–1900* (Taunton, UK: Somerset and Dorset Family History Society, 1998), 40

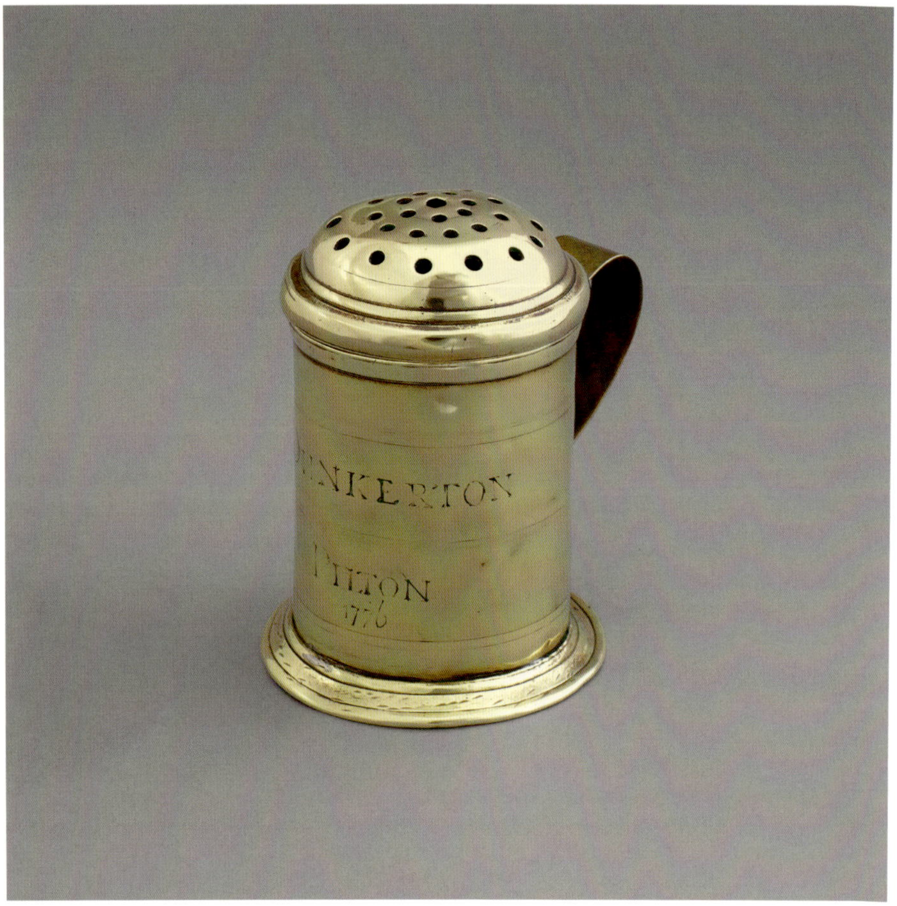

1768

Carving steel
Steel, ivory, silver
$14\frac{1}{4} \times 1\frac{1}{2}$ in. (36.2 × 3.8 cm)

Inscription: W^{M} MARRIMAN; 1 7 6 8

In the eighteenth century, as an important ceremonial component of proper dining, the host carved the meal's roasted meat at the table. Part of the theater included the blade's being sharpened on the steel beforehand and the fork's being impaled into the meat. The knife, fork, and carving steel would have been acquired as a set, although this steel is the only implement to have survived. It may have belonged to William Marriman, Esquire, who married Mary Batley on November 28, 1773, at St John Timberhill in Norwich.

PROVENANCE
bt. John H. Bryan of Lake Bluff, IL (1936–2018) from Lindsay Antiques, London, June 29, 1992

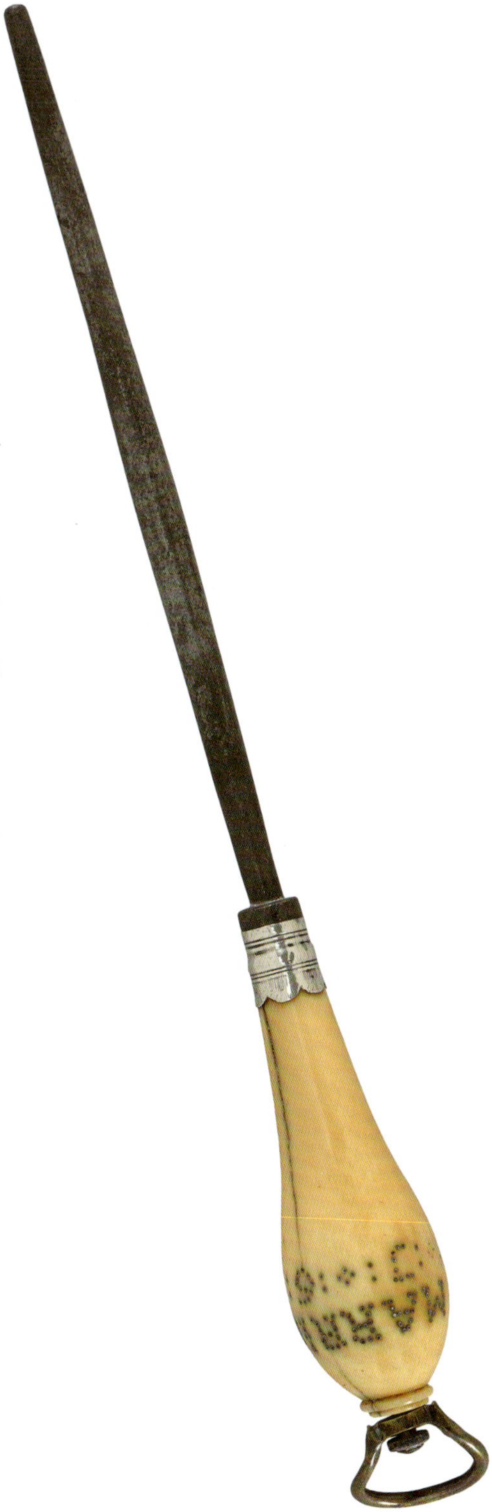

1658

Spoon
Brass
$10\frac{3}{8} \times 2 \times \frac{3}{4}$ in.
(26.4 × 5.1 × 1.9 cm)

Inscription: 1658

Too large for an eating utensil, this big spoon was presumably used for cooking. The inscription on the underside of the stem is badly worn, and the date can only just be made out.

PROVENANCE
bt. John H. Bryan of Lake Bluff, IL (1936–2018) from Rupert Gentle, Pewsey, Wiltshire, October 25, 1982

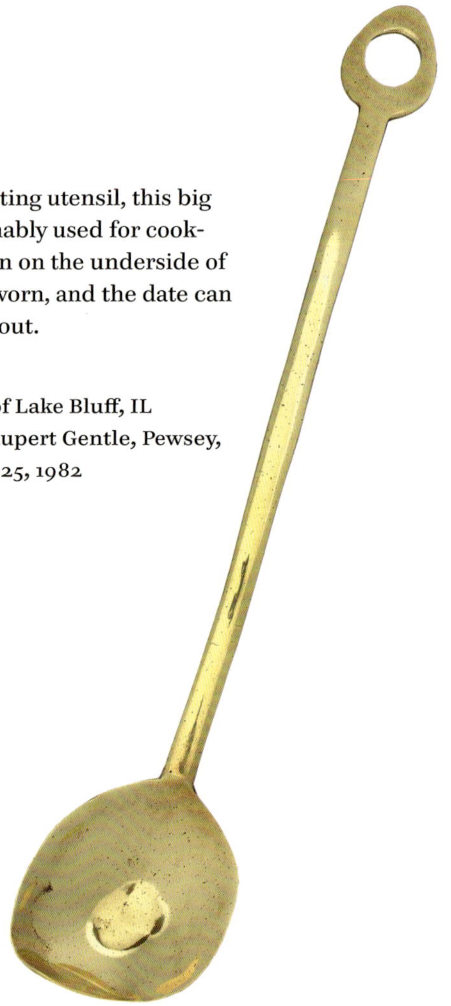

1683

Skimmer
Brass
$23\frac{1}{8} \times 7\frac{3}{4} \times 2\frac{1}{4}$ in.
(58.7 × 19.7 × 5.7 cm)

Inscription: F.R.S. 1683.

Also known as a "scummer," a skimmer lifted fat and solids from the surface of boiling soups in a pot; the clear broth would run back into the pot through the perforated blade. The implement might also be used to skim ripened cream from the settling pan at a dairy.

PROVENANCE
bt. John H. Bryan of Lake Bluff, IL (1936–2018) from Rupert Gentle, Pewsey, Wiltshire, November 3, 1996

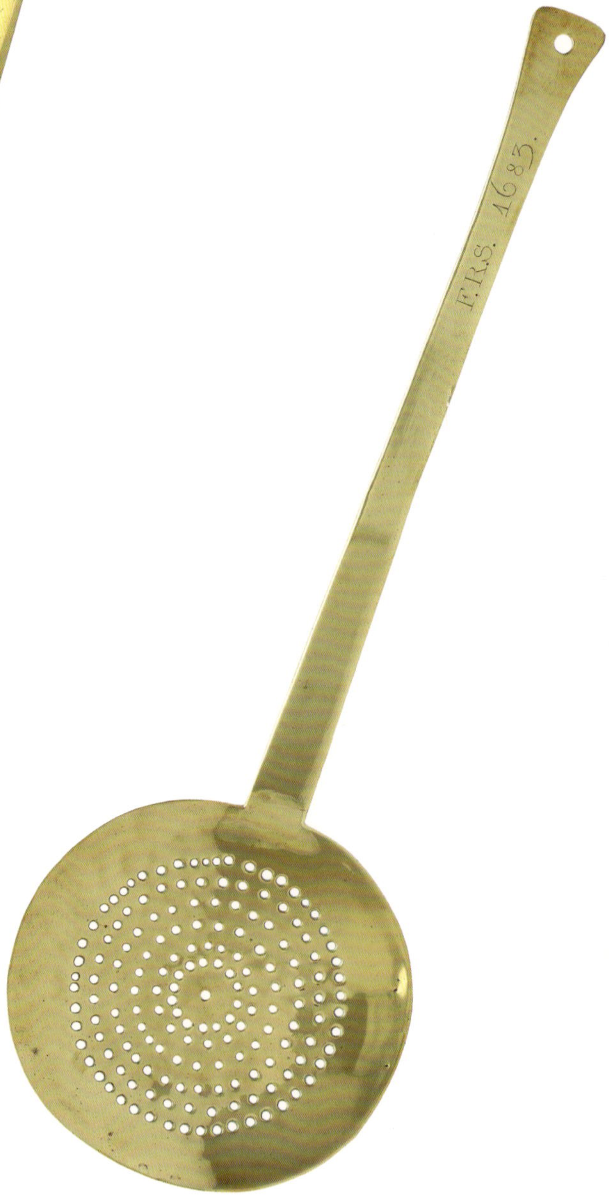

1701

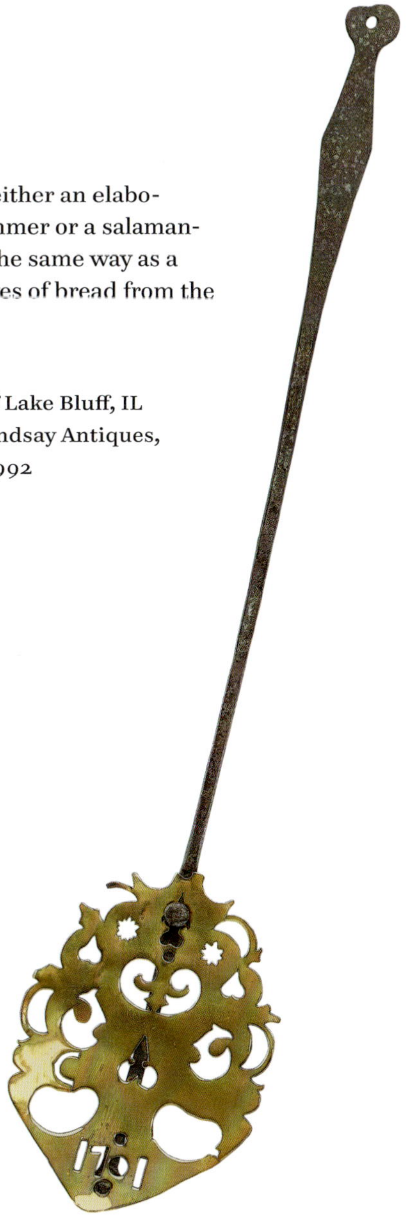

Skimmer or salamander
Brass, iron
21 × 4$\frac{1}{8}$ × $\frac{1}{4}$ in.
(53.3 × 10.5 × .64 cm)

Inscription: 1701

This implement is either an elaborately wrought skimmer or a salamander, a tool used in the same way as a peel to remove loaves of bread from the bake oven.

PROVENANCE
bt. John H. Bryan of Lake Bluff, IL (1936–2018) from Lindsay Antiques, London, August 7, 1992

1732

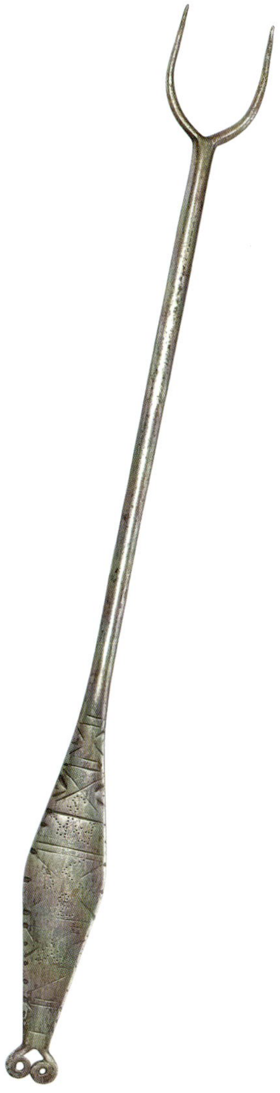

Fork
Steel
18$\frac{3}{4}$ × 1$\frac{3}{4}$ × $\frac{5}{8}$ in.
(47.6 × 4.5 × 1.59 cm)

Inscription: F|R|S; 1732

The heft and durability of this long fork suggests that it might have been used to remove meat from a boiling cauldron rather than simply to toast items in the hearth.

PROVENANCE
bt. John H. Bryan of Lake Bluff, IL (1936–2018) from Tobias Jellinek Antiques, London, October 18, 2017

1687

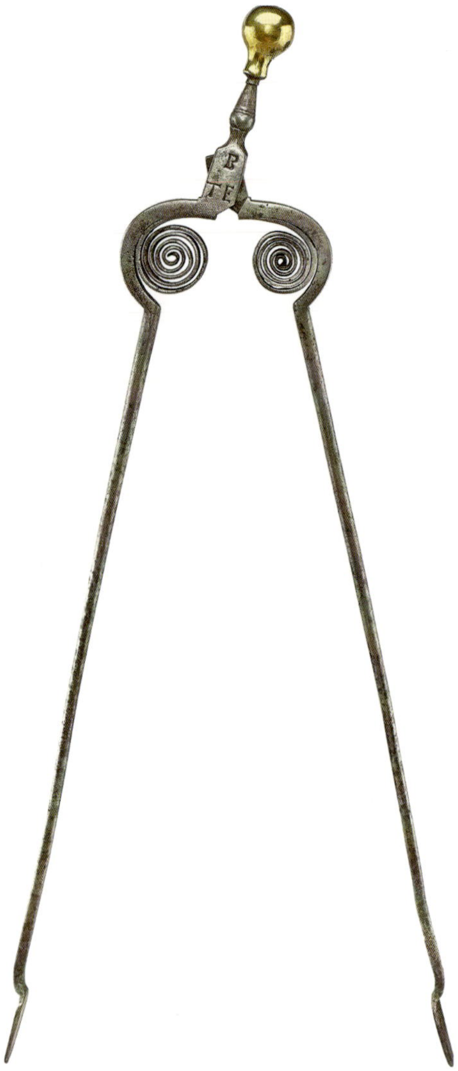

Fire tongs
Steel, brass
$26\frac{1}{4} \times 3\frac{7}{8} \times 1\frac{1}{8}$ in.
(66.7 × 9.8 ×
2.9 cm)

Inscription: T|B|E;
1687

Tongs like these would have been used to lift and position coals and embers within a fire or to transport them into portable braziers, dish warmers, and kettle stands. This implement belonged to a couple whose initials are too common to be discoverable.

PROVENANCE
bt. John H. Bryan of Lake Bluff, IL (1936–2018) from Tobias Jellinek, London, September 18, 2002

1698

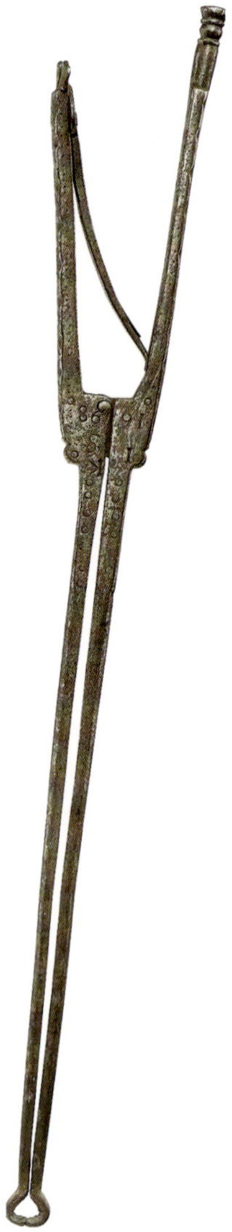

Ember tongs
Steel
$17\frac{1}{4} \times 2\frac{1}{4} \times 1$ in.
(43.8 × 5.7 × 2.5 cm)

Inscription: I|K|S|
16 98; P H|16 98

PROVENANCE
Bonhams, London, March 15, 2017 (lot 205); bt. John H. Bryan of Lake Bluff, IL (1936–2018) through Tobias Jellinek Antiques, London

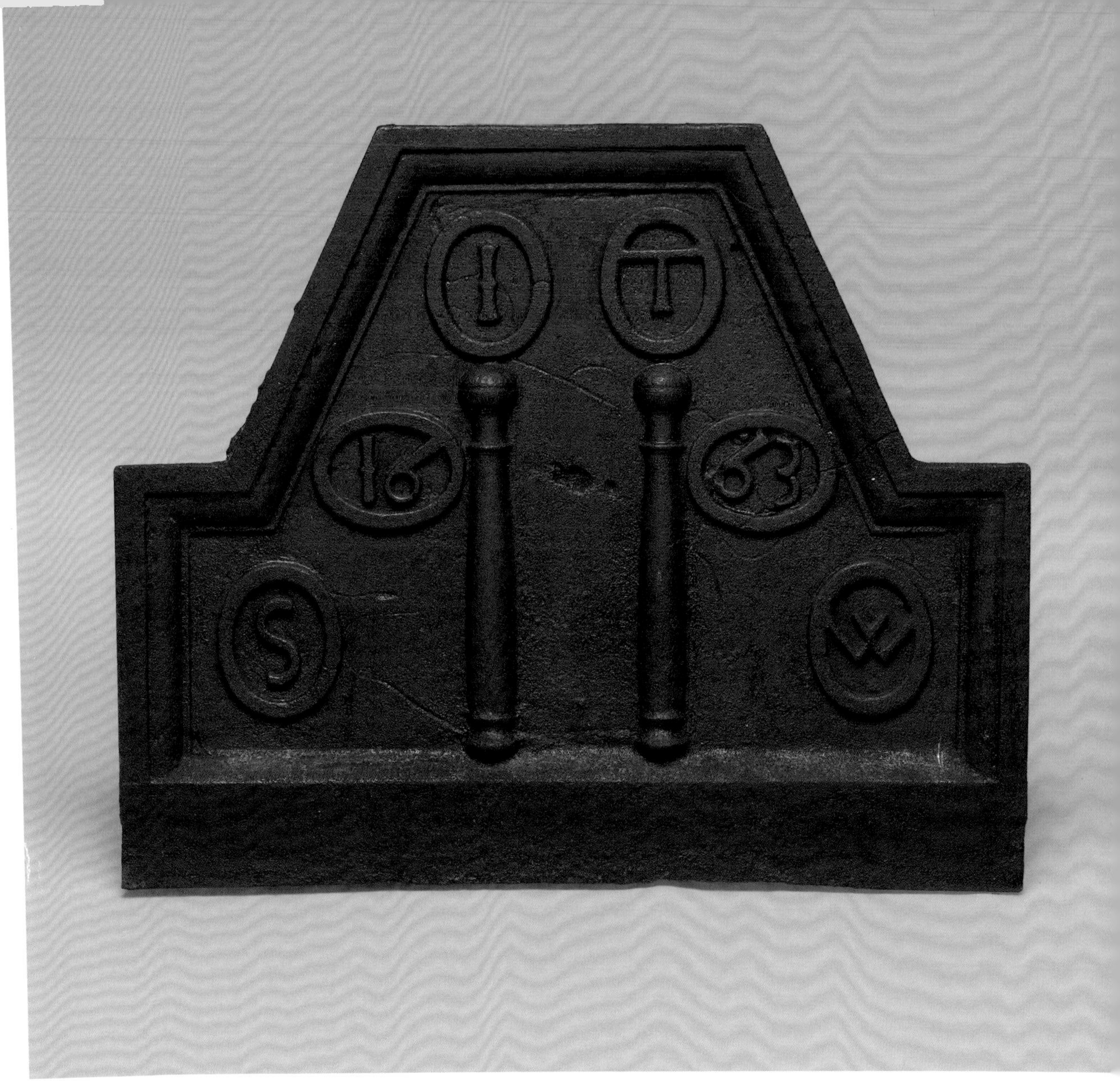

1663

Fireback

Iron

27 × 31½ × 1⅛ in. (68.6 × 80 × 2.9 cm)

Inscription: I T; 16 63; S W

This fireback came out of a house called Skipreed, a timber-framed structure probably of the sixteenth century. The house was located in Chiddingstone, Kent, close to the Sussex border, at one time the heart of the Wealdon iron industry. The decorative shapes, which imitate wool spindles, associate this fireback with the area around Cuckfield in Sussex, but its unusual form makes it an outlier.

PROVENANCE

bt. John H. Bryan of Lake Bluff, IL (1936–2018) from Alistair Sampson Antiques, London, February 24, 1989

LITERATURE

Jeremy Hodgkinson, "Pre-Restoration Iron Firebacks," *Journal of the Antique Metalware Society* 20 (2012): 2–15

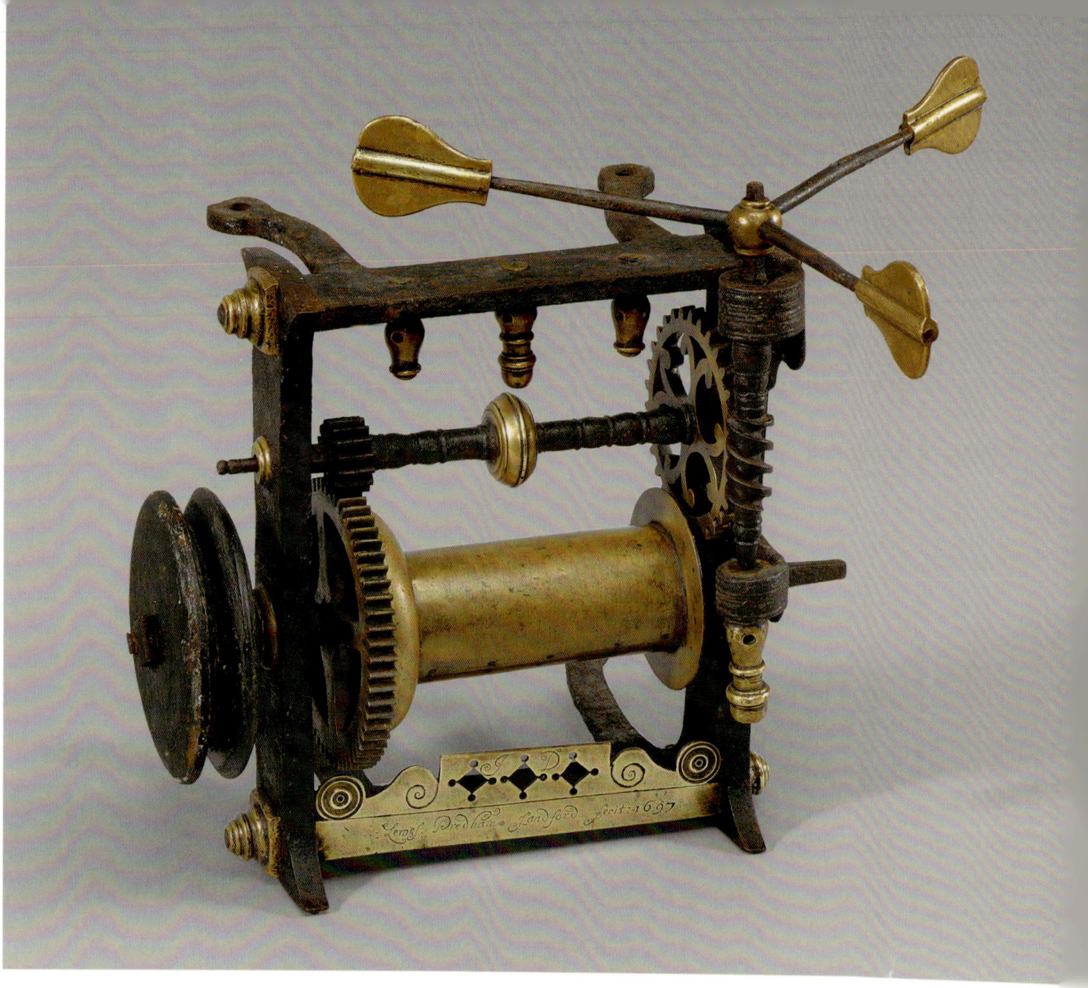

1697

Spit jack

Brass, iron, wood

15½ × 18 × 16 in. (39.4 × 45.7 × 40.6 cm)

Lewes Predham (active 1697; d. 1749)

Inscription: J D | Lewes Predham Sandford fecit: 1697

On the hearth, meat and stews cooked more evenly if all their surfaces faced the fire in turn, which necessitated rotating the food on a spit. Turning such a device by hand, however, took considerable time and effort. Mechanical clockwork alternatives like this one eliminated the manual work. Driven by a weight that lowered over time, its descent slowed by the three-finned regulator (or flywheel) above the device, this jack would have turned the spit by itself, demanding only periodic attention from the cook. This jack was made by the clockmaker Lewes Predham, or Pridham (d. 1749), of Sandford, Devon. Predham made clocks for the churches of All Saints, Dulverton (1708); St Michael, Minhead (1725); and the church at Bampton (1728). The latter two have been replaced; the clock at Dulverton is still in use and is wound each day. A lantern clock by Predham is at the Royal Ontario Museum, Toronto.

PROVENANCE

bt. John H. Bryan of Lake Bluff, IL (1936–2018) from Alistair Sampson Antiques Ltd., London, June 10, 1993

LITERATURE

Jack Kenneth Bellchambers, *Somerset Clockmakers* (London: Antiquarian Horological Society, 1968)

Jeanne Schinto, "The Clockwork Roasting Jack, or How Technology Entered the Kitchen," *Gastronomica* 4, no. 1 (2004): 33–40

Tony Weston, "English Roasting Jacks, Part 1: Weight-Driven Spit Jacks of the Seventeenth Century," *Journal of the Antique Metalware Society* 17 (2009): 14–26

1701 | 1781

Bellows
Elm, leather, iron
23¼ × 11¼ × 4½ in.
(59.1 × 28.6 × 11.4 cm)

Inscription:
The Ward of Aldersgate | Giuein by The Inquest | m^{r} Jacob Sutton Foreman | 1701; Repaird 1781 | Edw Kirby Foreman

This pair of bellows was first presented in 1701 to an inquest jury in the London ward of Aldersgate by their elected foreman, Jacob Sutton. The jury comprised twelve local men whose role it was to investigate and take action against petty lawbreakers in the ward. Juries were well known for doing themselves rather well, courtesy of the ratepayers, during their "ward mote" meetings. This behavior drove the Court of Common Council to issue Orders in 1694, and annually thereafter, to curb the "Extravagant Expenses at the Meeting of their Wardmote-Inquests, in Eating and Drinking, and also sending abroad Wine and Ale to the Inhabitants of the several Wards." Instead, inquests were to adjourn so that "Inquestmen may refresh themselves at their own houses" and at their own expense. Whether these bellows warmed the jurymen at one of their own houses (or perhaps the foreman's) or in a public meeting place, it is impossible to say, but the fact that they were mended and presented again by another foreman eighty years later suggests a prolonged history of making jurymen comfortable.

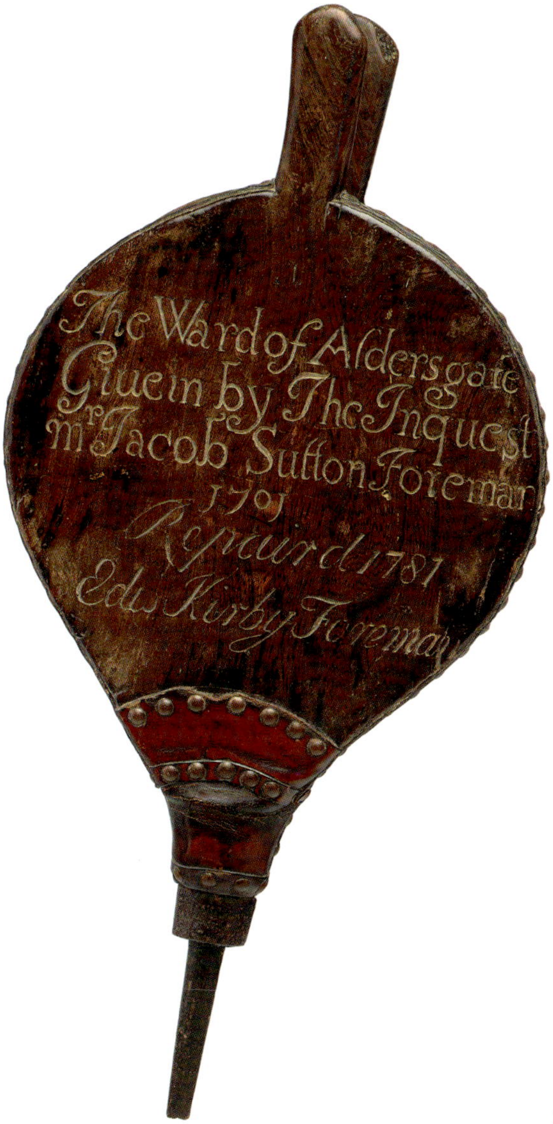

Jacob Sutton the Elder, who in his will described himself as "Brass ffounder Citizen and Loriner of London," was born on July 18, 1671 (TNA PROB 11/672/36). He gained his freedom from apprenticeship in 1696. His son, Jacob, was made free in 1713 and took over part of the business. Father and son lived within the vicinity of Loriners' Hall on London Wall. The word *loriner*, which is derived from the Latin *lorum*, means strap or harness, so the two men would have made metal components for bridles, harnesses, and spurs. Jacob the Elder was buried in the vault of St Anne and Agnes, Aldersgate, London, on June 18, 1734. Edward Kirby, the other foreman whose name is inscribed on the bellows, was a resident of Aldersgate Street; he was buried on January 19, 1814, at the impressive age of ninety-two.

PROVENANCE
bt. John H. Bryan of Lake Bluff, IL (1936–2018) from Alistair Sampson Antiques Ltd., London, March 31, 1992

1615

Warming pan

Brass, iron

$43\frac{1}{4} \times 12\frac{1}{4} \times 4$ in. (109.9 × 31.1 × 10.2 cm)

Inscription: THE : RIGHT : HON : THE : EARLE : OF : ESSIX : HIS : ARMES; 1615

Made for a man whose bed was likely never warm, this brass-and-iron pan would have been filled with red embers picked out of the fire. The hinged lid opens so the embers can be scooped into the bowl. After the lid was closed, the pan would be swiftly and skillfully passed up and down the length of the bed, under the bedcovers, its surface polished so as not to scratch the sheets. Once the pan had cooled sufficiently, it could be left in the bed to warm it further.

The lid on this warming pan was decorated with repoussé work depicting a reindeer, part of the Devereux family's coat of arms. In 1615, this family was headed by Robert Devereux, third Earl of Essex (1591–1646), who was then not yet twenty-five but was well on his way toward a life of misery. Twice cuckolded, frequently poisoned, and perpetually humiliated, Devereux was an angry, insecure man who was obsessed with reclaiming the honor his family had lost after his father was executed for treason in 1601. In 1604, James I restored Devereux's titles and made him a page to Prince Henry. The two men are depicted in a large portrait of about 1605 by Robert Peake in the Royal Collection, London (RCIN 404440).

Devereux's first wife was the much-admired court beauty Frances Howard (1590–1632), daughter of the Earl of Suffolk. He was fourteen, and she only eight months older. At about this time, he suffered what was a severe and surely

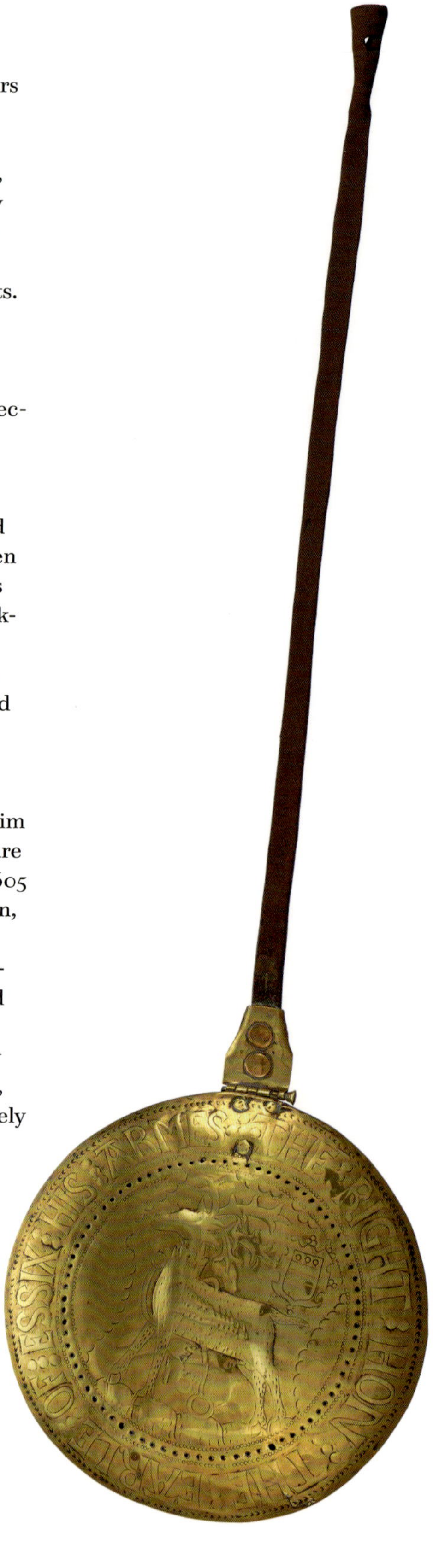

disfiguring bout of smallpox. Fallings-out with Prince Henry in 1610 and 1611 finally led Devereux to strike the prince with a tennis racquet, drawing blood. At Chartley, Staffordshire, Devereux's wife made her first attempt to poison him. In 1613, her family persuaded the king to establish a commission to grant Frances an annulment of the marriage, citing her husband's impotence (which was either physical or emotional) as the ground for separation, granted in September 1613. Devereux returned to Chartley and five years later to London to observe the trial and conviction of his former wife and her new husband, the Earl of Somerset, who had murdered Sir Thomas Overbury, finishing him off with a poisoned enema. Devereux went on to become the first captain general of the Parliamentary army, a role in which he served for thirty months and which he fulfilled with characteristic mediocrity. He died without an heir, and thus his title became extinct and his estates divided among his sisters and cousins.

By the nineteenth century, the warming pan was at Cassiobury Park, Hertfordshire, the seat of the Capel-Coningsby family, created Earls of Essex in 1661. It is not clear how this object arrived there, but it is possible it was acquired by George Capel-Coningsby, fifth Earl of Essex (1757–1839), antiquary, to whom the inscription would have appealed.

In the collection of the Victoria and Albert Museum, there is a punch-decorated lid with the arms of Devereux, dated 1630 (314-1880). There is also a similar warming pan, dated 1622, with the inscription, "Will searve ye Lord and my Hows." This object is illustrated in Gentle and Field.

PROVENANCE

Possibly acquired by George Capel-Coningsby, fifth Earl of Essex of Cassiobury, Hertfordshire (1757–1839); by descent to Adele Grant, Countess of Essex (d. 1922); her sale, Knight, Frank and Rutley at Cassiobury Park, Hertfordshire, 1922 (lot 619); bt. John H. Bryan of Lake Bluff, IL (1936–2018) from Christopher Clarke, Stow-on-the-Wold, Gloucestershire, May 27, 1992

LITERATURE

Rupert Gentle and Rachel Field, *Domestic Metalwork: 1640–1820*, revised and enlarged by Belinda Gentle (Woodbridge, UK: Antiques Collectors' Club, 1994), 377, no. 1.

G. B. Hughes, "Old English Warming Pans," *Country Life*, November 18, 1949, 1502–4.

Angela McShane, "Subjects and Objects: Material Expressions of Love and Loyalty in Seventeenth-Century England," *Journal of British Studies* 48, no. 4 (October 2009): 871–86.

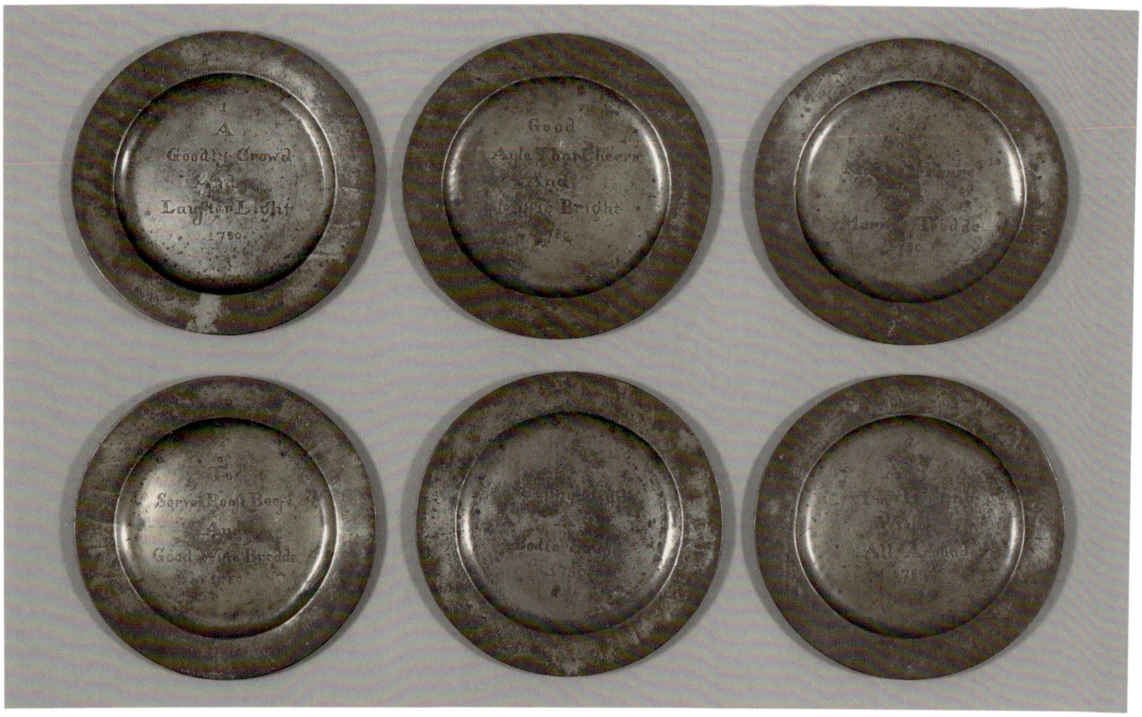

1750

Plates

Pewter

Each: $9\frac{1}{4} \times \frac{1}{2}$ in. (23.5 × 1.27 cm)

Charles Buckley Maxey (active 1750–1757)

Inscriptions:
1|A|Goodly Crowd|And|Laufter Light|1750

2|Good|Ayle That Cheers|And|Music Bright|1750

3|Some|Maidens Fayre|With|Merrie Tredde|1750

4|To|Serve Roast Beefe|And|Good Wyte Bredde|1750

5|A|Happie Mind|A|Bodie Sounde|1750

6|And|I'm Content|With|All Arounde|1750:

Merryman plates typically came in sets of six and included a light-hearted, six-line verse to cheer dinner guests as they enjoyed the hospitality of their host. This set bears the touchmark of a pelican vulning herself and standing on a globe, issued to Charles Buckley/Puckle Maxey in 1750. Charles Buckley Maxey, who shared his father's name, was apprenticed to Richard King, Pewterer, in 1743, which makes this set of plates one of the first of his commissions as an independent craftsman.

PROVENANCE

bt. John H. Bryan of Lake Bluff, IL (1936–2018) from Leo van Dongen in or before 1984

LITERATURE

Howard Herschel Cotterell, *Old Pewter, Its Makers and Marks in England, Scotland and Ireland: An Account of the Old Pewterer & His Craft* (repr., Rutland, VT: Charles Tuttle, 1963), 225

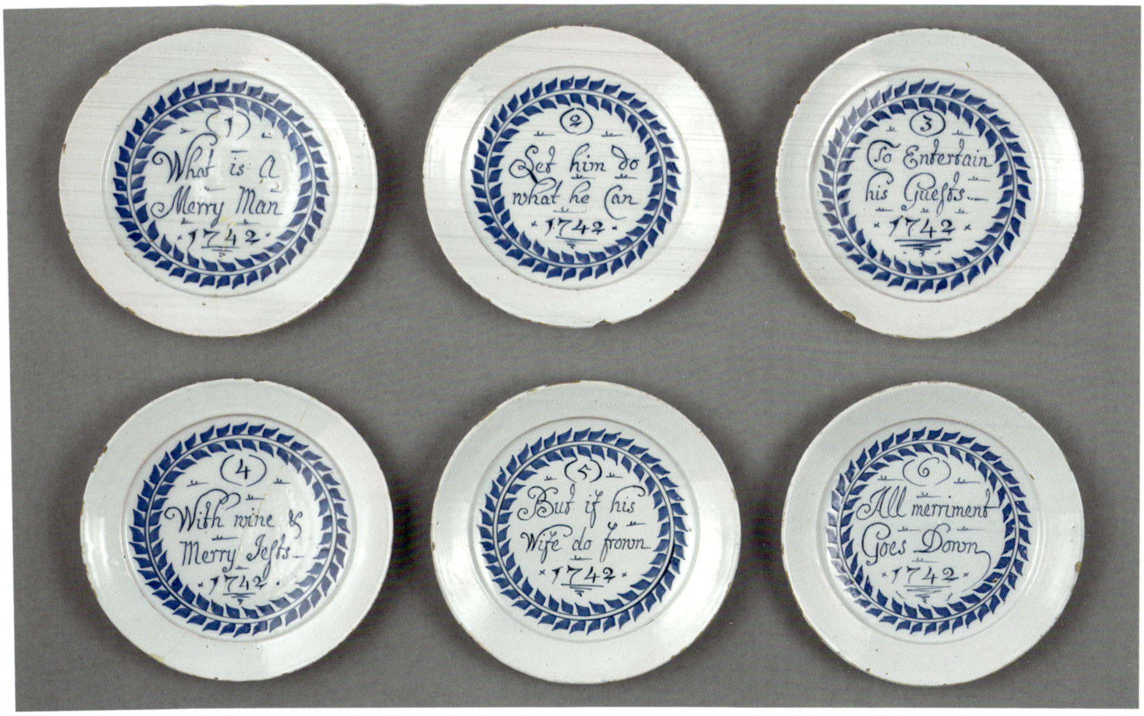

1742

Plates

Delftware

Each: 8 ¾ × 1 in. (22.2 × 2.54 cm)

Inscriptions:
(1)|What is A|Merry Man|1742

(2)|Let him do|what he Can|1742

(3)|To Entertain|his Guests|1742

(4)|With wine &|Merry Jests|1742

(5)|But if his|Wife do frown|1742

(6)|All merriment|Goes Down|1742

The fashion for sets of plates inscribed with mottoes came from Holland in the mid-seventeenth century. By the turn of the century, enthusiasm for this type of plate had waned in Holland, but it continued unabated in England well into the next century. Lipski and Archer list forty-nine examples of Delftware Merryman plates ranging from 1682 to 1752. This exact set of inscriptions is found on six plates of 1727 at the Victoria and Albert Museum, London (C.5-2016) and the Museum of London (A3837); an identical set dated 1752 was sold at Sotheby's, London, on December 3, 2013 (lot 30).

PROVENANCE

Simon Sainsbury; bt. John H. Bryan of Lake Bluff, IL (1936–2018) from Jellinek & Sampson Antiques, London, November 1991

LITERATURE

Louis L. Lipski and Michael Archer, *Dated English Delftware: Tin-Glazed Earthenware, 1600–1800* (London: Sotheby Publications, 1984), 110, nos. 482, 482A–E

1774

Plate
Delftware (Bristol or London Delft)
8 7/8 × 1 3/8 in. (22.5 × 3.5 cm)

Inscription: M. T| 1774

This plate, probably made in Bristol, is painted with the cracked-ice-and-fish decoration that imitated designs on Chinese porcelain.

PROVENANCE
Louis Gautier Collection; Puttick and Simpson, London, March 9, 1944 (lot 161); Sotheby's, London, February 24, 1987; bt. John H. Bryan of Lake Bluff, IL (1936–2018) from Alistair Sampson Antiques, London, March 2, 1987

LITERATURE
Louis L. Lipski and Michael Archer, *Dated English Delftware: Tin-Glazed Earthenware, 1600–1800* (London: Sotheby Publications, 1984), 149, no. 671A

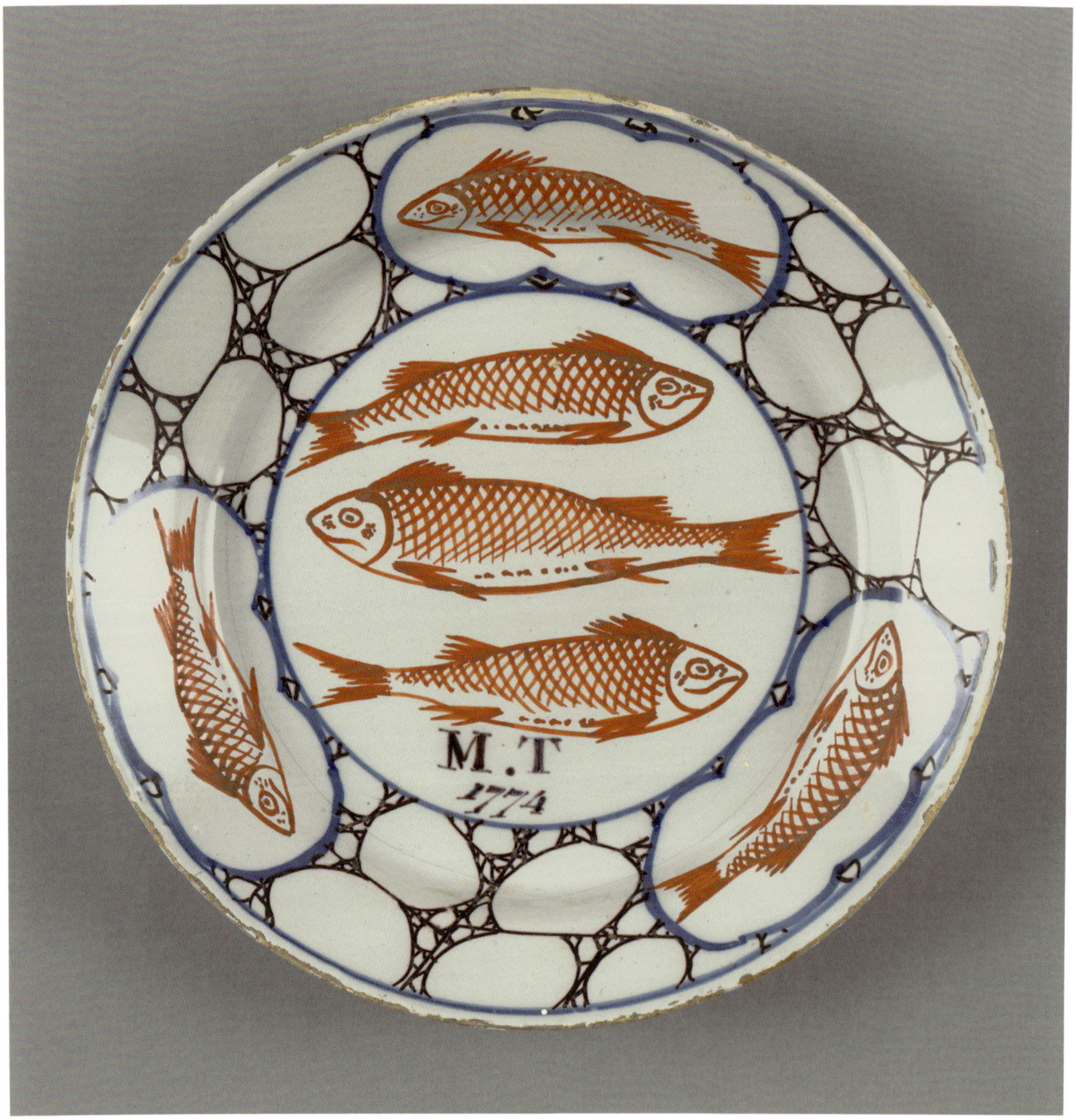

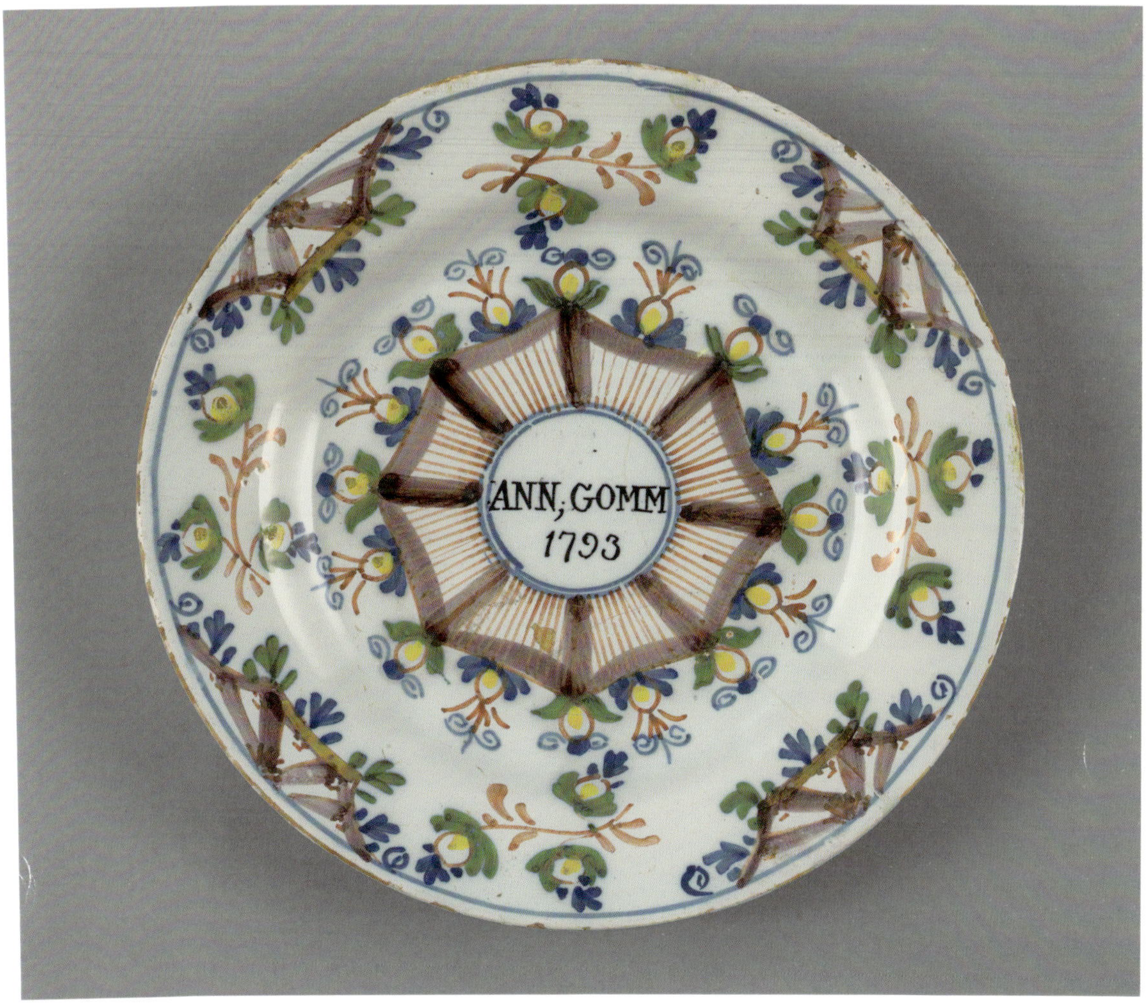

1793

Plate

Delftware (Lambeth Delft)

$8\frac{7}{8} \times 1\frac{1}{8}$ in. (22.5 × 2.9 cm)

John Brayne (active 1780; d. 1793)

Inscription: ANN;GOMM|1793

Painted decoration of this type has come to be synonymous with Ann Gomm, whose name appears on three plates dated 1793 but who remains unidentified.

Shards of this pattern have been found on the Lambeth High Street site of the pottery of John Brayne (active at that site 1789–1793). Brayne, "brownstone potter," made his will, in good health and mind, in 1791. He left his business to his sons, John and Benjamin, to be "carried on by them jointly," although the latter was given permission to pursue another profession if he chose to do so. The will was proved January 16, 1793 (TNA PROB 11/1227/119).

PROVENANCE

F. H. Garner of Wymeswold, Leicestershire (1893–1964); Sotheby's, London, March 2, 1965 (lot 57); John Philip Kassebaum of Mount Pleasant, SC (1932–2016); Sotheby's, London, October 1, 1991 (lot 64); bt. John H. Bryan of Lake Bluff, IL (1936–2018) through Lindsay Antiques, London

LITERATURE

Frank Britton, *London Delftware* (London: Jonathan Horne, 1987), 167

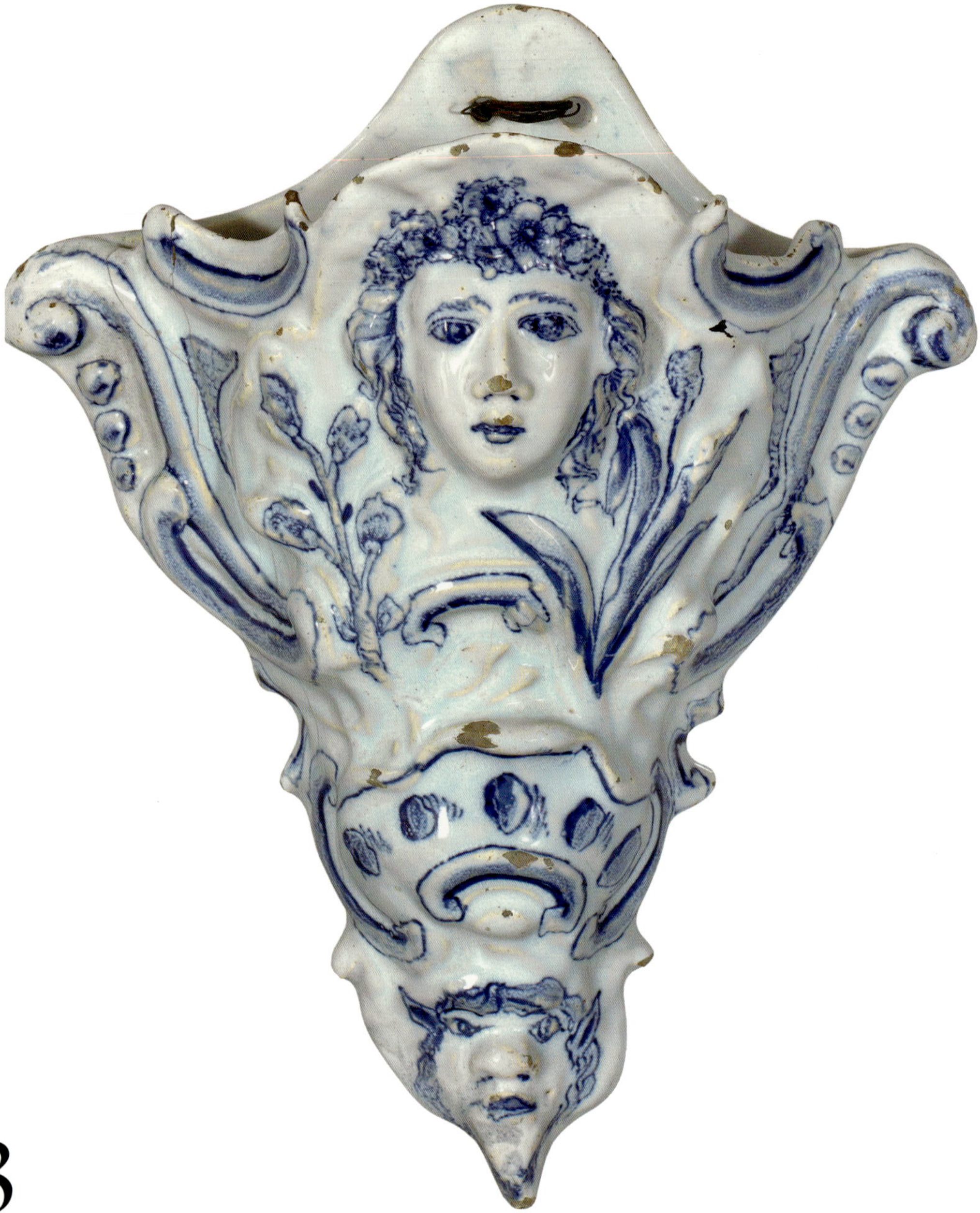

1753

Wall pocket
Delftware
$8\frac{1}{2} \times 6\frac{5}{8} \times 3$ in.
(21.6 × 16.8 × 7.6 cm)

Inscription:
M·S.|1753

Wall pockets were decorative wall vases that held small bouquets of flowers. Two of a similar design, sold at Olympia Auctions, London, on November 24, 2005 (lot 21), look to be from the same mold but decorated by different painters. Another of this design is in the Fitzwilliam Museum, Cambridge (C.2880-1928), where it is noted that wall pockets of this general shape are attributed to Liverpool. Four comparable examples can be found in the Bristol Collection.

PROVENANCE

bt. John H. Bryan of Lake Bluff, IL (1936–2018) from Lindsay Antiques, London, October 3, 1991

LITERATURE

Michael Archer, *Delftware in the Fitzwilliam Museum* (London: Philip Wilson Publishers, 2013), 312

Frank Britton, *English Delftware in the Bristol Collection* (London: Sotheby Publications, 1982), 102

1661

Plate
Delftware
$7\frac{1}{2} \times 7\frac{1}{4} \times \frac{3}{4}$ in.
(19.1 × 18.4 × 1.9 cm)

Inscription:
Weilcom my|
Friends|1661

Even by the standards of the day, this inscription displays eccentric spelling, suggesting it was painted by a potter who did not have English as his first language. It has also led to the suggestion that these plates were made in Holland for the export market. A wine bottle in the City Museum and Art Gallery, Birmingham, bears an identical inscription. The inscription on this plate is also very similar to that on a bottle with the same date and in the same collection: "You & i are|Earth|1661." This inscription also appears on a number of round and octagonal plates, suggesting a common place of manufacture for all.

PROVENANCE

bt. John H. Bryan of Lake Bluff, IL (1936–2018) from Jonathan Horne, London, March 23, 1996

LITERATURE

Leslie B. Grigsby, *The Longridge Collection of English Slipware and Delftware* (London: Jonathan Horne Publications, 2000), 1:110

Louis L. Lipski and Michael Archer, *Dated English Delftware: Tin-Glazed Earthenware, 1600–1800* (London: Sotheby Publications, 1984), 43, nos. 112–112C; 36, nos. 1488, 1489

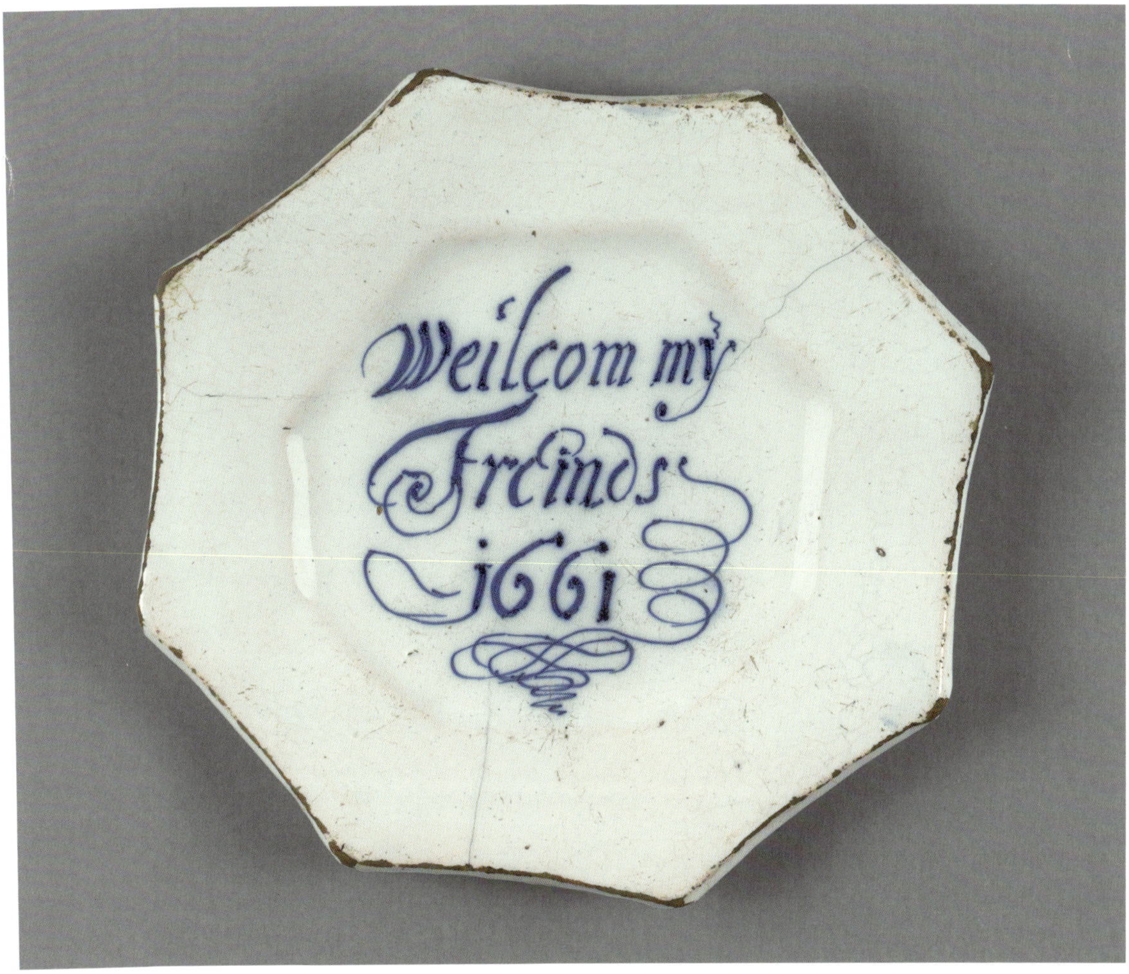

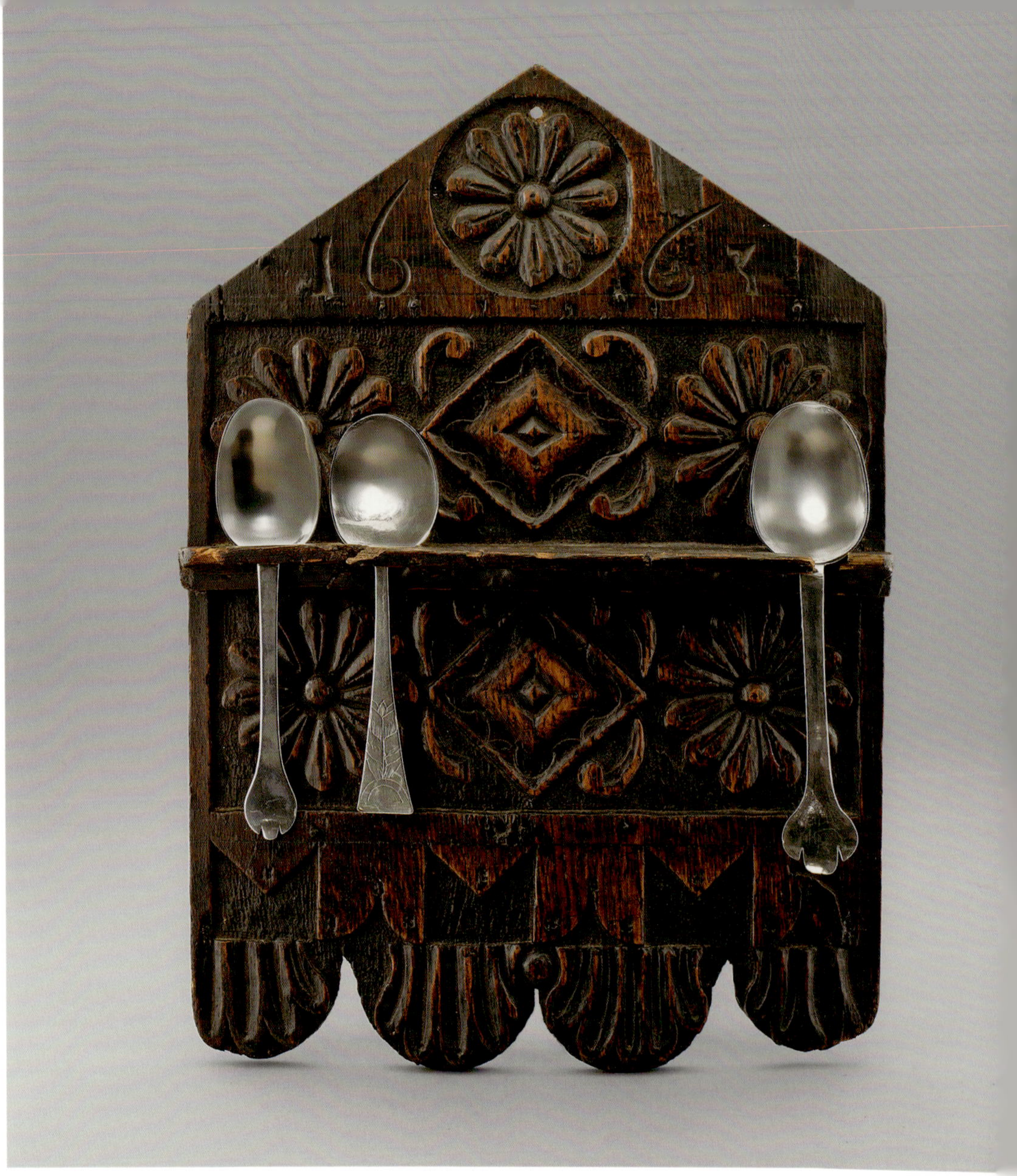

1663

Spoon rack

Oak

17½ × 12 × 1¾ in. (44.5 × 30.5 × 4.5 cm)

Inscription: 16 63

Prior to the introduction of forks, spoons were the most predominant eating utensils. Spoons were often given as gifts for baptisms and as bequests in wills. The giving of spoons also played an important role in the ceremonies of apprenticeship.

Owing to their inherent fragility, spoon racks of such a demonstrably early date are extremely rare. The board was fitted with fretted holding strips for spoons. Extant examples from the eighteenth century incorporate cutlery trays below the spoon holders.

PROVENANCE

John Vane, eleventh Baron Barnard of Raby Castle, County Durham; Christie's, London, October 10–11, 1993 (lot 122); John L. Boonshaft of Las Vegas, NV (1950–1997); Sotheby's, New York, January 23, 1998 (lot 81); bt. John H. Bryan of Lake Bluff, IL (1936–2018) through Tobias Jellinek Antiques, London

LITERATURE

Victor Chinnery, *Oak Furniture: The British Tradition, A History of Early Furniture in the British Isles and New England* (Woodbridge, UK: Antique Collectors' Club, 1979), 409

1677

Spoon
Silver
$6\frac{7}{8} \times 1\frac{7}{8} \times \frac{5}{8}$ in.
(17.5 × 4.8 × 1.59 cm)

Inscription: 1677; W R; W R

Until the mid-eighteenth century, cutlery was laid face down on the table. So, when this spoon was first in use, the double set of initials would have been visible while the spoon sat on the table, but the decorative tulip motif would have been hidden until the spoon was held in the hand.

PROVENANCE
John L. Boonshaft of Las Vegas, NV (1950–1997); Sotheby's, New York, January 23, 1998; bt. John H. Bryan of Lake Bluff, IL (1936–2018) through Tobias Jellinek Antiques, London

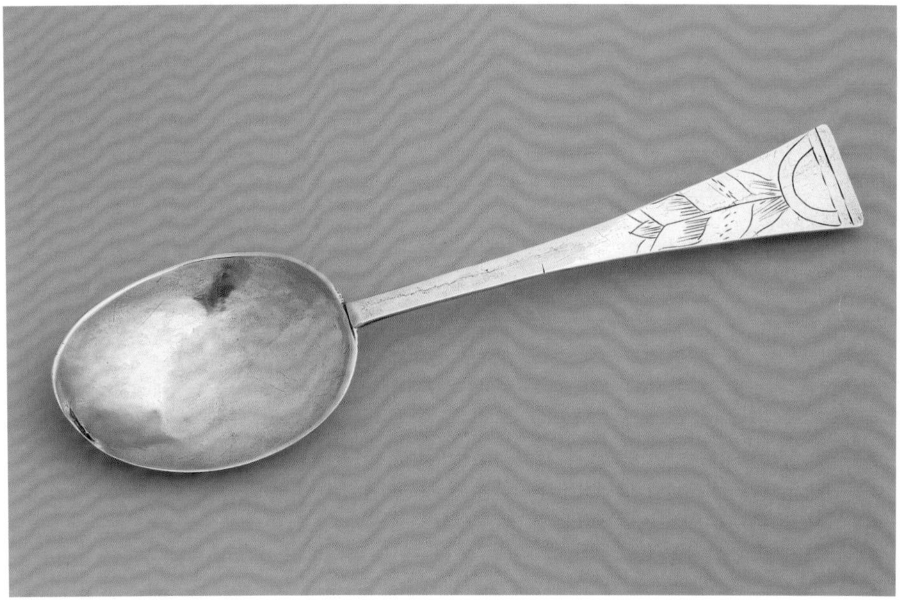

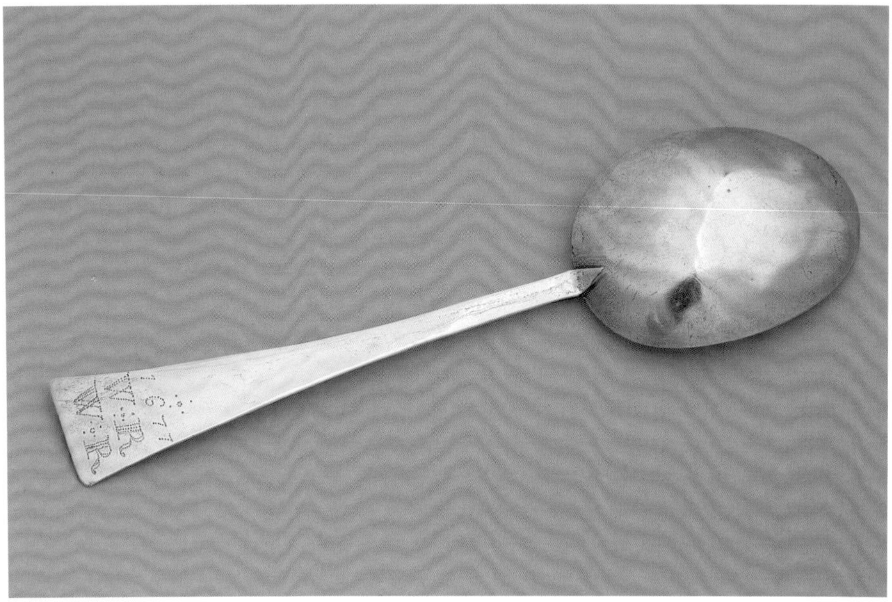

1681

Spoon
Silver
8 × 2 × 7/8 in. (20.3 × 5.1 × 2.22 cm)

Inscription: 1681; G D; T S

This spoon bears its inscription on the underside of the bowl and displays marks for Exeter on the stem.

PROVENANCE
John L. Boonshaft of Las Vegas, NV (1950–1997); Sotheby's, New York, January 23, 1998; bt. John H. Bryan of Lake Bluff, IL (1936–2018) through Tobias Jellinek Antiques, London

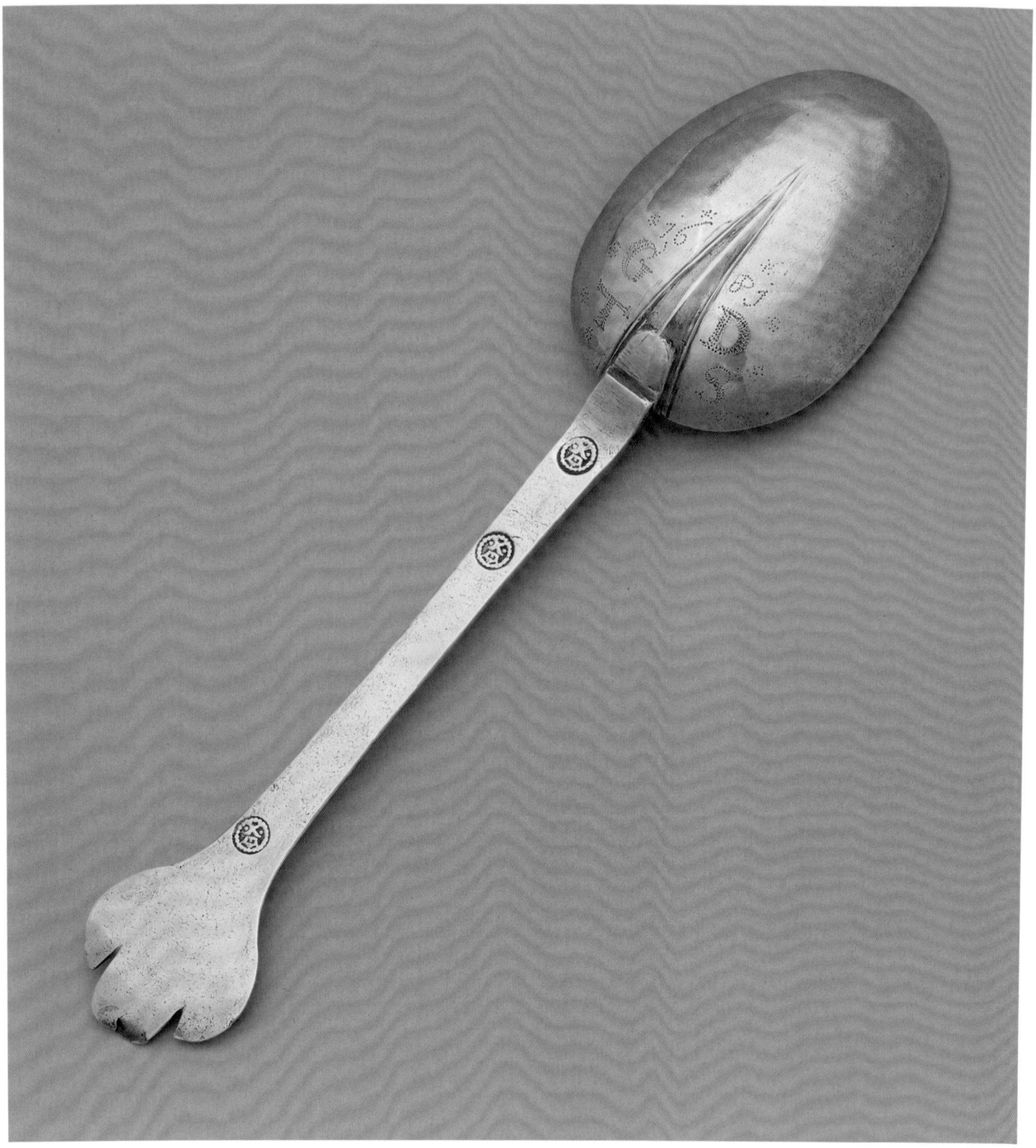

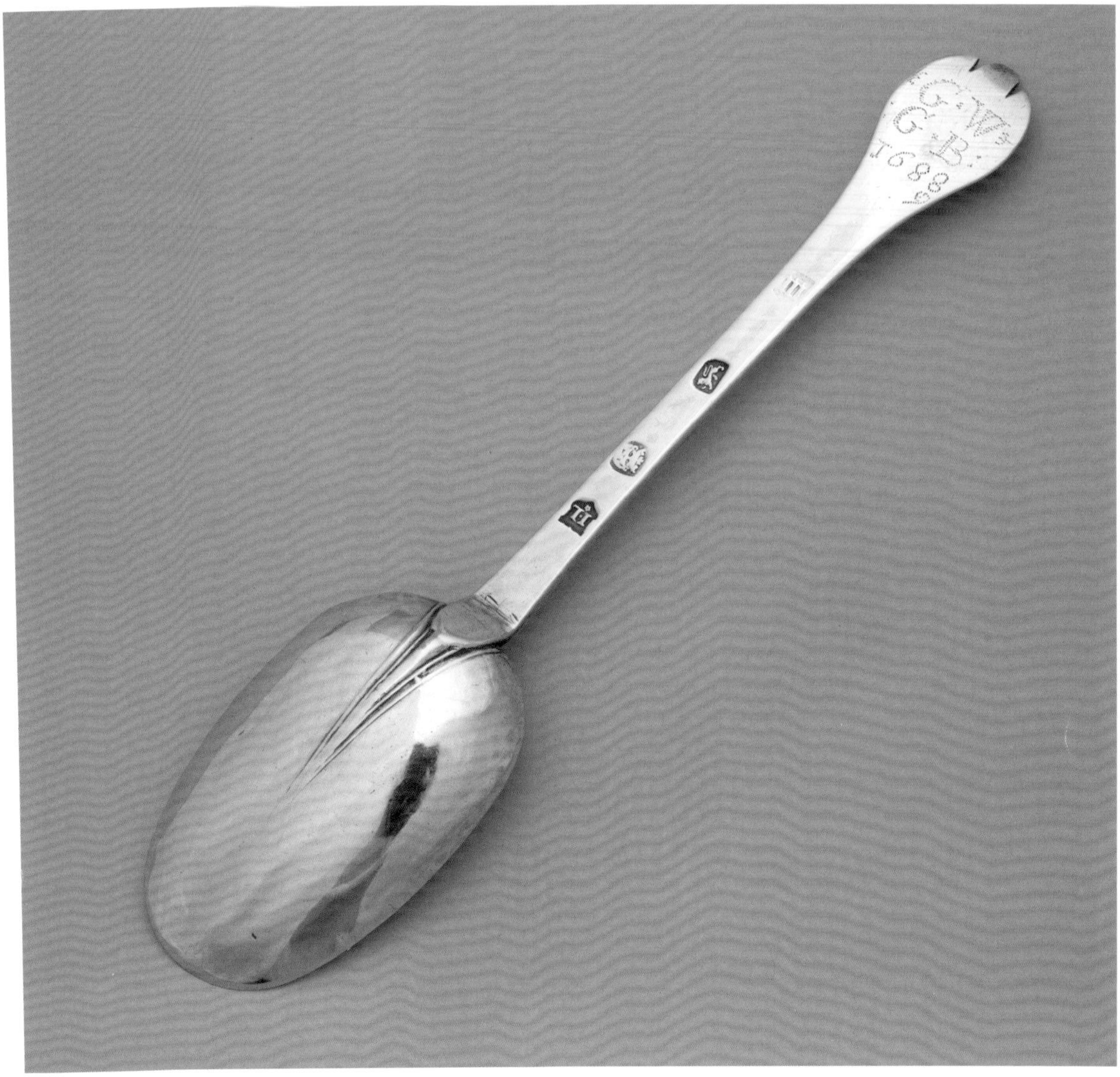

1688/9

Spoon
Silver
$7\frac{5}{8} \times 1\frac{3}{4} \times \frac{3}{4}$ in.
(19.4 × 4.5 × 1.9 cm)

Inscription: G W;
G B; 1688|9

The date inscribed on this spoon is a bit of a puzzle, as it seems to suggest a date of 1688/9 (that is, prior to Lady Day, 1689). Yet it bears the 1689 hallmark for London, which changed on the day the wardens of the Goldsmiths' Company were elected, which, after 1660, would have been May 29—by which point the turn of the new year, on March 25, had already occurred. Perhaps instead the inscription commemorates an event that had happened earlier that year.

The mark, bearing the initials "I. L." within a shield, with a mullet below, has not been identified but it appears on other spoons dating from 1691 to 1696. Jenkin Lewis (d. 1713) is a candidate for the maker.

PROVENANCE
John L. Boonshaft of Las Vegas, NV (1950–1997); Sotheby's, New York, January 23, 1998; bt. John H. Bryan of Lake Bluff, IL (1936–2018) through Tobias Jellinek Antiques, London

1764

Knitting sheath
Wood
$15\frac{1}{4} \times \frac{3}{4} \times \frac{5}{8}$ in.
(38.7 × 1.9 × 1.59 cm)

Inscription: J J; 1764; E·T

Knitters used a knitting sheath to hold one needle and support the knitting in progress. They knit with the other needle, "throwing" the working yarn with the free hand. Knitting was a vital activity in early modern households. Families required a variety of clothing items made from wool, including caps, hats, stockings, and hose. Women could also supplement their incomes by making and selling knitted goods—in a shop, from a market stall, or perhaps from their doorsteps to neighbors or passersby. This example, which is probably Welsh, has an integral chain and a hook for carrying the length of finished fabric. Both the tool—domestically carved from boxwood and fruitwoods—and the handwork it helped to produce were very much items of the everyday.

PROVENANCE
bt. John H. Bryan of Lake Bluff, IL (1936–2018) from Alistair Sampson Antiques Ltd., London, June 22, 1998

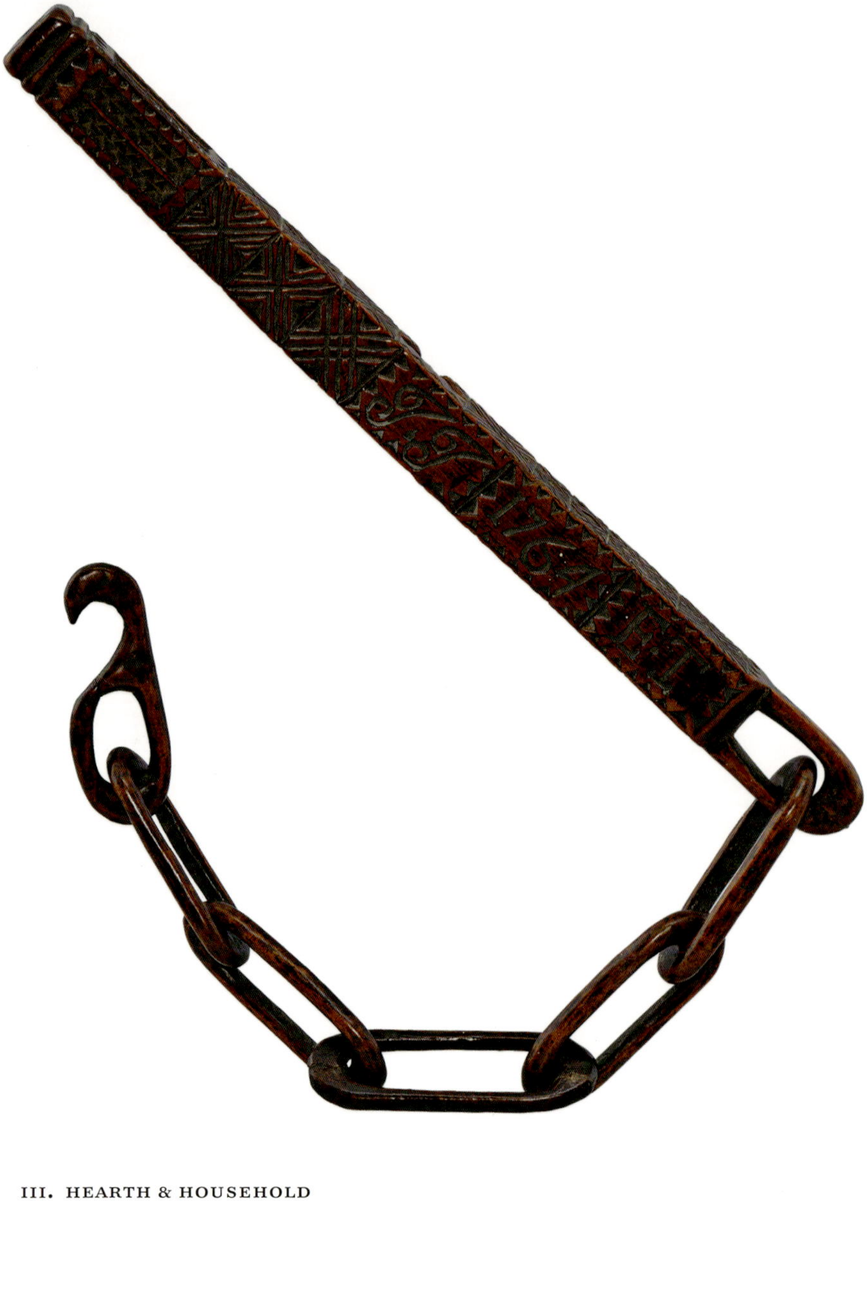

1786

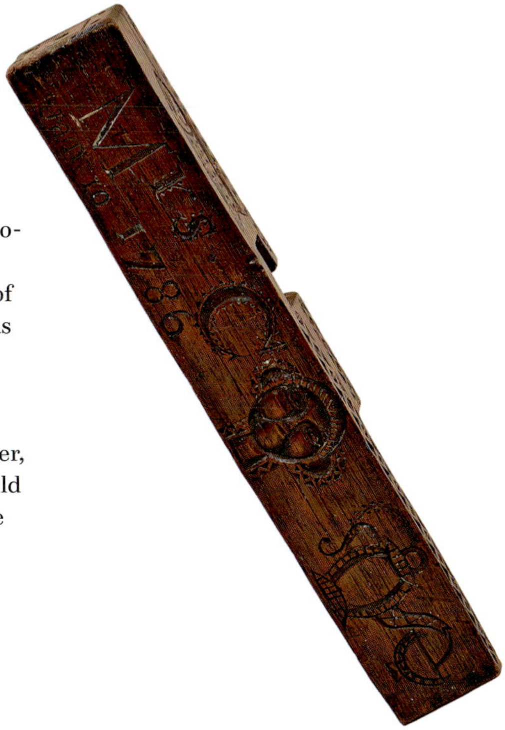

Knitting sheath
Mahogany
8 × 1¼ × ¾ in.
(20.3 × 3.2 ×
1.9 cm)

Inscription: M^rs C;
Jan 10 1786

Sometimes the decoration on knitting sheaths provides some sense of the people to whom they were presented as gifts. This particular sheath, redolent of affection for a characterful woman, was given to "M^rs C" on January 10, 1786. It is covered in clues about her likes and loves: tea, flowers, smoking, and fish—but the inevitable moral reminder, the hourglass, was a warning she should not waste too much time with her pipe and cup.

PROVENANCE
bt. John H. Bryan of Lake Bluff, IL (1936–2018) from Alistair Sampson Antiques Ltd., London, June 7, 1996

1790

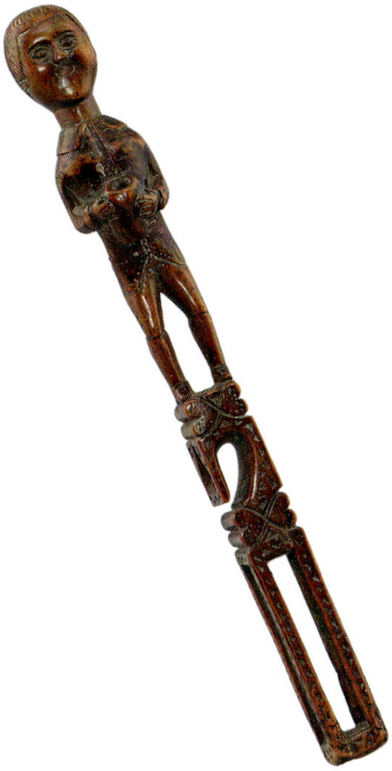

Knitting sheath
Spindle Tree
7½ × 1 × 1 in. (19.1 ×
2.5 × 2.5 cm)

Inscription: 1790|
W L

PROVENANCE
bt. John H. Bryan of Lake Bluff, IL (1936–2018) from Lindsay Antiques, London, June 29, 1992

1782

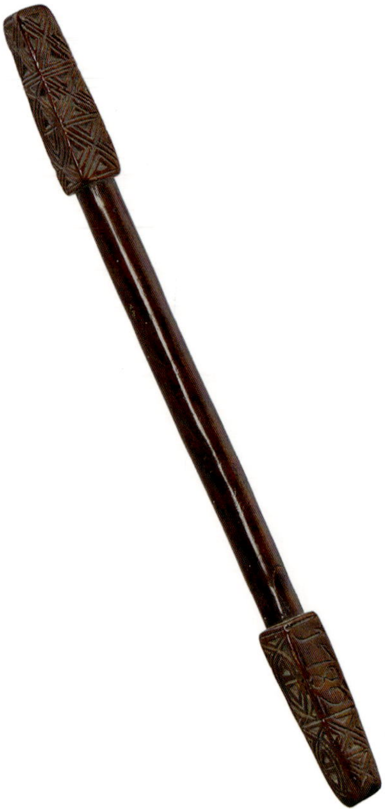

Knitting sheath
Wood
$7\frac{1}{2} \times 1 \times 1$ in. (19.1 × 1.27 × 1.27 cm)

Inscription: MJ; 1782

This knitting sheath has a very unusual form. Most knitting sheaths have a hole on one end, into which to insert the end of the stable needle. This sheath has a hole at each end, presumably for a different-sized needle, thus making the sheath more versatile.

PROVENANCE
bt. John H. Bryan of Lake Bluff, IL (1936–2018) from Alistair Sampson Antiques Ltd., London, April 22, 1999

1799

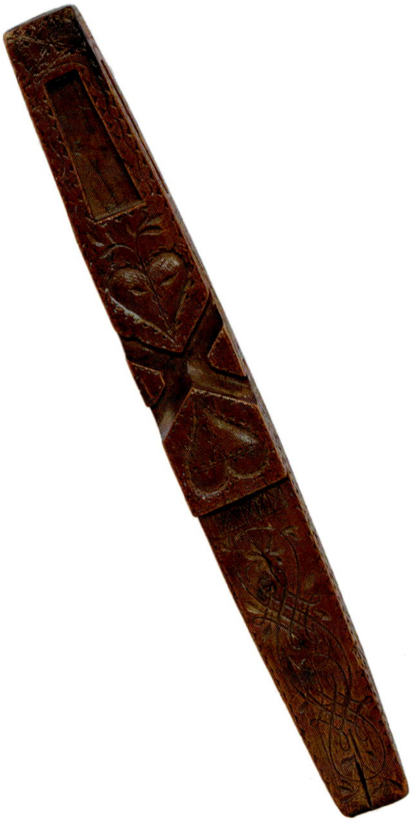

Knitting sheath
Boxwood
$7 \times 1 \times \frac{5}{8}$ in. (17.8 × 2.5 × 1.59 cm)

Inscription: 17 Mary Ridlen Moralee 99

One Mary Moralee was buried at St Andrew, Lamesley, in the county of Durham, on October 4, 1808, at age sixty-three.

PROVENANCE
Wilkinson's Auctioneers, Doncaster, December 3, 2017 (lot 118); bt. John H. Bryan of Lake Bluff, IL (1936–2018)

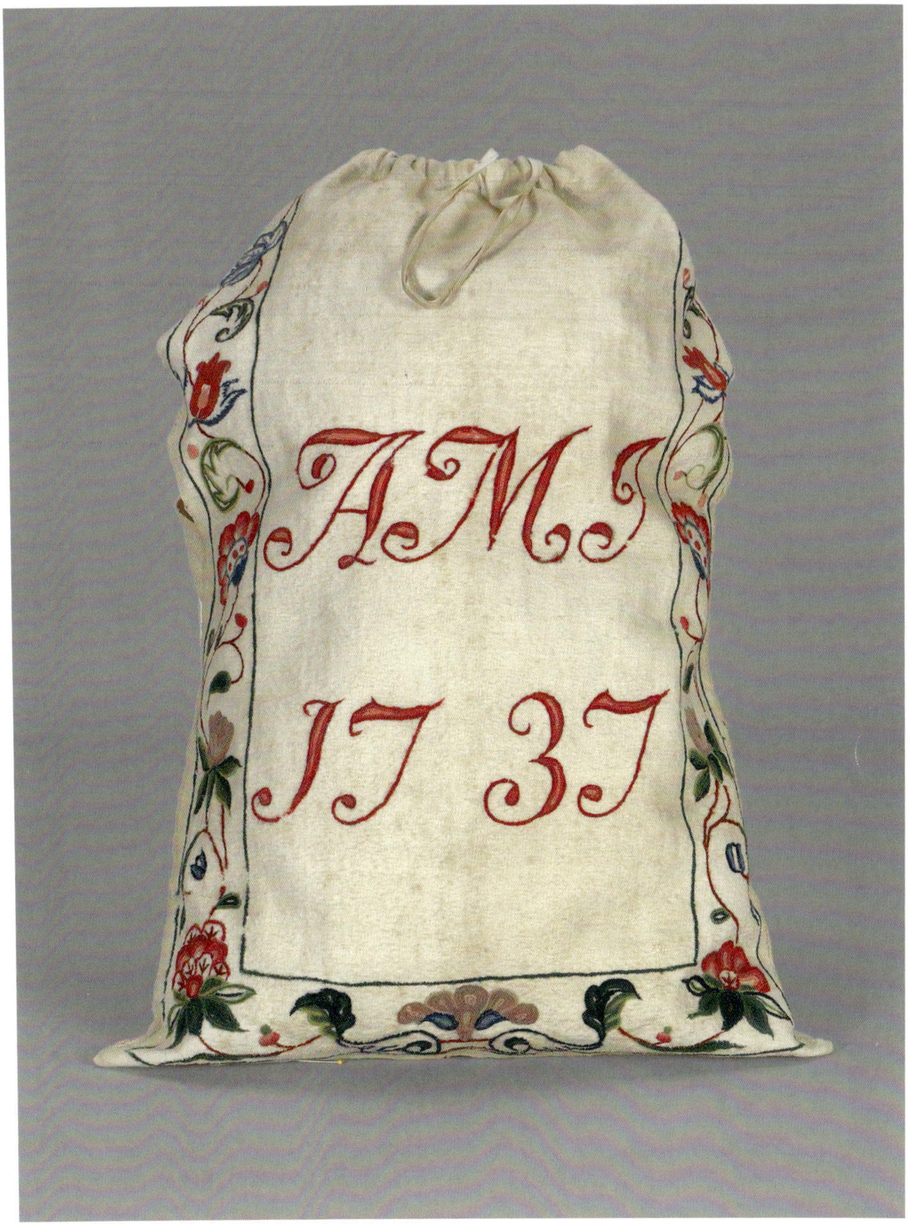

1737

Workbag

Cotton with wool thread

24½ × 15½ in. (62.2 × 39.4 cm)

Inscription: AMJ | 17 37

Workbags like this one were important domestic accessories of the eighteenth century. The designs in this example draw inspiration from painted Indian textiles, imports of which had been banned in England by this date.

Such bags were usually sufficiently large so as to hold the substantial hanks of colored worsted wool that young girls—from about the age of ten—used in their crewelwork. The girls often inscribed dates upon their finished work. An example in the Victoria and Albert Museum, London (T.166-1984) consists of two stitched panels with different dates, which suggests that the panels were considered individual works and that each was dated on its completion.

PROVENANCE

bt. John H. Bryan of Lake Bluff, IL (1936–2018) from Cora Ginsburg Inc., New York, August 3, 1994

1706

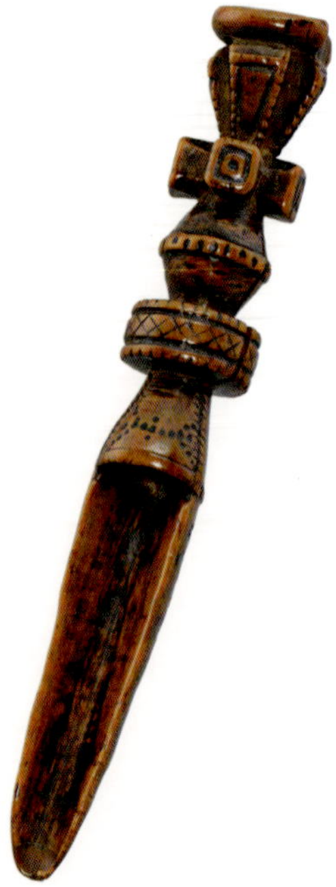

Apple corer
Fruitwood
$5\frac{1}{8} \times \frac{7}{8}$ in. (13 × 2.22 cm)

Inscription: R S | I G; 1706

It is difficult to distinguish between an apple corer and a cheese scoop. Both objects were designed to bore through a surface and extract a portion of the contents. This implement is simple in its carving and design and, like similar utensils of this period, is made of fruitwood.

PROVENANCE
bt. John H. Bryan of Lake Bluff, IL (1936–2018) from Alistair Sampson Antiques Ltd., London, September 9, 1995

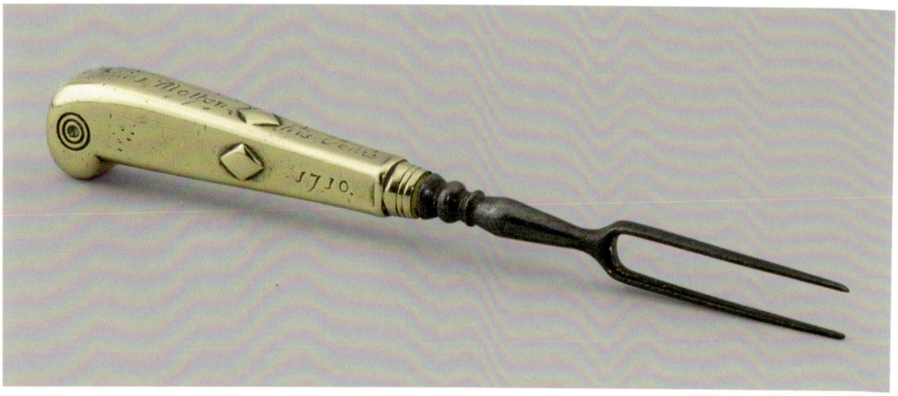

1710

Fork
Brass, steel
$6\frac{1}{4} \times 1 \times \frac{5}{8}$ in. (15.9 × 2.5 × 1.59 cm)

Inscription: Jonas Molten his Forke; 1710

Who Jonas Molten was remains a mystery, but this was his fork. The name may have been an approximation of "Malden," and, if so, it is possible that this fork belonged to an apothecary by that name, who was also of Malden in Essex. On August 13, 1713, a man by that name married Mary Robient in the nearby parish of Little Leighs—perhaps having impressed her with his mastery of what was then still a relatively novel eating utensil.

PROVENANCE
bt. John H. Bryan of Lake Bluff, IL (1936–2018) from Tobias Jellinek Antiques, London, February 17, 2017

1772

Apple corer
Bone, boxwood
$5\frac{5}{8} \times 1\frac{1}{2} \times 1\frac{1}{4}$ in.
(14.3 × 3.8 × 3.2 cm)

Inscribed: 1772

PROVENANCE
bt. John H. Bryan of Lake Bluff,
(1936–2018) from Alistair Samp
Antiques, London, April 9, 198

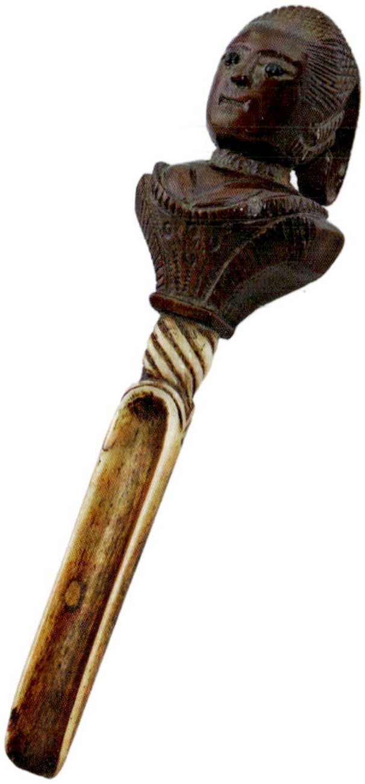

1690

Nutcracker
Boxwood
$3 \times 1\frac{7}{8}$ in. (7.6 × 4.8 cm)

Inscription: M H; 1690

Nutcrackers were used at the dining table for opening a variety of small, edible nuts, which were an important part of the eighteenth-century diet. Boxwood was the common material because of its tight grain and resistance to fracturing. With this style of nutcracker, the nut was cracked through pressure applied with the thumbscrew.

PROVENANCE
bt. John H. Bryan of Lake Bluff, IL (1936–2018) from Lindsay Antiques, London, June 29, 1992

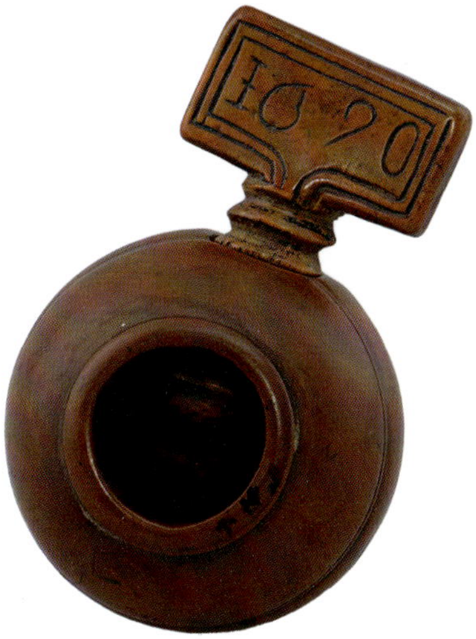

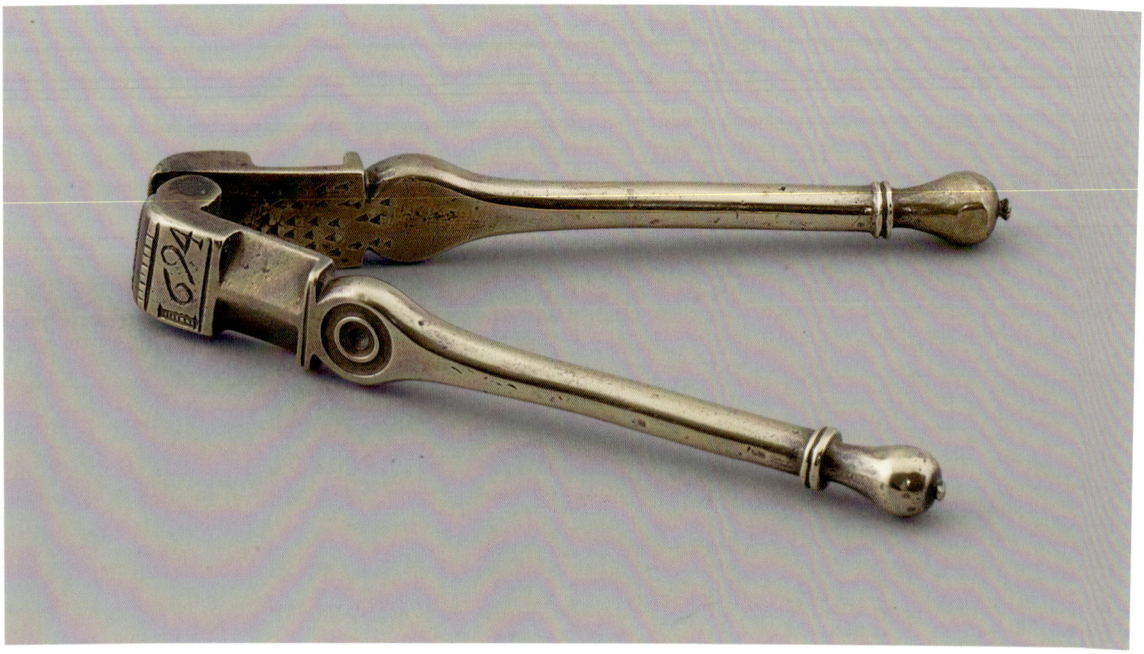

1694

Nutcracker
Brass
$\frac{5}{8} \times 4\frac{1}{4} \times \frac{3}{4}$ in. (1.59 × 10.8 × 1.9 cm)

Inscription: 1694

PROVENANCE
bt. John H. Bryan of Lake Bluff, IL (1936–2018) from Lindsay Antiques, London, November 11, 1991

1719

Nutcracker
Boxwood
$7 \times 1\frac{1}{2} \times 2$ in. (17.8 × 3.8 × 5.1 cm)

Inscription: ER| 1719; EMANVEL; ROGERS

The man with a movable jaw was a popular design for nutcrackers of the early eighteenth century. The nut was placed inside his mouth, and when the handles were tightened to close his mouth, the nut would crack.

PROVENANCE
bt. John H. Bryan of Lake Bluff, IL (1936–2018) from Alistair Sampson Antiques Ltd., London, August 8, 1995

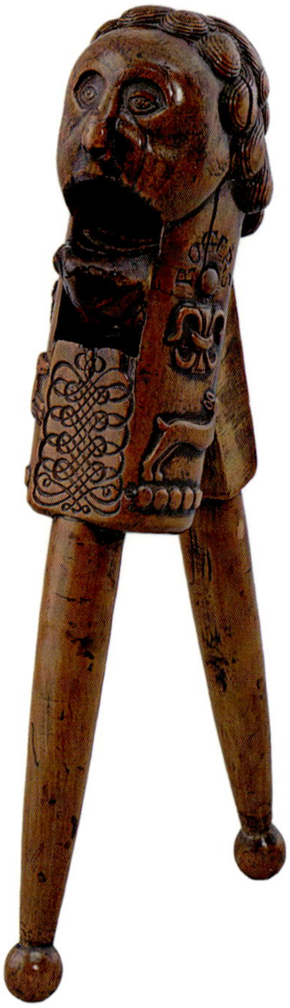

1765

Candle box
Brass
$7\frac{1}{4} \times 11\frac{1}{4} \times 4\frac{7}{8}$ in.
(18.4 × 28.6 × 12.4 cm)

Inscription:
R. Bown|1765

When hung from the wall, by means of the tapered brass straps, the contents of this candle box would have been safe from the advances of floor-dwelling rodents, which might otherwise access the edible contents.

PROVENANCE
bt. John H. Bryan of Lake Bluff, IL (1936–2018) from Alistair Sampson Antiques Ltd., London, February 21, 1995

LITERATURE
Rupert Gentle and Rachel Field, *Domestic Metalwork: 1640–1820*, revised and enlarged by Belinda Gentle (Woodbridge, UK: Antiques Collectors' Club, 1994), 172, no. 2

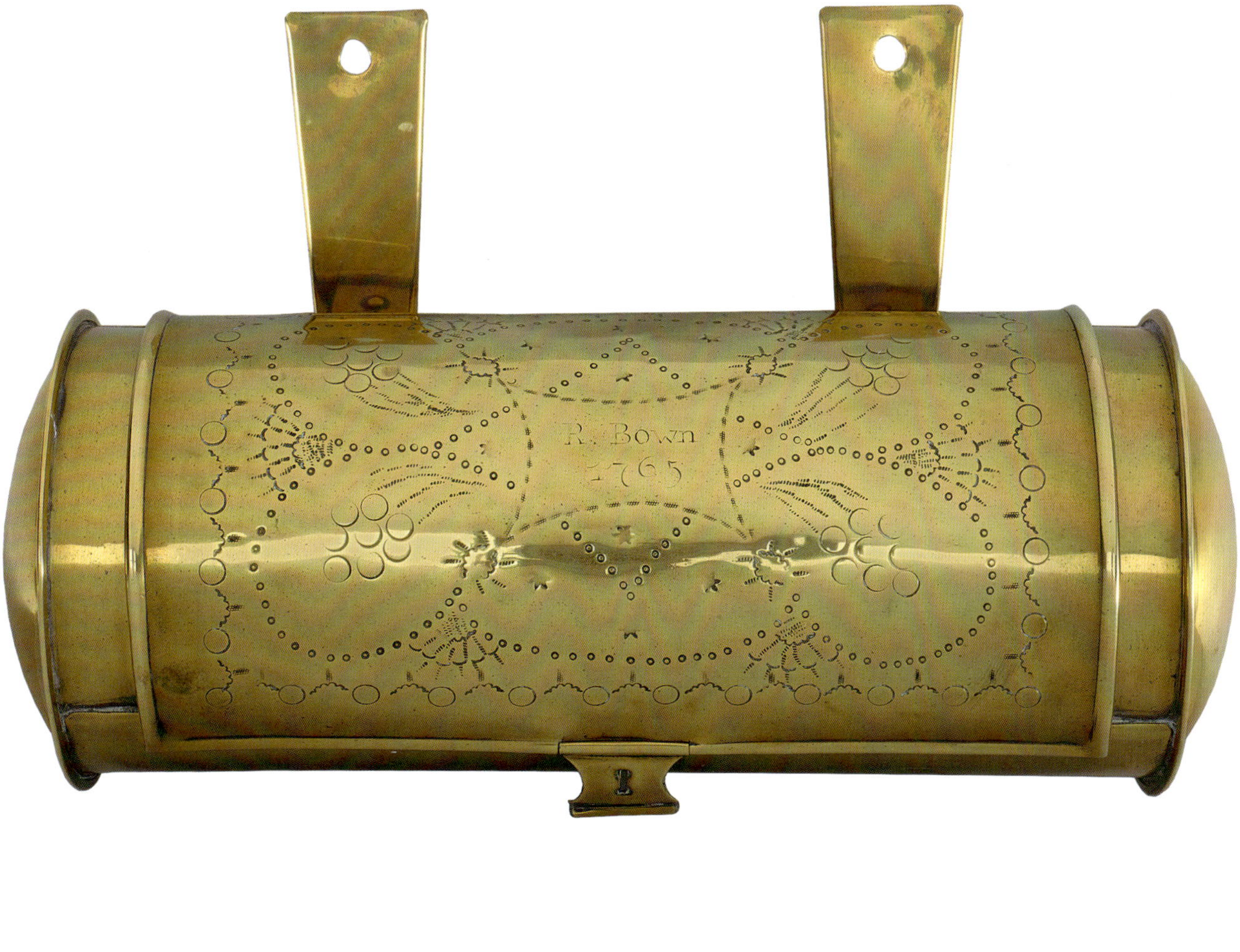

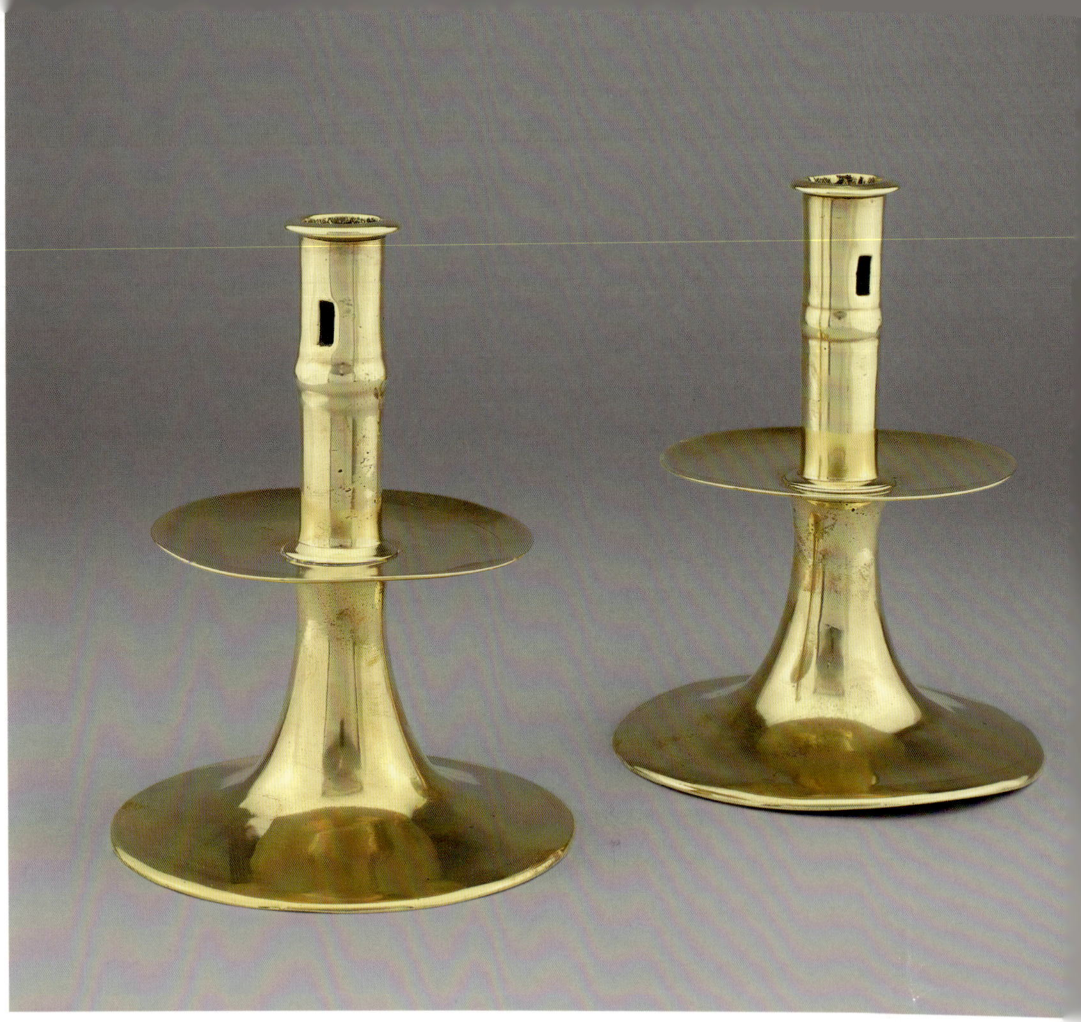

1702

Candlesticks

Brass

8 × 6 in. (20.3 × 15.2 cm)

Inscription, on both: Thomas Mendham| 1702 Briston| Norfolk

These candlesticks, with their flared brass bases, are known as trumpet candlesticks. Where the form appears in silver, hallmarks date this design from 1650 to 1663. Brass examples date to 1682, which makes these candlesticks a little late, but they may have begun life in domestic use before being gifted to a church, upon which occasion they would have been inscribed. The date is the year the church tower at Briston collapsed; it was never rebuilt. The apertures in the stems, for ejecting the candle inside, are later additions.

Mendham was a well-established name in East Anglia. The owner of these candlesticks was Thomas Mendham, probably the son of Elizabeth and Robert Mendham, baptized at Briston on June 24, 1677. In turn, that man's son may be the energetic evangelical preacher who answered the poet Robert Potter's 1785 pamphlet, *Observations on the Poor Laws and on Houses of Industry*, in which he described the cruelty of overseers at workhouses.

PROVENANCE

bt. John H. Bryan of Lake Bluff, IL (1936–2018) from Tobias Jellinek Antiques, London, November 17, 1998

LITERATURE

Jeremy Gregory and Jeffrey S. Chamberlain, *The National Church in Local Perspective: The Church of England and the Regions, 1660–1800* (Woodbridge, UK: Boydell Press, 2003), 178

Eloy Koldeweij, *The English Candlestick: 500 Years in the Development of Base-Metal Candlesticks, 1425–1925* (London: Christie's Books, 2001), 46, 50, no. 12

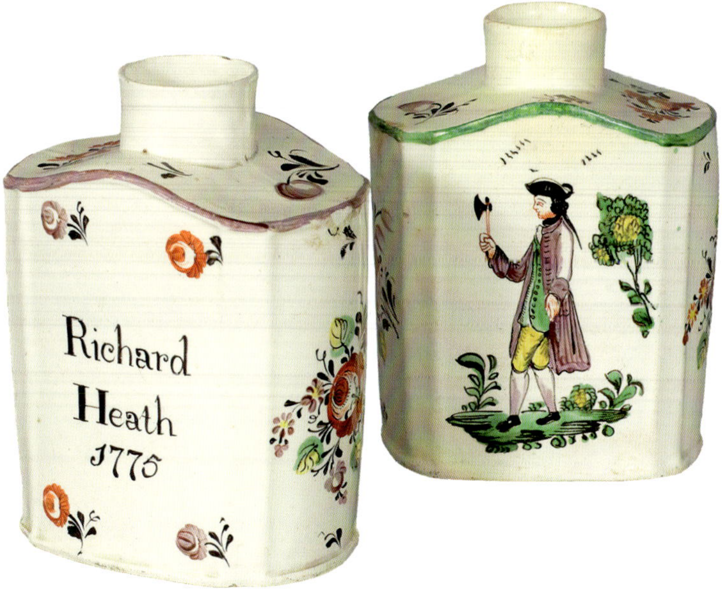

1775

Tea canisters

Creamware

Left: $4\frac{1}{4} \times 3 \times 2\frac{1}{4}$ in. (10.8 × 7.6 × 5.7 cm)

Right: $4\frac{1}{4} \times 3 \times 2$ in. (10.8 × 7.6 × 5.1 cm)

Inscription, on both: Richard| Heath|1775

The colors on these painted creamware canisters may have indicated their contents: green for green tea and purple for Bohea, a type of brownish-red black tea from China.

PROVENANCE

bt. John H. Bryan of Lake Bluff, IL (1936–2018) from Garry Atkins, London, February 21, 2017

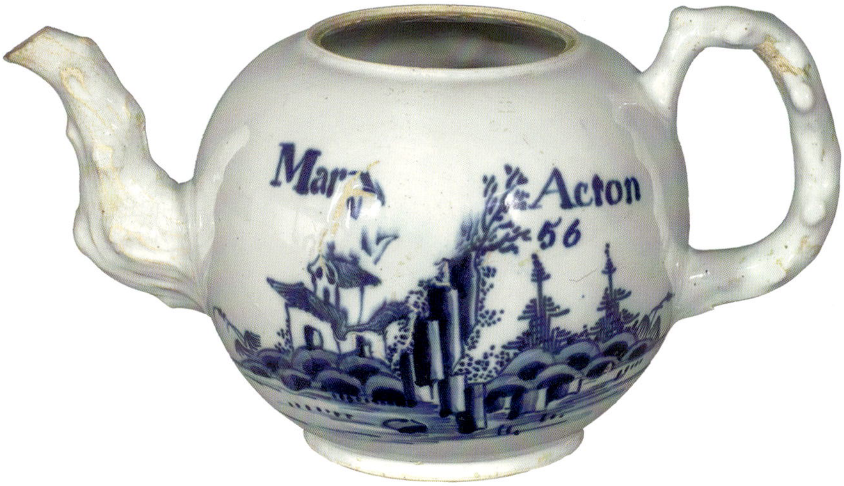

1756

Teapot

Tin-glazed stoneware

$4\frac{7}{8} \times 8\frac{3}{8} \times 4\frac{3}{4}$ in. (12.4 × 21.3 × 12.1 cm)

Inscription: Mary Acton; 17 56

Made in Liverpool, this teapot has rustic "crabstock" handles and spouts that mimic those of Chinese red stoneware.

PROVENANCE

bt. John H. Bryan of Lake Bluff, IL (1936–2018) from Jonathan Horne, London, March 10, 1987

1761

Mug

Salt-glazed stoneware with "scratch blue" decoration

$5\frac{1}{8} \times 5 \times 3\frac{5}{8}$ in. (13 × 12.7 × 9.2 cm)

Inscription: Martha Mason June 23rd 1761

Staffordshire and Devon were the two centers of production for this type of decorated stoneware, popular in the 1760s and 1770s. Devon's Bovey Tracey pottery produced a series with owners' names and dates incised in the manner seen here; shards found on America's eastern coast evince the existence of an export market for these products. Martha Mason, who has evaded discovery, clearly was a woman of independent means who had her name and the date marked on this lightweight and elegant mug.

PROVENANCE

Mr. John Eliot Hodgkin (1829–1912); Sir Victor and Lady Gollancz (1893–1967); The Troy Chappell Collection; Skinner Auctions, European Furniture and Decorate Arts sale, July 15, 2016 (lot 318); bt. John H. Bryan of Lake Bluff, IL (1936–2018) from Martyn Edgell Antiques Limited, Cambridgeshire, February 5, 2017

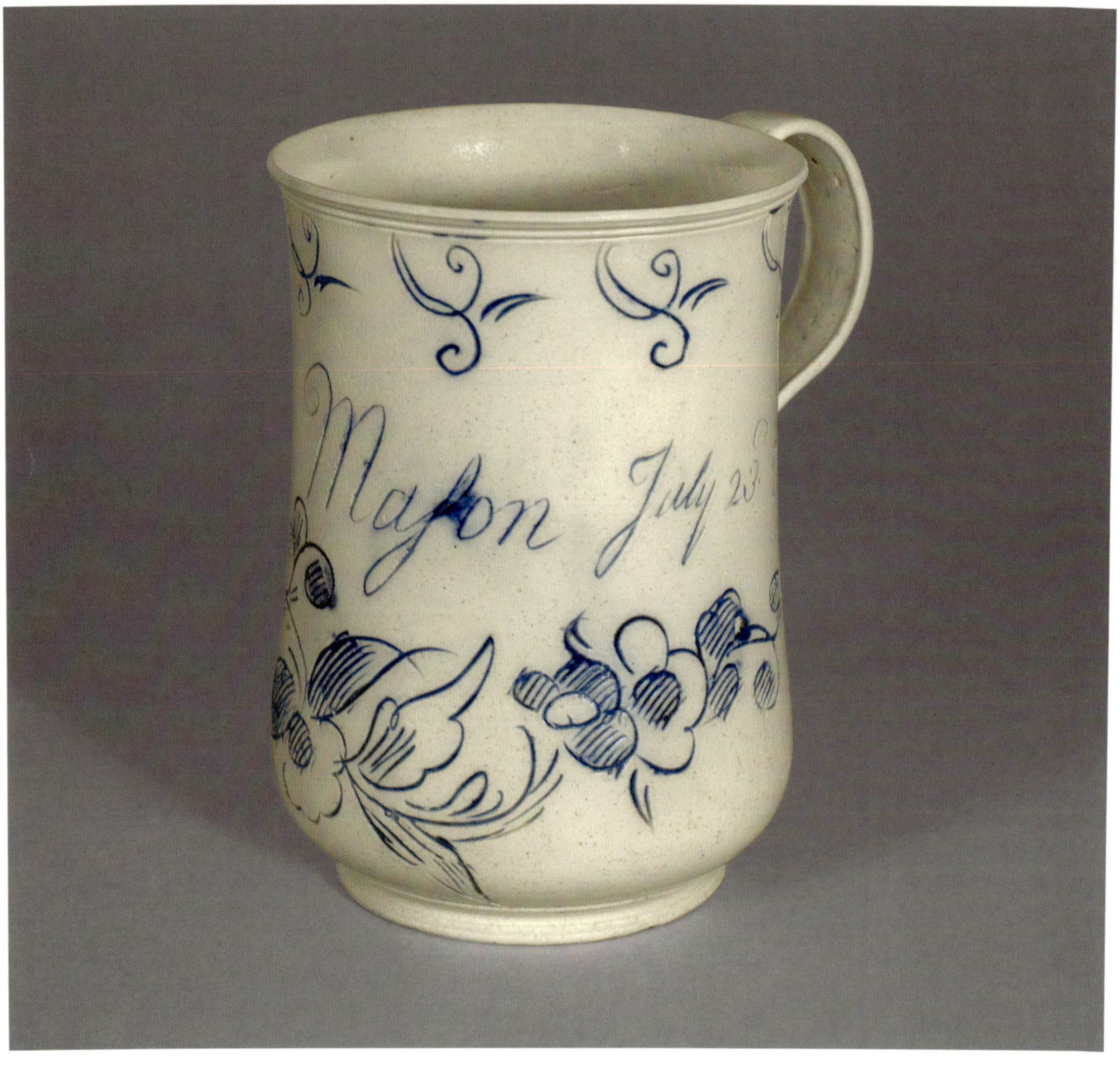

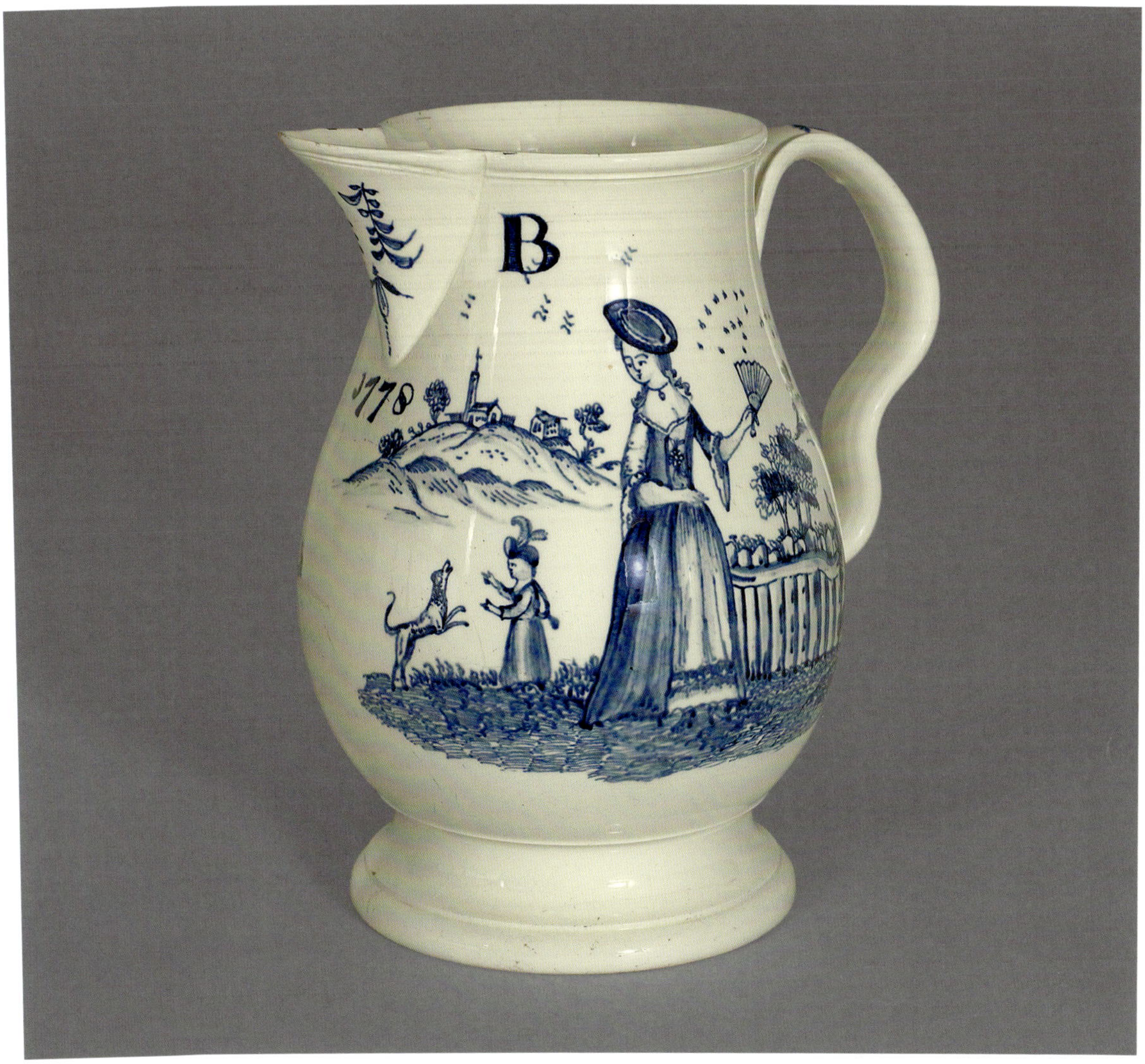

1778

Jug
Creamware
$8\frac{1}{2} \times 7\frac{5}{8}$ in. (21.6 × 19.4 cm)

Inscription: WB 1778

Creamware is a type of earthenware that has a lead glaze over a pale-colored body. It was developed by Josiah Wedgwood over the course of the middle decades of the eighteenth century. Thought to have been made in Yorkshire, this creamware jug would have been used—appropriately enough—as a container for cream. By 1778, when this jug was made, the fashion for the style had reached its apogee.

Like all Western wares that sought to imitate porcelain, this creamware bears chinoiserie decoration that is a crude attempt to imitate the painted scenes on Chinese exports.

PROVENANCE

bt. John H. Bryan of Lake Bluff, IL (1936–2018) from Martyn Edgell Antiques Limited, Cambridgeshire, February 5, 2017

1639

Bible box

Oak, painted

4 × 14$\frac{1}{8}$ × 9$\frac{7}{8}$ in. (10.2 × 35.9 × 25.1 cm)

Inscription: 1639 I M

It is extremely rare to find a Bible box and a Bible that have remained together, but, judging from the dates on this object and the next, this was the case here. The box is decorated with paintwork in blue, white, red, and orange—vestiges of which are original—but the Prussian blue pigment identified during paint analysis indicated that the polychrome scheme was refreshed at some point in the eighteenth century. The marginalia in the book suggests that both box and Bible changed hands a number of times through the centuries, yet they somehow stayed together.

PROVENANCE

bt. John H. Bryan of Lake Bluff, IL (1936–2018) from Seligmann Antiques, London, June 10, 1997; bt. John H. Bryan III of Nashville, TN, September 5, 2013

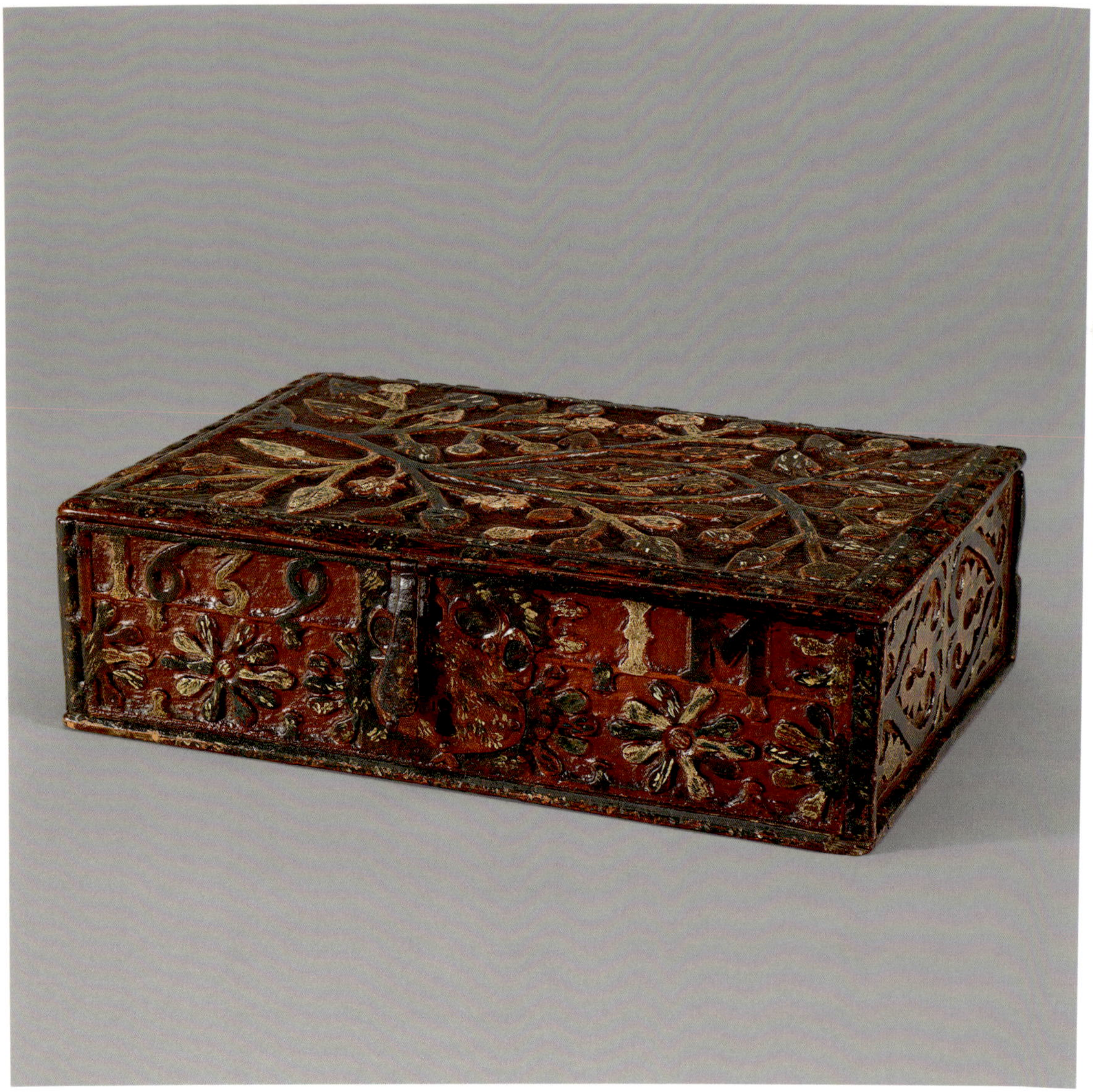

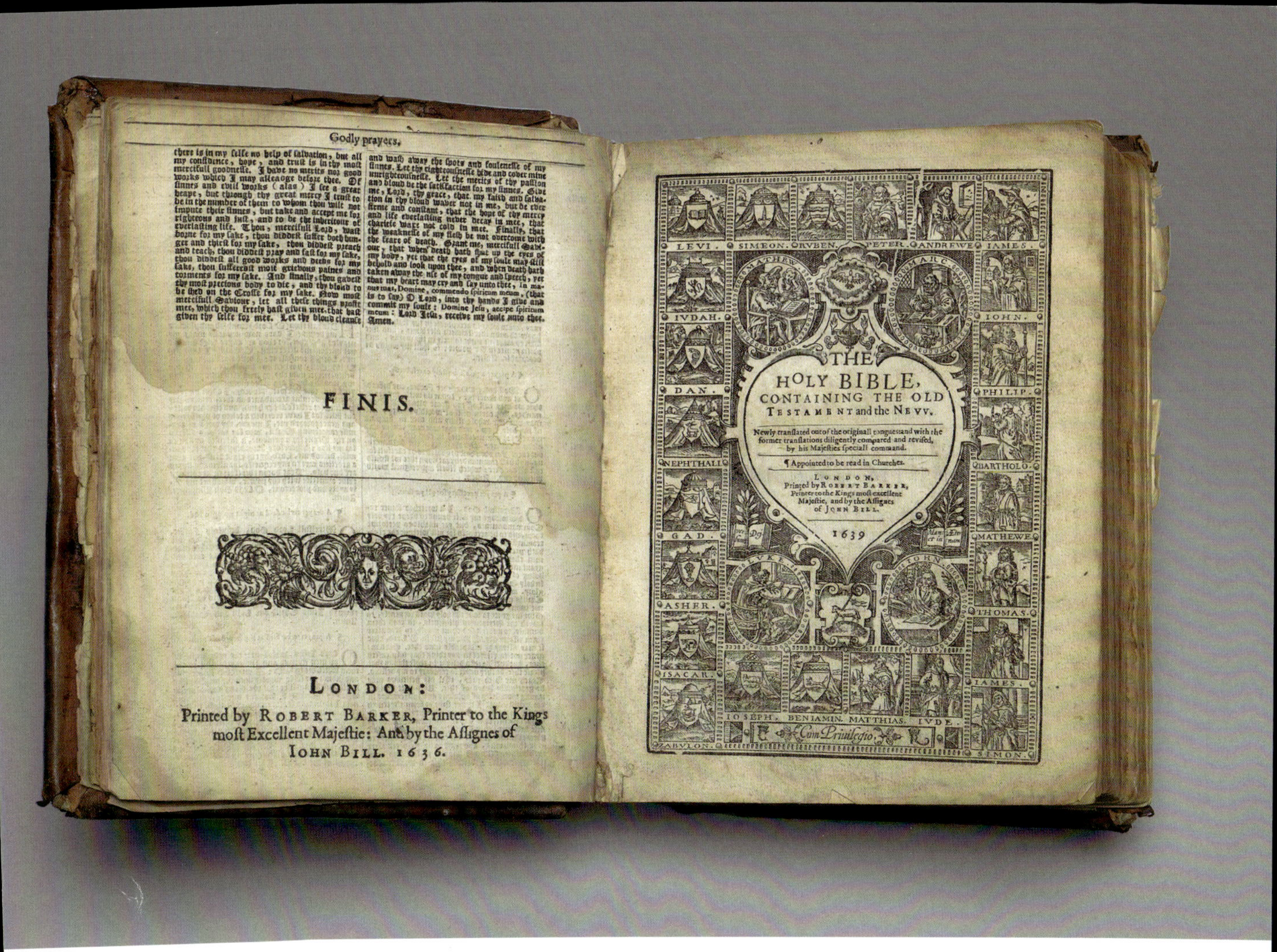

1636 | 1639 | 1641 | 1693 | 1872

Bible

Paper, board, leather

$2\frac{3}{4} \times 7 \times 9\frac{1}{2}$ in. ($7 \times 17.7 \times 24$ cm)

Inscription: LONDON: Printed by ROBERT BARKER, Printer to the Kings|most Excellent Majestie: And by the Assignes of|IOHN BILL 1636; Thomas Hirons|Brandon 1872; I have mad the fomer verses|unto vereses of almite god and JOHN Lay|Tomas Whight|I Remembe|ar my love|to youe ho|pinge that|youe are| in good|health as|I thanke god I a|me will|iam lee his bo|oke|good|give him gra|ce|th John Lea : John; Tomus Corties|I Remember|my Love unto|youe hopinge|that you are| in good heall|the at this ti|me but I am not|at this ti|me willia|m lee his|book good|give him [____]|[____] thereon|to looke|John Lea; Samwell whight|I Remembear| my Love unto|you hoping|John Lea|Samwell whight; John Lea

The whole BOOKE OF PSALMES, Collected into English Meeter, by Thomas Sternebold, Iohn Hopkins,| and others, conferred with the Hebrew, with|apt Notes to singe them withall.|Set forth and allowed to be sung|in Churches, of all the people together be-|fore and after Morning and Evening Prayer; as also|before and after Sermons; and moreover in pri-|vate houses, for their godly solace and comfort, laying|apart all ungodly Songs and Ballads, which tend only to the nourishment of Vice, and corrupting of Youth. [____] LONDON|Printed by E. G. for the Company of|Stationers 1641.

John Clemans|Newton|4 plac'd; Cause mouartall|men will not re|pent the hevens in shoures will not|re lent|will Leigh; cause mouretall|will not|[____]pent; John Leigh|shut not up my sule|with them in sine that take thear|fille nor yet my life among|those men that ske much|blood to spille|John Leigh; Richard harries; My shepherd is the living Lord northing therefore I need in pastures faire with waters calem he|set me for to feed he div cover and glade my soule and brought my minde in frame to walk|in paths of rightousnesse for his most holy name yea though I walk in valc of death yet will I|[____] none ill thy rod thy stafe do comfort me, and thou art with me still and in thy presence [____] my taboule though shalt spread thous shalt Lord fill full my cup and [____] John Leigh; Thomas clement 1693|Rit this; The sword|shall them|devours eache|one; I saide I wolde loke to my wayse|for feare I should goe ronge; Bee it known to all men|By this present|that I John Lea doe, bounde; Josph Stretton His Booke; pacem| mine|John Lea; Robert Wisd[____]| finis; I [____] out|thye [____]; Cnoment|Covenante; William Neale

THE|NEW TESTAMENT|OF OUR|LORD AND SAVIOUR|JESUS CHRIST.|Newly translated out of the originall Greek: and with the|former translations diligently compared and revised|by his Majesties speciall command.|Appointed to be read in Churches.|LONDON,|Printed by ROBERT BARKER, Printer to the Kings most excellent|Majesteis, and by the Assignes of JOHN BILL.|1639

This book contains three texts: a Book of Common Prayer of 1636, a King James Bible of 1639, and a book of Psalms of 1641, bound together in a single volume. The names in various inscriptions place the book in Oxfordshire for much of its history and, later, in Brandon, Warwickshire, when in 1872 it was the property of a carpenter named Thomas Hirons.

One of the first individuals to annotate this book was a man called John Leigh (or Lay or Lea), who seems to have taken great delight in adding to the margins small notes of remembrance to his friends. He had a strange habit of funneling his text into lines of decreasing length, to the extent that they eventually become barely legible. In one of these notes, he wished his friends good health and remembered his love to them but also recorded that he was himself, at that time, not in good health. Elsewhere, Leigh transcribed the words of Psalm 23, which may have been a comfort to him, particularly during a period of illness.

PROVENANCE

bt. John H. Bryan of Lake Bluff, IL (1936–2018) from Seligmann Antiques, London, June 10, 1997; bt. John H. Bryan III of Nashville, TN, September 5, 2013

1767

Mug
Salt-glazed stoneware
$5\frac{1}{2} \times 5\frac{1}{8} \times 6\frac{3}{4}$ in.
(14 × 13 × 17.1 cm)

Inscription:
William Bennit|
june 24,, 1767

One of the characteristics of Nottingham ware from this period is that it often carried inscriptions with names, dates, and the place of manufacture. Often this information claimed authorship for the maker, but in this instance, the inscription appears to refer to the owner, who has yet to be identified.

PROVENANCE
bt. John H. Bryan of Lake Bluff, IL (1936–2018) from Martyn Edgell Antiques Limited, Cambridgeshire, February 5, 2017

LITERATURE
Frank L. Wood, *The World of British Stoneware: Its History, Manufacture and Wares* (Kibworth Beauchamp, UK: Matador, 2014), 7–8

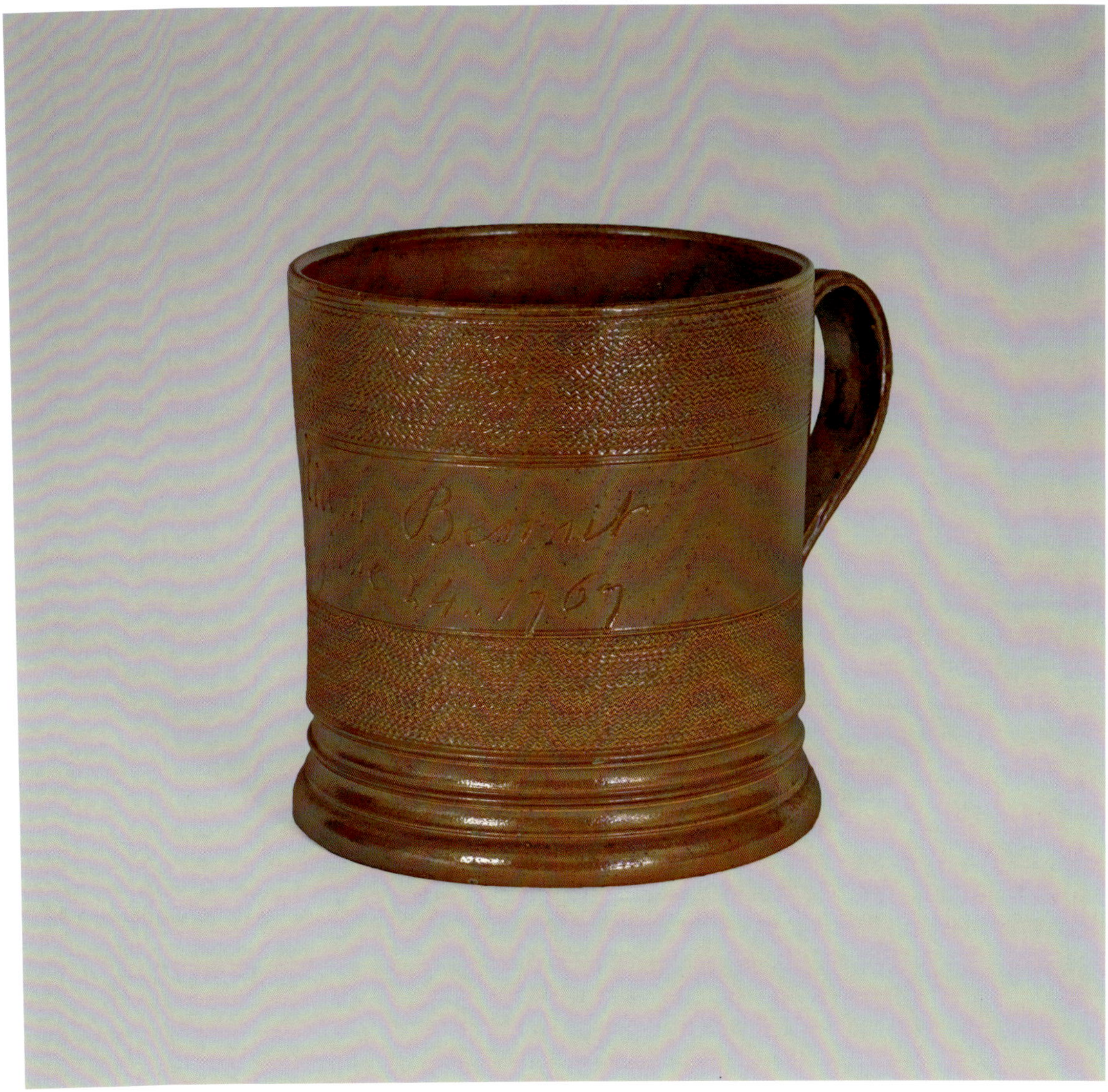

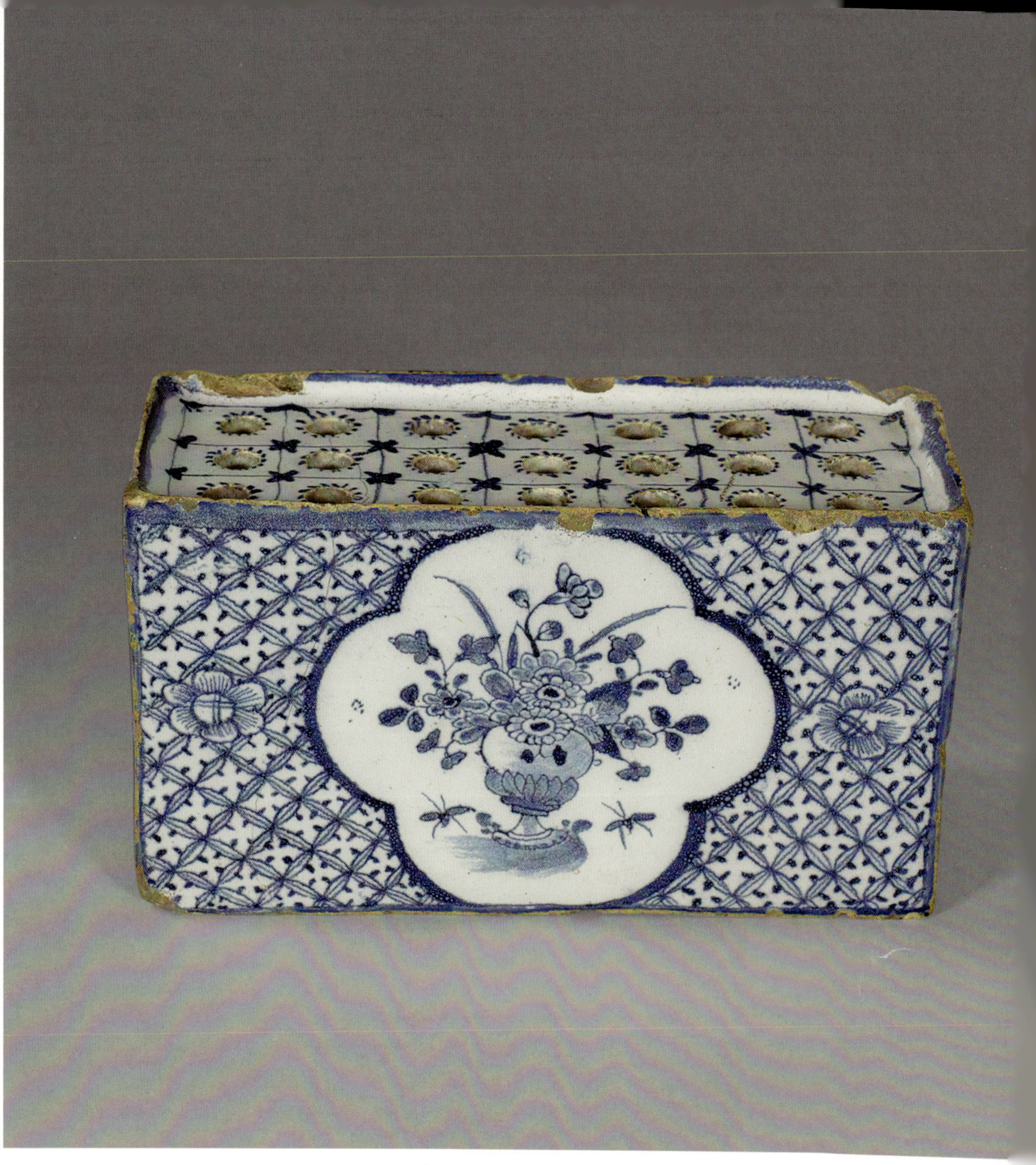

1737

Flower brick

Delftware

$3\frac{1}{2} \times 6\frac{1}{8} \times 2\frac{1}{4}$ in. (8.9 × 15.6 × 5.7 cm)

Inscription: E|1|7|3|7

This brick is one of a pair; the other sold to John Philip Kassebaum and later became part of the Longridge Collection. These sets were not uncommon. The bricks have a type of decoration that was popular on English Delftware throughout most of the eighteenth century. Here, the decoration gestures at the object's use: the bouquet in a vase mirrors what would be the contents of the brick—multiple stems of freshly cut flowers.

PROVENANCE

Woolley and Wallis, Salisbury, March 11, 1982 (lot 307); bt. John H. Bryan of Lake Bluff, IL (1936–2018) from Jellinek & Sampson Antiques, London, July 26, 1982

LITERATURE

Leslie B. Grigsby, *The Longridge Collection of English Slipware and Delftware* (London: Jonathan Horne Publications, 2000), 1:414

Louis L. Lipski and Michael Archer, *Dated English Delftware: Tin-Glazed Earthenware, 1600–1800* (London: Sotheby Publications, 1984), 365, nos. 1591, 1591A

1720

Model dogs
Delftware
Each: 6 × 5¾ × 2⅞ in. (15.2 × 14.6 × 7.3 cm)

Inscription, on both: 1720

There are no other known examples of painted Delftware figures made from this mold. This pair of model dogs, like other Delftware models, was purely decorative, made to adorn a mantelpiece or shelf.

PROVENANCE
bt. John H. Bryan of Lake Bluff, IL (1936–2018) from Lindsay Antiques, London, February 27, 1984

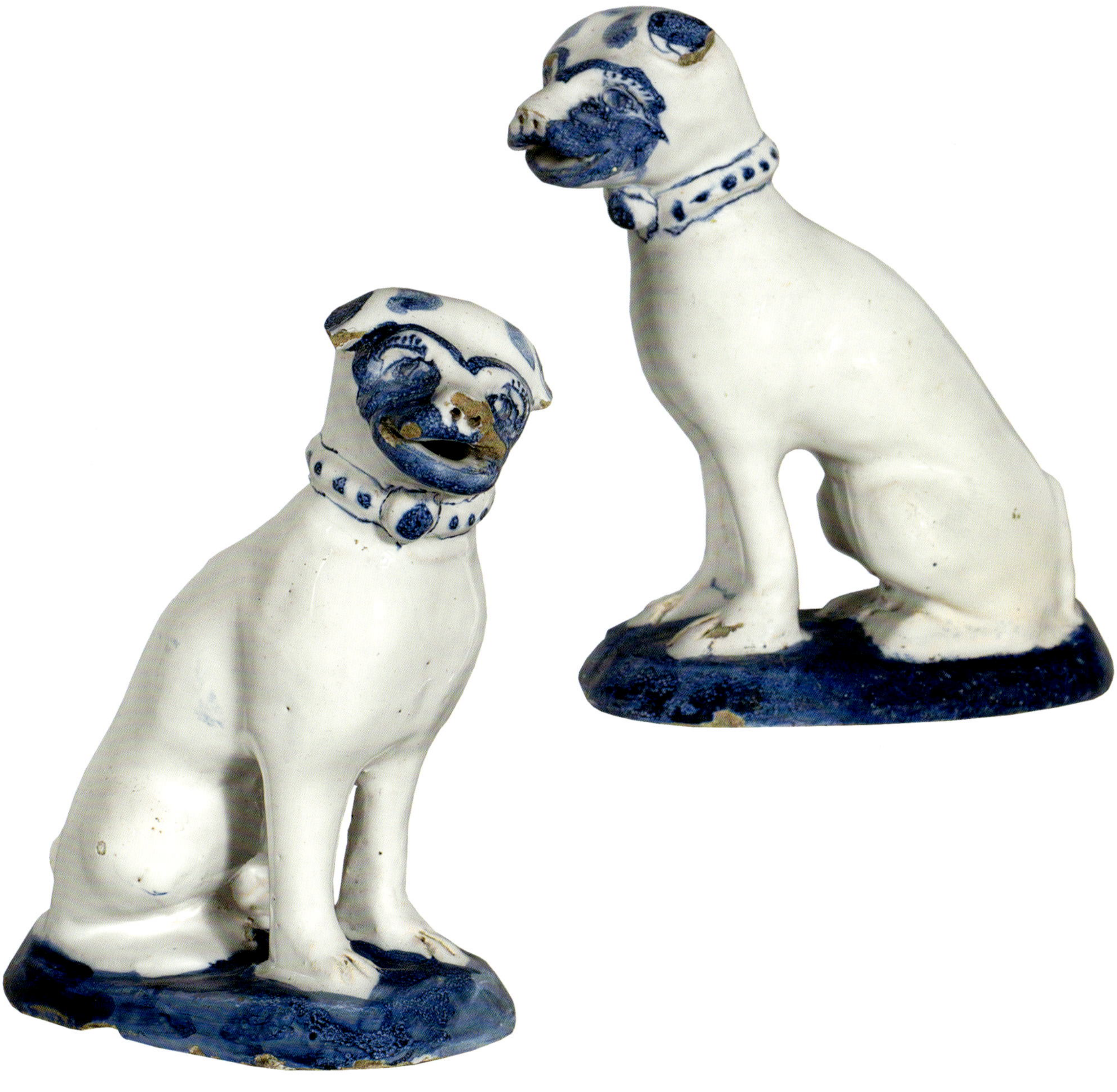

1721

Dog bowl
Brass
$4\frac{1}{8} \times 7$ in. (10.5 × 17.8 cm)

Inscription:
Will|1721

"Will" is surely the name of the dog that drank from this bowl, not the name of the bowl's owner.

PROVENANCE
bt. John H. Bryan of Lake Bluff, IL (1936–2018) from Mallett Antiques, New York, June 17, 2013

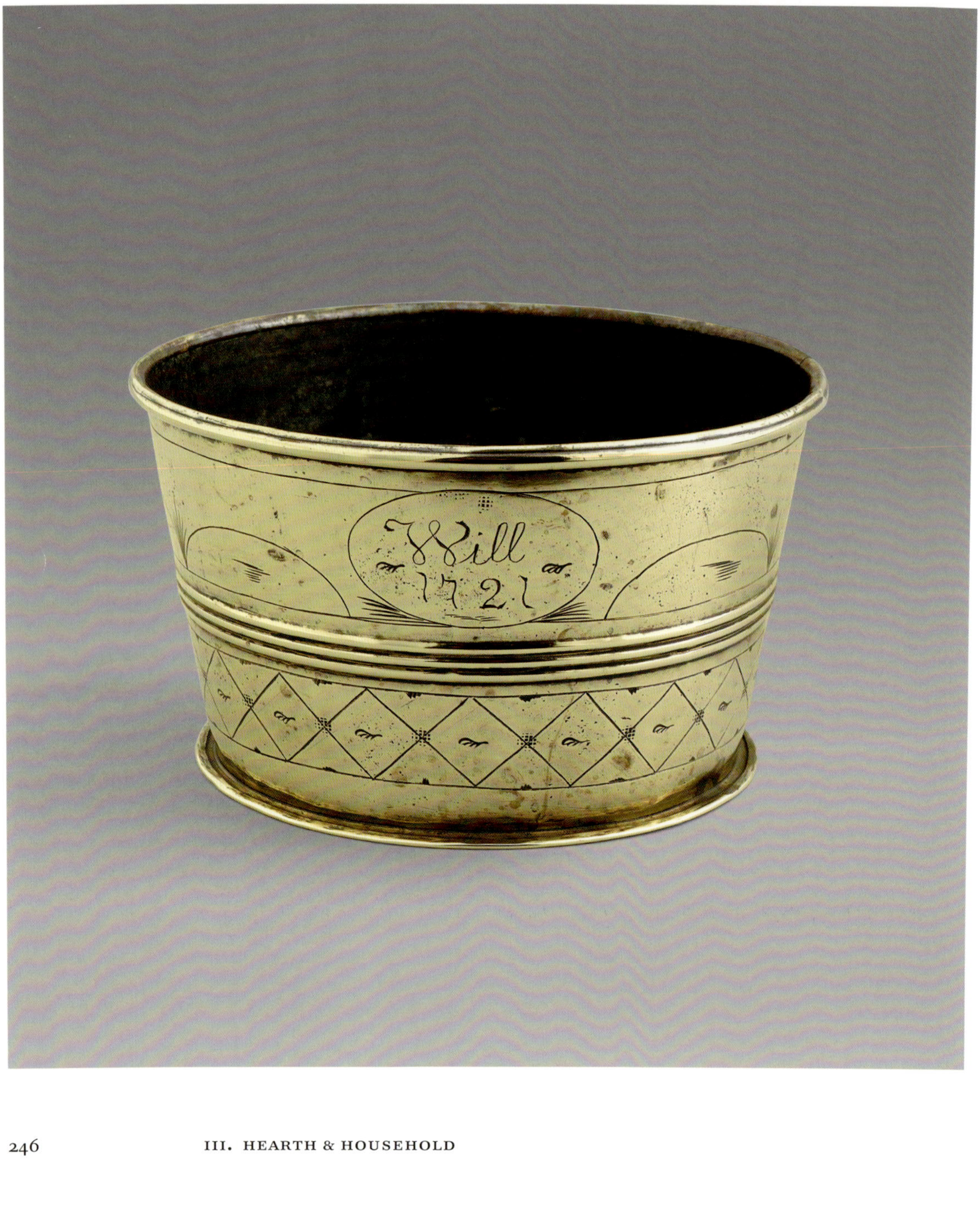

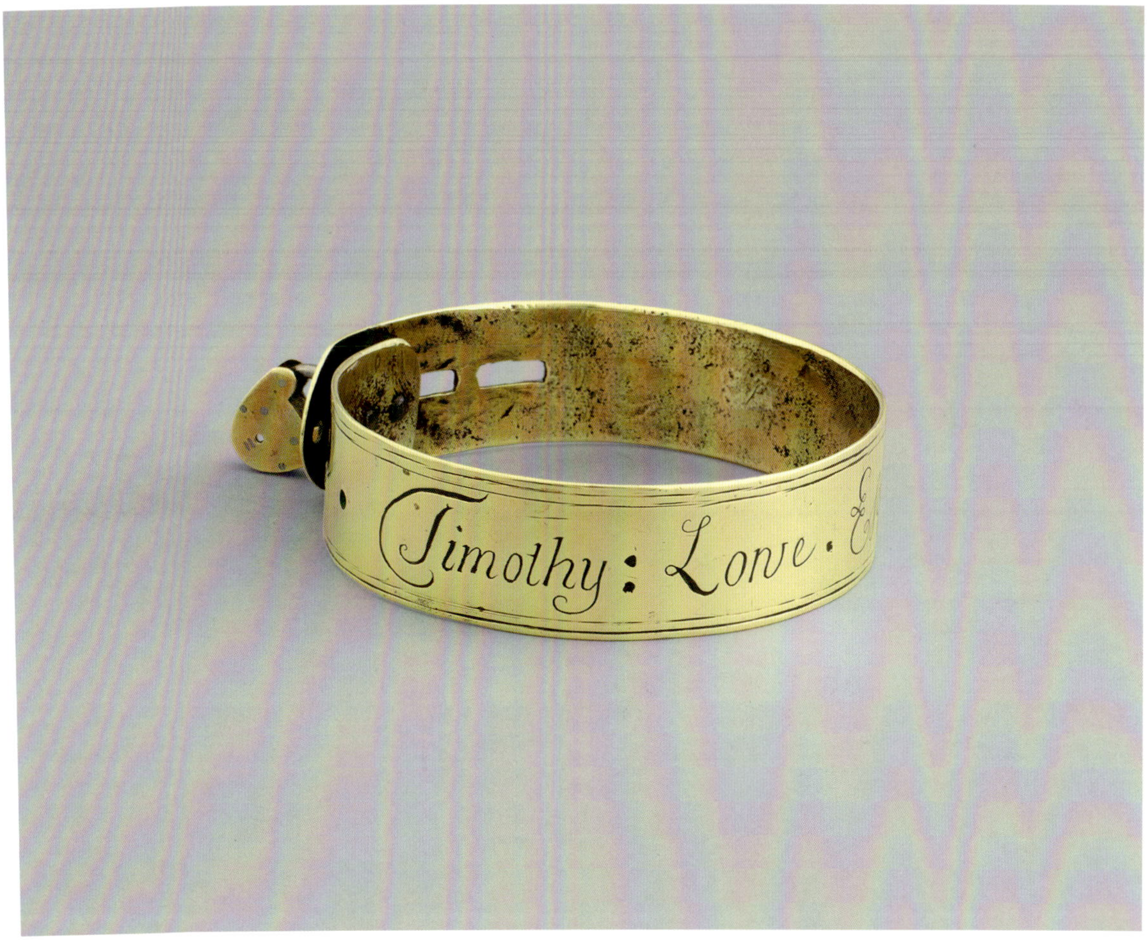

1721

Dog collar

Brass, steel

4 × 7 in. (10.2 × 17.8 cm)

Inscription:
Timothy: Lowe. Esq. of New Work 1721

At one time, dog collars were designed to protect the dog: broad iron collars lined with spikes shielded the animal's throat from attacks by other animals. In the eighteenth century, decorative dog collars became more common. The simplest were made with a brass or copper ring secured with a padlock, like this one. Collars typically bore the name of the owner rather than the name of the dog. They also indicated a place to which to return the animal should it run away or get lost.

The dog that wore this collar was property of Timothy Lowe of Newark Park in Ozleworth, Gloucestershire, son and heir of Sir Gabriel Lowe. In centuries prior, the Lowe family had been London merchants. They enlarged the hunting lodge built in the mid-sixteenth century by the Henrican courtier Sir Nichlas Poyntz (1510–1556), converting it into a home for permanent habitation. The Lowes owned Newark Park until a year after this dog collar was inscribed; the property is now cared for by the National Trust.

PROVENANCE

Bonhams, London, September 28, 2016 (lot 78); bt. John H. Bryan of Lake Bluff, IL (1936–2018) through Tobias Jellinek Antiques, London

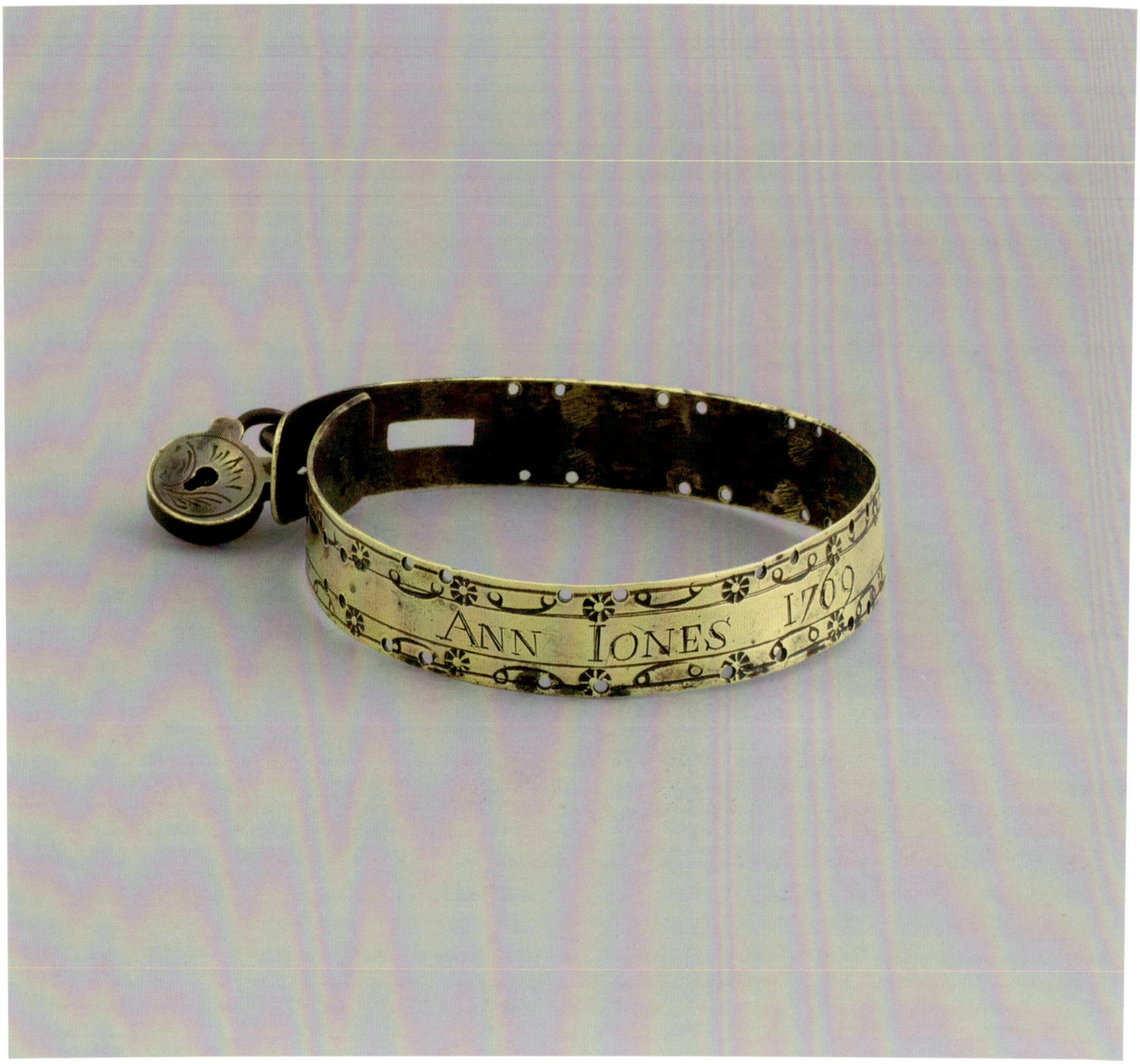

1769

Dog or monkey collar

Brass, steel

$\frac{1}{2} \times 2\frac{3}{4}$ in. (1.27 × 7 cm)

Inscription: ANN JONES 1769

Adjustable to two positions, this collar would originally have had a leather lining to reduce chafing. It is not clear whether it was made for a small dog or for a pet monkey. Monkeys were commonly kept by those who could afford such exotic pets.

PROVENANCE

Bonhams, London, September 28, 2016 (lot 76); bt. John H. Bryan of Lake Bluff, IL (1936–2018) through Tobias Jellinek Antiques, London

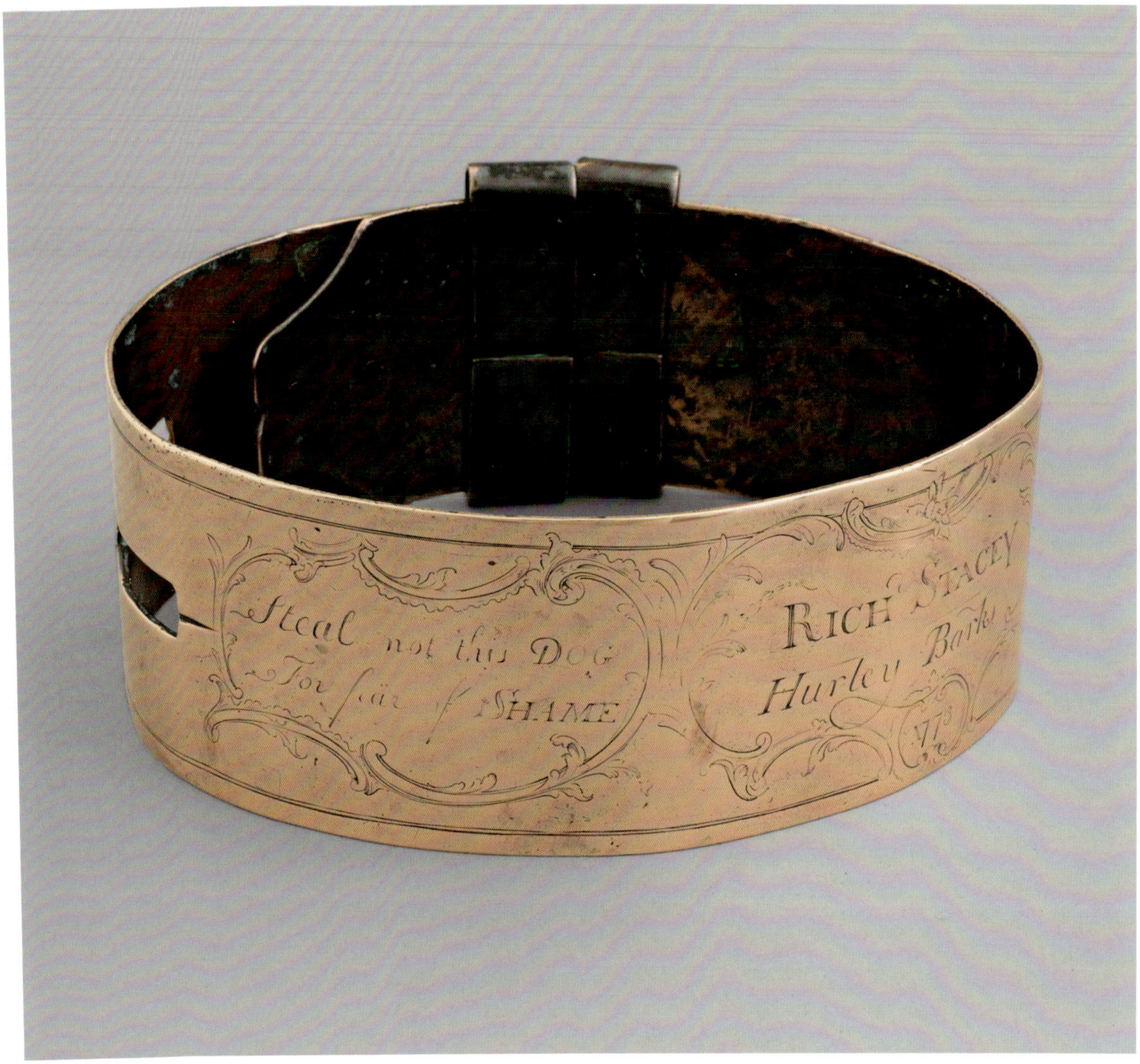

1773

On July 23, 1772, Richard Stacey, "Innholder," married Susannah Lawrence in the ancient church of St Mary the Virgin, Hurley, Berkshire, by the banks of the Thames and not far from Bisham Abbey. Both he and his bride were twenty-five years of age. The bondsman was James Sale, "Gent." A year later, Stacey had this collar made for his dog, which would have helped him keep order at his inn. His establishment may have been the Olde Bell, a timber-framed coaching inn that is still in business today, just a stone's throw from the church.

PROVENANCE
bt. John H. Bryan of Lake Bluff, IL (1936–2018) from Christopher Bangs, London, August 10, 1995

Dog collar

Copper

$2\frac{1}{2}$ × 6 in. (6.4 × 15.2 cm)

Inscription: Steal not this DOG|For fear of SHAME; RICH[d] STACEY| Hurley Barks|1773 For hear you See| The owners NAME

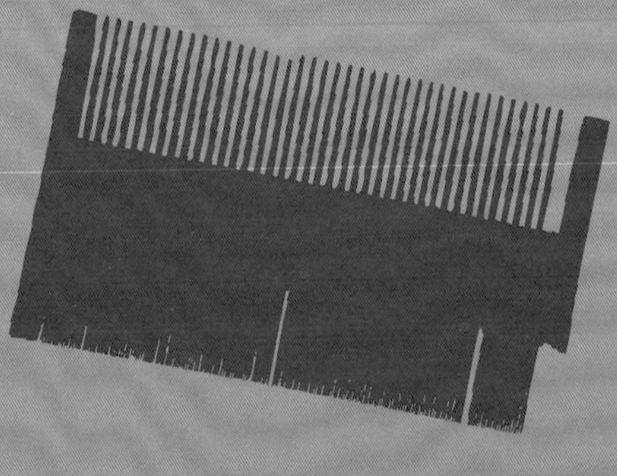
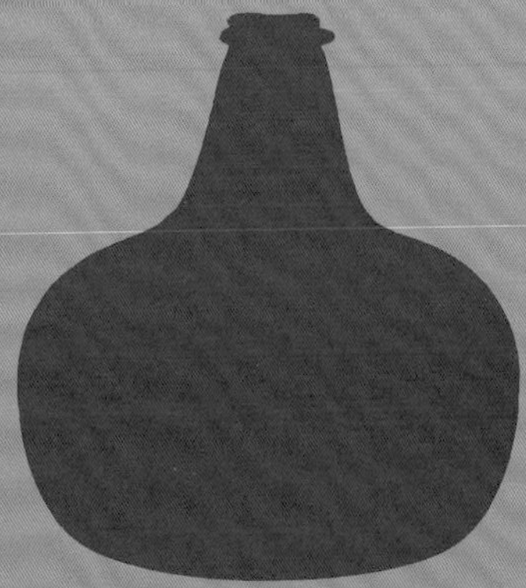

IV. FASHION & FRIENDSHIP

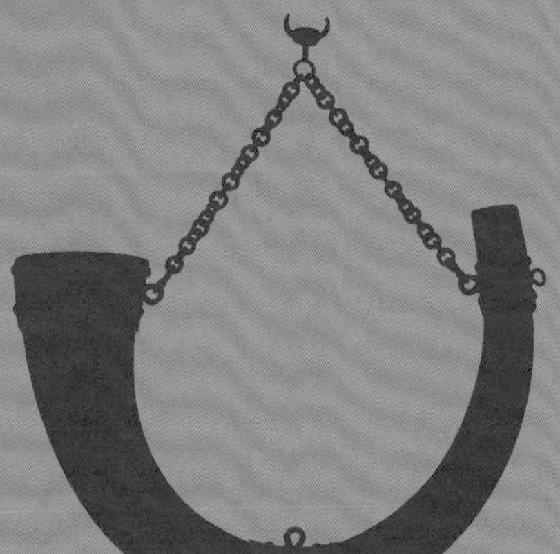
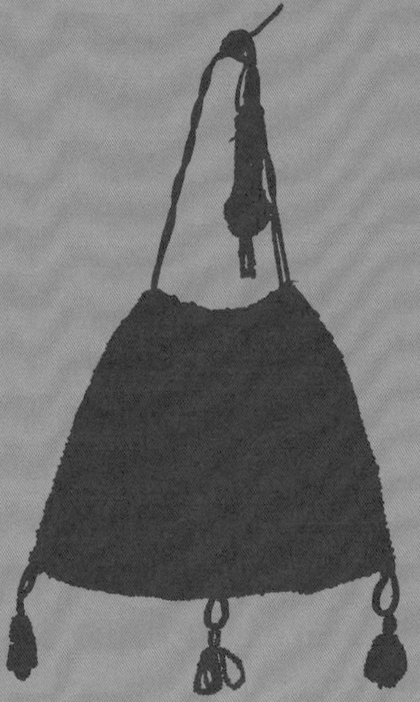
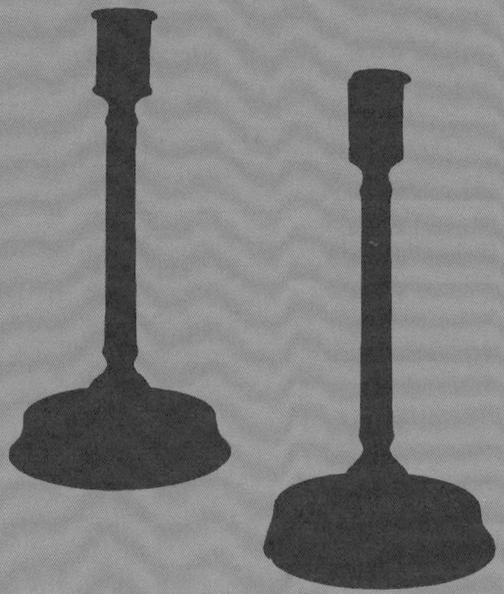

COFFEEHOUSES, INNS, TAVERNS, AND OTHER HOSTELRIES WERE important sites for the consumption of newly introduced commodities, many of them intoxicants containing alcohol, nicotine, or caffeine. All inspired their own craze. Whether the sugar in the punch or the tobacco in the pipe, these new substances were the products of Britain's colonial enterprise—and, with that, of a violent and exploitative system of ensnarement, dislocation, and enforced labor, upon which the slave economy depended. But, by the time these commodities arrived in Britain and were dissolved into tea, thrown down the throat, or fired up the nose, the hours, days, and months of grueling work that had gone into their production and transportation had been obfuscated: a lump of sugar, nipped from a cone, carried no indication of where, how, or by whom it had been produced.

Along with habits in intoxication, fashions in dress changed rapidly as Britain became ever more entrenched as a consumer society. Hats, wigs, walking sticks, and other accoutrements shifted in shape and style, reaching elaborate and often gravity-defying heights in some metropolitan circles. Men who sported these extravagant accessories were pejoratively dubbed "Macaroni" (named for the newly fashionable Italian pasta) and lampooned as affected, effeminate, and "Italianate." Clothes themselves were, as indexes of changing fashion, almost never embroidered with dates, although small items like buckles and baldrics, which were particularly vulnerable to theft, were sometimes marked with names and dates.

In a period when even the choice of one's drink could signal political allegiance, communal drinking was a serious business. By modern standards, people drank a lot and enjoyed doing it; white, red, and fortified wines were drunk in formidable quantities, beginning at the start of the day. Communal toasts, regularly made in both the tavern and the home, helped reinforce shared loyalties and common values. A mug or tankard that bore a particularly auspicious date provided those gathered with a call to raise their glasses in memory of a monarch or a moment in a monarch's reign—and a call for another round of drinks.

Interpersonal relationships and friendships were begun, galvanized, and maintained through the practices of gift giving and sharing. Many receptacles for tobacco, for instance, bear inscriptions that speak to these systems of generosity and reciprocity. In the seventeenth and eighteenth centuries, the term *friend* applied to spouses, members of the wider family, trade colleagues, important patrons, and close acquaintances. One depended on "friends" to speak on your behalf during times of trouble, to put your name forward when you sought new opportunities, and to help out when times were hard. As such, to claim someone as a friend was a vital social and economic asset in early modern Britain.

The transfer of property, income, and influence, however great or small, could make or break an individual—especially a woman or a younger son, who perhaps had been brought up without a trade or had failed to thrive in one and had no claim on family support. Consequently, maintaining family relationships through regular visits, offers of hospitality and entertainment, and the giving of gifts was particularly important.

When friendship could not be expressed with the offer of a pinch of snuff, it could be sustained through correspondence. It is thus unsurprising to see dates on objects associated with writing, situating the correspondence within a temporal framework as the writer recorded events from the past and anticipated events to come.

1674

Ale bottle
Glass
8 × 6 in. (20.3 × 15.2 cm)

Inscription:
THISTLE ~~ BOON ~~; Bydder 1674

Thought to be the second-earliest Welsh sealed bottle, this ale bottle is the earliest extant example that carries a name and date. It has a dry capacity of 10 ounces (800 milliliters). It belonged to Thomas Bydder (d. 1681), a yeoman farmer of Thistleboon, near the parish of Oystermouth, south of Swansea.

England exported dark glass bottles like these throughout the world. Many have applied seals that identify the owner, who was sometimes the proprietor of a tavern, as is true for the bottle on the facing page.

PROVENANCE
Roger Dumbrell; bt. John H. Bryan of Lake Bluff, IL (1936–2018) from Jonathan Horne, London, September 6, 1990

LITERATURE
Dwight P. Lanmon, *The Golden Age of English Glass: 1650–1775* (Woodbridge, UK: Antiques Collectors' Club, 2011), 300, no. 107

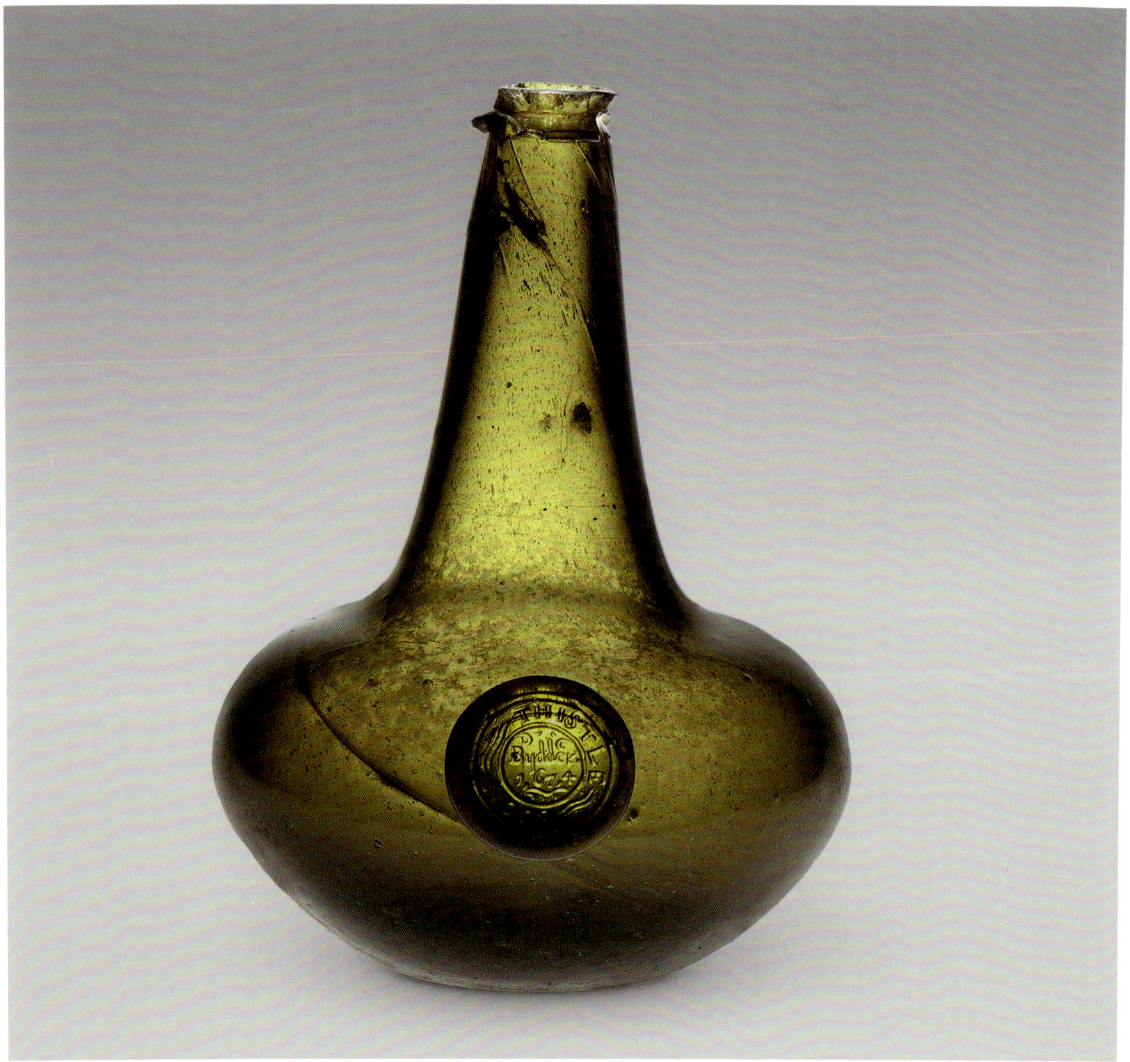

1684

A pint bottle made for Anne Morrell (d. 1696) and William Turton (her son from her first marriage) at the Crown Tavern in Oxford.

Until her death in 1696, Anne and her son managed the Crown Tavern, located on what is now 3 Cornmarket, Oxford, which she had leased from New College in 1665. They provided wine and bottles for the city council on Guy Fawkes Day in 1681. Despite earlier kerfuffles with the university over her license to serve wine, Anne died a wealthy and successful woman (TNA PROB 11/431/14).

There are several surviving sealed bottles from Crown Tavern, some with the ciphers of subsequent licensees. The earliest dated Morrell seal is from 1674 and the latest is from 1701, when the tavern was licensed to Anne's granddaughter, Joan Morrell.

PROVENANCE

Roger Dumbrell; bt. John H. Bryan of Lake Bluff, IL (1936–2018) from Jonathan Horne, London, September 6, 1990

LITERATURE

Dwight P. Lanmon, *The Golden Age of English Glass: 1650–1775* (Woodbridge, UK: Antiques Collectors' Club, 2011), 504

E. T. Leeds, "17th and 18th Century Wine-Bottles of Oxford Taverns," *Oxoniensia* 6 (1941): 44–55

Ale or wine bottle

Glass

6 × 5 in. (15.2 × 12.7 cm)

Inscription: OX ON|EWAMOR [in monogram] 1684

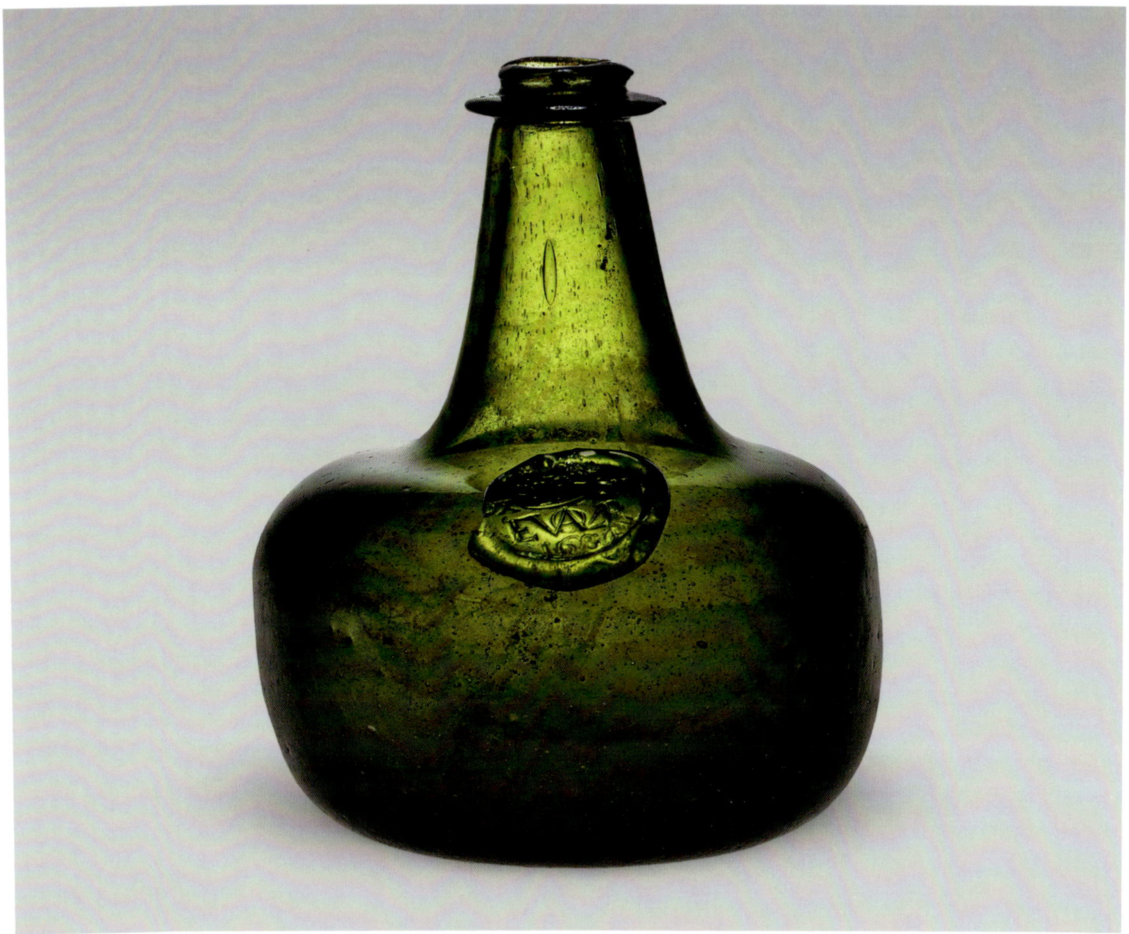

1691

Wine bottle
Glass
6½ × 6½ in. (16.5 × 16.5 cm)

Inscription:
R·N|1691

This wine bottle has a dry capacity of 950 milliliters (32 ounces). It is believed to have been dug up near Devizes in Wiltshire.

PROVENANCE
bt. John H. Bryan of Lake Bluff, IL (1936–2018) from Jonathan Horne, London, August 10, 1989

LITERATURE
Dwight P. Lanmon, *The Golden Age of English Glass: 1650–1775* (Woodbridge, UK: Antiques Collectors' Club, 2011), 300, no. 107

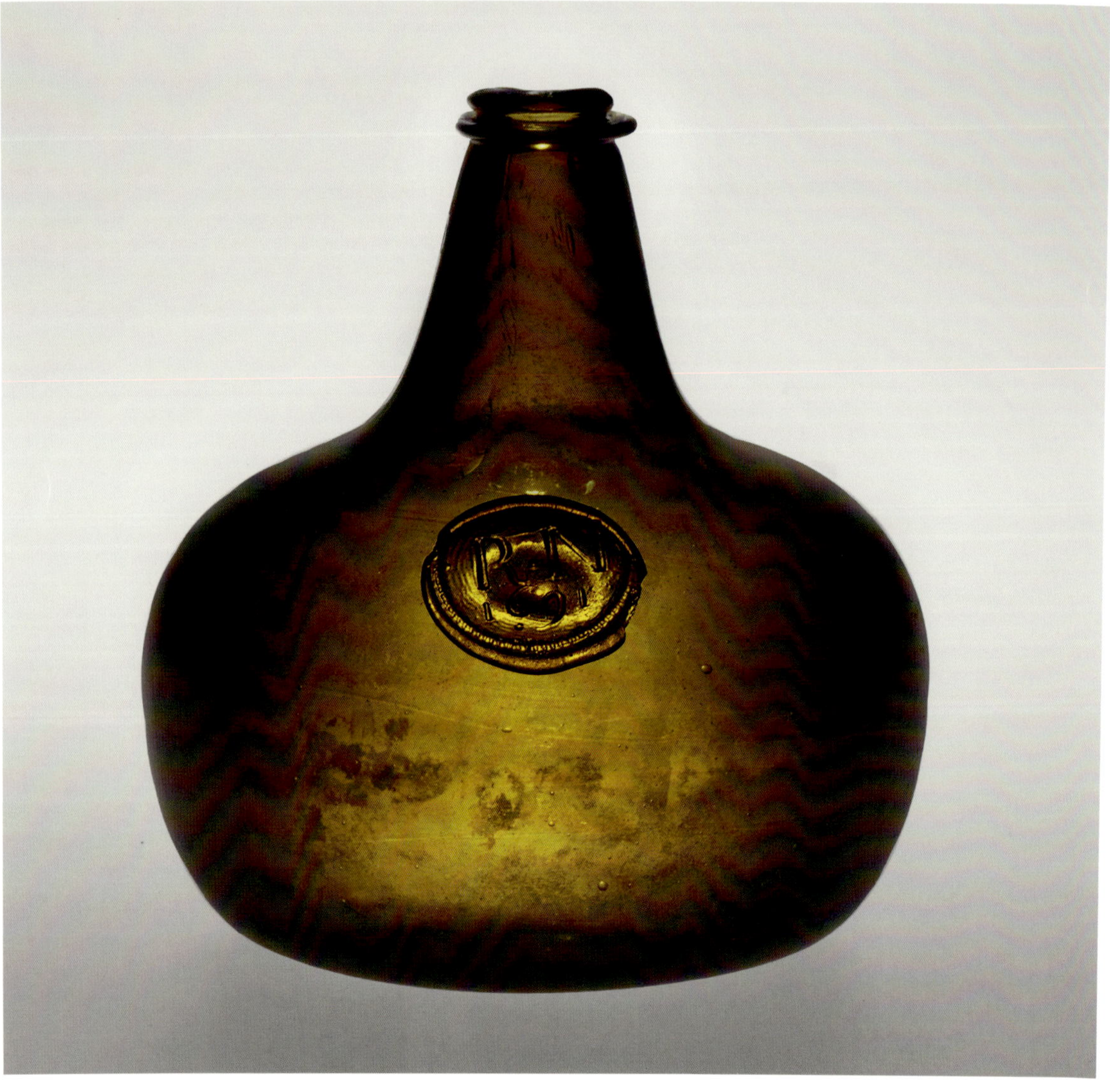

1705

This bottle, with a dry capacity of 1 liter (34 ounces), is ideally suited for beer. The initials of what were presumably a husband and wife have not been identified, but the survival of another bottle with the same initials in the collection of the Guernsey Museums and Galleries raises the possibility of a link with the Channel Islands.

PROVENANCE

Tom Floyd; his sale, Christie's, London, February 13, 1990 (lot 17); bt. John H. Bryan of Lake Bluff, IL (1936–2018) through Jonathan Horne, London

LITERATURE

Dwight P. Lanmon, *The Golden Age of English Glass: 1650–1775* (Woodbridge, UK: Antiques Collectors' Club, 2011), 299, no. 105

Ale bottle

Glass

7 × 7 in. (17.8 × 17.8 cm)

Inscription:
ID·I|1705

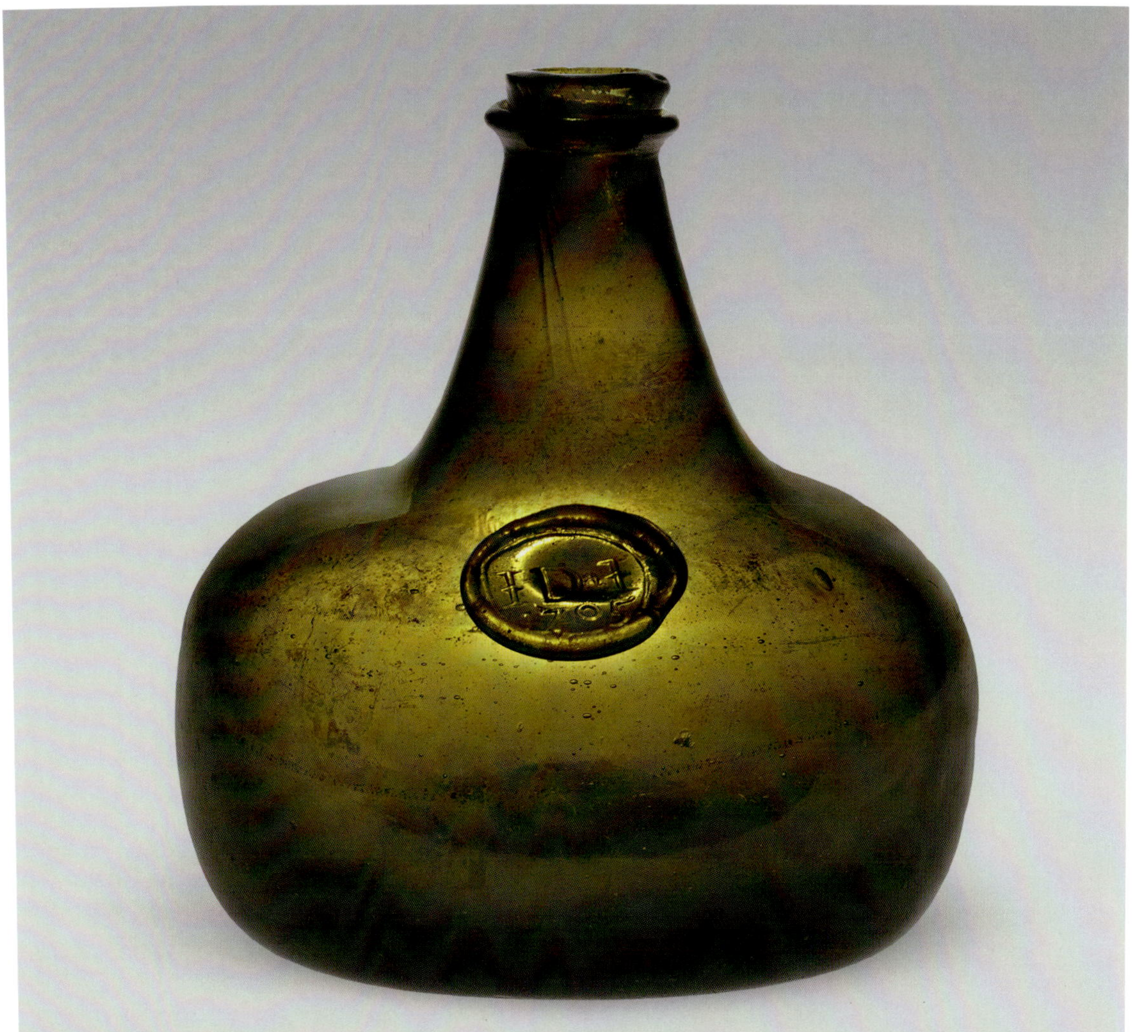

1712

Ale or wine bottle

Glass

6½ × 6½ in. (16.5 × 16.5 cm)

Inscription:
R|Reynolds|·1712·

This liter (quart) bottle for ale or wine was made for R. Reynolds.

PROVENANCE

bt. John H. Bryan of Lake Bluff, IL (1936–2018) from Jonathan Horne, London, January 4, 1987

LITERATURE

Dwight P. Lanmon, *The Golden Age of English Glass: 1650–1775* (Woodbridge, UK: Antiques Collectors' Club, 2011), 302, no. 111

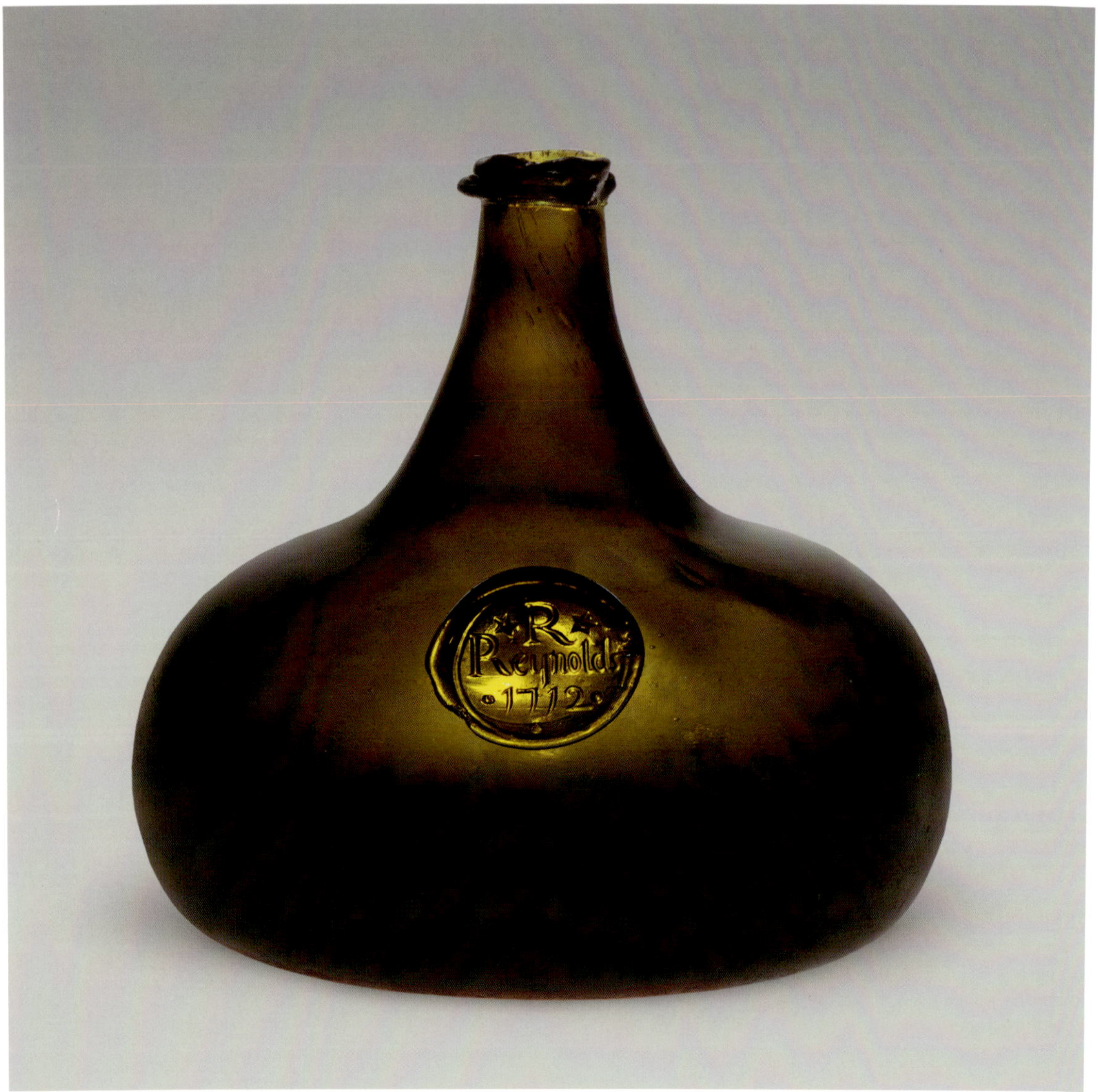

1723

This bottle, for ale or wine, has a dry capacity of 470 milliliters (16 ounces).

PROVENANCE
Finlater, Mackie, Tood and Co. Ltd., London, Collection; Sotheby's, London, July 21, 1980; Jim Galloway Collection; Tom Floyd; his sale, Christie's, London, February 13, 1990 (lot 26); bt. John H. Bryan of Lake Bluff, IL (1936–2018) through Jonathan Horne, London

LITERATURE
David Burton, *Antique Sealed Bottles 1640–1900 and the Families Who Owned Them* (Woodbridge, UK: Antiques Collectors' Club, 2015), 602

Dwight P. Lanmon, *The Golden Age of English Glass: 1650–1775* (Woodbridge, UK: Antiques Collectors' Club, 2011), 303, no. 112

Ale or wine bottle
Glass
6 × 5 × 4 in. (15.2 × 12.7 × 10.2 cm)

Inscription: LS|1723

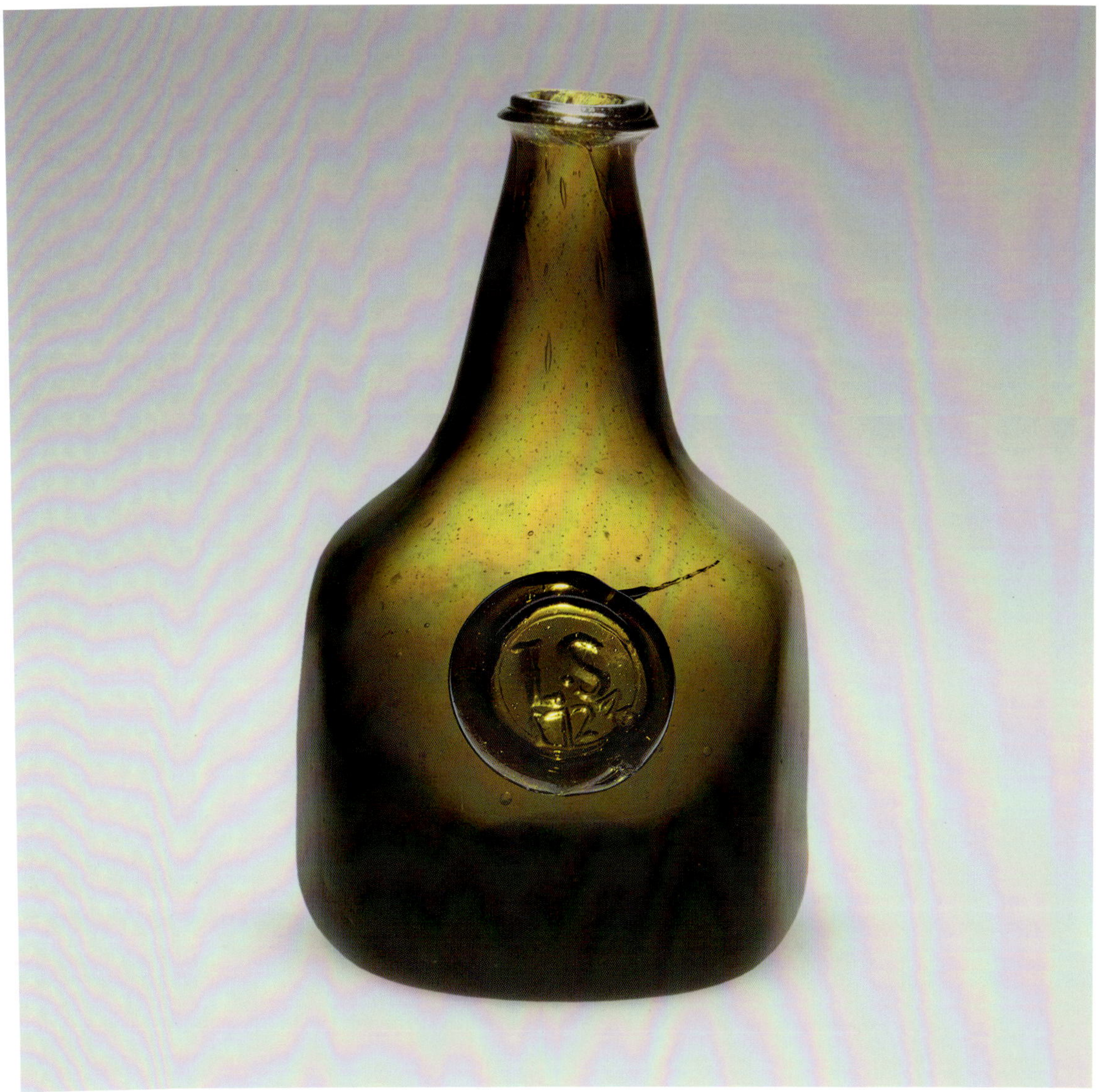

1723

Wine bottle
Glass
$6\frac{1}{2} \times 4\frac{1}{2}$ in. (16.5 × 11.4 cm)

Inscription:
R|Newman|1723

This pint bottle, with a dry capacity of 470 milliliters (16 ounces), was dredged from the mud in Dartmouth harbor during a search for coal that colliers had dropped there. Robert Newman (1676–1739) was a merchant in Dartmouth, Devon, whose family had traded with Newfoundland in codfish and port wine for two hundred fifty years. By tradition, a Newman ship loaded with port was pursued while en route to London and blown far off course, forcing the captain to seek the Newman estate at St. John's, Newfoundland. The wine was kept there in storage until brought back to London the following year, whereupon it was discovered that it had improved considerably. Thenceforth, wine from Oporto, Portugal, was matured in Newfoundland and sold in Britain, perhaps in bottles such as this one.

In his will (TNA PROB 11/701/188), Newman left his second wife, Elizabeth (née Holdsworth), two hundred pounds and also the dwelling house and malt house in Dartmouth, where presumably this bottle had been in use. Newman was buried at St Petrox, Dartmouth, on May 7, 1739.

PROVENANCE
Probably Robert Newman of Dartmouth, Devonshire (1676–1739); listed as being in a Newman family member collection in 1949; Bob Metselaar Collection; BBR auction, March 8, 1992; bt. John H. Bryan of Lake Bluff, IL (1936–2018) from Sheppard Cooper Ltd., London, June 15, 1992

LITERATURE
David Burton, *Antique Sealed Bottles 1640–1900 and the Families Who Owned Them* (Woodbridge, UK: Antiques Collectors' Club, 2015), 280, 600–601

Dwight P. Lanmon, *The Golden Age of English Glass: 1650–1775* (Woodbridge, UK: Antiques Collectors' Club, 2011), 303, no. 113

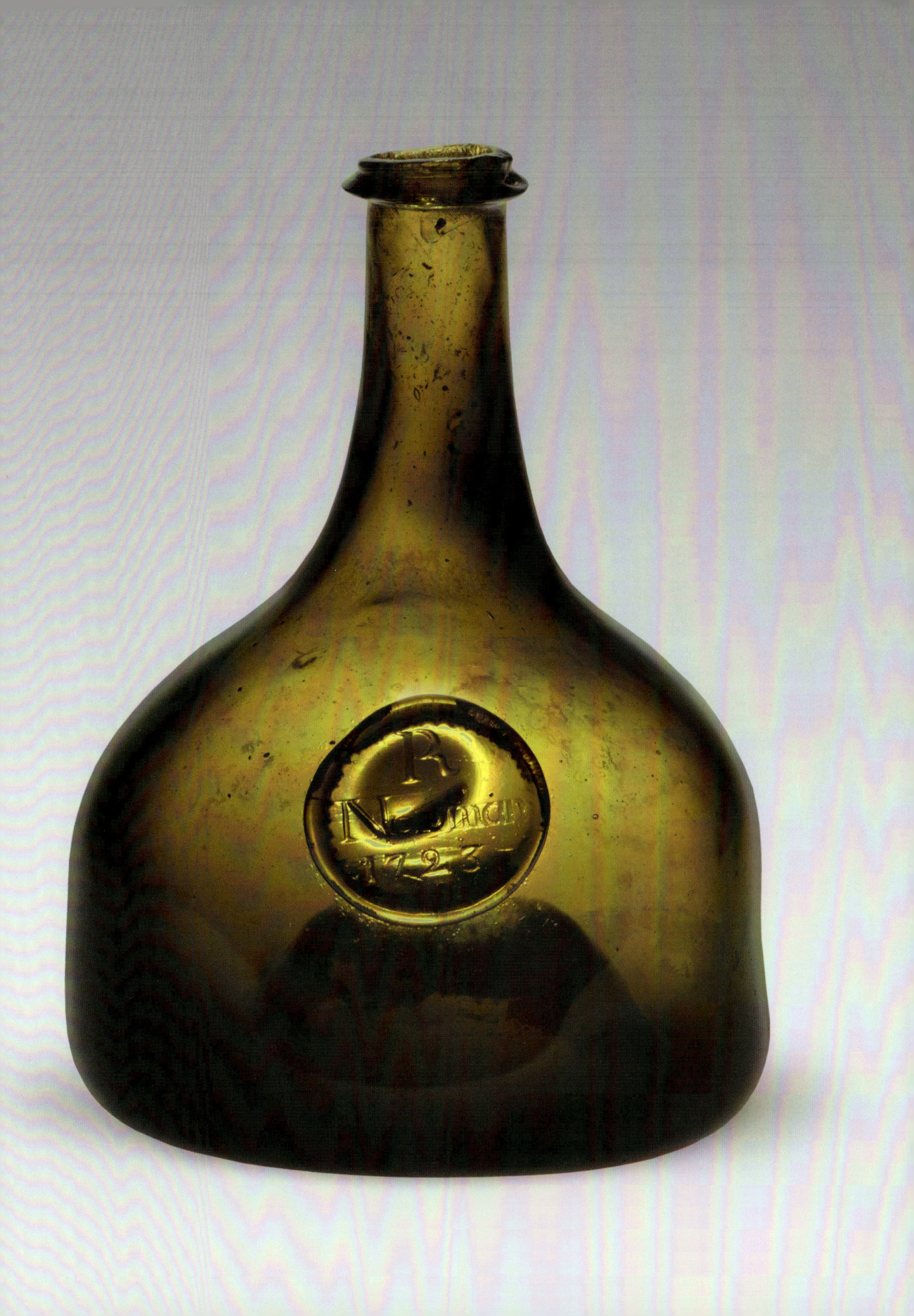
R
1723

1725

Ale or wine bottle
Glass
$5\frac{1}{2} \times 4\frac{1}{2}$ in. (14 × 11.4 cm)

Inscription: Wm| Allen|Looe|1725

This half-pint ale or wine bottle with a dry capacity of 300 milliliters (10 ounces) belonged to William Allen, who was probably the man buried at St Mary-by-Looe on December 9, 1754. This parish is immediately east of the town of Looe in South Cornwall.

PROVENANCE
Tom Floyd; his sale, Christie's, London, February 13, 1990 (lot 26); bt. John H. Bryan of Lake Bluff, IL (1936–2018) through Jonathan Horne, London

LITERATURE
David Burton, *Antique Sealed Bottles 1640–1900 and the Families Who Owned Them* (Woodbridge, UK: Antiques Collectors' Club, 2015), 609

Dwight P. Lanmon, *The Golden Age of English Glass: 1650–1775* (Woodbridge, UK: Antiques Collectors' Club, 2011), 302, no. 110

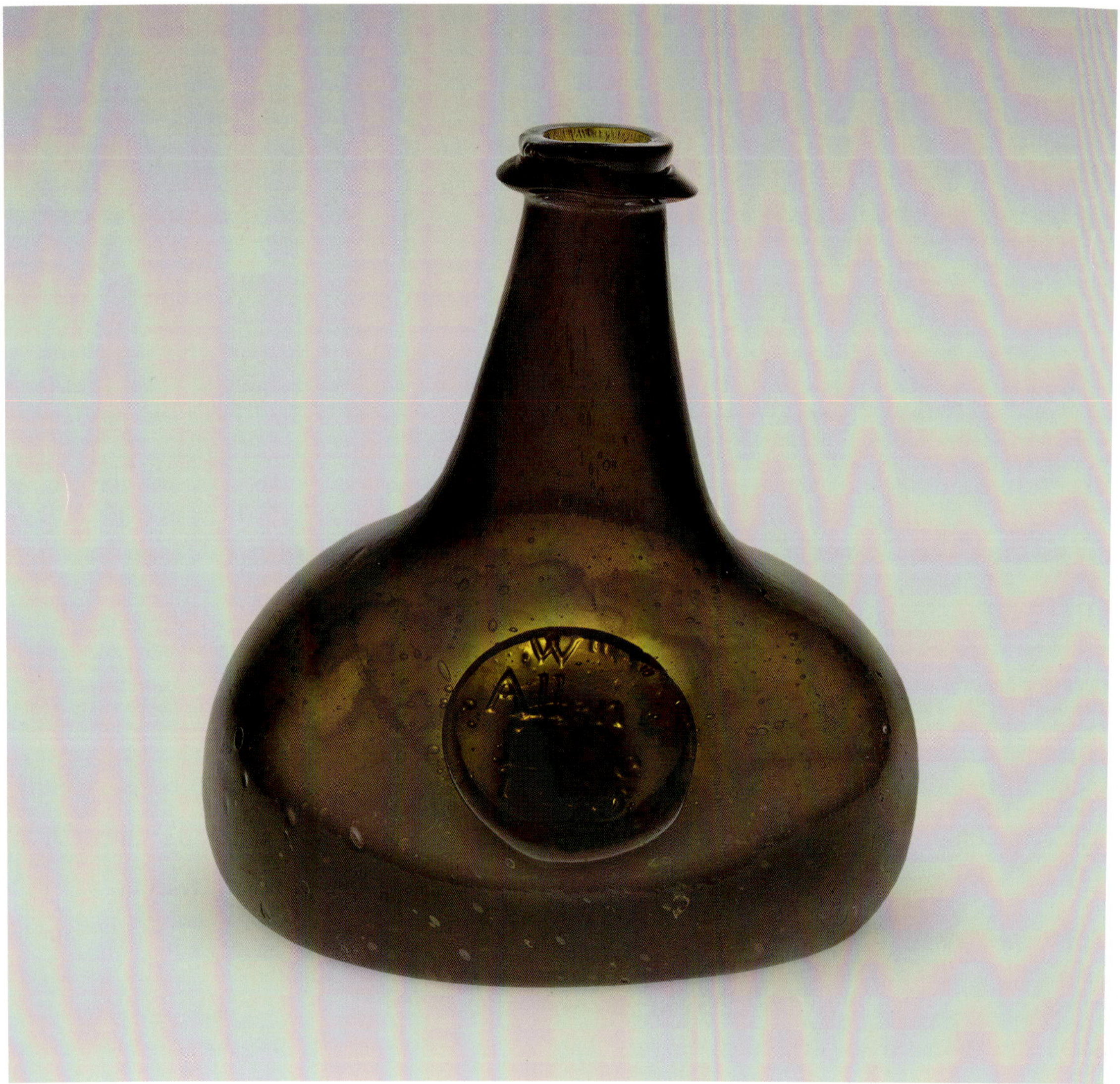

1731

Another ale bottle like this one, which has a capacity of 1130 milliliters (38 ounces), can be found in the Museum of London (NN24329). The owner of this bottle might be Roger Oxenham of Tiverton, who had his will proved in 1744 at the Archdeaconry Court of Exeter.

PROVENANCE

bt. John H. Bryan of Lake Bluff, IL (1936–2018) through Jonathan Horne, London, January 4, 1987

LITERATURE

David Burton, *Antique Sealed Bottles 1640–1900 and the Families Who Owned Them* (Woodbridge, UK: Antiques Collectors' Club, 2015), 53, 636–37

Dwight P. Lanmon, *The Golden Age of English Glass: 1650–1775* (Woodbridge, UK: Antiques Collectors' Club, 2011), 309, 117

Ale bottle

Glass

8 ½ × 5 in. (21.6 × 12.7 cm)

Inscription: Roger| Oxenham|1731

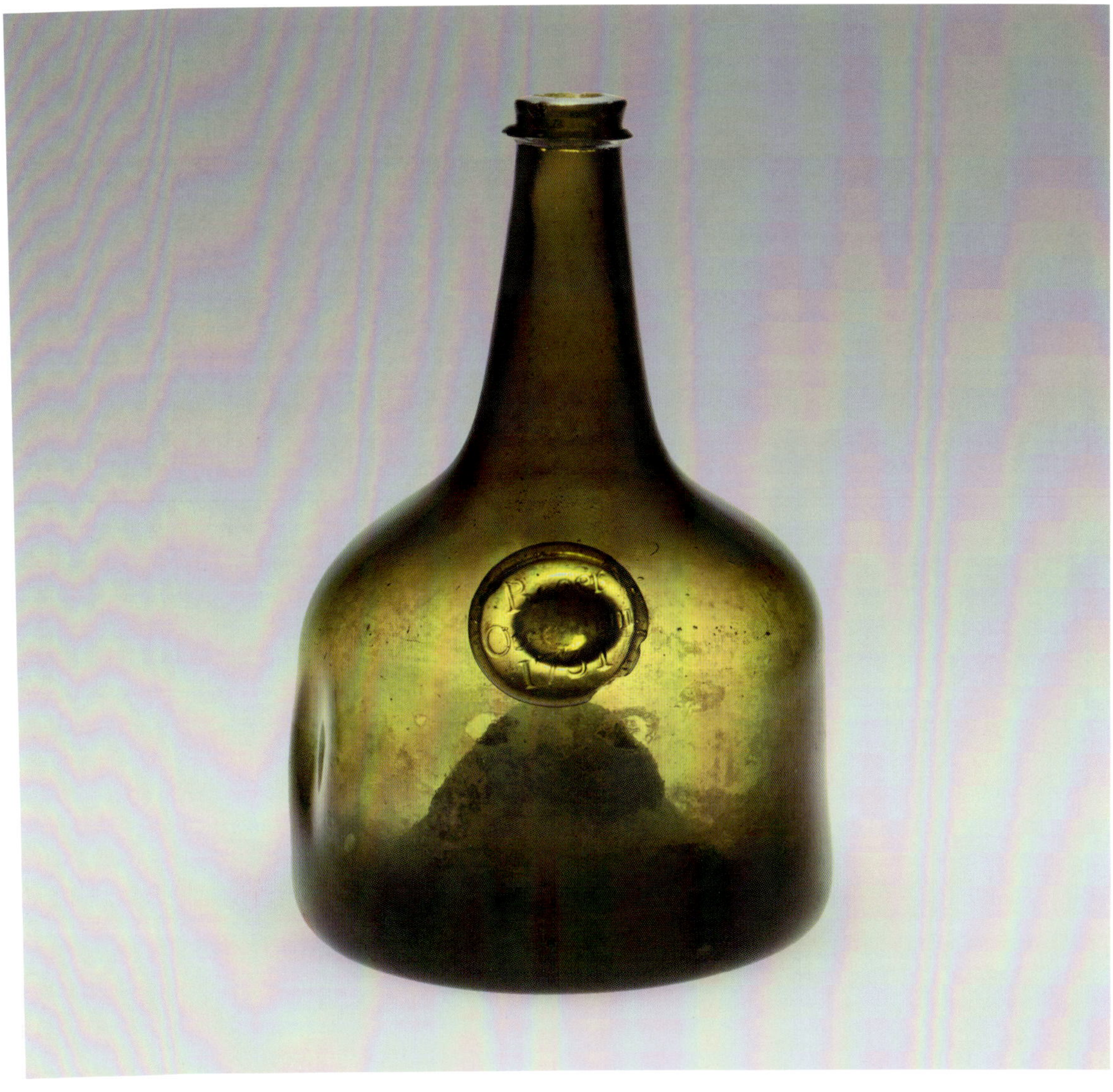

1739

Wine bottle
Glass
$7\frac{1}{4}$ × 6 in. (18.4 × 15.2 cm)

Inscription: John Littler 1739

This wine bottle, which retains its original cork, is engraved overall. There are depictions of flowers; a woman with arms raised and a cup in one hand and flowers in the other; a man beside a table, drinking and smoking; a child riding a dog; a carnation, rose, and tulip emerging from an elaborate urn; and a thistle and tulip growing from the ground. A bottle engraved in the same fashion was exhibited at the County Museum, Truro, in 1976 (no. 50).

PROVENANCE

Robin Butler; bt. John H. Bryan of Lake Bluff, IL (1936–2018) from Jonathan Horne, London, January 31, 1992

LITERATURE

David Burton, *Antique Sealed Bottles 1640–1900 and the Families Who Owned Them* (Woodbridge, UK: Antiques Collectors' Club, 2015), 668

Dwight P. Lanmon, *The Golden Age of English Glass: 1650–1775* (Woodbridge, UK: Antiques Collectors' Club, 2011), 304, no. 114

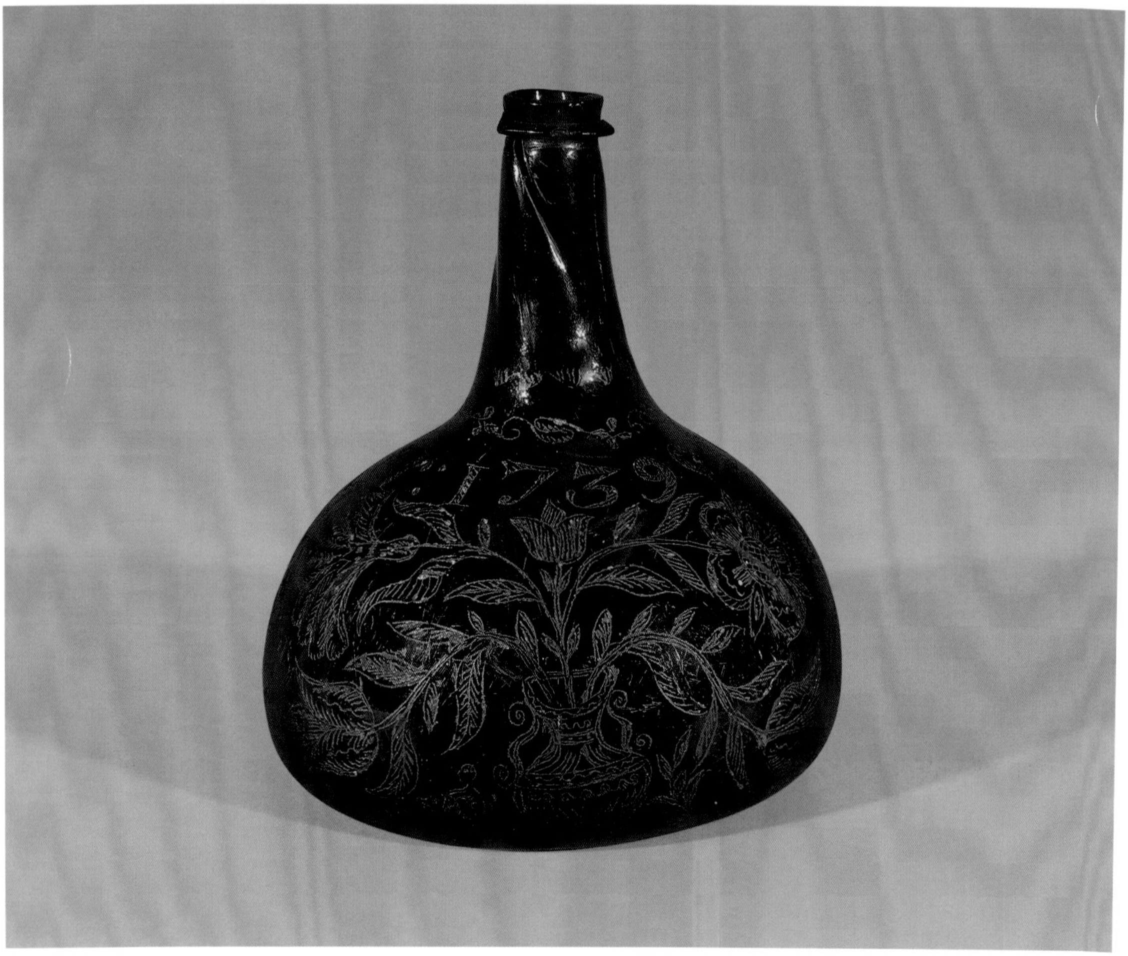

1745

This ale bottle has a dry capacity of 1050 milliliters (36 ounces).

PROVENANCE

bt. John H. Bryan of Lake Bluff, IL (1936–2018) from Jellinek & Sampson, London, May 30, 1980

LITERATURE

David Burton, *Antique Sealed Bottles 1640–1900 and the Families Who Owned Them* (Woodbridge, UK: Antiques Collectors' Club, 2015), 61, 686

Dwight P. Lanmon, *The Golden Age of English Glass: 1650–1775* (Woodbridge, UK: Antiques Collectors' Club, 2011), 309, 118

Ale bottle

Glass

8 ¾ × 4 ½ in. (22.2 × 11.4 cm)

Inscription:
H|lLoyd|1745

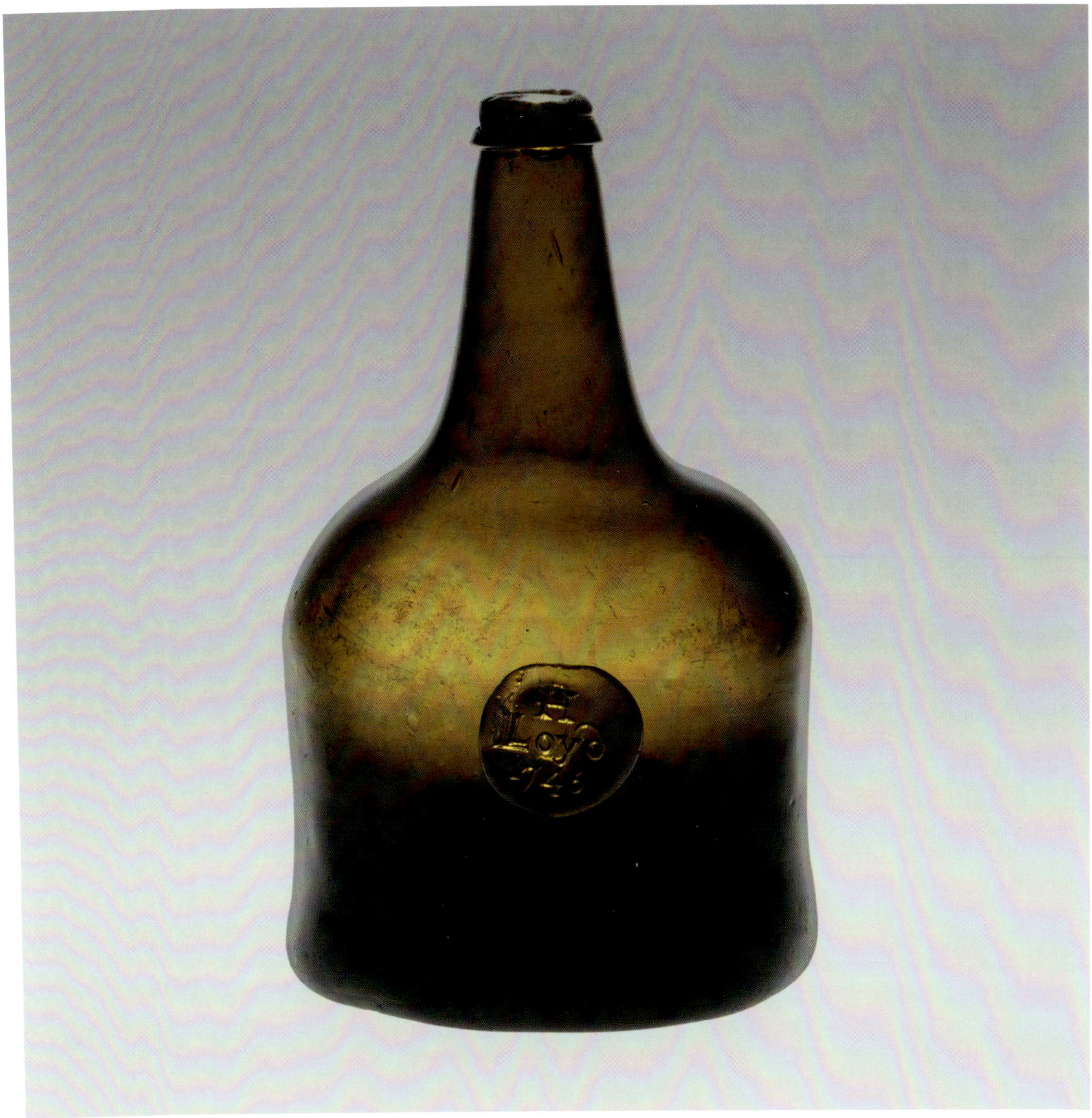

1751

Wine bottle
Glass
$9\frac{1}{2} \times 4\frac{1}{2}$ in. (24.1 × 11.4 cm)

Inscription:
H|Jarman|1751

This wine bottle has a dry capacity of 950 milliliters (32 ounces).

PROVENANCE
Sotheby's, Billingshurst, Sussex, March 23, 1989; bt. John H. Bryan of Lake Bluff, IL (1936–2018) from Jonathan Horne, London, August 10, 1989

LITERATURE
David Burton, *Antique Sealed Bottles 1640–1900 and the Families Who Owned Them* (Woodbridge, UK: Antiques Collectors' Club, 2015), 696

Dwight P. Lanmon, *The Golden Age of English Glass: 1650–1775* (Woodbridge, UK: Antiques Collectors' Club, 2011), 309, no. 119

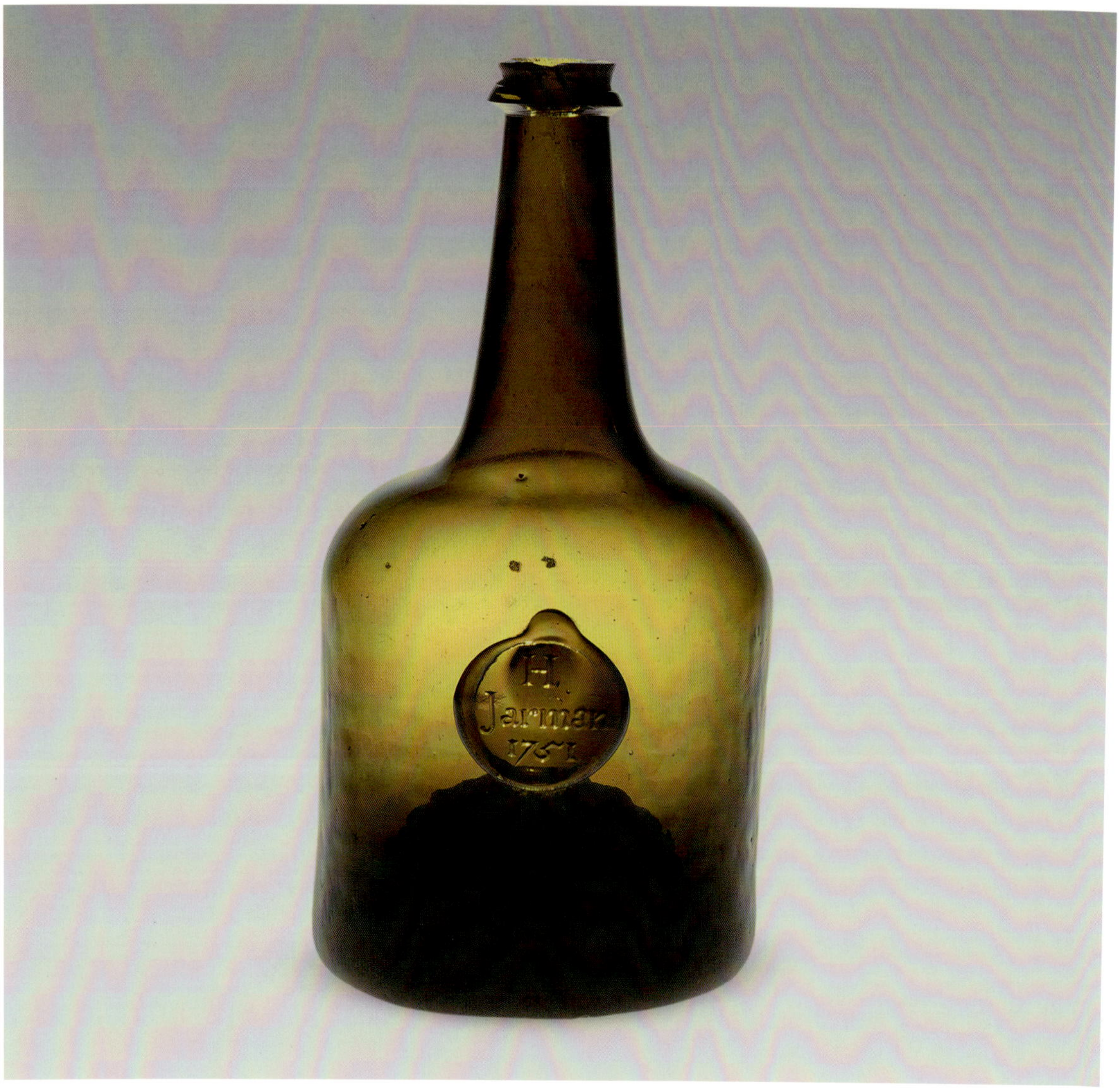

1751

The unusual style of this bottle, with its short, stubby neck and flared lip, suggests it was a storage jar, possibly used to ferment or preserve fruits. Thomas Jemmitt of London, "Fruiterer," son of Ezekiel, was a fishmonger. He served as master of the company in 1742 and in 1746 he took on an apprentice named Thomas Duppa, charging an eighty-pound premium. The premium was one hundred pounds for Edmond Chamberline the following year. There was also a sugar refiner of Lambeth by the same name, who made his will in 1762 (TNA PROB 11/879/247).

PROVENANCE

bt. John H. Bryan of Lake Bluff, IL (1936–2018) from Alistair Sampson Antiques Ltd., London, October 7, 1991

LITERATURE

David Burton, *Antique Sealed Bottles 1640–1900 and the Families Who Owned Them* (Woodbridge, UK: Antiques Collectors' Club, 2015), 697

Jar

Glass

$10\frac{3}{8} \times 7\frac{1}{4}$ in. (26.4 × 18.4 cm)

Inscription: THO^S.| JEMMITT|1751

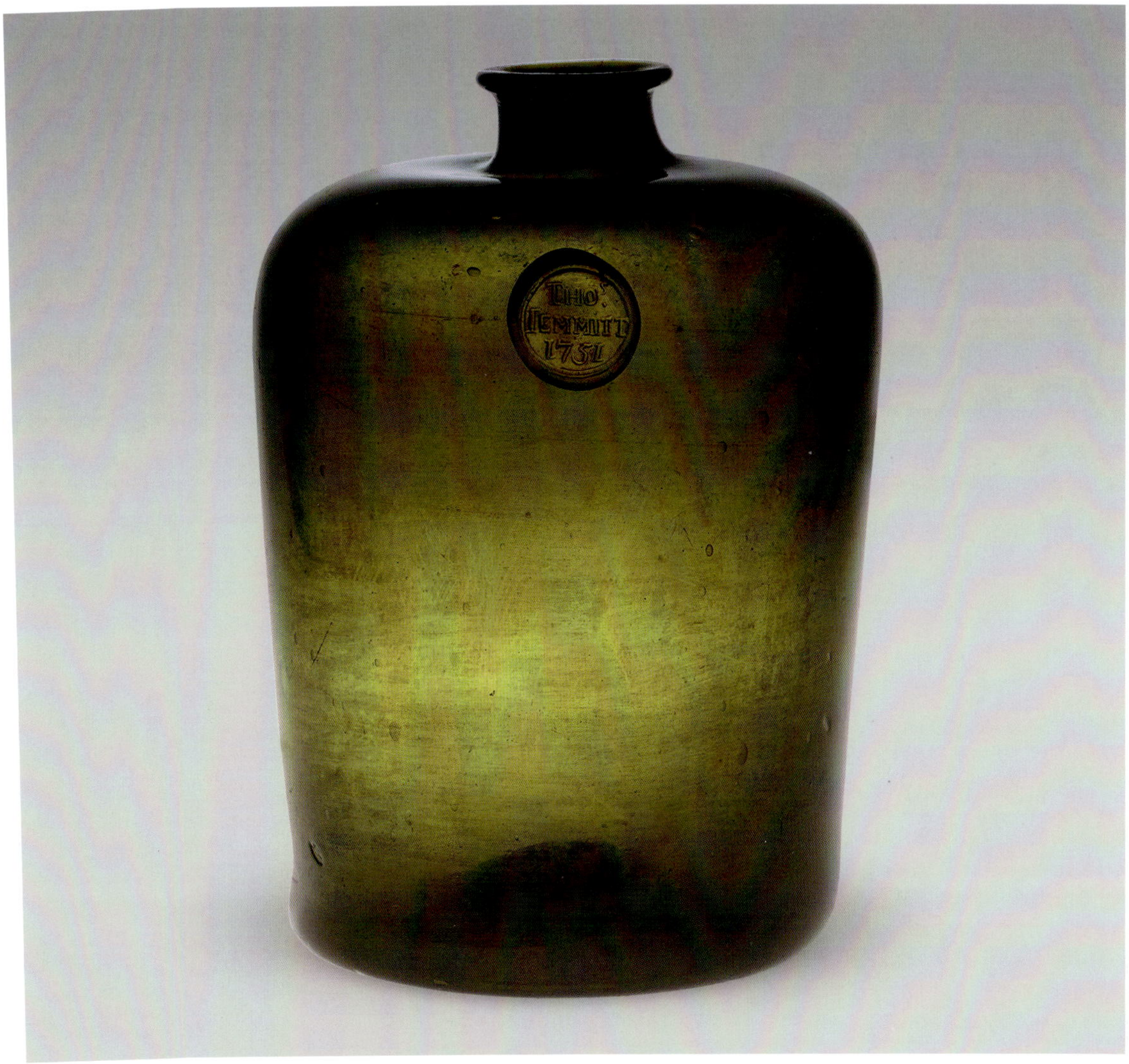

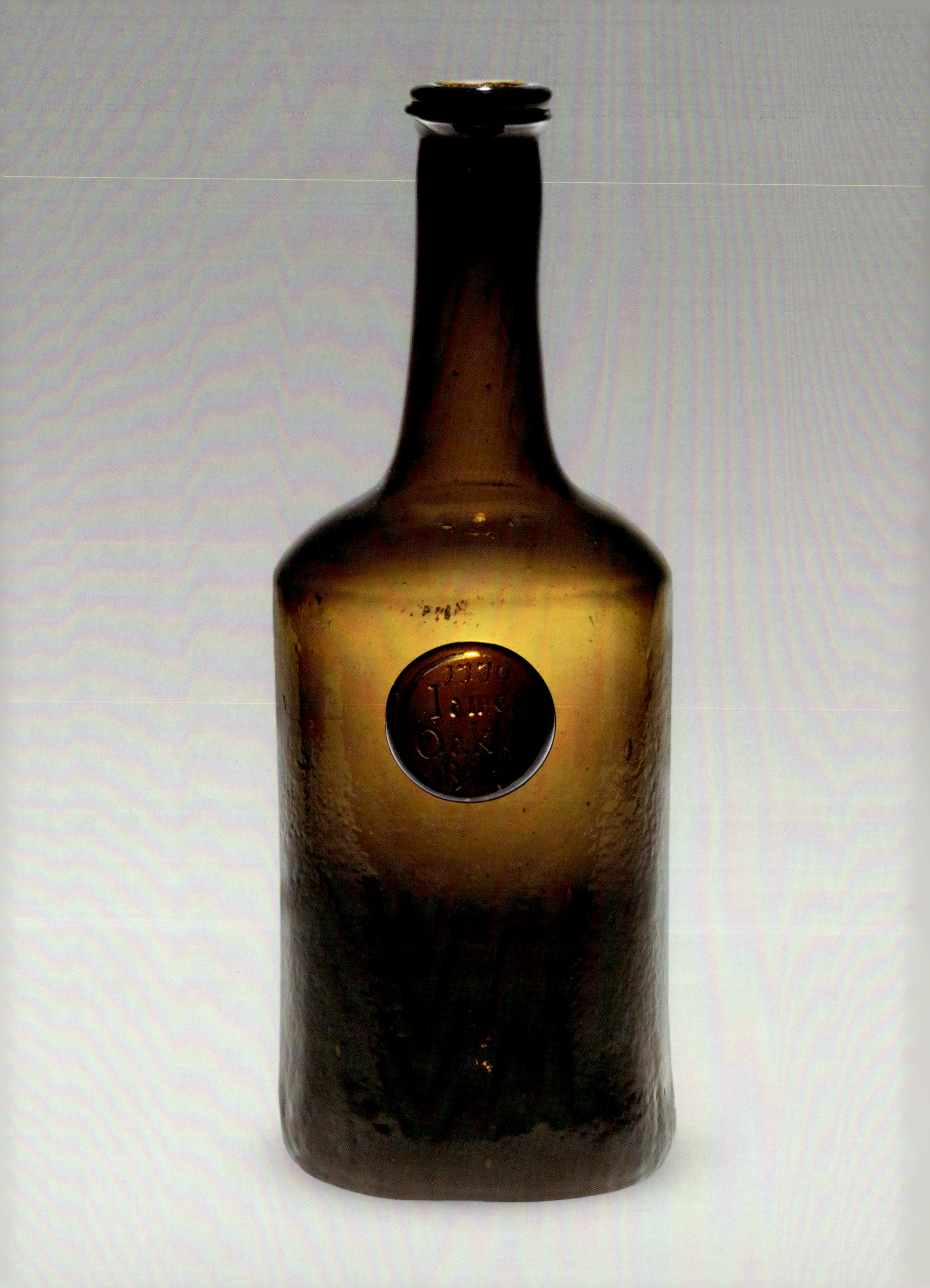

1770

Wine bottle
Glass
$10\frac{1}{4} \times 3\frac{1}{2}$ in. (26 × 8.9 cm)

Inscription: 1770|Iames|Oakes|Bury

James Oakes was one of the wealthiest merchants in Bury St Edmunds, the second-largest town in Suffolk. His seal appears on a number of bottles, including another in the Bryan Collection. This bottle has a dry capacity of 700 milliliters (24 ounces).

Educated at the Edward VI Grammar School, Oakes was apprenticed to his uncle, Orbell Ray, and entered into his yarn-making industry, which he inherited in 1768. He became one of the town's wealthiest merchants, expanding into selling coal and hops and, later, into banking. Oakes's diaries—which begin in November 1778 and end in October 1827, when he was nearly eighty-six years old—chart his life as a provincial businessman.

Oakes was the Duke of Grafton's political agent in the town and, through the Duke's influence, became receiver general of the Land Tax for the Western Division of Suffolk. He was alderman of the town five times and served as a justice of the peace for forty-five years. His house at 81 Guildhall Street in Bury St Edmunds had additions to the designs of John Soane between 1789 and 1791. The building retains a portion that originally served as part of the yarn warehouses.

In 1784 Oakes purchased the Green Dragon Public House on Guildhall Street, adjacent to his existing property. In doing so, he cannily caught the business of the combers who worked in his yarn-making manufactories, thus returning their wages to his pocket. Oakes died wealthy, with a fortune that ran to the tens of thousands of pounds.

PROVENANCE

James Oakes of Bury St Edmunds, Suffolk (1741–1829); bt. John H. Bryan of Lake Bluff, IL (1936–2018) from Jonathan Horne, London, January 4, 1987

LITERATURE

David Burton, *Antique Sealed Bottles 1640–1900 and the Families Who Owned Them* (Woodbridge, UK: Antiques Collectors' Club, 2015), 742

Jane Fiske, ed., *The Oakes Diaries: Business, Politics and the Family in Bury St Edmunds, 1778–1827* (Woodbridge, UK: Boydell Press for the Suffolk Records Society, 1990), 32:47

Dwight P. Lanmon, *The Golden Age of English Glass: 1650–1775* (Woodbridge, UK: Antiques Collectors' Club, 2011), 313, no. 123

1769

Ale bottle
Glass
$10\frac{3}{8} \times 7\frac{1}{4}$ in. (24.1 × 12.1 cm)

Inscription: J | Dunstan | 1769

PROVENANCE
bt. John H. Bryan of Lake Bluff, IL (1936–2018) from Alistair Sampson Antiques Ltd., London, March 30, 1990

LITERATURE
David Burton, *Antique Sealed Bottles 1640–1900 and the Families Who Owned Them* (Woodbridge, UK: Antiques Collectors' Club, 2015), 736

Dwight P. Lanmon, *The Golden Age of English Glass: 1650–1775* (Woodbridge, UK: Antiques Collectors' Club, 2011), 309, no. 120

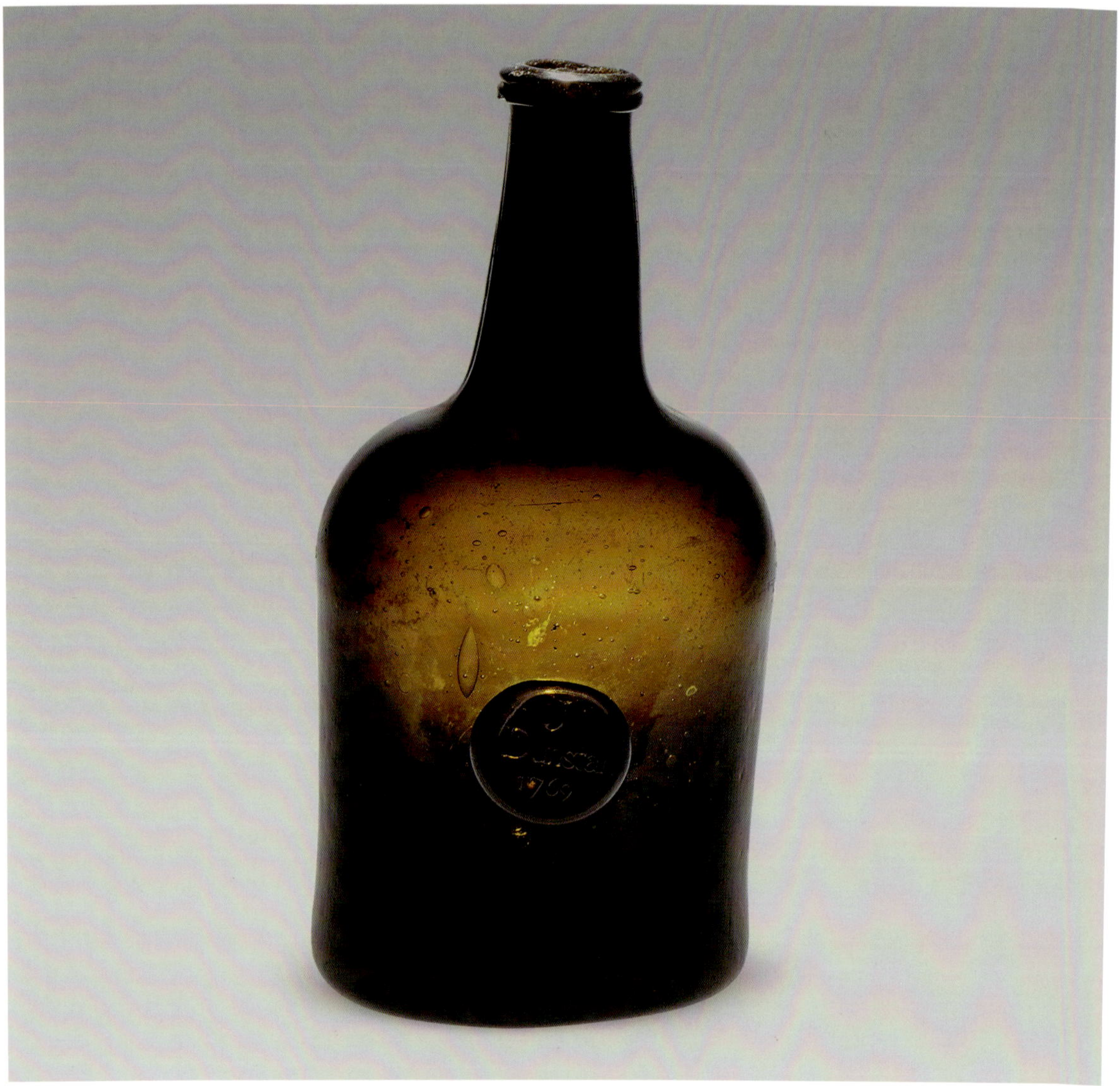

1785

This wine bottle was made soon after James Oakes purchased the Green Dragon Public House in Bury St Edmunds, Suffolk. It has a dry capacity of 700 milliliters (24 ounces), which is a wine pint and a half.

PROVENANCE

James Oakes of Bury St Edmunds, Suffolk (1741–1829); Gordon Renny Collection; bt. John H. Bryan of Lake Bluff, IL (1936–2018) from Alistair Sampson Antiques Ltd., London, March 30, 1990

LITERATURE

David Burton, *Antique Sealed Bottles 1640–1900 and the Families Who Owned Them* (Woodbridge, UK: Antiques Collectors' Club, 2015), 784

Dwight P. Lanmon, *The Golden Age of English Glass: 1650–1775* (Woodbridge, UK: Antiques Collectors' Club, 2011), 313, no. 124

Wine bottle

Glass

10¼ × 3½ in. (26 × 8.9 cm)

Inscription: JA^{S}| OAKES|BURY|1785

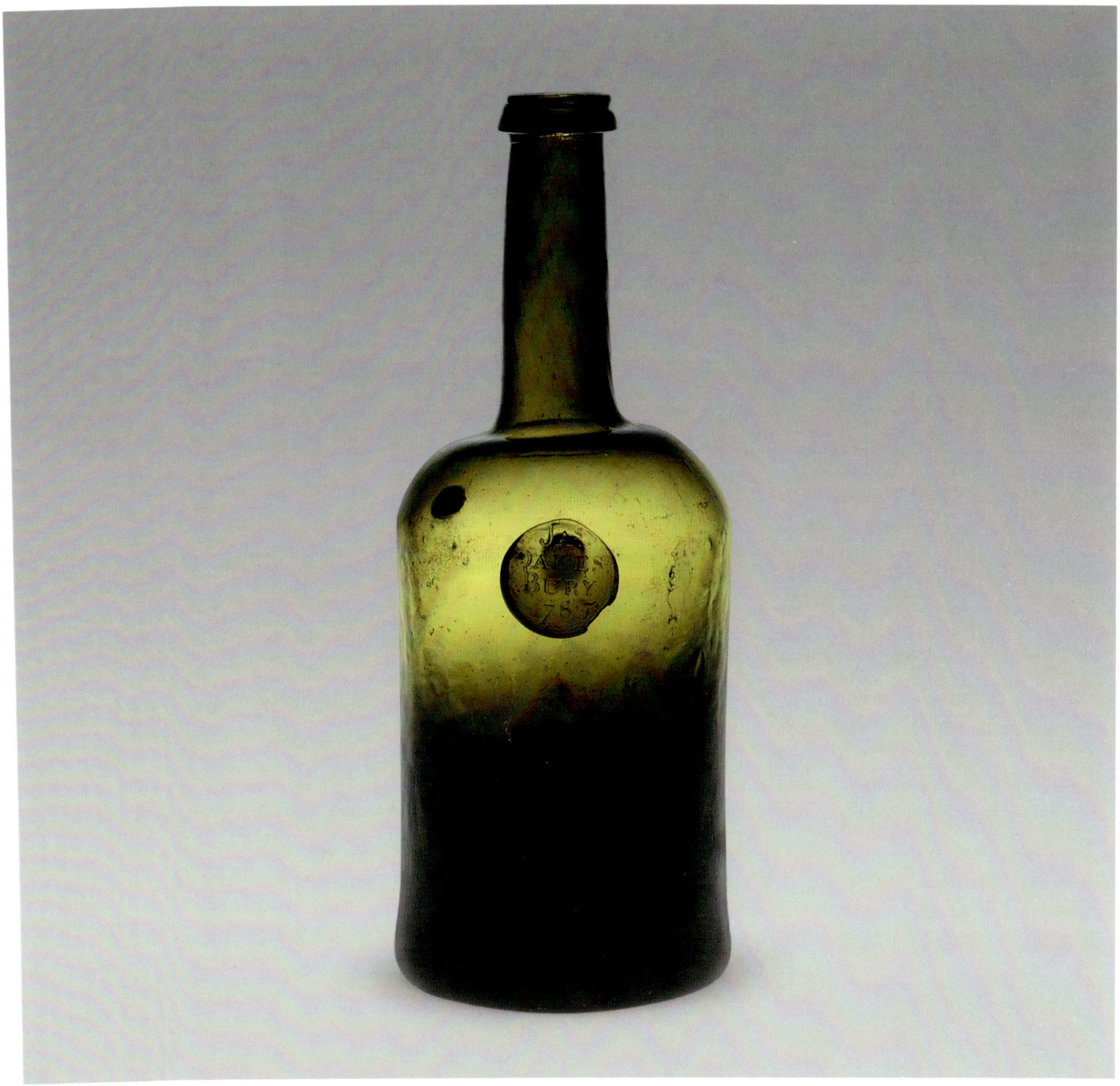

1793

Ale bottle
Glass
12 × 4½ in. (30.5 × 11.4 cm)

Inscription:
T·F|1793

This bottle may possibly have a connection to Thomas Fox, of Falmouth, Cornwall. His family members were prominent merchants, shipping agents, and consuls in the town. This bottle has a dry capacity of 1675 milliliters (57 ounces).

PROVENANCE
bt. John H. Bryan of Lake Bluff, IL (1936–2018) from Jonathan Horne, London, August 3, 1998

LITERATURE
David Burton, *Antique Sealed Bottles 1640–1900 and the Families Who Owned Them* (Woodbridge, UK: Antiques Collectors' Club, 2015), 805

Dwight P. Lanmon, *The Golden Age of English Glass: 1650–1775* (Woodbridge, UK: Antiques Collectors' Club, 2011), 313, no. 125

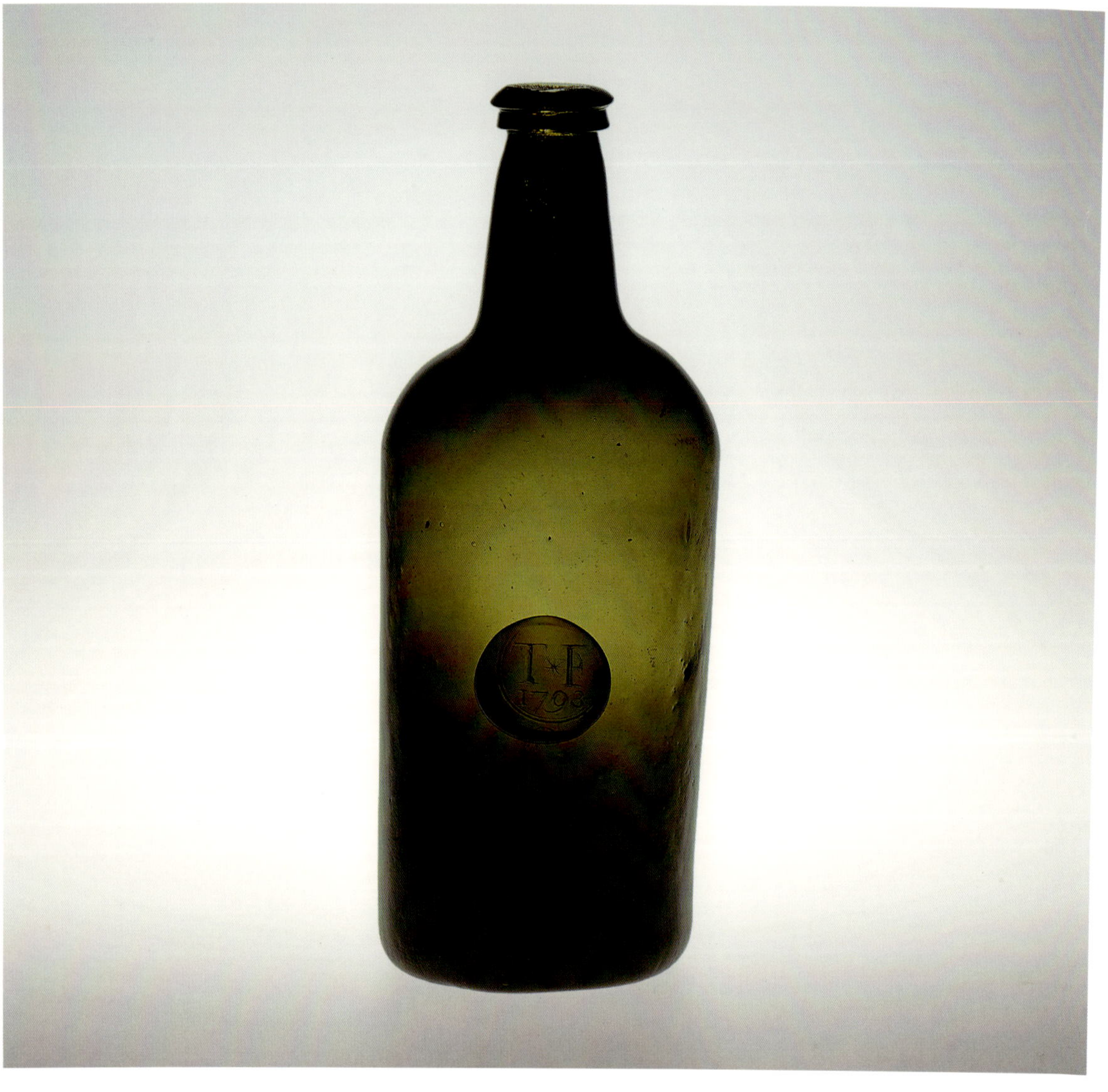

1801

In the collection of the Hereford Museum, there is another example of this seal on a bottle associated with Tom Gosell, market gardener from Tarrington, Herefordshire. It is a "magnum," which is to say that it holds two standard bottles of liquid. It has been suggested that the bottle was made to hold champagne, a popular drink among members of fashionable society.

PROVENANCE

bt. John H. Bryan of Lake Bluff, IL (1936–2018) from Alistair Sampson Antiques Ltd., London, December 17, 1998

LITERATURE

David Burton, *Antique Sealed Bottles 1640–1900 and the Families Who Owned Them* (Woodbridge, UK: Antiques Collectors' Club, 2015), 821

Dwight P. Lanmon, *The Golden Age of English Glass: 1650–1775* (Woodbridge, UK: Antiques Collectors' Club, 2011), 315, no. 127

Wine bottle

Glass

10 ½ × 5 ¾ in. (26.7 × 14.6 cm)

Inscription: TG|1801

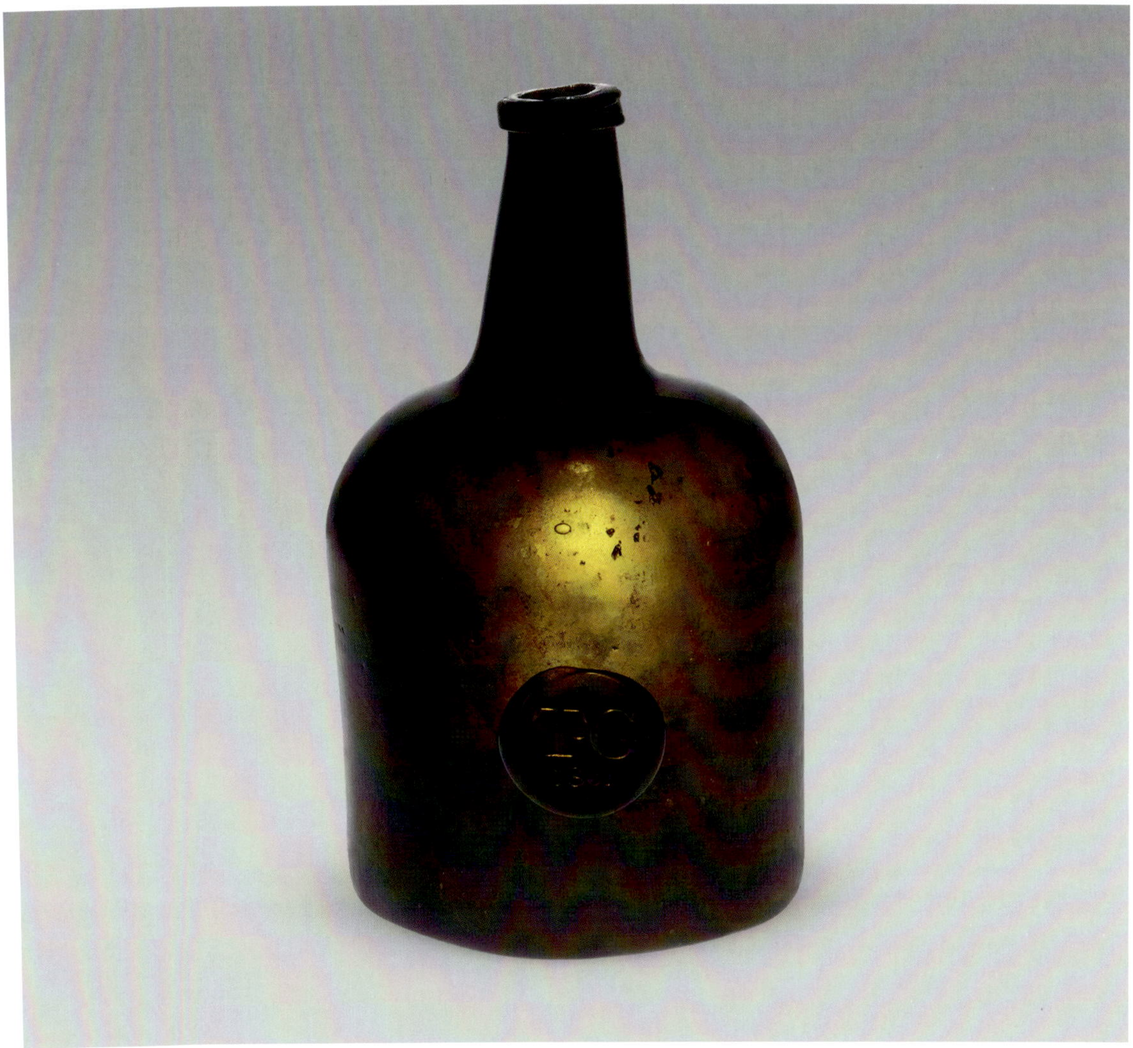

1693

Cabinet

Oak

$20 \times 38\frac{1}{2} \times 10$ in. (50.8 × 97.8 × 25.4 cm)

Inscription: Maids I advise you have a Care|Glasses & Lasses are Brittle ware:|1693

Drinking glasses were expensive commodities in the seventeenth century, so they were kept safe in a wall-hung case like this one, known as a glass case, cupboard, frame, or perch. This oak case is fitted internally with a single shelf. The front composes as three open-fretwork panels: the panel at left displays the Rose of England, the central panel (which is the door) displays the Harp of Ireland, and the panel at right the Thistle of Scotland.

The painted inscription is a common joke of the period, which required a northern accent to rhyme *glass* with *lass*.

PROVENANCE

Collection of John Fardon, Birmingham; Sotheby's, London, September 28, 2004 (lot 61); bt. John H. Bryan of Lake Bluff, IL (1936–2018) through Tobias Jellinek Antiques, London

LITERATURE

Victor Chinnery, *Oak Furniture: The British Tradition, A History of Early Furniture in the British Isles and New England* (Woodbridge, UK: Antique Collectors' Club, 1979), 339, no. 3:322

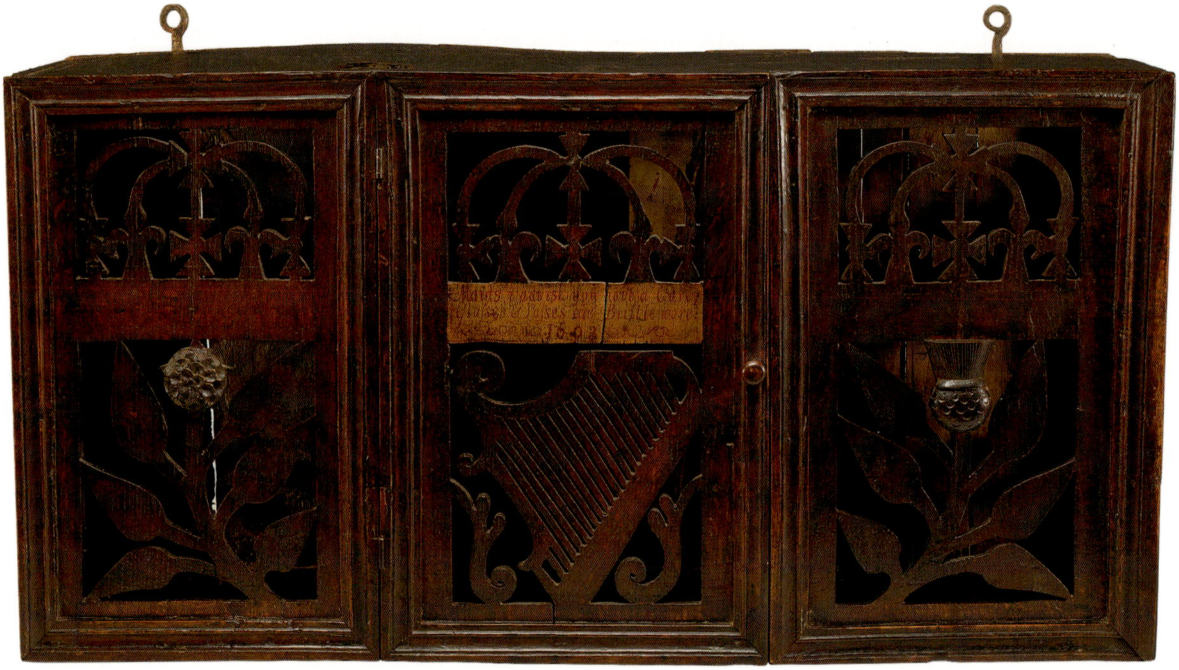

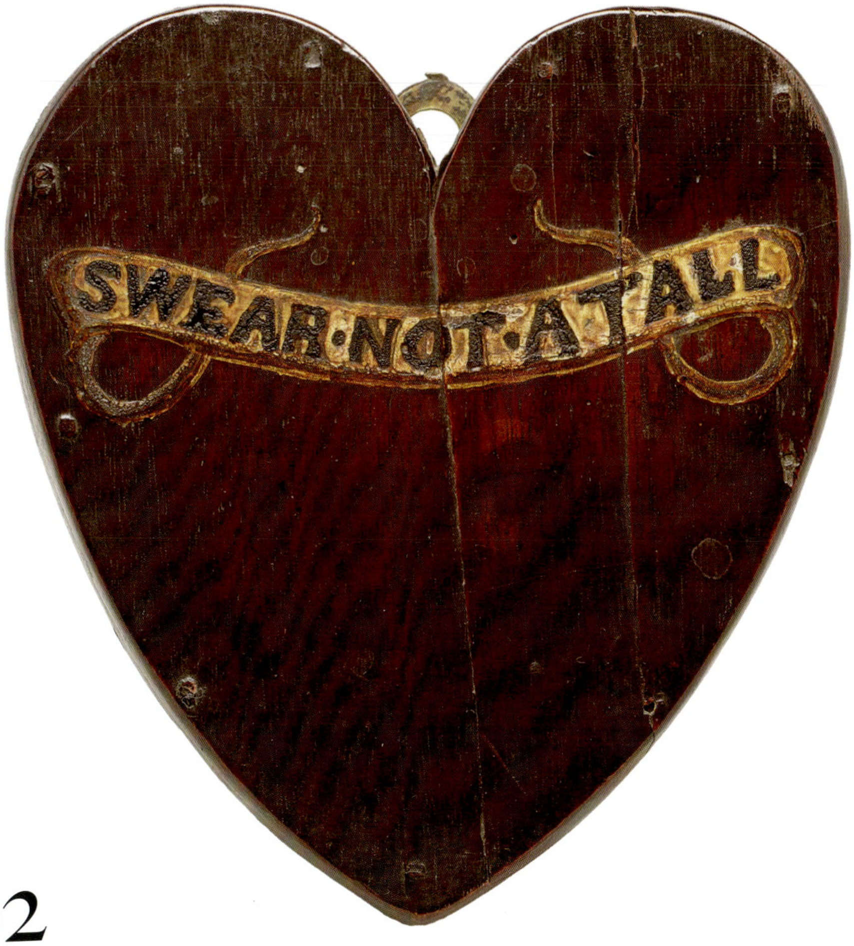

1762

Swear box

Mahogany and elm, painted

$6\frac{7}{8} \times 6 \times 2\frac{3}{4}$ in. (17.5 × 15.2 × 7 cm)

Inscription: SWEAR·NOT·AT ALL; James [illegible] | 1762

The phrase is derived from Matthew 5:34 in the King James Bible, which reads: "But I say unto you, 'Swear not at all; neither by heaven; for it is God's throne.'" Ecclesiastical debates went back and forth on whether this passage prohibited oath making, but the words painted on this mahogany box warn against profanity, intended not only to extort good behavior from family and neighbors but also fines from those who forget themselves.

PROVENANCE

The Arthur Golding Barrett Collection, Tennants of Yorkshire, April 26, 1986 (lot 518); John L. Boonshaft of Las Vegas, NV (1950–1997); Sotheby's, New York, January 21, 1998 (lot 121); bt. John H. Bryan of Lake Bluff, IL (1936–2018) through Tobias Jellinek Antiques, London

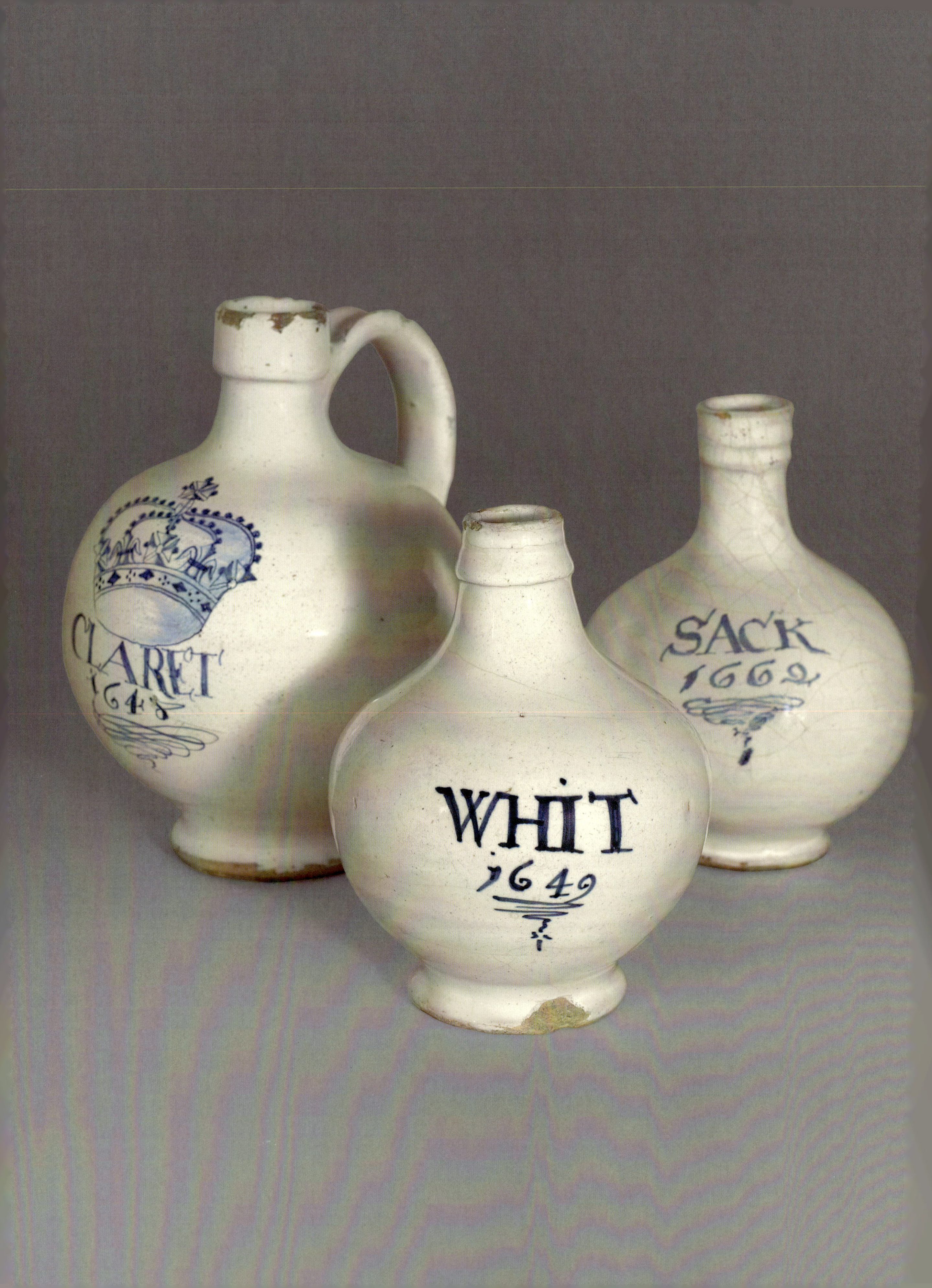
CLARET
SACK
1662
WHIT
1649

1648

Wine serving bottle

Delftware

$7\frac{1}{2} \times 5\frac{1}{2} \times 4\frac{3}{4}$ in. (19.1 × 14 × 12.1 cm)

Inscription:
CLARET|1648

In the seventeenth century, wine was drunk when young and consumed in great quantities, with little heed given to vintage. The popularity of wine as a New Year's present has led to the suggestion that the dates that routinely appear on these types of bottles relate to the gift rather than to the vintage. Dated examples—products of the potteries at Pickleherring, Rotherhithe, and possibly also Brislington—range from 1621 to 1719. This wine bottle is prominently decorated with a crown, a show of loyalty to a monarch who was soon to lose his head.

PROVENANCE

Rous Lench Court, Worcestershire; May 29, 1990 (lot 15); bt. John H. Bryan of Lake Bluff, IL (1936–2018) through Alistair Sampson Antiques Ltd., London

LITERATURE

Louis L. Lipski and Michael Archer, *Dated English Delftware: Tin-Glazed Earthenware, 1600–1800* (London: Sotheby Publications, 1984), 324, no. 1365

1649

Wine serving bottle

Delftware

$6 \times 4\frac{3}{8}$ in. (15.2 × 11.1 cm)

Inscription:
WHIT|1649

The word "WHIT" painted in blue on the surface of this bottle means "white," indicating that it was a container for white wine.

PROVENANCE

bt. John H. Bryan of Lake Bluff, IL (1936–2018) from Lindsay Antiques, London, June 10, 1988

1662

Wine serving bottle

Delftware

$6\frac{1}{8} \times 4\frac{1}{2}$ in. (15.6 × 11.4 cm)

Inscription:
SACK|1662

This wine bottle, with a dry capacity of 470 milliliters (16 ounces), contained Canary "sack," a white wine that was made in Gran Canaria and Málaga.

PROVENANCE

T. Scholes; Sotheby's, London, January 16, 1962 (lot 14); Joseph V. Vizcarra, Chicago, IL; Sotheby's, New York, April 12, 1984 (lot 376); bt. John H. Bryan of Lake Bluff, IL (1936–2018) from Alistair Sampson Antiques, London, April 17, 1984

LITERATURE

Louis L. Lipski and Michael Archer, *Dated English Delftware: Tin-Glazed Earthenware, 1600–1800* (London: Sotheby Publications, 1984), 336–37, no. 1490

1670 | 1671

Goblet
Glass, silver
8 × 3¼ in. (20.3 × 8.3 cm)

This goblet contains two silver threepenny coins of Charles II, dated 1670 and 1671, but the goblet itself probably dates from the last decade of the seventeenth century or the first decade of the eighteenth century. The style of production would suggest that this was a so-called Tory Cup, made for drinking to the health of the Stuart dynasty, probably during the reign of Queen Anne.

PROVENANCE
John L. Boonshaft of Las Vegas, NV (1950–1997); Christie's, London, October 9, 1979 (lot 132); Christie's, London, November 23, 1993 (lot 78); bt. John H. Bryan of Lake Bluff, IL (1936–2018) from Sheppard and Cooper Ltd., London, November 24, 1993; Art Institute of Chicago, IL, gift of Mr. and Mrs. John H. Bryan, 2012 (2012.869)

LITERATURE
Dwight P. Lanmon, *The Golden Age of English Glass: 1650–1775* (Woodbridge, UK: Antiques Collectors' Club, 2011), 110–11, no. 26

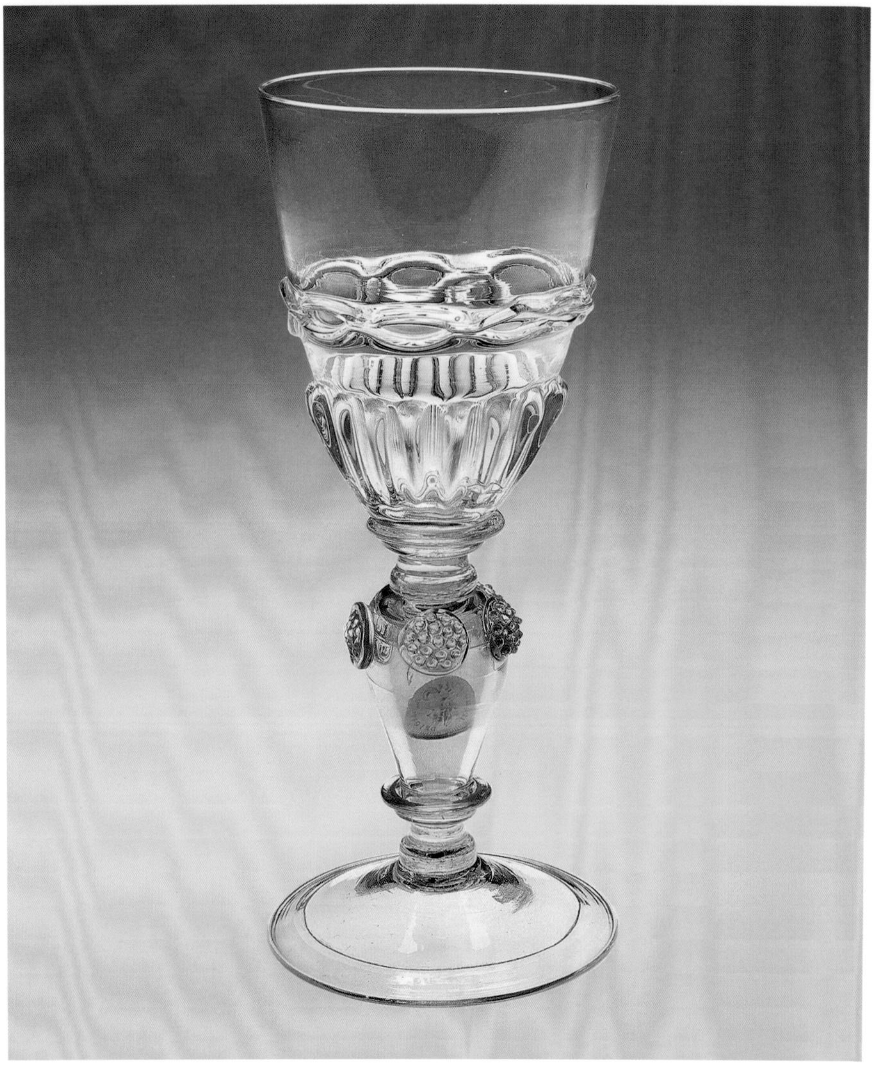

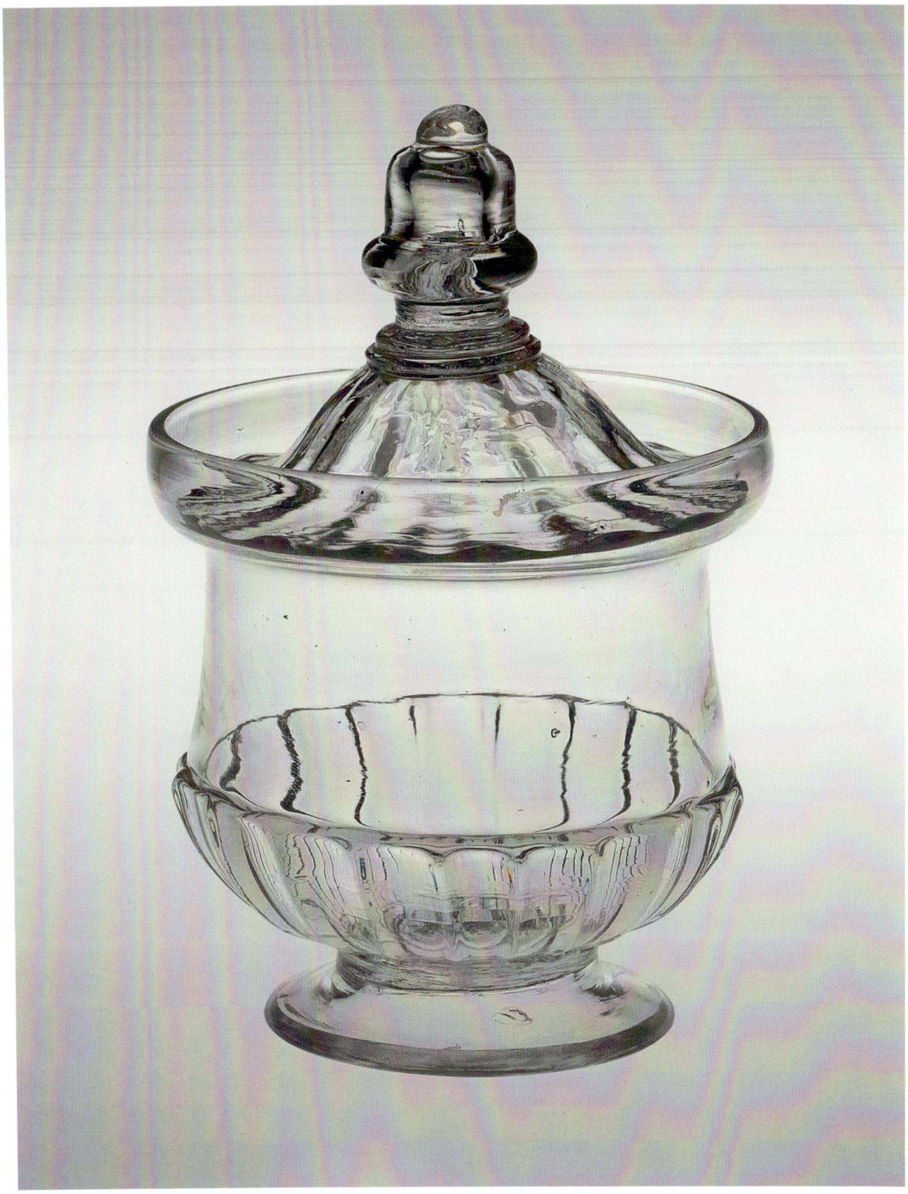

1681

Jar

Glass, silver

6 × 3⅞ in. (15.2 × 9.8 cm)

This covered sugar bowl, or preserves jar, has an apparently uncirculated silver coin of Charles II, dated 1681, encased in the bottom. The form suggests, however, that the jar was made a century or more later, possibly for a Jacobite dining club at one of the universities. Sugar was not available in granular form in the eighteenth century. It was bought in large paper cones, about 18 inches (45.7 centimeters) long. The sugar was broken off the cone with the aid of a hammer and "nippers," pliers with sharp blades. Once nipped, the sugar was transported to the table in a vessel like this.

PROVENANCE

bt. John H. Bryan of Lake Bluff, IL (1936–2018) from Mallet and Son Antiques Ltd., London, December 31, 1998

LITERATURE

Dwight P. Lanmon, *The Golden Age of English Glass: 1650–1775* (Woodbridge, UK: Antiques Collectors' Club, 2011), 94–95, no. 17

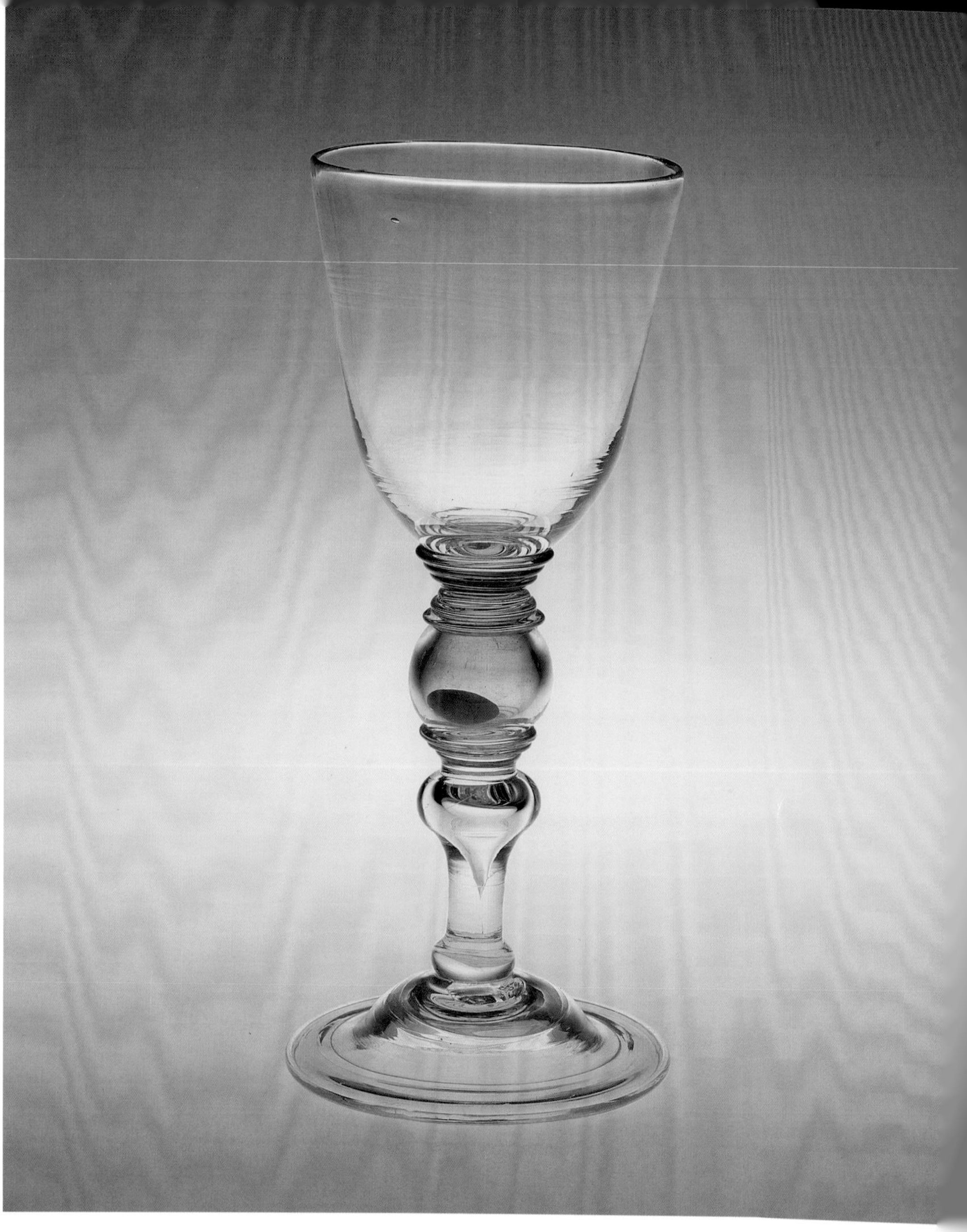

1723

Goblet

Glass, silver

9 ½ × 4 in. (24.1 × 10.2 cm)

This glass goblet contains a coin that is dated to 1723. The form of the vessel suggests, however, that the goblet was made about 1730, some years after the coin was minted.

PROVENANCE

Maureen Thompson, London; bt. John H. Bryan of Lake Bluff, IL (1936–2018) from Sheppard and Cooper Ltd., London, June 15, 1992

LITERATURE

Dwight P. Lanmon, *The Golden Age of English Glass: 1650–1775* (Woodbridge, UK: Antiques Collectors' Club, 2011), 110–11, no. 25

1724

Half-pint wine bottles like this one could be carried in the pocket so the owner could enjoy a drink while out and about. In 1772, the *Lady's Magazine* announced the death of "Mr. James Gouldsbury, Esquire," of King Street, Bloomsbury, on March 11.

PROVENANCE
Christie's, London, June 2, 1987 (lot 32); bt. John H. Bryan of Lake Bluff, IL (1936–2018) through Delomosne and Son Ltd., London

LITERATURE
Dwight P. Lanmon, *The Golden Age of English Glass: 1650–1775* (Woodbridge, UK: Antiques Collectors' Club, 2011), 280–81, no. 96

Wine bottle
Glass
$7\frac{1}{2} \times 4 \times 1\frac{1}{2}$ in. (19.1 × 10.2 × 3.8 cm)

Inscription: James Gouldsbury|1724

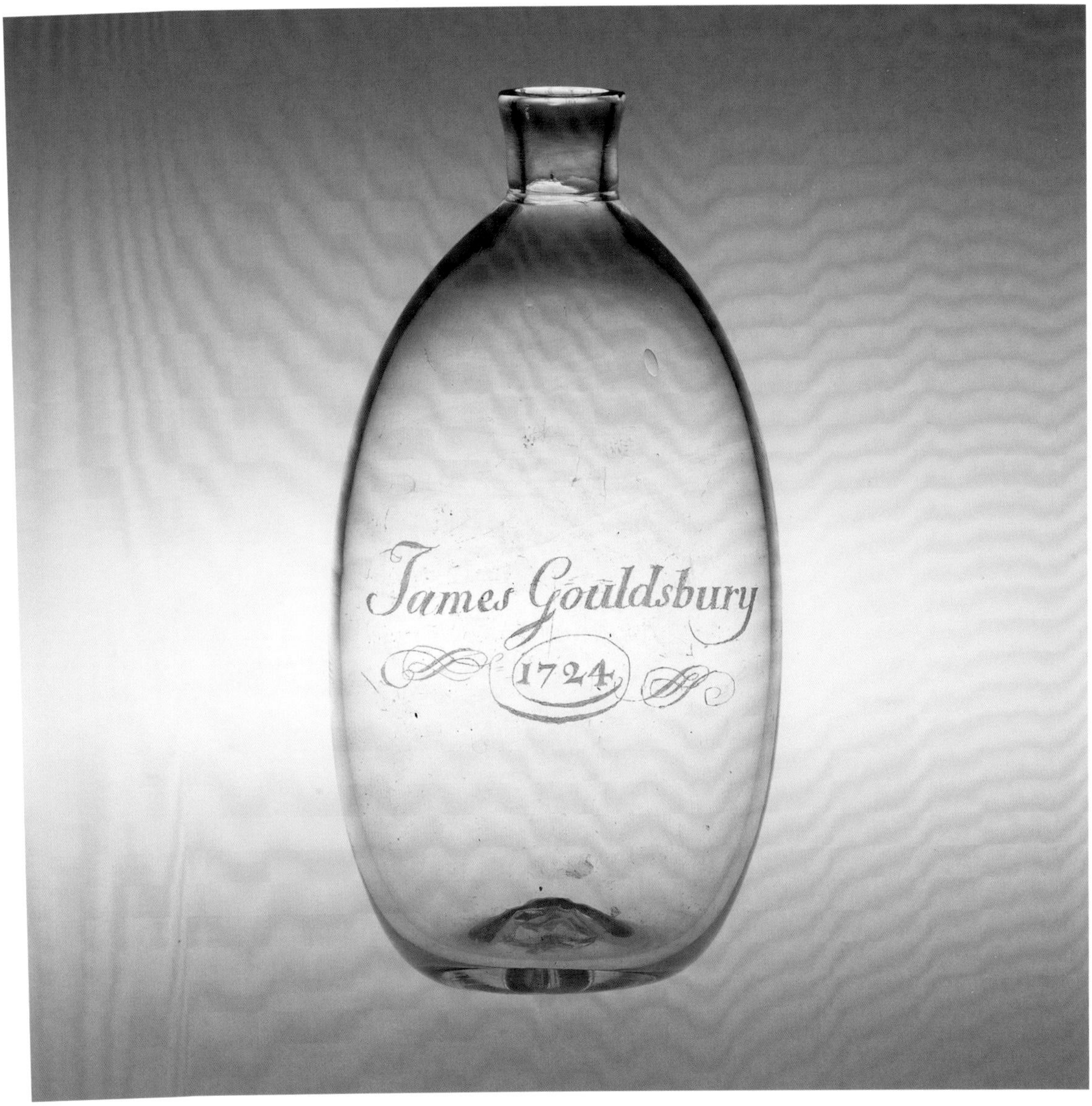

1659

Wine costrel
Leather
$9 \times 7\frac{1}{2} \times 8\frac{1}{2}$ in.
(22.9 × 19.1 ×
21.6 cm)

Inscription: IW
NW|RD; RD|1659;
W C

This leather costrel, designed for carrying liquids, was suspended by straps slung over the carrier's shoulder. The several initials punched into it remain a puzzle. This costrel has a dry capacity of 2240 milliliters (76 ounces), which was precisely the right capacity to hold a gallon of wine. Because costrels were portable, workers could wear them into the fields. As a popular ballad put it,

> A Leather Bottle is
> good,
> far better than Glasses
> or Cans of Wood
> For when a man is at
> work in the Field
> poor Glasses and Pots
> to comfort will yield:
> Then a good Leather
> Bottle standing him by,
> he may drink always
> when he is dry
> It will revive the spirits,
> and comfort the brain,
> wherefore let none this
> Bottle refrain.
>
> —"A Pleasant New Song, in Praise of a Leather Bottle," 1678

PROVENANCE
John Fardon Collection; Christie's, London, May 1, 1996 (lot 150); Christie's, London, July 8, 2014 (lot 194); bt. John H. Bryan of Lake Bluff, IL (1936–2018) from Tobias Jellinek Antiques, July 11, 2014

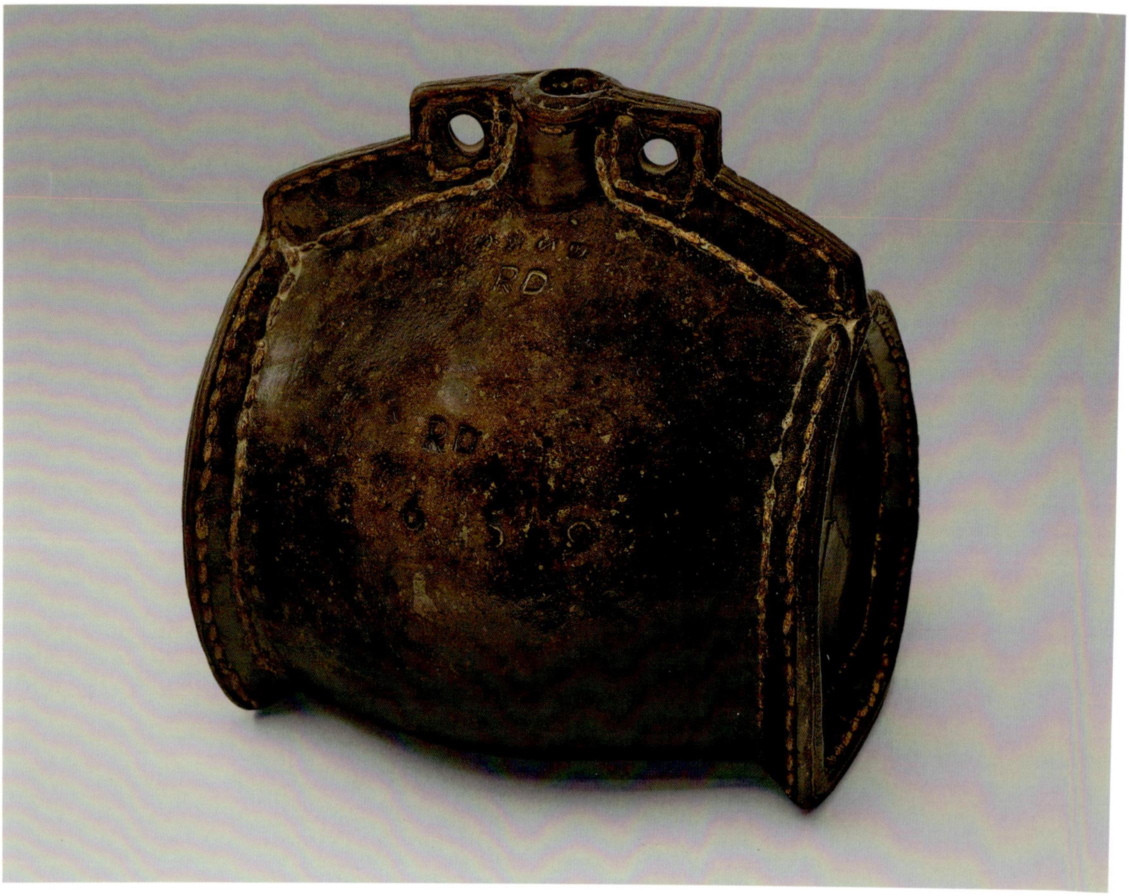

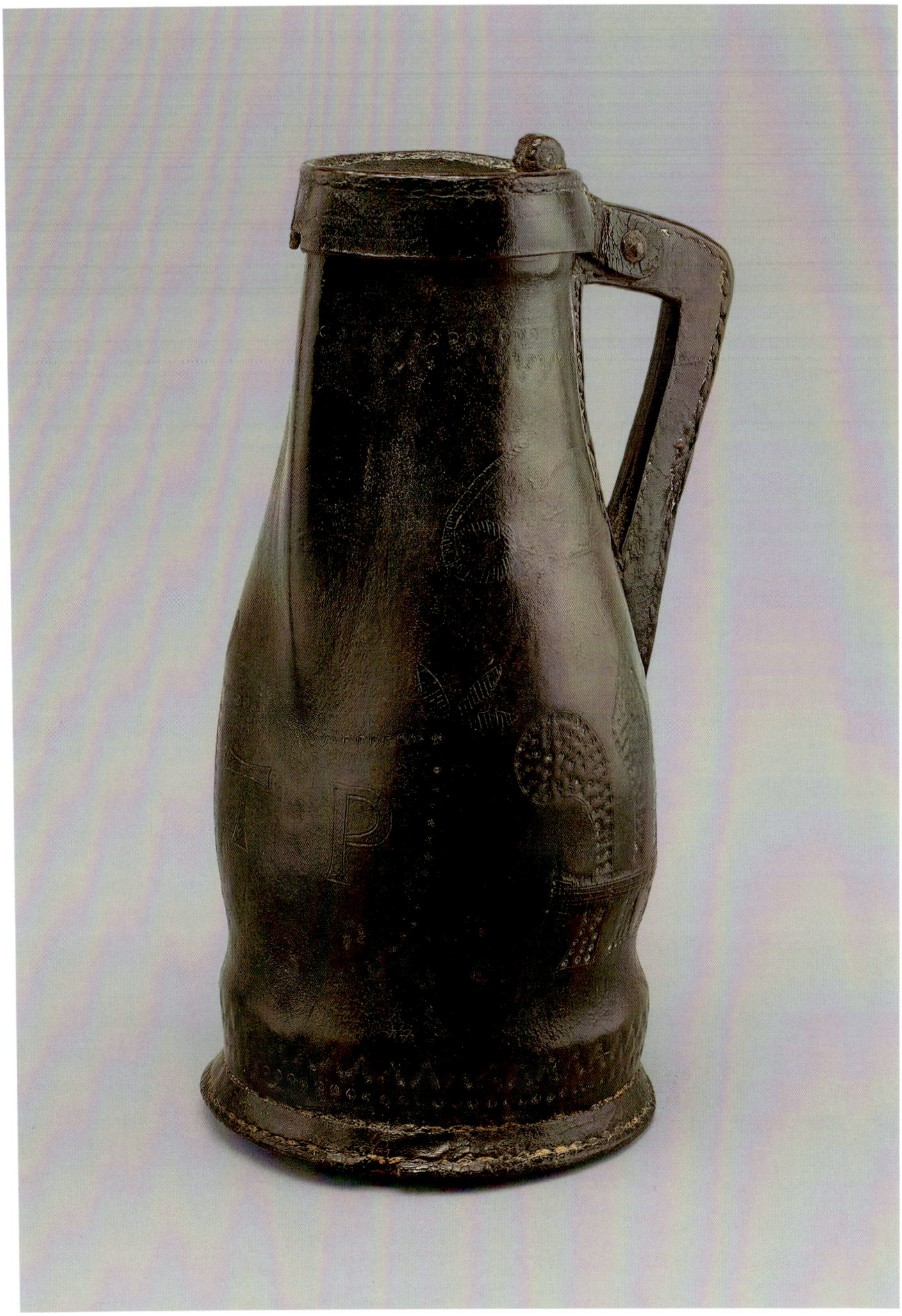

1662

Ale jug
Leather
9 × 7½ × 8½ in.
(38.7 × 19.1 × 21.6 cm)

Inscription: 16 62; TP

This jug is dated 1662, the year of Charles II's wedding to Catherine of Braganza. With a capacity of 2240 milliliters (76 ounces), it held a gallon of ale or beer. Also known as a bombard or black-jack, this style of jug was made from boiled leather (often referred to by the French term *cuir bouilli*). The hot leather was shaped while malleable and then stitched together. This jug has a hinged lid to keep the liquid contents sealed within the vessel.

PROVENANCE
bt. John H. Bryan of Lake Bluff, IL (1936–2018) from Jonathan Horne, London, March 15, 1988

1708

Ale pistol flask
Leather
$18\frac{3}{4} \times 4\frac{3}{4} \times 3\frac{1}{4}$ in.
(47.6 × 12.1 ×
8.3 cm)

Inscription: 1708;
1708

Like a pistol, this style of flask could be hung from the waist and worn against the leg, and its contents enjoyed on the hoof.

PROVENANCE
bt. John H. Bryan of Lake Bluff, IL (1936–2018) from Christopher Bangs Ltd., Felsham, Suffolk, March 26, 2018

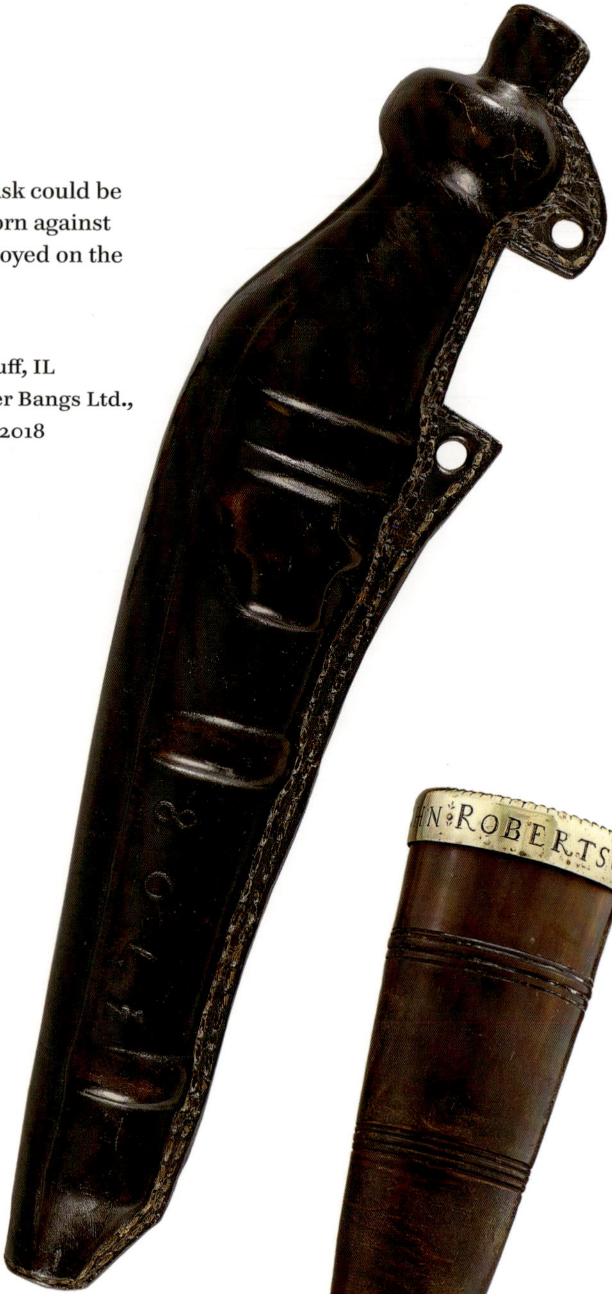

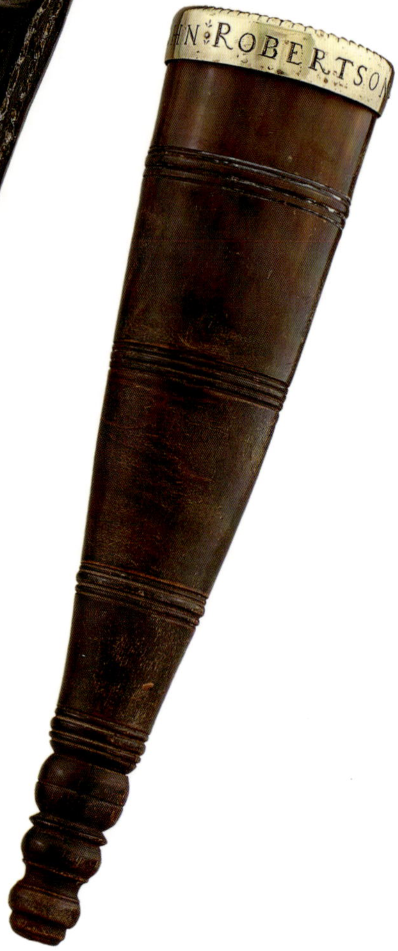

1752

Stirrup cup or powder flask
Horn, brass
$9\frac{3}{4} \times 2\frac{3}{8}$ in. (24.8 ×
6 cm)

Inscribed: IOHN
ROBERTSON 1752

A stirrup cup was a "parting cup," presented to departing guests who were traveling on horseback. The serrated top, however, might suggest this object is a powder flask for shooting that has lost its top.

PROVENANCE
bt. John H. Bryan of Lake Bluff, IL (1936–2018) from Tobias Jellinek Antiques, London, February 8, 1996

1776

Plate
Delftware
$9\frac{1}{4}$ × 1 in. (23.5 × 2.5 cm)

Inscription: MARY HUNT; 1776

Plates with an approximate date and design in Lipski and Archer depict similar pastoral scenes: refreshments being brought out to visitors arriving at an inn, two men in a sawpit, and a lady and gardener in a formal garden. It is difficult to know whether such scenes had particular relevance to the plate owners or whether they were merely pleasant evocations of rural hospitality and cheer.

Archer describes one plate, dated 1772, as having a noticeably greenish glaze that pools on the back. The reddish-brown rim and a wide elaborate border are also characteristics of late eighteenth-century wares made in Lambeth.

PROVENANCE

Dreweatt Neate, Donnington Priory, Berkshire, February 18, 1998; bt. John H. Bryan of Lake Bluff, IL (1936–2018) from Alistair Sampson Antiques Ltd., London, March 9, 1998

LITERATURE

Michael Archer, *Delftware in the Fitzwilliam Museum* (London: Philip Wilson Publishers, 2013), 153, no. B.109

Frank Britton, *English Delftware in the Bristol Collection* (London: Sotheby Publications, 1982), 299, no. 19:24

Louis L. Lipski and Michael Archer, *Dated English Delftware: Tin-Glazed Earthenware, 1600–1800* (London: Sotheby Publications, 1984), 149, no. 668; 151, no. 678

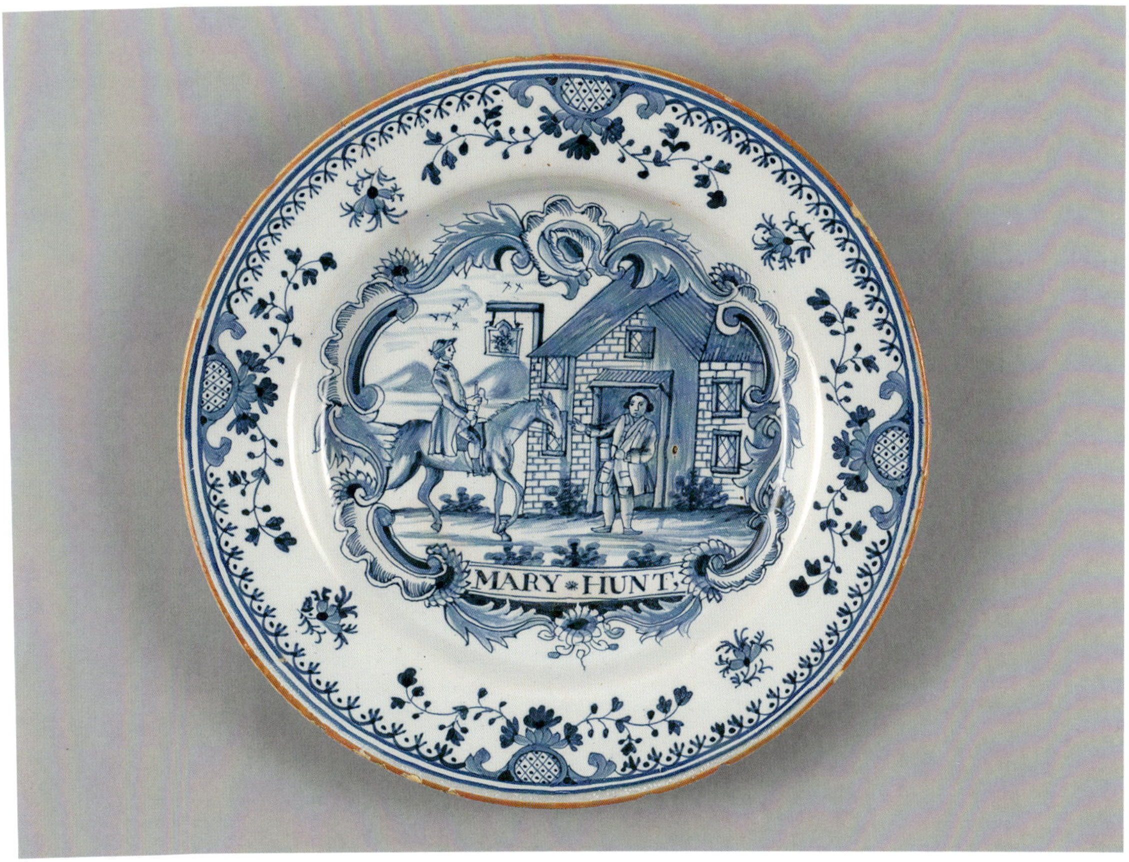

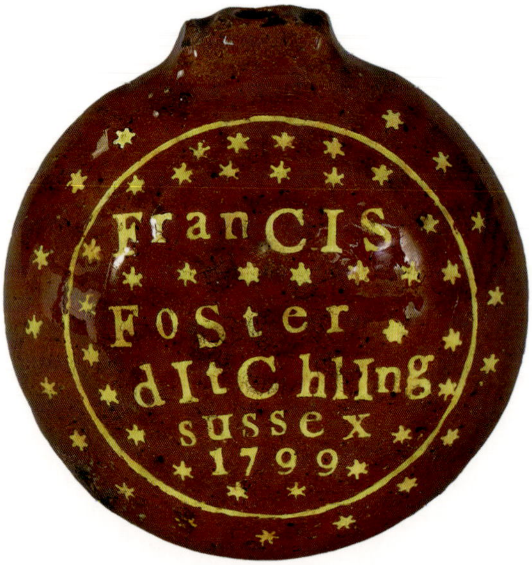

1799

Flask

Lead-glazed earthenware

$3\frac{1}{8} \times 3 \times 1\frac{3}{4}$ in. (7.9 × 7.6 × 4.5 cm)

Inscription:
FrancIS|FoSter|dItChlIng|sussex|1799 When|heare Is|A dram|you may drInk It If you|Can

The minute size of these Sussex-made flasks has led to skepticism as to whether they were ever used. The inscription here suggests that perhaps they were, although it might also be a joke. The charming jumps between upper- and lowercase letters in the inscription were produced by pushing metal type into the clay before firing. The indentations were then filled with white clay and scraped clean before the lead glaze was applied and fired—a technique developed by Sussex potteries.

A very similar style of inscription is found on a tobacco jar inscribed "ANN REVES JUNE 18 1811" in the Victoria and Albert Museum, London (C.71&A-1947). A flask of nearly identical form, also in that collection, made in Chailey, East Sussex, bears the inscription "JANUARY 17 1797" (C.574-1925).

The owner of this flask was an agricultural laborer, born in about 1781, who at the time of the 1841 census was working as a brickmaker in Wivelsfield. He was probably the same Francis Foster who married Ann Carter at Wivelsfield in 1804.

PROVENANCE

bt. John H. Bryan of Lake Bluff, IL (1936–2018) from Martyn Edgell Antiques Ltd., Cambridgeshire, February 5, 2017

LITERATURE

Robin Hildyard, *English Pottery 1620–1840* (London: V&A Publications, 2005), 23–24

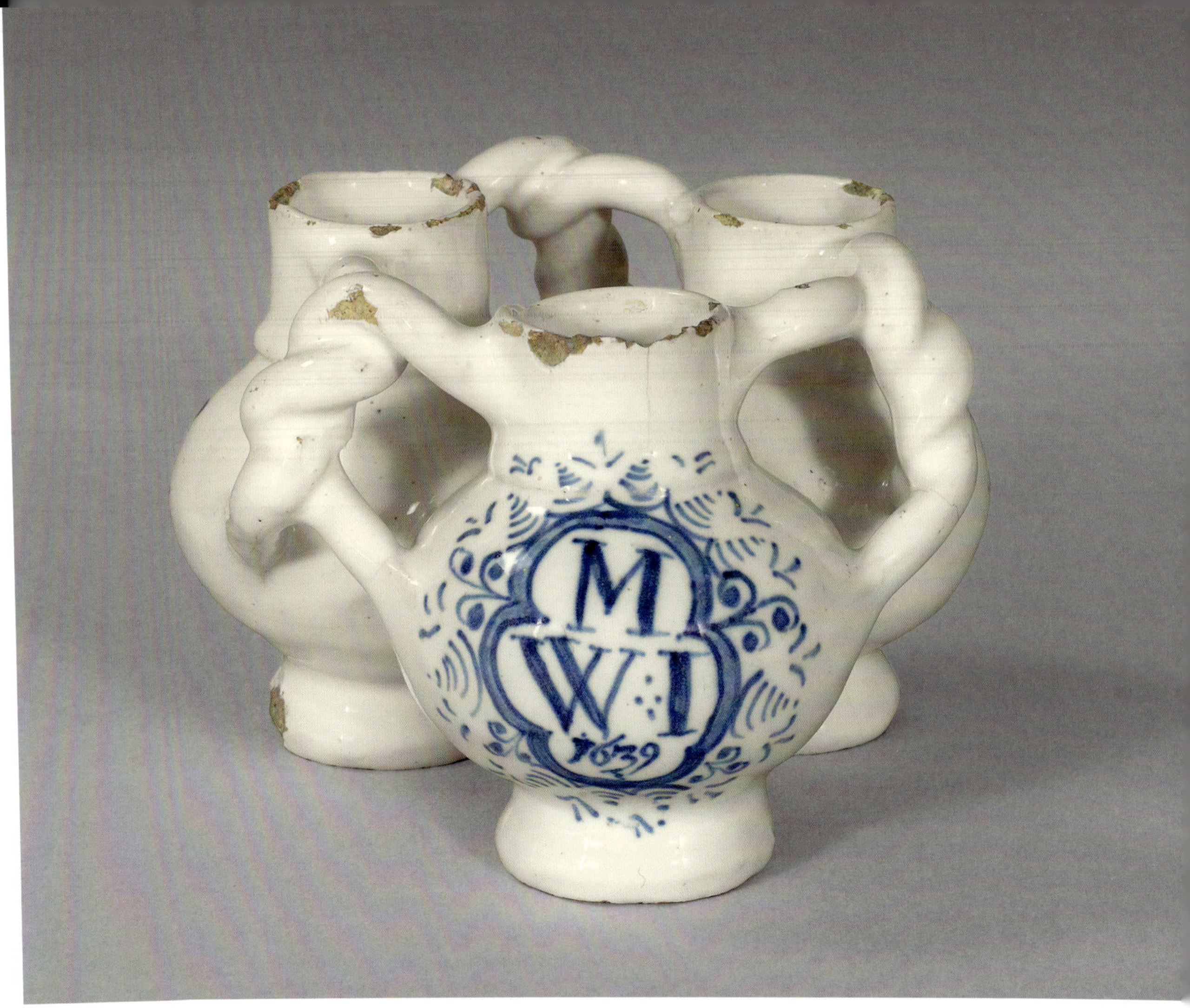

1639

Fuddling cup

Delftware

$3\frac{1}{4} \times 4\frac{3}{4}$ in. (8.3 × 12.1 cm)

Inscription:
W|M|I|1639

Fuddling cups tricked drinkers into consuming three, sometimes four times the amount they thought they were imbibing, thus quickening the time it took for them to get fuddled. According to Lipski, who does not cite his source, a different alcoholic drink was poured into each vessel, and the cup was shaken so that all the liquids mixed together. The drinker was then asked to taste the mixture and guess the beverage.

Dated fuddling cups range from 1633 to 1649, indicating that the fashion for them was relatively short-lived; however, based upon their style of decoration, a very small number can be dated to the early eighteenth century, suggesting that the form persisted after the initial craze had passed.

PROVENANCE

D. Howard; Dolz Collection, Alexandria, VA; Sotheby's, London, October 20, 1981 (lot 44); Nick Narganes; bt. John H. Bryan of Lake Bluff, IL (1936–2018) from Lindsay Antiques, London, June 10, 1988

LITERATURE

Louis L. Lipski and Michael Archer, *Dated English Delftware: Tin-Glazed Earthenware, 1600–1800* (London: Sotheby Publications, 1984), 198–99, no. 880

1740

Mug
Leather, silver
5¾ × 3⅞ × 5⅛ in.
(14.6 × 9.8 × 13 cm)

Inscription: Tho[s]
Legg 1740

This leather mug has a silver rim that is later than the mug itself. The rim is engraved with the name "Thomas Legg" and the date "1740." It is also stamped with the maker's mark, "D. C.," possibly belonging to the London silversmith Daniel Cowald (1723–1778). It has a dry capacity of 450 milliliters (15 ounces).

PROVENANCE
Watergate Antiques, Chester, April 17, 1998; Christie's, London, July 8, 2014 (lot 203); bt. John H. Bryan of Lake Bluff, IL (1936–2018) through Tobias Jellinek Antiques, London

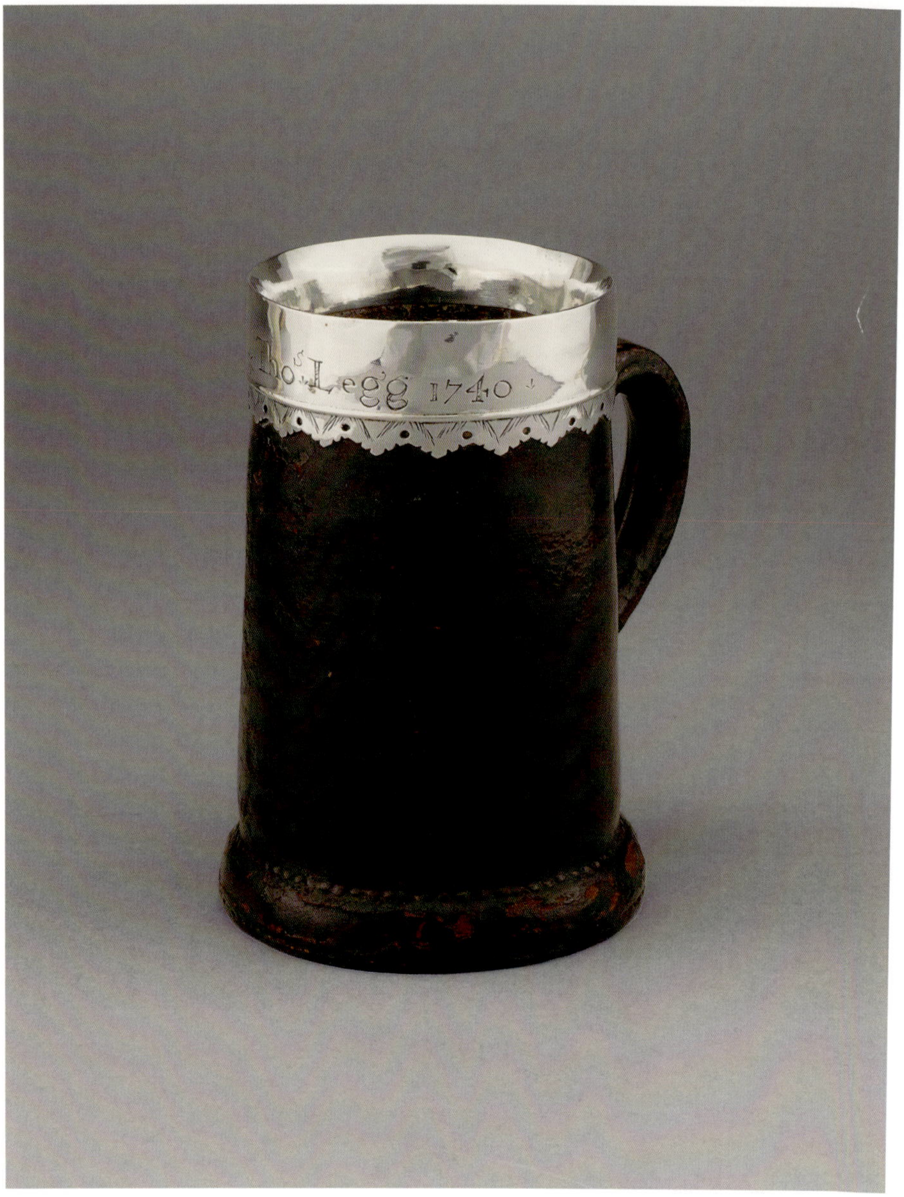

1754

This heavy stoneware mug bears a "G. R." excise mark, for "George Rex," which indicates Bristol rather than London as the place of manufacture and, with that, a possible connection to the factory of Thomas Frank at Redcliff Back. A similar tankard, dated 1739, is in the collection of the Victoria and Albert Museum, London (414:855-1885), signed by John Harwell while apprenticed to Joseph and Sarah Taylor at Thomas and Richard Frank's Redcliff Back pottery. That tankard and this one share similar molded and applied decorations. In particular, the same huntsman figure, blowing a horn, appears on both. In fact, the letter forms are sufficiently similar to suggest that the Bryan mug may also be the work of Harwell. About the time this mug was made, Harwell set up his own pottery on Redcliff Back. He died in or before 1774.

PROVENANCE

Burt Latham; Jonathan Horne Ltd., London; Irvin and Anita Schorsch; Sotheby's, New York, January 20, 2016; bt. John H. Bryan III of Nashville, TN, from Tim Gleason, February 12, 2016

LITERATURE

Robin Hildyard, *Browne Muggs: English Brown Stoneware* (London: Victoria and Albert Museum, 1985)

Mug

Salt-glazed stoneware

$7\frac{3}{4} \times 5\frac{1}{2} \times 7\frac{3}{4}$ in. (19.7 × 14 × 19.7 cm)

Inscription:
William Robbins 1754|I wish you all merry|& fill it once more

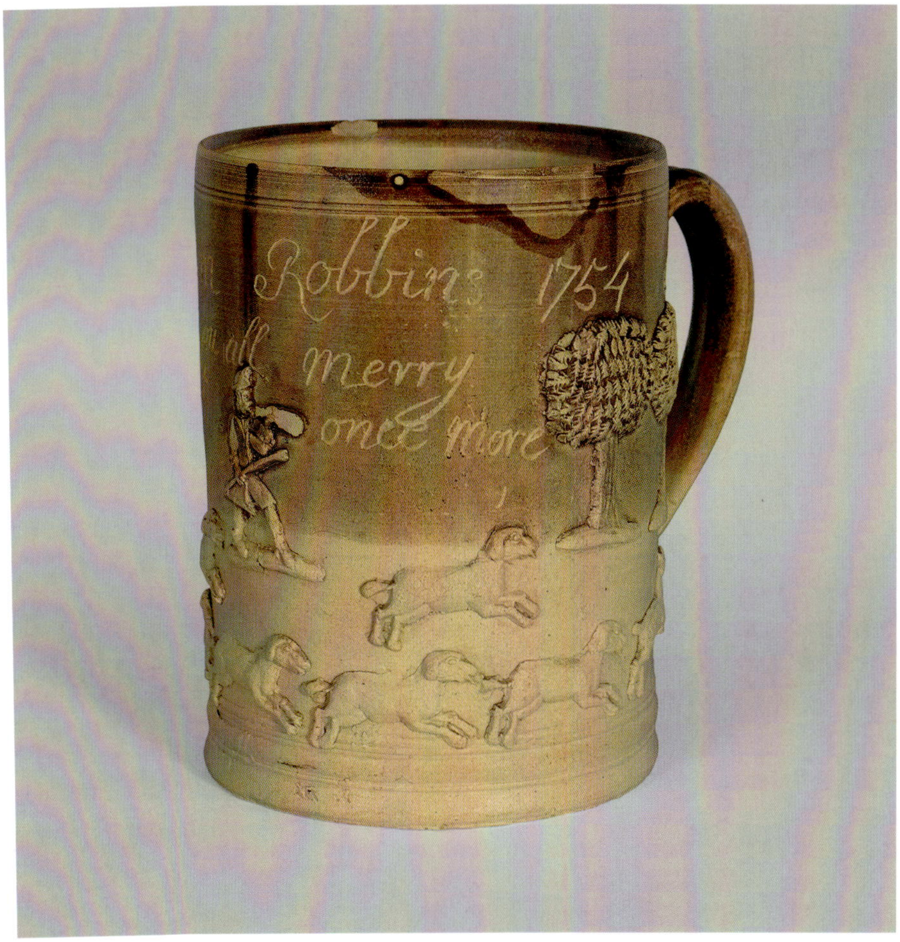

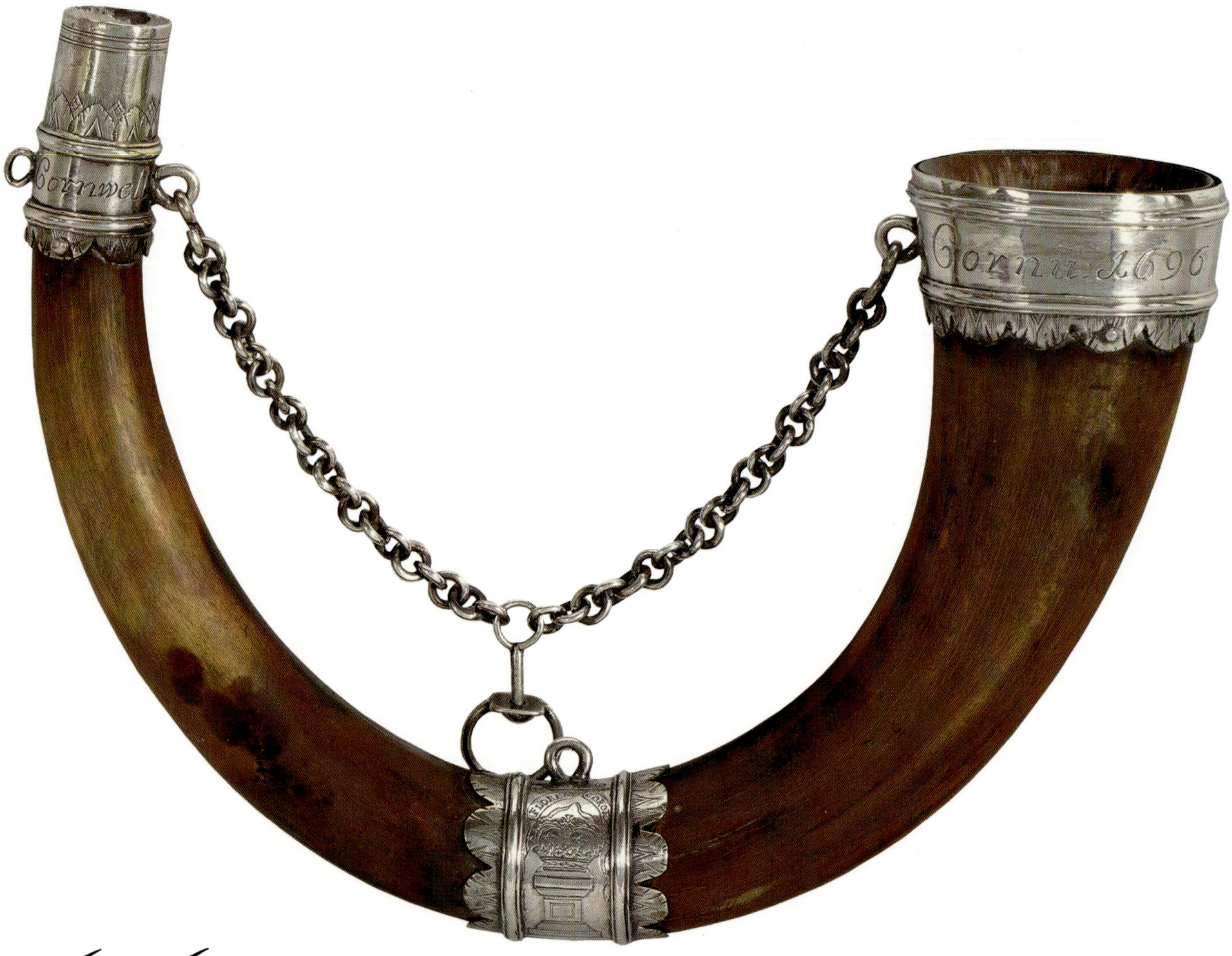

1696

Hunting horn

Horn, silver

$8\frac{1}{2} \times 7\frac{1}{4} \times \frac{3}{4}$ in.
(21.6 × 18.4 ×
4.5 cm)

Inscription: Cornū de Cornwell; Floreat virtūte Cornū 1696| FLOREAT CORNU CORONE; CORNUSALUTI

The identity of the owner of this hunting horn is a puzzle. Cornwell Manor is in Oxfordshire and was home to the Penistone family, but this coat of arms is not theirs.

PROVENANCE

bt. John H. Bryan of Lake Bluff, IL (1936–2018) from Peter Cameron, London, December 5, 2016

1746

Cup
Horn, silver
$4\frac{1}{4} \times 2\frac{1}{2}$ in. (10.8 × 6.4 cm)

Inscription: Rob: Farmerly : Huntsman to His Grace the Duke of Leeds Keefton 1746

The ancestral home of the Duke of Leeds, Kiveton Hall, Yorkshire, was demolished in 1812. Robert Farmerly, "Huntsman to His Grace the Duke of Leeds," was interred February 7, 1750/1, at Harthill, Yorkshire. The duke he served was Thomas Osborne, fourth Duke of Leeds (1713–1789), who, like his huntsman, is buried at All Hallows, Harthill.

PROVENANCE
bt. John H. Bryan of Lake Bluff, IL (1936–2018) from Tobias Jellinek Antiques, London, October 18, 2017

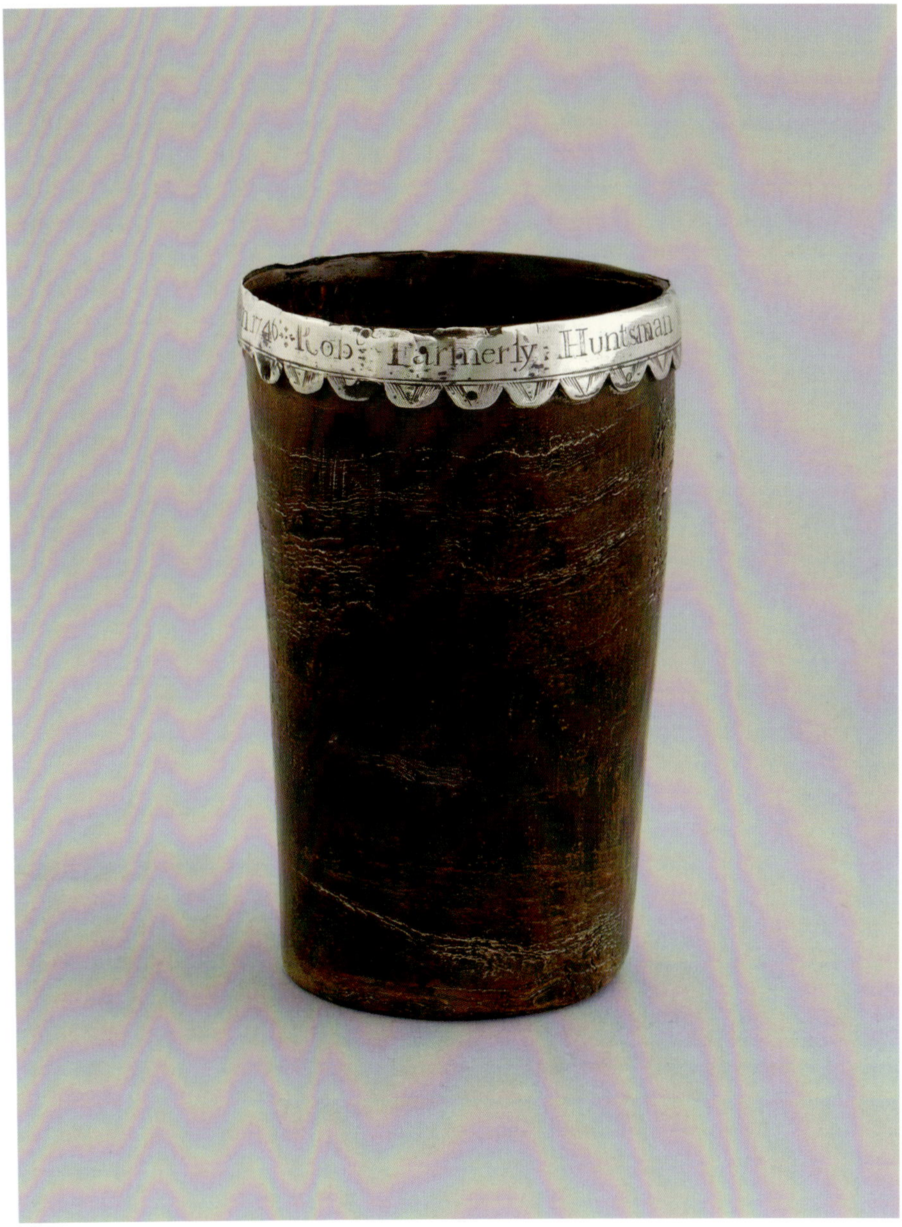

1776

Fishing reel
Brass, iron
$3\frac{1}{2} \times 2 \times 2\frac{1}{2}$ in.
(8.9 × 5.1 × 6.4 cm)

Inscription: 1776
JN^O TOMBS.

This fishing reel was possibly the property of John Tombs. He was buried, in 1778, aged fifty-two, at St Andrew Ombersley, Droitwich, Worcestershire, which is just off the River Severn, where he may have fished.

PROVENANCE
bt. John H. Bryan of Lake Bluff, IL (1936–2018) from Jellinek & Sampson Antiques, London, May 19, 1981

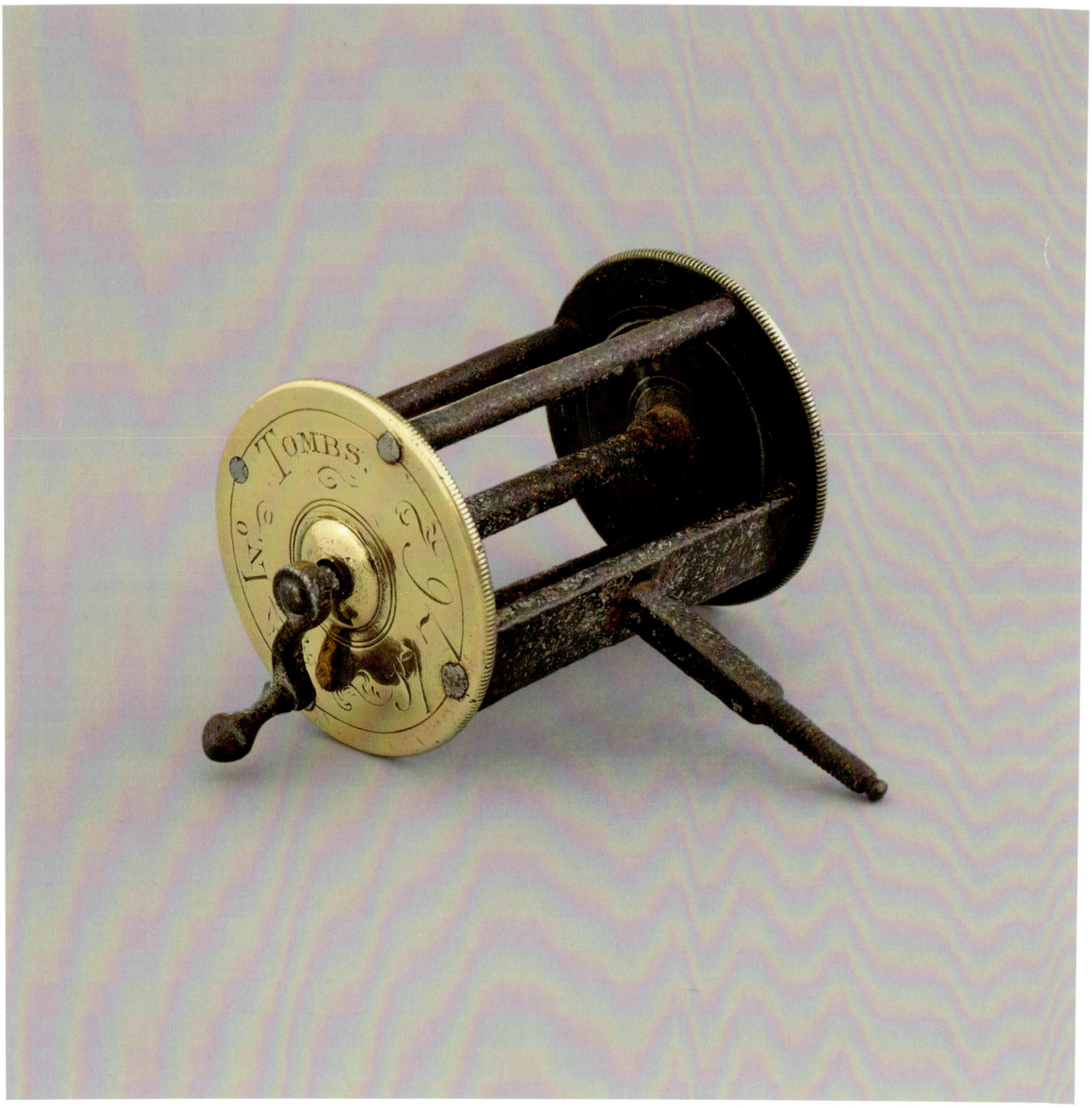

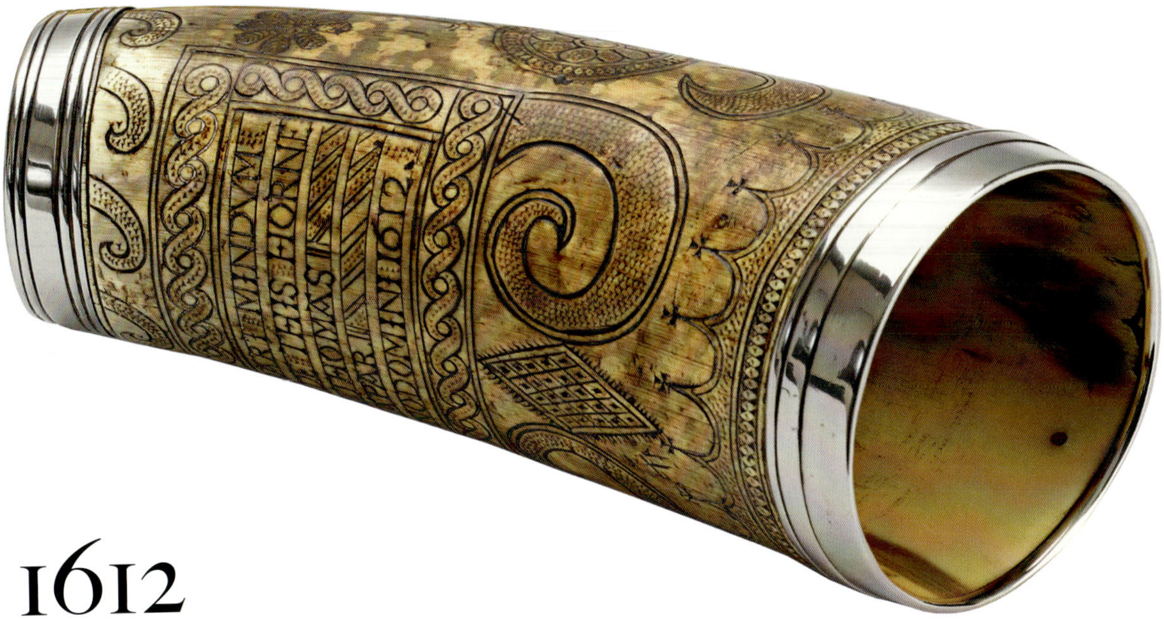

1612

Powder horn

Horn, silver

$3\frac{5}{8} \times 5\frac{3}{4} \times 2\frac{1}{2}$ in. (9.2 × 14.6 × 6.4 cm)

Robert Mindum (active 1593; d. 1615?)

Inscription:
ROBART MINDVM| MADE THIS HONRE|FOR THOMAS|DRAPAR| ANNO DOMINI|1612

This object is one of only two known powder horns made by Robert Mindum; the other is in the Museum of London, made by Mindum for William Morris in 1601. To judge by the style of the silver mounts, this example was converted into a drinking vessel at some point in the eighteenth century.

PROVENANCE

Collection of Stuart Kirby of Long Valley, NJ (1952–2019); bt. John H. Bryan III of Nashville, TN, from Adam Ambros, New York, July 22, 2019

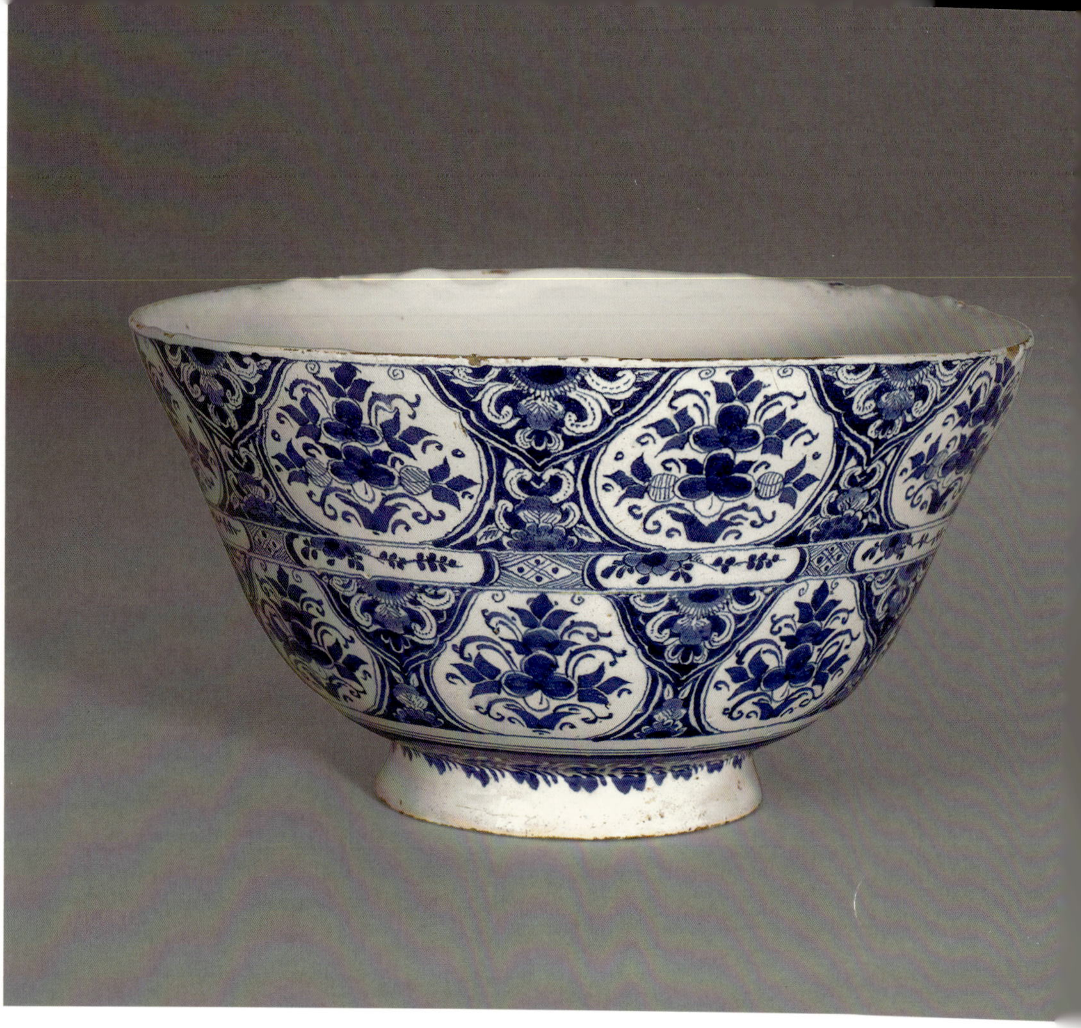

1729

Punch bowl
Delftware
8 1/8 × 14 5/8 in. (20.6 × 37.1 cm)

Inscription: The Gift of Benjamin| Richards to Thomas|Gwynn Esq[r] 1729

Ceramic punch bowls were made in Europe from the 1680s onward. Smaller ones were generally for individual use; larger ones might have been made for public houses and communal drinking. Punch arrived in Europe in the seventeenth century via the East India trade. It originally contained arrack, a rum-like spirit from Goa and Jakarta, mixed with lemon juice, water, sugar, and spices, but by the eighteenth century, there were many different recipes in circulation.

Thomas Gwyn of Llanarthney in Carmarthenshire, Gentleman, had his will proved at the Prerogative Court of Canterbury on July 3, 1766 (TNA PROB 11/920/163). However, there was another Thomas Gwyn of Gwempa, Carmarthenshire (1691–1752), in which case it would follow that the gift giver, Benjamin Richards, was the mercer of Carmarthen, Carmarthenshire, by that name who died in November 1749 and was buried near the monument to his wife at St Peter's Church, Carmarthen (TNA PROB 11/775/399).

PROVENANCE
Cecil Baring, third Baron Revelstoke (1864–1934); Puttick and Simpson, London, April 26, 1937 (lot 167); Louis Gautier; Puttick and Simpson, London, March 9, 1944 (lot 132); Christie's, London, July 12, 1982 (lot 107); bt. John H. Bryan of Lake Bluff, IL (1936–2018) from Jellinek & Sampson Antiques, London, September, 1982

LITERATURE
Louis L. Lipski and Michael Archer, *Dated English Delftware: Tin-Glazed Earthenware, 1600–1800* (London: Sotheby Publications, 1984), 249, no. 1083

1665

The Boar's Head Tavern on Knightrider Street in London was at the junction with Do Little Lane. Its proprietor, John Stillington, was the son of William, of Earith, Huntingtonshire, Gentleman, who had been apprenticed to Thomas Robinson, Vintner, on March 3, 1656/7. The owner's decision to inscribe his name on these candlesticks would have deterred light-fingered customers from liberating them from the tavern.

PROVENANCE
bt. John H. Bryan of Lake Bluff, IL (1936–2018) from Alistair Sampson Antiques Ltd., London, March 9, 1995

LITERATURE
Eloy Koldeweij, *The English Candlestick: 500 Years in the Development of Base-Metal Candlesticks, 1425–1925* (London: Christie's Books, 2001), 75, no. 42

Candlesticks
Brass
8 × 3⅞ in. (20.3 × 9.8 cm)

Inscription, on both: John Stillington at y^e Boares head|(and Dragon)|Taverne in Knight Rider Streete 1665

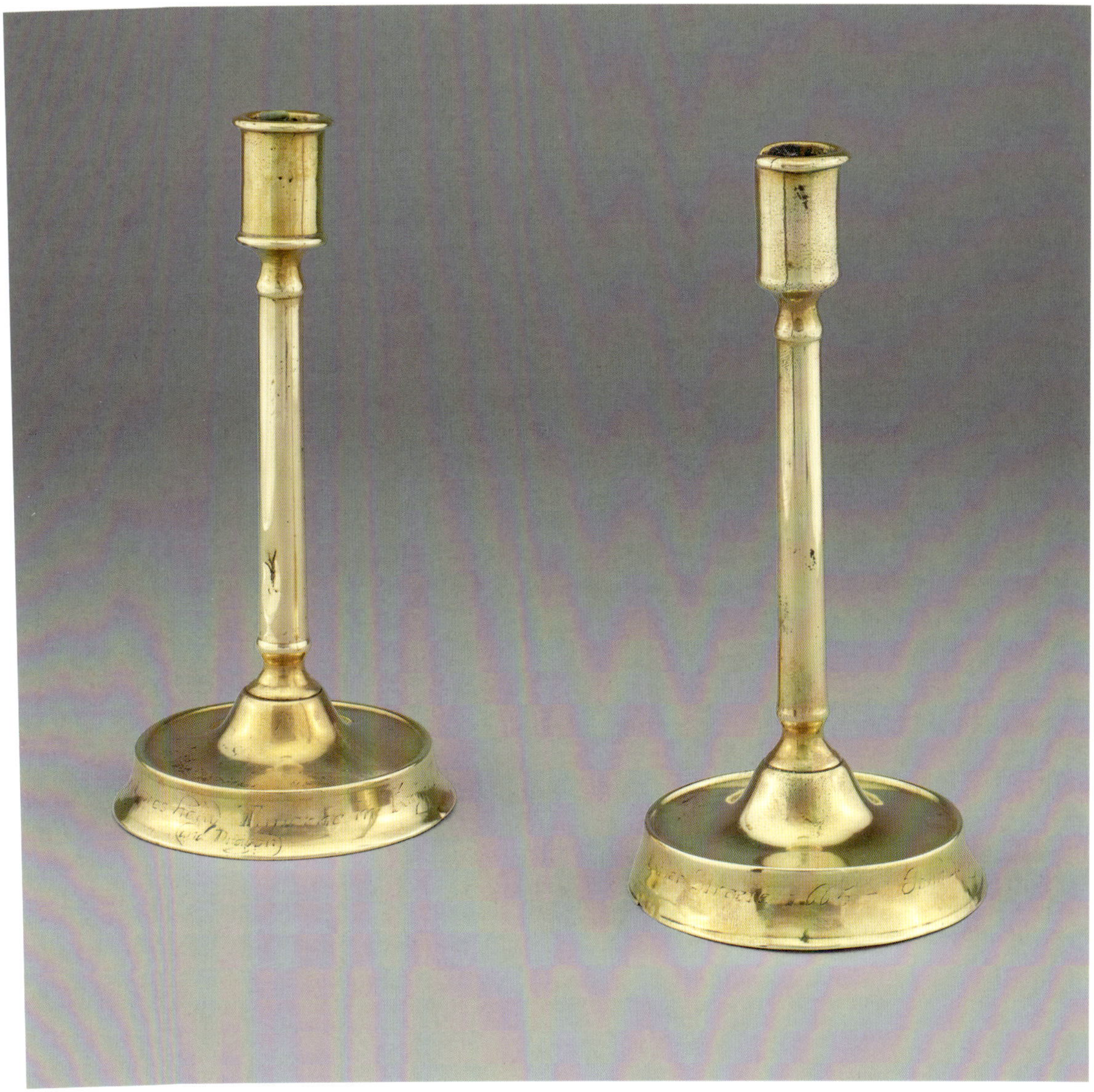

1709

Candlestick

Brass

$7\frac{7}{8}$ × 4 in. (20 × 10.2 cm)

Inscription: Gen. Constantine at y^e Grecians Coffe house in Deveroux Court in y^e Strand by Temple barr 1709

This simple candlestick, unremarkable in its design and typical in its manufacture, bore witness to some of the most significant scientific, political, and philosophical debates of the age. It belonged to the Grecian, a coffeehouse just off the Strand, named for the nationality of its famous and long-lived proprietor, George Constantine (d. 1728). Constantine had moved to this location on Devereux Court about a decade after his previous business on Threadneedle Street had burned to the ground in the Great Fire of London. There he had sold the imported intoxicants—coffee, chocolate, sherbet, and tea—that had caused a craze in coffeehouse culture. By the early part of the eighteenth century, there were more than eight thousand coffee shops in the capital city alone.

Initially called the Turks Berry-Drink, coffee was deemed to be a cure for indigestion, tiredness, and headaches; it was also believed to prevent a number of ailments. One contemporary advertisement claimed that "it will prevent Drowziness, and make one fit for Business. It is observd, that in Turkey, where it is generally drunk, that they are not troubled with the Gout, Stone, Dropsy, or Scurvy, and that their Skins are exceedingly clear and white.... It is neither laxative nor restringent."

Coffeehouses were important meeting places for various groups to share knowledge and contest ideas. Newspapers and auction catalogues were sold there, too; it is no surprise that the Grecian is mentioned in the first edition of the weekly magazine the *Spectator*. Its patrons included writer and journalist Richard Steele, writer Oliver Goldsmith, and the notorious Samuel Foote, the "one-legged Georgian comedian" whose characters Lady Pentweazle in *Taste* and Mother Cole in *The Minor* are the prototypes for the modern-day "pantomime dame." The Grecian was also frequented by members of the Royal Society, including Sir Isaac Newton, Sir Hans Sloane, Edmond Halley, and James Douglas, the antiquarian William Stukeley, and the poet and statesman Joseph Addison. When patrons gathered together, the light from the candles and the caffeine from the coffee would have sustained their conversation long after sunset, whether they were debating the merits and dangers of freethinking or, as they did in 1712, observing the dissection of a dolphin.

Many coffeehouses attracted a specific clientele that shared political affinities. The Grecian was no exception; it hosted a secret club called The Old or Real Whigs, whose members were otherwise known as the Commonwealthmen, a group that resisted the existence of a standing army, which they saw as a direct threat to liberty. The group promoted republicanism and subsequently had a great influence on Republicanism in the United States. The Grecian stayed open until 1843, after which the coffeehouse was converted into a pub.

PROVENANCE

bt. John H. Bryan of Lake Bluff, IL (1936–2018) from Alistair Sampson Antiques Ltd., London, February 4, 1993

LITERATURE

Brian Cowan, *The Social Life of Coffee: The Emergence of the British Coffeehouse* (New Haven, CT: Yale University Press, 2005)

James Douglas, *A Supplement to the Description of the Coffee-Tree* (London: Printed for Thomas Woodward, 1727)

Markman Ellis, ed., *Eighteenth-Century Coffee-House Culture* (London: Pickering and Chatto, 2006), 4:219, 253, 254

Jonathan Harris, "The Grecian Coffee House and Political Debate in London, 1688–1714," *London Journal* 25 (2000): 1–13

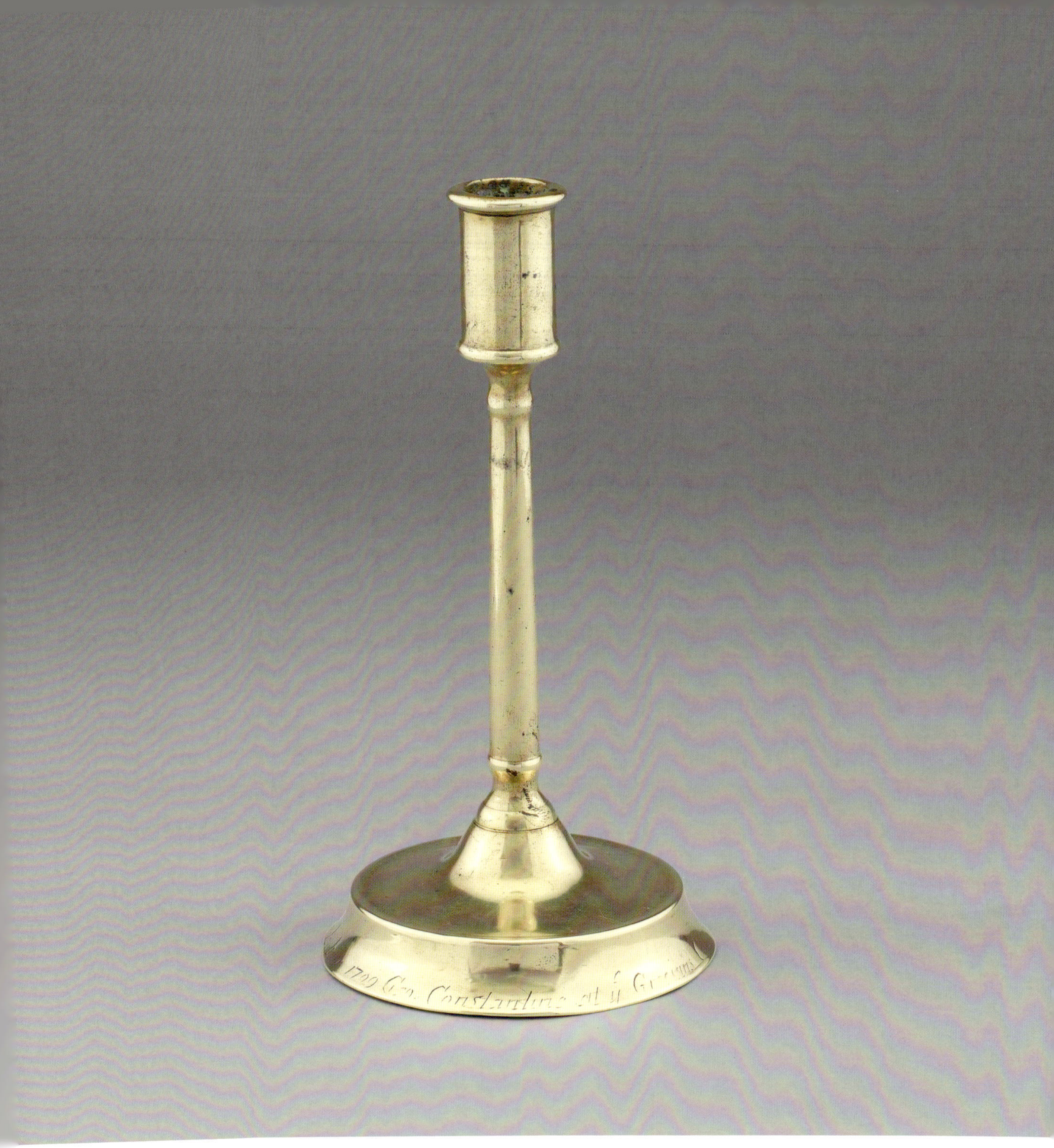

C. S. Knighton, ed., *Calendar of State Papers: Domestic Series of the Reign of Anne Preserved in the Public Record Office* Vol. 4, *October 1705–December 1706* (Woodbridge, UK: Boydell Press; London: The National Archives, 2006), 376

Eloy Koldeweij, *The English Candlestick: 500 Years in the Development of Base-Metal Candlesticks, 1425–1925* (London: Christie's Books, 2001), 74, no. 41

1628

Wine bottle

Delftware (Southwark)

8 ¼ × 6 in. (21 × 15.2 cm)

Inscription: 1628

This Delftware wine bottle is nearly identical to a small number of other bottles dated 1628. All are painted with blue paint, depicting scenes with rocks, plants, flowers, and birds. These bottles, along with a mug dated 1629 and another dated 1630, are among the dated objects that have the bird-on-rocks decoration based on Chinese porcelain from the late Ming dynasty. This form of decoration occurs frequently on Christian Wilhelm's Pickleherring pottery, although pottery shards with decoration of this type have also been found in the other Southwark potteries at Montague Close and at Rotherhithe.

The year 1628 was of great significance to Wilhelm—it was the year he was granted a royal patent that gave him a monopoly on the manufacture of gallyware (tin-glazed earthenware). His petition to Charles I boasts of his invention of "makeinge of white earthen potts and vessells glazed both wthin and wthout wch shewes as faire as China dishes" (TNA SP 16/408, fol. 19).

Wilhelm had arrived in England in late September 1604 and took over part of the property called Pickleherring on the bank of the Thames, north of Horsley Down, Southwark. The property name—which lived on for centuries as the Pickle Herring Stairs, leading down to the waterfront—derived from the nickname of Peter van Durant, alias Pickle Herring, who had operated a brewery on that site. Initially, Wilhelm (or Wilhelmson, as he was also known) manufactured the synthetic blue pigment smalt, made from powdered glass mixed with cobalt oxide, but he had been frustrated by the grant of a monopoly to another manufacturer of smalt.

In about 1612, Wilhelm converted his manufactory into a pottery, which employed forty men. On his death, it passed to his son-in-law, Thomas Townsend, and then, into the first quarter of the eighteenth century, into the hands of other owners outside the family.

In his will, made on March 8, 1629/30, and proved on April 9 (TNA PROB 11/157/361), Wilhelm described himself as a "gallipot maker." He cited property in the Palatinate region of Germany and—as a "great adventurer into Virginia" and other parts "beyond the seas"—stocks he had acquired that concerned "sope ashes and pott ashes" (potashes used in the making of smalt).

PROVENANCE

Manfield, 1913; Thomas George Burn, Rous Lench Court, Worcestershire; Christie's, London, May 29, 1990 (lot 13); bt. John H. Bryan of Lake Bluff, IL (1936–2018) from Alistair Sampson Antiques Ltd., London, May 31, 1990

LITERATURE

Michael Archer, *Delftware in the Fitzwilliam Museum* (London: Philip Wilson Publishers, 2013), 220

Frank Britton, *London Delftware* (London: Jonathan Horne, 1987), 108

Isabel Davies, "Seventeenth-Century Delftware Potters in St Olave's Parish, Southwark," *Surrey Archaeological Collections* 66 (1969): 11–31

Leslie B. Grigsby, *The Longridge Collection of English Slipware and Delftware* (London: Jonathan Horne Publications, 2000), 1:244–45

Louis L. Lipski and Michael Archer, *Dated English Delftware: Tin-Glazed Earthenware, 1600–1800* (London: Sotheby Publications, 1984), 309, no. 1252

Hugh Tait, "Southwark (Alias Lambeth) Delftware and the Potter Christian Wilhelm" (part 1), *Connoisseur* 146, no. 587 (August 1960): 36–42; (part 2), *Connoisseur* 147, no. 591 (March 1961): 22–29

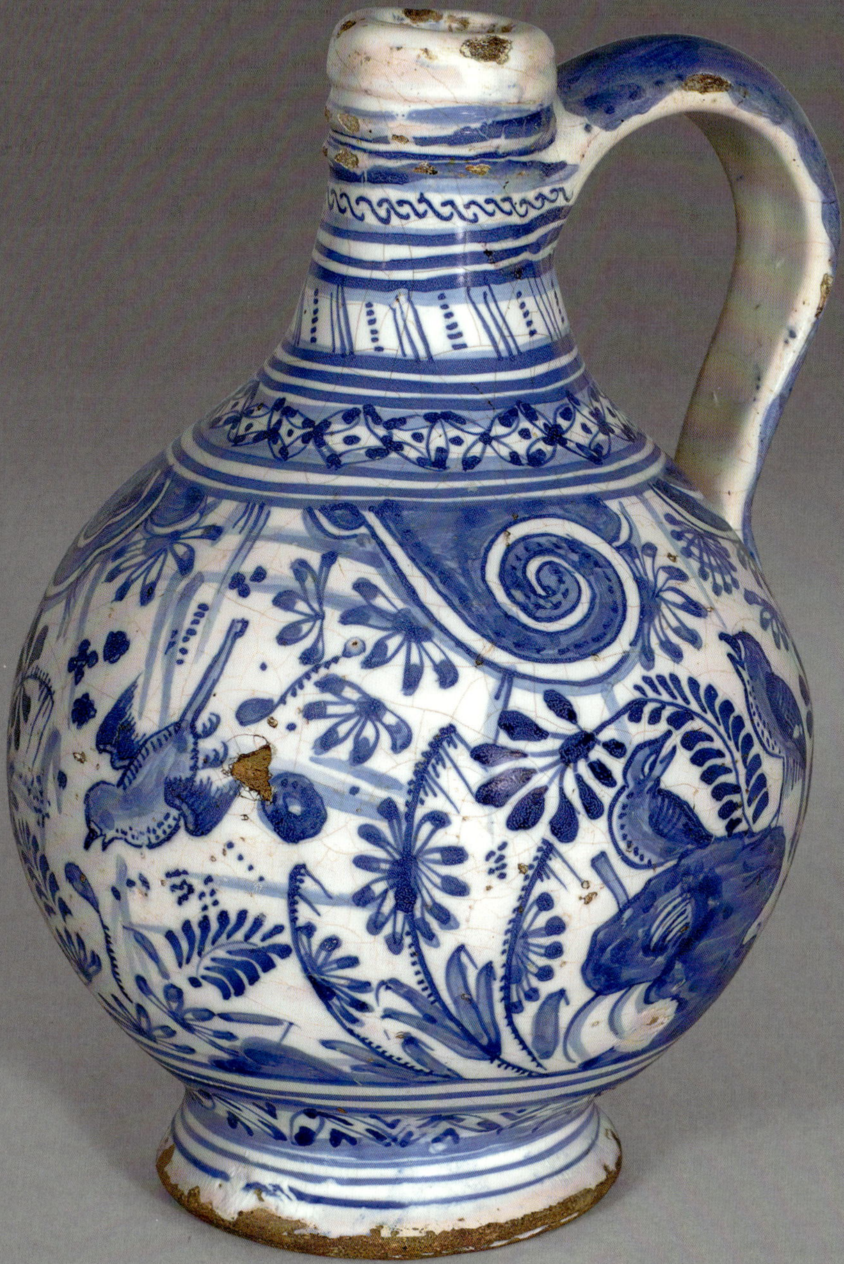

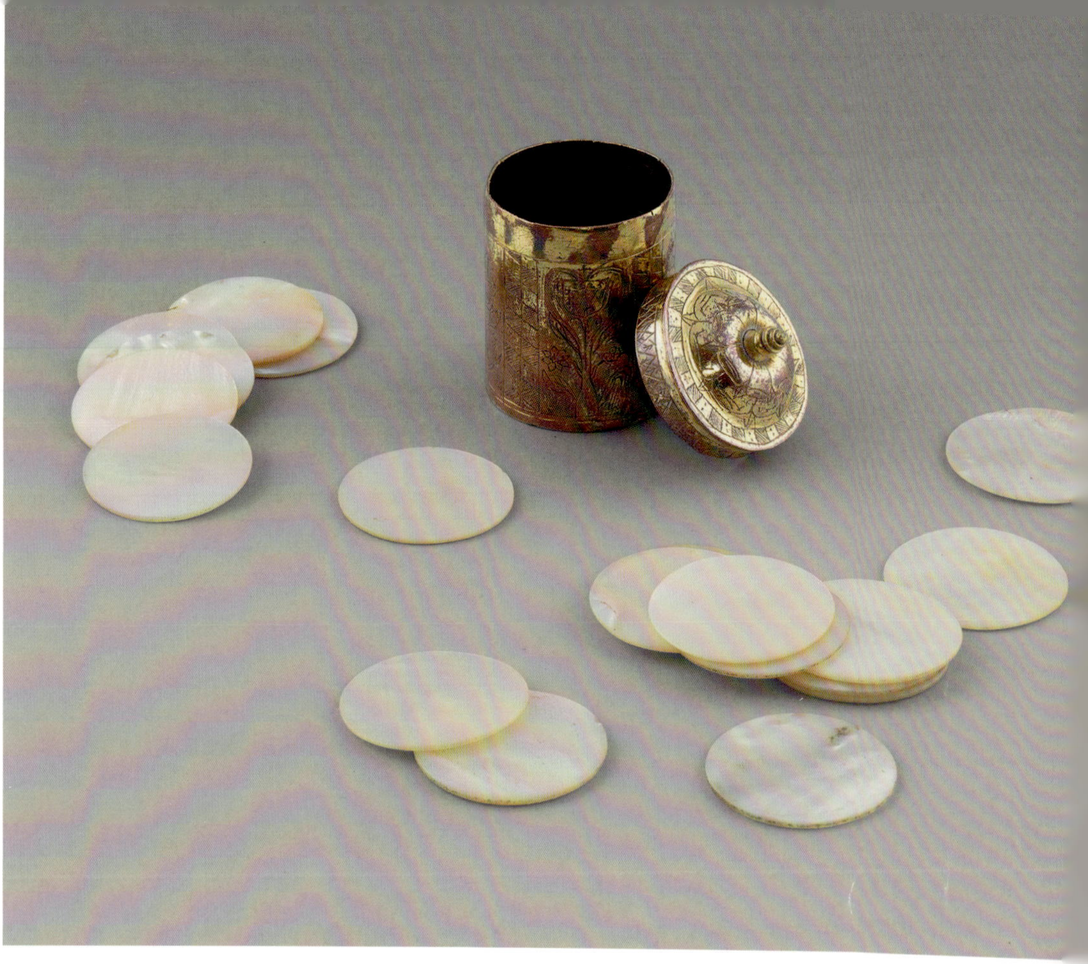

1625

Counter box
Gilded brass
2⅜ × 1⅜ in.
(6 × 3.5 cm)

Inscription:
1625|EW

Counter boxes like this one were made in England from the late sixteenth century. The mother-of-pearl counters are presumed to be later in date than the box.

PROVENANCE
bt. John H. Bryan of Lake Bluff, IL (1936–2018) from Alistair Sampson Antiques Ltd., London, June 22, 1998

1652

Case
Horn, brass
4⅞ × 2⅛ × ¼ in.
(12.4 × 5.4 × .64 cm)

Inscription: T L; 1652

PROVENANCE
bt. John H. Bryan of Lake Bluff, IL (1936–2018) from Tobias Jellinek Antiques Ltd., London, October 18, 2017

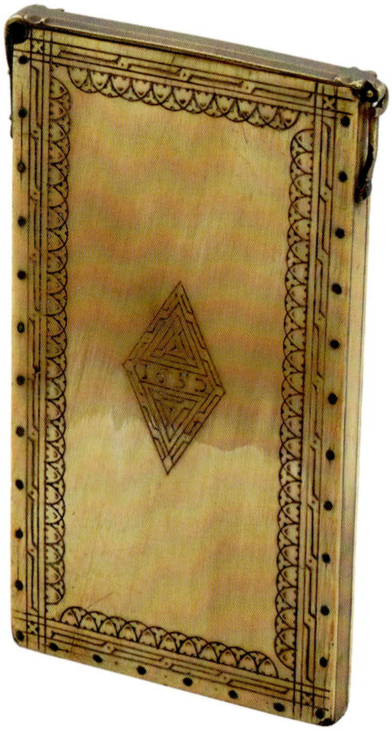

1683/4

Box

Silver

$\frac{7}{8} \times 1\frac{3}{4}$ in. (2.22 × 4.5 cm)

Inscription, on verso: Bought on y^{e}|Frozen River of|Thames Feb. 1st 1683/4|Elizabeth Foche

Frost fairs were held on the River Thames between 1608 and 1814. The frost fair of 1684—the year of the most severe freeze in the river's history—began in mid-January. Depictions of the frost fair that year show the booth from which trinkets such as this silver box were sold, along with scenes of bull baiting and, remarkably, of carriages driving across the river. There was also sledding, skittles games, and ox roasts before the ice melted in early February.

This silver box—a memento of the frost fair, dated February 1, 1683/4—was bought by Elizabeth Foche, née Horsey (d. 1693), a member of London's citizen elite. She was the second wife of the phenomenally wealthy Sir John Foche, Draper of St Margaret Lothbury. The two married on December 20, 1681, at All Hallows Staining, just off Fenchurch Street and near London Bridge. They had three children together: Meriel (baptized October 28, 1690), William (January 21, 1690/1), and Samuel (June 27, 1693). Samuel's birth proved fatal for Elizabeth; she died on June 13, 1693, aged thirty-two.

The box is one of just five silver, or silver-mounted, articles known to have been purchased at the 1684 fair. The Museum of London has a trifid spoon with an inscription dating its purchase as January 29 and a leather mug with silver rim purchased on January 28. The Victoria and Albert Museum, London, has a miniature glass mug with an engraved silver rim, bought on January 27 (C.156-1997). Another trifid spoon, a lace back, appeared at auction at Lawrence of Crewkerne, on January 19, 2016 (lot 92), inscribed with a January 28 date of purchase and "E. W.," the owner's initials. This filigree box in the Bryan Collection is, therefore, the only piece to have an identifiable owner.

PROVENANCE

bt. John H. Bryan of Lake Bluff, IL (1936–2018) from Peter Cameron, London, December 3, 2016

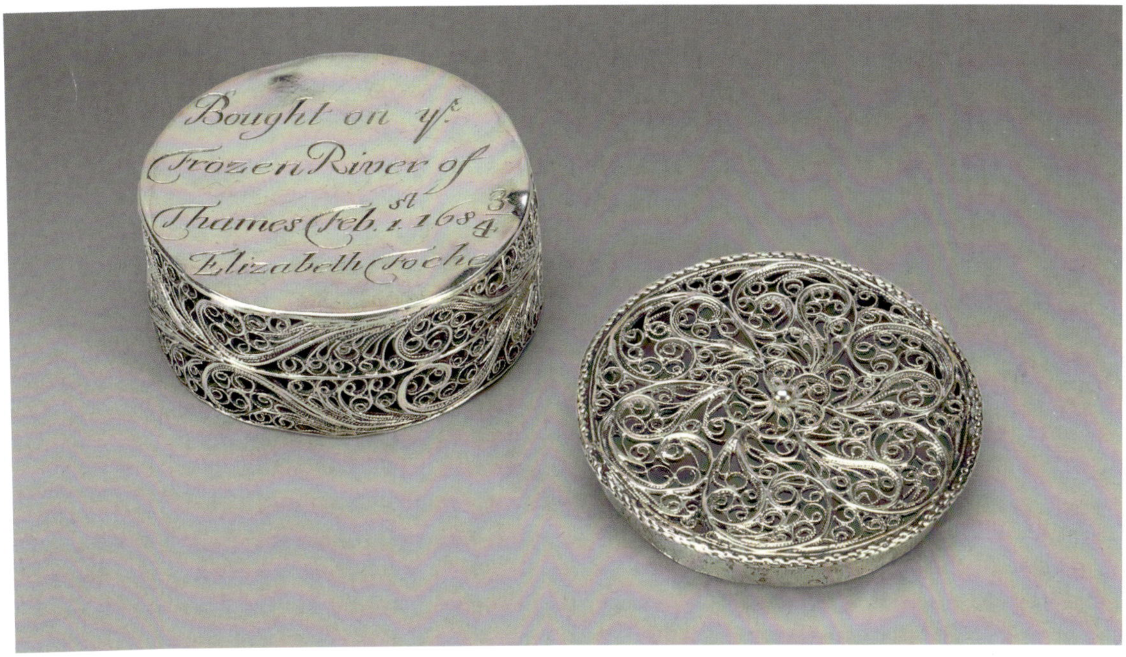

1734

Cribbage board
Fruitwood
$10\frac{1}{4} \times 1\frac{3}{8} \times \frac{1}{2}$ in.
(26 × 3.5 × 1.27 cm)

Inscription:
Richard Maswood
Decr y^{e}. 31 1734

The antiquary and biographer John Aubrey (1626–1697) attributed the invention of cribbage to the gamester, gallant, amorist, and sometime poet John Suckling (1609–1641?). The rules of the game have remained unchanged since Suckling's day. Cribbage is played with cards, and players accumulate points by creating matching pairs or a "run," a sequence of cards of consecutive value. The points are tallied on distinctive cribbage boards of the type seen here, which have two sets of sixty holes, one set for each player. Score is kept by advancing pegs, or "spillikins," which in these boards have all been lost.

Plausibly, Richard Maswood was the cordwainer buried on June 1, 1742, at St James Piccadilly, London. On July 21, 1739, he had, in clandestine fashion at the Fleet, married a widow named Frances Bowyer, also of St James Piccadilly.

PROVENANCE
bt. John H. Bryan of Lake Bluff, IL (1936–2018) from Alistair Sampson Antiques Ltd., London, October 21, 1999

1761

Cribbage board
Wood, bone, brass
$10\frac{3}{4} \times 1\frac{3}{4} \times \frac{3}{4}$ in.
(27.3 × 4.5 × 1.9 cm)

Inscription:
Hannah · Reay · y^{e} ·
Novr 5 · 1761

PROVENANCE
bt. John H. Bryan of Lake Bluff, IL (1936–2018) from Martyn Edgell Antiques Ltd., Cambridgeshire, August 10, 2017

1771

Cribbage board
Brass, mahogany
$11\frac{1}{4} \times 2\frac{3}{8} \times \frac{5}{8}$ in.
(28.6 × 6 × 1.59 cm)

Inscription:
W.|O.|T.|
Decem^{r}. 20 1771

PROVENANCE
bt. John H. Bryan of Lake Bluff, IL (1936–2018) from Price Glover Antiques Inc., New York, January 31, 1984

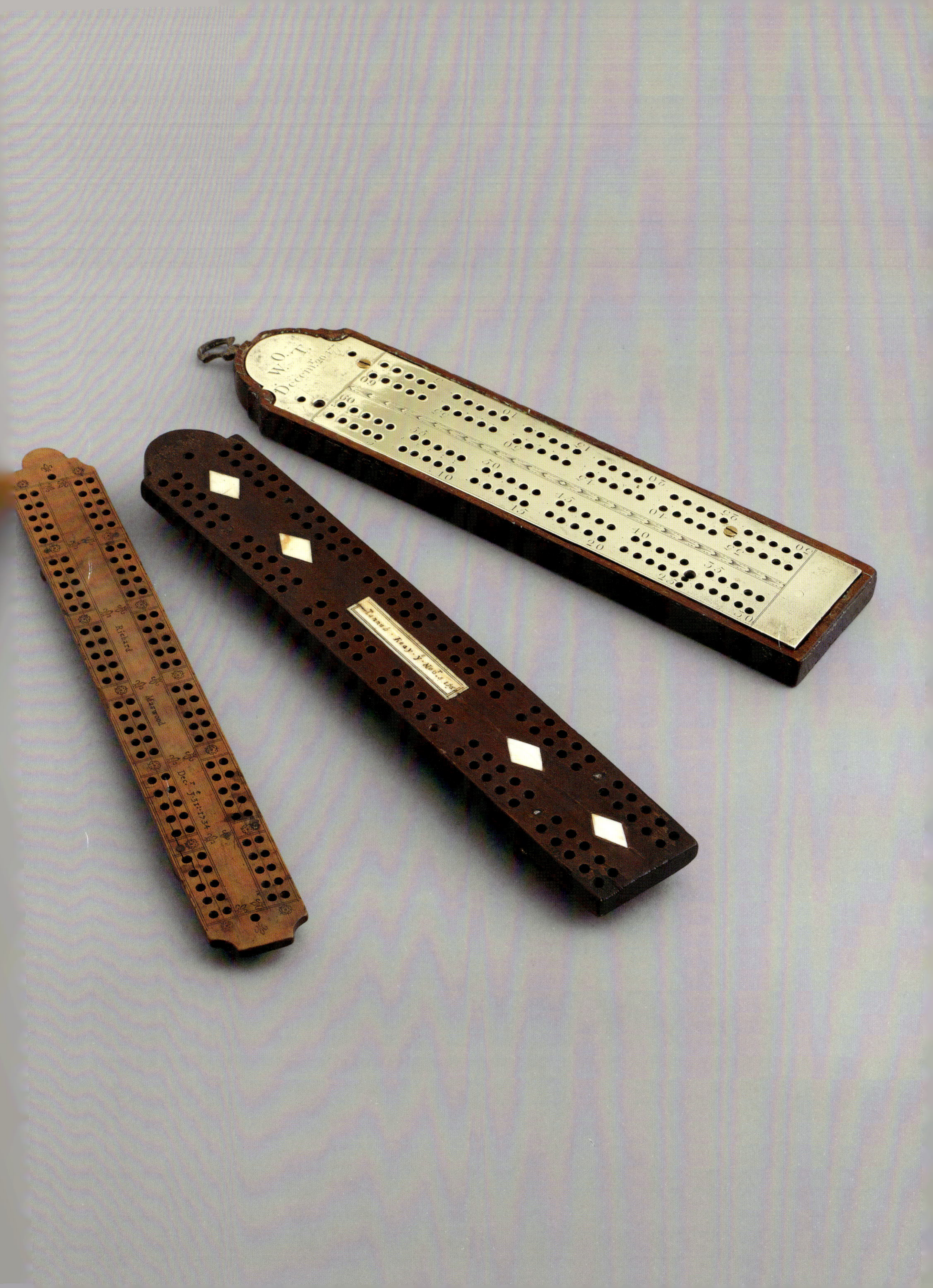

1664

Pipe tamper
Boxwood, silver
$3\frac{3}{8} \times 1 \times \frac{3}{4}$ in. (8.6 × 2.5 × 1.9 cm)

Inscription: VI.VIL. LAY.ROY; 16 64 R G

Soon after its introduction to the British Isles in the late sixteenth century, tobacco became a subject for popular debate. Many critics associated the plant, which originated in the New World, with paganism and incivility. Among them was King James VI and I, who, in an anonymously published anti-tobacco treatise in 1604, described smoking as "a custome lothsome to the eye, hatefull to the Nose, harmefull to the braine, dangerous to the Lungs." But proponents of tobacco, including many contemporary physicians and, most famously, Sir Walter Raleigh, won out, and by 1614, a great number of tobacco shops were operating in or near London.

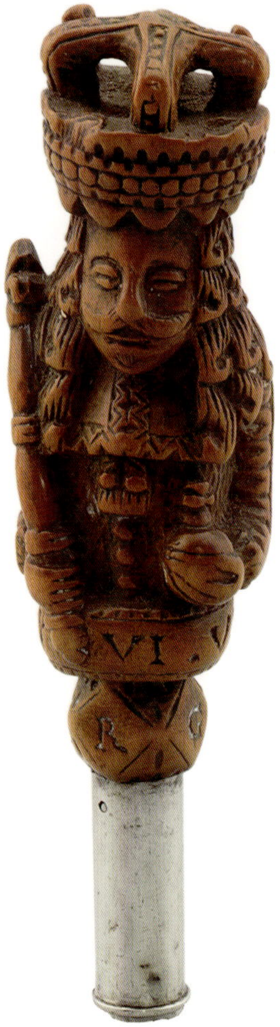

The explosion of pipe smoking in the seventeenth century led to the invention of many small smoking accessories, including pipe tampers like this one. For the perfect drag, the pipe smoker packed the bowl with tobacco very carefully—not too loose, not too tight. The pipe tamper, a simple tool made of wood or bone, served as a better alternative to one's fingertip for packing down tobacco. Tampers were commonly a product of amateur carvers and often depicted popular contemporary figures. Famous examples include depictions of Shakespeare carved from the mulberry trees grown in his garden in Stratford-upon-Avon and, in the nineteenth century, comic representations of the Duke of Wellington, an outspoken critic of the smoking habit.

This tamper, dated 1664, depicts Charles II (the grandson of the tobacco-hating James VI and I) just four years after the end of the Interregnum and his Restoration to the throne. A year after the tamper's creation, the Great Plague of 1665 caused a spike in smoking, as tobacco usage was seen as a way to immunize against the disease. Children were taught to smoke, and Eton College made smoking at breakfast compulsory for its students.

PROVENANCE

Philips, London, July 6, 1998; bt. John H. Bryan of Lake Bluff, IL (1936–2018) through Alistair Sampson Antiques Ltd., London

LITERATURE

Owen Evan-Thomas, *Domestic Utensils of Wood: XVIth to XIXth Century* (London: Owen Evan-Thomas, 1932)

1751

The tamper from 1751, engraved with the name "Hignell Fulker," shows a coursing greyhound catching a hare. The design was very popular in the eighteenth century, and numerous tampers with similar compositions survive. By the time of this tamper's creation, pipe smoking had been supplanted in popularity by snuff taking, seen by many at the time as a more elegant and refined means of taking tobacco.

One "Hignello Fulker" was baptized at Kingsclere, Hampshire, in 1751; his mother's name was Elizabeth. In 1766 he was apprenticed to Oliver Beck, wheelwright of Hartley Wintney.

PROVENANCE

bt. John H. Bryan of Lake Bluff, IL (1936–2018) from Peter Cameron, London, December 3, 2016

Pipe tamper

Gilded wood, silver

4 × 2 × $\frac{7}{8}$ in. (10.2 × 5.1 × 2.22 cm)

Inscription:
HIGNELL
FULKNER|1751

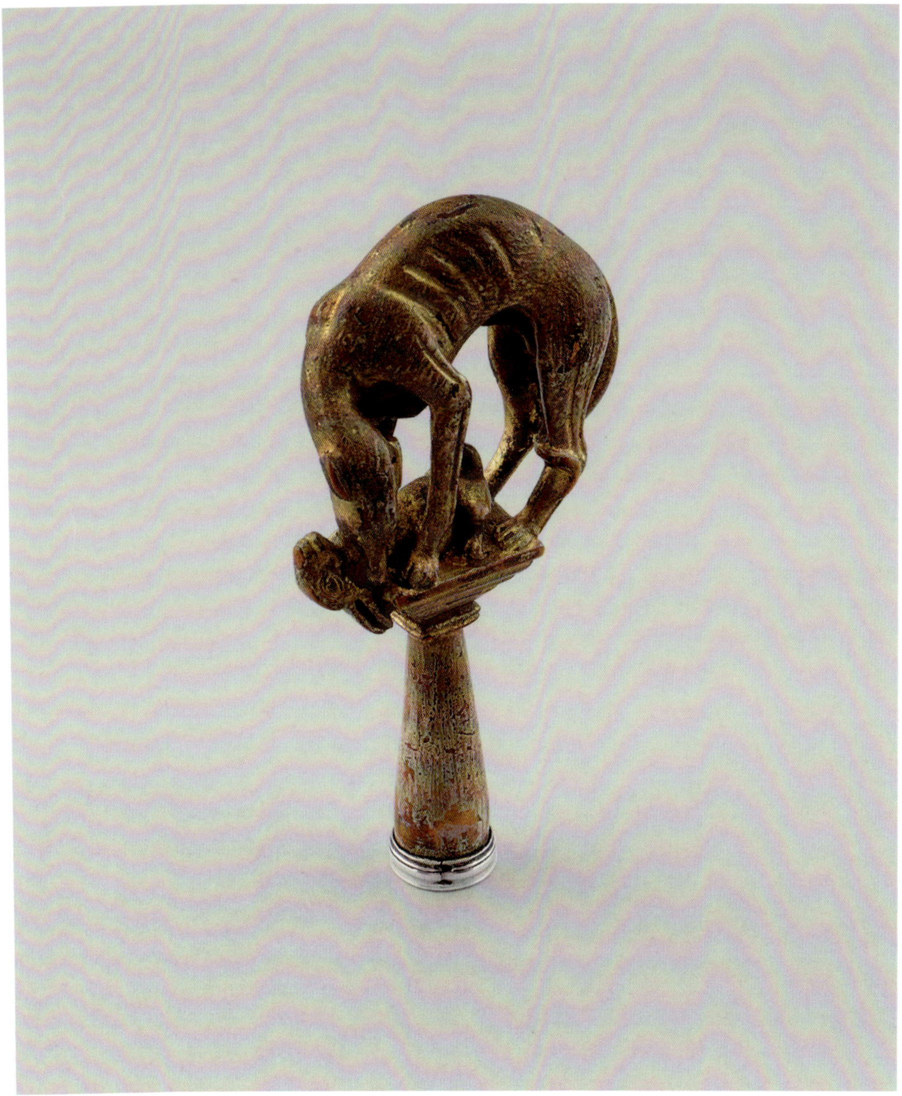

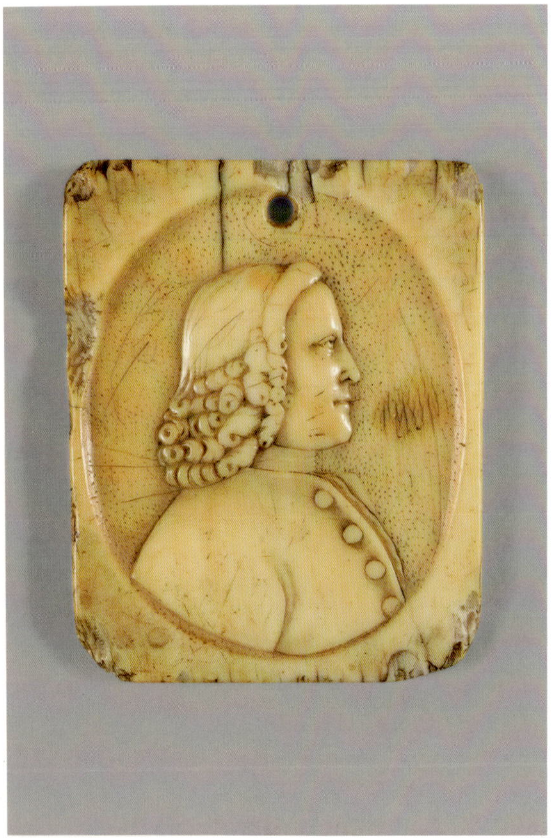

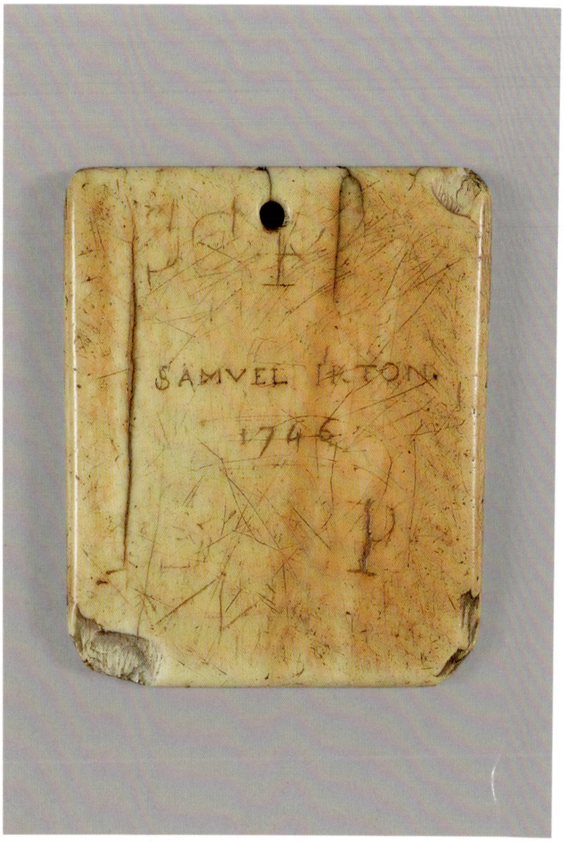

1746

Portrait
Ivory
2¾ × 2⅛ × ¼ in. (7 × 5.4 × .64 cm)

Inscription, on verso: SAMVEL IRTON|1746

In 1764, the London merchant Samuel Irton (1714–1766) put together a manuscript book about his ancestors, the Irtons of Irton Hall in Cumberland. He was assisted in his research by "Mr. Warburton," Somerset Herald, and "Mr. Jones," who together produced what could best be described as a hopeful hodgepodge of a genealogy. Samuel was proud of both his ancestry and the contents of his home. He made a careful inventory of each room, ending with this statement: "All which being exceeding good, I purchased and I beg may never be parted from the house."

Samuel was the youngest son of George Irton and his wife, Elizabeth Poole, and he inherited Irton Hall upon his brother George's death. On November 13, 1752, Samuel married Frances Tubman, daughter and heiress of Robert Tubman of Cockermouth. They had eight children and were succeeded at Irton by their son George. By the early nineteenth century, the estate was heavily mortaged, and, in 1866, the last direct decendant (also Samuel) died, after which the estate passed through a number of hands. It is now a hotel.

PROVENANCE

bt. John H. Bryan of Lake Bluff, IL (1936–2018) from Lindsay Antiques, London, June 29, 1992

LITERATURE

S. Taylor, "The Irtons of Irton Hall," *Transactions of the Cumberland and Westmorland Antiquarian and Archaeological Society* 41 (1941): 72–123

1610 | 1661

Tobacco box

Boxwood, horn, brass

1 × $2\frac{3}{4}$ × $2\frac{3}{8}$ in.
(2.5 × 7 × 6 cm)

Inscription: THE YEARE OF OVR LORD GOD|1661; LET THE FREE MAN COME FILL BUT LET THE; CLOSE MAN STAND STILL BUY AS WEL AS I; 1610|FRANCIS CLARKE HIS AGE IVLY 8

Francis and his twin brother, George, sons of John Clarke, were baptized in Othery, Somerset, on July 8, 1610. The same parish register records the marriage of their father to Joane Avoks on May 10, 1596. If this is the same Francis Clarke referred to in this inscription, then it seems likely that he was the man by that name who was buried in the neighboring village of High Ham in Somerset on November 6, 1681.

PROVENANCE

Golding Barett Collection; Bonham's, London, March 15, 2017 (lot 200); bt. John H. Bryan of Lake Bluff, IL (1936–2018) through Tobias Jellinek Antiques, London

LITERATURE

Angela McShane, "The New World of Tobacco," *History Today* 67, no. 4, (April 2017): 41–47

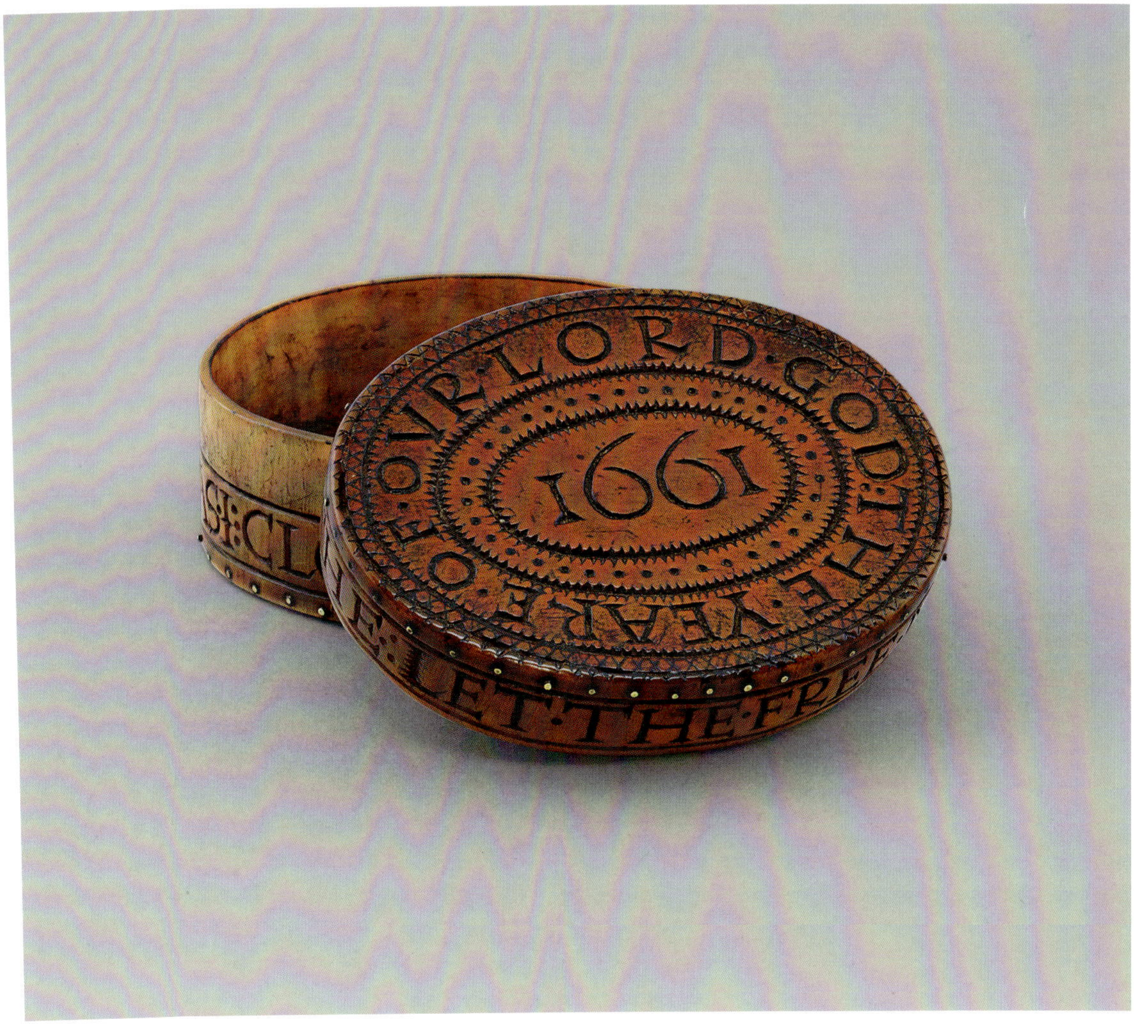

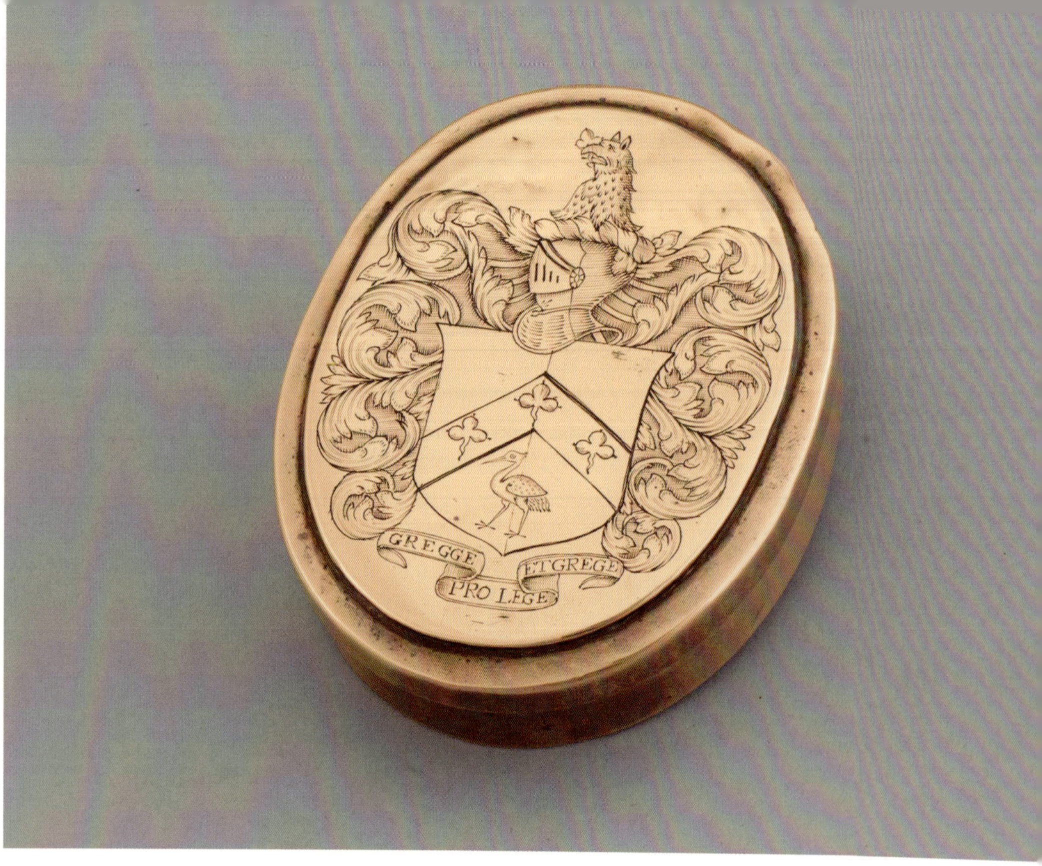

1662

Tobacco box

Copper

$\frac{3}{4} \times 3\frac{5}{8} \times 2\frac{3}{4}$ in.
(1.9 × 9.2 × 7 cm)

Inscription:
GREGGE|
PRO LEGE|ET
GREGE; [on verso]
Aperpetuall|
Almanack; Samuel
Gregge vrgeas|
legem An° regni
Car:|II: XIIII An°
dom 1662

For Samuel Gregge, law was a family business. As the motto on his tobacco box denotes, Gregge was "for the law and for the people"—although to judge from the paper record, the family was mainly in it for themselves. The Gregges of London were a cadet branch of an old Cheshire family whose claim to fame was a marriage to a coheiress of Robert Starkey of Stretton, to whom the fifteenth-century judge, Humphrey Starkey, chief baron of the Exchequer, seems to have been related. This connection was clearly a source of pride for the Gregges; although it might have made some heralds wince, Samuel had the image of a stork, the heraldic symbol for the Starkey family, incorporated into his coat of arms. A more accurate display appears on the carved stone that bears the Gregge coat of arms, once part of the Old Bell in Holborn Hill (a property acquired by the family in 1679) and now in the collections of the Museum of London.

Samuel was born on May 2, 1644. He was the son of Elizabeth (daughter of Robert Foot, Grocer) and her husband, Ralph (d. 1680), who had married on March 21, 1636/7, at Holy Trinity, Clapham. Their child was baptized on May 4 at the parish church of St Augustine, Watling Street, in the shadow of old St Paul's Cathedral. Ralph Gregge was a member of Clement's Inn, one of the various Inns of Court to the west of the city, within easy walking distance of his home. Samuel and his brother John entered Clement's Inn in 1664 with their father as surety. Their brothers later followed in their footsteps: Ralph in 1669, Thomas in 1676, Joseph in 1680, and Francis in 1683. Their eldest sibling, Robert, apprenticed to a pewterer called Thomas Gregge, likely a relative, in 1660.

So, when this tobacco box was made, Samuel Gregge would have been about eighteen years old, two years away from admission to Clement's Inn. The perpetual almanac on the back of the box would have helped him manage his time.

To consult the almanac, Gregge would find the year in the second or tenth column. Leap years have two adjacent letters and other years one. The leap year

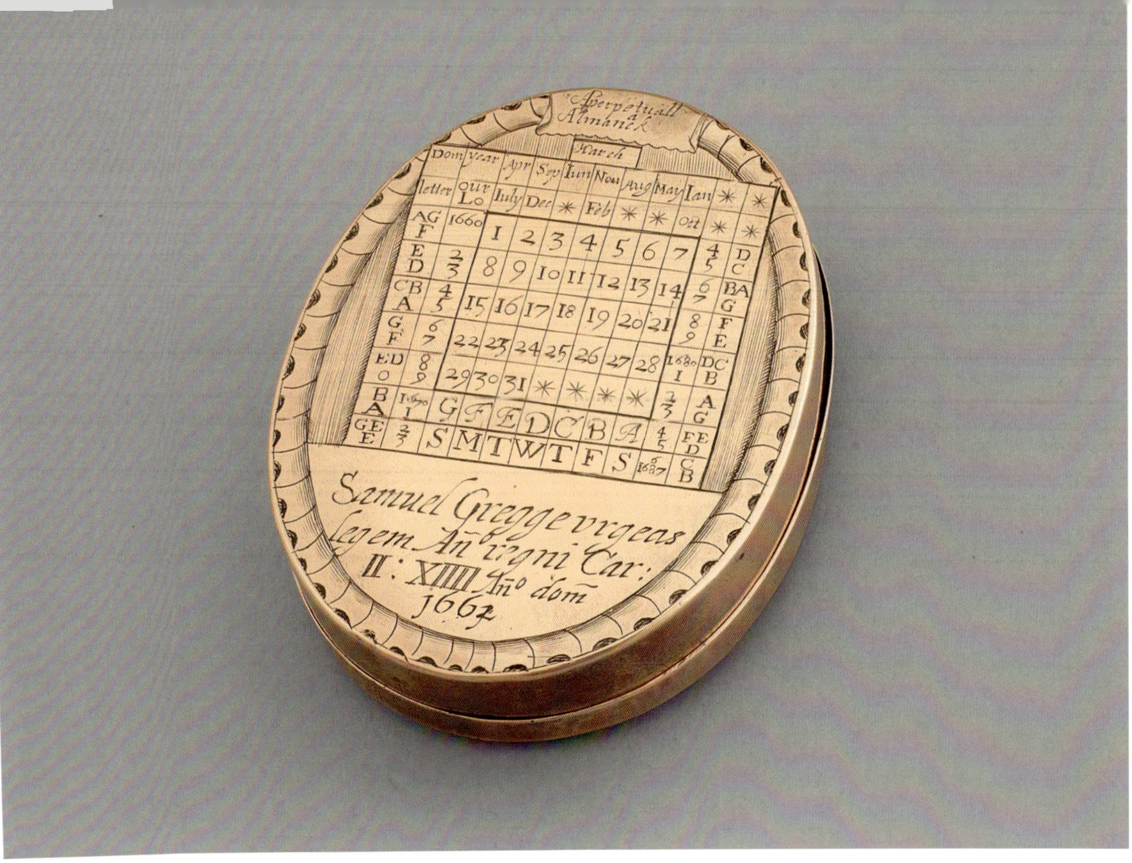

1660 corresponds to the letters *A* and *G*. To determine the day of the week for the first of April and the first of July (shown at the top of the third column), Gregge would locate the letter *G* in the top alpha row at the bottom of that column (the *A* applies only to January and February). The *S* in the row below indicates that both days fell on a Sunday that year. As Gregge read down the column, he would learn the dates for all the Sundays in April and July. He could then easily determine dates for subsequent days of the week. In 1661, which corresponds to *F*, those same dates fell on Mondays. With this system, Gregge could look into the future and predict that, because the letter code for 1670 is *B*, the first, eighth, fifteenth, twenty-second, and twenty-ninth of April and July would be Fridays.

Neither his father nor his brother Ralph mentioned Samuel in their wills. The father wrote his will from his home in Hammersmith in March 1673 and, when he died in 1680, bequeathed to his son Thomas "my president [precedent] Books in writing for the proceedings of the Court of Kings Bench" and half of his law books. Francis was left all the precedent books for the Court of Common Pleas and the remaining half of the law books. Ralph gave his son Benjamin "my watch (which I usually weare in my Pockett), with my walking staff with a silver head and mine other staves to my sonnes Robert Joseph and Francis." As described in his will, in his rooms at Clement's Inn, he had a watch and "an allarme."

Samuel's disappearance from the records remains a mystery. Perhaps he was the black sheep of the family. He could have been the Samuel Gregg buried at St James, Bath, on August 28, 1679. It is possible he had fallen sick and traveled to the spa for a remedy, only to die there.

PROVENANCE

bt. John H. Bryan of Lake Bluff, IL (1936–2018) from Tobias Jellinek Antiques, London, October 15, 2015

LITERATURE

Cecil Carr, *Pension Book of Clement's Inn* (London: Bernard Quaritch for the Selden Society, 1960), 78:255

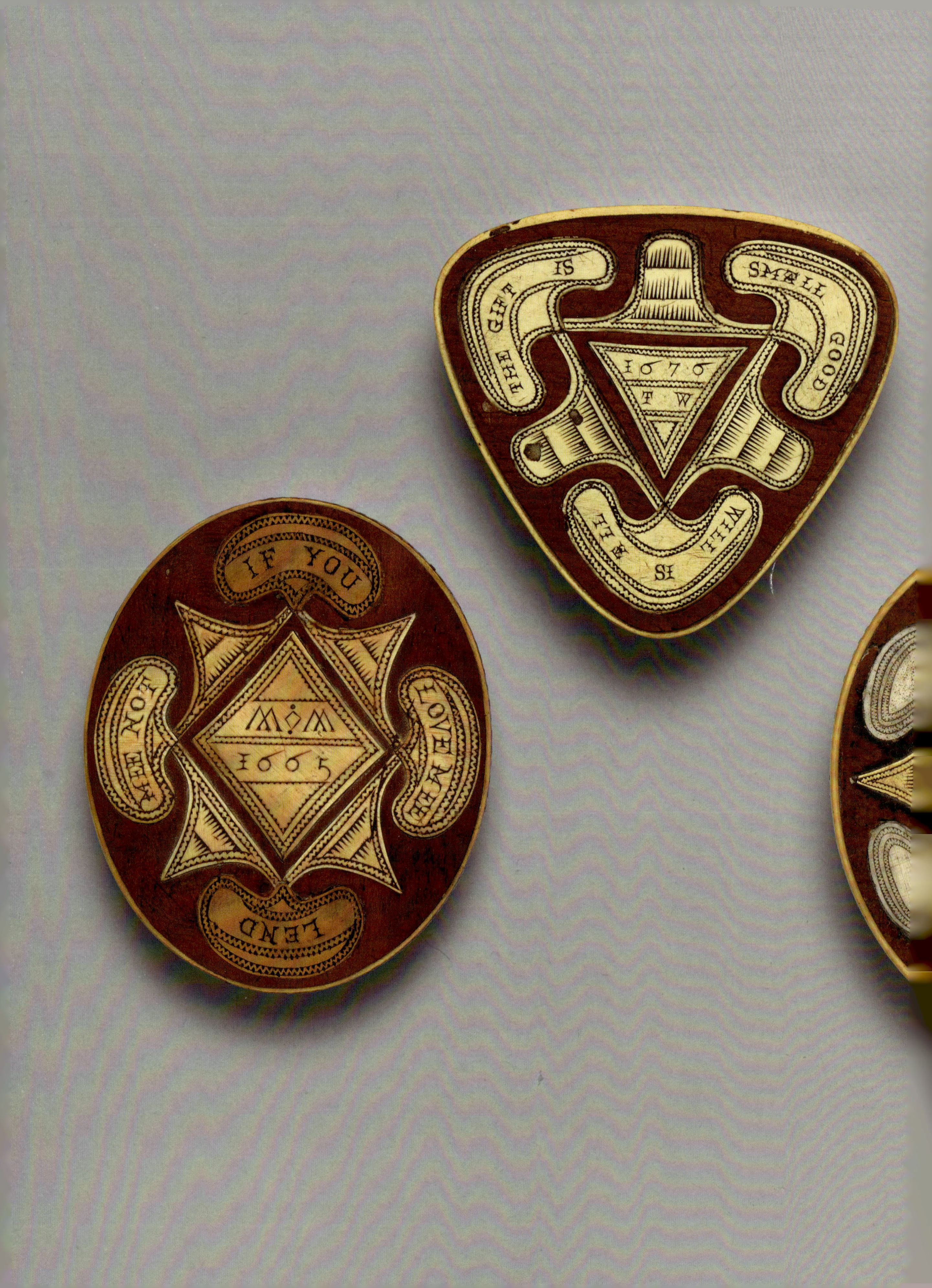
THE GIFT IS SMALL GOOD WILL IS ALL
1676
T W
IF YOU LOVE ME LEND ME NOT
M M
1665

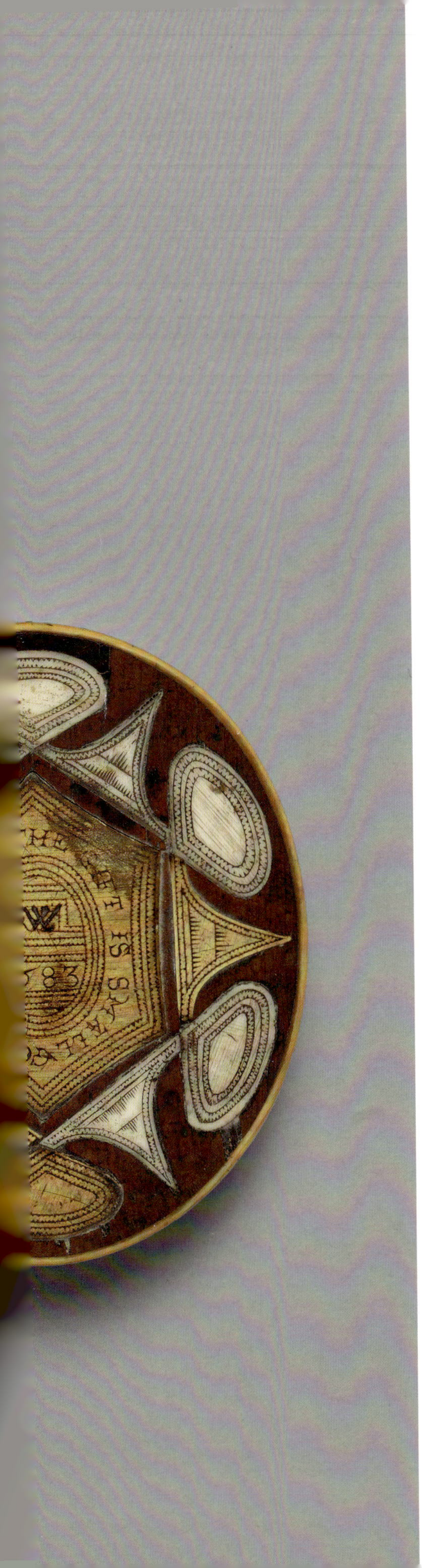

1665

Tobacco box
Cedar, horn, bone
$1 \times 3 \times 3\frac{1}{2}$ in. ($2.5 \times 7.6 \times 8.9$ cm)

Inscription: IF YOU LOVE MEE LEND ME NOT; W W|1665

These three boxes are part of a larger group, all made with a similar design and bearing appropriate inscriptions and dates ranging from 1657 to 1710, which suggests that there may have been successive generations of makers working in the same idiom during these six decades.

PROVENANCE
Gift to John H. Bryan of Lake Bluff, IL (1936–2018)

1676

Tobacco box
Cedar, horn, bone
$1 \times 3\frac{1}{2} \times 3\frac{1}{4}$ in. ($2.5 \times 8.3 \times 9.5$ cm)

Inscription: THE GIFT IS SMALL GOOD WILL IS ALL; 1676|T W

PROVENANCE
bt. John H. Bryan of Lake Bluff, IL (1936–2018) from Lindsay Antiques, London, June 29, 1992

1683

Tobacco box
Cedar, horn, bone
$1\frac{1}{4} \times 3\frac{1}{4} \times 3$ in. ($3.2 \times 8.3 \times 7.6$ cm)

Inscription: THE GIFT IS SMALL GOOD WILL IS ALL; R W; 1683

PROVENANCE
bt. John H. Bryan of Lake Bluff, IL (1936–2018) from Halycon Days, London, June 22, 1994

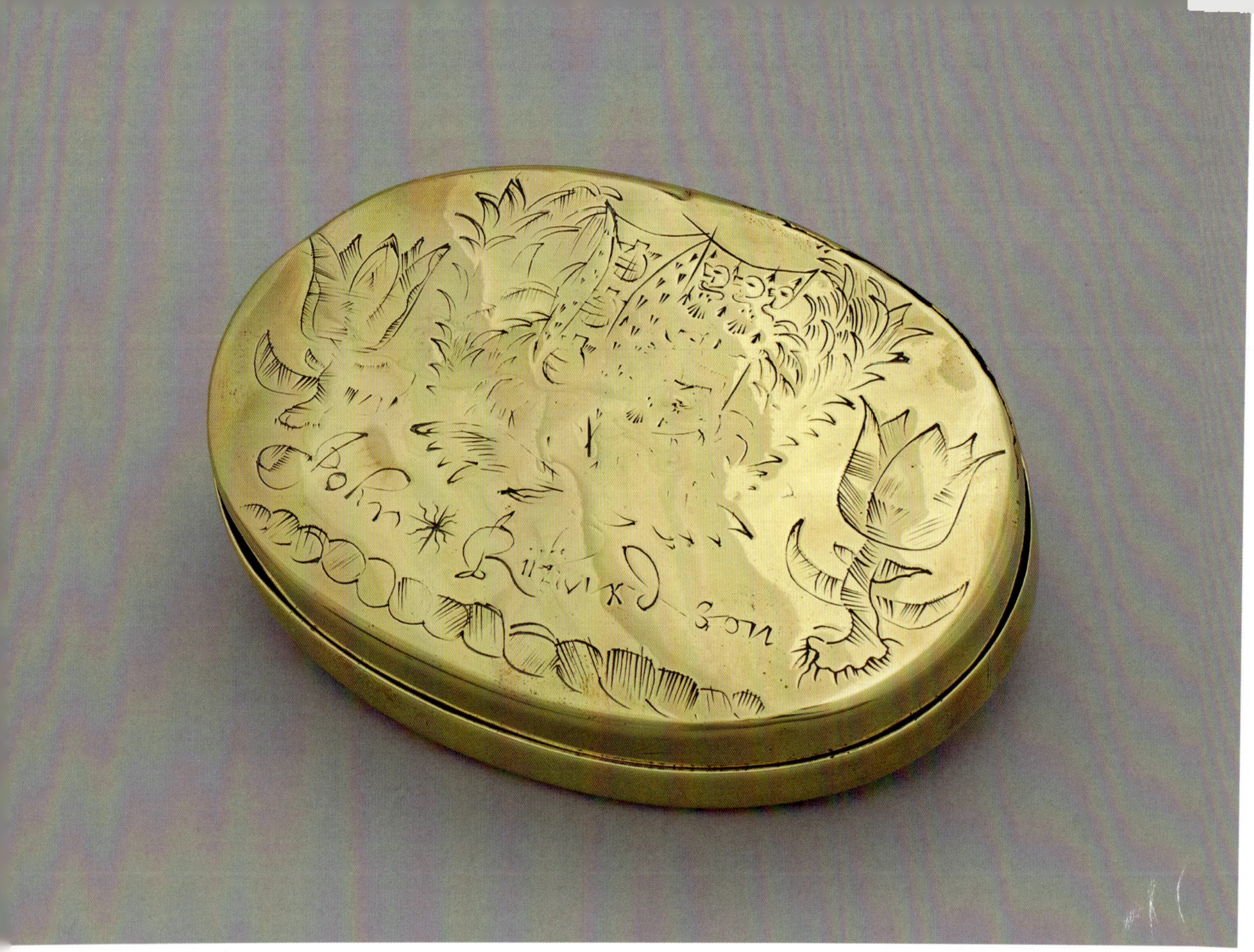

1725

Tobacco box
Brass
1 × $3\frac{7}{8}$ × $2\frac{7}{8}$ in.
(2.5 × 9.8 × 7.3 cm)

Inscription:
John Pinder|York
Owner|1725; John
Richardson; T S

In 1716, John Pinder (d. 1758), then aged fourteen, left home to attend Mr. Dowbiggin's school at Thornton in West Yorkshire. At age eighteen, he entered St John's College, Cambridge, and in 1725 he graduated and began his career as a clergyman. He celebrated the date of his first appointment by inscribing onto his modest, brass snuffbox his name, the name of the diocese in which he worked, and the word "owner" (should the box go missing). Perhaps he kept the box as a memento of a friend or mentor. Or perhaps he was attracted to it as an object in need of a loving owner.

In any case, his choice of snuffbox tells us something of a man who preferred to take his tobacco from an object marked with the patina of time, friendship, and conviviality rather than from one that was merely fashionable and new. Perhaps Pinder knew the previous owner, John Richardson. A clergyman by that name lived nearby.

PROVENANCE

bt. John H. Bryan of Lake Bluff, IL (1936–2018) from Tobias Jellinek Antiques, London, February 8, 1996

LITERATURE

Angela McShane, "Belonging and Belongings: Identity, Emotion and Memory Stored in a Tobacco Box," in Dagmar Freist, Sabine Kyora, and Melanie Unseld, eds., *Transkulturelle Mehrfachzugehörigkeit als kulturhistorisches Phänomen: Räume, Materialitäten, Erinnerungen* (Bielefeld, Germany: Transcript, 2019), 59–82

1726

Tobacco box

Brass

$1\frac{5}{8} \times 5\frac{1}{8} \times 3\frac{1}{8}$ in.
(4.1 × 13 × 7.9 cm)

Inscription: John Craford Hamer-Man in Crafords-dike his Box 1726; Woman and Wine will now|and then decive of Some wise|Men

This brass snuffbox has been personalized with a date but also with the address and trade details of the owner and, on the base, a witty rhyme. John Craford was a blacksmith of Crawford's Dike. In 1700, the town had unsuccessfully lobbied the Scottish Parliament for a new harbor. Not to be denied, Crawford's Dike and the connected town of Greenock voluntarily imposed a local tax on beer in order to raise the money themselves. The effort was successful, and the new harbor was begun in 1707. By 1740, not only was the harbor complete, the debt incurred in the construction was paid off. Metalworkers were notoriously heavy beer drinkers because their hot work made them thirsty, so John Craford is likely to have paid more in beer taxes than most during his working life—but the outcome was a matter of considerable pride. There would be no shortage of work for this new-made "Hamer-Man" in his self-made port town.

PROVENANCE

bt. John H. Bryan of Lake Bluff, IL (1936–2018) from Tobias Jellinek Antiques, London, December 19, 1994

LITERATURE

Angela McShane, "Belonging and Belongings: Identity, Emotion and Memory Stored in a Tobacco Box," in Dagmar Freist, Sabine Kyora, and Melanie Unseld, eds., *Transkulturelle Mehrfachzugehörigkeit als kulturhistorisches Phänomen: Räume, Materialitäten, Erinnerungen* (Bielefeld, Germany: Transcript, 2019), 59–82

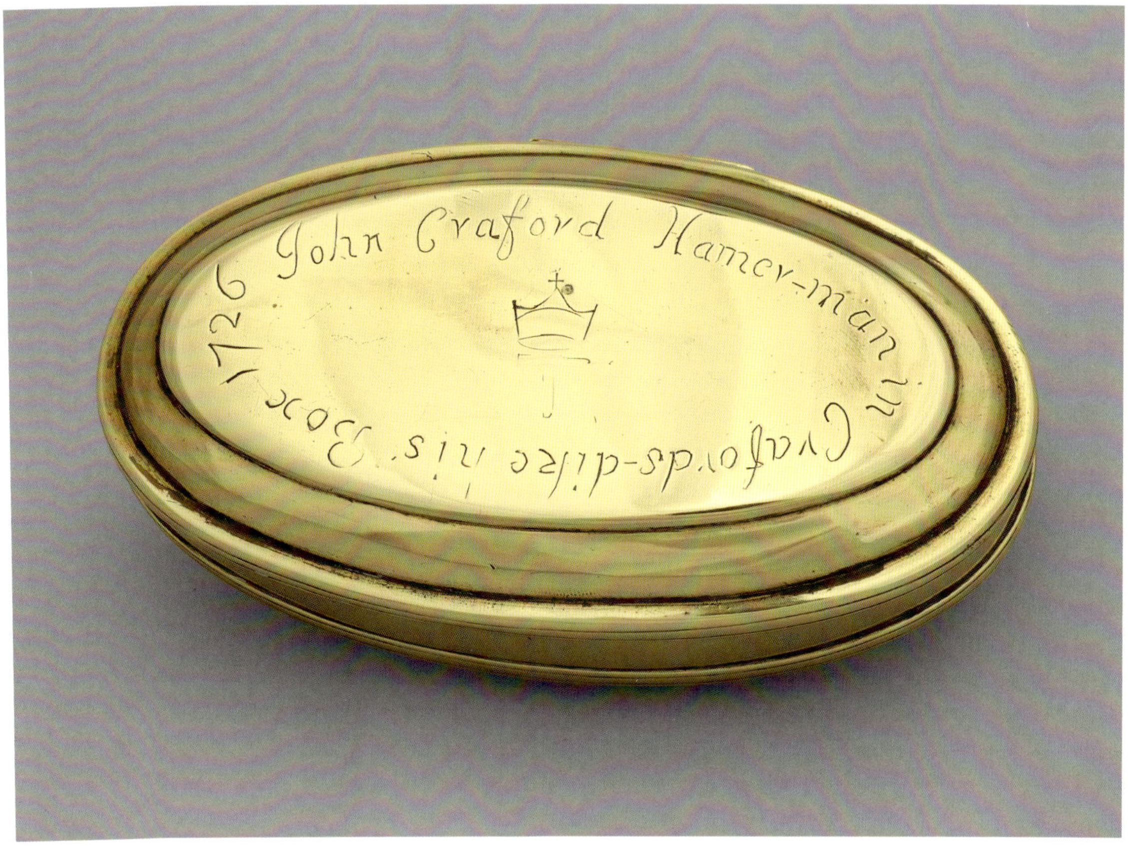

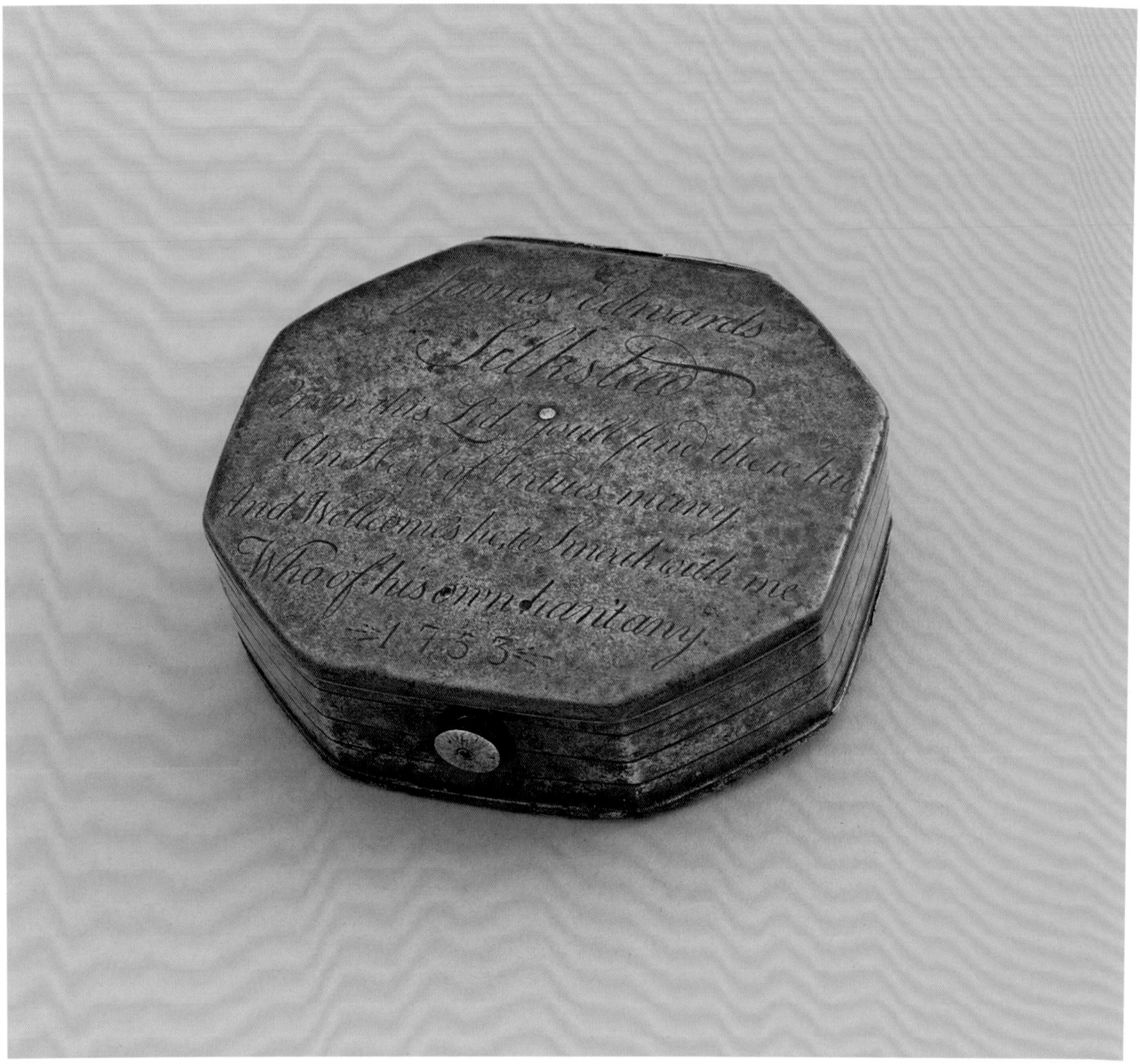

1753

Tobacco box

Steel

$1\frac{1}{4} \times 3\frac{1}{4} \times 3$ in.
(3.2 × 8.3 × 7.6 cm)

Inscription: James Edwards|Silksteed| Open this Lid, you'll find there hid|An Herb of Virtues many|And Wellcome's he, to Smoak with me|Who of his own han't any|1753

James Edwards, yeoman, was born in Shalfleet on the Isle of Wight in 1724. He was still a bachelor when this hospitable inscription was inscribed upon his tobacco box. By this time, he had moved inland and was living in Silkstead, Hampshire. Two years later, at age thirty, he married Elizabeth Mersh of Broughton, age twenty-five, on October 25, 1755. A remarkably similar tobacco box dated 1736 is in the collection of Colonial Williamsburg, Williamsburg, VA.

PROVENANCE

Gillespie Collection; bt. John H. Bryan of Lake Bluff, IL (1936–2018) from Christopher Bangs, London, November 15, 1990

LITERATURE

Rupert Gentle and Rachel Field, *Domestic Metalwork: 1640–1820*, revised and enlarged by Belinda Gentle (Woodbridge, UK: Antiques Collectors' Club, 1994), 425

1756

Tobacco box or boot-powderer

Steel

$2\frac{3}{8} \times 5\frac{1}{8} \times \frac{7}{8}$ in. (6 × 13 × 2.22 cm)

Inscription: Will[m]. Skreen|Canterbury 1756; SUCCESS|TO TRADE

The perforated holes in this object suggest that the contents would be dispersed by shaking—as would boot powder, which was shaken onto boots to help the wearer pull them on and off. The sliding aperture, however, was also employed in steel tobacco boxes, many of which were made in Birmingham.

William Skreen was a pattern maker and a Quaker of Canterbury. He was apprenticed to John Love in 1719 and later took on his own apprentices. He died on May 9, 1779, and was buried at Canterbury on May 13.

PROVENANCE

bt. John H. Bryan of Lake Bluff, IL (1936–2018) from Christopher Bangs Ltd., Felsham, Suffolk, August 6, 2018

LITERATURE

Donald L. Fennimore, "Birmingham Steel Tobacco Boxes," *Journal of the Antique Metalware Society* 3 (1995): 1–6

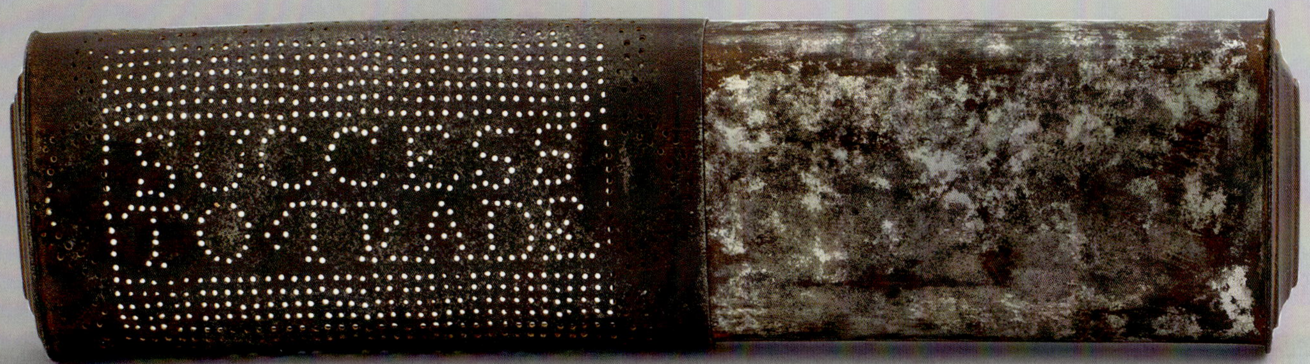

1720 | 1745

Tobacco box
Horn
$1\frac{3}{4} \times 2\frac{7}{8} \times 3\frac{3}{7}$ in. (4.5 × 7.3 × 9.8 cm)

Inscription:
WO | 1745; 1720

The three tobacco boxes shown here are all by the same unknown maker. Their engraved decoration depicts scenes of agricultural and domestic labor. This box, which has two dates, was probably made in 1745, possibly to celebrate an event that had occurred a quarter of a century earlier. The engraving is an illustration of a farmer driving a hand plow into the soil. Also depicted are two blades for field work.

PROVENANCE
bt. John H. Bryan of Lake Bluff, IL (1936–2018) from Alistair Sampson Antiques, London, June 16, 1999

LITERATURE
Angela McShane, "The New World of Tobacco," *History Today* 67, no. 4 (April 2017): 41–47

1767

Tobacco box
Horn
$1\frac{3}{8} \times 3$ in. (3.5 × 7.6 cm)

Inscription:
DP | 1767

This box and the adjacent one, described below, show women at work: in one a woman is spinning yarn, and in the other a woman is seated at a table, knitting. It is likely that these two tobacco boxes were made for and owned by women, who would have dipped into them for a pinch of snuff.

PROVENANCE
bt. John H. Bryan of Lake Bluff, IL (1936–2018) from Tobias Jellinek Antiques, London, July 11, 2017

LITERATURE
Angela McShane, "The New World of Tobacco," *History Today* 67, no. 4 (April 2017): 41–47

1767

Tobacco box
Horn, with replaced wooden bottom
$1\frac{1}{4} \times 3$ in. (3.2 × 7.6 cm)

Inscription:
AB | 1767

PROVENANCE
bt. John H. Bryan of Lake Bluff, IL (1936–2018) from Tobias Jellinek Antiques, London, July 11, 2017

LITERATURE
Angela McShane, "The New World of Tobacco," *History Today* 67, no. 4 (April 2017): 41–47

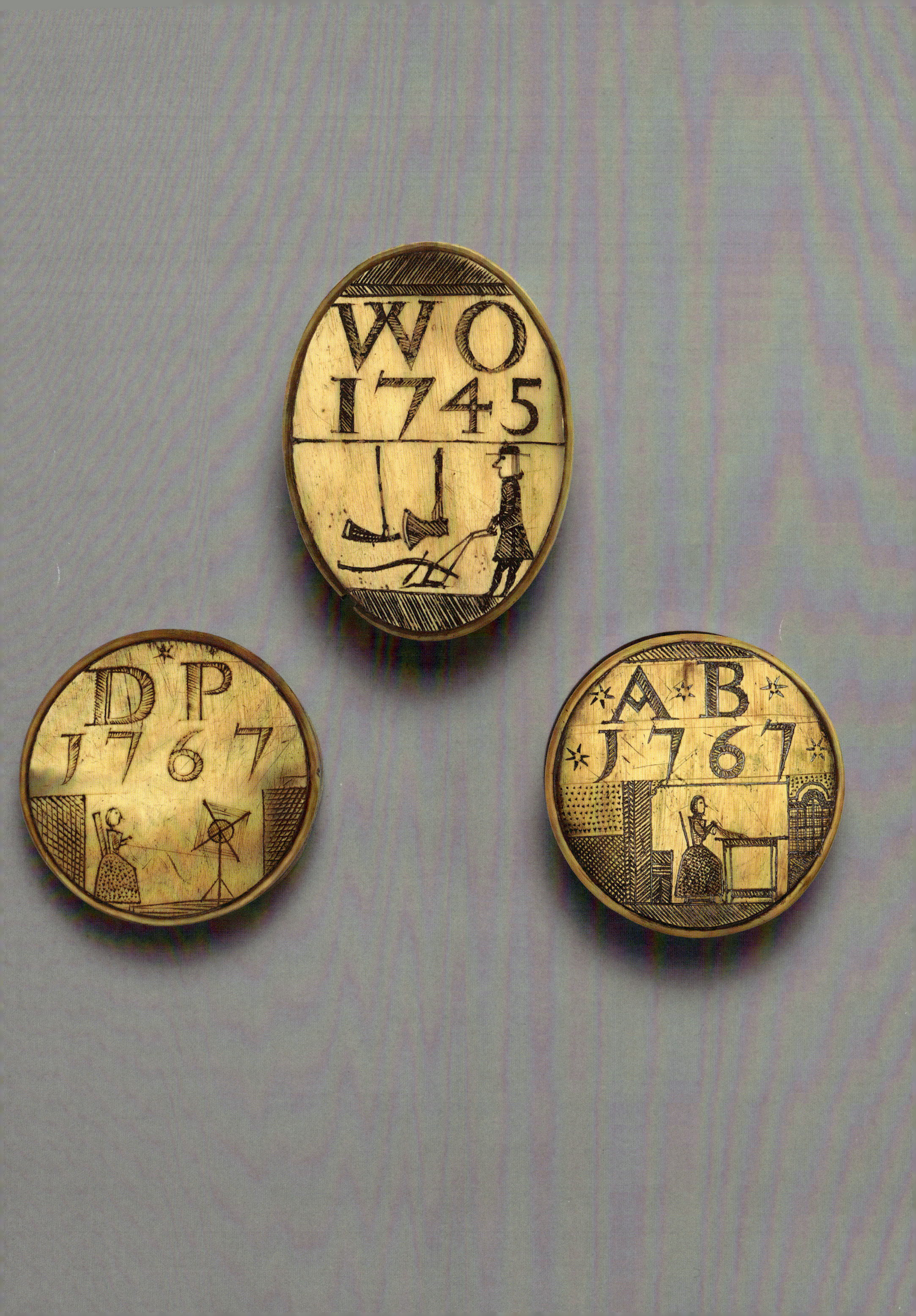
WO
1745
DP
1767
AB
1767

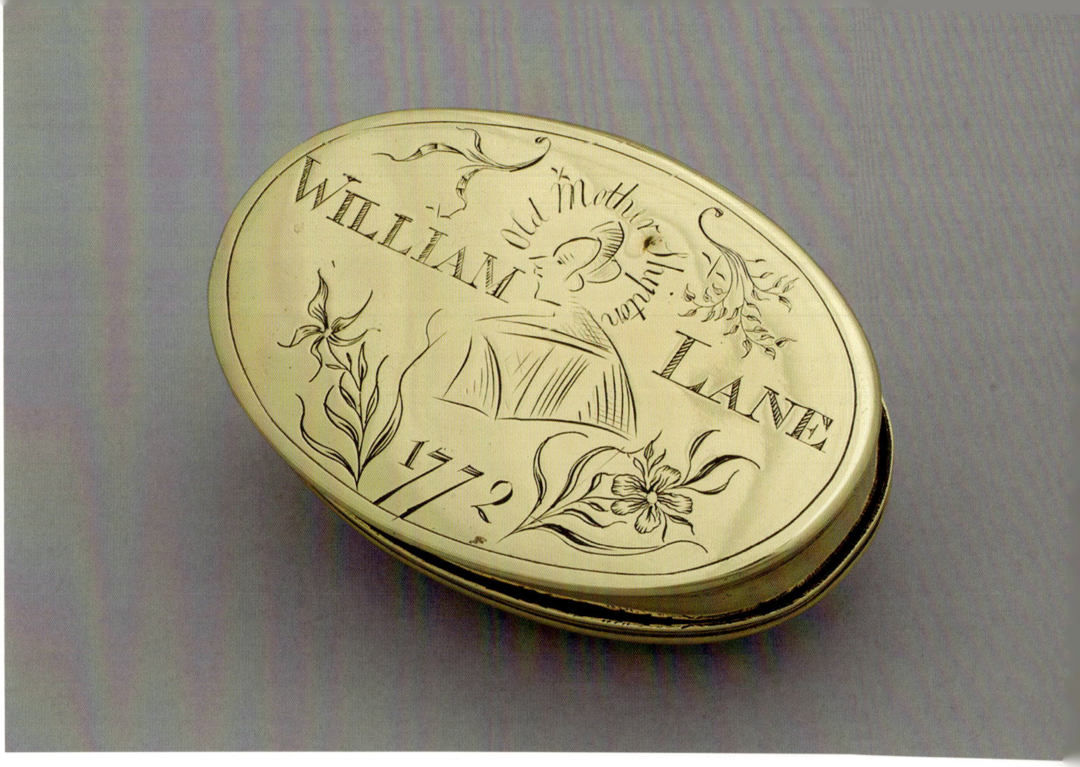

1772

Tobacco box

Brass

$1\frac{1}{8} \times 3\frac{1}{4} \times 2\frac{1}{4}$ in.
(2.9 × 8.3 × 5.7 cm)

Inscription: Old Mother Shipton [clockwise around head] | WILLIAM LANE | 1772

Old Mother Shipton, "Our English Sibyl," was a legendary soothsayer and prophetess. Believed to have been born in the late fifteenth century in Knaresborough, Yorkshire, and raised in a cave that still bears her name, she maintained a place in the popular imagination for centuries after her death. Mother Shipton was the subject of plays and, according to a sales catalogue of March 1690/1 at the King's Arms Tavern, she was also the subject of a small painting by Brueghel (lot 251). Her prophecies appeared in popular print in the seventeenth century, and the accompanying illustrations depicted her with warts and a grotesque nose and chin, hunched over a walking stick.

The prophecies, often recorded as two-line nonsense rhymes, were reissued regularly and thought to have ongoing relevancy, even if the supporting explanation required some creative thought. For example, the rhyme "When the cow doth ride the Bull; Then Priest beware the Skull" was understood to be prescient of the actions of Henry VIII. As the Earl of Richmond, a title he continued to hold as king, he "gave the cow his arms" and rode the "Bull" (Anne Boleyn, whose father's coat of arms contained a bull's head as its crest). Their union precipitated the Dissolution of the Monasteries, leading to injury to the priesthood. Perhaps William Lane repeated Old Mother Shipton's sayings, or proffered his own, as he pinched tobacco to his nose or dragged on his pipe.

PROVENANCE

bt. John H. Bryan of Lake Bluff, IL (1936–2018) from Tobias Jellinek Antiques, London, February 17, 2017

LITERATURE

Anonymous, *Fourteene strange Prophesies: besides Mother Shiptons and Mr. Saltmarsh, predicting wonderfull events to betide these yearse of calamity, 1648* (London: Printed for R. Harper, 1648)

Anonymous, *The Strange and Wonderful HISTORY of Mother Shipton: Plainly setting forth Her prodigious Birth, Life, Death, and Burial* (London: Printed for W. H., 1686)

Rupert Gentle and Rachel Field, *Domestic Metalwork: 1640–1820*, revised and enlarged by Belinda Gentle (Woodbridge, UK: Antiques Collectors' Club, 1994), 430, no. 28

1773

Tobacco box

Horn

$1\frac{3}{4}$ × 3 × 4 in. (4.5 × 7.6 × 10.2 cm)

Inscription: R · WP [in monogram] | 6 · 9 | 1773

This box is possibly from the Humpherson Factory at Bewdley, Worcestershire.

PROVENANCE

bt. John H. Bryan of Lake Bluff, IL (1936–2018) from Lindsay Antiques, London, June 29, 1992

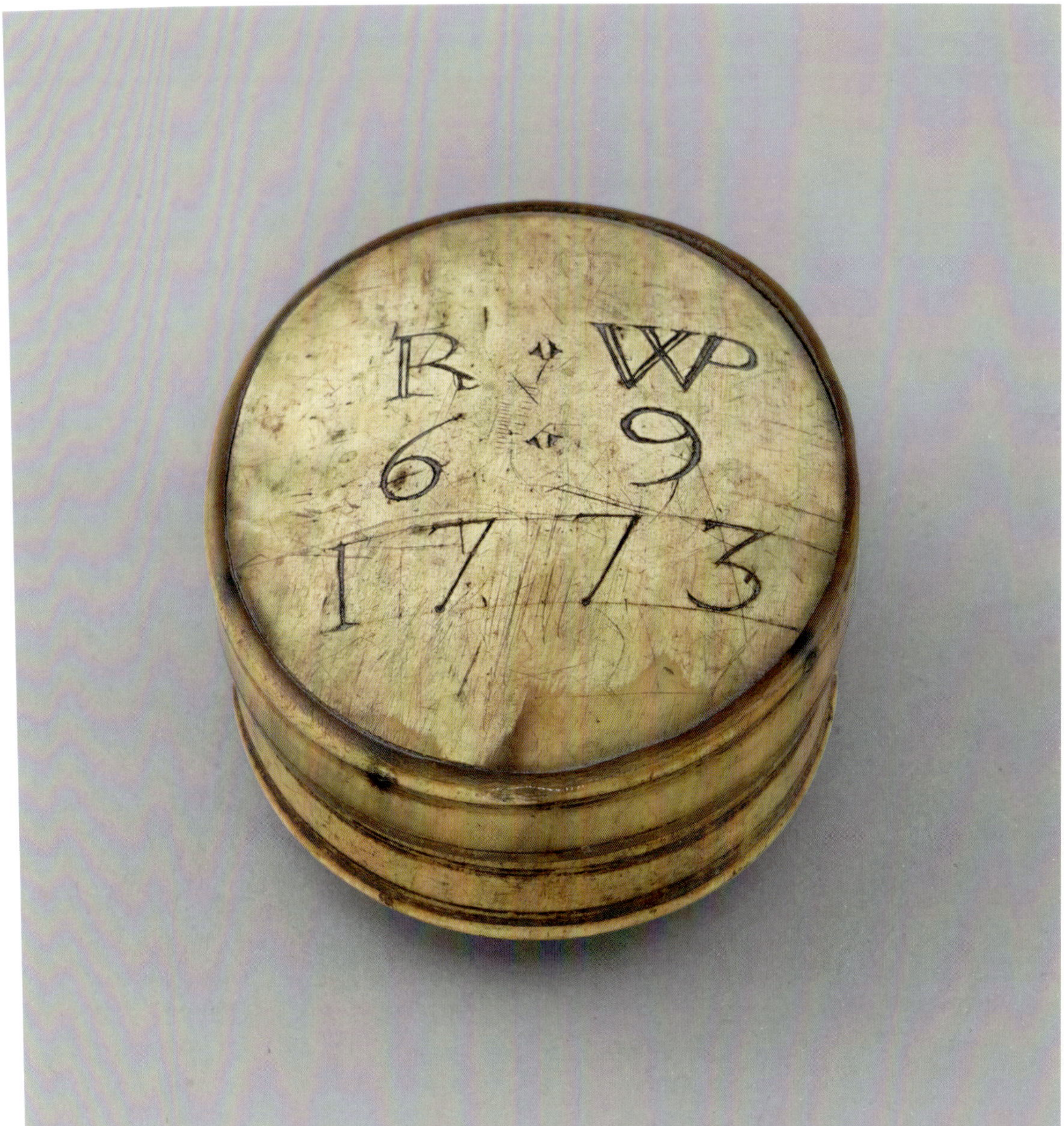

1689|1739

Tobacco box or pounce pot

Brass

$1\frac{1}{4} \times 3\frac{1}{4}$ in. (3.2 × 8.3 cm)

Inscription: Edward Greaves|1689; John Barlow|1739

This small perforated pot may have been used to distribute pounce, or sand, over inked writing on paper. The pounce helped dry the ink and prevented it from smudging.

PROVENANCE

Collection of Michael Gillingham; bt. John H. Bryan of Lake Bluff, IL (1936–2018) from Alistair Sampson Antiques Ltd., London, December 7, 1993

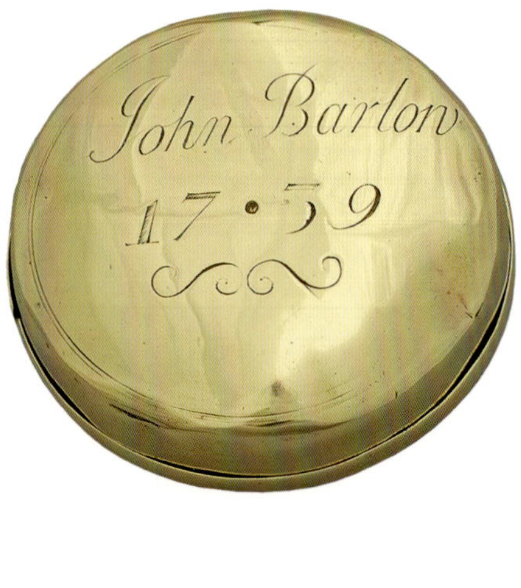

1724

Tobacco box

Brass, lead

$1\frac{1}{8} \times 4\frac{1}{2} \times 2\frac{5}{8}$ in. (2.9 × 11.4 × 6.7 cm)

Inscription: Nicho Vipond|1724

Nicholas Vipond was an attorney of the Court of Common Pleas, Norwich, and auditor of Norwich Cathedral. On February 24, 1711/2, he married Mary Andrews at St Michael Coslany, Norwich, where he lived until his death in 1743.

PROVENANCE

bt. John H. Bryan of Lake Bluff, IL (1936–2018) from Alistair Sampson Antiques Ltd., London, July 29, 1999

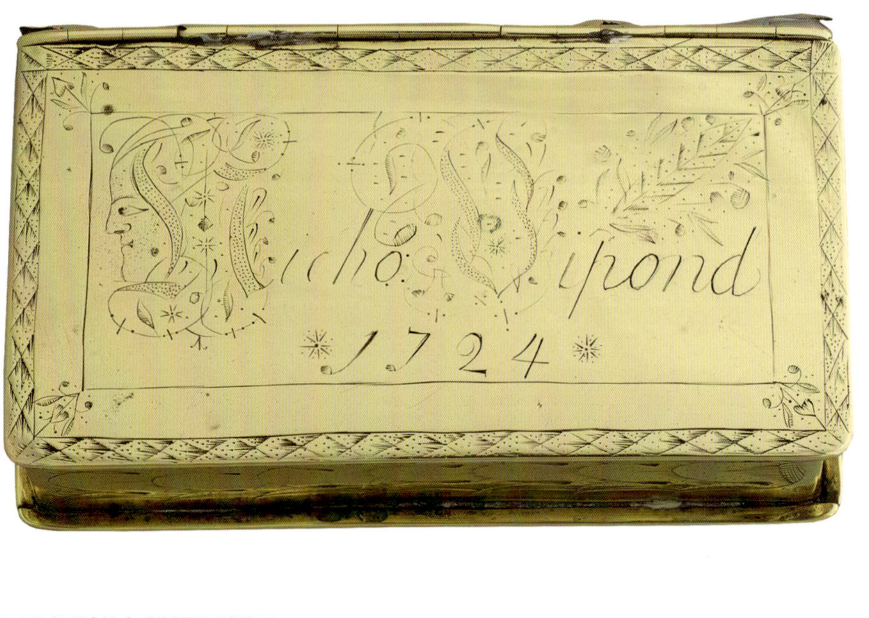

1730

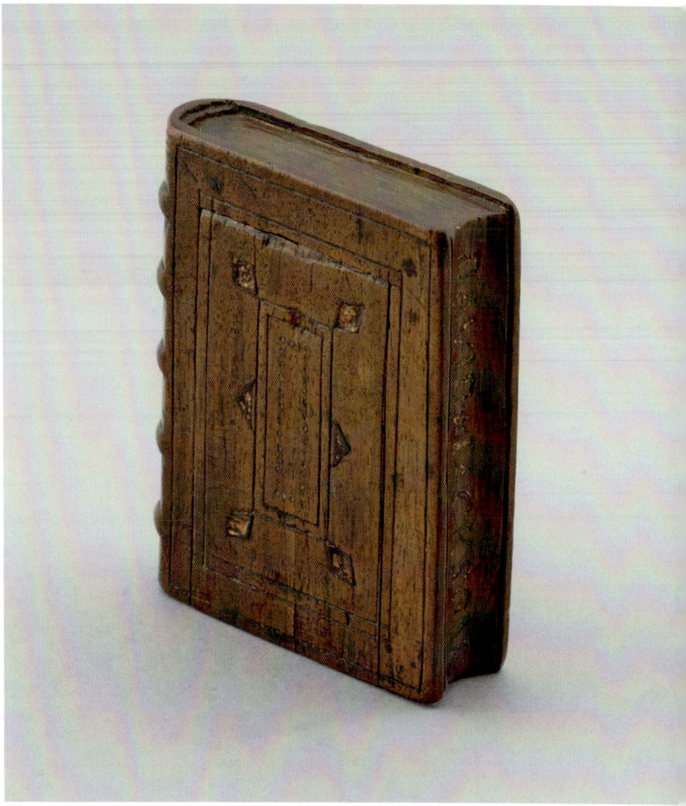

Tobacco box
Walnut, silver
$3\frac{1}{4} \times 2\frac{1}{2} \times \frac{3}{4}$ in.
(8.3 × 6.4 × 1.9 cm)

Inscription: DAVID JACKSON; 1730

The lid of this wooden tobacco box, which takes the shape of a book, slides open to give the user access to the contents.

PROVENANCE
bt. John H. Bryan of Lake Bluff, IL (1936–2018) from Jellinek & Sampson Antiques, London, February 1979

1785

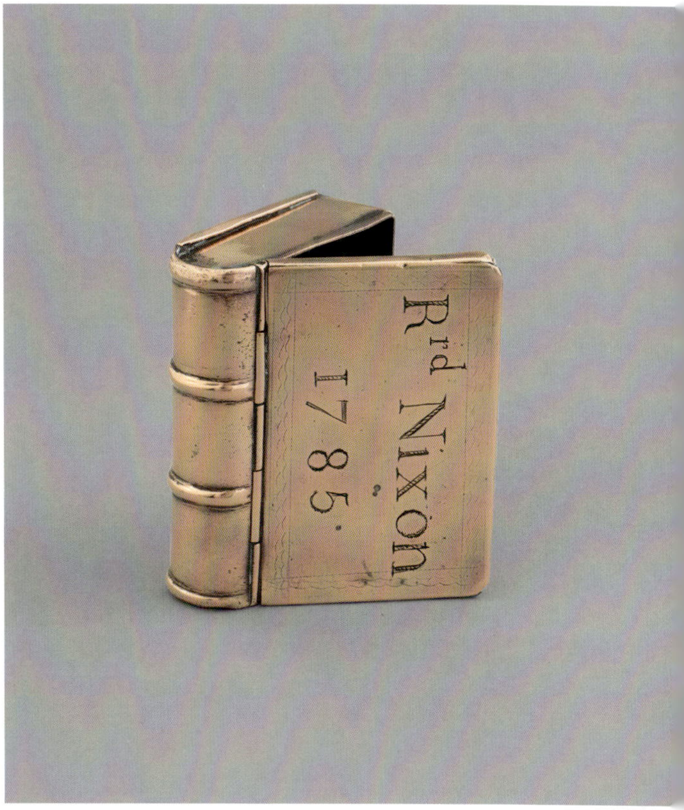

Tobacco box
Copper
$3\frac{1}{4} \times 2\frac{5}{8} \times 1$ in.
(8.3 × 6.7 × 2.5 cm)

Inscription: R^{rd} Nixon. 1785

The cover of this book-like box is hinged on the spine to reveal the contents.

PROVENANCE
bt. John H. Bryan of Lake Bluff, IL (1936–2018) from Tobias Jellinek Antiques, London, October 15, 2015

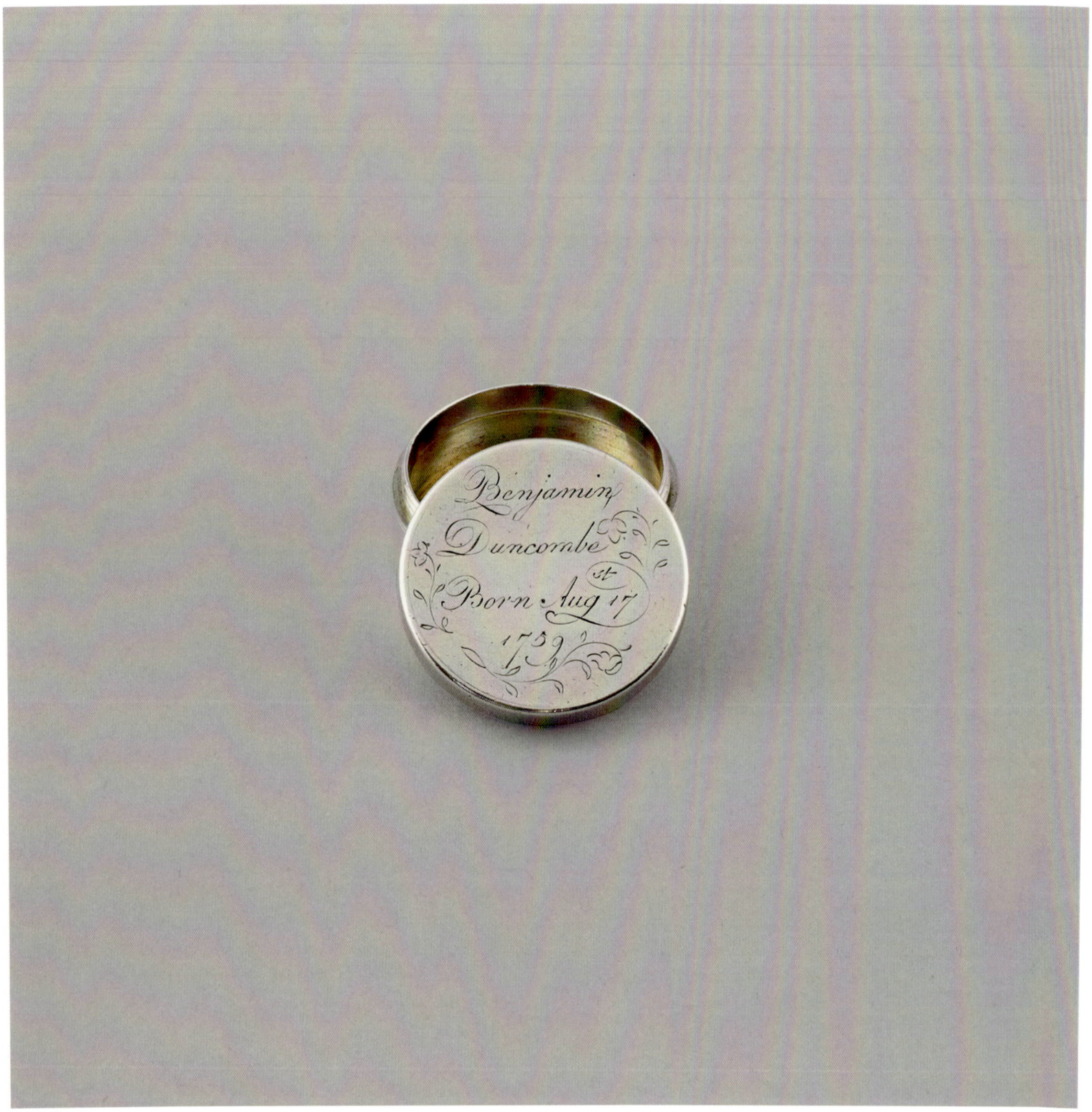

1759

Box
Silver
$\frac{5}{8} \times 1\frac{1}{8}$ in. (1.59 × 2.9 cm)

Inscription:
Benjamin|
Duncombe|Born
Augst 17|1759

The small size of this box suggests that it is possibly a patch box—a box that held small cosmetic patches that were used to cover unsightly blemishes on the skin. It might instead have contained comfit sweets or was a snuff box, presented as a gift to a newborn.

PROVENANCE
bt. John H. Bryan of Lake Bluff, IL (1936–2018) from Peter Cameron, London, December 3, 2016

1778

Etui
Steel
4 × $\frac{7}{8}$ × $\frac{5}{8}$ in. (10.2 × 2.22 × 1.59 cm)

Inscription:
T · H|Oliver|1778; GUEST

This tool-kit-cum-manicure-set enabled the owner (who in 1778 was T. H. Oliver) to trim his beard, pluck his nose hair, clean his cuticles, gouge wax from his ears, and measure a small distance with the ruler and dividers—all while on the move. The small knife could also be screwed into a handle for ease of use. It has been suggested that the inscription "GUEST" on the knife blade was the name of a precursor of the company that became Guest, Keen and Nettleford. Alternatively, there was also a foundry called Guest in Dowlais, Wales.

PROVENANCE
bt. John H. Bryan of Lake Bluff, IL (1936–2018) from Jim Marshall, Pen & Pencil Gallery, The Wirral, Merseyside, May 10, 2017

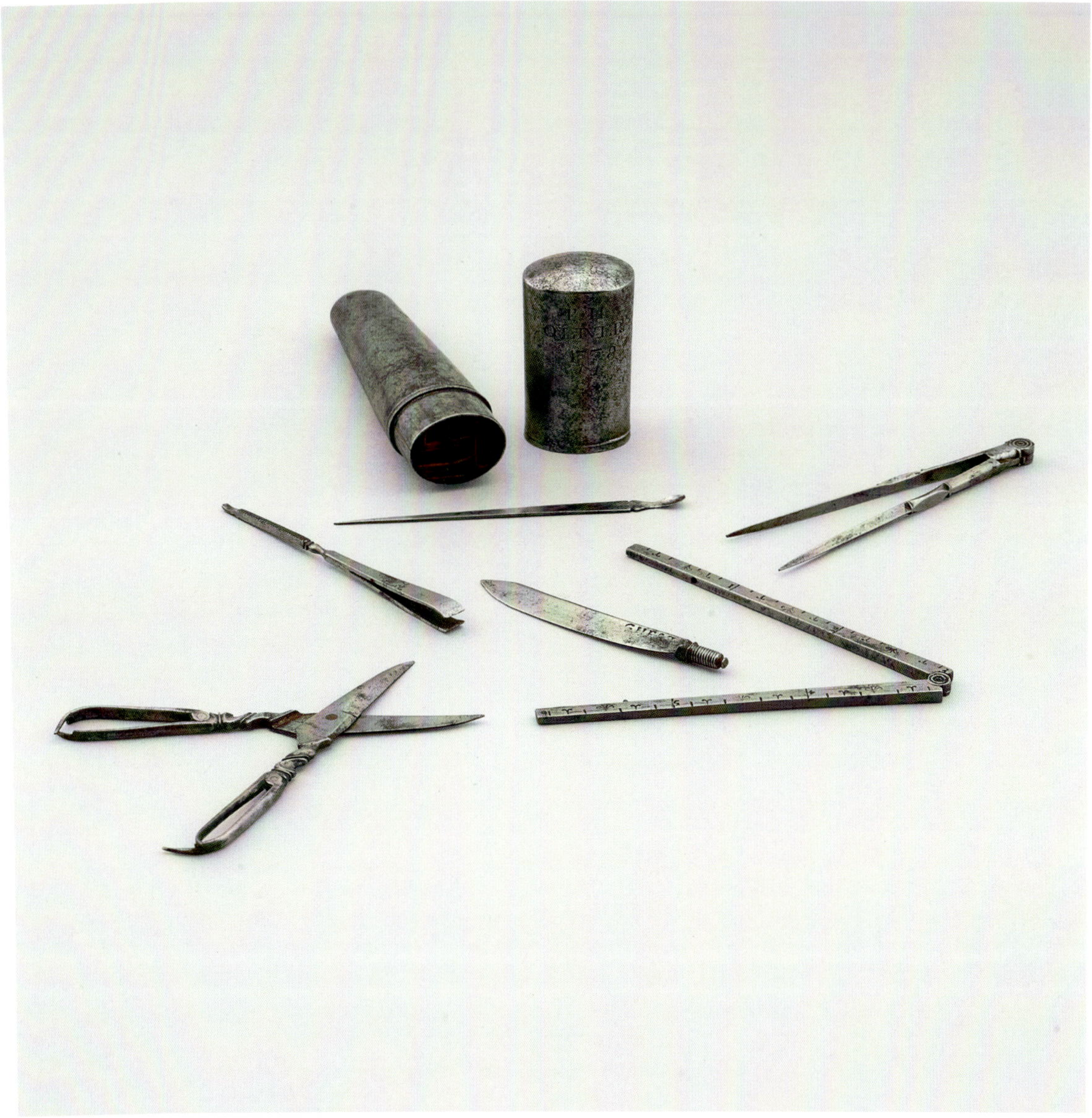

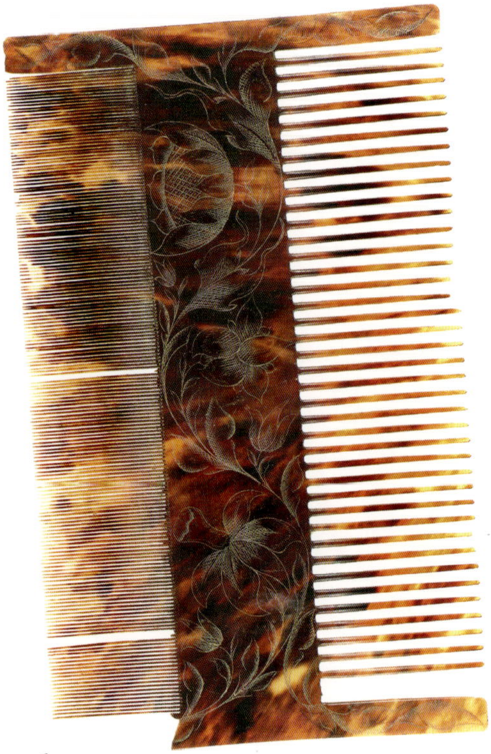
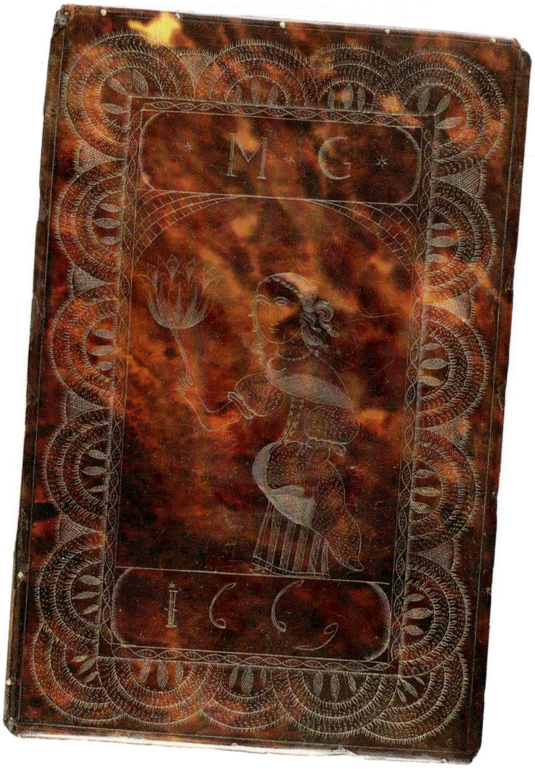

1669

Comb and case

Tortoiseshell

Comb: 7¾ × 4¾ × ⅛ in. (19.7 × 12.1 × .32 cm); case: 8¼ × 5½ × ½ in. (21 × 13.3 × 1.27 cm)

Inscription:
M G|1669

Prized for its color, translucence, and shine, tortoiseshell was a luxury commodity in the seventeenth and early eighteenth centuries. Artisans in the West Indies collected tortoiseshell from native hawksbill turtles and fashioned the material to make objects to sell to wealthy consumers in Europe.

The comb has two rows of teeth: the wider-gauge teeth would be used to shape and smooth a periwig; the finer teeth would remove lice. This set is a particularly early example of a group of combs made in Jamaica during the latter half of the seventeenth century; other dated examples range from 1671 to 1692, the year in which an earthquake decimated Port Royal. Of these, only two combs have the owners' names inscribed: "Sarah Henly, Jamaica 1670" (Bonham's, London, November 21, 2018 [lot 133]), and "Elizabeth Fleming, 1678." Elizabeth was likely the sister of John Fleming (d. in or before 1692), a merchant of Port Royal and a plantation owner in St. Elizabeth, Jamaica.

From what little we know about these combs, it seems that they were commissioned by female members of the plantocracy, who sometimes sent them as gifts. This fact raises the possibility that the initials inscribed on this comb belong to Margaret, wife of Daniel Gosling of Port Royal Jamaica, who made his will in 1684 (TNA PROB 11/425/397).

PROVENANCE

Gift of Titi Halle, Cora Ginsburg LLC, to John H. Bryan of Lake Bluff, IL (1936–2018), October 1, 2014

LITERATURE

Donald F. Johnson, "Combing the Roots of Colonialism: Jamaican Tortoiseshell Combs and Social Status, 1655–1692," *Winterthur Portfolio* 43, no. 4 (Winter 2009): 313–34

Geoffrey Wills, "Jamaican Engraved Tortoiseshell Wig-Combs," *Connoisseur Yearbook* (1957): 76–77

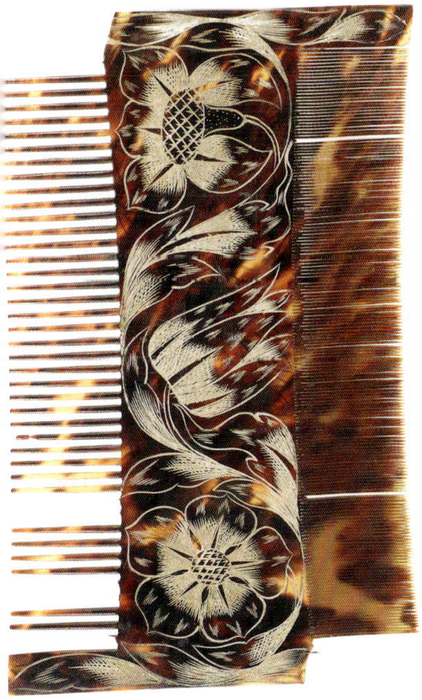
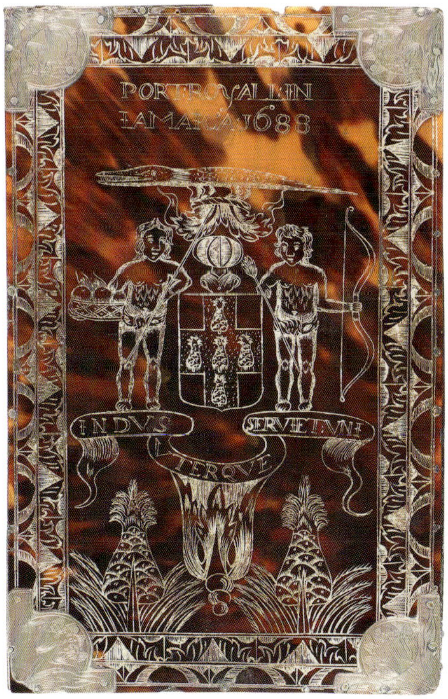
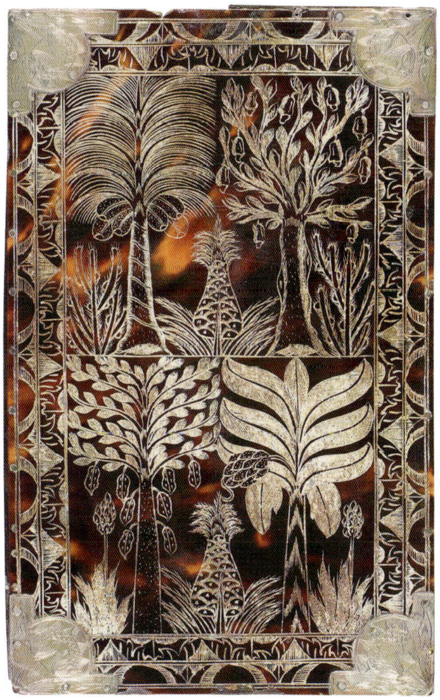

1688

Comb and case

Tortoiseshell, silver

Comb: $7\frac{1}{4} \times 4\frac{1}{4} \times \frac{1}{8}$ in. (18.4 × 10.8 × .32 cm); case: $7\frac{7}{8} \times 4\frac{7}{8} \times \frac{3}{8}$ in. (20 × 12.4 × .95 cm)

Inscription: PORT ROYAL IN | IAMAICA 1688; INDVS VTERQVE SERVIET VNI

Dated 1688, this comb case was made in Port Royal, Jamaica, which was founded in 1656, a year after the conquest of the island and seizure from the Spanish by Oliver Cromwell. Prior to its almost total destruction in an earthquake of 1692, Port Royal was Jamaica's principal port and the center for all contraband trade with Spanish America. The comb bears the Jamaican coat of arms, devised by the archbishop of Canterbury, which displays a crocodile and five pineapples within a shield and flanked by two native figures standing on a banderole. A very similar example, with the same coat of arms, inscribed "A CASHU TREE" ("cashew") on the verso but lacking a date, is in the Museum of Fine Art, Boston (60.1173a-b); another comb with the same inscription, dated 1688, sold at Cheffins, Cambridge, November 20, 2016 (lot 570).

The Boston comb shares the same style mounts, thought to have been added in London, which suggests there was a substantial export market from Jamaica. That being said, the Institute of Jamaica has eleven of these combs in their collection, indicating that many of these objects remained in the country and were made to satisfy local demand from the mercantile elite and plantocracy.

Other dated examples that bear the Jamaican coat of arms and the same silver mounts are inscribed 1690 and 1692, respectively, and appear to be by the same maker as this comb and case. Both are in private collections. The only named comb maker active in Jamaica was a man called Paul Bennett, recorded as being in Port Royal in 1692, about whom almost nothing is known.

PROVENANCE

Percival D. Griffiths of Sandridgebury, St. Albans, Hertfordshire (1862–1938); Price and Isobel H. Glover, New York; Sotheby's, New York, January 18, 2018 (lot 601); bt. John H. Bryan of Lake Bluff, IL (1936–2018) through Christian Jussel, Stonington, CT

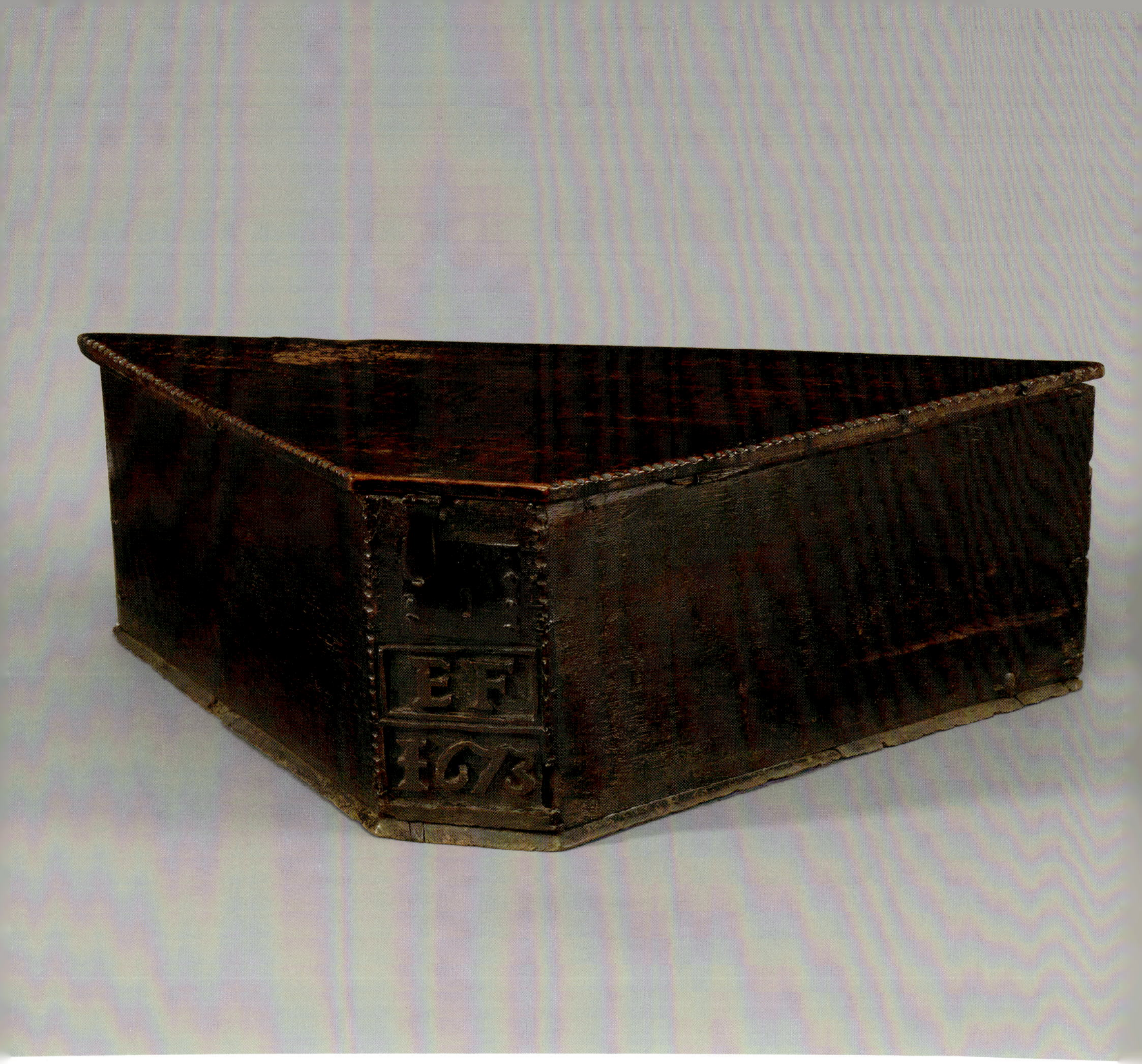

1673

Hatbox
Oak, iron
8 ¾ × 28 × 15 ½ in.
(22.2 × 71.1 × 39.4 cm)

Inscription:
E F|1673

This sturdy tricorn hatbox of oak and iron was ideally suited for travel in a rickety coach or rolling sea cabin. It would have protected the owner's precious headwear from the depredations of long journeys, ensuring that the traveler would look at his best upon arrival at his destination.

PROVENANCE
Collection of John Fardon, Birmingham; Christie's, London, October 24, 1991 (lot 44); bt. John H. Bryan of Lake Bluff, IL (1936–2018) from Robert Hischhorn, London, August 23, 1999

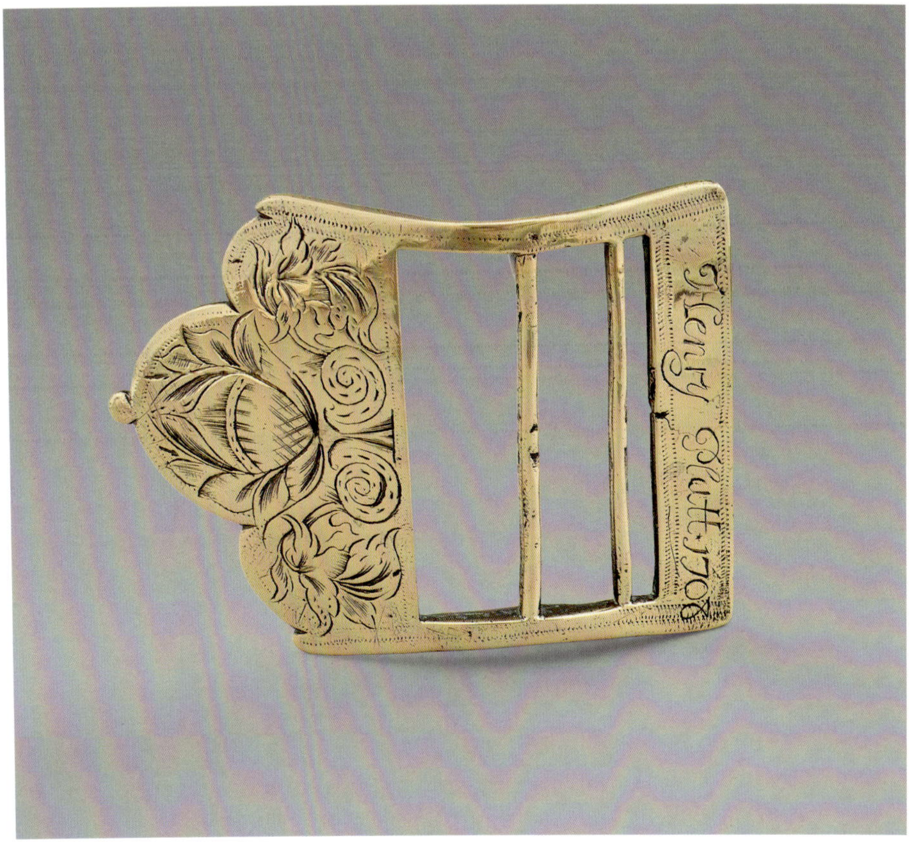

1708

Buckle

Brass

4 × 5⅜ in. (10.2 × 13.7 cm)

Inscription: Henry Platt·1708

This brass buckle was made to secure a baldric, an ornamental belt that was worn across the chest to support and transport a weapon, bugle, or drum.

PROVENANCE

bt. John H. Bryan of Lake Bluff, IL (1936–2018) from Alistair Sampson Antiques Ltd., London, January 10, 2001

1596

Shoehorn

Horn

$6\frac{1}{2} \times 1\frac{5}{8} \times 1\frac{1}{4}$ in.
(16.5 × 4.1 × 3.2 cm)

Robert Mindum
(active 1593;
d. 1615?)

Inscription: 1596;
THIS IS ROBART
HEND/ART MINDVM

During the sixteenth and seventeenth centuries, the appearance of one's footwear and legs (usually covered by fine stockings) were important markers of status. Throughout the sixteenth century, innovations in leatherworking resulted in the production of shoes that perfectly conformed to the desires of the fashionable: tight fitting and supple, with soft backs that would not damage expensive stockings. Because this new style of shoe was difficult to put on, men and women began to use shoehorns, made from naturally curved oxhorn. The depiction of a fashionably attired woman on this shoehorn suggests its owner was female.

Twenty-six shoehorns and two powder horns made by Robert Mindum are known to exist, bearing dates that range from 1593 to 1613 (pages 291 and 328). They are inscribed with Mindum's name, the names of the owners, and, in one instance, the names of the individuals who commissioned the shoehorn as a gift. Despite ongoing interest among collectors and antiquarians, nothing is known about Mindum himself. Yet clues are to be found in the names inscribed upon his shoehorns, even though some of those individuals remain unidentified. There is sufficient evidence to show that Mindum was operating in a small corridor of Cambridgeshire around the Isle of Ely, a part of the country then surrounded by fenland. This idea is supported by the provenance of two shoehorns that appear to have been in East Anglia for much of their history (at Saffron Walden, Essex, and at Rushbrook, Suffolk) and that of the powder horn now in the Museum of London (formerly in the collection of Rev. W. F. Creeny of Norwich), which again points to an East Anglian origin.

There is one shoehorn, however, about which somewhat more is known: the earliest surviving shoehorn, dated 1593 and now in the Museum of London, was made for Hamlet Radesdale (or Rotchedale), "THE COVPER OF LONDAN." This shoehorn provides additional information both about its owner and about Mindum—in that the common spelling of London was apparently unfamiliar to him. Hamlet Radesdale was indeed a Cooper of London; he lived in the extramural parish of St Giles-without-Cripplegate from about 1575 to 1605. It is not clear what may have brought Radesdale into rural Cambridgeshire, but it is possible that he, too, worked in horn and had a professional relationship with Mindum. Or perhaps he simply chanced upon Mindum on a journey to the coast at King's Lynn, Norfolk.

Many of the other names inscribed on Mindum's dated shoehorns can be linked to individuals domiciled in parishes along a ten-mile corridor between Cambridge and Ely: Jane Ayers, Witchford (shoehorn dated 1593); Robert Gotobed, Landbech and Stretham (1595); Richard Crab, Little Thetford (1595); John Gibson, Soham (1597); Ambrose Buckle, Ely (1598); Francis Hinson, Fordham (1600); Benjamin Gibson, Ely (1603); Bridget Dearsley, Soham (1605); and Thomas Draper, Cambridge (1612). The two outliers are Anthony Violet, Foudlen, and Hillary Forby, Saham Toney, but these locations in Norfolk are still within the general area. None of these individuals appears to have been particularly wealthy, and many were from families who had been and remained in the area for centuries.

With all this in mind, it seems sensible to suggest that Mindum was the "Robert Mendum" of Littleport, Cambridgeshire, whose will was proved at the Consistory Court of Ely in 1615. Littleport is directly north of Ely and on the road toward Downham Market and King's Lynn.

PROVENANCE

Collection of Stuart Kirby of Long Valley, NJ (1952–2019); bt. John Bryan III of Nashville, TN, from Adam Ambros, New York, July 22, 2019

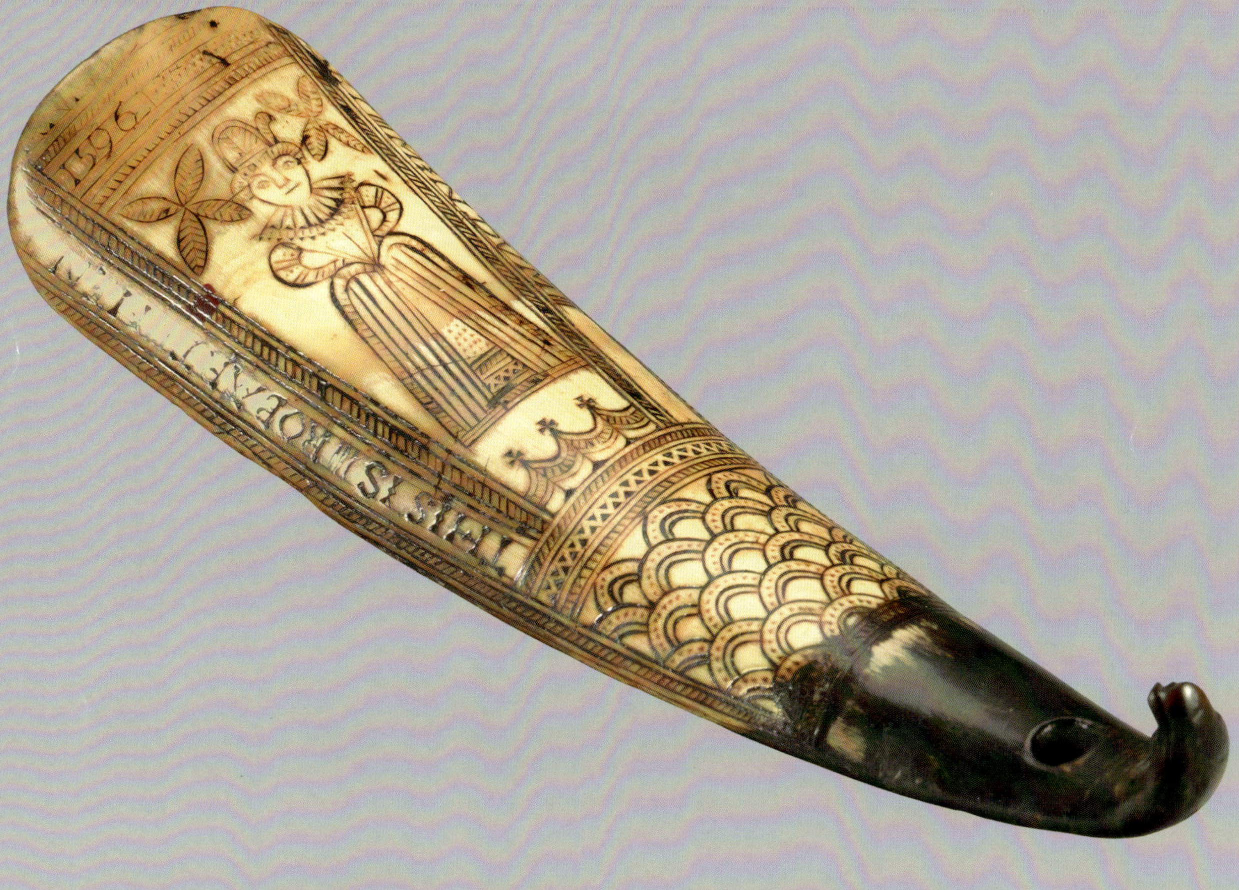

1597

Shoehorn
Horn
$5\frac{1}{2} \times 1\frac{1}{2} \times 1\frac{3}{8}$ in.
(14 × 3.8 × 3.5 cm)
Robert Mindum
(active 1593;
d. 1615?)

Inscription: 1597 ;
THIS IS WILL'YAM
S ; OF ROBART
MINDVM

LITERATURE
Joan Evans, "Shoe-Horns and a Powder Flask by Robert Mindum," *Burlington Magazine for Connoisseurs* 85, no. 500 (1944): 282–84

PROVENANCE
Christie's, London, July 12, 2005, (lot 104); bt. John H. Bryan of Lake Bluff, IL (1936–2018) from Tobias Jellinek Antiques, London, August 5, 2005

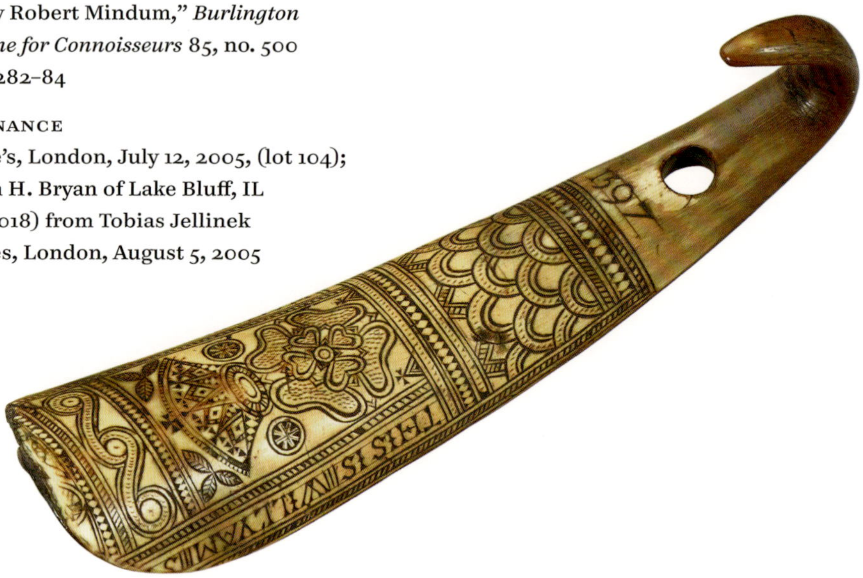

1598

Shoehorn
Horn
$6\frac{1}{2} \times 2 \times 1\frac{1}{4}$ in.
(16.5 × 5.1 × 3.2 cm)
Robert Mindum
(active 1593;
d. 1615?)

Inscription:
ROBART MINDVM
MADE THIS SHOING
HORNE FOR ROSE
FALES ANNO
DOMINI 1598

PROVENANCE
Collection of Stuart Kirby of Long Valley, NJ (1952–2019); bt. John Bryan III of Nashville, TN, from Adam Ambros, New York, July 22, 2019

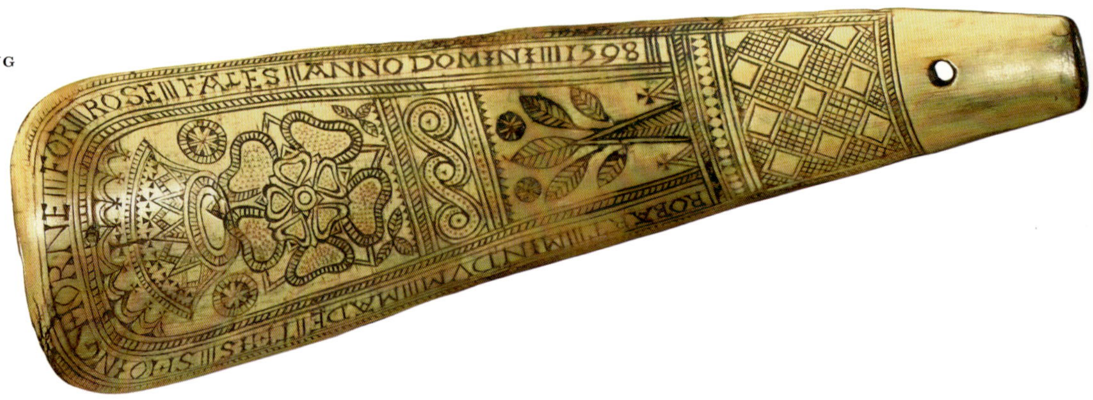

1694

The owner of this brass shoehorn was probably Edward Spurrier, who was buried at St Paul's, Hammersmith, on January 11, 1719/20.

PROVENANCE

bt. John H. Bryan of Lake Bluff, IL (1936–2018) from Jellinek & Sampson Antiques, London, October 17, 1980

Shoehorn

Brass

$5\frac{1}{4} \times 1\frac{3}{4} \times 1\frac{1}{2}$ in.
(13.3 × 4.5 × 3.8 cm)

Inscription:
EDVVARD
SPVRYER|1694

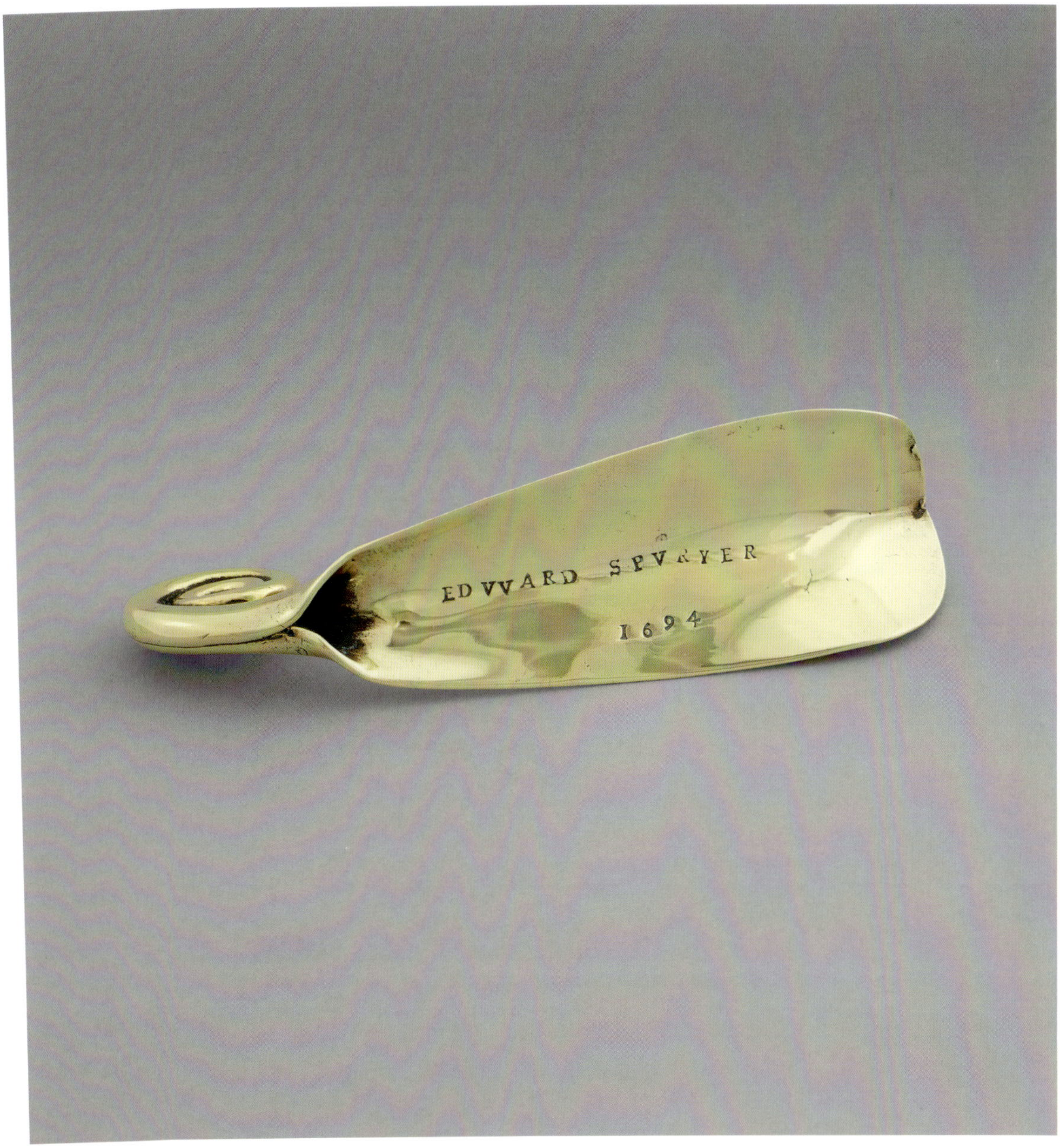

1685

Brush

Horn, brass, animal hair

$1\frac{1}{2} \times 2\frac{1}{4} \times 4\frac{1}{4}$ in. (3.8 × 5.7 × 10.8 cm)

Inscription: 1685

Clothes brushes were used for brushing dried mud and dirt from outer garments, and were, therefore, essential to the routine cleaning of woolen items, which were not washed in water due to risk of shrinkage and felting.

PROVENANCE

bt. John H. Bryan of Lake Bluff, IL (1936–2018) from Alistair Sampson Antiques, Ltd., London, December 17, 1998

1698

Brush

Horn, brass, animal hair

$1\frac{5}{8} \times 3\frac{3}{8} \times 2$ in. (4.1 × 8.6 × 5.1 cm)

Inscription: R B 1698

PROVENANCE

bt. John H. Bryan of Lake Bluff, IL (1936–2018) from Tobias Jellinek Antiques, London, September 11, 1999

1774

Brush

Fruitwood, animal hair

$2 \times 2\frac{3}{4} \times 7\frac{1}{2}$ in. (5.1 × 7 × 19.1 cm)

Inscription: I C|1774

PROVENANCE

Tristram Jellinek of London (1933–1995); Sotheby's, London, May 10, 1996 (lot 799); bt. John H. Bryan of Lake Bluff, IL (1936–2018) through Tobias Jellinek Antiques, London

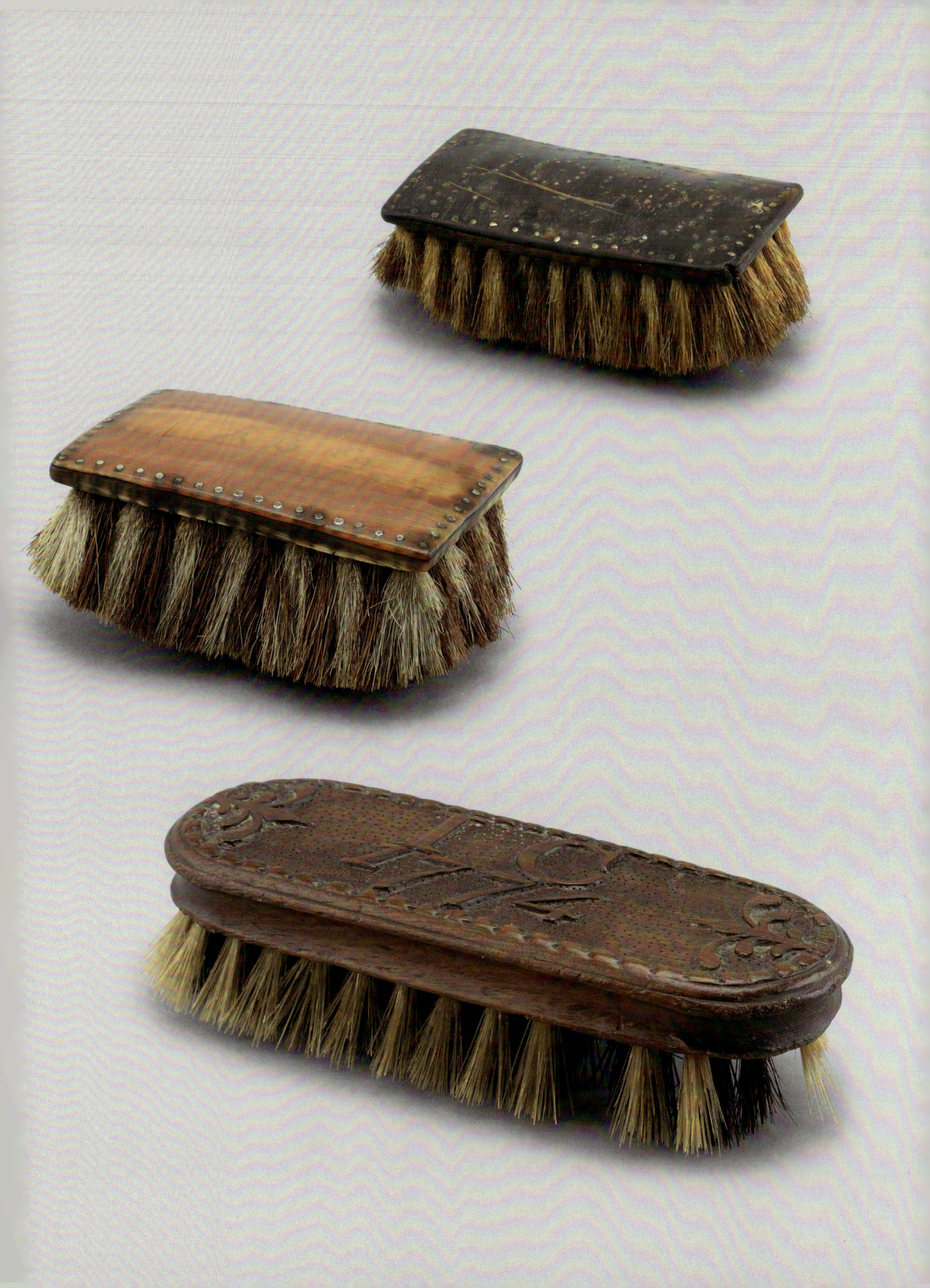
1774

1601

Walking stick

Malacca cane, whale tooth, silver, brass, steel

35 × 4 × 1¼ in. (88.9 × 10.2 × 3.2 cm)

Inscription: ·I·B 1·6·0·1·

The fashion for walking sticks grew over the course of the seventeenth century and reached fever pitch in the 1660s. If the date on this cane is to be taken at face value, it may be one of the very earliest of its type. Alternatively, the date could have been added retrospectively and might record the birth date of its owner, "I B."

Fashioned from Malacca cane from Sumatra and finished with a whale-tooth handle, this walking stick incorporated materials imported from around the world.

Although walking sticks could be extremely luxurious—with costs ranging from a few shillings to several pounds—they were not exclusively the property of the urban rich. For example, in 1664, the Lancashire apprentice shopkeeper Roger Lowe was delighted when, as a reward for his excellent work, his master promised him a "steak" (stick) and a new "hatt" as part of his attire in the new year. Simply owning a fine new walking stick was, perhaps, itself sufficient reason to add a celebratory date to its decoration, and the addition of the owner's name or initials guarded against its being mistaken, when laid aside, for one of the hundreds of others that walked the streets.

PROVENANCE

bt. John H. Bryan of Lake Bluff, IL (1936–2018) from Michael C. German, London, March 9, 1998

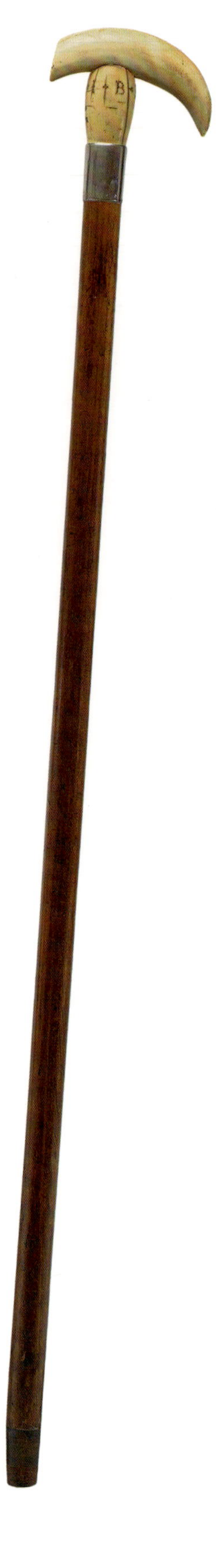

1729

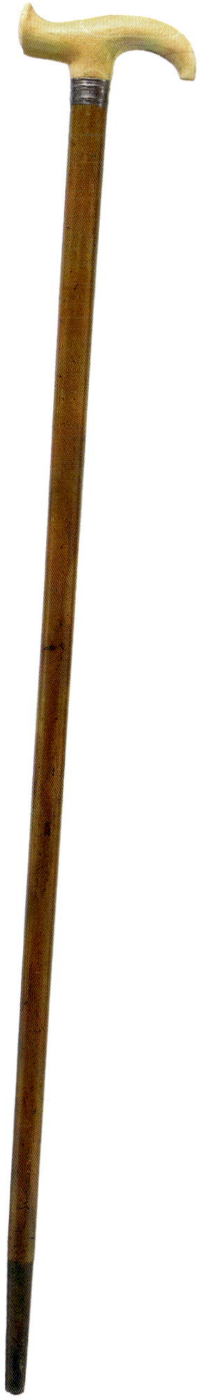

Walking stick
Malacca cane, horn, silver, brass, steel
$36\frac{1}{8} \times 4\frac{3}{4} \times 1\frac{1}{8}$ in. (91.8 × 12.1 × 2.9 cm)

Inscription: AMOS LINDLEY YARME 1729

Yarm was a cosmopolitan town in North East England. It boasted a port and was an important hub for coach travel. As a well-to-do master merchant, with apprentices of his own, Amos Lindley had his own take on the modern fashion for walking sticks. He preferred the solidity of wood to rattan but was able to afford an "opera" style ivory handle, which, with its curvilinear shape, would offer real support. The richness of his walking stick was topped off with a silver collar on which "1729" was engraved. We can only speculate about the significance of this date for Amos Lindley, but perhaps it commemorated the event at which his walking stick was first "worn." Lindley was an important figure in his local society, and 1729 was the year in which the first horse races were staged at the new Yarm Racecourse. It seems highly likely that Lindley attended this very public event, along with all the other great and good of the town.

PROVENANCE
bt. John H. Bryan of Lake Bluff, IL (1936–2018) from Michael German Antiques Ltd., London, January 7, 2003

1746

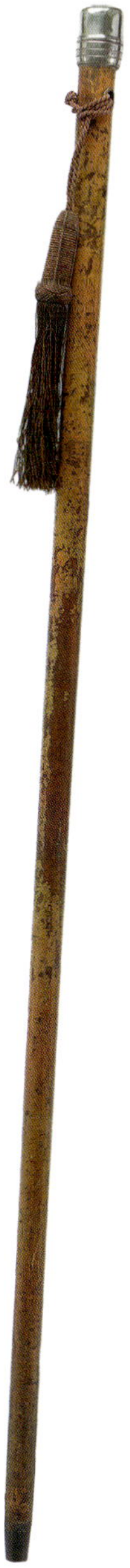

Walking stick
Wood, silver, brass, steel
$36\frac{3}{8} \times 1\frac{1}{8}$ in. (92.4 × 2.9 cm)

Inscription: Culloden Moor| April 16th|1746.

The inscription on this walking stick marks the Hanoverian victory at the Battle of Culloden on April 16, 1746—the last of the great Jacobite risings to restore the House of Stuart to the British throne. After Culloden, Charles Edward Stuart (known as Bonnie Prince Charlie and as the Young Pretender) returned to exile in France for the remainder of his life. The inscription does not make clear which side the owner of this stick might have been on, although it has been described as a Scottish walking stick, which might suggest that it belonged to a Jacobite supporter; it is unclear, however, how long after the battle the inscription might have been made.

PROVENANCE
Christie's, Glasgow, June 12, 1996 (lot 109); bt. John H. Bryan of Lake Bluff, IL (1936–2018) from D. S. Lavender Antiques Ltd., London, July 16, 1996

1761

Inkstand

Delftware

$1\frac{1}{2} \times 2\frac{7}{8}$ in. (3.8 × 7.3 cm)

Inscription:
S.. B|august|26|1761

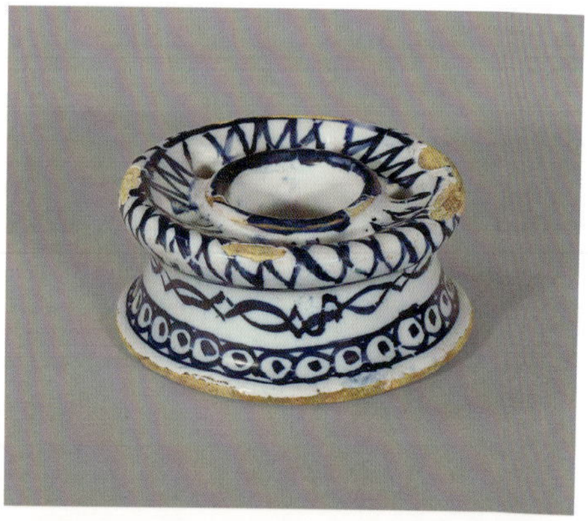

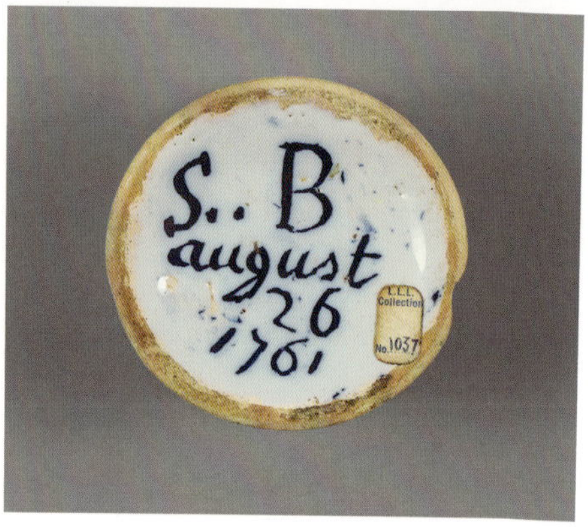

PROVENANCE

Mrs. Radford; Sotheby's, London, November 3, 1943 (lot 23); T. W. G. Thomas; Sotheby's, London, June 13, 1961 (lot 11); Louis L. Lipski, London; Sotheby's, London, March 10, 1981 (lot 127); bt. John H. Bryan of Lake Bluff, IL (1936–2018) from Jellinek & Sampson Antiques, London, March 12, 1981

LITERATURE

Louis L. Lipski and Michael Archer, *Dated English Delftware: Tin-Glazed Earthenware, 1600–1800* (London: Sotheby Publications, 1984), 363, no. 1586

1765

Inkstand

Brass, lead

$1\frac{1}{2} \times 2\frac{1}{4} \times 2\frac{1}{4}$ in. (3.8 × 5.7 × 5.7 cm)

Inscription:
J A|1765

This inkstand, or well, is one of three extant examples that have engraved brass tops and heavy lead bases. The weight of the base kept the inkstand stable in the event that it was accidentally knocked over, which avoided the calamity of ink being spilled everywhere. The other two examples are inscribed, respectively, "T S|1766" and "F R|1768" and appear to be by the same maker. They, too, have four smaller holes in the corners to hold quill pens.

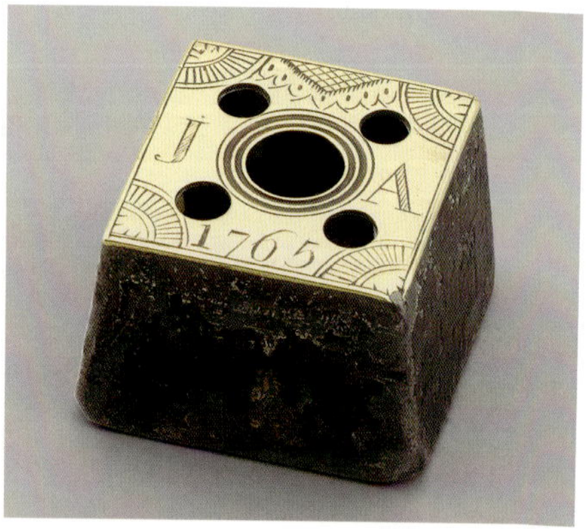

PROVENANCE

bt. John H. Bryan of Lake Bluff, IL (1936–2018) from Alistair Sampson Antiques Ltd., London, November 9, 1989

LITERATURE

Rupert Gentle and Rachel Field, *Domestic Metalwork: 1640–1820*, revised and enlarged by Belinda Gentle (Woodbridge, UK: Antiques Collectors' Club, 1994), 363, nos. 7, 8

1767

Quill holder
Delftware
$1\frac{7}{8} \times 5 \times 2\frac{3}{4}$ in.
(4.8 × 12.7 × 7 cm)

Inscription: I C 1767

PROVENANCE
Mrs. Vincent E. Morris, Connecticut; bt. John H. Bryan of Lake Bluff, IL (1936–2018) from Earle D. Vandekar, New York, March 17, 1992

LITERATURE
Louis L. Lipski and Michael Archer, *Dated English Delftware: Tin-Glazed Earthenware, 1600–1800* (London: Sotheby Publications, 1984), 365, no. 1590

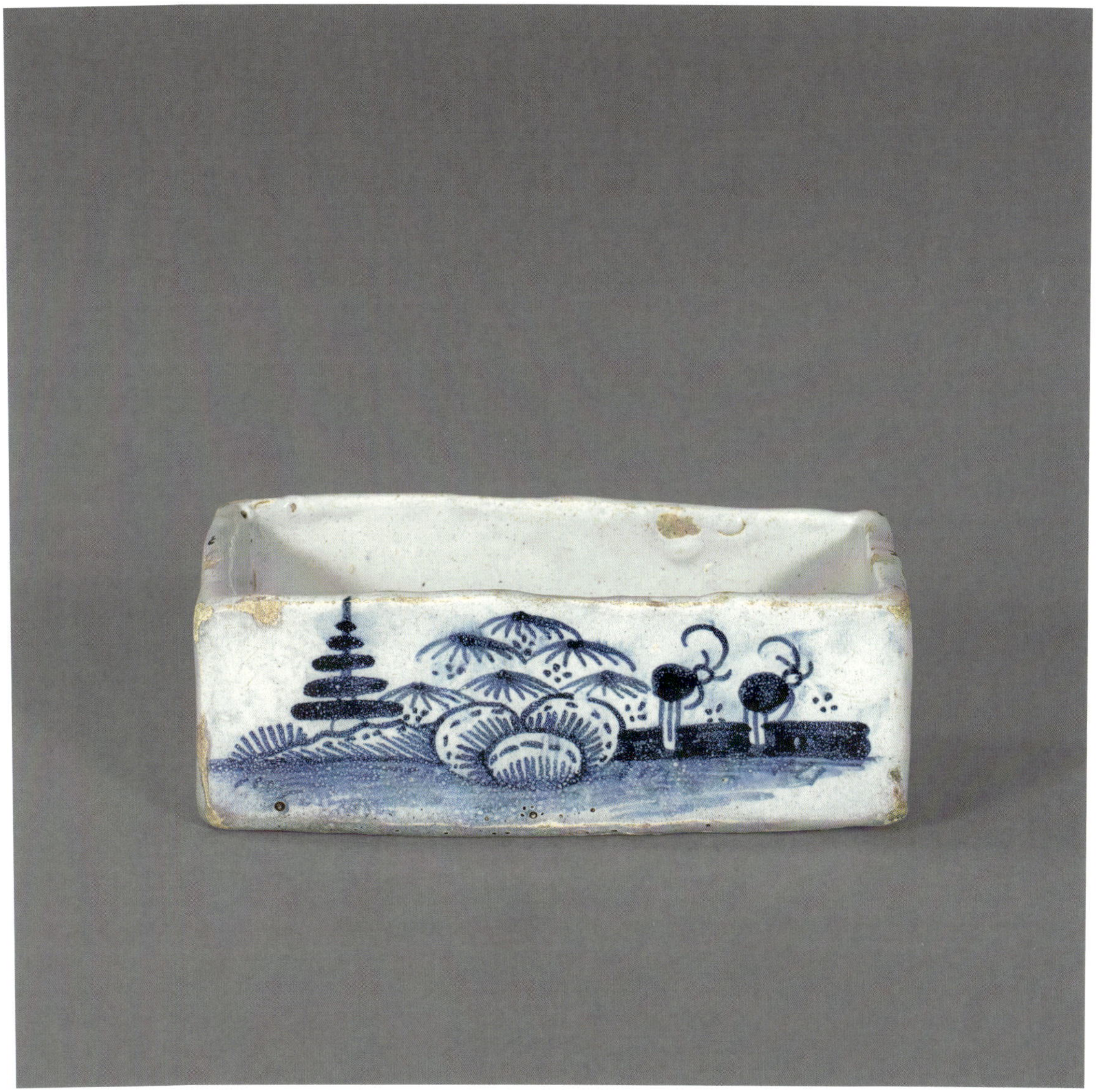

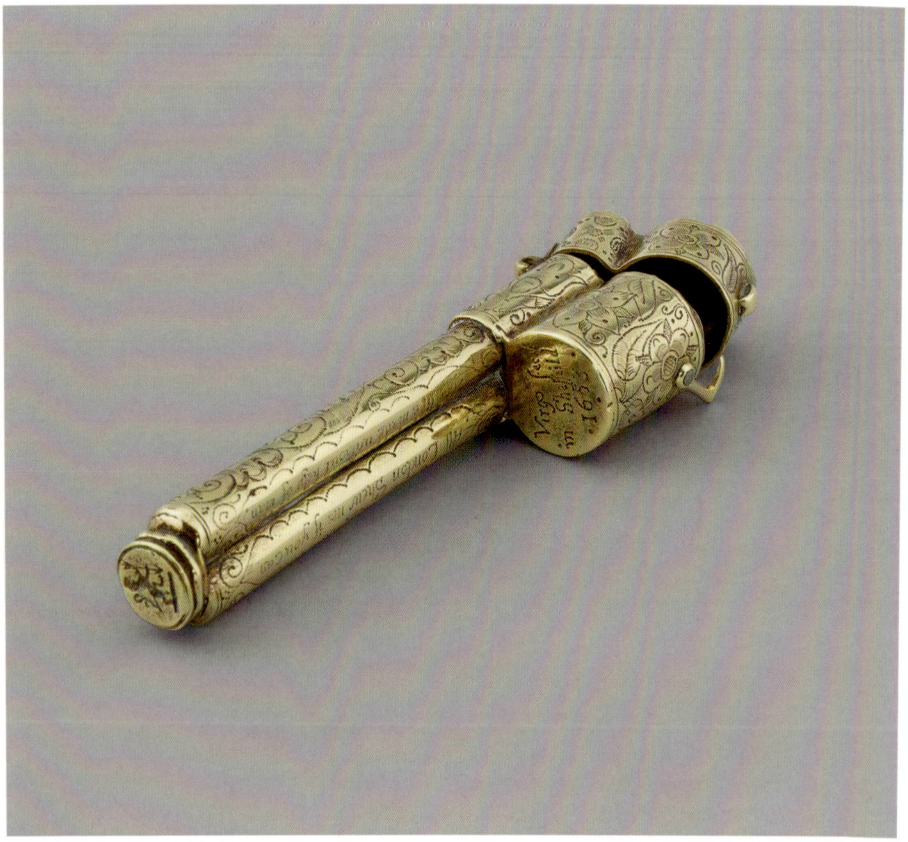

1653

Penner

Brass

$4\frac{1}{4} \times 1\frac{1}{4} \times \frac{3}{4}$ in. (10.8 × 3.2 × 1.9 cm)

Thomas Madin (active 1652; d. 1670?)

Inscription: All London Shew me if you can|The like made without help of man; Virgo fecit|in Sheffield 1653

A penner is a pen case, and this example was made by an elusive craftsperson working in Sheffield between 1652 and 1657 who went by the moniker Virgo. This maker made brass writing cases and implements; a small ivory, brass, and horn box in the Victoria and Albert Museum, London (W.1-1956); and possibly also the pastry jigger in the Bryan Collection (page 207). This penner has a double-barrel form to hold two pens and a round ink reservoir, in the same configuration as the example dated 1652 in the collection of the Victoria and Albert Museum (M.201&2-1914) and one dated 1654 in the Museum of London (46.78/587a). The writer used the penner in conjunction with a wax seal kept in a case—another item from the Madin workshop (facing page). The writer endorsed a document by impressing his or her seal (in this case, a matrix of a crest of a heraldic lion rampant, carrying an arrow or spear) and then sealed the letter by the same means.

Virgo seems to have been a punning pseudonym for *Maden*, or *Madin*, which was the name of a family of cutlers working in Sheffield in the late seventeenth century, and it has been suggested that the maker of this object was one of that family's forebears. Nevertheless, the maker's true identity remains an enigma: a Beatrice Mathion, widow, was buried in Sheffield in 1665, and a Thomas Mathion was buried at the Cathedral Church of St Peter and St Paul, Sheffield, on January 28, 1669/70, but it is impossible to say with any certainty whether they were from the family in question.

PROVENANCE

bt. John H. Bryan of Lake Bluff, IL (1936–2018) from Jonathon Trace, Rifton, NY, February 6, 2006

LITERATURE

Michael Finlay, *Western Writing Implements in the Age of the Quill Pen* (Wirral, UK: Pen & Pencil Gallery, 1990), 140

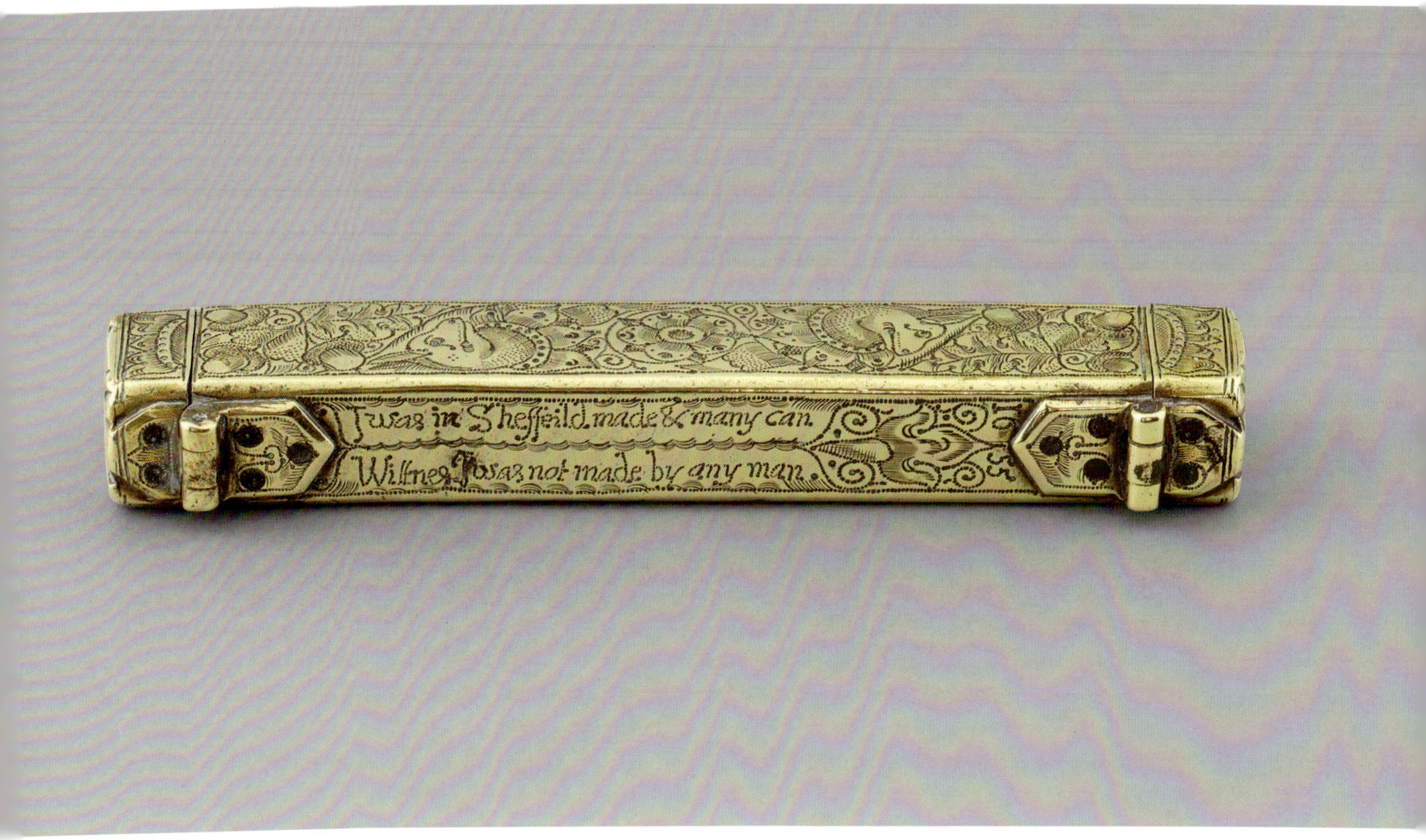

1655

Sealing wax case
Brass
$\frac{7}{8} \times \frac{3}{4} \times 5$ in. (2.22 × 1.9 × 12.7 cm)
Thomas Madin (active 1652; d. 1670?)

Inscription: I was in Sheffield made & many can|Witnes I was not made by any man; 1655

This sealing wax case is of the same design as two others, both at the Victoria and Albert Museum, London, and dated 1656 (M.149-1992) and 1657 (M.118-1930).

PROVENANCE

bt. John H. Bryan of Lake Bluff, IL (1936–2018) from Michael & Jane Dunn Antiques, Claverack, NY, February 16, 1988

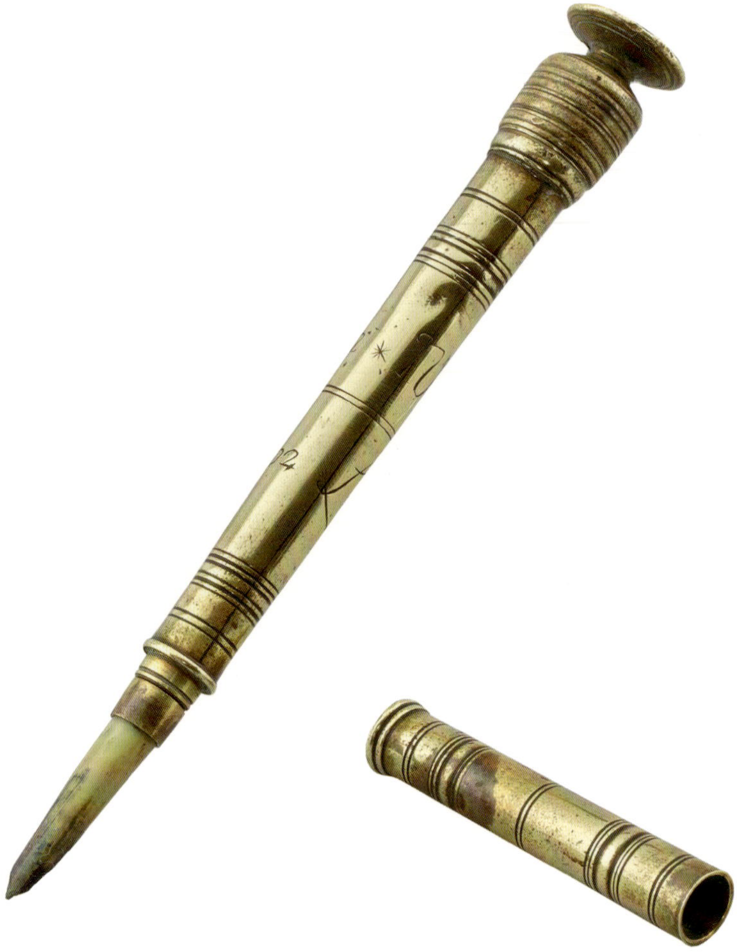

1702

Fountain pen

Brass, quill

$5\frac{1}{4} \times \frac{5}{8}$ in. (13.3 × 1.59 cm)

Inscription:
William Line|1702

This is the earliest dated fountain pen, and it was owned by the aptly named William Line, who might be identified as the man by that name who lived his entire life, from 1680 to 1732, in Warwickshire. At its terminal, the pen has a rebus seal of a lion; the brass tube that forms the body serves as the ink reservoir, from which ink percolates to the quill nib through a horsehair, flax, or cotton plug. The ink flow was adjusted by compressing the plug. The cap has a protruding rod, which sealed the pen when not in use.

PROVENANCE

bt. John H. Bryan of Lake Bluff, IL (1936–2018) from Jim Marshall, Pen & Pencil Gallery, The Wirral, Merseyside, May 14, 2017

1761

The coat of arms (Argent, a chevron between three griffins passant sable) at the terminal of this sealing wax case can be identified as belonging to the Finch family, Earls of Nottingham and Winchelsea and Earls of Aylesford. The mullet is a mark of cadency, which identifies the third son, who might be identified as John Finch (d. 1763) of Bushey, Hertfordshire. Prior to his marriage, Finch had a daughter named Elizabeth with an actress named Elizabeth Younger. This daughter may have been the giver of this gift in 1761. Famously, Finch was stabbed by Sally Salisbury, a notorious courtesan, on the night of December 18, 1722, at the Three Tuns Tavern in Chandos Street, London.

It has been suggested that the mark on the underside of the small lid is for David Tanqueray of London, which, if so, would have been made about 1715. Tanqueray's wife, Anne, daughter of the Huguenot David Williaume, was a silversmith in her own right.

PROVENANCE

bt. John H. Bryan of Lake Bluff, IL (1936–2018) from Peter Cameron, London, December 3, 2016

Sealing wax case

Silver

$3\frac{7}{8}$ × 1 in. (9.8 × 2.5 cm)

Inscription:
M·F|Given her by|E·F|1761.

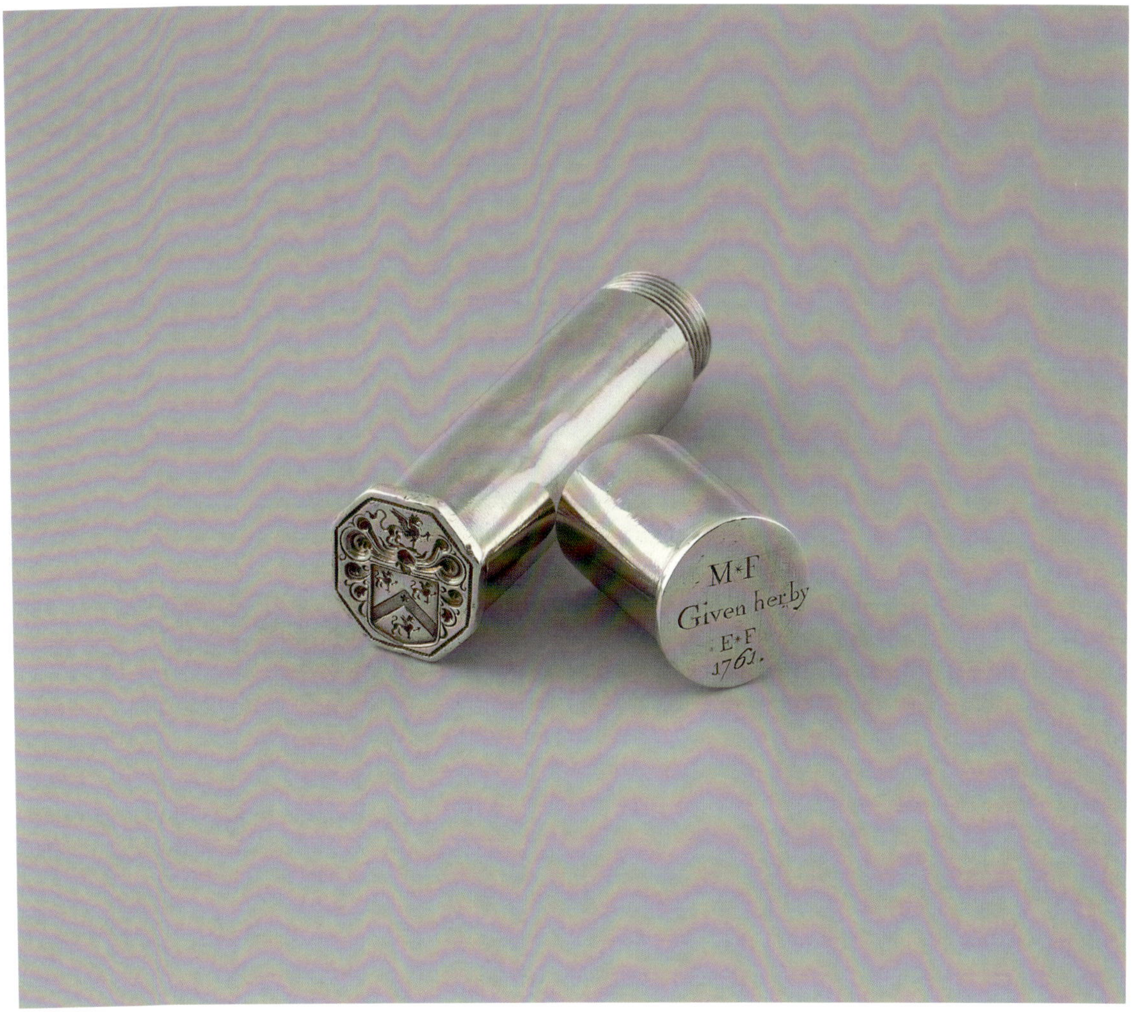

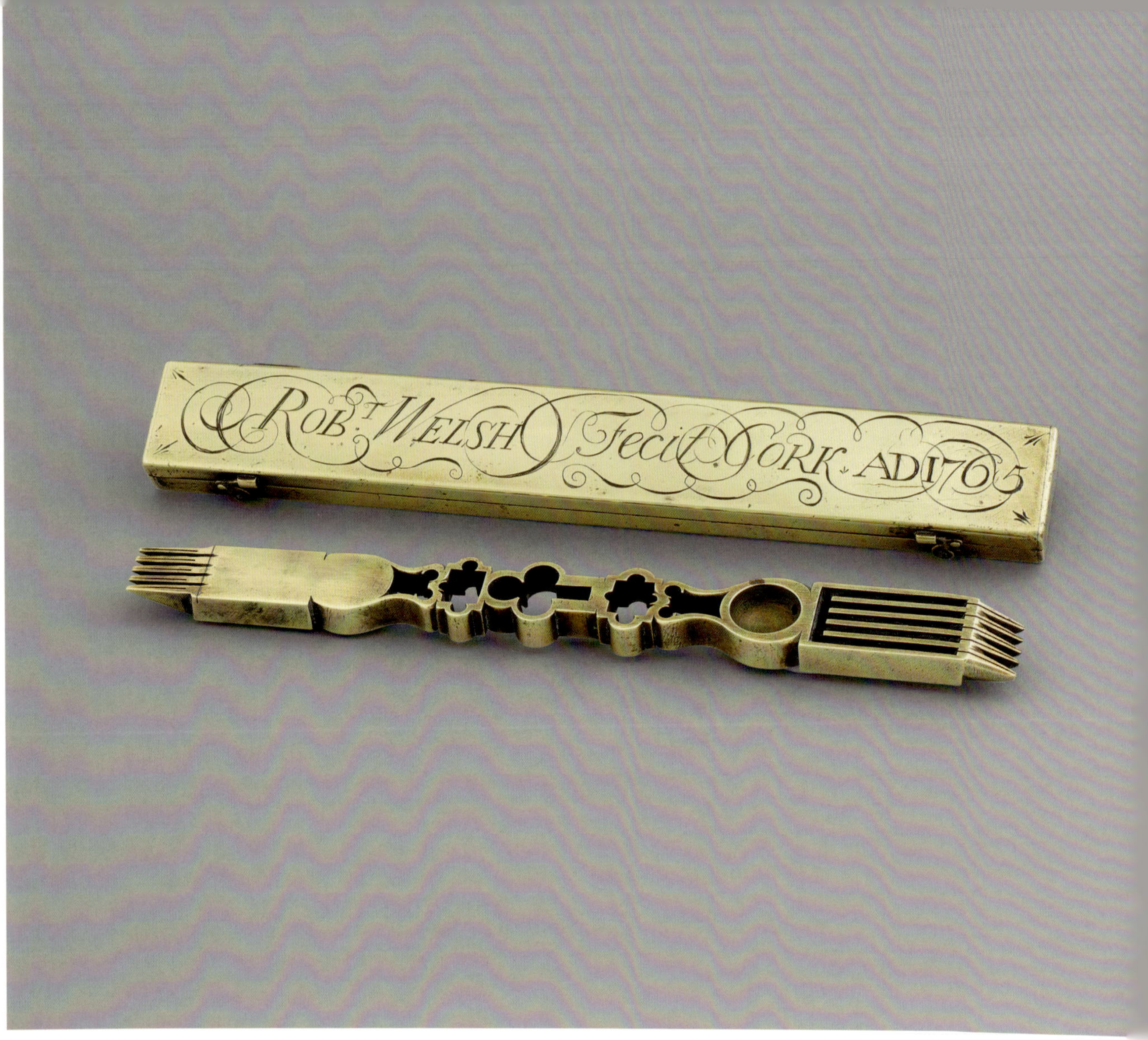

1765

Rastrum
Brass
½ × 6 × ⅞ in. (1.27 × 15.2 × 2.22 cm)

Inscription: ROB^T WELSH Fecit Cork AD 1765

Rastrum is the Latin word for "rake," and the instrument known by this name was used to make blank musical manuscript pages. Ink was poured into the dished reservoir, and as the rake was dragged along the paper, the ink flowed into five channels and then from the tips of the five projections. When the rake was tilted at about twenty degrees, it produced the five-line musical staff. The maker, Robert Welsh of Cork, Ireland, has not been traced.

PROVENANCE
M. S. Dudley Westropp of Dublin (1868–1964); Michael Finlay; bt. John H. Bryan of Lake Bluff, IL (1936–2018) from Jim Marshall, Pen & Pencil Gallery, The Wirral, Merseyside, May 14, 2017

1625

Sweet bag
Silk, glass beads
$6\frac{3}{8} \times 11 \times \frac{1}{2}$ in.
(16.2 × 27.9 × 1.27 cm)

Inscription:
THEGIFTOF;
AFREND1625

Sweet bags containing dried herbs and sweet-smelling flowers would ward away bad smells and miasmas; placed between linens, they would keep the fabrics fragrant and fresh. This example is made from polychrome glass beads threaded on silk thread and worked into a looped mesh. It is lined with yellow silk (in twill weave), which forms a casing for the plaited yellow drawstring at the top edge. The drawstring is decorated with a knob tassel covered with a stitched tube of glass beads. Along the lower edge are three stiffened thread loops oversewn with yellow silk in buttonhole stitch, from which hang looped tassels of strung beads.

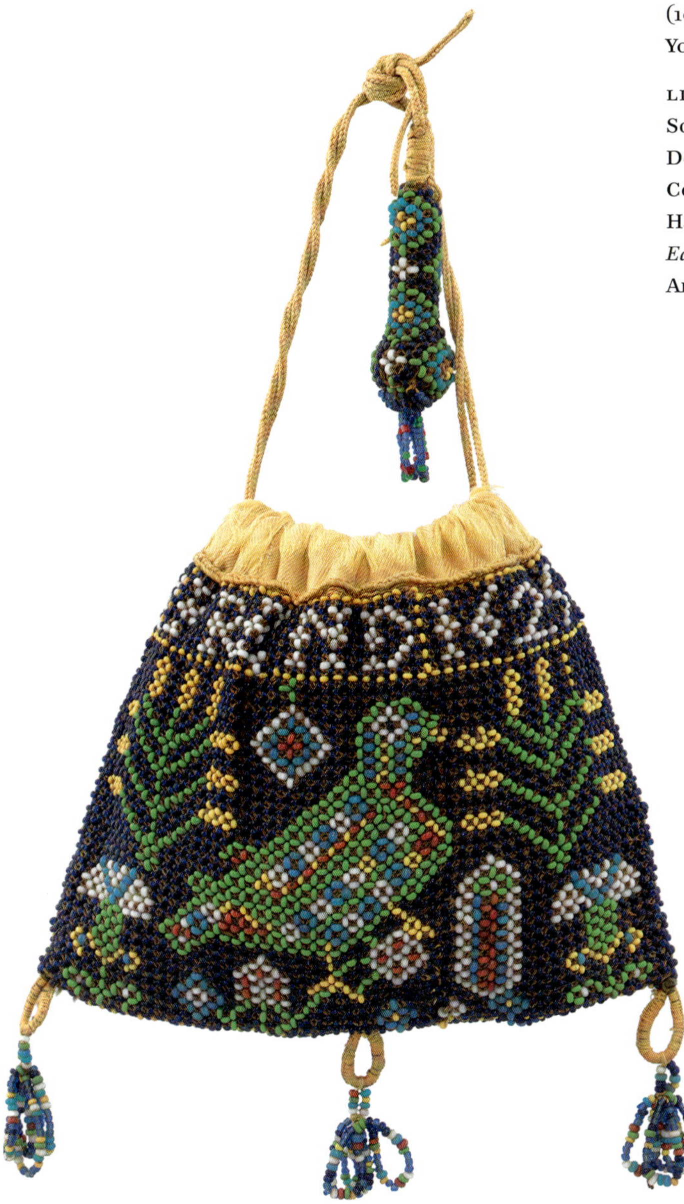

PROVENANCE

Sir Frederick Richmond (1873–1953); Christie's, London, June 23, 1987 (lot 69); Alistair Sampson Antiques, London; bt. John H. Bryan of Lake Bluff, IL (1936–2018) from Cora Ginsburg, New York, February 5, 1990

LITERATURE

Sophie Cope, "Women in the Sea of Time: Domestic Dated Objects in Seventeenth-Century England," in Merry E. Wiesner-Hanks, ed., *Gendered Temporalities in the Early Modern World* (Amsterdam: Amsterdam University Press, 2018), 47–68

1672

Snuff flask
Gourd
$2\frac{3}{8} \times 2\frac{1}{2}$ in. (6 × 6.4 cm)

Inscription:
·GILLIAN·FORAHM· ANNO·1672; ·A friend at neede doth gold excede · ·

This decorated gourd snuff flask was made for a woman named Gillian Forham. The globular body is carved with a rosette, the royal coat of arms, and flanking panels that contain scrolls and trees.

PROVENANCE
bt. John H. Bryan of Lake Bluff, IL (1936–2018) from Tristram Jellinek, London, June 29, 1992

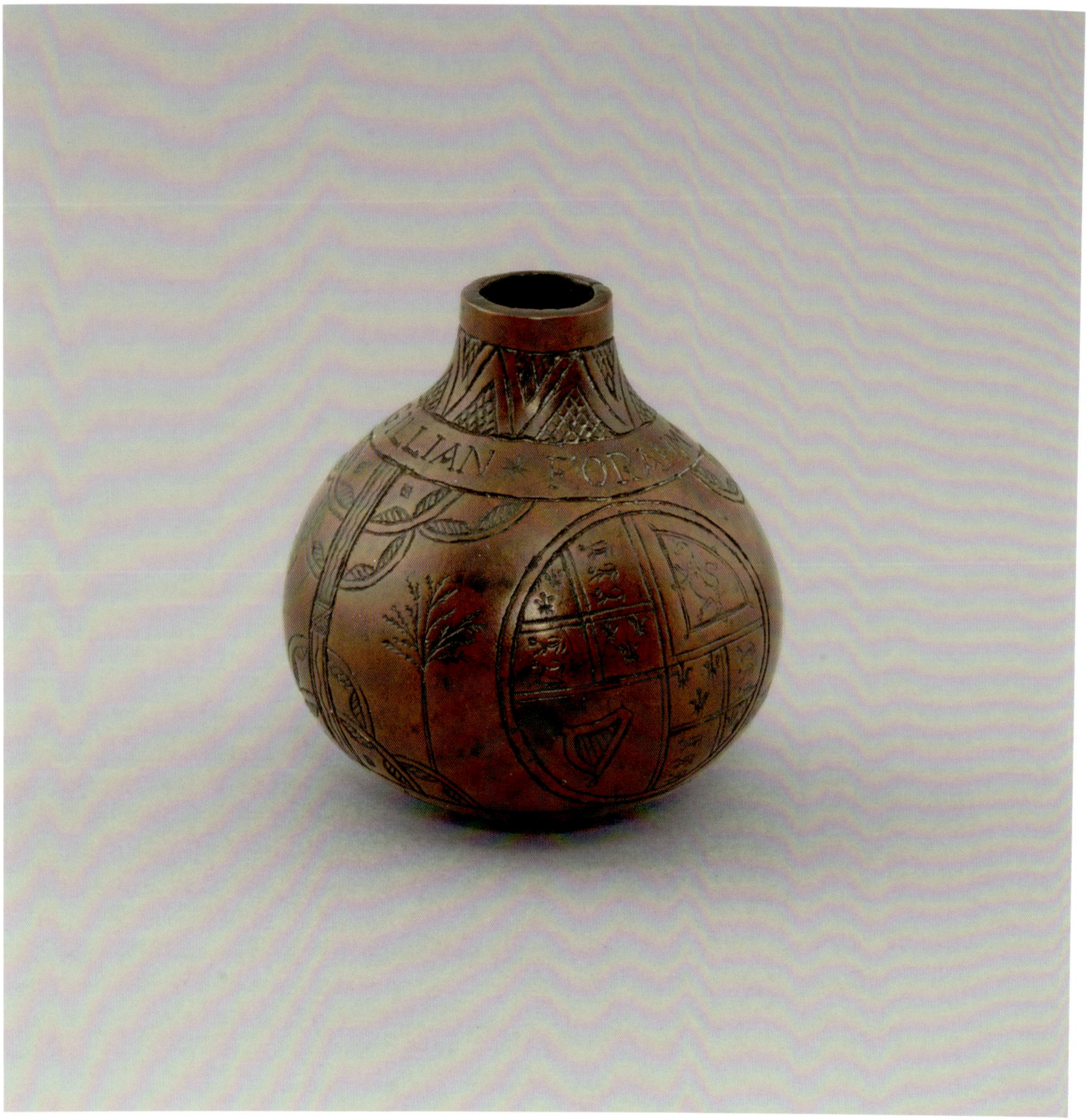

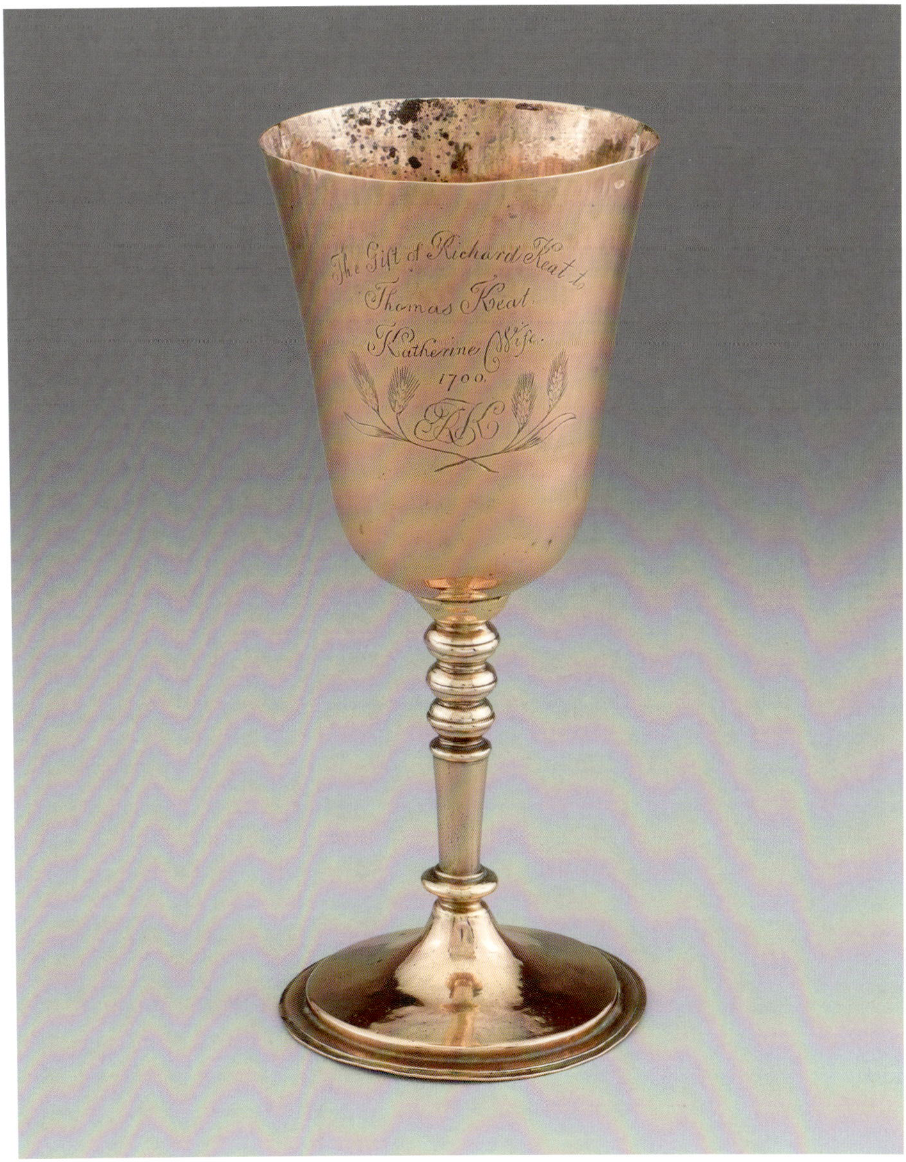

1700

Standing cup

Copper, brass

8 ¾ × 3 ⅝ in. (22.2 × 9.2 cm)

Inscription: The Gift of Richard Keat to|Thomas Keat|Katherine Wife|1700|RK [in monogram]

Originally gilded, this bell-metal standing cup was probably given by Richard Keat as a betrothal gift to his relative Thomas Keat and Thomas's bride-to-be, Katherine. The cup could be related to the marriage of the Thomas Keats and Catherine Owen who were granted license to marry by the bishop of London in 1701.

PROVENANCE

Sotheby's, Billinghurst, Sussex, June 22, 1993 (lot 652); bt. John H. Bryan of Lake Bluff, IL (1936–2018) through Christopher Bangs, London

1754

Pocket globe and case

Paper, card, sharkskin

$3\frac{1}{4}$ in. (8.3 cm)

Nathaniel Hill (1708–1768)

Inscription: A NEW|Terrestrial|GLOBE|by Nath Hill|1754

Nathaniel Hill was a prominent London maker of globes and instruments in the mid-eighteenth century. In 1730, he began as an apprentice to Richard Cushee (1696–1733), a Merchant Taylor by guild, but a land surveyor, engraver, and mapmaker and globe maker by trade. Cushee's shop was at the sign of the Globe and Sun on Chancery Lane, where one could also purchase copies of the astronomer Joseph Harris's *The Description and Use of Globes, and the Orrery* (London: Printed for Thomas Wright, 1732), whose introduction contained a brief account of the solar system. After Cushee's death in 1733, Hill and Cushee's wife took over the running of the business. This small pocket globe, which came with a sharkskin case that contained charts of the Northern and Southern skies, was one of four available sizes advertised in a trade card of the 1740s.

PROVENANCE

bt. John H. Bryan of Lake Bluff, IL (1936–2018) from Trevor Philips & Sons Ltd., London, June 10, 1998

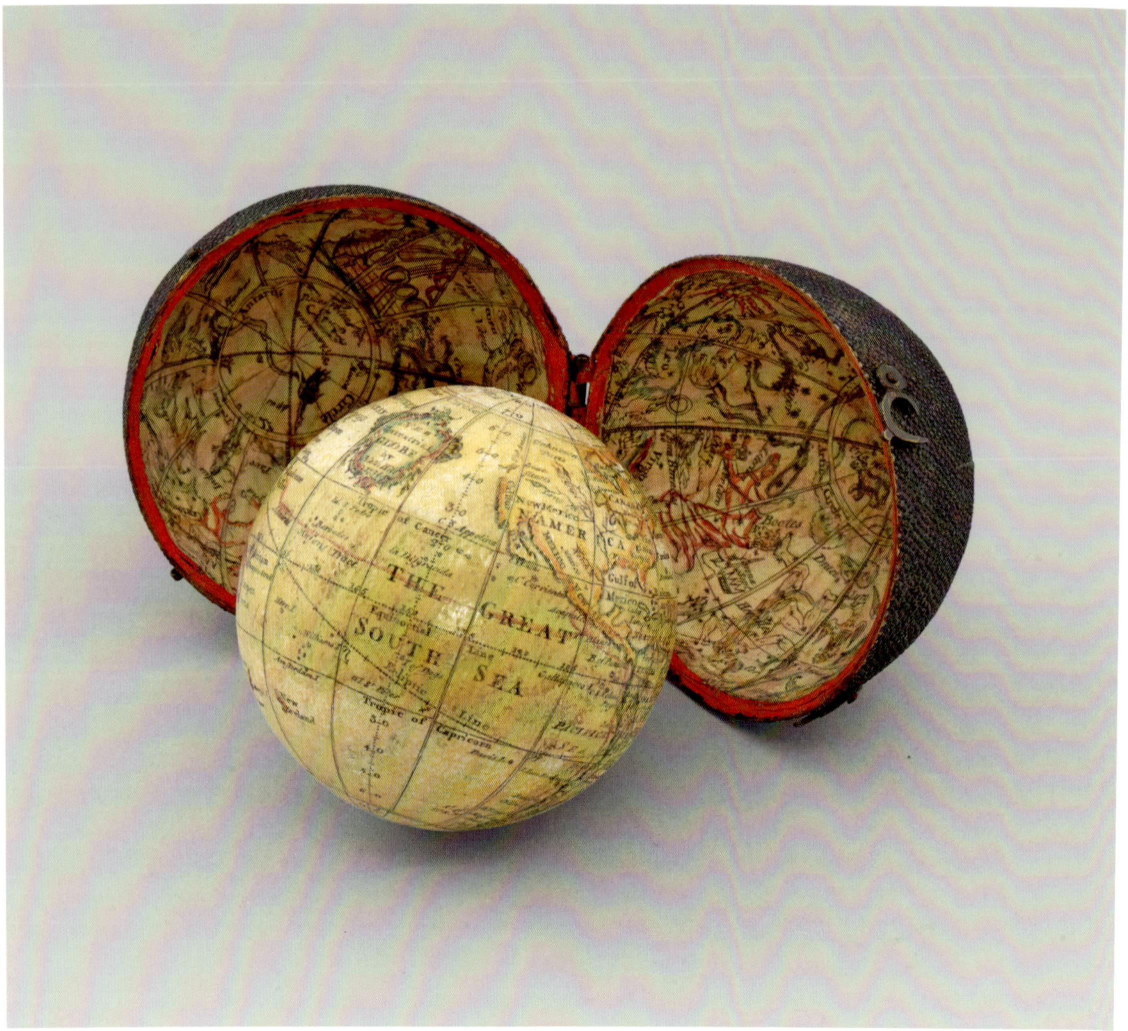

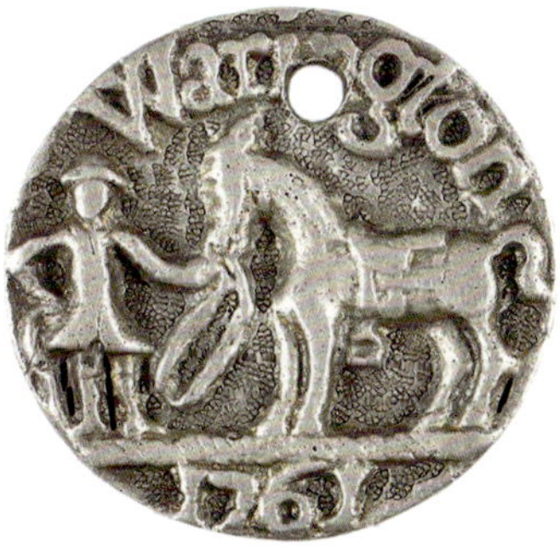

1761

Horse-racing pass
Silver
$\frac{1}{8} \times 1\frac{3}{8}$ in. (.32 × 3.5 cm)

Inscription:
Warington|1761

Horse racing took place at Latchford Heath, a village a mile south of Warrington, Cheshire, where events had been held since the first half of the century. Other racecourse passes like this one, also dated 1761, are made of copper rather than silver and have stamped rather than raised numbers. A racing pass conferred free access to the racecourse and to the grandstand, either for a prolonged period or for life. Unfortunately for the owner of this pass, racing at Latchford ceased soon after this pass was issued.

PROVENANCE
Dixon Noonan Webb, London, December 12, 2016 (lot 3226); bt. John H. Bryan of Lake Bluff, IL (1936–2018) from Peter Cameron, London, December 19, 2016

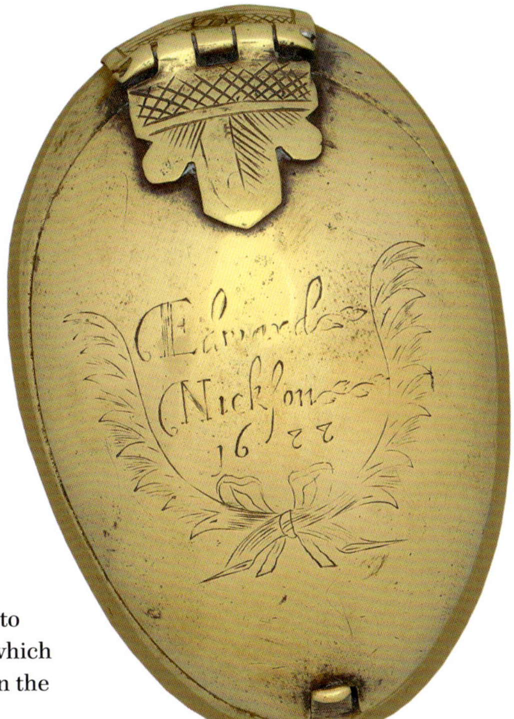

1688

Hand warmer
Brass
$3\frac{1}{4} \times 2\frac{1}{4} \times 2\frac{1}{4}$ in. (8.3 × 5.7 × 5.7 cm)

Inscription:
Edward|Nickson|16 88

The lid of this hand warmer lifts to receive a piece of hot charcoal, which would then smolder safely within the closed capsule, emitting heat.

PROVENANCE
bt. John H. Bryan of Lake Bluff, IL (1936–2018) from Alistair Sampson Antiques Ltd., London, December 7, 1993

1773

Magnifying glass
Brass, glass, wood
$3\frac{1}{2} \times 2\frac{3}{4} \times \frac{3}{8}$ in.
(8.9 × 7 × .95 cm)

Inscription:
T:Gill|CARNABY|1773

This magnifying glass appears to have belonged to the laborer Thomas Gill. He died at age seventy-seven and was buried on July 10, 1804, in Carnaby, Yorkshire.

PROVENANCE
Jack Casimir Ltd., London; Michael Gillingham of London (1933–1999); bt. John H. Bryan of Lake Bluff, IL (1936–2018) from Alistair Sampson Antiques Ltd., London, December 7, 1993

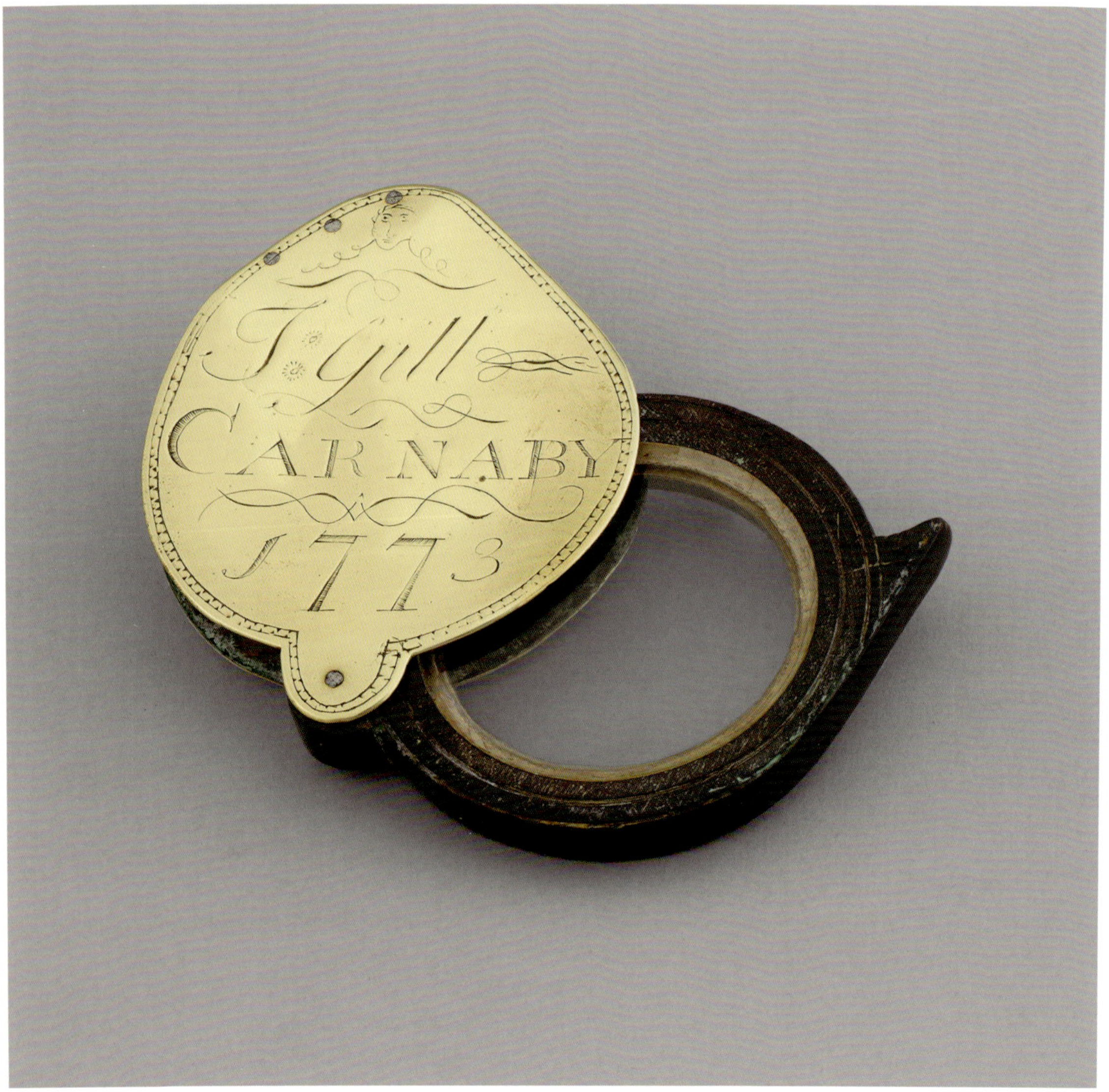

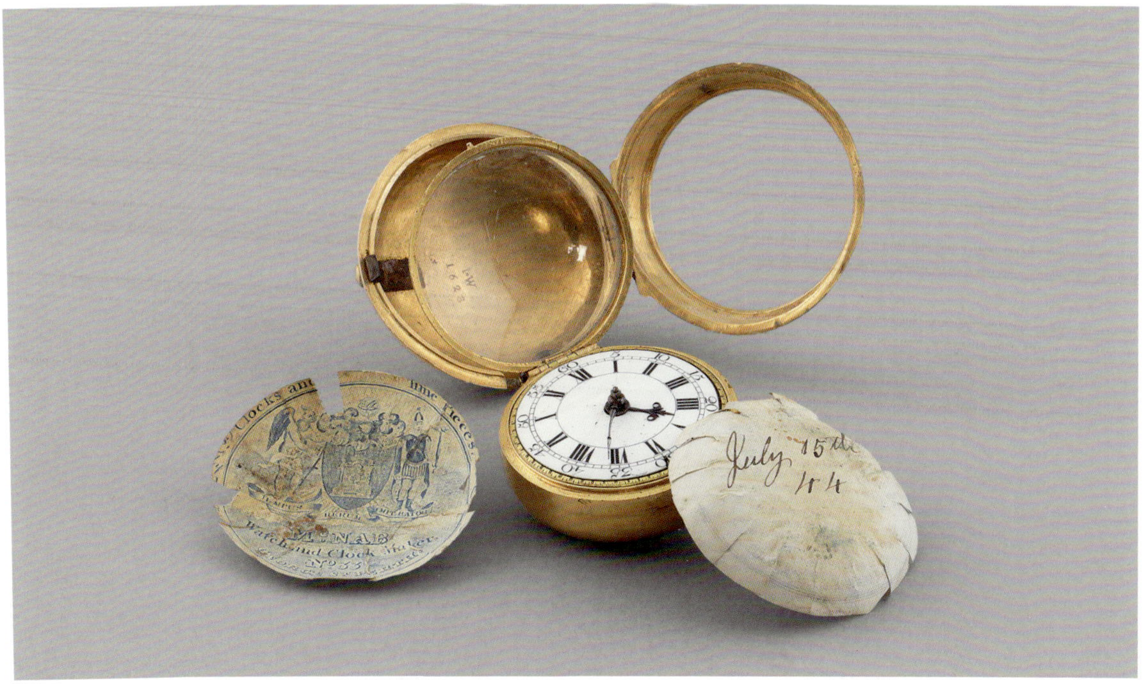

1723 | 1740 | 1744

Pocket watch and case

Gold, enamel, paper

$2\frac{5}{8} \times 2\frac{1}{8} \times 1\frac{1}{4}$ in. (6.7 × 5.4 × 3.2 cm)

Andrew Dunlop (active 1701; d. 1735)

Inscription: I W|1628; I W|1628; And^r Dunlop 1628|LONDON; June 17th |40; July 15th |44

Beautifully worked pocket watches like this one were often decorated inside and out, encouraging their owners to appreciate the internal mechanism as much as the external watch face. By the 1700s, watchmakers had developed new systems of authentication that combined several methods of recording time, including dates. The workings of this watch are inscribed with the name of the watchmaker, Andrew Dunlop, clock- and watchmaker of St Martin-in-the-Fields, London, whose will was proved on June 10, 1735 (TNA PROB 11/671/324).

The inside of the watchcase is stamped with "I W," the goldsmith's initials, and with the London gold assay hallmark for the year 1723. The inner and outer cases and the workings are all marked with the serial number, "1628," which places the watch within the sequence of Dunlop's output. Two enclosed watch papers record the care taken by its Scottish owner, who, in 1740 and 1744, sent it for maintenance to a specialist firm called McNabb's in George Street, Perth. The systematic features of name and serial number offered guarantees of authenticity and identified the watch as unique. This information would also be useful if the watch were to go missing, as in a case reported in the *Caledonian Mercury*, April 26, 1744:

> Lost on Saturday, April the 21st, betwixt the Gray's Close and Miln's Square, betwixt 8 and 9 o'clock at night, a large gold watch with Andrew Dunlop's name, with two gold cases and a green silk string with a small seal and an inscription on it, Love for Love.

PROVENANCE

Christie's at Fingask Castle, Glasgow, April 26–28, 1993 (lot 694); bt. John H. Bryan of Lake Bluff, IL (1936–2018) through Alistair Sampson Antiques Ltd., London

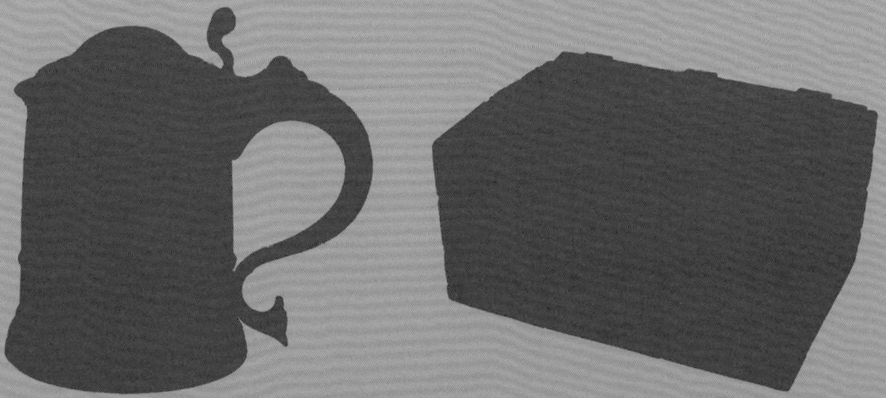

V. CROWN & CHURCH

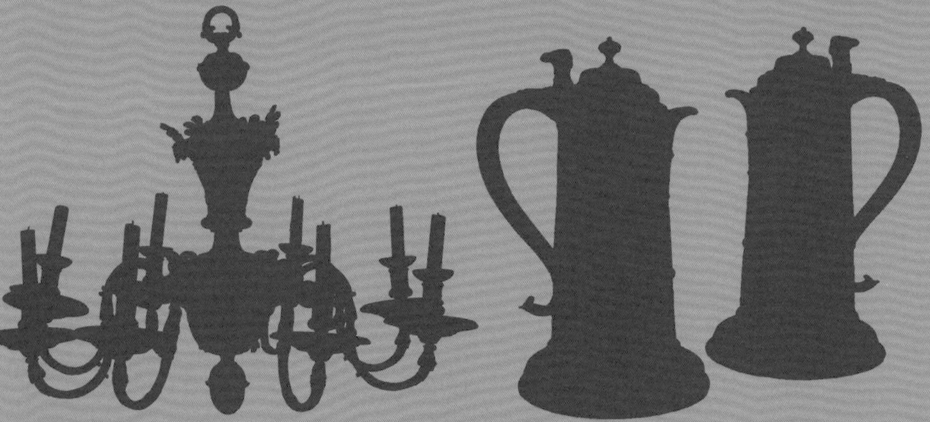

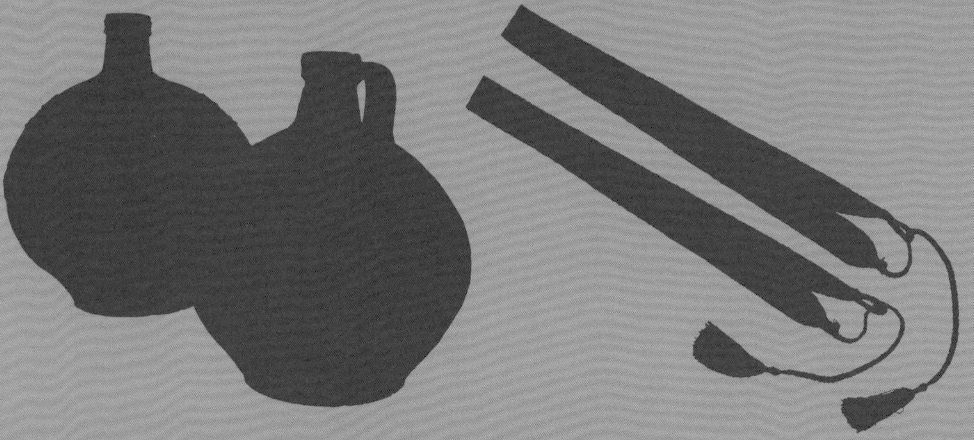

AS THE GOVERNING BODY WITH JURISDICTION OVER THE RITUALS associated with all the significant moments in the early modern life cycle—baptism, marriage, and death—the Churches of England and Ireland and their Presbyterian counterparts in Scotland loomed large. With the monarch as the head of the Church of England, the institutions of church and crown were inextricably bound. The mandate enacted by Thomas Cromwell in 1538 required that all parishes keep records of baptisms, marriages, and burials, thus making British society one of the most closely surveyed and monitored of all its Northern European counterparts. Parish officials stored these records in parish chests, like the one included in this catalogue (page 352).

In 1549—the earliest date inscribed on an object included in this catalogue—the Reformed Church of Edward VI issued the Book of Common Prayer, adherence to which was compulsory across England and Wales. The Book of Common Prayer provided complete forms of service for daily worship in English (and later also in Welsh), which had a homogenizing effect, eroding both regional dialects and local practices. And yet, religious belief was far from uniform. The seventeenth century saw a sharp rise in nonconformity and—considering the objects in this catalogue as an index—a rise particularly in Quakerism. Such groups eschewed the Church of England and worshipped in their own meeting places, while families who had maintained the Catholic faith had to keep their religious observance secret, often living in fear of discovery and arrest.

Gifts of plate to churches to support the sacramental service of administering communion in the Anglican Church were common forms of bequest among the pious of early modern Britain. As various objects in this section of the catalogue prove, objects regularly moved from private ownership within the home to the church, and it was often upon this occasion that the object was marked with an inscription recording and celebrating the gift (page 359).

Dated objects that displayed loyalty to current or previous regimes were found throughout the country, both within the home and within the halls of livery companies, taverns, and churches. Three monarchical dynasties and one commonwealth republic held power in Britain during this time, which entailed long periods of continuity and relative stability but also a series of violent changes in religious and dynastic regime: the back-and-forth between Protestantism and Catholicism in the mid-Tudor period; the arrival of the Stuart monarchy in 1603; the outbreak of civil war in 1642; the execution of Charles I in 1649; the establishment of the Protectorate in 1653 and its subsequent demise; the Restoration of the Stuart monarchy in 1660; the removal of James II from the throne in 1688; and, finally, the Hanoverian succession in 1714.

James II's preferment of Catholic religion and Catholic subjects created a schism in Britain that reverberated for more than half a century. As objects in this catalogue attest (pages 373 and 378), a large number of people owned objects that commemorated the discovery of the Popish Plot, the failure of the Duke of Monmouth's Rebellion, and the Hanoverian victory at the Battle of Culloden—the final and decisive pitched battle of the Jacobite rebellion that curtailed the aspirations of James II's descendants in their bid to reclaim the throne. The Hanoverian dynasty was secure and, with it, the ancient institutions of the law, the monarchy, and Parliament, which since then have engendered a strong but not entirely justified sense of continuity in British history.

1549

Portrait

Oil on panel

35 × 27½ × 2¼ in. (88.9 × 69.9 × 5.7 cm)

Attributed to William (Guillim) Scrots (active 1537–1553)

Inscription: EDVARDI SEXTI ANGLIE FRNCIE ET/HIBERNICE REGIS VERA EFFIGIES EO PRIM/ TEMPORE QVO REGIA CORONA/ EST INSIGNITVS, ÆTATIS SVE, 10 ANO, 1549.

The date on this frame is incongruous with facts. Edward VI was born on October 12, 1537, at Hampton Court Palace, and therefore he would not have been in his tenth year in 1549, as the inscription suggests. This portrait seems to show Edward as a boy of about eight, which would have been in about 1545. One explanation is that the picture was made in 1549, a copy of an earlier picture, creating confusion about the true age of the king—but this reasoning feels unsatisfactory, particularly given that there are a number of pentimenti in the picture, suggesting that it was not a slavish copy of a preexisting likeness.

Another possibility is the inscription refers to Edward's succession on January 28, 1547, in the tenth year of his life, and that the frame, or at least the inscription, was added at a later date. Another portrait of Edward VI in private ownership (formerly in the collection of the Roberts family of Glassenbury, Kent) bears an inscription on what appears to be the original frame: "EDVARDI SEXTI ANGLIE, FRNCIE ET/HIBERNICE REGIS VERA EFFIGES EO PRIV/TEMPORE QVO REGIA CORONAEST INSIGNTIVS ÆTATIS SVE, 10, ANO, 1547." The similarity of these two inscriptions is striking and points to this treatment as an established convention for frames on portraits of the young king.

Although the background has lost the depth of its original copper-green glaze, the painting is in good condition, made with expensive quality pigments like azurite blue and red lake. A series of changes to the position of both hands, made in a dry drawing medium, suggest that this portrait may have been the primary version, raising the likelihood that it is the work of the salaried court painter William Scrots. Scrots's portrait of Edward VI of about 1547, in the Royal Collection in London, shows the prince with a more mature face than in the Bryan painting, and that work presumably postdates this one by at least a year.

For a number of centuries, this painting was owned by the More family of Loseley Park, but it can only be traced back in inventories to 1777. At some point in this painting's subsequent history, Edward's face was intentionally scratched with a sharp edge about ten times. This type of vandalism is fairly common on portraits of this age, particularly on images of monarchs.

PROVENANCE

Probably either Sir William More (1520–1600) of Loseley Park, Surrey, or his son, Sir George More (1553–1632), and by descent in the More family at Loseley Park, where it is first recorded in an inventory of 1777; Christie's, London, June 9, 2004 (lot 6); bt. John H. Bryan of Lake Bluff, IL (1936–2018) through Gurr Johns, London; bt. John H. Bryan III of Nashville, TN, September 5, 2013

LITERATURE

Gervase Jackson-Stops, ed., *The Treasure Houses of Britain: Five Hundred Years of Private Patronage and Art Collecting* (Washington, DC: National Gallery of Art; New Haven, CT: Yale University Press, 1985), 84

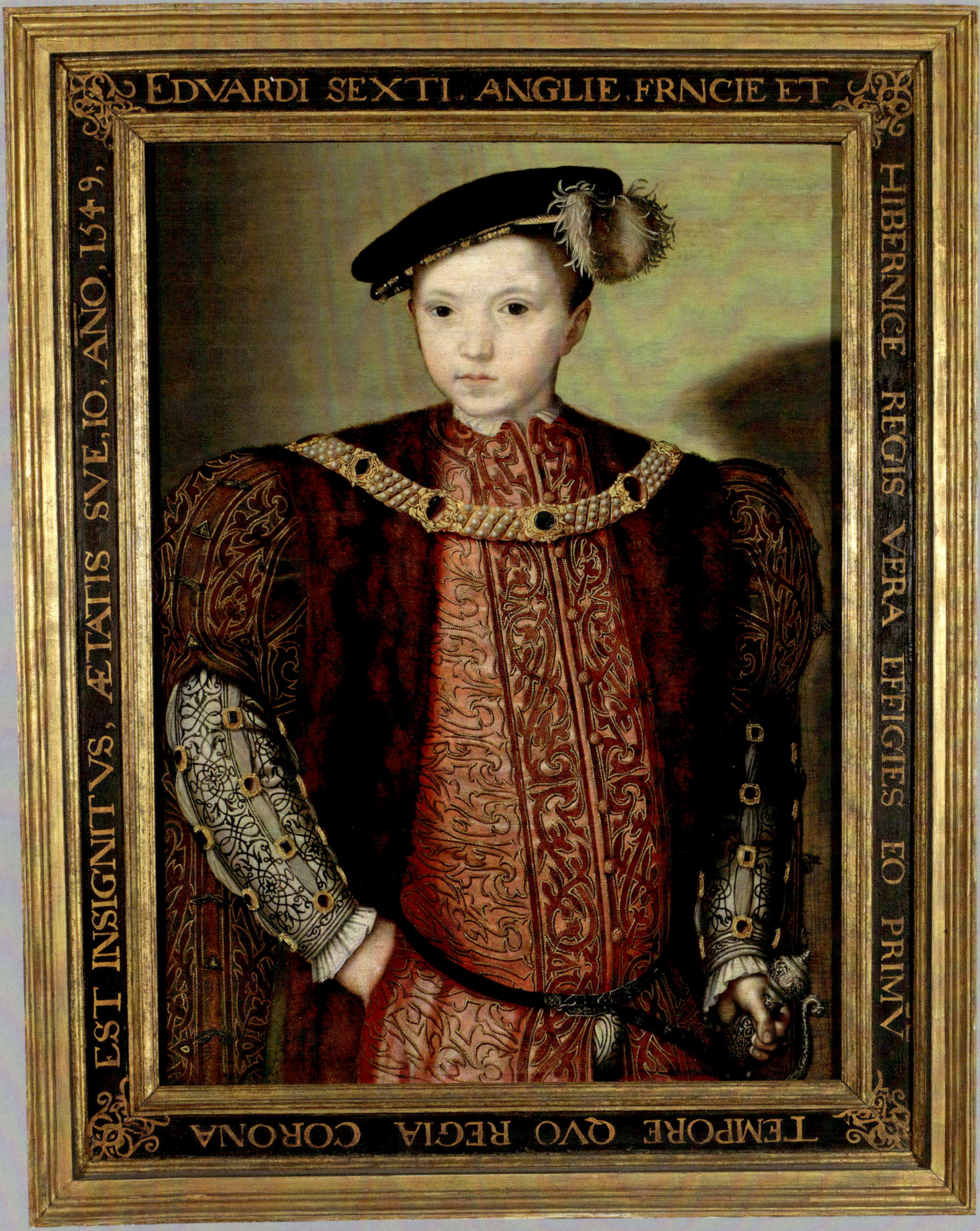
EDVARDI SEXTI. ANGLIE. FRNCIE ET
HIBERNICE REGIS VERA EFFIGIES EO PRIMV
TEMPORE QVO REGIA CORONA
EST INSIGNITVS, ÆTATIS SVE, 10, AÑO, 1549,

1717

Parish chest

Oak, iron

19 × 31½ × 23 in. (48.3 × 80 × 58.4 cm)

Inscription:
W BLANN|I AYLEE; 1717

Presided over by the parish priest, with support from elected and nominated churchwardens, this chest was secured with the use of three locks. Keys were distributed among the serving officials so as to ensure that no one person had sole access to its contents. The characters nailed into this parish chest are difficult to read, making it impossible to identify the vestrymen who inscribed the date and their names.

Along with the plate used to administer the sacrament, the parish chest held the church's records of vestry business and the distribution of alms money to the poor. It also contained the parish registers, from which much of the biographical information in this book is derived. In 1538, Thomas Cromwell mandated the keeping of records of registration of births, marriages, and deaths. Sixty years later, by royal decree, these records were copied into books of parchment. This parliamentary bill stipulated that there should be copies of earlier registers dating from the beginning of the mandate, "but especially since the first year of her Majesty's [Elizabeth I's] reign." For this reason, many parish registers only survive from 1558, the transcriber taking this directive as tacit permission to forgo copying the first twenty years of registers.

PROVENANCE

Mr. and Mrs. Bruce Howard of Taunton, Somerset; Victor Chinnery (1944–2011); bt. John H. Bryan of Lake Bluff, IL (1936–2018) from Victor Chinnery, Quercus Consultants, Wiltshire, July 8, 2004

LITERATURE

W. E. Tate, *The Parish Chest: A Study of the Records of Parochial Administration in England* (Cambridge: Cambridge University Press, 1969), 44–45

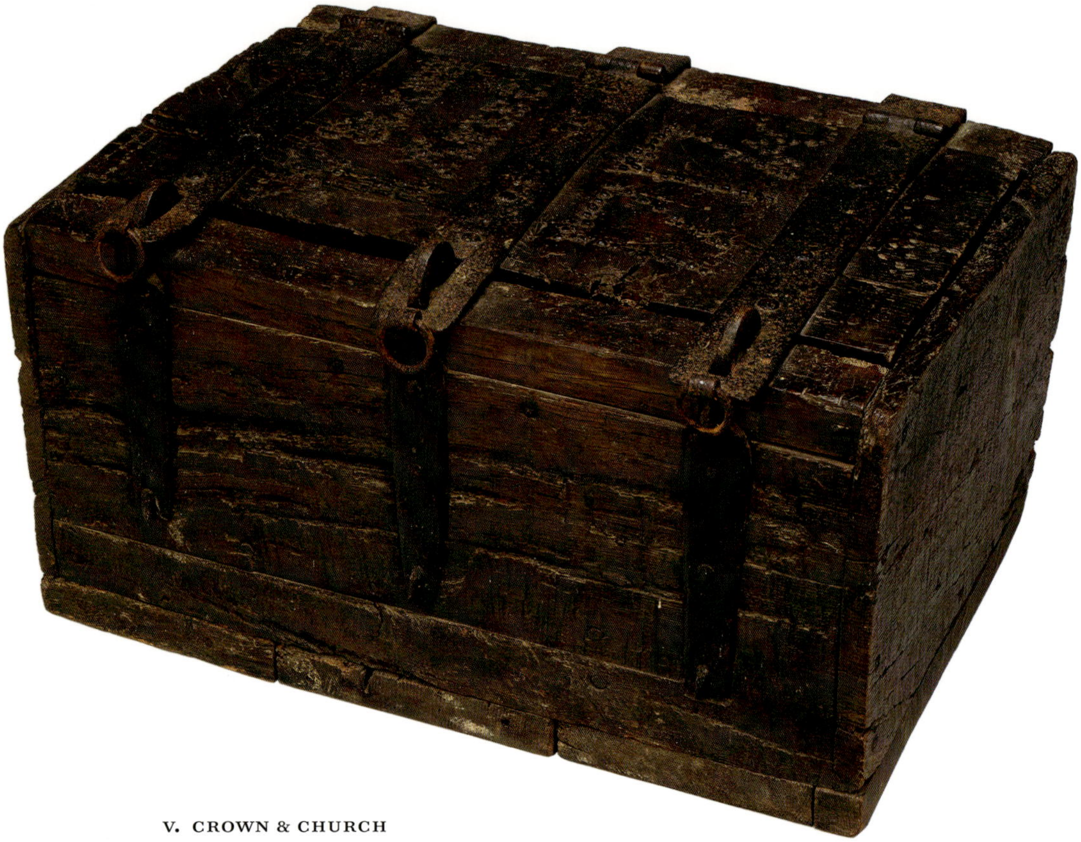

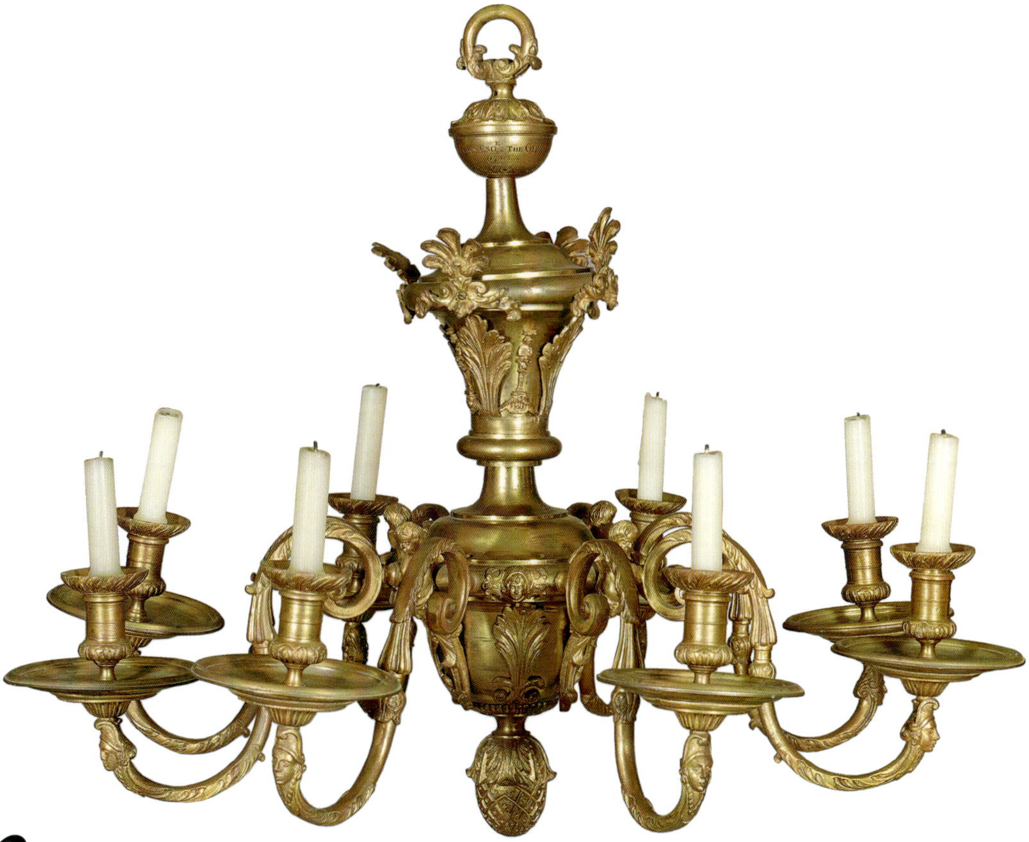

1762

Chandelier
Gilded brass
25 × 31 in. (63.5 × 78.7 cm)

Inscription: THE GIFT OF M[RS]. ANN PEERS 2[d]. WIFE OF CHA[S]. ESQ[R].|1762

Made in the early eighteenth century, this gilded-brass chandelier was a gift to the church of St Katherine of Chiselhampton from Ann, second wife of Charles Peers of Olney, Buckinghamshire. In 1748, Peers purchased the Chiselhampton estate in Oxfordshire from Sir Thomas Doyley. The Peerses were newcomers to Oxfordshire but soon made their presence known with a series of large building projects, including Chiselhampton House, which still stands. In 1763, they rebuilt the chapel of St Mary at Chiselhampton, which, after its completion, was rededicated to St. Katherine and, for the first time in its history, established its own graveyard. Today, the church retains its box pews and is still lit with candle candelabras. The church is famous for its service, featuring silver-gilt chalices and patens and an enormous pair of tankard flagons. Initially, these objects had been gifts of the French Church and Dutch Church to Sir Charles Peers (1661–1737). Peers was a director of the East India Company and the Bank of England, a common councilman for Aldgate Ward, and lord mayor of London. His descendants gave these objects to the church that the family had built at Chiselhampton.

PROVENANCE

Mr. and Mrs. Glenn C. Randall; M. Turpin Antiques, London; bt. John H. Bryan of Lake Bluff, IL (1936–2018) from G. Randall Inc., Washington, DC, August 14, 1992

LITERATURE

Reinier Baarsen and others, *Courts and Colonies: The William and Mary Style in Holland, England and America* (New York: Cooper-Hewitt Museum, 1988), 169, no. 125

Peter Schiffer, Nancy Schiffer, and Herbert Schiffer, *The Brass Book: American, English and European, Fifteenth Century through 1850* (Exton, PA: Schiffer Publishing, 1978), 223

1655

Chalice
Silver
$8\frac{3}{4} \times 5$ in. (22.2 × 12.7 cm)
Attributed to J. Burgis of Mitchel

Inscription: The guift of Ionathan Rowley Gent: who deceased Anno 1655 18^{oz} 18^{dwt}

In 1653, Jonathan Rowley, Draper, who died in 1655, was mayor of Shrewsbury. This chalice is one of a pair marked for 1620, with the initials "JB" for the maker Burgis of Mitchell and engraved with Rowley's arms (Rowley of Wyken and Rowley, Shropshire). Rowley's will, made in February 1654, specified the bequest to St Chad's, Shrewsbury, of "Twentie pounds for two guilt Bowles" made for "the administration of the Sacrament." If the value of the two bowls was less than that amount, Rowley asked that the remainder be spent on a diaper or damask cloth for the Communion table. The ministers of the parish were also bequeathed five pounds apiece (TNA PROB 11/250/509).

PROVENANCE
Sotheby's, Paris, April 12, 2018 (lot 251); bt. John H. Bryan of Lake Bluff, IL (1936–2018) from Peter Cameron, London, April 13, 2018

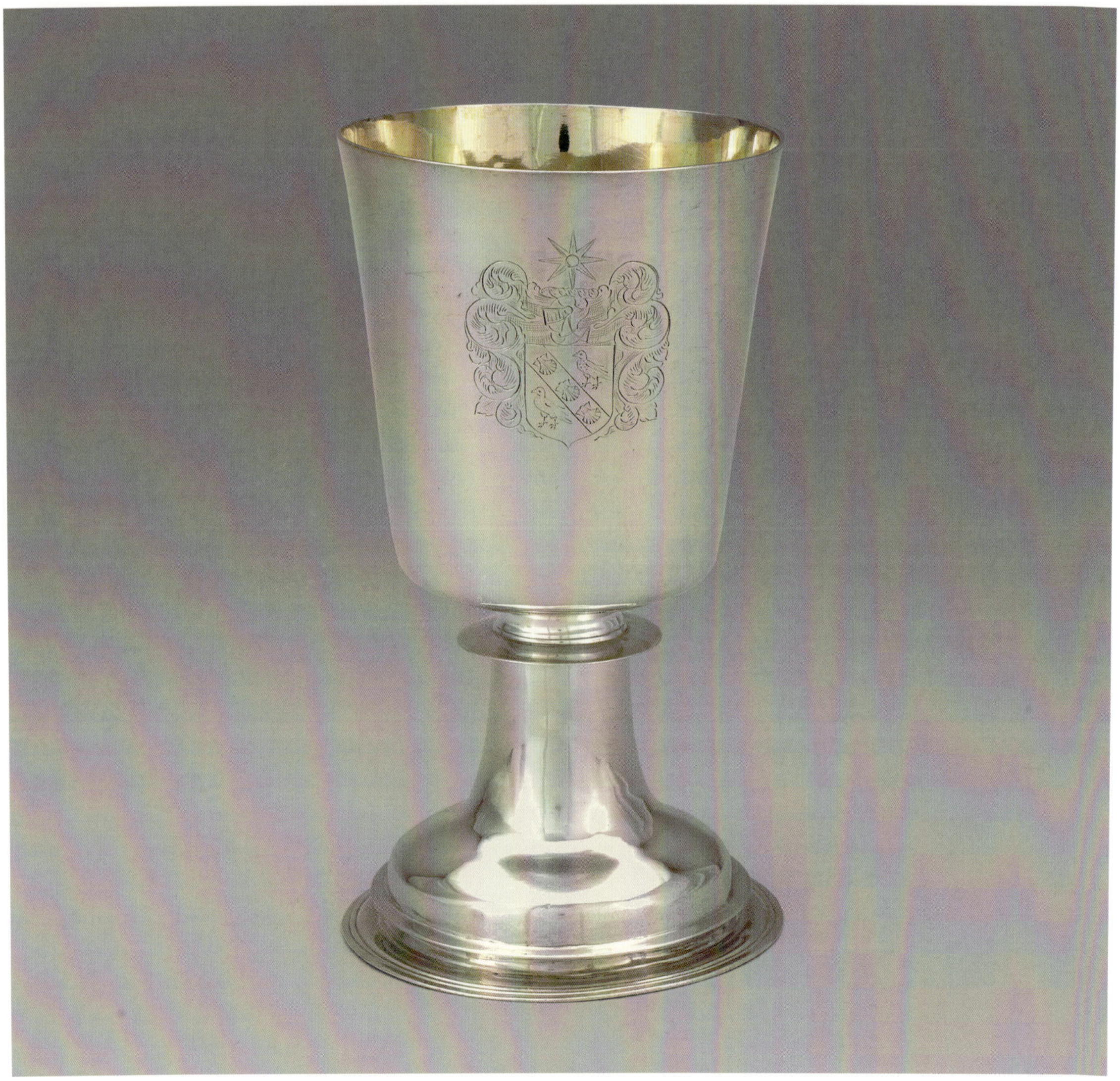

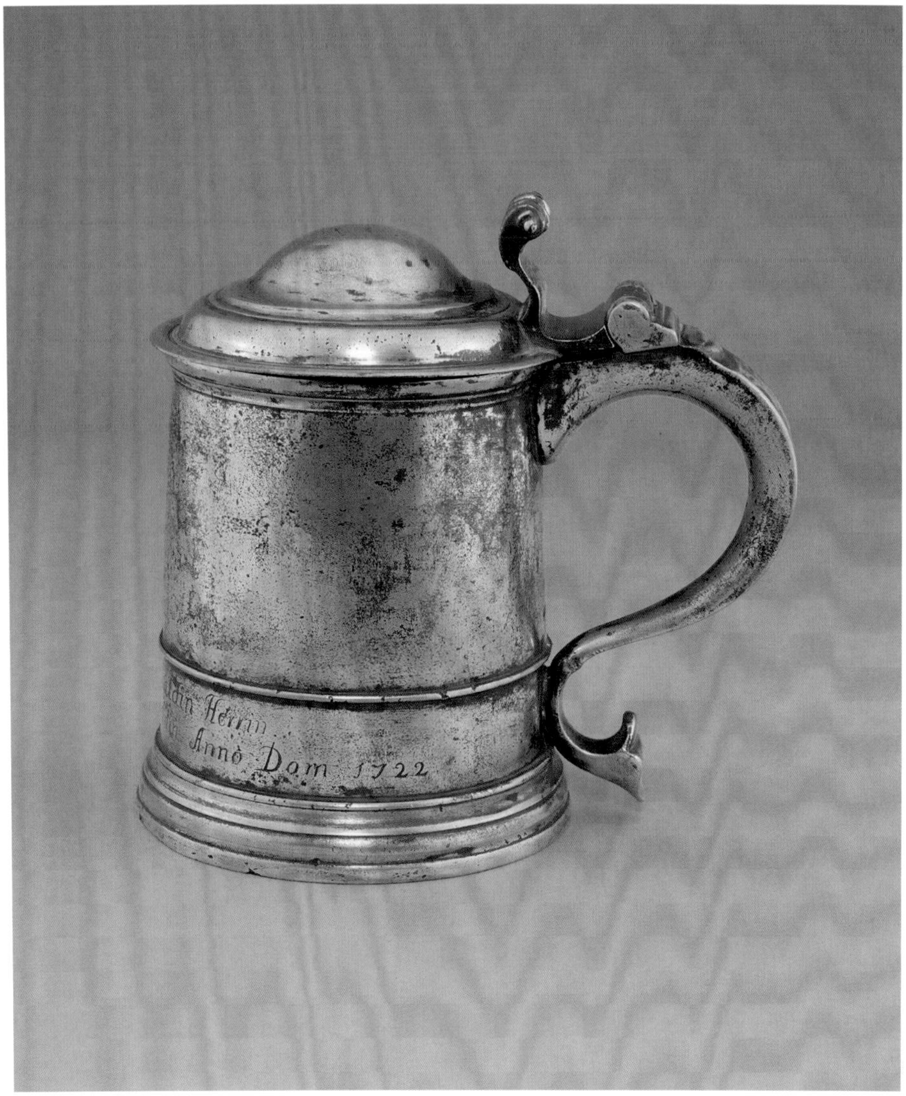

1722

Tankard

Pewter

$7\frac{1}{8} \times 7$ in. (18.1 × 17.8 cm)

Attributed to Thomas Carpenter (active 1713–1740)

Inscription: In usum Ecclesiae Chaldin Herrin | Samuel Cherry Church Wardin Anno Dom 1722

As the inscription states, this tankard was for the use of the church of Chaldon Herring, near Weymouth, Dorset. It was the gift of Samuel Cherry, church warden, probably the man by that name buried in the nearby Bere-Regis on April 20, 1746. The tankard has a dry capacity of 39 ounces (1150 milliliters), thus containing one and a half ale pints or nearly three wine pints (a standard size for Communion tankards). The hallmarks beneath the rim are attributed to the pewterer Thomas Carpenter, who was active in London between 1720 and 1740.

PROVENANCE

A. H. Isher & Son, Cheltenham, February 27, 1953; Frank Holt; his sale, Philips, Chester, October 20, 1998 (lot 71); Bonhams, London, February 24, 2016 (lot 34); bt. John H. Bryan of Lake Bluff, IL (1936–2018) through David Moulson, Alcester, Warwickshire

LITERATURE

Ronald F. Homer, "Thomas Carpenter," *Journal of the Pewter Society* 9 (Spring 1993): 18–19

Christopher Peal, *Pewter of Great Britain: For Pleasure and Investment* (London: John Gifford, 1983), 104

1695

Communion flagons

Brass

Each: 14 × 6 ¾ × 9 ½ in. (31.1 × 17.1 × 24.1 cm)

Inscription, on both: 1695

This pair of Communion flagons have the date "1695" engraved into their undersides, presumably recording the year the gift was given. The flagons have a dry capacity of 41 ounces (1225 milliliters), or three wine pints.

PROVENANCE

bt. John H. Bryan of Lake Bluff, IL (1936–2018) from Brian Beet, London, May 7, 2002

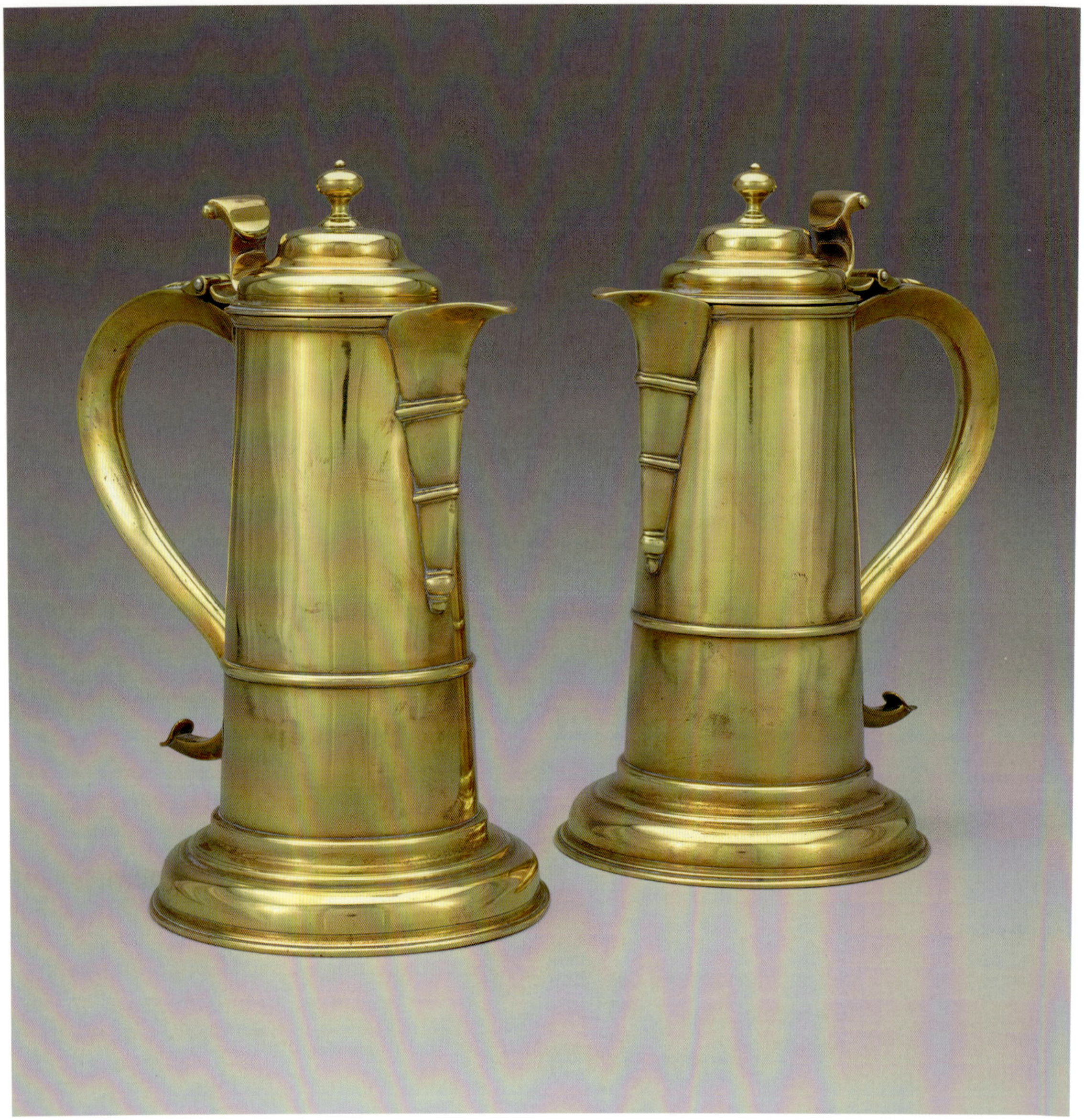

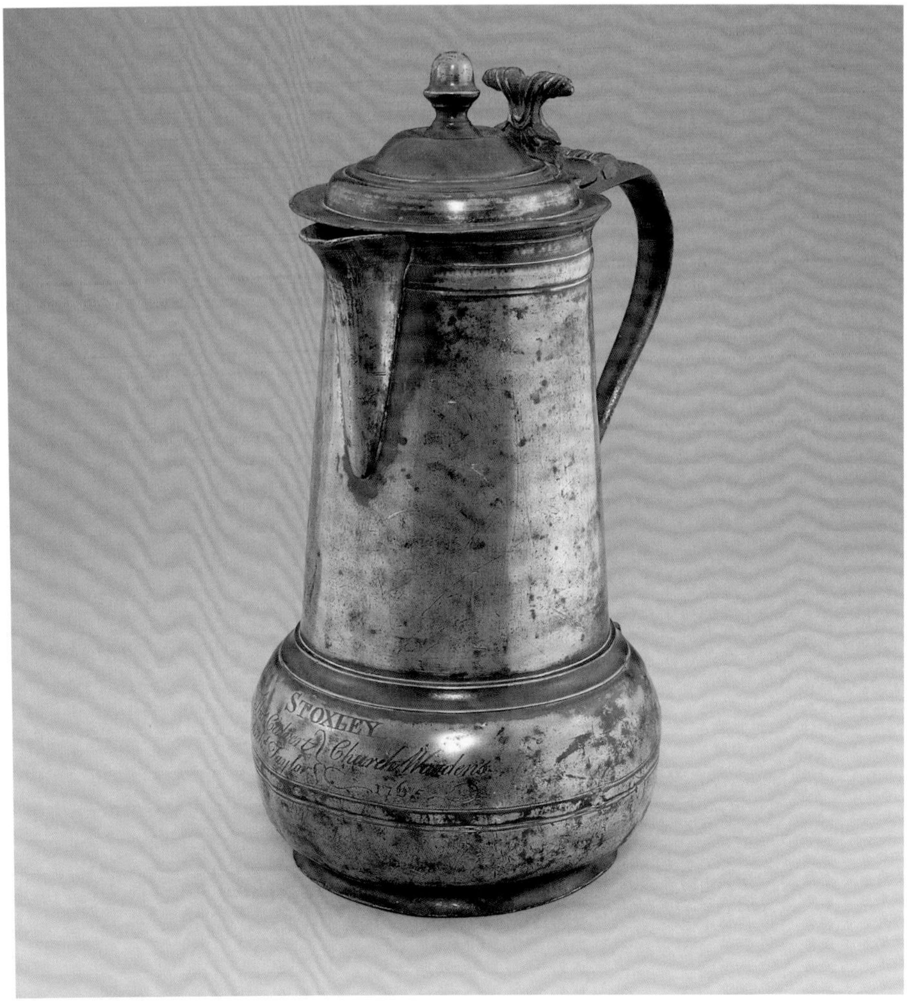

1765|1968

Communion flagon

Pewter

14 × 7 × 9½ in. (35.6 × 17.8 × 24.1 cm)

Inscription: STOXLEY | Thos Calvert Church Wardens|Richd Taylor 1765; 153| Stokesley|Holt|1968

This flagon could hold a gallon of wine (131.9 ounces, or 3900 milliliters) for the Communion service. It was formerly the property of the church of St Peter and St Paul, Stokesley, North Yorkshire.

PROVENANCE

Frank Holt; his sale, Philips, Chester, October 20, 1998 (lot 33); Bonhams, London, February 24, 2016 (129); bt. John H. Bryan of Lake Bluff, IL (1936–2018) through David Moulson, Alcester, Warwickshire

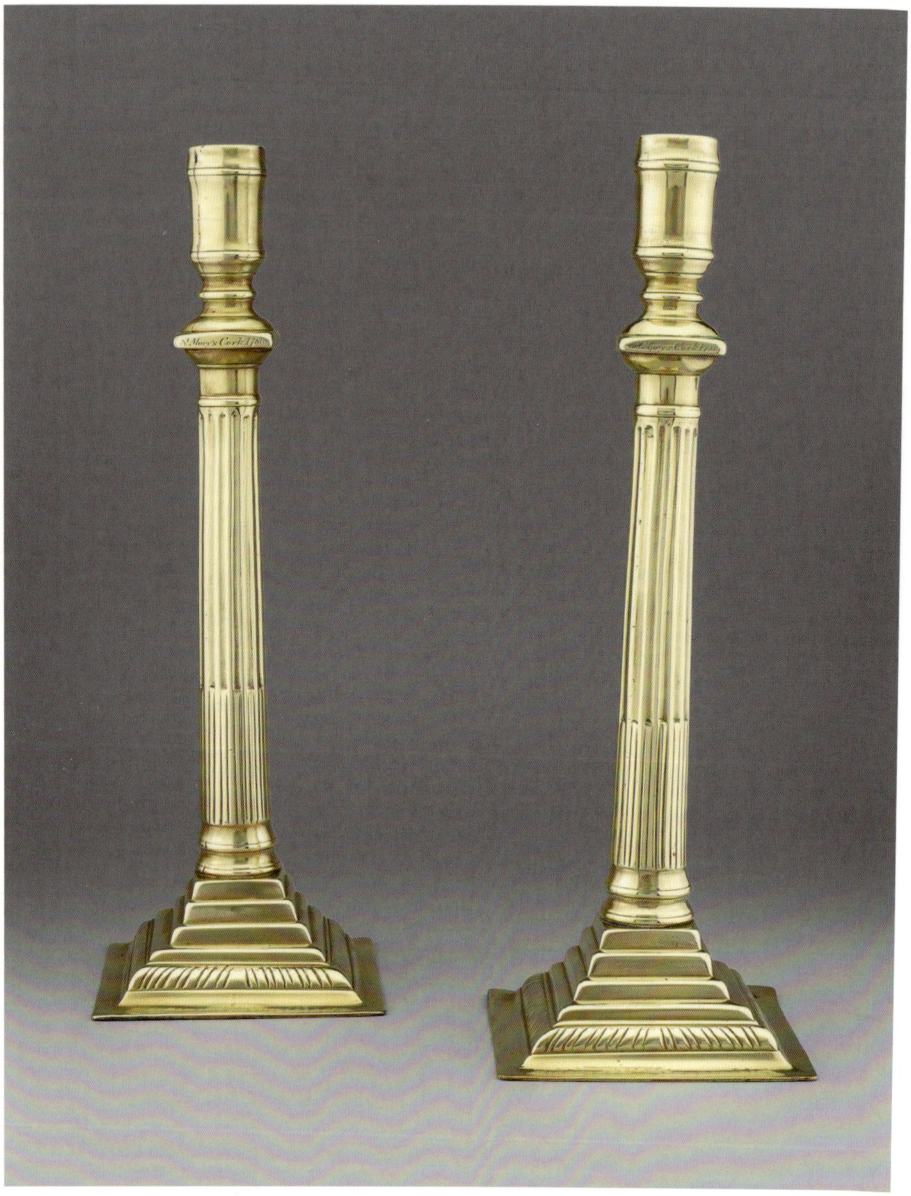

1786

Candlesticks

Brass

Each: 20 3/8 × 7 in. (51.8 × 17.8 cm)

Inscription, on both: St Mary's Cork 1786

Candlesticks for the altar were frequently donated to the church, and this style of candlestick became very popular in the late 1760s. They are of a neoclassical design, characterized by a stem that has the form of a straight column and a socket that is shaped like a Corinthian or Doric capital. In 1786, they were donated to St Mary's Cork, Ireland.

PROVENANCE

bt. John H. Bryan of Lake Bluff, IL (1936–2018) from Alistair Sampson Antiques, London, August 11, 1983

1705

Communion paten

Silver

$2\frac{3}{8} \times 8\frac{1}{2}$ in. (6 × 21.6 cm)

John Bodington (active 1697–1727), 1697

Inscription: Ad usus sacros Ecclesia Nun Etonensis in agro Warwici hanc patinam deo dicavit Anna Trotman 1705

The inscription on this Communion paten, or tazza, indicates that Anne Trotman gave it to the church at Nuneaton in Warwickshire, presumably in memory of her parents. Her father, Anthony, had died in 1703, and his wife, Abigail (Anne's mother), in 1705. Anthony, who had been born in Bishopston in Wiltshire, bequeathed moiety (or half share) in his household goods and plate to his wife (TNA PROB 11/472/280). In turn, Abigail's bequests to Anne in 1705 of her silver punch bowl and her largest silver salver suggest she had a substantial collection of plate (TNA PROB 11/482/102). After making a series of bequests of cash to her other children, Abigail stipulated that the rest of her possessions should go to her daughters Anne and Frances, who later bequeathed to the church patens that presumably had served as salvers in the Trotman household, to be repurposed for ecclesiastical use. The silver trays were made by John Bodington (active 1697–1727) of the Mitre, Foster Lane, who was beadle of the Goldsmiths' Company from 1714 to 1727. Bodington also produced flagons for Ashford, Kent; a paten for Woodchurch (1707); a Communion flagon for North Cerney in Gloucestershire; and a cup and cover, called the Boyle Cup, at Trinity College, Cambridge.

PROVENANCE

Bonhams, London, April 5, 2017 (lot 90); bt. John H. Bryan of Lake Bluff, IL (1936–2018) through Peter Cameron, London

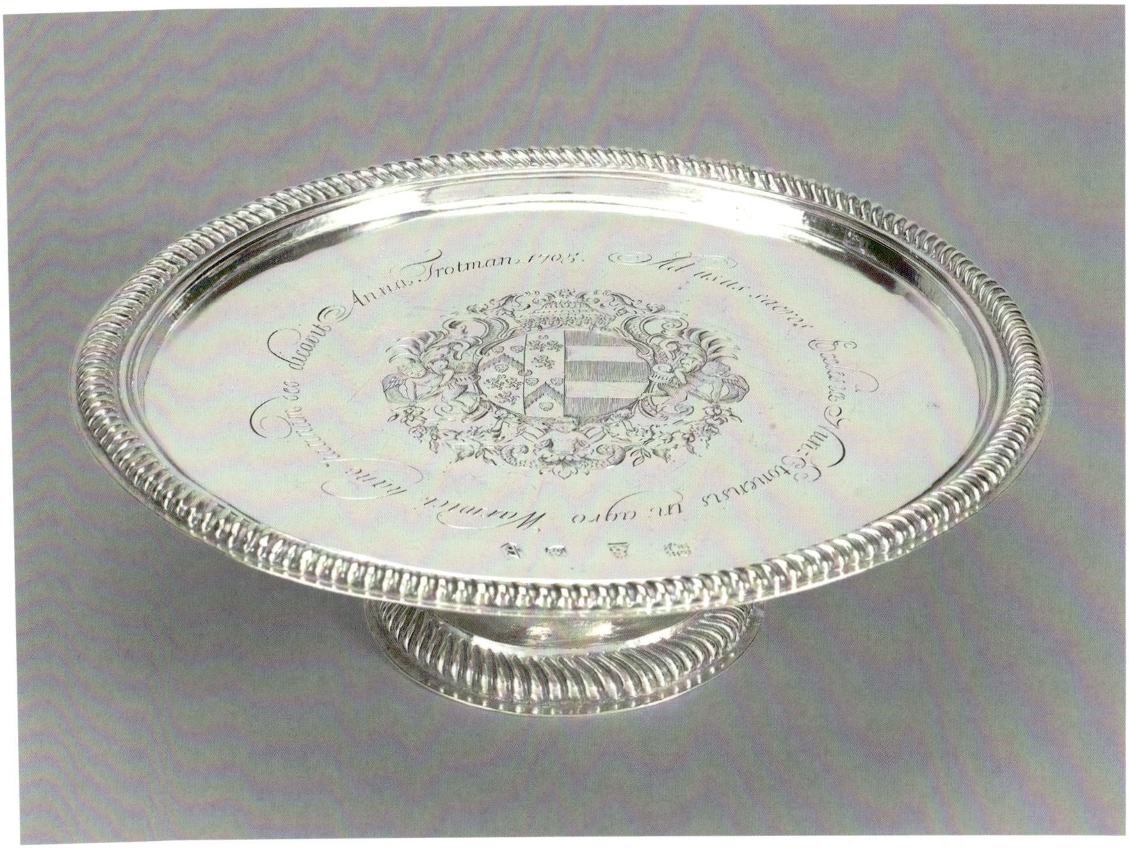

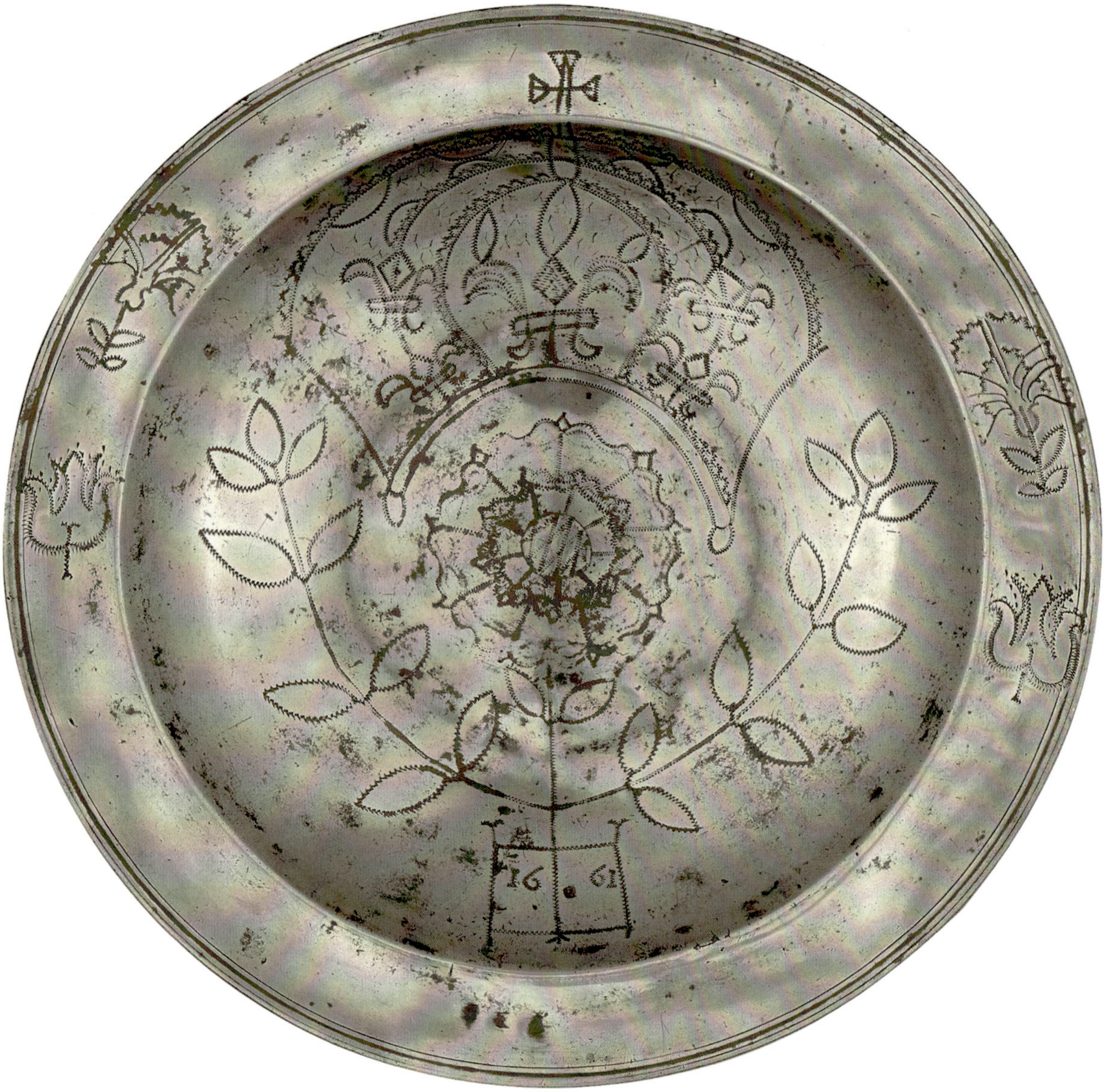

1661

Dish
Pewter
$1\frac{7}{8} \times 14\frac{1}{2}$ in. (4.8 × 36.8 cm)

Inscription: 16 61

The date on this pewter dish commemorates the coronation of Charles II on St. George's Day (April 23), 1661. The Stuart device of a crowned Tudor rose upon a stem alludes to the event.

PROVENANCE

Sotheby's, London, July 31, 1981 (lot 48); bt. John Bryan III of Nashville, TN, from David Moulson, Alcester, Warwickshire, October 25, 2016

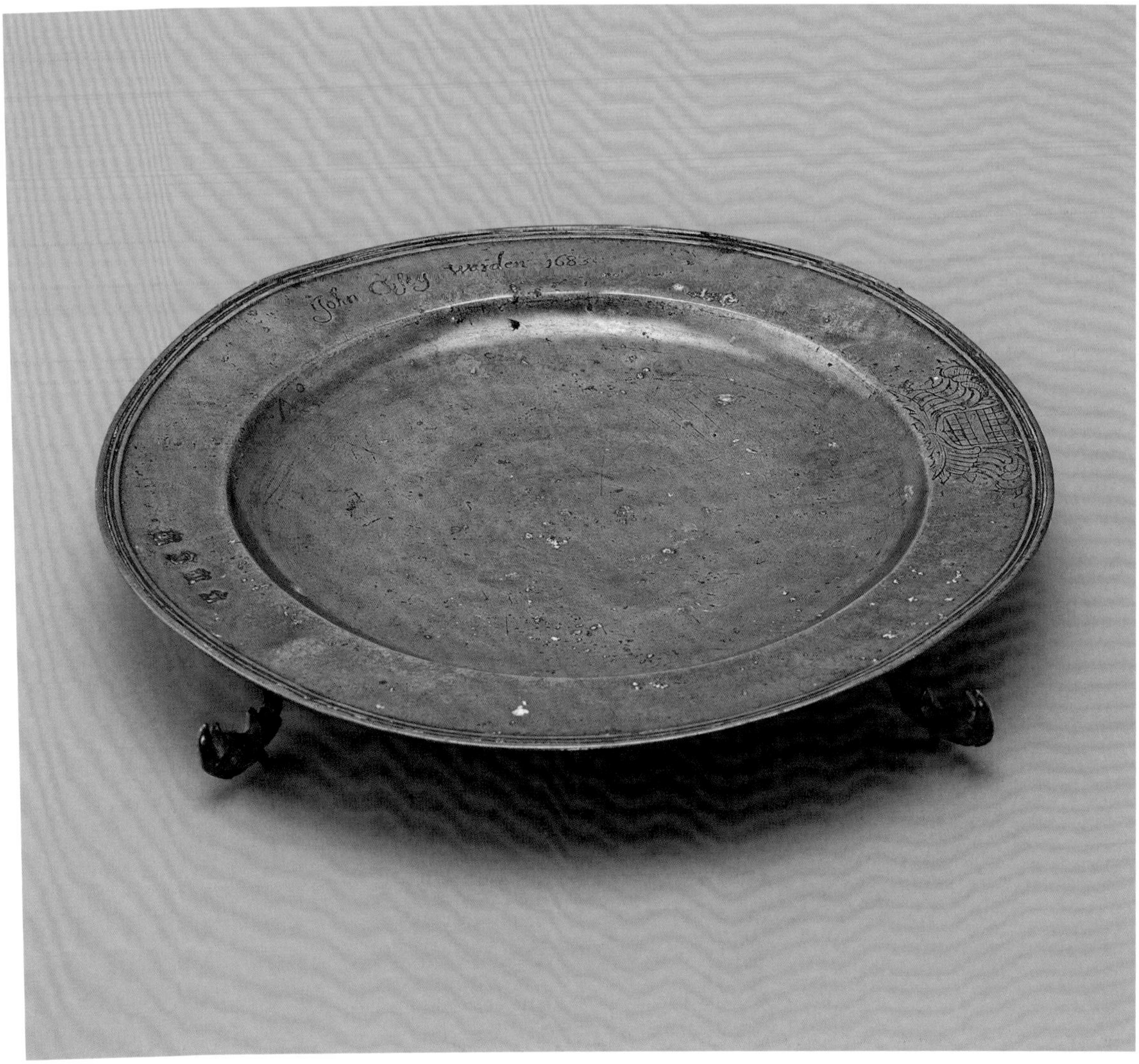

1683

Alms dish

Pewter

$1\frac{1}{2} \times 10\frac{5}{8}$ in. (3.8 × 27 cm)

Inscription: John Cosey Warden 1683

Alms dishes were used in churches to collect money for the poor, which was then administered by the church wardens of the parish. This alms dish was the gift of the Citizen and Pewterer John Cowsey, or Cosey or Coursey (d. 1710), of Drury Lane and St-Giles-in-the-Fields, and may have been of his own making. Between 1662 and 1678, Cowsey took on a series of apprentices. In 1681, he was one of the witnesses who gave evidence in the inquest surrounding the Popish Plot. His will was proved at the Prerogative Court of Canterbury on May 26, 1710 (TNA PROB 11/515/253).

PROVENANCE

Isher Collection; bt. John H. Bryan of Lake Bluff, IL (1936–2018) from Alistair Sampson Antiques, London, August 6, 1987

LITERATURE

Historical Manuscripts Commission, *The Manuscripts of the House of Lords 1678–1688* (Her Majesty's Stationery Office: London, 1887), 46, 48

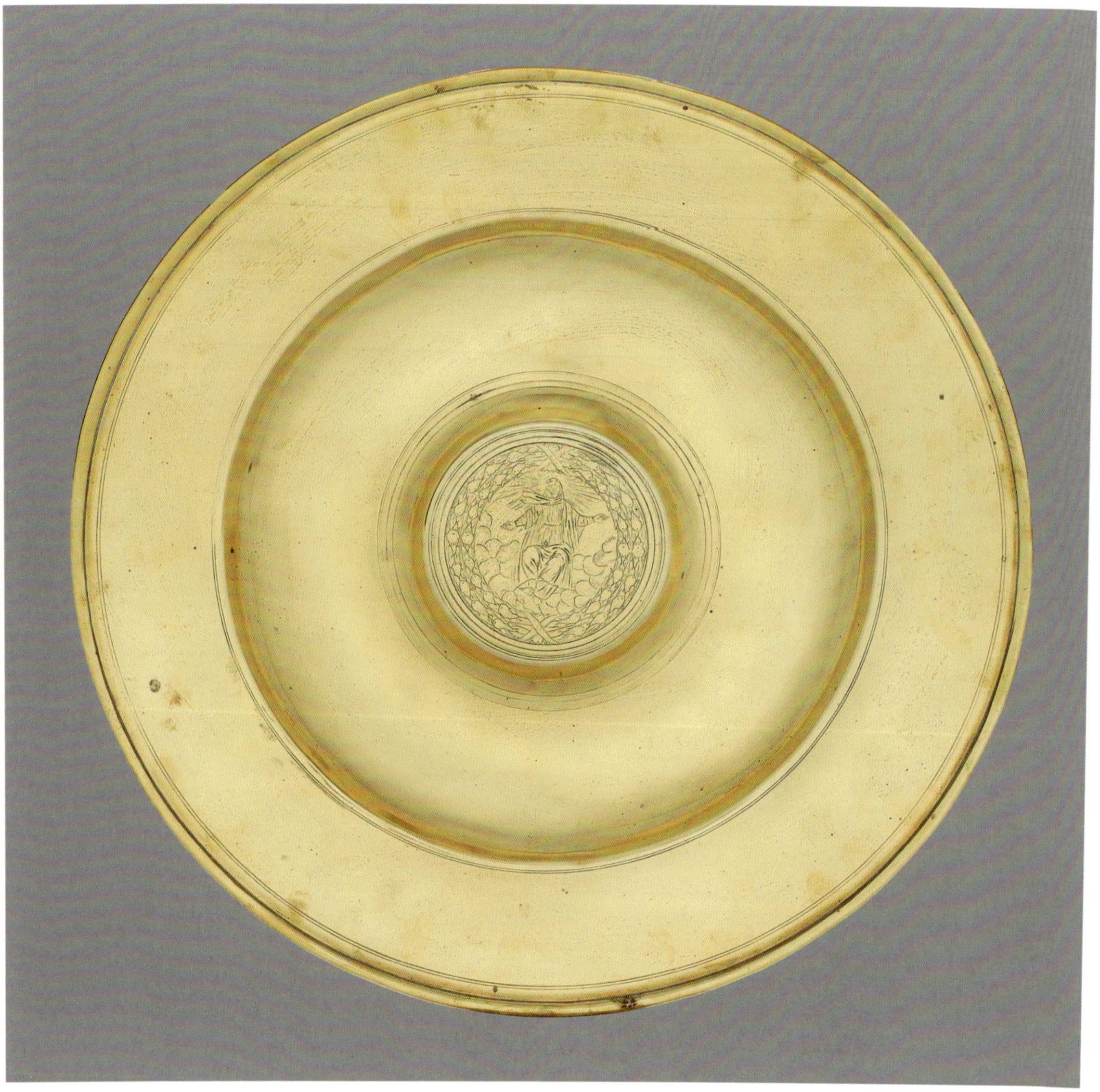

1710

Alms dish

Brass

$1\frac{3}{4} \times 19$ in. (4.5 × 48.3 cm)

Inscription, on verso: ANNO DOMINI|MDCCX

The raised center of this impressively large alms dish is engraved with a depiction of the Virgin Mary seated in the clouds within a ribbon-tied laurel wreath. Originally silvered, the inscription giving the date is on the verso.

PROVENANCE

bt. John H. Bryan of Lake Bluff, IL (1936–2018) from Rupert Gentle, Pewsey, Wiltshire, January 7, 1983

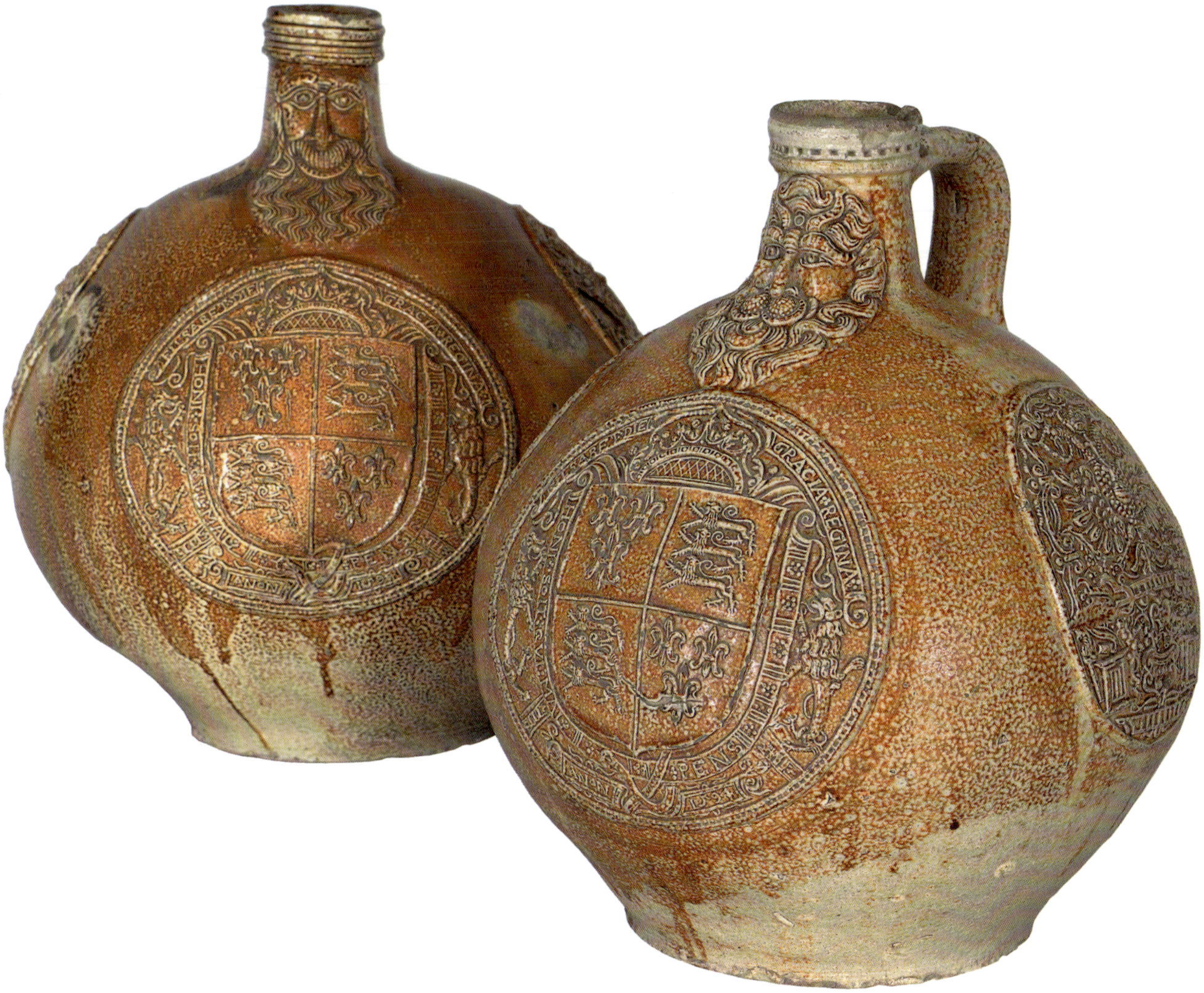

1594

Bellarmine bottles

Stoneware

Left: 13¾ × 10½ in. (34.9 × 26.7 cm)

Right: 15¾ × 12 in. (40 × 30.5 cm)

Capacity (each): 24 pints (11250 ml)

Inscription: ELISABET DIE GRACIA REGINA; ANON 1594; HONI SOIT QVI MAL PENS

Robust German stoneware bottles of this design survive in considerable numbers, but very few of the bearded figures adorning their necks have such jolly expressions on their faces. As to how this style of bottle got its curious name, one theory is that the bottles were intended to ridicule the Counter-Reformation priest Cardinal Bellarmine (1542–1621). Skeptics of this theory instead call them "Bartmann jugs," from the German term *Bartmänner*. These examples bear the queen's coats of arms and that of the Holy Roman Emperor. Molds for similar coats of arms have been discovered in Raeren, now part of modern Belgium.

PROVENANCE

Sotheby's Contents of Littlecote House, Wiltshire, November 20–22, 1985 (lots 4 and 5); bt. Jonathan Horne, London; Private Collection; bt. John H. Bryan of Lake Bluff, IL (1936–2018) from Jonathan Horne, London, December 19, 2000

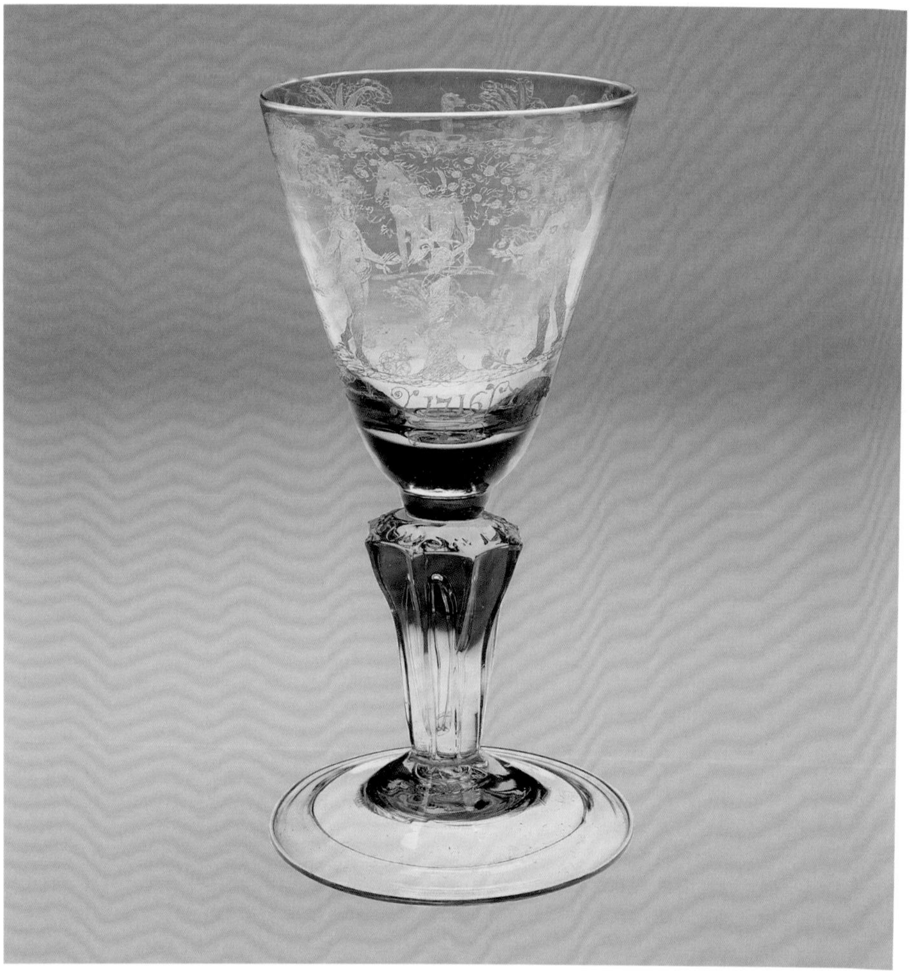

1716

Glass

$8\frac{1}{2} \times 3$ in. (21.6 × 7.6 cm)

Inscription:
GOD SAVE KING GEORGE|1716

This scratch-engraved goblet commemorating the coronation of King George I is made of blown and blow-molded, colorless lead glass. The diamond-point engraving depicts Adam and Eve in the Garden of Eden. A crowned serpent is entwined around the tree trunk. The moment shown is the Fall, when Adam accepts from Eve the forbidden fruit from the Tree of the Knowledge of Good and Evil. This is a typical depiction of this scene, although Adam and Eve are usually shown on opposite sides of the tree. Below the scene, which features eleven animals (including a camel, stag, lion, spotted ox, boar, fox, dog, squirrel, doe, and crouching rabbit), is engraved the date.

PROVENANCE

Dr. Burd before 1857; The Misses E. M. C. and E. N. A. Atkinson and Mrs. M. K. Gedge, Christie's London, October 2, 1974 (lot 107); Richard Emanuel, London; Eila Grahame, London; Alastair McAlpine, West Green House, Hartley Wintney, Hampshire; Anonymous sale, Sotheby's, London, September 15, 1992 (lot 36); bt. John H. Bryan of Lake Bluff, IL (1936–2018) from Sheppard and Cooper Ltd., London, September 15, 1992; Art Institute of Chicago, IL, gift of Mr. and Mrs. John H. Bryan, 2012

LITERATURE

Dwight P. Lanmon, *The Golden Age of English Glass: 1650–1775* (Woodbridge, UK: Antiques Collectors' Club, 2011), 148–53, no. 45

1666

Mortar

Bronze

$4\frac{1}{2} \times 5\frac{3}{4}$ in. (11.4 × 14.6 cm)

Edward Neale (active 1641; d. 1695)

Inscription: E N 1666

Edward Neale, the founder who made this mortar, was probably the son of the Henry Neale, who had a foundry in Somerford Keynes and cast a bell dated 1637. Subsequently, Henry Neale moved his business to Burford, where this mortar would have been made.

There are at least twenty extant bells by Edward Neale, including those for the Church of St James the Great, Fulbrook, Oxfordshire, in 1649 and 1662. Like many bell founders, Neale also made mortars; another example, dated 1659 and made for Francis Keble, is now in the Tolsey Museum, Burford.

Neale made his will on November 17, 1695, and "very sick and weake in body," he signed his name with a trembling rendition of his initials. He left his grandson Henry Wills all the tools, bell metal, and bells "belonging to my shop and workehouse." An inventory of his goods made on November 25, 1695, valued his wearing apparel and the money contained within it at three pounds. The contents of the workshop included a bell, six pounds of pewter, thirty-two pounds of "fruffe," and sixty-eight pounds of lead, anvils, bellows, and tools. Presiding over all those was a clock described in the inventory as "old," suggesting that at the time of Neale's death it may already have been in the shop for some time.

PROVENANCE

Robin Wigington (1932–2002); Woolley & Wallis, Salisbury, April 12, 2017 (lot 65); bt. John H. Bryan of Lake Bluff, IL (1936–2018)

LITERATURE

Henry Thomas Ellacombe, *The Church Bells of Gloucestershire* (Exeter, UK: William Pollard, 1881), 9

Alfred Heneage Cocks, *The Church Bells of Buckinghamshire: Their Inscriptions, Founders, and Uses, and Traditions* (London: Jarrold and Sons, 1897)

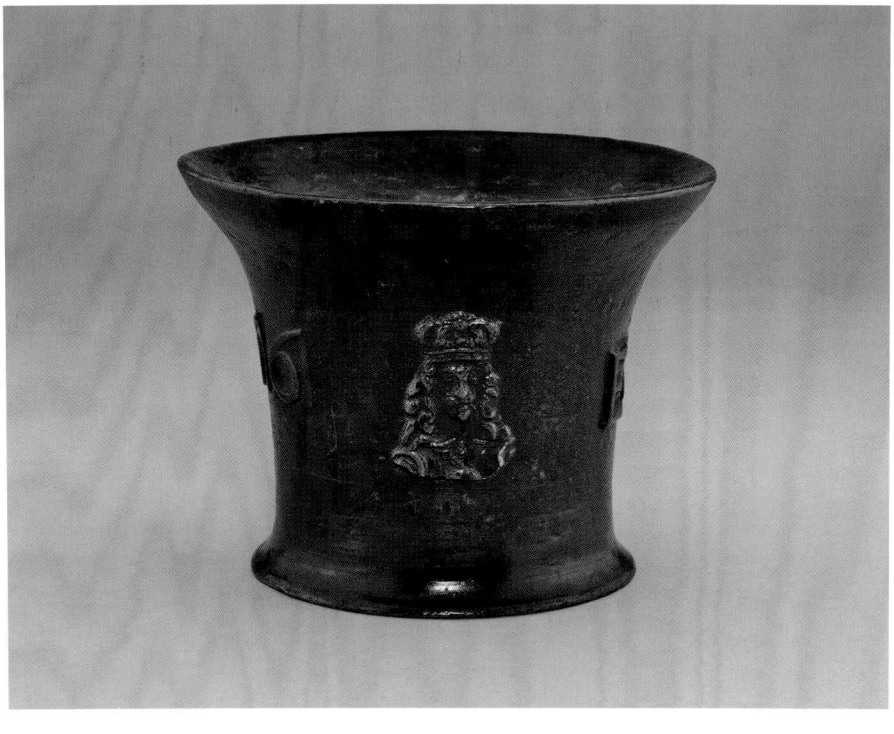

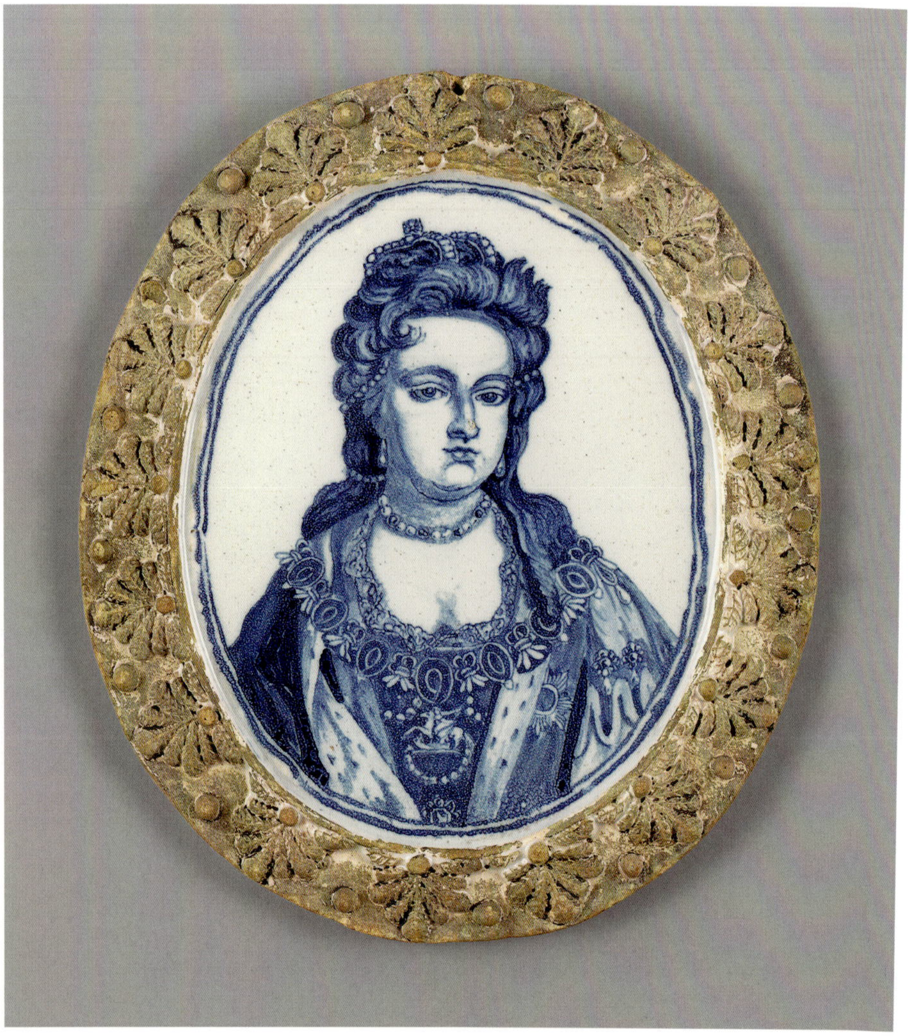

1704

Plaque
Delftware
9 ¼ × 8 × ¼ in.
(23.5 × 20.3 × .64 cm)

Inscription, on verso: 1704

This blue-and-white painted plaque, which is dated 1704 on the verso, portrays Queen Anne in her robes of state. The likeness is most probably taken from a mezzotint made after 1702 by John Smith, after Sir Godfrey Kneller. Another example, dated 1704 and almost certainly by the same hand, is in the Cuckfield Museum Trust (C.48). There is also a very similar portrait plaque of Anna Sophia of Hanover in the Fitzwilliam Museum, Cambridge (C.1356-1928). It, too, is dated 1704 on the verso and is also presumably by the same maker.

It seems the biscuit frame was perhaps gilded at one time, but traces of black enamel suggest it was repainted on Queen Anne's death in 1714.

PROVENANCE

Anonymous sale, Sotheby's, London, June 4, 1968 (lot 134); Mrs. Coleman Curtis, Virginia; Price Glover, New York; Christie's, London, June 14, 1988 (lot 9); Sainsbury sale, Christie's, London, June 18, 2008 (lot 141); bt. John H. Bryan of Lake Bluff, IL (1936–2018) through Sampson & Horne Antiques, London

LITERATURE

Michael Archer, *Delftware in the Fitzwilliam Museum* (London: Philip Wilson Publishers, 2013), 343–44

Louis L. Lipski and Michael Archer, *Dated English Delftware: Tin-Glazed Earthenware, 1600–1800* (London: Sotheby Publications, 1984), 403, no. 1684

1728

The date on this embroidery has led to the suggestion that this picture depicts George II, Queen Caroline, and Frederick, Prince of Wales, but it more likely depicts Queen Anne and her consort, Prince George of Denmark, Duke of Cumberland, accompanied by a young figure possibly meant to represent their son, Prince William, Duke of Gloucester (1689–1700). But what would have been known by 1728 is that there were no surviving children, despite Anne's seventeen pregnancies. After her death in 1714, she passed the throne to her second cousin, George of Hanover.

PROVENANCE

bt. John H. Bryan of Lake Bluff, IL (1936–2018) from Mallett & Son Antiques Ltd., London, May 25, 2006

Embroidered picture

Silk with silk thread

11 ¼ × 15 ¼ in. (28.6 × 38.7 cm)

Inscription:
E 1728 A

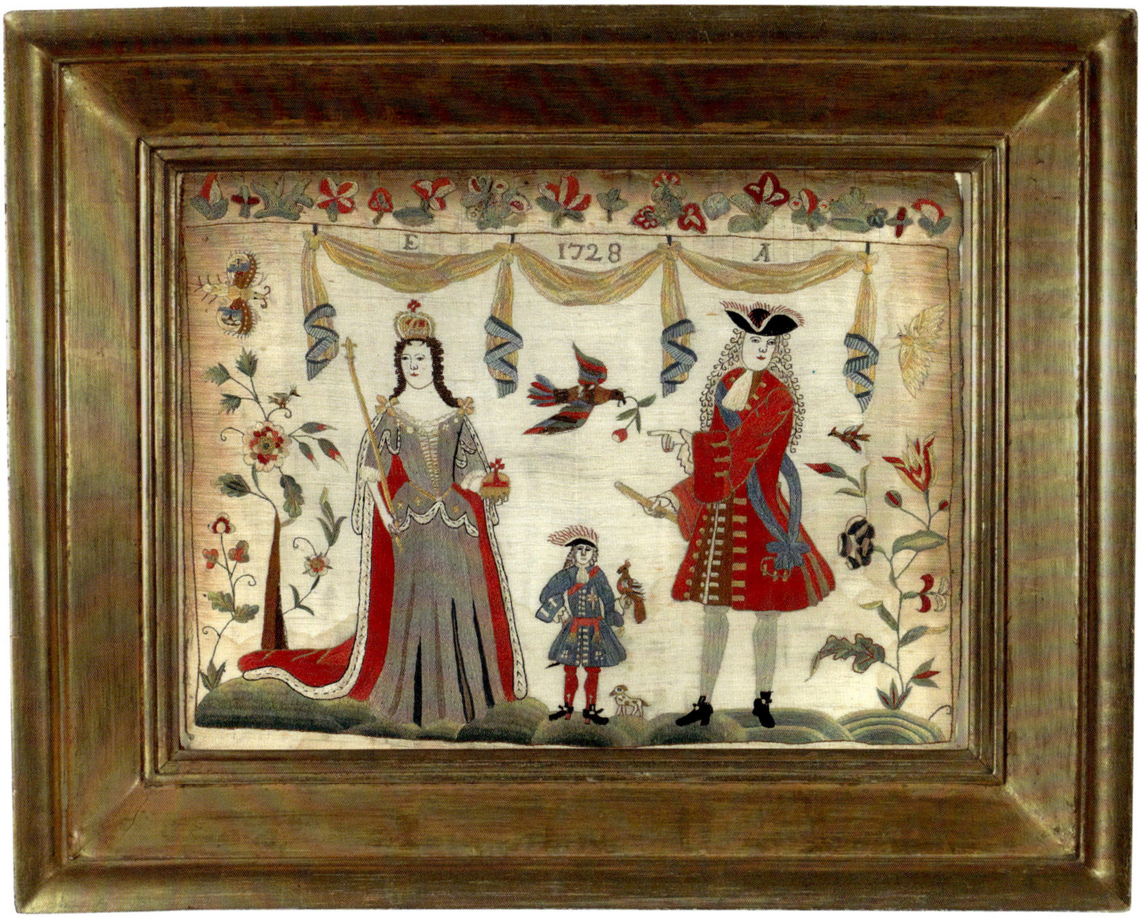

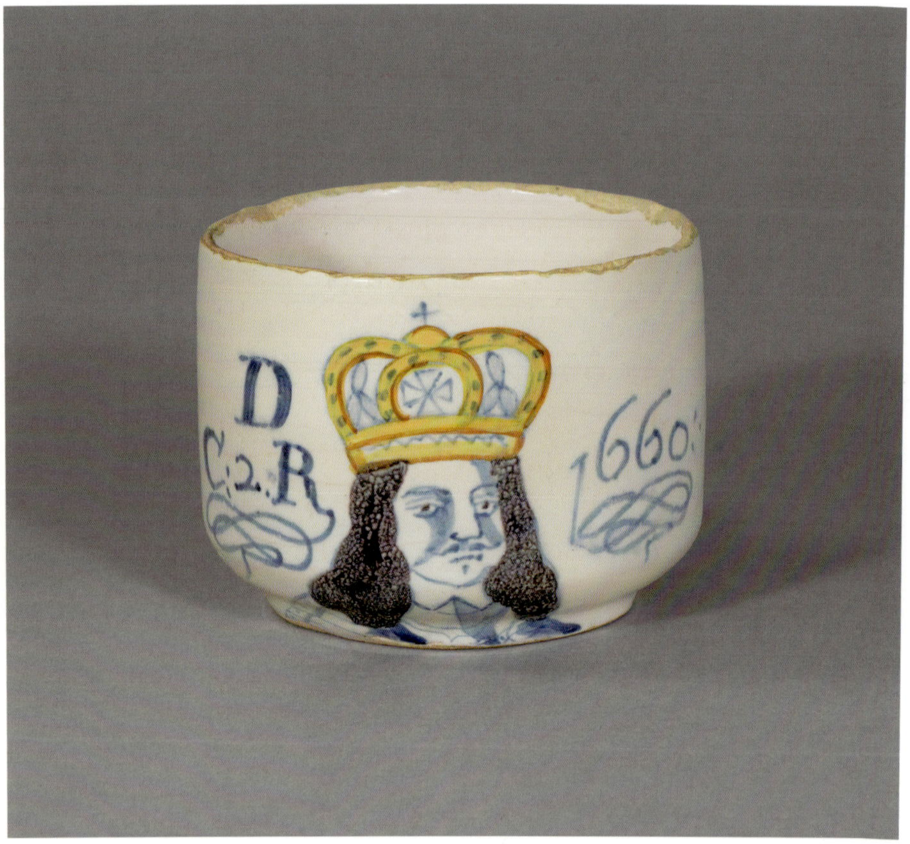

1660

Cup
Delftware
$2\frac{5}{8} \times 3\frac{3}{8} \times 4\frac{5}{8}$ in.
(6.7 × 8.6 × 11.7 cm)

Inscription:
C2|D|R; 1660

This cup was made in the year that Charles II returned from exile and was restored to the throne. By his estimation, he had never left and so considered 1660 to be the twelfth year of his reign.

PROVENANCE
bt. John H. Bryan of Lake Bluff, IL (1936–2018) from Alistair Sampson Antiques Ltd., March 19, 1999

1673

Although chargers were occasionally used as part of the table setting, their size and pictorial potential made them particularly well suited for display in demonstrative shows of allegiance to the crown.

PROVENANCE
Christie's, London, May 7, 1999 (lot 501); bt. John H. Bryan of Lake Bluff, IL (1936–2018) through Alistair Sampson Antiques Ltd., London

Charger
Delftware
3 × 15½ in. (7.6 × 39.4 cm)

Inscription: C^2 R^x; 1673

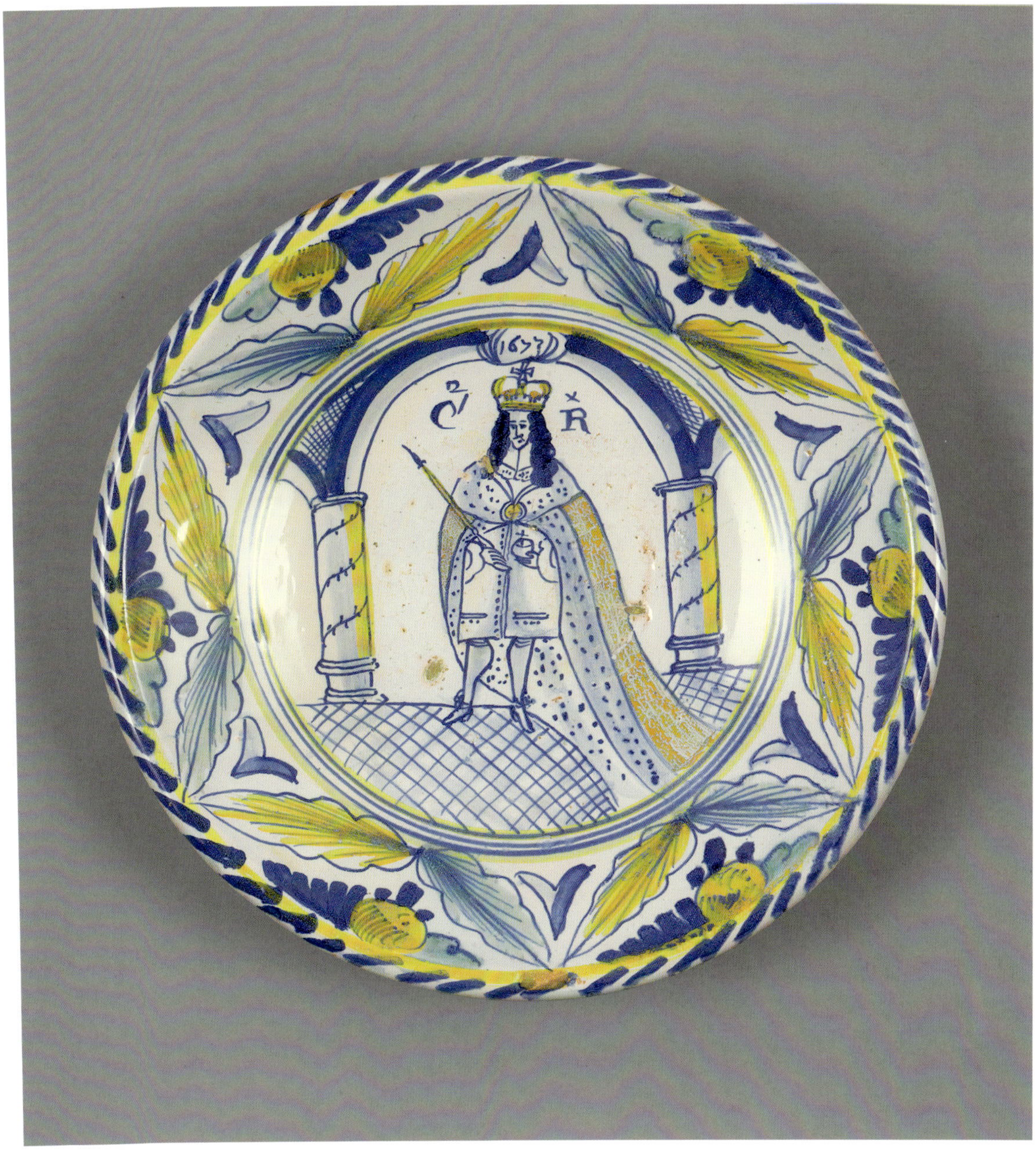

1691

Plate
Defltware
1 × 8½ in. (2.5 × 21.6 cm)

Inscription:
W|M|R; 16 M W 91

The letters on this plate stand for "William" and "Mary, Regibus," with the triad of initials giving precedence to Mary as the monarch and William as her consort. The two ruled together from Mary's accession to the throne in 1689 until her death in 1694. This design was a popular choice for painted Delftware plates, sometimes incorporating the initials of the owners.

PROVENANCE
W. A. H. Harding Trust; Christie's, London, October 10, 1983 (lot 28); bt. John H. Bryan of Lake Bluff, IL (1936–2018) through Alistair Sampson Antiques Ltd., London

LITERATURE
Kieron Tyler, Ian Betts, and Roy Stephenson, *London's Delftware Industry: The Tin-Glazed Pottery Industries of Southwark and Lambeth* (London: Museum of London Archaeology Service, 2008)

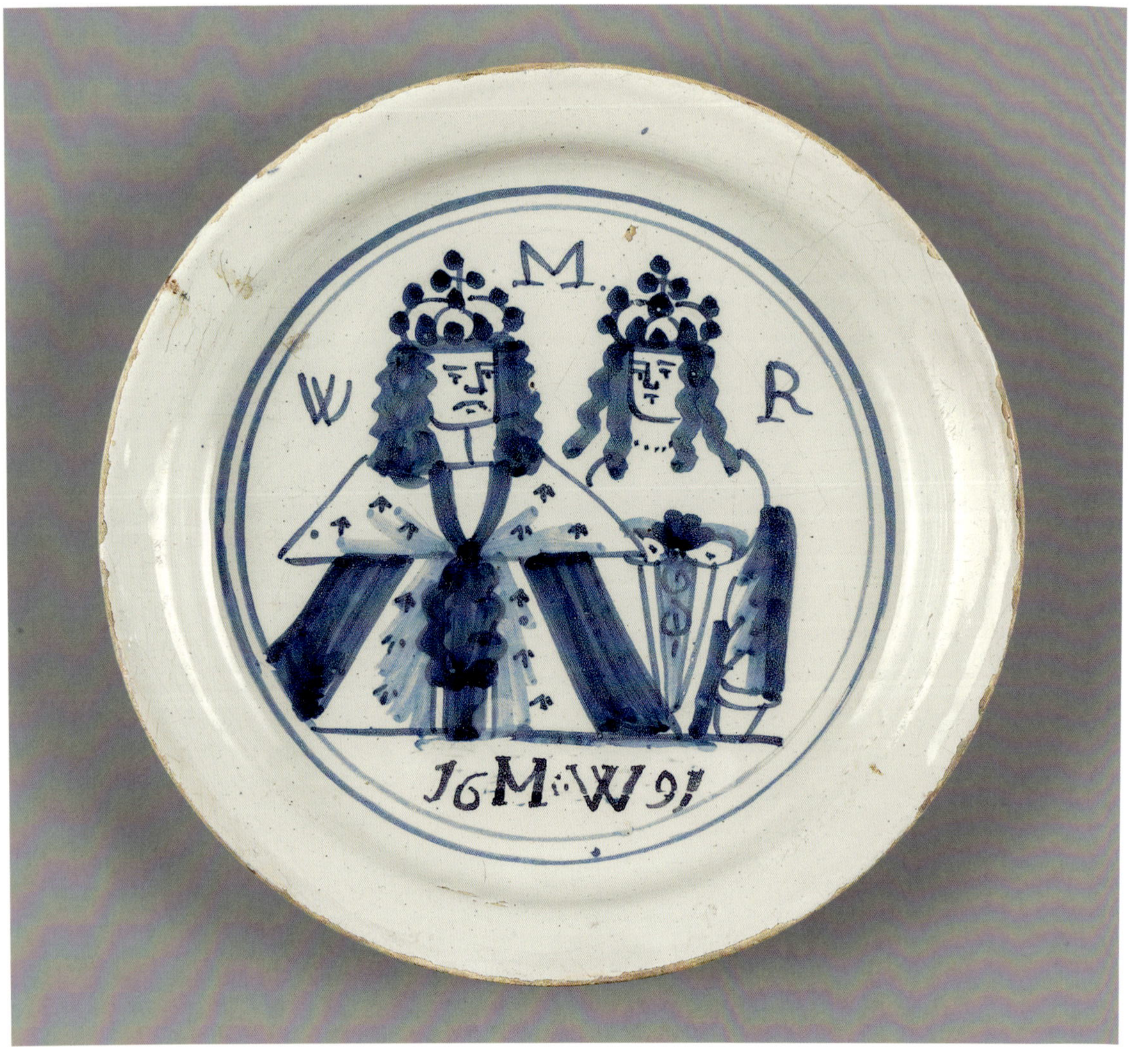

1717

This plate is one of a series of which a number survive. All have the same design and are inscribed with the same date. The other plates, in the Victoria and Albert Museum, London (C.83-1931), are inscribed: "To the Pious Memory of Queen Anne," "In Memory of Queen Anne 1717," and "God save King George 1737."

PROVENANCE
Joseph V. Vizcarra, Chicago, IL; Sotheby's, New York, April 12, 1984 (lot 393); John Philip Kassebaum of Mount Pleasant, SC (1932–2016); Sotheby's, London, October 1, 1991 (lot 101); bt. John H. Bryan of Lake Bluff, IL (1936–2018) from Lindsay Antiques, London, October 1, 1991

Plate
Delftware
$1\frac{1}{8} \times 8\frac{7}{8}$ in. (2.9 × 22.5 cm)

Inscription:
God Save|King George|1717

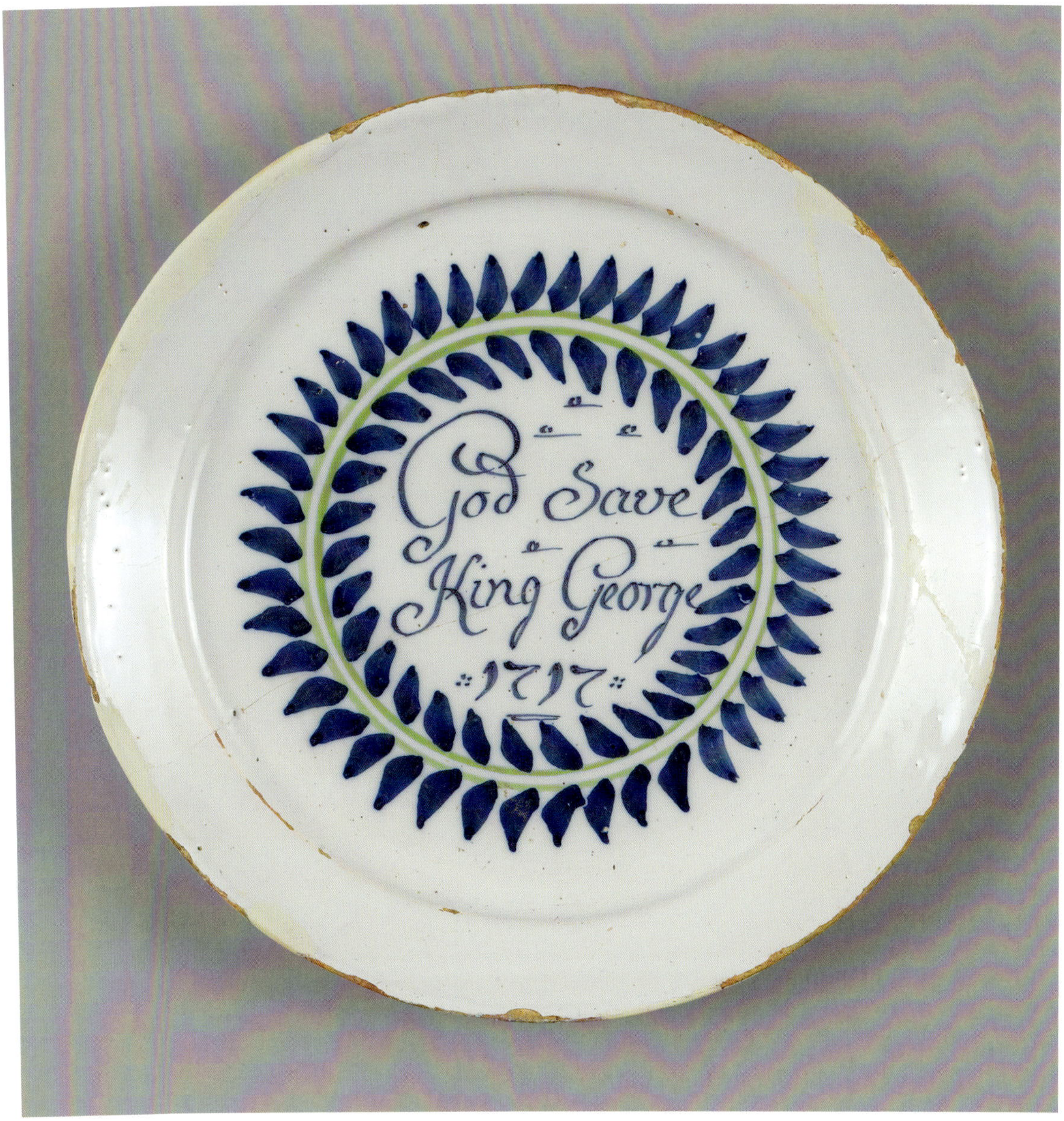

1738

Plate
Delftware
1 × 8 in. (2.5 × 20.3 cm)

Inscription: To y^{e} Pious Memory|of Queen Caroline|1738

Caroline of Brandenburg-Ansbach (1683–1737) was the queen consort of George II. This plate was made a year after her death, which was widely mourned throughout Britain.

PROVENANCE
Herbert E. Ward, Black Bass Hotel, Lumberville, PA; Bonhams, London, September 8, 1990; bt. John H. Bryan of Lake Bluff, IL (1936–2018) from Alistair Sampson Antiques Ltd., London, April 28, 1992

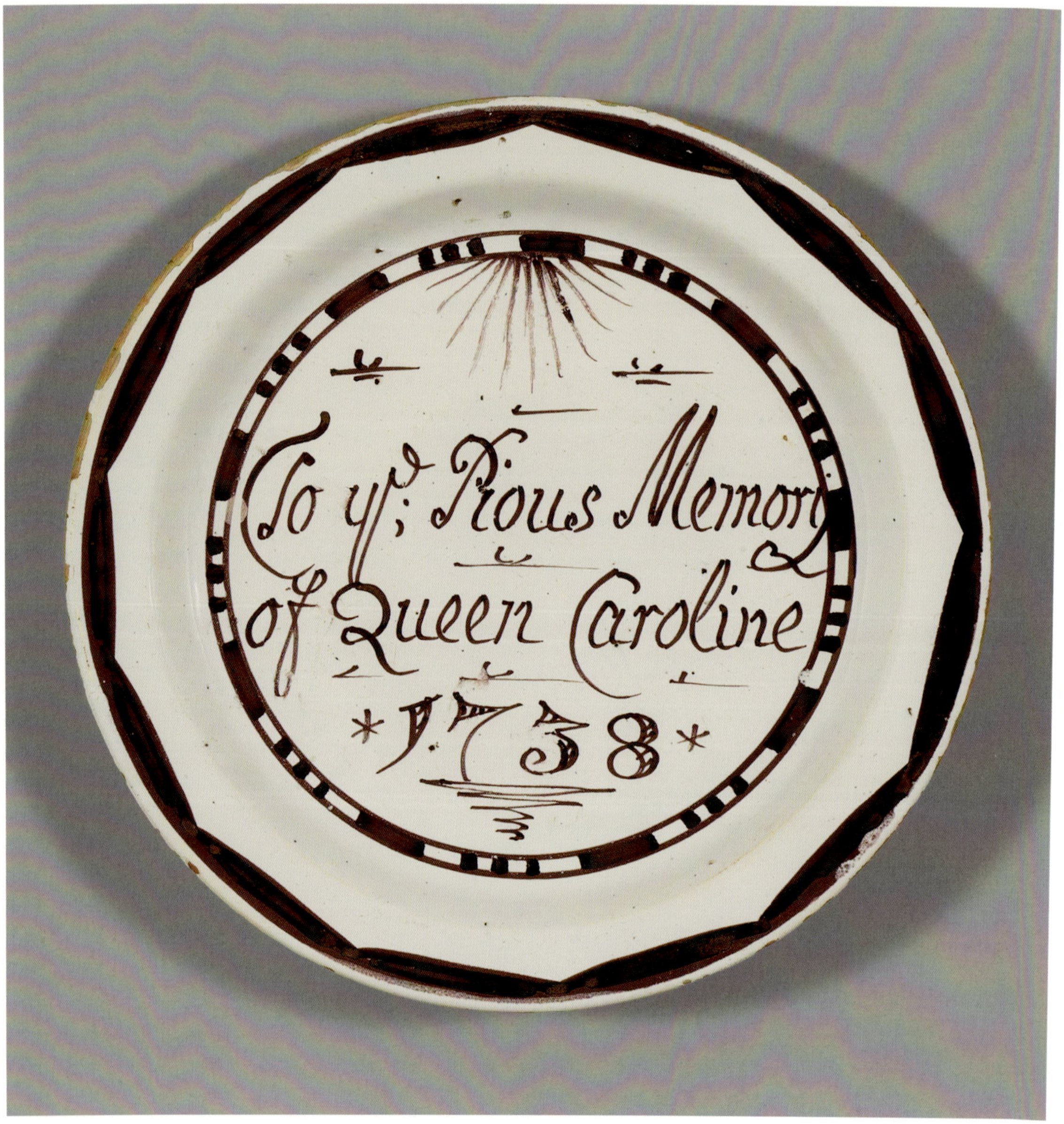

1746

Possibly made in Liverpool, this plate depicts William Augustus, Duke of Cumberland (1721–1765), as victor at the Battle of Culloden in 1746.

PROVENANCE
Longridge Collection; Private Collection, London; bt. John H. Bryan III of Nashville, TN, from John Howard at Heritage, Oxfordshire, January 12, 2017

LITERATURE
Leslie B. Grigsby, *The Longridge Collection of English Slipware and Delftware* (London: Jonathan Horne Publications, 2000), 1:85

Plate
Delftware
1 × 8 3/8 in. (2.5 × 21.3 cm)

Inscription:
1746|Duke William|For Ever.

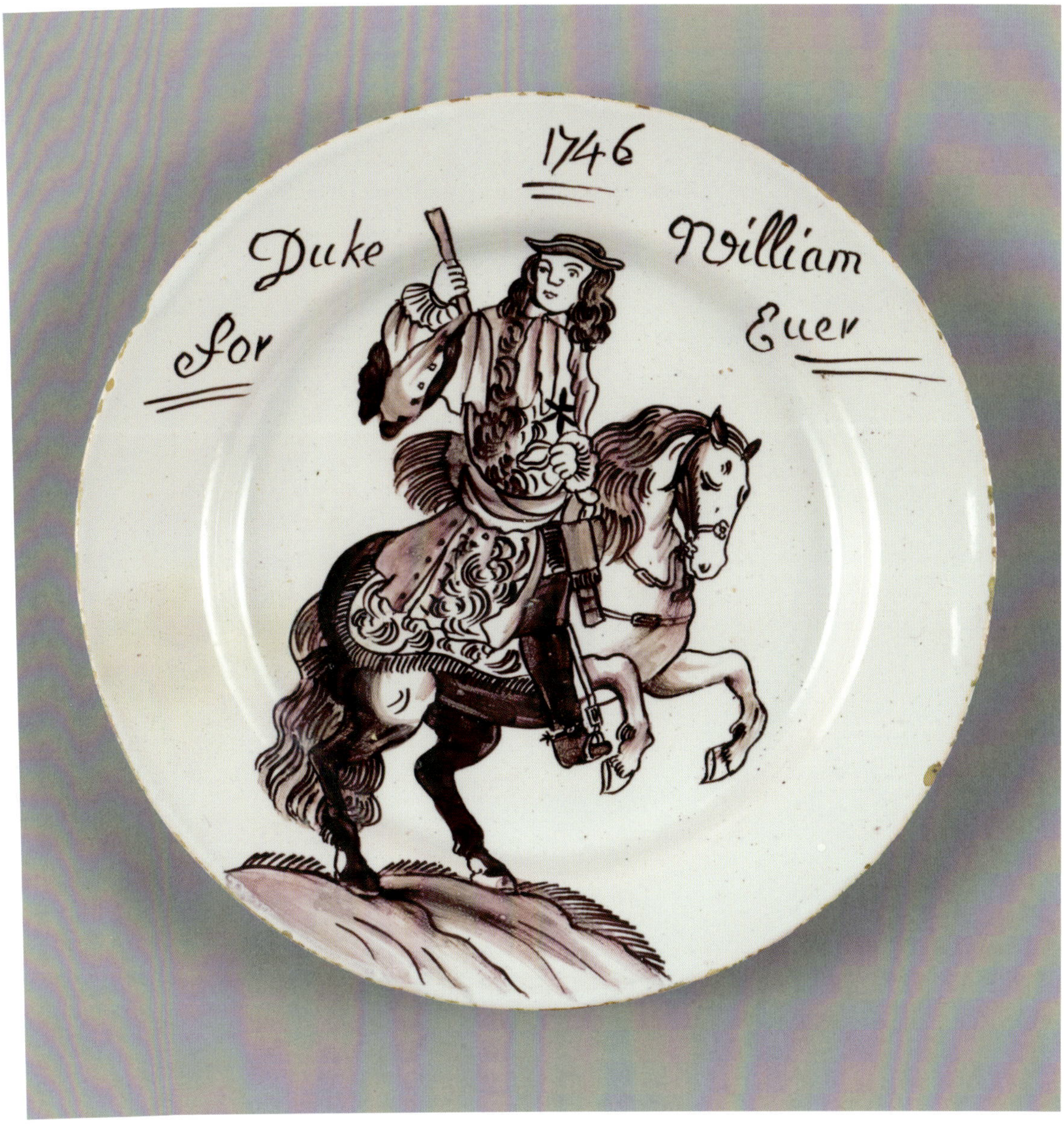

1739

Plate

Delftware with powdered manganese decoration

$1\frac{3}{8}$ × 8 in. (3.5 × 20.3 cm)

Inscription:
I G|1739

The biblical scene depicted on this plate is described in the Book of Genesis account of the life of Joseph, in which Joseph is sold to Potiphar as a slave. Potiphar's wife attempts to seduce Joseph, and angry at his resistance, she falsely accuses him of rape.

It is thought this plate might be from the Wincanton pottery near Bristol, which was founded by Nathaniel Ireson (1686–1769) and Thomas Lindslee (1725–1750). A lilac manganese-powder border with reserved panels decorated in blue, like the one shown here, is often attributed to Wincanton.

PROVENANCE

Anonymous sale, Christie's, London, August 10, 1993 (lot 16); bt. John H. Bryan of Lake Bluff, IL (1936–2018) from Lindsay Antiques, London, August 17, 1991

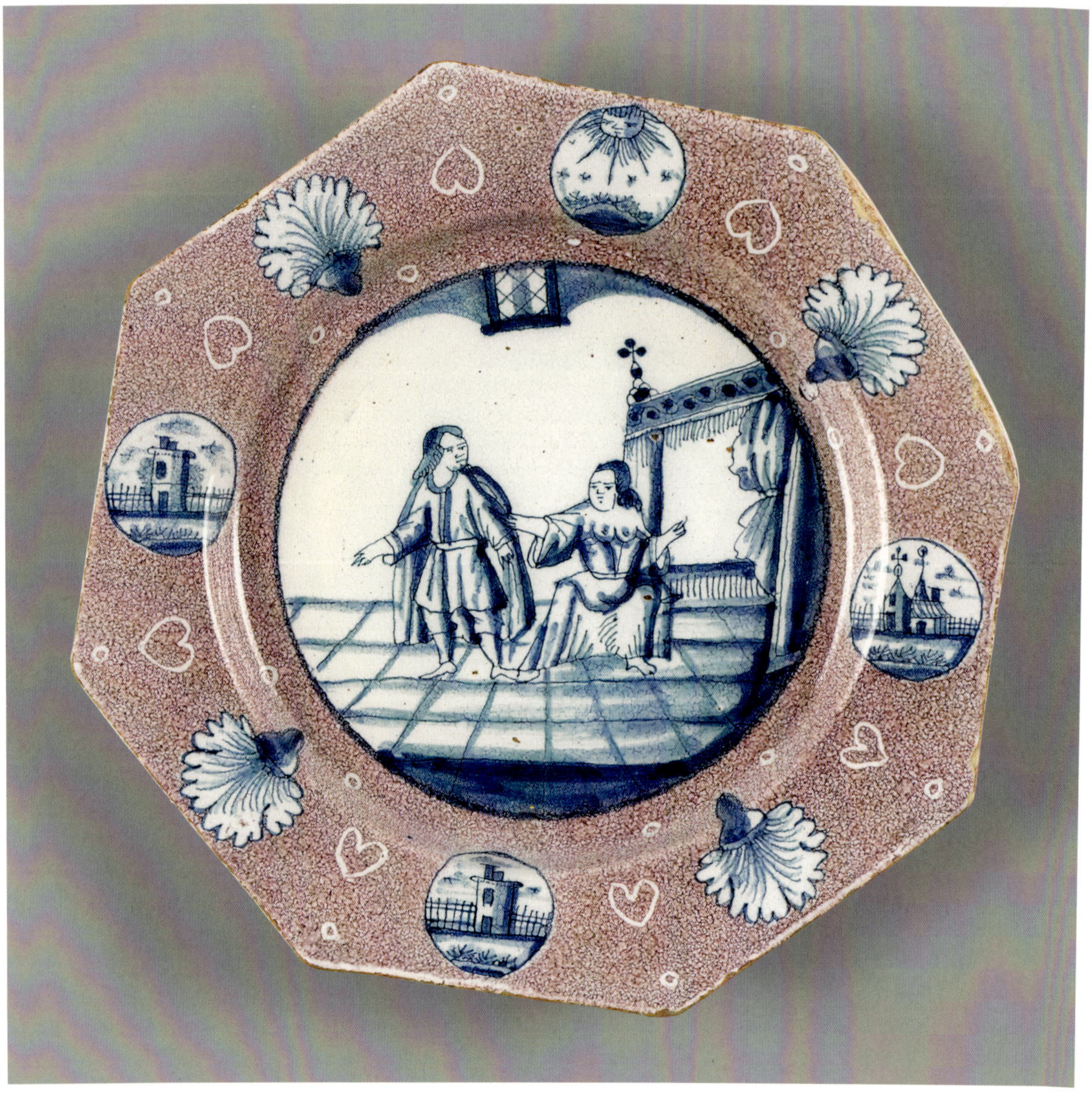

1704

Portrait

Watercolor and gouche on vellum

$2\frac{3}{4} \times 2\frac{1}{8}$ in. (7 × 5.4 cm)

Anne Chéron (ca. 1663–1718)

Inscription: pient a St Germaine|en Laye par anne|Cheron femme|de Belle de 1704

This miniature of the Old Pretender was made by Anne Chéron (ca. 1663–1718). Her husband, Alexis Simon Belle (1674–1734), whom she married on November 12, 1701, made the portrait from which this likeness is derived. Belle was the principal painter at the exiled Jacobite court at Saint-Germain-en-Laye. Together, he and his wife were responsible for disseminating the Pretender's portrait in both small and large formats.

PROVENANCE

Horace Walpole Collection; bt. John H. Bryan of Lake Bluff, IL (1936–2018) from D. S. Lavender Antiques Ltd., London, March 18, 1998; gifted June 6, 2017, to John H. Bryan III of Nashville, TN

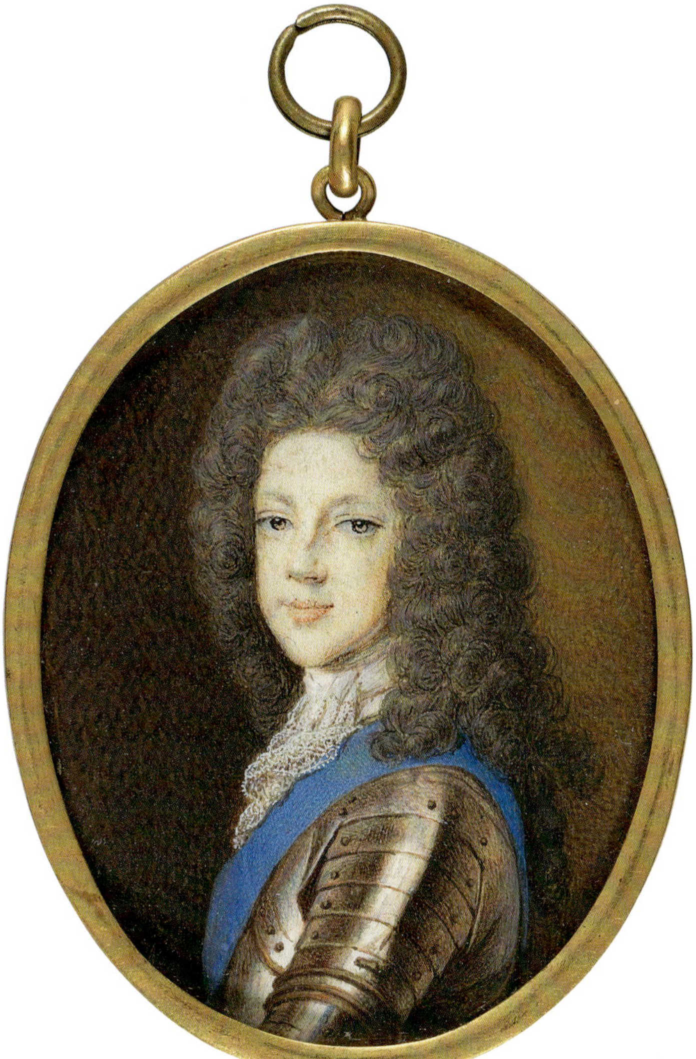

1758

Portrait

Graphite on vellum

$4\frac{1}{2} \times 3\frac{1}{2}$ in. (11.4 × 8.9 cm)

Jo. Winder

Inscription: L:t
Jo: Winder Fecit
Chester 1758

Nothing is known about the Lieutenant John Winder who made this graphite drawing on vellum, known as a plumbago. The subject is Prince William Augustus (1721–1765), Duke of Cumberland, a younger son of King George II, vanquisher of the Jacobites, who is suitably attired in faux classical armor.

PROVENANCE

bt. John H. Bryan of Lake Bluff, IL (1936–2018) from D. S. Lavender Antiques Ltd., London, March 23, 1999; gifted June 6, 2017, to John H. Bryan III of Nashville, TN

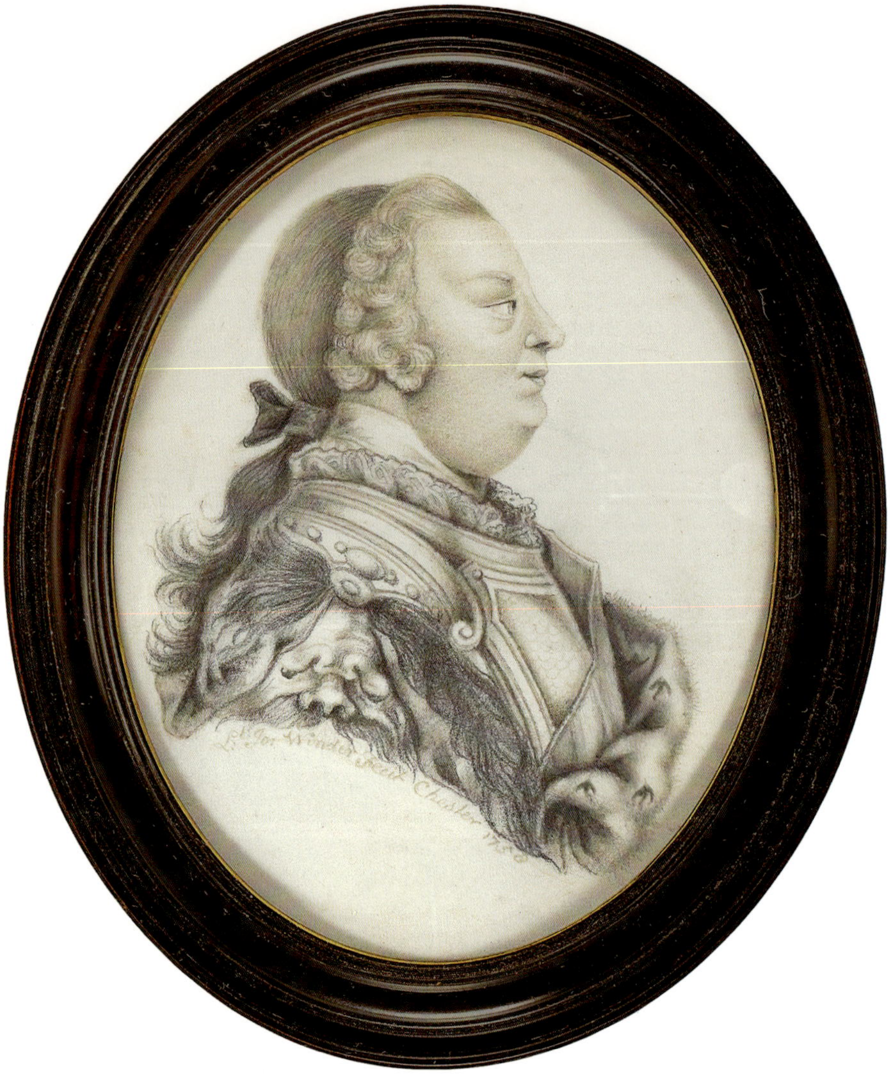

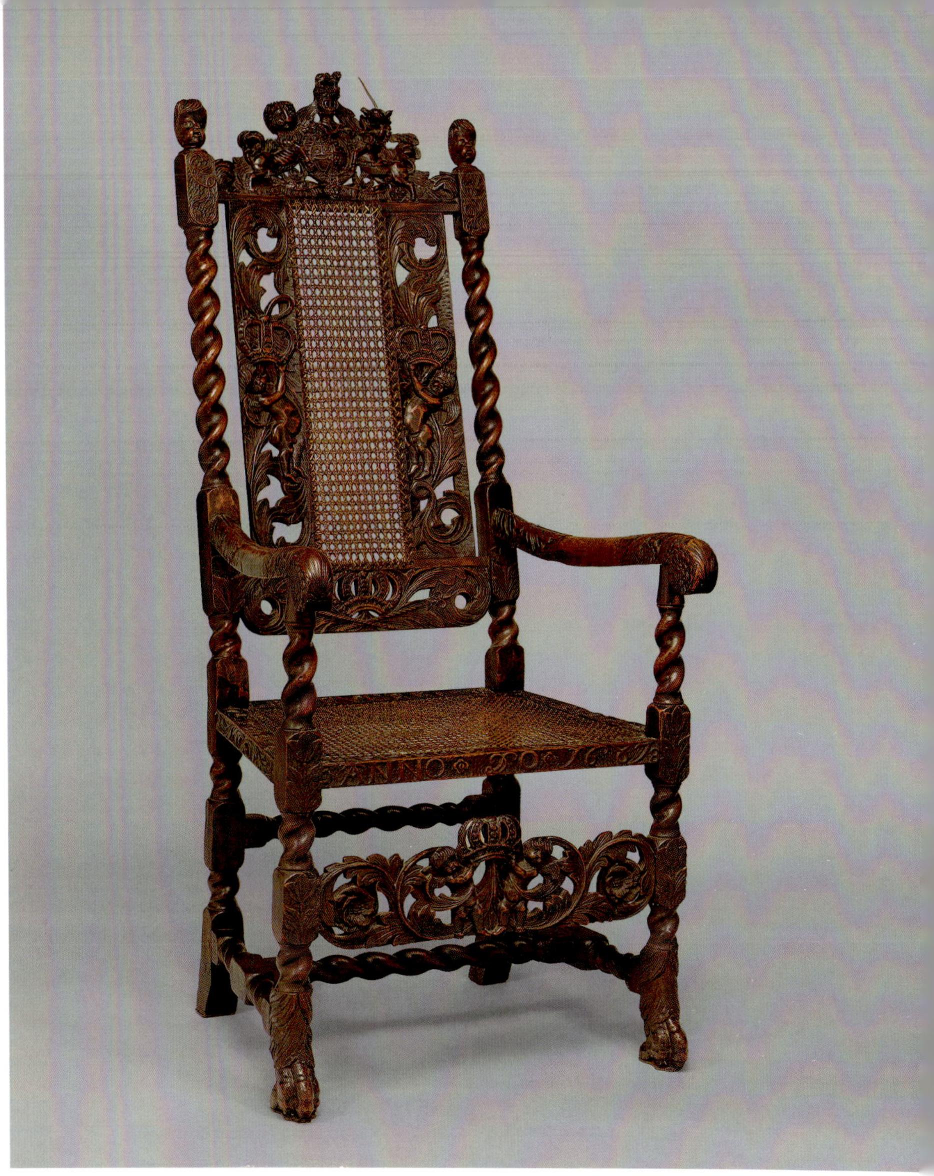

1687/8

Armchair

Walnut, cane, lead

$53\frac{1}{2} \times 23\frac{1}{8} \times 22\frac{3}{4}$ in. (135.9 × 58.7 × 57.8 cm)

Inscription: HONI SOIT QVI MAL Y PENSE; DIEVET M N DROIT' FEBVERY y^{e} 20 GEORGE LEWIS ANNO DO 1687_{8}

The significance of the date in the inscription on this chair is impossible to work out. The most likely interpretation is that this chair was made for George Lewis or Louis (1660–1727), at the time elector of Hanover and later George I of Britain. He had visited the court of Charles II in December 1680 and stayed into the following year. The chair may have been sent subsequently as a gift, but the motivation remains a mystery.

PROVENANCE

George Weare Braikenridge of Brislington, Bristol (1775–1856); Christie's, London, June 25, 1931 (lot 53); S. W. Wolsey, London; Mr. and Mrs. J. Edgar Stone, 1966; Sotheby's, New York, December 8, 1989 (lot 472); bt. John H. Bryan of Lake Bluff, IL (1936–2018) from Ronald A. Lee (Fine Arts) Ltd., London, June 15, 1990

LITERATURE

Adam Bowett, *English Furniture from Charles II to Queen Anne, 1660–1714* (Suffolk, UK: Antique Collectors' Club, 2002), 92

Adam Bowett, *100 British Chairs* (Suffolk, UK: Antique Collectors' Club, 2015), 28

Geoffrey Beard and Judith Goodison, *English Furniture, 1500–1840* (Oxford, UK: Phaidon-Christie's, 1987)

1689

Garters

Silk and metal thread

$43\frac{3}{4} \times 1\frac{1}{4}$ in. (111.1 × 3.2 cm)

Inscription, on both: ELIZABETH BISHOP IERVSALEM 1689

Jerusalem garters were associated with pilgrimages to the Holy Land, but Jewish weavers in North Africa and Europe seem also to have made them for export. This pair was made for Elizabeth Bishop. Her name and the date are woven into the fabric, which was made with the simple, ancient technique of tablet (or card) weaving. The inscriptions are the same on each garter in the pair.

PROVENANCE

Loaned by Cora Ginsburg LLC, New York, to John H. Bryan of Lake Bluff, IL (1936–2018)

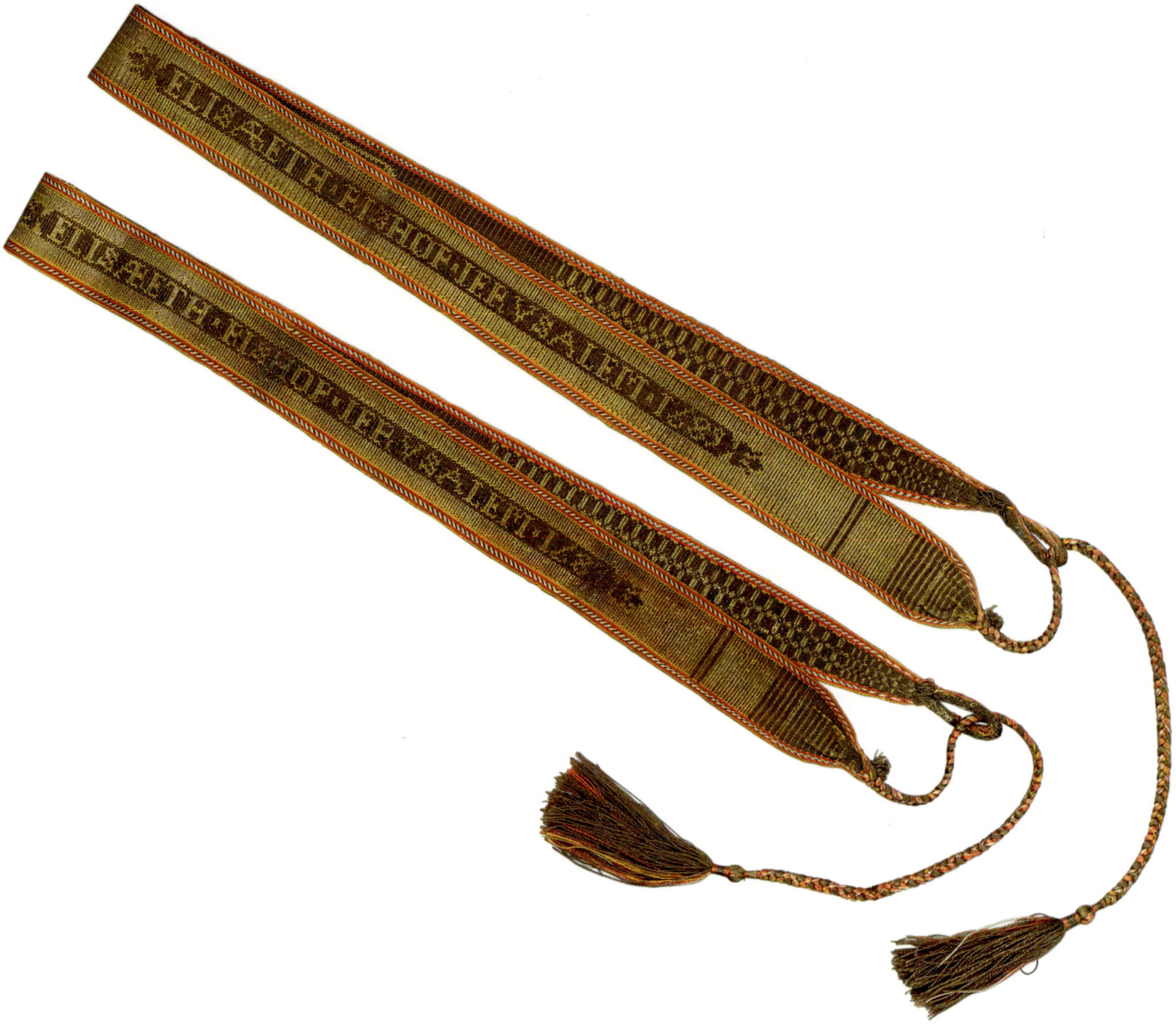

1698

Girdle
Silk thread
68 ½ × ½ in. (174 × 1.27 cm)

Inscription:
FAVOVR IS
DECEITFVLL AND
BEAVTY IS VAIN BVT
A WOMAN THAT
FEARETH THE LORD
SHALL BE PRAISED
GIVE HER OF THE
FRVIT OF HER
HANDS AND LET
HER OWN WORKS #
LORD MY SOVL LIKE
NOAHS WEARY DOVE
CAN FIND NO REST
BVT IN THY ARK
A BOVE REBECCA
EGGELTON HER
GIRDEL 1698

Braided into this girdle is a partial text from Proverbs 31, which describes a woman's role as a faithful wife, mother, and keeper of the household. If not truncated, the tenth line of the inscription would continue, "praise her in the gates." The girdle was made using a complex double-braiding technique that requires two people to work together to create the patterns. Instructions for the technique are found in extant pattern books, which include selections of verses to work into the braids.

This girdle was worn around the waist by Rebecca Eggelton. On June 17, 1702, she married Christopher Truss of St Alban, Wood Street, at the Quaker meeting place at the Bull and Mouth Inn, not far from his parish in Aldersgate. The couple made a public declaration of their union on September 17 later that year. Truss was the son of Christopher Truss, a wax chandler late of Reading; Rebecca, who was born on June 22, 1680, was the daughter of the deceased Thomas Eggleton of Nightingale Lane, Chandler. Truss also was a citizen and member of the Wax Chandlers' Company of London. By 1713, if not before, he was working as a cabinetmaker, and the couple was living in his parish of St Alban, Wood Street. Truss died on November 17, 1733, aged sixty-seven. Nothing more is known of Rebecca.

PROVENANCE
Loaned by Cora Ginsburg LLC, New York, to John H. Bryan of Lake Bluff, IL (1936–2018)

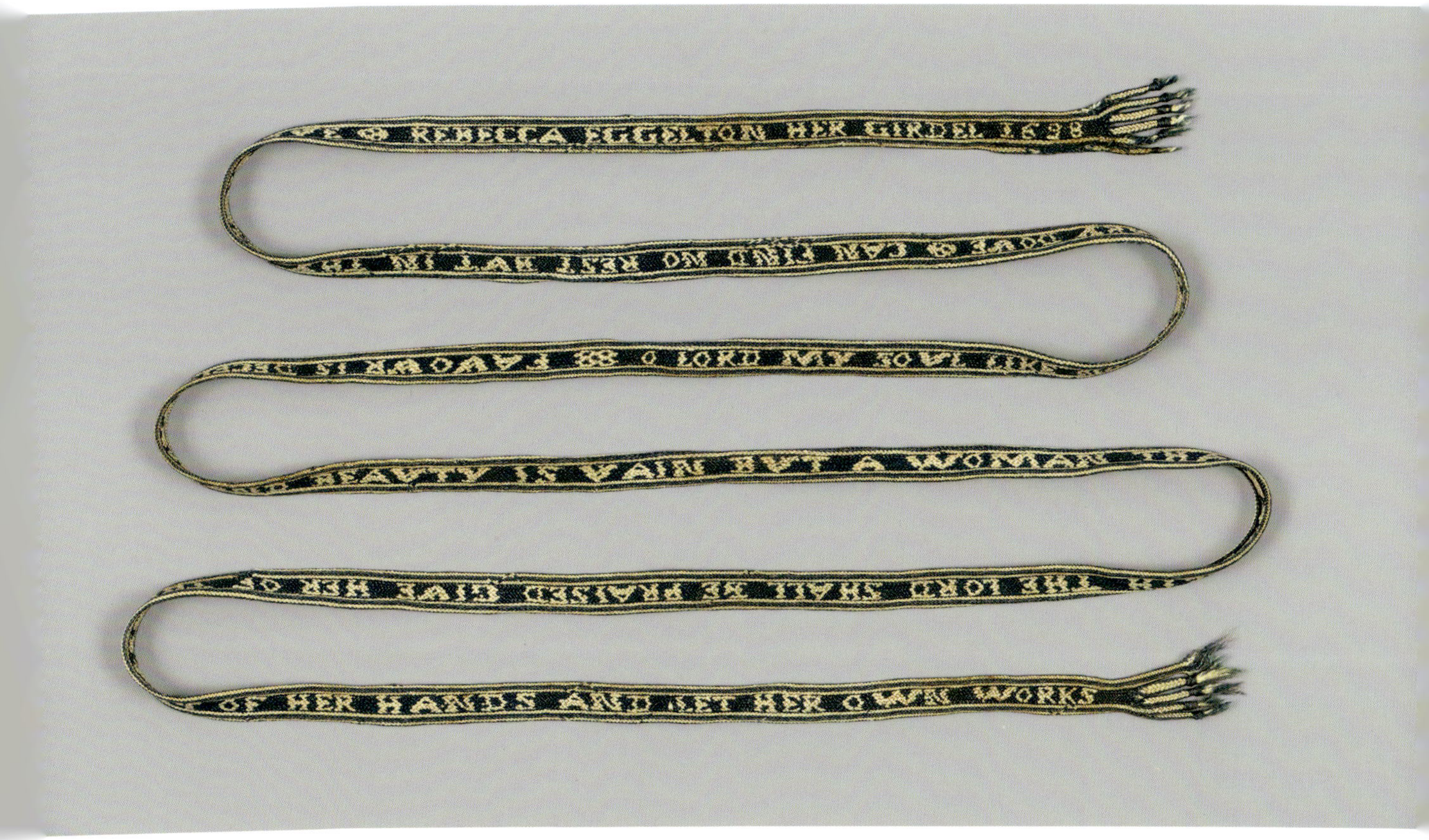

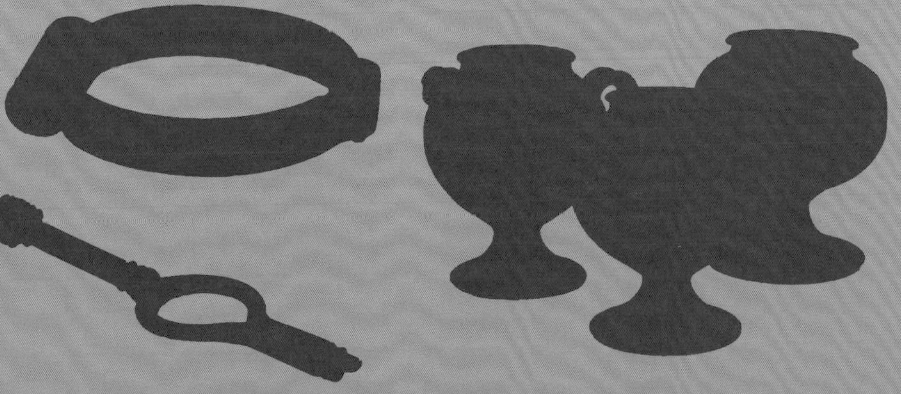

VI. POWER & DOMINION

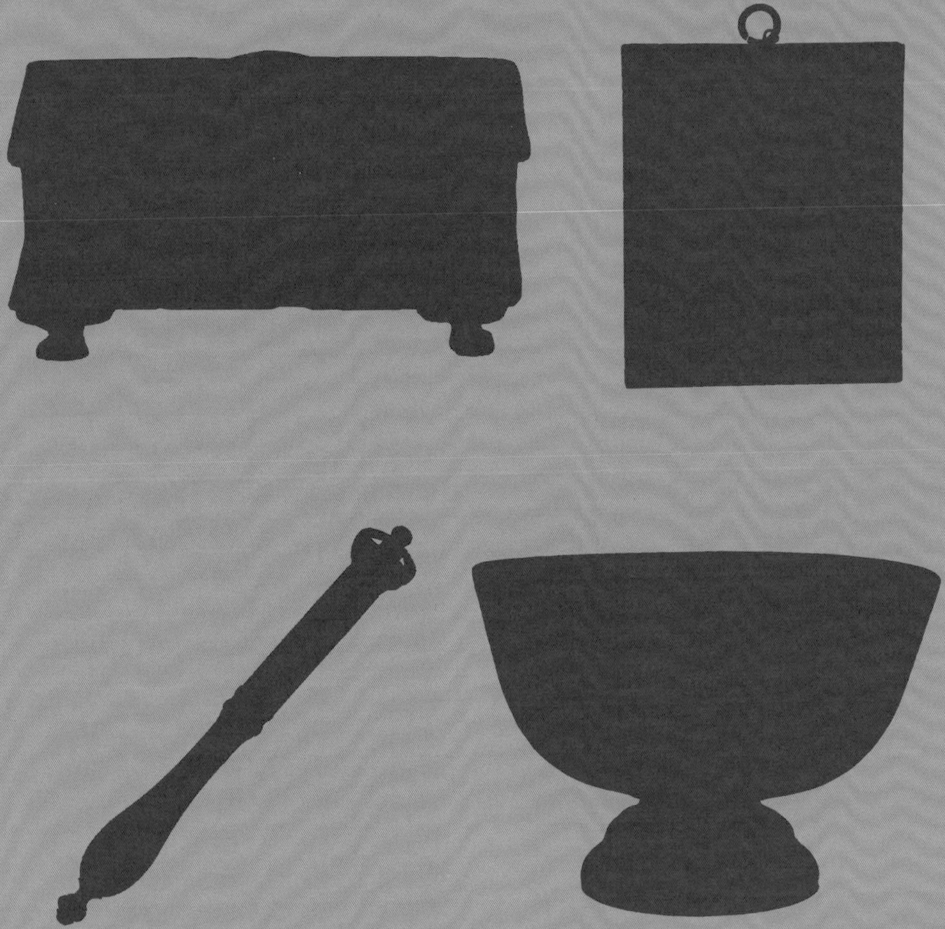

MUCH LIKE THE PERFECT GEORGIAN SQUARES AND CHURCHES OF London, the objects in this book could present a distorted sense of this period of British history as one of untroubled peace and prosperity. Furthermore, sentimental objects may provide a misleading degree of distance from the violence of slavery that was generating much of the wealth that was to make London a great city and that also provided the nation's army and navy with the resources to triumph in its many theaters of war.

In the seventeenth and eighteenth centuries, Britain made its appearance on the global stage with a series of decisive victories, both at home against the Jacobite threat and abroad against Spain and France, adversaries that vied for maritime supremacy. Both motivating and underpinning this aggression was Britain's steadily increasing involvement in the transatlantic slave trade, which between 1600 and 1775 had resulted in the ensnarement and dislocation of more than 1.8 million people.

Initially, more than half of the immigrants arriving in Britain's North American colonies were indentured servants from Britain and Ireland. However, once the carnage of the English Civil War (1642–51) had depleted the supply of ready labor, this system could no longer meet demand. Britain therefore looked with increasing frequency to the coast of West Africa for the labor force required to work the plantations. Thus began the so-called triangular trade, whereby enslaved labor was cleaved from Africa and transported across the Atlantic under horrific conditions in what became known as the Middle Passage. Enslaved Africans were brought to the Caribbean and American colonies to harvest goods—sugar and tobacco—that were then sent to Europe to be consumed and harvest woods to be made into furniture, staves, and truncheons (pages 390–92). Some manufactured objects, be they garments or guns, were then shipped from Britain to the west coast of Africa and traded for more slaves.

In the same ships that brought materials from the colonies to the burgeoning port cities of Bristol and Liverpool, there were sometimes men, women, and children of African descent who were leaving lives of slavery in the colonies only to enter a new period of bondage in Britain. Although English and Scottish laws were unclear about the status of slavery within the shorelines of Britain, a small but significant number of people of African heritage were sought for purchase, advertised for sale, or kept in physical bondage as household servants. The ankle iron dated 1733 (page 384) bears an inscription that declares it and the person restrained by it to be the property of John Deverell of Corn Street, Bristol. This object is a stark reminder of this complex and too easily forgotten aspect of Britain's involvement in the history of slavery.

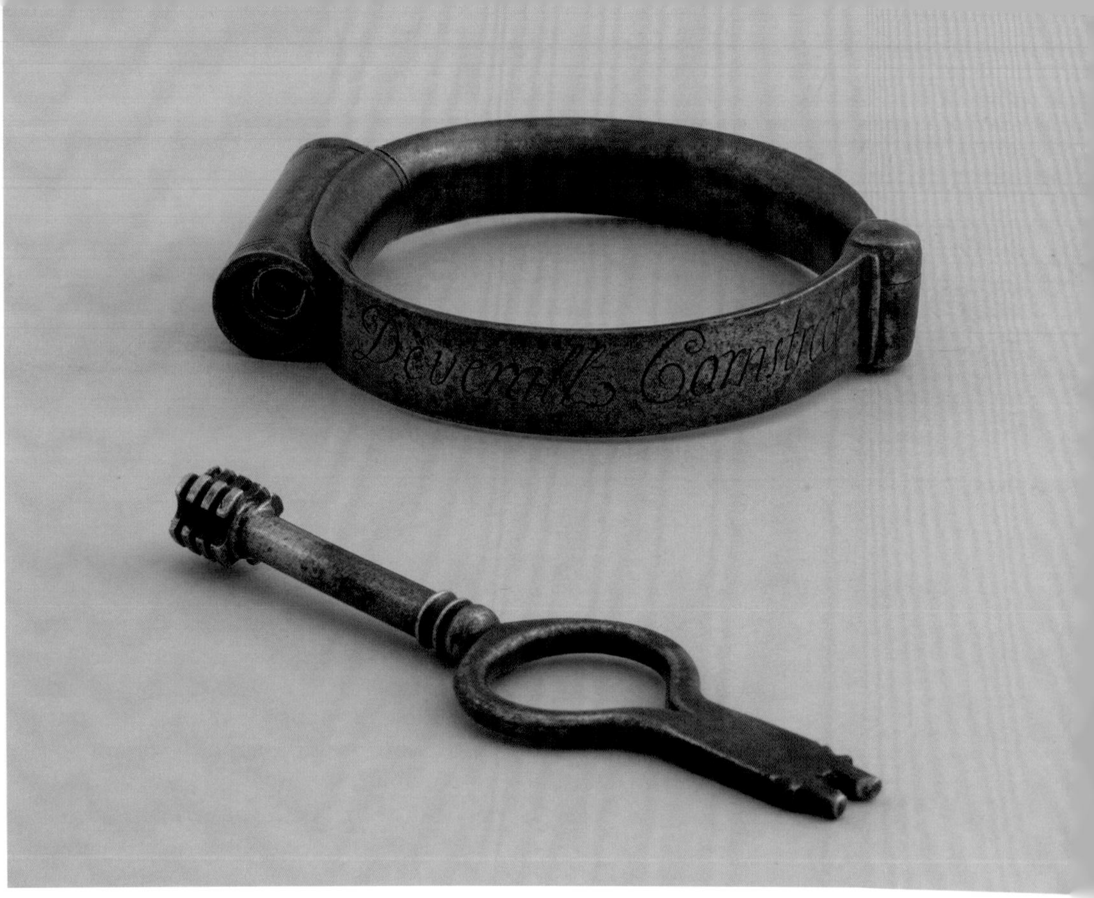

1733

Ankle iron and key

Steel

Iron: $1\frac{1}{4} \times 3\frac{1}{2}$ in. (3.2 × 8.9 cm); key: $4\frac{1}{4}$ in. (10.8 cm)

Inscription:
Deverall Corn street
Bristoll, 1733

This device offers a poignant reminder of the violent history and invidious nature of slavery in Britain. The scant scholarship and euphemistic language describing the small number of similar objects suggests a resistance—bordering on reluctance—to revisit a period of British history in which people of African heritage were brought into Britain as enslaved servants.

The combination of the durable steel and sophisticated screw-key locking system helped this object achieve its callous function: binding the person's ankle and possibly tethering the limb to chains and other shackles. The inscription furthered the containment by providing the name of the owner, "Deverell"; his address, "Corn Street, Bristol"; and the date, "1733." There is no record of who wore this shackle. The maker is also unknown, but the barrel lock and hinge resemble those on another shackle in the National Maritime Museum, Greenwich (ZBA2474), which carries the name Samuel Bosanquet (1700–1765), of Leyton, Essex, and the date "1746," described in the past as a "Slave ownership bracelet" and "identification band." Both irons resemble others made after 1780 by the Birmingham manufacturer Hiatt, one of which is in the collection of the Bristol Museums (TB3299).

The 831 newspaper advertisements (collated to date) published between 1700 and 1780 by individuals seeking to recover runaway slaves (mostly young and male) provide a sense of the prevalence of slave ownership within the British Isles before legal rulings in the late 1770s addressed the confusing and contradictory slavery laws. Several advertisements appeared in Bristol newspapers, a city with a significant number of people of African origin. One notice, published in the *Daily Courant* of July 23, 1723, was paid for by Paul Moon, a soap boiler. He offered a two-guinea reward for the return of an unnamed "Negro young Man, about 20 Years of Age, who ran away from his Master the 7th Instant; he is of a short Stature, full

Face, Scars about both his Eyes, wears a dark Duroy Coat and Waistcoat, and a light-colour Pair of Plush Breeches; with a Steel Collar about his Neck, with his Master's Name ingrav'd upon it." Neck collars, more often made of silver or brass, also provided the name of the "owner." Five years after publishing his notice, Moon was bankrupt and ordered to surrender himself to the commissioners at Cooke's Coffeehouse, which was, coincidently, on Corn Street. Possibly less of a coincidence is that Paul's father—also named Paul and a member of a Quaker family of St Philip and St Paul Bristol—had his will witnessed by Jeremiah Deverell in 1726.

The latter is one of two possible owners of this object: Jeremiah (1668–1751) and John Deverell (d. 1760), father and son, both surgeons of Bristol. Wills of various Deverells proved in the Consistory Court of Bristol suggest that this family had been domiciled in the city for the best part of a century. They may also have been related to John Deverell, Barber-Surgeon of London, who died aboard the *Happy Entrance* off the east coast of India in November 1670 (TNA PROB 11/345/347). Benjamin Deverell, Jeremiah's brother, was a Bristol merchant who died in Virginia in about 1717 (TNA PROB 11/558/113); John's first wife was the daughter of the Bristol mariner Hugh Rainstorp (d. 1718). Thus, there were multiple ways by which an enslaved person might have arrived on Corn Street in 1733. There were also other slave-holding members of Bristol society: the probate inventory of John Weaver, mariner, of 1725 includes one "Negro Boy" valued at nine pounds sterling; an advertisement in *Felix Farley's Bristol Journal* of September 24, 1757, gave notice of a runaway called John, sought by Captain Ezekiel Nash, whose address was Blake and Palmer's, again in Corn Street.

Of Jeremiah and his son, John, only the latter can be definitely associated with Corn Street. The residence was on the south side of the street, at the intersection with Cock Lane, in the parish of St Werburgh. Jeremiah Deverell had held the lease since 1704 but in 1720 assigned it to John, upon John's marriage to Sarah Rainstorp. By 1734, if not before, Jeremiah was voting in St James parish and he is associated with various other addresses in this period but not with Corn Street. This information gives the clearest indication that the object belonged to John and not his father. There is no immediate record of a black presence in St Werburgh. However, the parish register of St Augustine, where John Deverell moved to in about 1746, records the baptism of Mary Reece, a child of African heritage, in 1733 and her burial on January 22, 1736/7. Likewise, Sarah Lewis, "a Negro," was baptized in that parish on May 7, 1747. Other records of baptisms of persons described as of African descent in St Augustine include Phillis (October 3, 1724), Samuel (February 20, 1727), Charles Dyer (August 2, 1728), Elxander (October 23, 1730), Richard Smith (October 3, 1734), Elizabeth (April 20, 1735), and Thomas Dorset (August 31, 1735).

John Deverell may have served as a surgeon on slaving voyages, although the paper record suggests that he remained domiciled in Bristol throughout his life. As the second-most important person on slaving vessels after the master, the surgeon had the right to trade one enslaved male and one enslaved female during the voyage for earnings over and above his pay. Deverell might have then used this shackle to distinguish which African he considered his property. Surgeons received "head money" for each slave who arrived in the colony alive and was sold. The other, more likely possibility is that he placed the iron upon an enslaved servant in Bristol to prohibit escape and also to provide proof of "ownership" in the event the servant was abducted to be resold or ran away seeking freedom.

Attrition rates aboard slavers in the first half of the eighteenth century were as high as ten percent, as unsanitary conditions provided an environment that allowed infectious diseases, particularly dysentery and fevers, to thrive. Thomas Aubrey's *The Sea-surgeon, or The Guinea man's vade mecum* (1729) describes the unspeakable deprivations suffered by ensnared Africans on the Middle Passage. Aubrey recalled: "Very often they are abused by Sailors, who beat and kick them to that Degree, that sometimes they never recover, and then the Surgeon is blamed for letting the Slaves dye, when they are murthered, partly by Strokes, and partly famished." Although such texts perpetuated prejudicial stereotypes of West African cultures, testament from Bristol surgeons in the latter half of the century—brought as evidence to the House of Lords in an investigation into the slave trade—helped build the case for abolition. Deverell's experience and expertise in venereal disease, as evinced by mention of his work in John Douglas's *A Dissertation on the Venereal Disease* (1737), may have come by way of treating victims and perpetrators of sexual exploitation aboard slave ships and in the plantations.

John Deverell made his will on January 8, 1759, and was buried at St Werburghs on February 1, 1760. The next day, *Felix Farley's Bristol Journal* duly reported the death of Deverell, an "eminent surgeon" of St Augustine's Back. Three years earlier, the same journal had advertised Captain Nash's offer of a reward for an enslaved servant named John, if returned to Corn Street.

Note: The editors thank Professor Jonathan Barry, Professor Madge Dresser, Dr. Jonathan Horner, Professor Simon P. Newman, and Dr. Christine Whyte for their assistance with this entry. The information about runaway slaves is from https://runaways.gla.ac.uk.

PROVENANCE

bt. John H. Bryan of Lake Bluff, IL (1936–2018) through Lindsay Antiques, London, June 29, 1992

LITERATURE

Madge Dresser, *Slavery Obscured: The Social History of the Slave Trade in an English Provincial Port* (London: Continuum, 2001), 74–75

Stephen Farrell, Melanie Unwin, and James Walvin, eds., *The British Slave Trade: Abolition, Parliament and People* (Edinburgh: Edinburgh University Press for the Parliamentary History Yearbook Trust, 2007), 280–81

Edwin George and Stella George, eds., *Bristol Probate Inventories: Part 3, 1690–1804* (Bristol Record Society, 2008), 60:113–14

Douglas Hamilton and Robert J. Blyth, eds., *Representing Slavery: Art Artefacts and Archives in the Collections of the National Maritime Museum* (Aldershot, UK: Lund Humphries, 2007), 157

Roger H. Leech, *The Topography of Medieval and Early Modern Bristol: Part I* (Bristol, UK: Bristol Record Society, 1997), 64–65

Simon P. Newman, "Freedom-Seeking Slaves in England and Scotland, 1700–1780," *English Historical Review* (October 3, 2019), https://doi.org/10.1093/ehr/cez292

Richard Brinsley Sheridan, "The Guineas Surgeons on the Middle Passage: The Provision of Medical Services in the British Slave Trade," *International Journal of African Historical Studies* 14, no. 4 (1981): 601–25

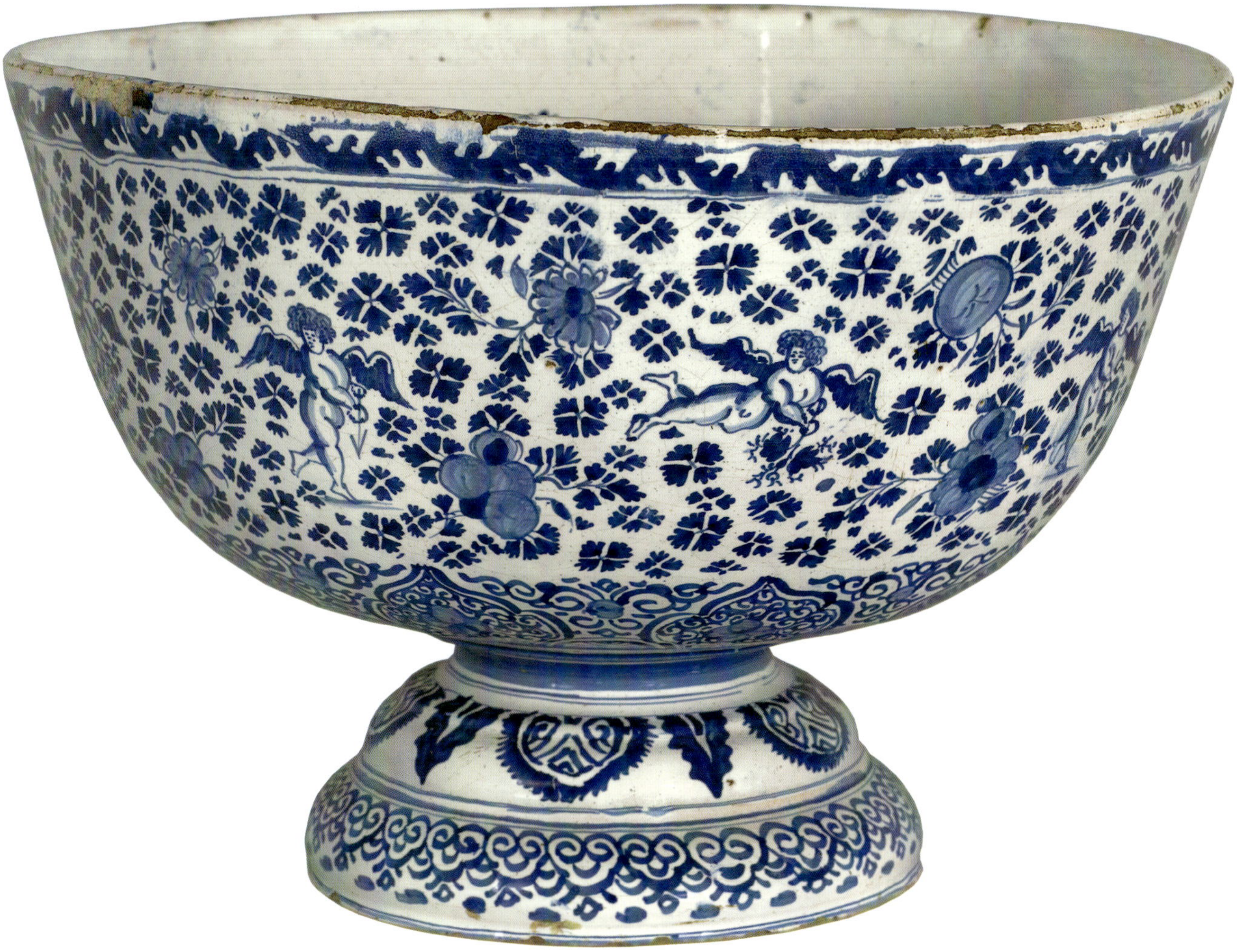

1701

Punch bowl
Delftware
$9\frac{1}{4} \times 13\frac{3}{4}$ in. (23.5 × 34.9 cm)

Inscription: G|R|E; 1701

The coat of arms on this ceremonial punch bowl is that of the Saddlers' Company, whose motto is, "Hold fast, sit sure," which could be easily slurred to comic effect after a glass or two of punch.

PROVENANCE
J. M. Furniss Collection; Cornhill & Job, Liverpool, March 13, 1956 (lot 881); Thomas George Burn, Rous Lench Court, Worcestershire; Sotheby's, London, July 1, 1986 (lot 34); bt. John H. Bryan of Lake Bluff, IL (1936–2018) from Lindsay Antiques, London, July 1, 1986

LITERATURE
Louis L. Lipski and Michael Archer, *Dated English Delftware: Tin-Glazed Earthenware, 1600–1800* (London: Sotheby Publications, 1984), 238, no. 1044

1714

Blunderbuss pistols

Brass, steel, iron, walnut, flint

14½ × 3 × 6 in. (36.8 × 7.6 × 15.2 cm)

James Turton

Inscription, on both: William Foord|May the 1/14 1714; Ja Turton Fecit

With their short, high-caliber barrels, this pair of English flintlock blunderbuss pistols would have been deadly at short range. These were often the weapon of choice for both highwaymen and those charged with repelling them. Nothing is known of the gunsmith James Turton or William Foord, for whom this brace of pistols was made.

PROVENANCE

William Keith Neal Collection; bt. John H. Bryan of Lake Bluff, IL (1936–2018) from Peter Finer Ltd., Warwickshire, January 20, 2018

LITERATURE

W. Keith Neal and D. H. L. Back, *Great British Gunmakers 1540–1740* (Norwich, UK: Historical Firearms, 1984)

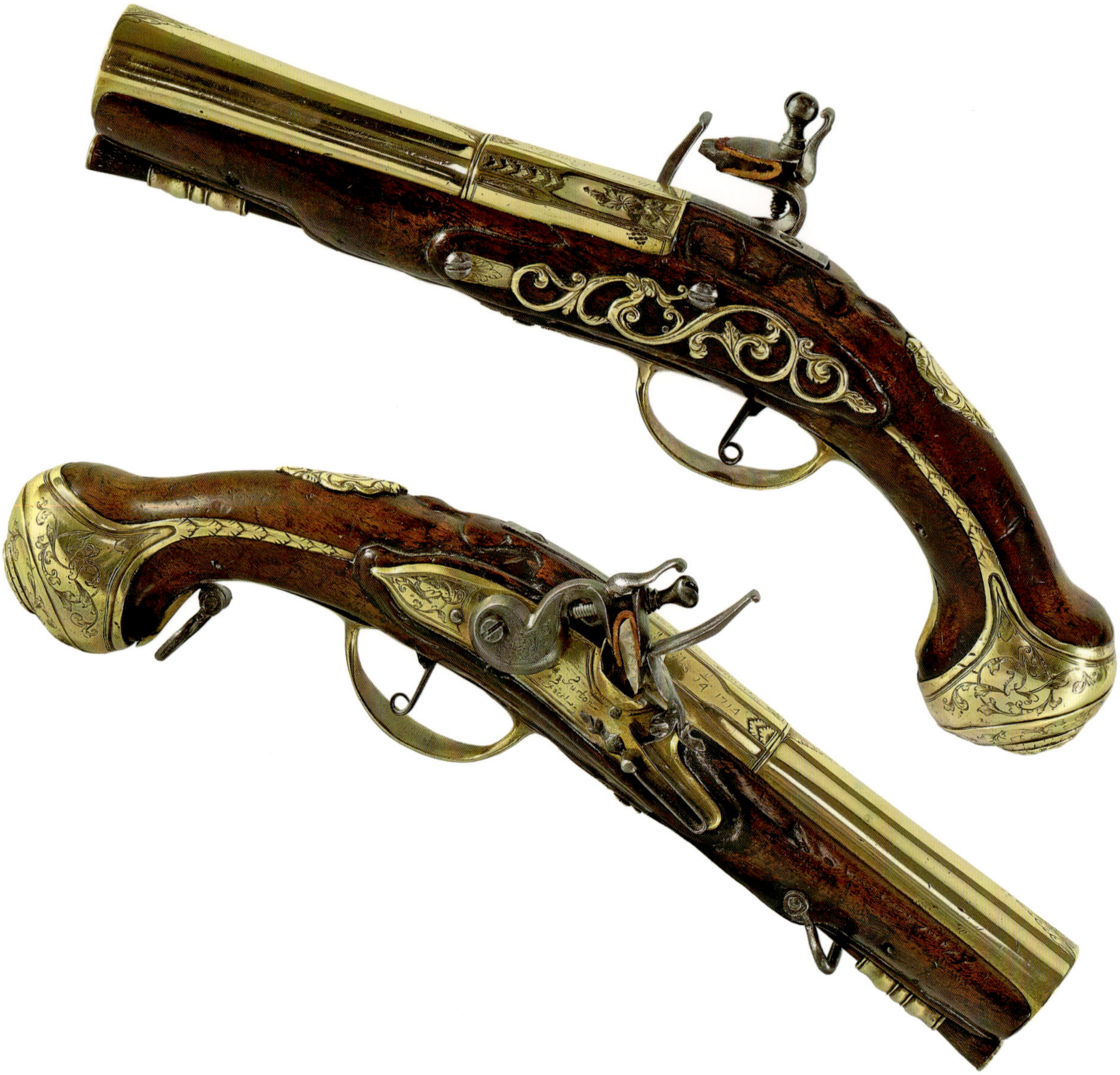

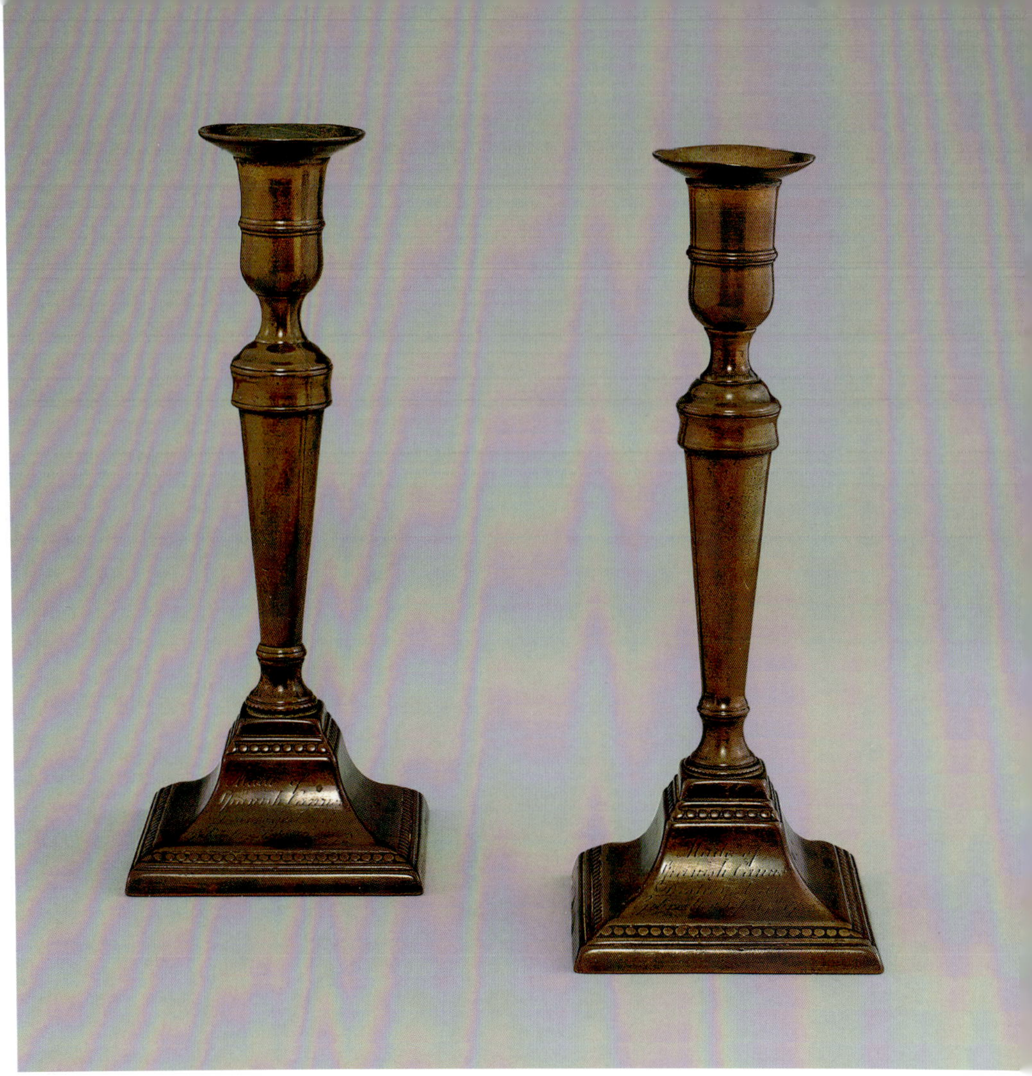

1782

Candlesticks

Gunmetal

$9\frac{7}{8} \times 3\frac{7}{8}$ in. (25.1 × 9.8 cm)

Tomson Warner (active 1738–1816)

Inscription (left): MADE OF/ SPANISH CANNON/ DESTROYED AT/ GIBRALTER SEP^{R} 14th 1782; T. WARNER

Inscription (right): MADE OF SPANISH/CANNON DESTROY'D/ AT GIBRALTER/ SEP^{R}:14·1782

These candlesticks commemorate the repulse of the Spanish siege of Gibraltar on September 14, 1782, the day after the Spanish and their Bourbon allies had launched their final attack. The British held back the advancing line of Spanish artillery batteries by peppering their wooden defenses with red-hot shot, nicknamed "hot potatoes." This pair of candlesticks was among several that Tomson Warner made from Spanish cannons. Both carry the same inscription.

Warner, who was born in Watford, was from a Quaker family of brass founders. He served his apprenticeship under an ironmonger in Ampthill, Bedfordshire, and later set up business in London with his brother John. In 1806, a William Glover was indicted for stealing five pounds in weight of copper, property of Tomson Warner, John Warner (his son), and Robert Warner. Glover was found guilty and transported for seven years.

At the time of his death, Tomson Warner was living in Uxbridge. His will was proved at the Prerogative Court of Canterbury on August 2, 1816 (TNA PROB 11/1583/32).

PROVENANCE

bt. John H. Bryan of Lake Bluff, IL (1936–2018) from Alistair Sampson Antiques Ltd., London, October 1, 1998

LITERATURE

Eloy Koldewij, *The English Candlestick 1425–1925: 500 Years in the Development of Base-Metal Candlesticks* (London: Christie's Books Limited, 2001), 190, no. 180

1756

Staff
Mahogany, brass
$19\frac{1}{4} \times 2$ in. (48.9 × 5.1 cm)

Inscription:
Hartington|J L 1756

"Hartington" might refer to the town in Northumberland in the North of England or to Hartington, Derbyshire, in the Midlands. This staff or truncheon probably served a commemorative purpose rather than a physical one. In 1756, the mahogany for this staff almost certainly came from Jamaica. Under British colonial rule, the wood would have been found, harvested, and transported by enslaved African people.

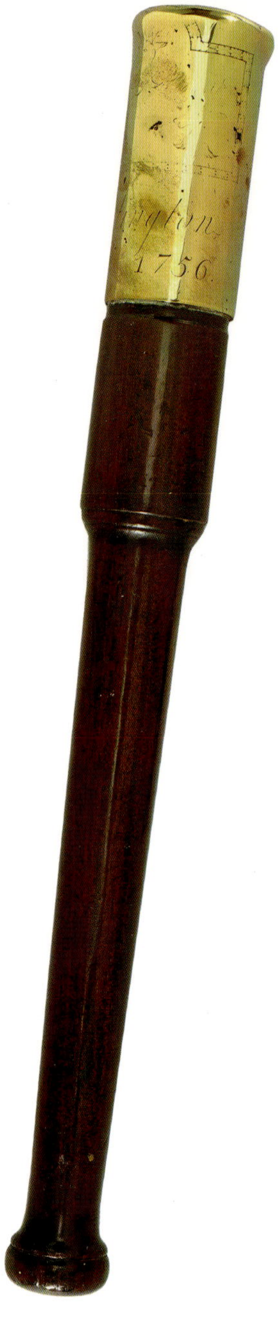

PROVENANCE

bt. John H. Bryan of Lake Bluff, IL (1936–2018) from Tobias Jellinek Antiques, London, January 26, 1995

LITERATURE

Jennifer Anderson, *Mahogany: The Costs of Luxury in Early America* (Cambridge, MA: Harvard University Press, 2012)

E. R. H. Dicken, *The History of Truncheons* (Ilfracombe, UK: Arthur H. Stockwell, 1952)

1789

Staff

Lignum vitae, brass

21 × $1\frac{1}{4}$ in. (53.3 × 3.2 cm)

Inscription: BR; 1789 | Edwd Markland Esq | MAYOR

After a career as a merchant, Edward Markland (1749–1832) was elected mayor of Leeds in 1789 and 1807. This sleek staff commemorated the first of his two elections: the etched date, incorrectly engraved initially as 1799, has scratch marks where the year was changed to 1789. Ornamental in purpose but suggesting the violence of a club or a truncheon, the staff would have served as a visual reminder of Markland's political power and social prestige. By the late eighteenth century, the exportation of hardwood from Jamaica had stalled, but the lignum vitae in this staff may nonetheless have come from Jamaica or another Caribbean island, where it was harvested and processed by enslaved African laborers.

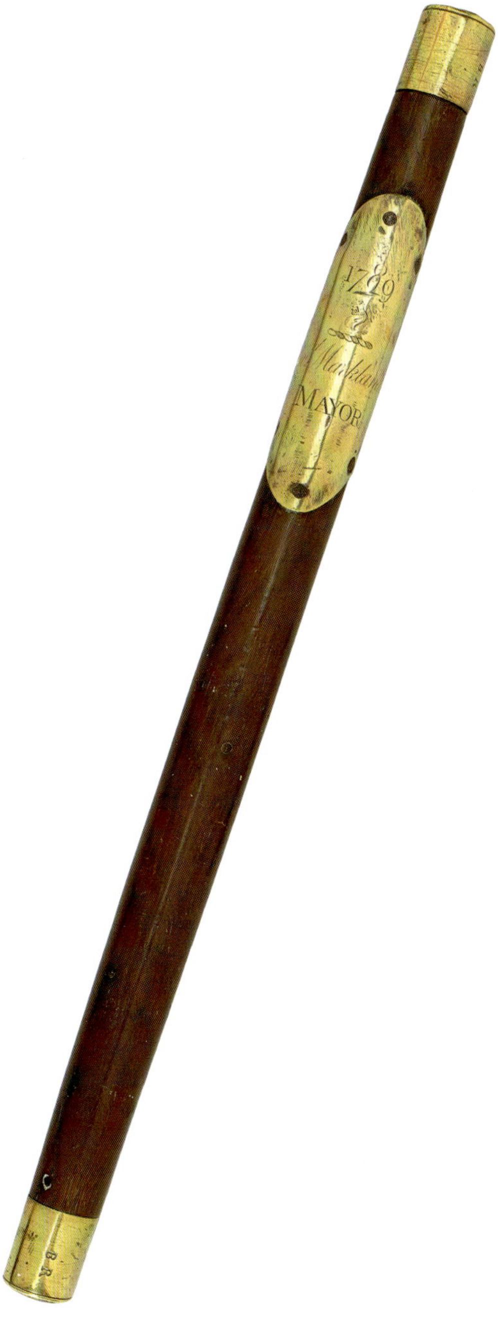

PROVENANCE

bt. John H. Bryan of Lake Bluff, IL (1936–2018) from Tobias Jellinek Antiques, London, February 8, 1996

LITERATURE

E. R. H. Dicken, *The History of Truncheons* (Ilfracombe, UK: Arthur H. Stockwell, 1952)

Rev. R. V. Taylor, *The Biographia Leodiensis: Or, Biographical Sketches of the Worthies of Leeds and Neighbourhood* (London: Simpkin, Marshall, 1865)

1769

Tipstaff

Lignum vitae, brass

8 ½ × 1 ¼ in. (21.6 × 3.2 cm)

Inscription: Daniel Munton|Inspector of|the Pavement|of the City of|London| 1769

Although the phrase *inspector of the pavement* eventually became a slang reference to a person locked in a pillory, it here means something both more practical and more prestigious. Daniel Munton, a member of the Worshipful Company of Paviors, which paved and maintained London's streets, served an important role in the guild. He took an apprentice and was instrumental in the appointment of a new clerk to the company in 1789. Although Munton acted in this official capacity for his company, he described himself as a stone mason of Aldgate in his will, which was proved on March 20, 1797 (TNA PROB 11/1287/140).

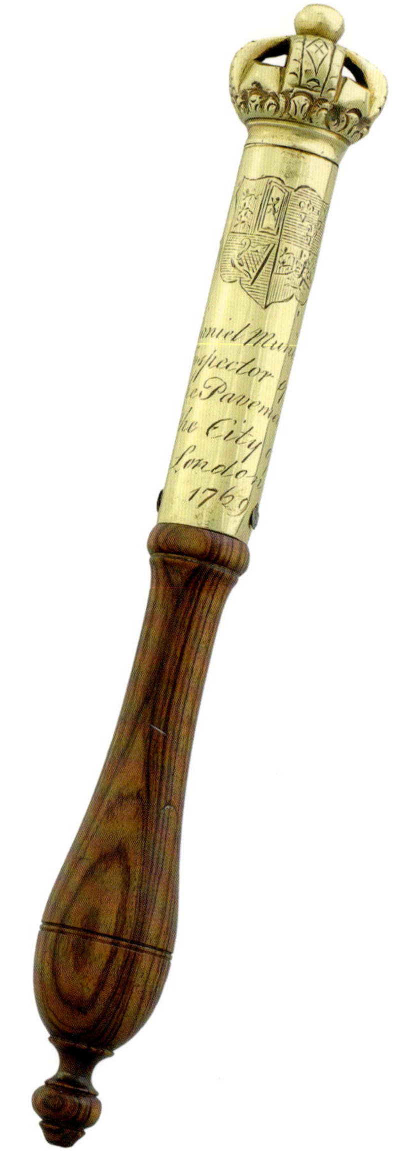

PROVENANCE

Tristram Jellinek of London (1933–1995); bt. John H. Bryan of Lake Bluff, IL (1936–2018) from Alistair Sampson Antiques, London, August 6, 1997

LITERATURE

E. R. H. Dicken, *The History of Truncheons* (Ilfracombe, UK: Arthur H. Stockwell, 1952)

Charles Welch, *History of the Worshipful Company of Paviors of the City of London* (London: The Worshipful Company of Paviors, 1909)

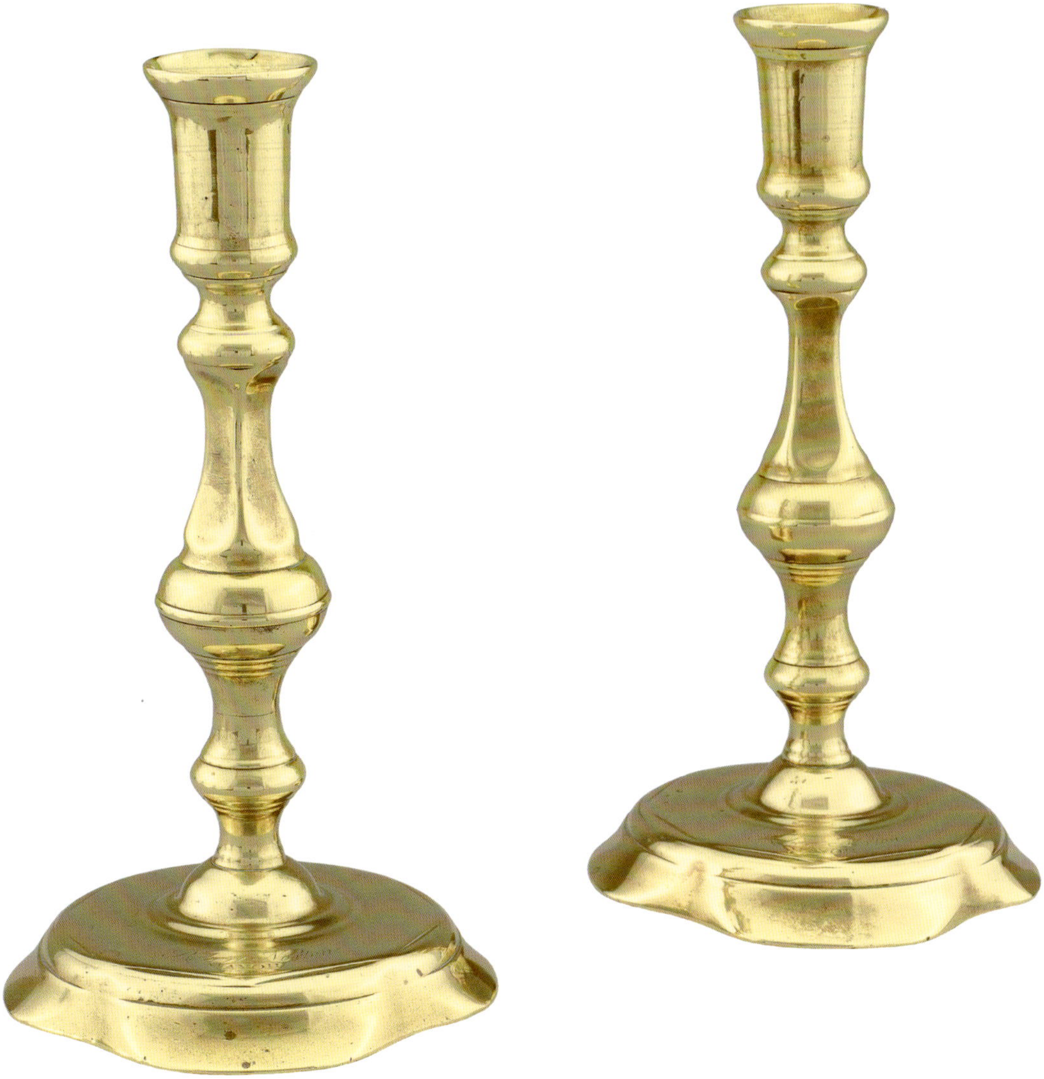

1734

Candlesticks

Brass

10 × 5¼ in. (25.4 × 13.3 cm)

Inscription, on both: Robert Norton|MAYOR|1734

This pair of candlesticks was made to celebrate Robert Norton's term as mayor of the city of Chichester, Sussex, in 1734. Norton lived between 1694 and 1768 and served as mayor on two other occasions, in 1747 and 1766.

PROVENANCE

Casimir Collection; bt. John H. Bryan of Lake Bluff, IL (1936–2018) from Alistair Sampson Antiques, London, June 10, 1983

1754

Plate
Delftware
$1\frac{1}{8}$ × 9 in. (2.9 × 22.9 cm)

Inscription:
Fortescue.& Harris.|
For Ever.|1754

This Delftware plate was made in support of the candidates John Fortescue and John Harris (1703–1768) on the occasion of the May 31, 1754, election for the Barnstaple constituency in Devon. The results were Harris, 217 votes; George Amyand (1720–1766), 205 votes; and Fortescue, 154 votes. Harris and Amyand were returned with clear majorities.

PROVENANCE
C. E. Roberts Chanter in 1914; Christie's, London, May 12, 1969 (lot 6); with Jonathan Horne, London; Syd Levethan: The Longridge Collection; Christie's, London, November 3, 2011 (lot 337); The Birkett Collection; Sotheby's, London, From Earth to Fire, May 4, 2017 (lot 88); bt. John H. Bryan of Lake Bluff, IL (1936–2018)

LITERATURE
Leslie B. Grigsby, *The Longridge Collection of English Slipware and Delftware* (London: Jonathan Horne Publications, 2000), 2:92

Louis L. Lipski and Michael Archer, *Dated English Delftware: Tin-Glazed Earthenware, 1600–1800* (London: Sotheby Publications, 1984), 129, no. 575

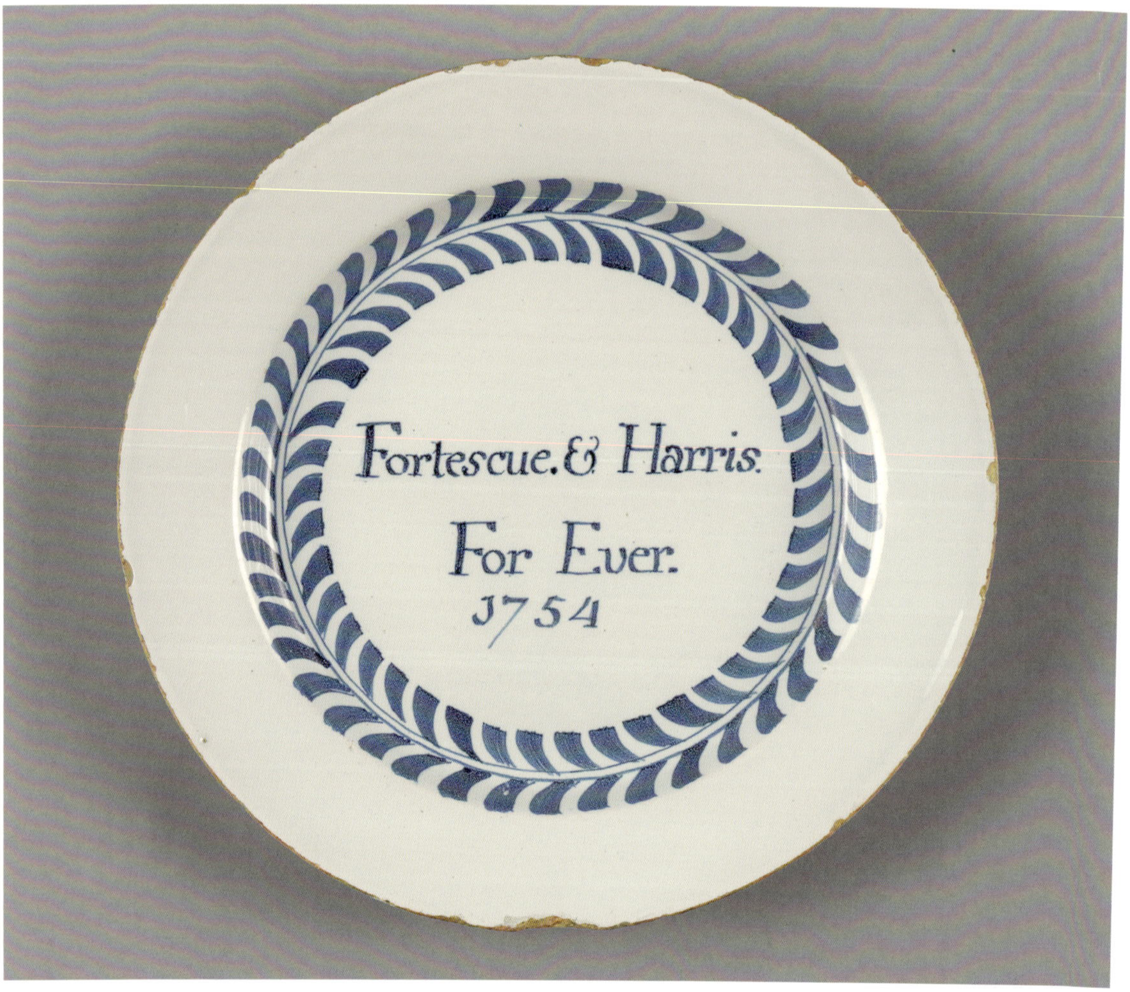

1761

This plate is part of a set of at least three extant plates, all with bianco-sopra-bianco ("white on white") decoration and made in Bristol. The painted decoration depicts a horse and rider identified as Abraham Bosavern, a customs officer from Polperro, Cornwall, patrolling the coast between Looe and Fowey. Bosavern was buried at St Mary by Looe on February 21, 1782.

PROVENANCE

Warren Baker, Indiana; Private Collection, London; bt. John H. Bryan of Lake Bluff, IL (1936–2018) from John Howard at Heritage, Oxfordshire, June 30, 2017

LITERATURE

Louis L. Lipski and Michael Archer, *Dated English Delftware: Tin-Glazed Earthenware, 1600–1800* (London: Sotheby Publications, 1984), 140, no. 622

Plate

Delftware

1 × 9 in. (2.5 × 22.9 cm)

Inscription:
A BOSAVERN LOOE.|1761

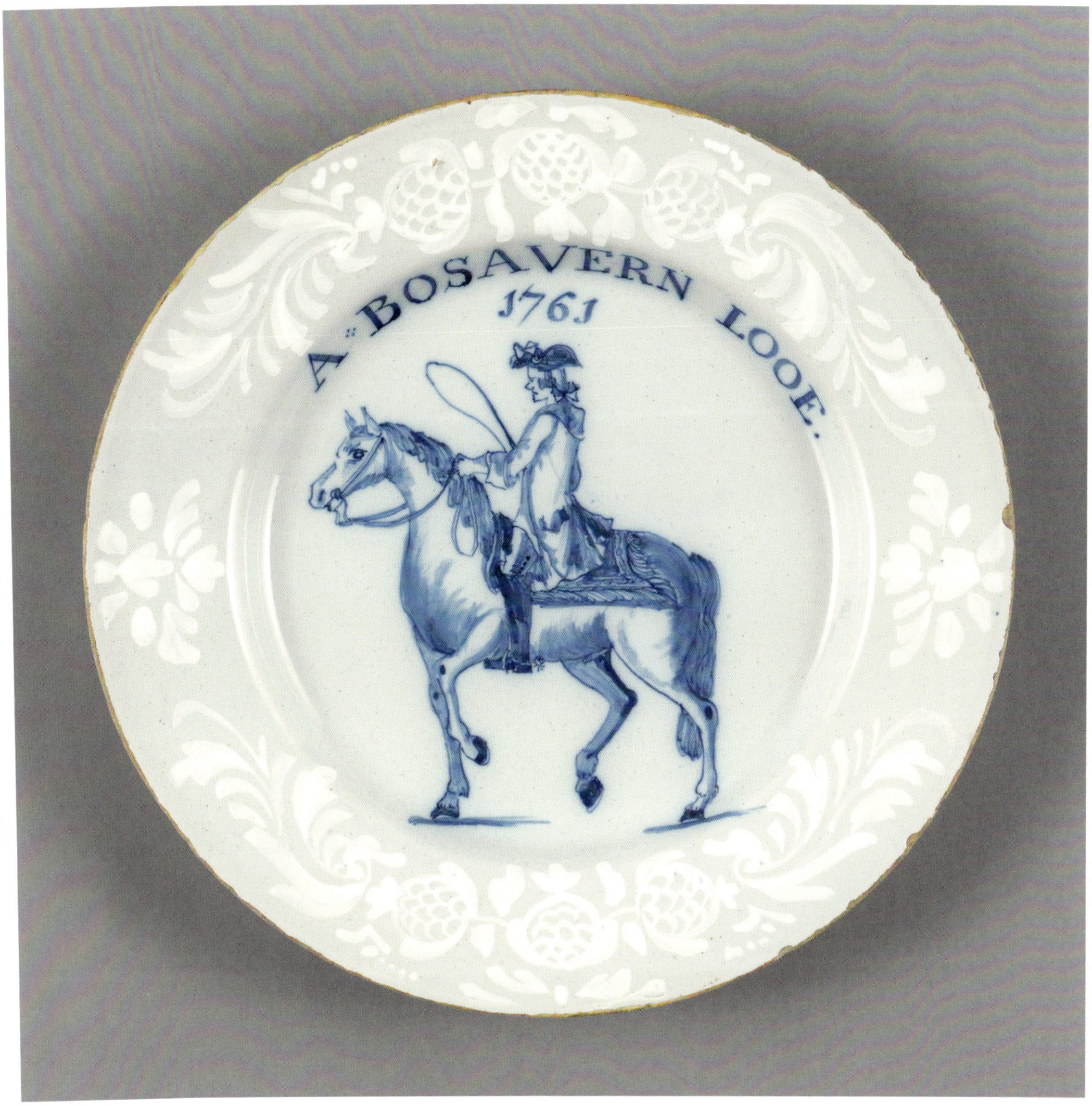

1767

Plate
Delftware
1 × 9¼ in. (2. 5× 23.5 cm)

Inscription: THO^S: SMITH : WOOL HOPE 1767.

This plate, for a Thomas Smith at Woolhope, Herefordshire, may well have commemorated him as the winner of a local race in 1767.

PROVENANCE
bt. John H. Bryan of Lake Bluff, IL (1936–2018) from Alistair Sampson Antiques Ltd., London, April 28, 1992

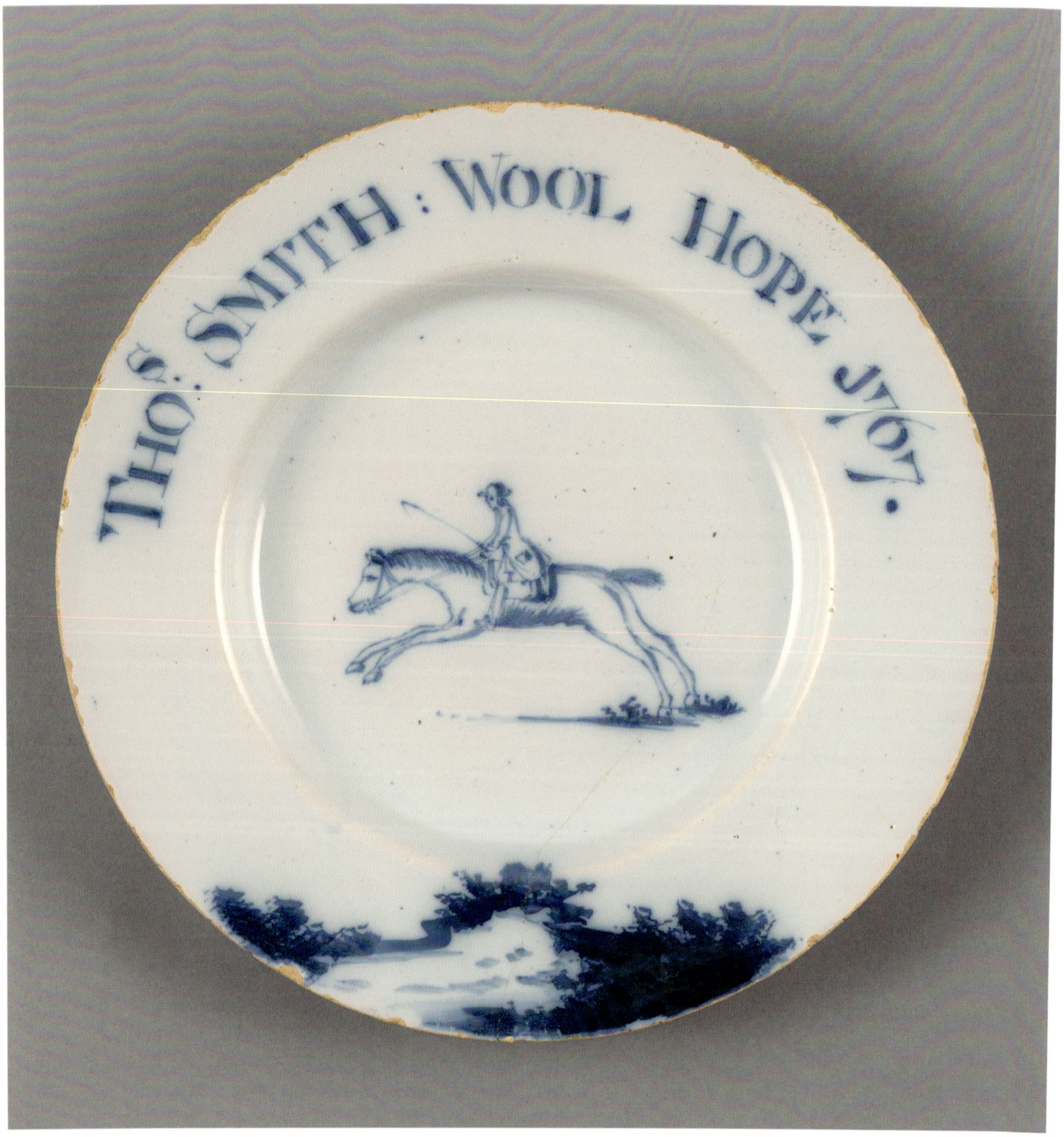

1770

The painting featured on this plate may be of a slave ship. The vessel is similar to that depicted on a ship bowl in the Liverpool Museum (50.60.13), which bears the inscription, "Success To The Dobson 1770."

PROVENANCE

L. A. Compton; Christie's, London, February 12, 1990 (lot 10); bt. John H. Bryan of Lake Bluff, IL (1936–2018) from Alistair Sampson Antiques Ltd., London, February 15, 1990

Plate

Delftware

$1\frac{1}{8} \times 9\frac{1}{4}$ in. (2.9 × 23.5 cm)

Inscription: P|1770

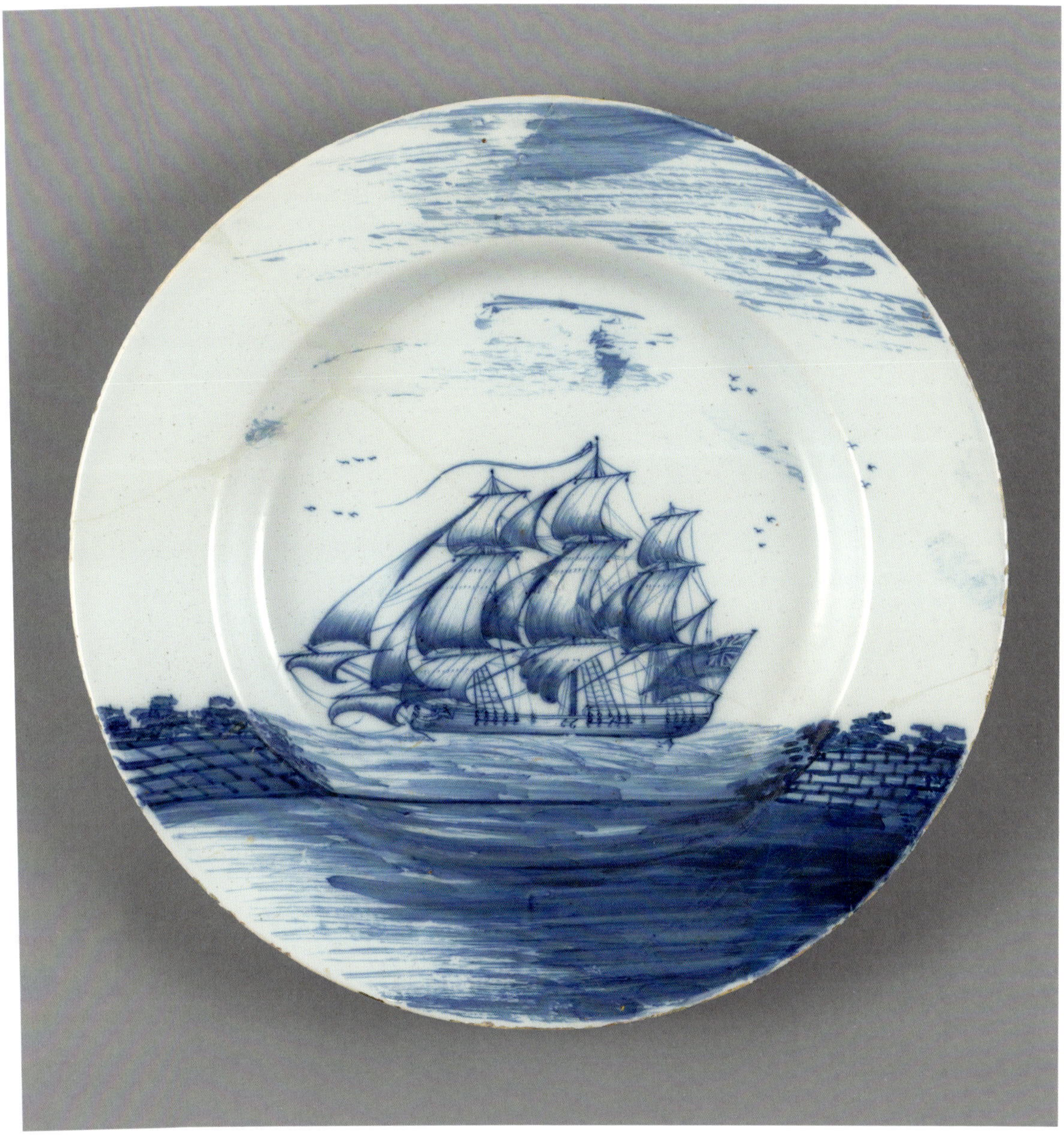

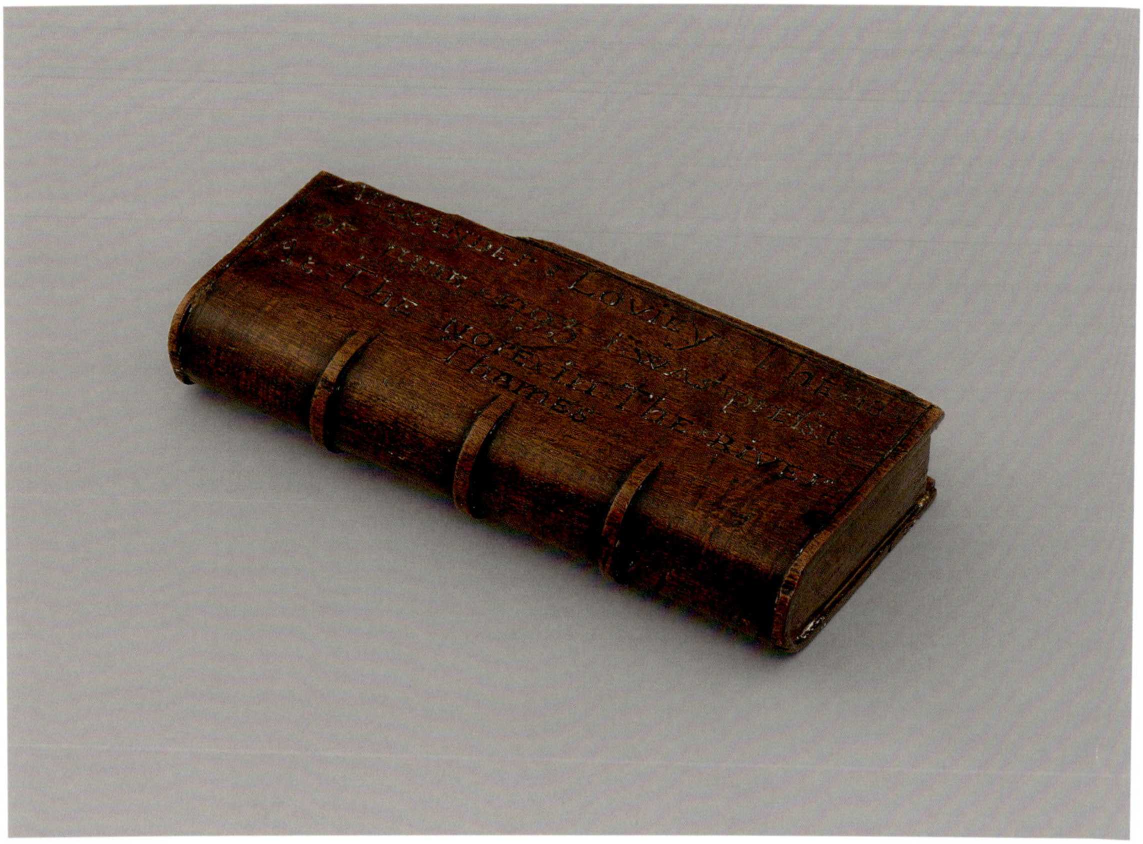

1793

Box
Wood
$1\frac{1}{4} \times 6\frac{7}{8} \times 3$ in.
(3.2 × 17.5 × 7.6 cm)

Inscription:
ThomAs ANthoNy WittoN WilsoN|I cAmE on BoArD ThE 7 oF|MArch At shEArnEst RivEr| ChATlhEM 1793; ALEXANDEr LOVIEY THE 18|OF JUnE 1793 i WAS PrEss,t|At ThE NorE in ThE RivEr|ThamEs

The Nore was an anchorage at the Thames Estuary, where, as this inscription records, Alexander Loviey was "presst" into the navy in June 1793. The word *press* (as in *press-gang*) was a corruption in regular use at the end of the eighteenth century of the word *prest*, an old French word meaning "a loan or advance." The organization at the ports charged with obtaining man power was known as the Impress Service, and once coerced or forced into service, recruits became "imprest" or "prest" men.

The Impress Service was limited to enlisting men who were "seamen," a word given a broad interpretation. Although age limits were set at eighteen to fifty-five years, they were frequently ignored. Anthony Thomas Witton Wilson was baptized on January 8, 1783, at St Sepulchre-without-Newgate, London; he would have just turned ten when he went aboard. Loviey would have been a little older; when he eventually made his will aboard HMS *Majestic* on November 23, 1794, he had a wife named Elizabeth living on King Street in South Shields. George Blagdon Wescott (1732–1798) was the captain of the *Majestic*, a seventy-four-gun, third-rate ship of line that fought at the Battle of the Nile, where Wescott was killed—and possibly Loviey, too, as his will was proved in March 1800.

PROVENANCE

bt. John H. Bryan of Lake Bluff, IL (1936–2018) from Tobias Jellinek Antiques, London, January 19, 2017

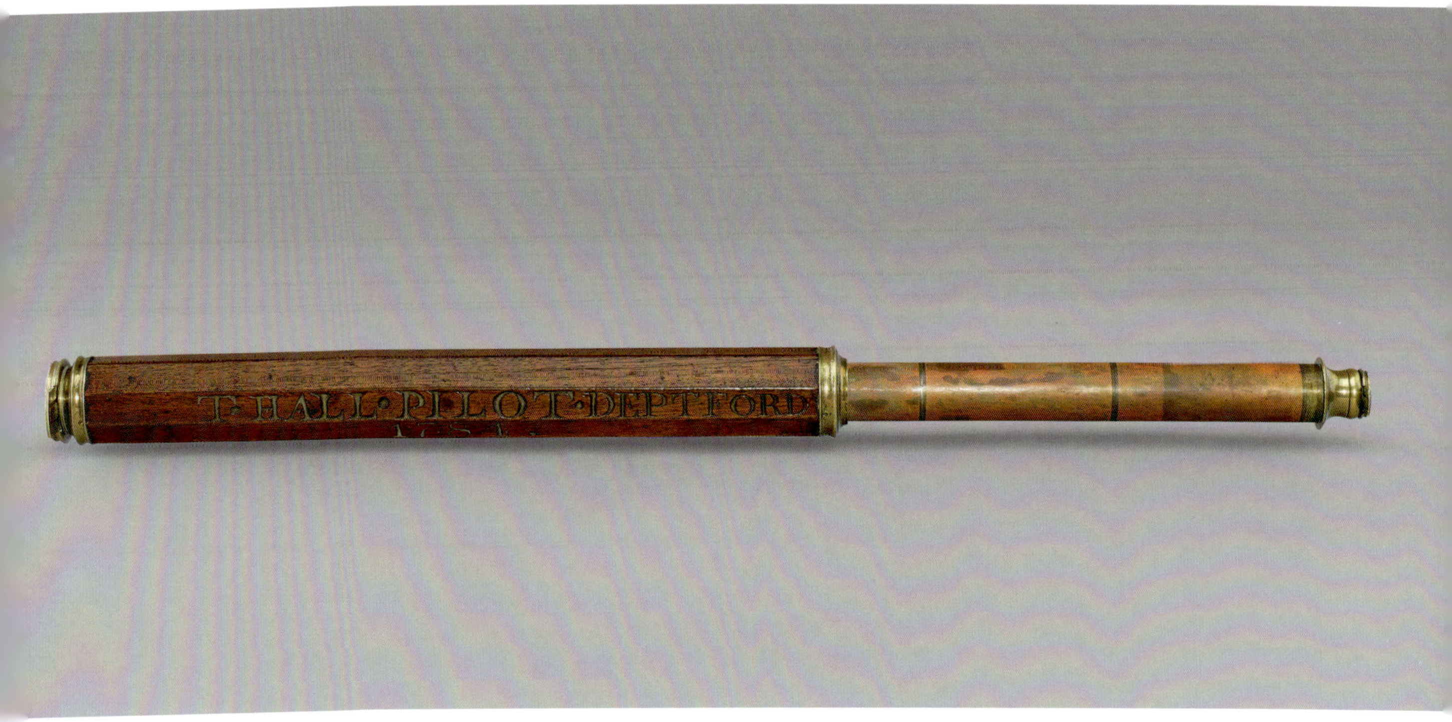

1784

Telescope

Brass, mahogany, glass

$13\frac{1}{2} \times 1\frac{1}{2}$ in. (34.3 × 3.8 cm)

Inscription: T·HALL·PILOT·DEPTFORD| 1784.

Ship pilots guided vessels through busy or dangerous waters, such as at harbors or the mouths of rivers. They were expert ship handlers with in-depth knowledge of the local waters. Pilots were often self-employed, meaning that they needed to have quick transport to get from port to the incoming ships, a necessity that eventually led to the development of the "pilot cutter," a fast boat that carried pilots to the larger vessels.

PROVENANCE

Tristram Jellinek of London (1933–1995); Sotheby's, London, The Tristram Jellinek Collection, May 10, 1996 (lot 695); bt. John H. Bryan of Lake Bluff, IL (1936–2018) through Tobias Jellinek Antiques, London

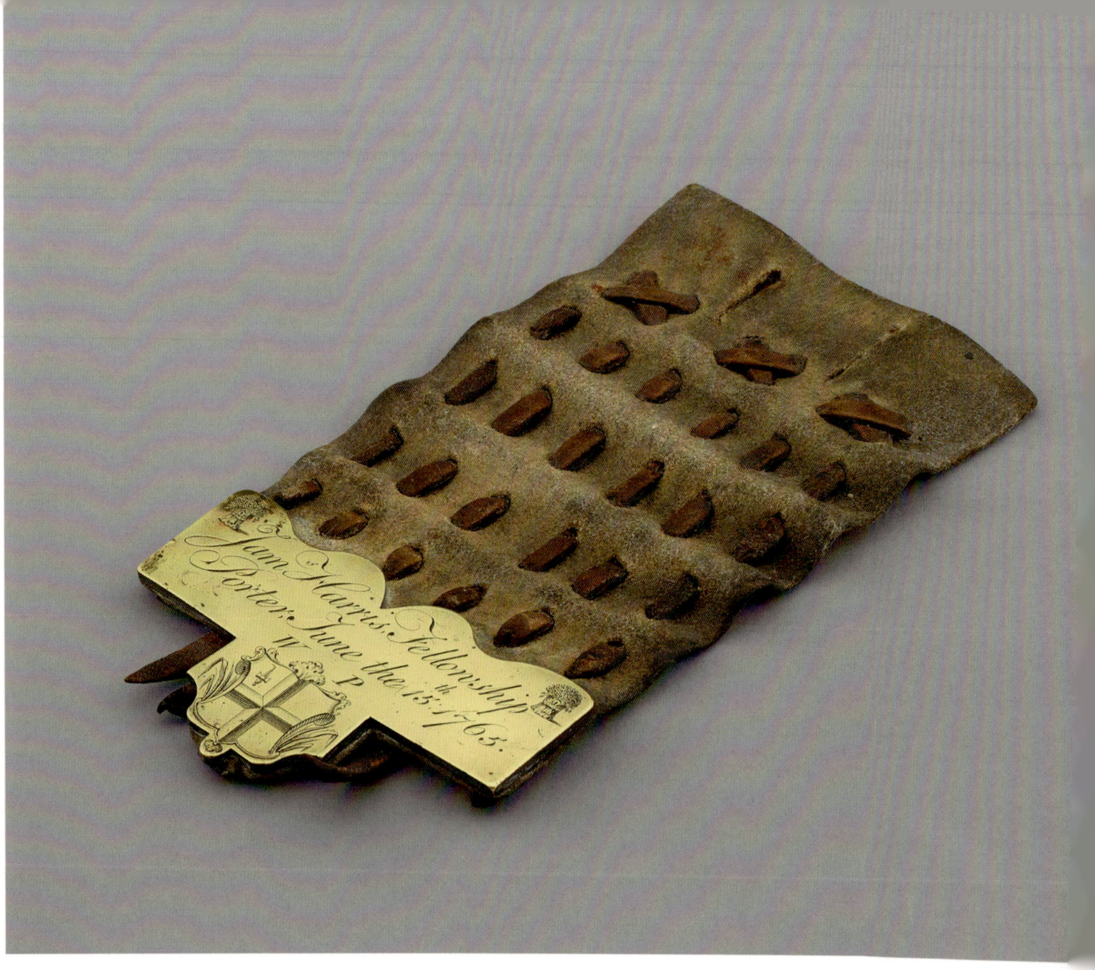

1765

Porter's tally

Brass, leather

$7\frac{1}{4} \times 3\frac{1}{2} \times \frac{1}{4}$ in. (18.4 × 8.9 × .64 cm)

Inscription:
Sam Harris, Fellowship|Porter, June the 15th 1765; W P

The Fellowship Porters comprised two groups: Tackle Porters and Ticket Porters. Their work involved unloading, landing, and housing various goods and transporting corn, salt, coal, fish, and fruits of all types. Each Ticket Porter wore a metal badge or ticket, inscribed with his name, his number as registered, and the date he was granted his freedom. The leather pieces held the tallies for loads carried—sticks with notches to mark numbers or payments—and were attached by two holes to the wearer's belt. A similar tally carrier and badge, with the inscription, "Robert Wiseman Fellowship Porter Oct 5 1785," was sold by Woolley & Wallis, Salisbury, January 23, 2011 (lot 409).

On December 29, 1742, Samuel Harris of St Martin-in-the-Fields, Porter, and Mary Mathars of the same parish were joined in a clandestine marriage at the Fleet.

PROVENANCE

bt. John H. Bryan of Lake Bluff, IL (1936–2018) from Alistair Sampson Antiques Ltd., London, June 7, 1996

1779

Coffer

Copper, iron, leather, wood

6¼ × 9 × 6½ in. (15.9 × 22.9 × 16.5 cm)

Inscription: BRIDGE PORTERS| For, Y^{E} City & Liberty of|WESTMINSTER| JAN,^{Y} 1,st 1779

The twenty-four members of the Society of Bridge-Porters enjoyed the privilege of transporting burdens to and from Queens Bridge in New Palace Yard in Whitehall and all the wharves of the city.

This box, which belonged to the society, has two lock plates and is bolted throughout. Inside, leather folds on either side of the coin slot ensure that, even if the box is turned upside down, the coins will not fall out.

PROVENANCE

bt. John H. Bryan of Lake Bluff, IL (1936–2018) from Tobias Jellinek Antiques, London, January 6, 2017

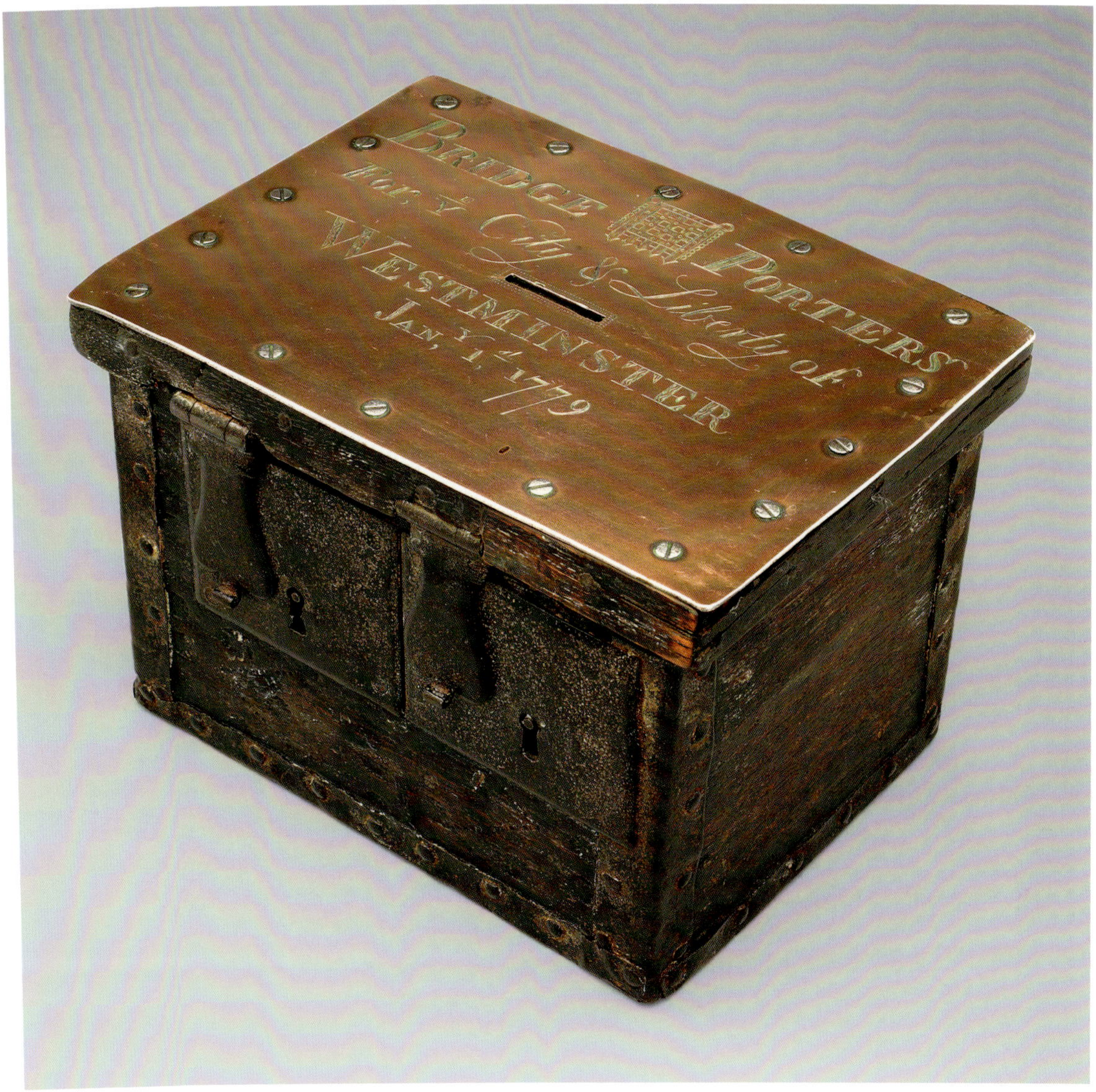

1671

Wet drug jar
Delftware
$7\frac{1}{4} \times 5 \times 7\frac{1}{2}$ in. (18.4 × 12.7 × 19.1 cm)

Inscription: S:VIOLARVM|B:B|1671

Drug jars were very much utilitarian items, made to be used but also, to a degree, to be displayed. The opening has a curved lip to allow the securing of a textile cover over the contents. Wet drug jars usually have a spout with a cork to allow the decanting of liquid mixtures.

As the inscription indicates, this jar contained a syrup derived from violets. It was mainly used as flavoring but offered the added bonus of having a mild laxative effect.

PROVENANCE
bt. John H. Bryan of Lake Bluff, IL (1936–2018) from Alistair Sampson Antiques Ltd., London, December 22, 1989

1671

Wet drug jar
Delftware
$7 \times 5 \times 7\frac{1}{4}$ in. (17.8 × 12.7 × 18.4 cm)

Inscription: S:AVRANTIO|B:B|1671

The "angel with outspread wings" design has been primarily seen on examples dating from 1659 to 1725. The initial letter of the inscription indicates the type of substance the jar contained. In this case, "S:AVRANTIO" stands for *syrupus aurantiorum*, or "syrup of orange." This syrup was primarily used as flavoring rather than for any specific medicinal purpose.

PROVENANCE
bt. John H. Bryan of Lake Bluff, IL (1936–2018) from Alistair Sampson Antiques Ltd., London, December 22, 1989

1677

Wet drug jar
Delftware
$7\frac{1}{2} \times 5\frac{1}{2} \times 8\frac{1}{4}$ in. (19.1 × 14 × 21 cm)

Inscription: S.E: SVCC:CITR|I.R.|1677

The inscription indicates this jar held *syrupus e succo citri*, a syrup of lemon juice, which was mainly used as a flavoring rather than as a medicinal remedy.

PROVENANCE
W. A. H. Harding Trust; Christie's London, October 10, 1983 (lot 1); bt. John H. Bryan of Lake Bluff, IL (1936–2018) from Alistair Sampson Antiques, London, October 11, 1983

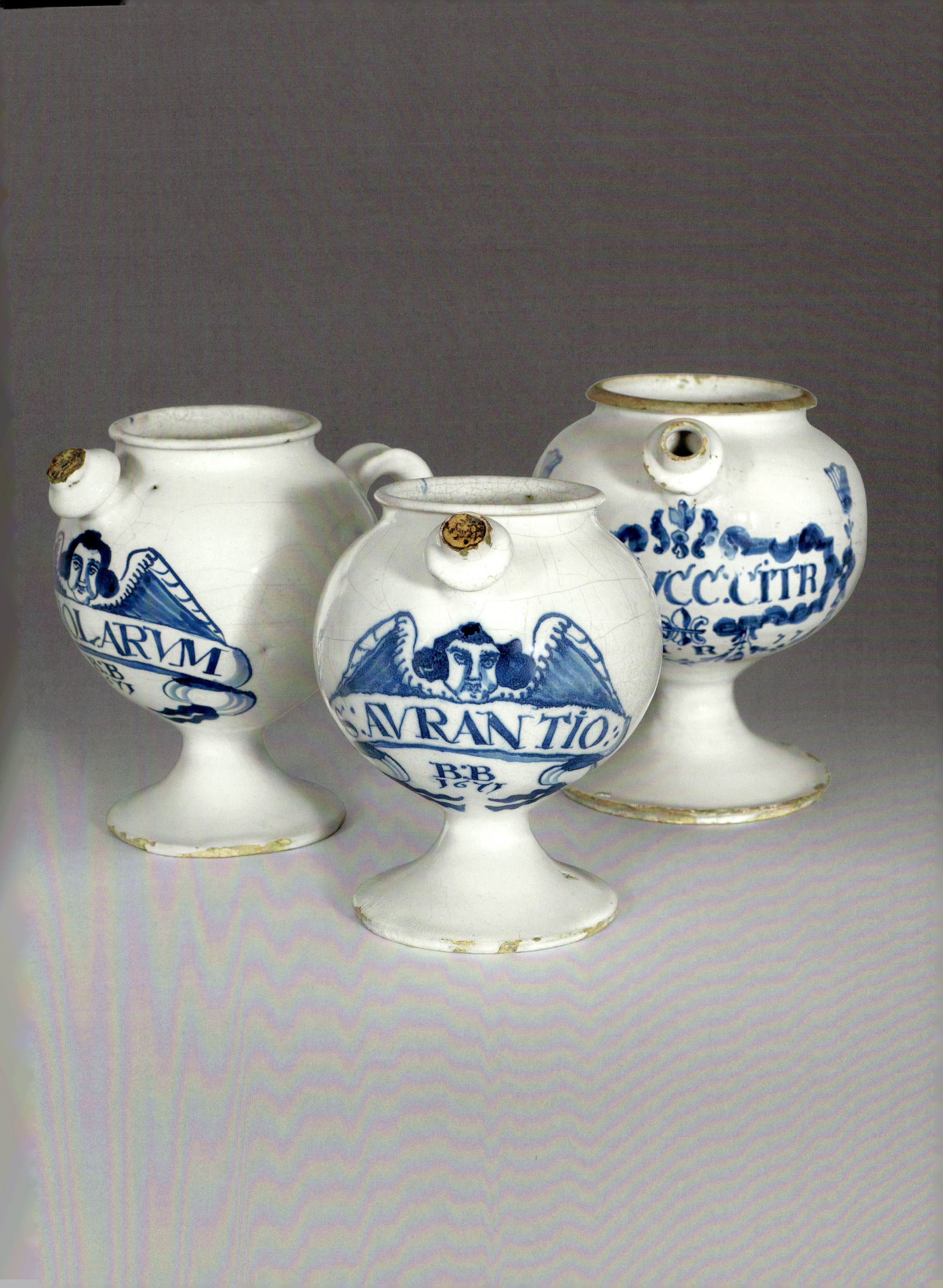
OLARVM
S:AVRANTIO:
B:B
VCC:CITR

1681

Dry drug jar
Delftware
7 × 5 in. (17.8 × 12.7 cm)

Inscription:
SALVIÆ; C 1681

"C" could stand for *confectio* ("confection"), *conserva* ("conserve") or *cortex* ("bark"). This jar contained *conserva salviae*, a conserve of sage leaves, to treat fevers. It was added to liquor, diet drinks, and medicinal ales, as a sweetener, and as a cleanser of the blood.

PROVENANCE
Sotheby's, London, October 1983; bt. John H. Bryan of Lake Bluff, IL (1936–2018) from Alistair Sampson Antiques, Ltd., London, November 17, 1983

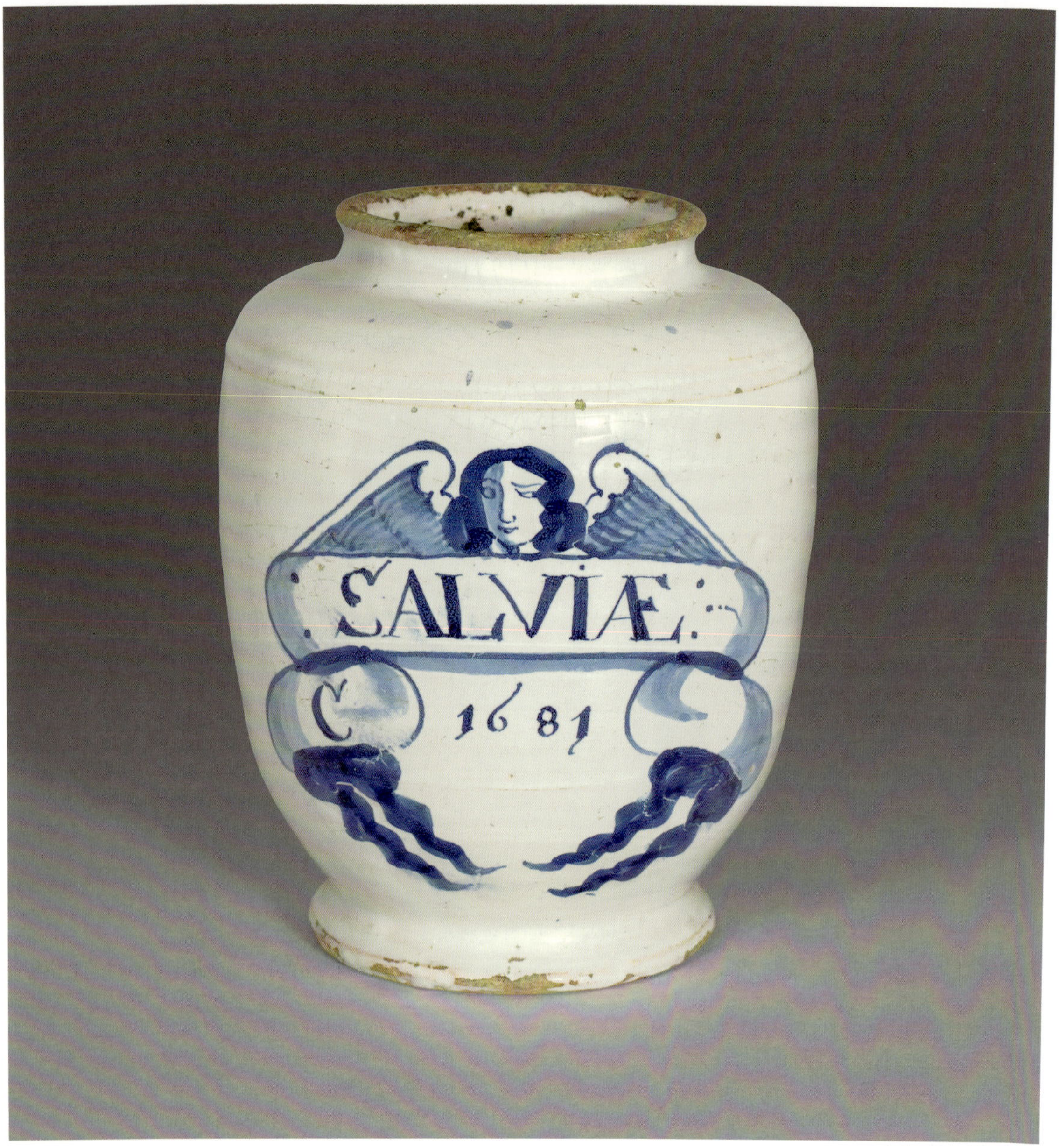

1757

Attributed to the Bow Porcelain Manufactory of New Canton, this mug is painted with the arms of the Company of Tylers and Bricklayers, which comprises two brick axes and a brush positioned between a chevron and a fleur-de-lys. The initials have not been identified.

PROVENANCE
Stacey's Auctioneers and Valuers, Rayleigh, Essex, September 17, 2017 (lot 800); bt. John H. Bryan of Lake Bluff, IL (1936–2018) from E. & H. Manners, London, September 29, 2017

Mug
Porcelain
5¾ × 6 in. (14.6 × 15.2 cm)

Inscription: IN GOD IS ALL OUR TRUST; M|W,|E|Ian^r,|1757

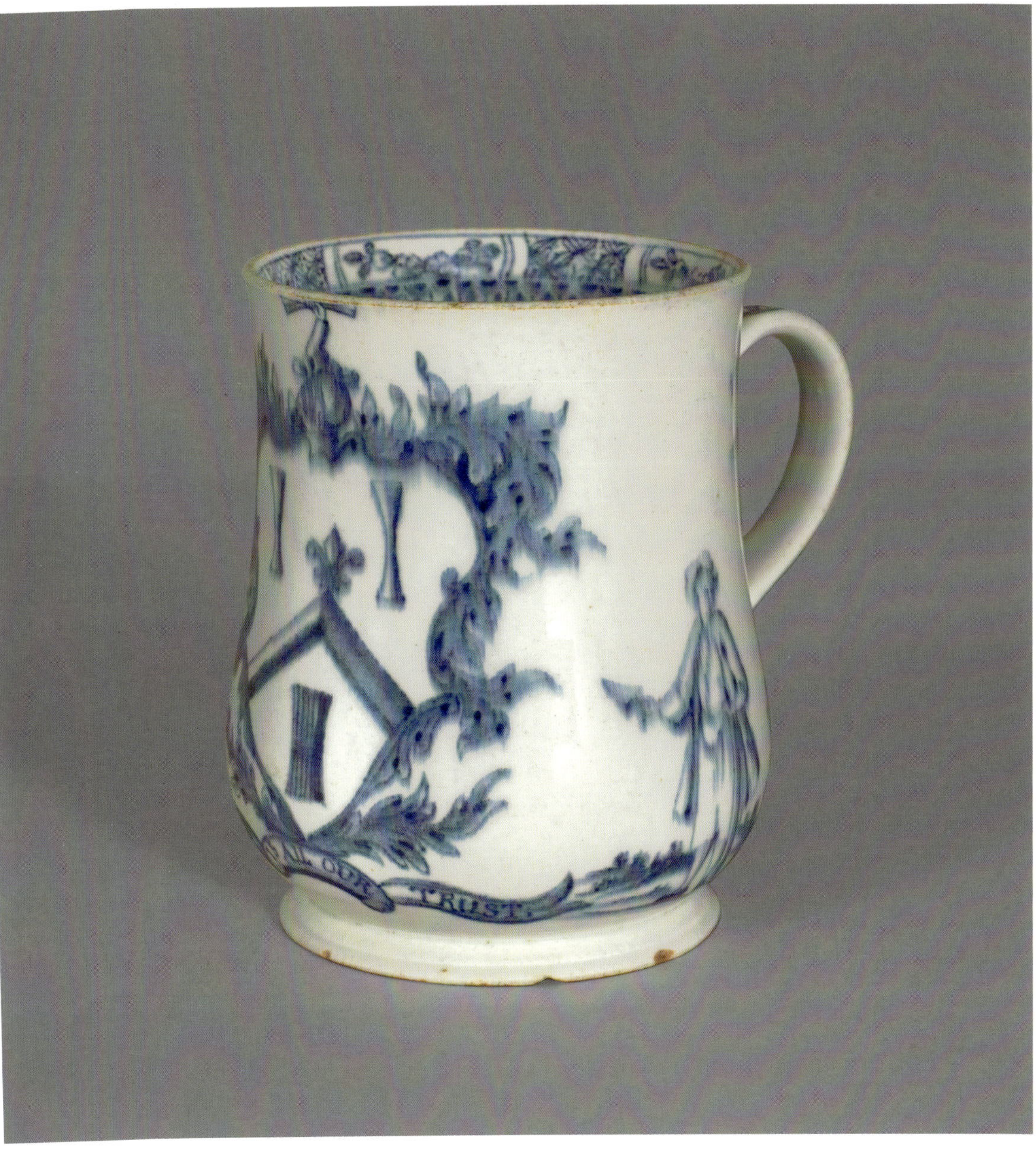

1726

Candlestick
Brass
$5\frac{3}{8} \times 5\frac{7}{8}$ in. (13.7 × 14.9 cm)

Inscription: S^{T} BARTHOLOMEWS HOSPITAL NOV.^{BR} 19TH 1726

This style of candlestick is known as a brass-socket "dwarf" or "desk" candlestick. Two similar candlesticks remain at St Bartholomew's Hospital in London.

PROVENANCE
bt. John H. Bryan of Lake Bluff, IL (1936–2018) from Alistair Sampson Antiques Ltd., London, October 1, 1998

LITERATURE
Eloy Koldewij, *The English Candlestick 1425–1925: 500 Years in the Development of Base-Metal Candlesticks* (London: Christie's Books Limited, 2001), 103, no. 70

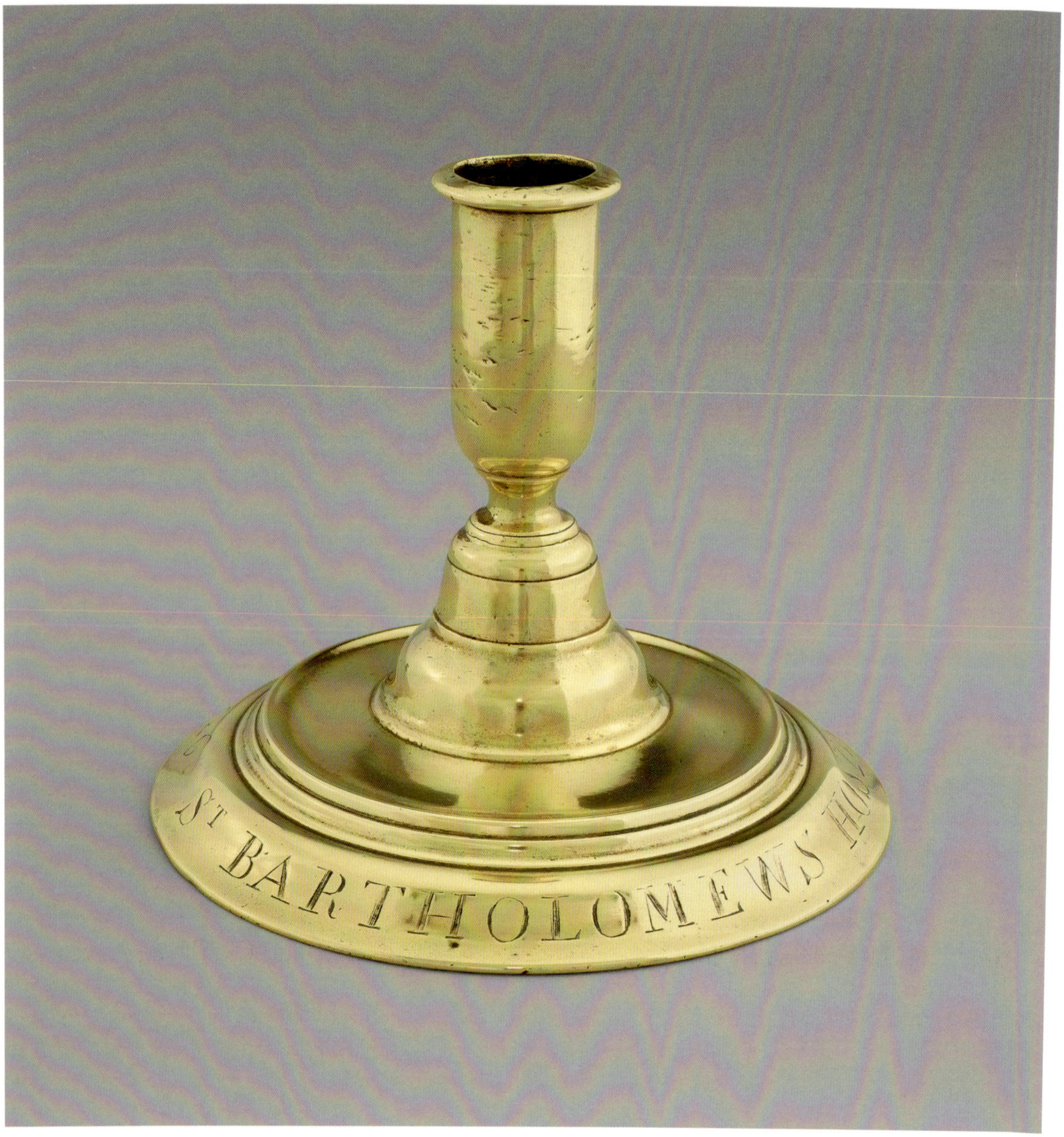

1745

Candlesticks

Gilded brass

Each: $9\frac{1}{4} \times 5\frac{3}{8}$ in. (23.5 × 13.7 cm)

John Bignell (active 1745) and Robert Clee (active 1711; d. 1773)

Inscription, on both: J^o^n Bignell cast this 1745 From|Corinthian Brass Roman Medals; Chasd by R Clee in London| Voucher to the Same

On the tops of their bases, these gilded-brass candlesticks are inscribed with the crest of Sir Thomas Bootle (1685–1753) of Lathom Hall, Lancashire. Bootle acquired his title in 1745, the year of the great Jacobite Rebellion, after delivering a loyal address in the House of Commons. He was also an important courtier, serving as chancellor to Frederick, Prince of Wales, between 1740 and 1751, and to George, Frederick's successor, from 1751 until Bootle's death in 1753.

The two makers were also acknowledged in two long inscriptions on the side of each base. The first tells how "J^o^n Bignell cast this 1745 From Corinthian Brass Roman Medals." John Bignell was apprentice to the candlestick maker Thomas Merry, who was active in London from 1718 to 1739. The second inscription reveals that the candlesticks had been "Chasd by R Clee in London Voucher to the Same." Unusually, the well-known engraver, Robert Clee, used his name and reputation to vouch for the historical origins of the brass. Clee, of Panton Street in St Martin-in-the-Fields, was an etcher, engraver, and printmaker whose will was proved on May 25, 1773 (TNA PROB 11/987/391).

PROVENANCE

bt. John H. Bryan of Lake Bluff, IL (1936–2018) from Jeremy Ltd., London, December 16, 1991

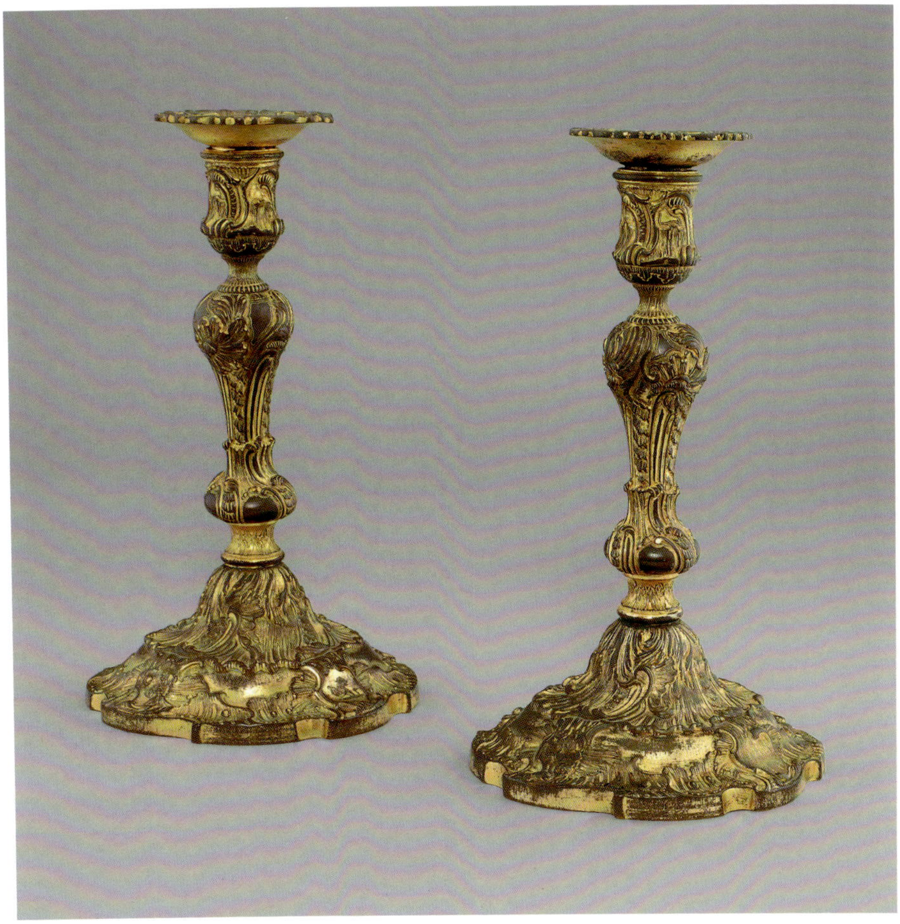

1757

Candlesticks
Brass
10 3/4 × 7 1/4 in. (27.3 × 18.4 cm)

Inscription, on both: WE ARE ONE; THE GIFT OF|M^R JN^O WHITE| MASTER|1757; W^M EASTWOOD|W^M SHEMELD| WARDENS

These candlesticks, whose bases seem to melt away like molten wax, are from a larger set of eight made for the master and wardens of the Armourers and Braisiers' Company. John White, master, took on apprentices between 1733 and 1758. At the time he made his will, which was proved in 1775, White was living in St Botolph-without-Bishopsgate (TNA PROB 11/1014/40).

PROVENANCE
The Worshipful Company of Armourers and Brasiers until 1838; Board Fine Arts Auctioneers, Haverhill, Suffolk, May 1986 (lot 225); bt. John H. Bryan of Lake Bluff, IL (1936–2018) from R. Jorgensen Antiques, Wells, ME, April 28, 1990

LITERATURE
Eloy Koldewij, *The English Candlestick, 1425–1925: 500 Years in the Development of Base-Metal Candlesticks* (London: Christie's Books Limited, 2001), 155, no. 141

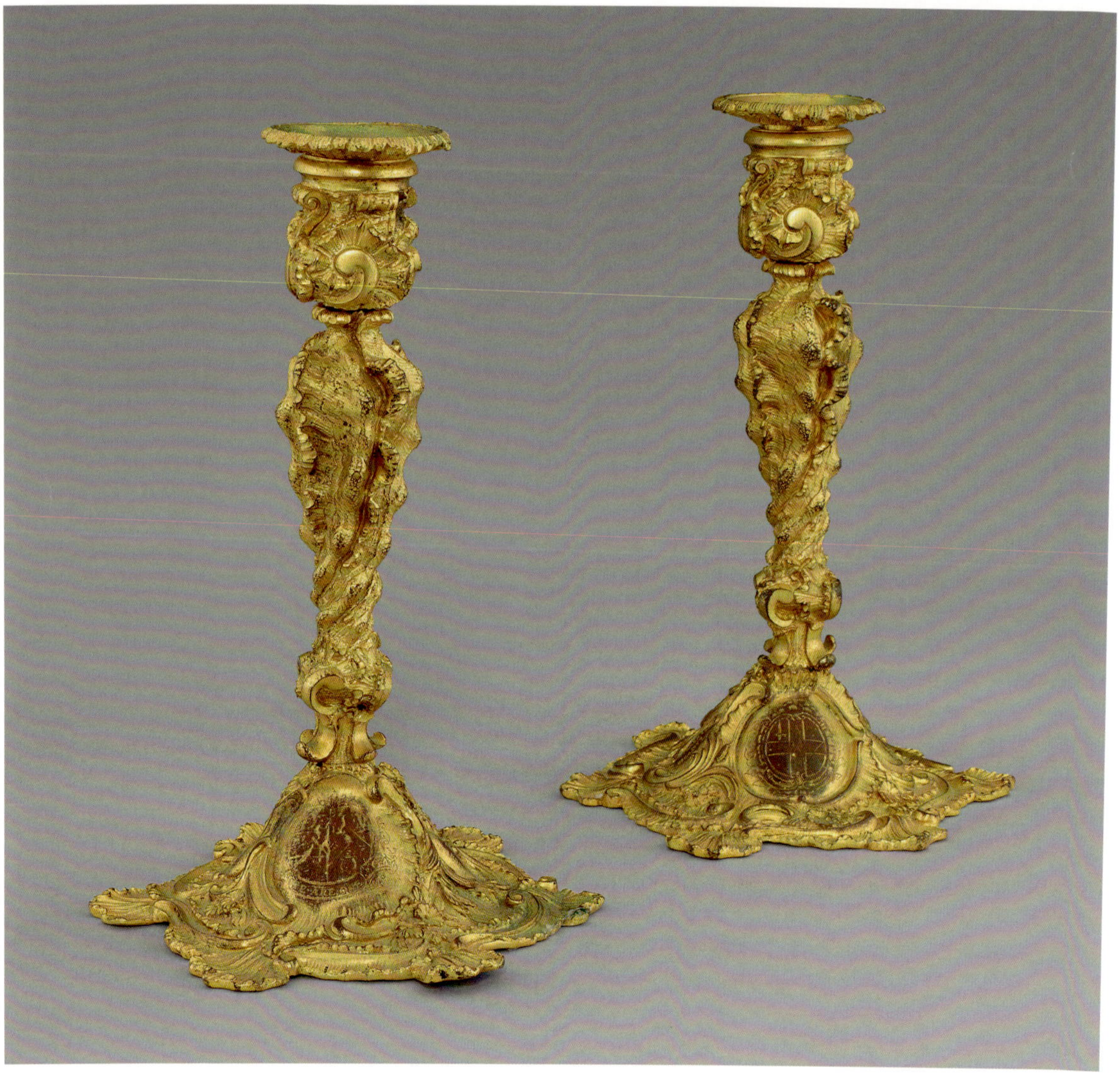

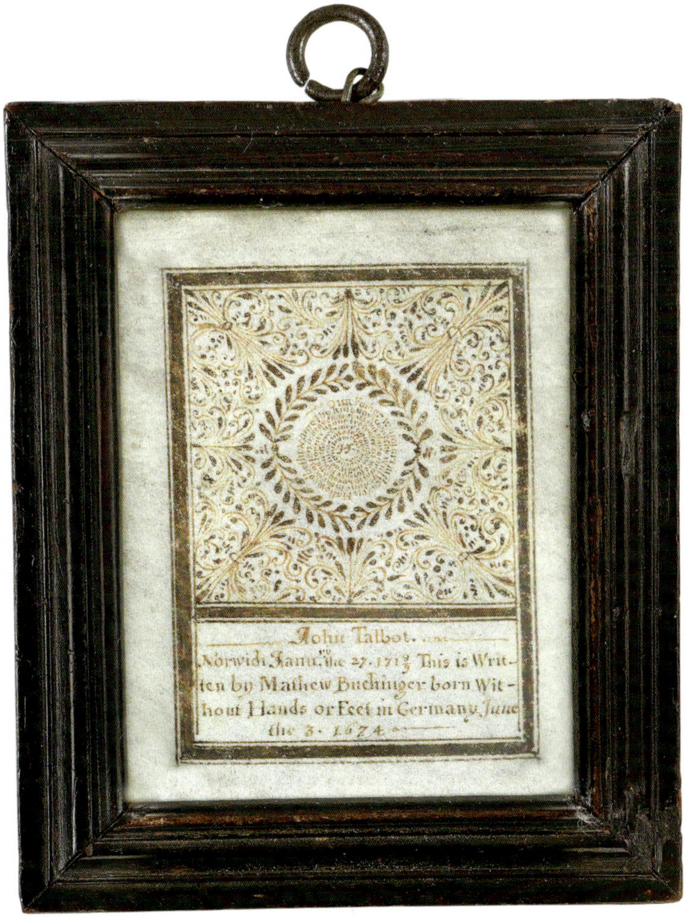

1674 | 1718/9

Calligraphic miniature

Ink on vellum

2¼ × 2 in. (7 × 5.1 cm)

Matthew Buchinger (1674–1739)

Inscription: John Talbot.|Norwich Janu^ry the 27. 1718/9 This is Writ-|ten by Mathew Buchinger born Wit-|hout Hands or Feet in Germany June|the 3. 1674

This small panel features a calligraphic miniature of the Lord's Prayer by the talented polymath Matthew Buchinger (1674–1739). A biographer reports that this particular talent of Buchinger's was well known: the antiquary Francis Grose reported that Buchinger once wrote the Lord's Prayer within the compass of a shilling. The fact that he was born without hands and legs did not inhibit him from becoming one of the best-known performers of the day. He played more than a dozen musical instruments at court both in England and Germany. He was also an accomplished dancer, magician, and bowler, proficient in making ships in bottles, and a sure shot with a pistol. Buchinger married four times and had eleven children.

An elegy written in 1722 (although, in fact, Buchinger did not die until 1739) remarked upon his many talents and virtues, including his musical abilities and the "very great Command" he had of his pen: "He practis'd Musick too, it did appear, / Tho' he no finger has, to have an Ear."

PROVENANCE

bt. John H. Bryan of Lake Bluff, IL (1936–2018) from Alistair Sampson Antiques Ltd., London, June 22, 1998

LITERATURE

Ricky Jay, *Matthias Buchinger: The Greatest German Living* (Los Angeles: Siglio, 2016)

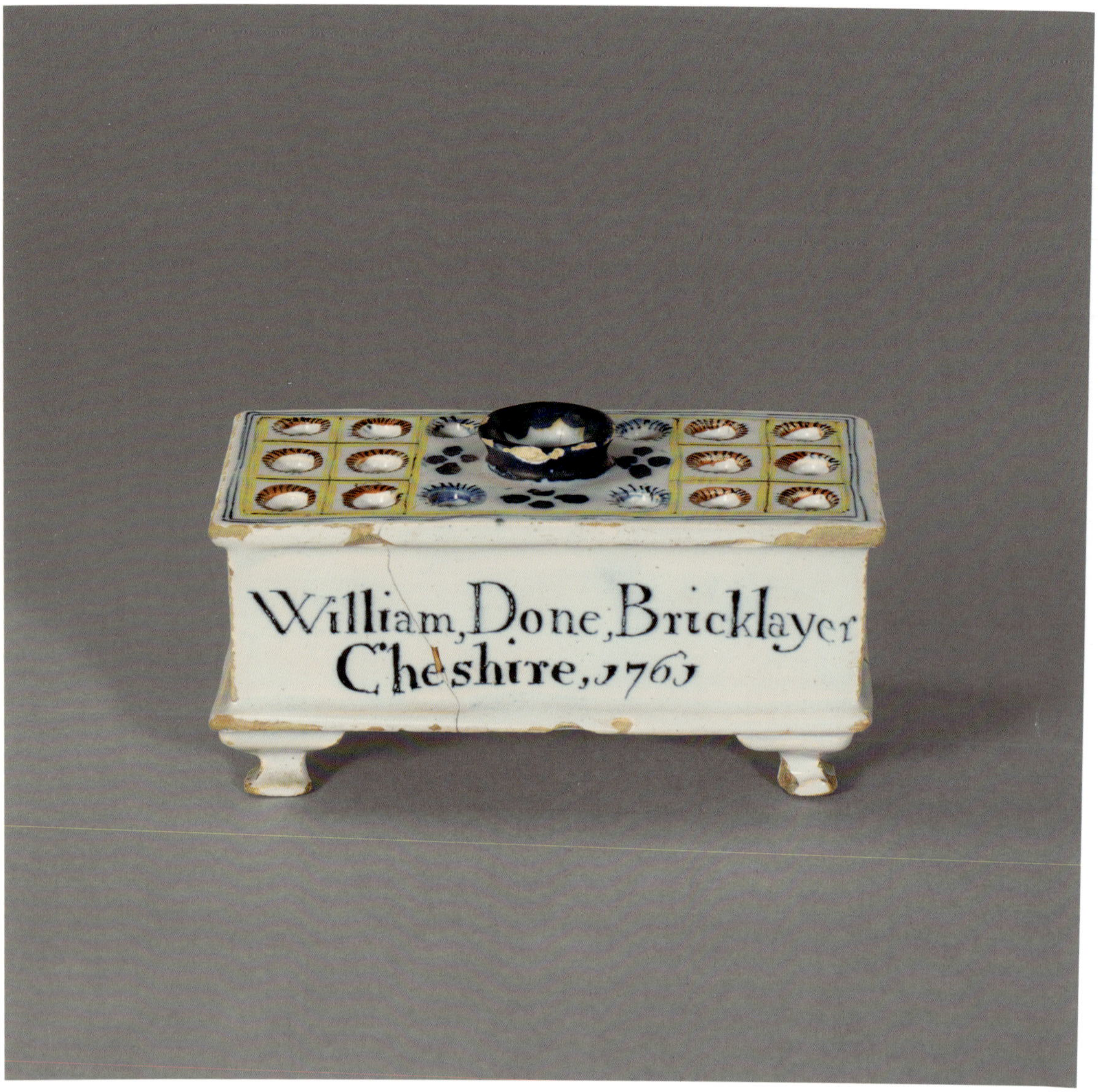

1761

Flower brick
Delftware
$2\frac{1}{4} \times 5 \times 2$ in. ($5.7 \times 12.7 \times 5.1$ cm)

Inscription: William,Done, Bricklayer| Cheshire,1761

A William Done was buried on May 28, 1777, at Eccleston, Cheshire, aged sixty-two. This fact and the colorful "Fazakerley" palette of the decoration—named for the village on the outskirts of Liverpool— would suggest that this is Liverpool Delft.

PROVENANCE
Joseph V. Vizcarra, Chicago, IL; Sotheby's, New York, April 13, 1984 (lot 432); bt. John H. Bryan of Lake Bluff, IL (1936–2018) through Alistair Sampson Antiques, London, April 17, 1984

LITERATURE
Louis L. Lipski and Michael Archer, *Dated English Delftware: Tin-Glazed Earthenware, 1600–1800* (London: Sotheby Publications, 1984), 364, no. 1587

1763

Brick fragment
Brick
$7\frac{3}{4} \times 13\frac{3}{4} \times 2\frac{1}{8}$ in.
(19.7 × 34.9 × 5.4 cm)

Inscription:
B|T·EE|1763

The bricks that make up this piece are not of the same depth as regular bricks of the period. These must have formed a veneer on the exterior of a building built in 1763. Some of the letters—perhaps all—may be the initials of the builders, but their identities remain unknown.

PROVENANCE
Woolley & Wallis, Salisbury, October 4, 2017 (lot 47); bt. John H. Bryan of Lake Bluff, IL (1936–2018) from Tobias Jellinek Antiques, London, October 12, 2017

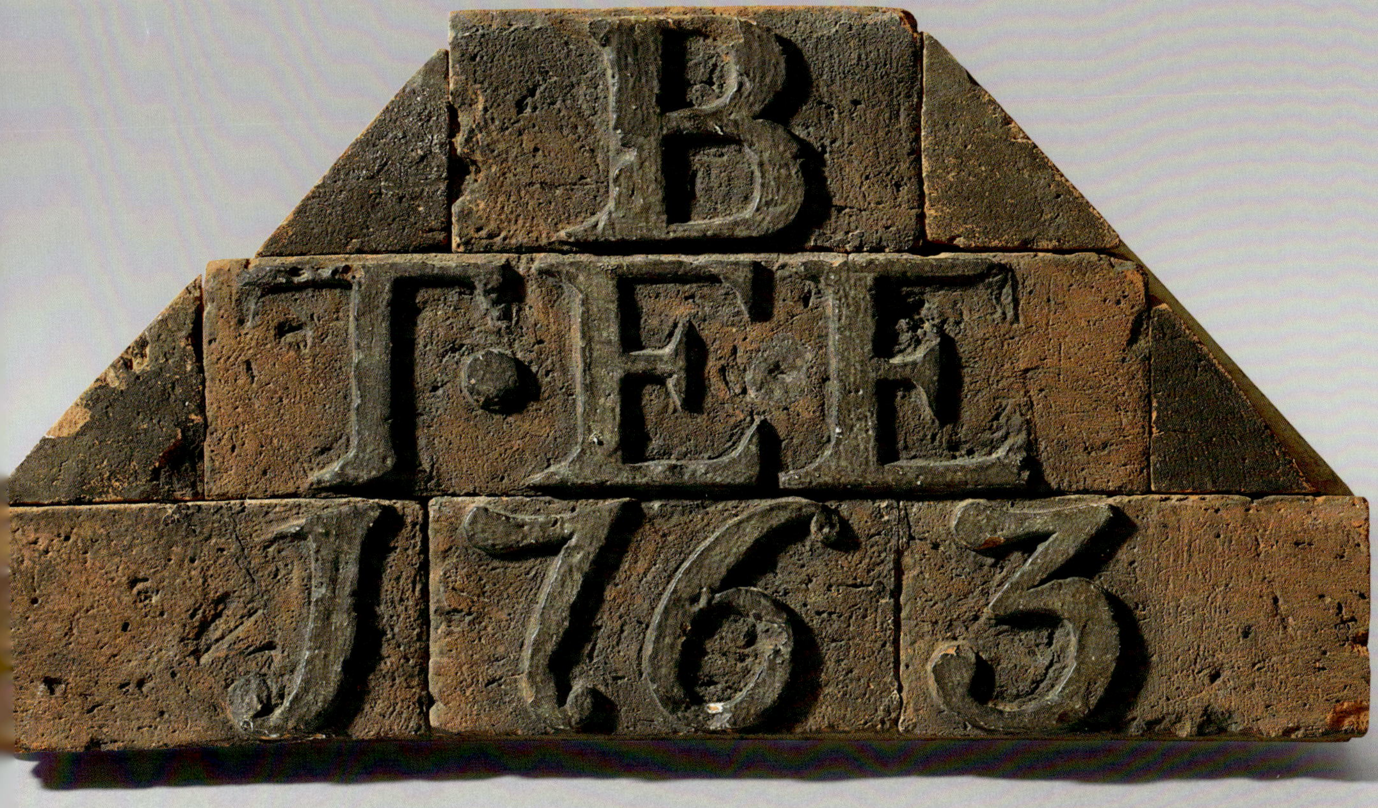

1691

Fountainhead
Lead
8 × 7½ × 2¼ in.
(20.3 × 19.1 × 5.7 cm)

Inscription: 1691

Fountainheads are also sometimes called "water quills." The word *quill* was used because the lead pipe that served as the spout was the size of a goose or swan's quill. Fountainheads were found both in public spaces and in private homes, emitting water from the town's water conduits. This example, in the form of a lion mask, is molded with a frilling mane. The lion's "paws" grasp a rectangular plate bearing the date, and the water spout issues from its mouth.

PROVENANCE
bt. John H. Bryan of Lake Bluff, IL (1936–2018) from Lindsay Antiques, London, December 12, 1994

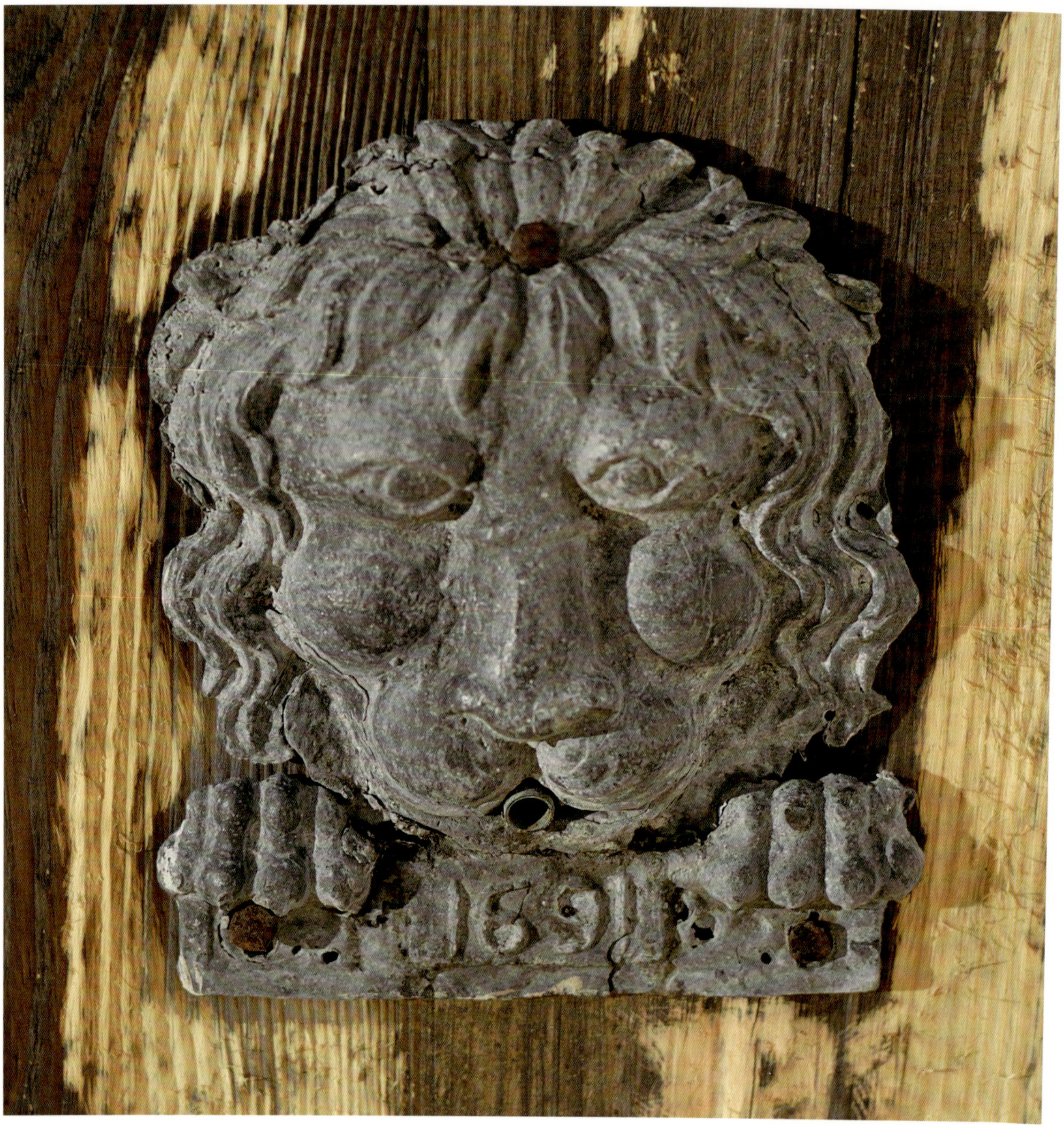

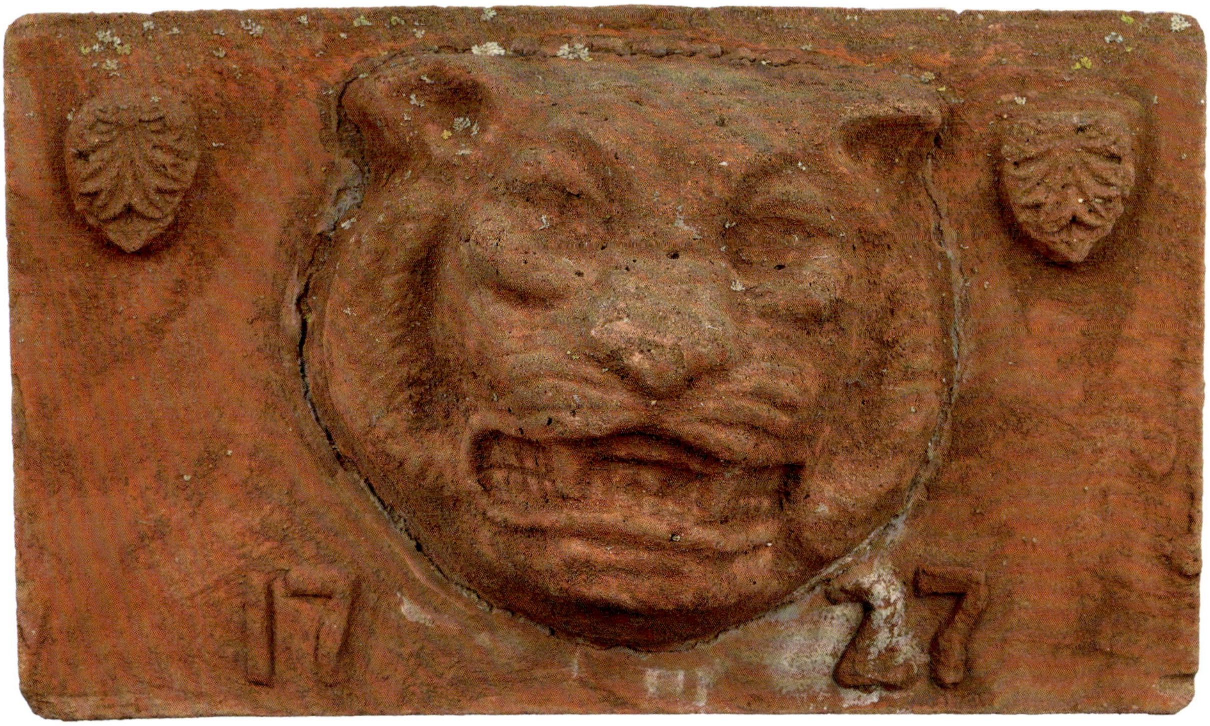

1727

Architectural fragment

Sandstone

8 ½ × 14 × 5 ½ in. (21.6 × 35.6 × 14 cm)

Inscription: 17 27

This grimacing leopard, baring its teeth above the date 1727, would have served as a particularly unwelcoming embellishment on a building or structure erected in that year.

PROVENANCE

Gift to John H. Bryan of Lake Bluff, IL (1936–2018)

1618

Portrait
Oil on panel
$21\frac{1}{8} \times 15\frac{1}{2}$ in. (53.7 × 39.4 cm)
William Larkin
(b. ca. 1580; d. 1619)

Inscription: Ætatis Suæ·56·|June·5·1618 ·

This portrait is remarkable for several reasons. The handling of the paint is instantly recognizable as that of the Jacobean portraitist William Larkin, although the choice of sitter is unusual for this artist, who otherwise painted court elites. Thomas Trayton (1562–1638) was a trusted member of the household of Richard Sackville, third Earl of Dorset (1589–1624), and an established member of the Sussex gentry, but he was not of the same social standing as Larkin's regular clientele. This painting is the only portrait by Larkin to contain a coat of arms and to cite the date in which the likeness was made, so perhaps either the sitter or artist anticipated that Trayton might not immediately be recognized and was keen to indicate his armigerous status.

By the time this painting was made, Trayton was in the employ of the so-called spendthrift earl, who, in under a decade, had decimated the fortune it had taken his family a century to accumulate. As Dorset's woodward, Trayton would have spent much time in Kent and Sussex, where the Sackville family had considerable land holdings. His responsibilities included the management of the coppiced woodlands that fueled the Wealdon iron industry of Kent and Sussex, upon which Dorset's fortune relied and his debts were underwritten. Trayton was one of the beneficiaries of Dorset's mismanagement and he acquired land in Sussex that was sold by his employer to service his debts. Trayton's son, Ambrose, married into the Chiddingly branch of the Sackville family and was buried at Withyham Church near the ancestral seat at Buckhurst in Sussex.

Buckhurst was Dorset's chosen residence, from which he launched his annual progress across Sussex and Kent, hunting and cavorting in the company of the largest retinue of gentlemen of any aristocrat of the day—all of whom enjoyed excessive salaries due to the largess of their patron. Meanwhile, at Knole in Kent, Dorset's wife, Lady Anne Clifford, languished in solitude, terrorized by the erratic behavior of her spouse, committing her misery to posterity in her remarkable diary, which survives in part for this period of her life. Thanks to the diary entries, and the inclusion of the unusually specific date on this painting, it is possible to provide the context for this picture's production.

Lady Anne Clifford's diary records that she and her daughter had their portraits made by William Larkin the summer of 1618, and the inscription on Trayton's picture marks the date as June 5, placing Larkin's visit to Knole in early June. With this in mind, it is possible to envisage that when William Larkin traveled to Knole to take in the likenesses of Lady Anne and her daughter Margaret, he also found time to make a portrait of one of Dorset's household. That the central board of the panel support was derived from a tree felled at some point after 1390 (and recycled this way) suggests that Larkin may not have planned in advance to paint this portrait and had to improvise a source for his panel support.

It is possible that Trayton's portrait may have been made at the behest of Dorset, who was clearly a lively and enthusiastic patron for pretty much anything for which he could part with his money. As mentioned, he was also devoted to his gentlemen servants. Among the pictures that hung at Buckhurst and were recorded in an inventory of 1619, there were a number of portraits of favorites, including the unknown "Mr. Singleton" and Mr. Gravenier (a gentleman usher to Dorset). Also inventoried was a portrait, described, remarkably, as of "Mr Larkine the picture maker"—the only reference to a self-portrait of the

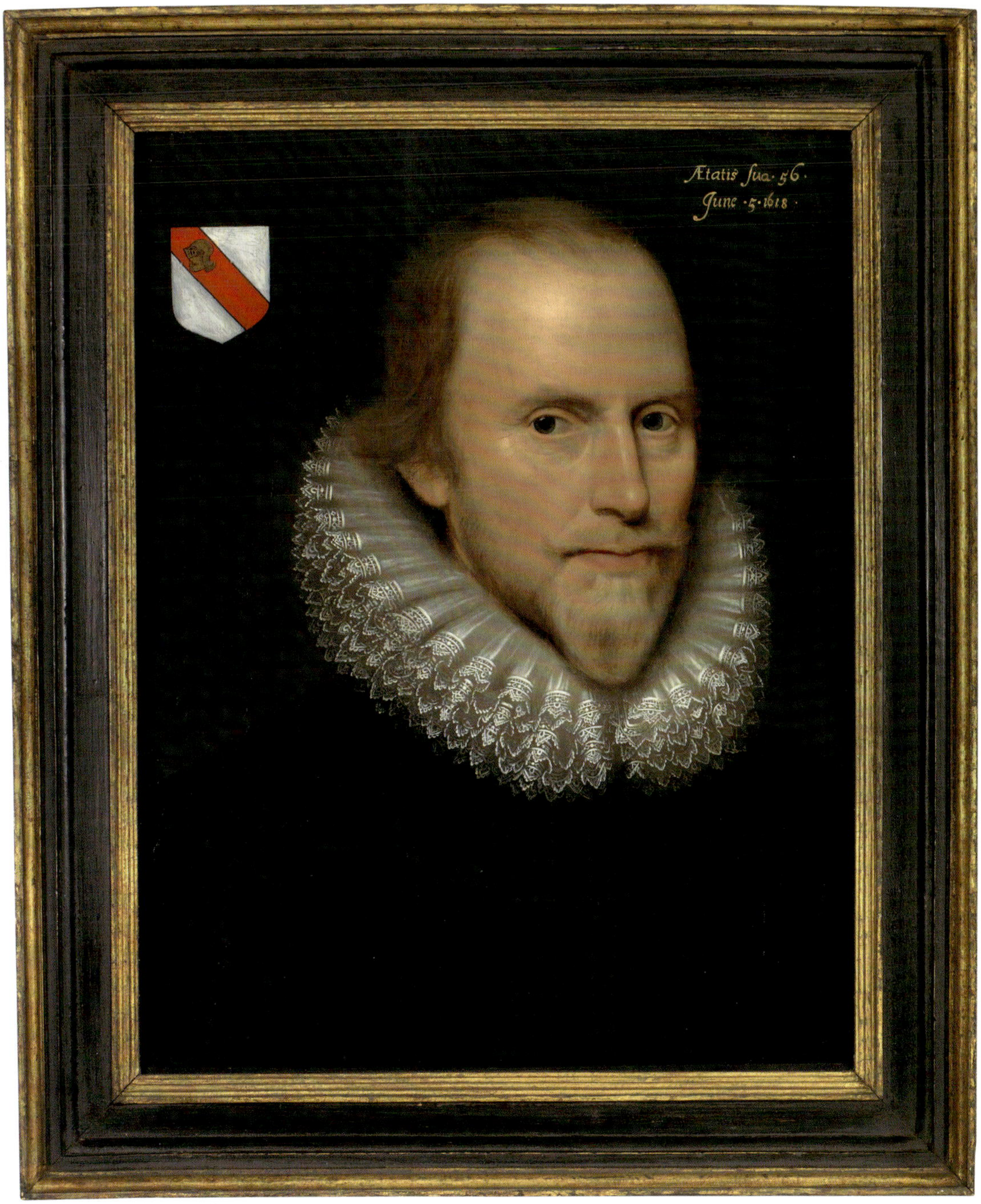

painter, although it is not known to have survived.

PROVENANCE

Private Collection, Spain, until 2016; bt. John H. Bryan III of Nashville, TN, through Gurr Johns, London

LITERATURE

Edward Town, "The Owl in the Desert: Lady Anne Clifford at Knole, 1609–1624," *National Trust Historic Houses and Collections Annual* (2018): 12–18

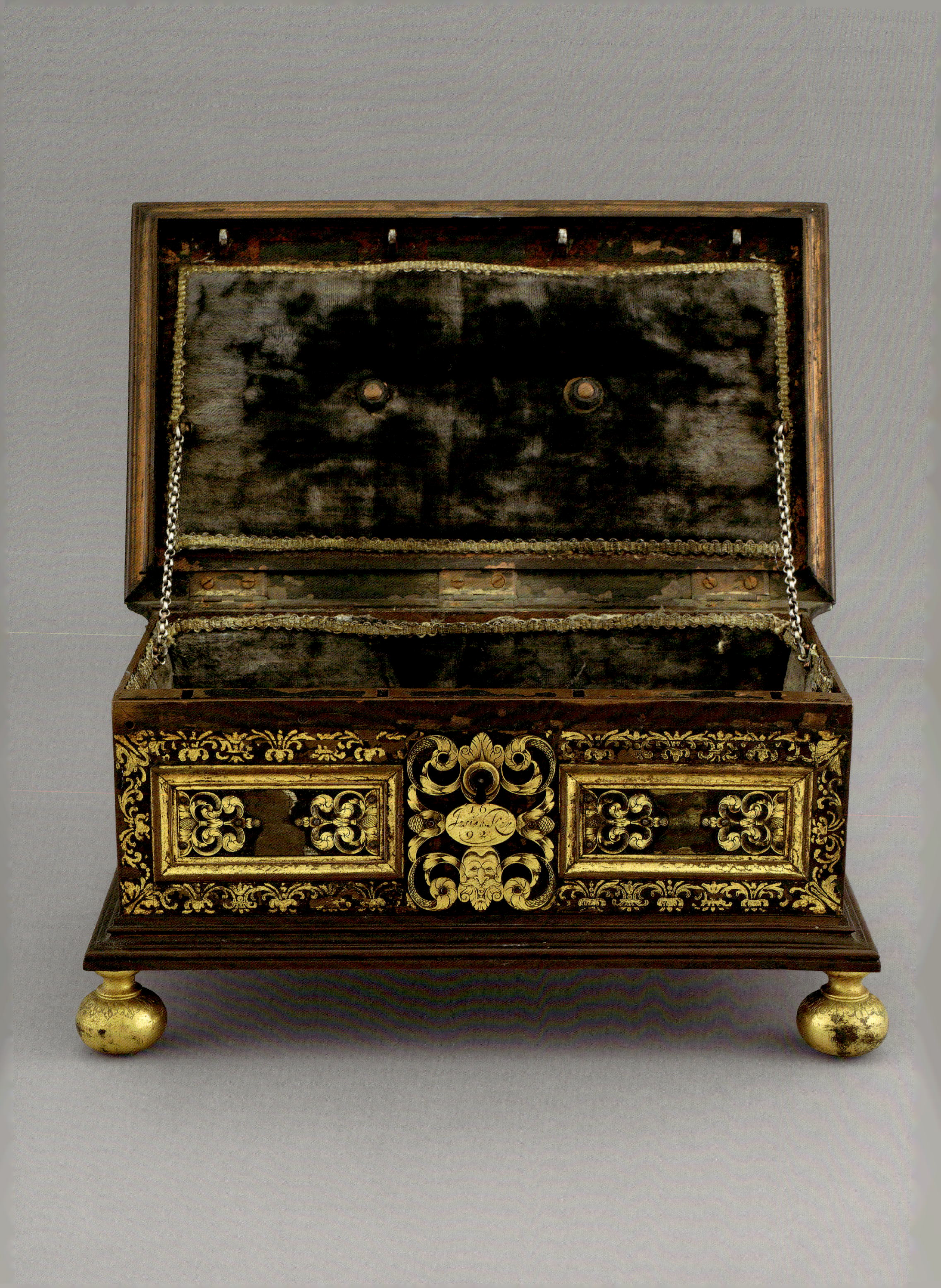

1692

Casket

Brass and parcel-gilt [partial-gilt] bronze

5 × 9½ × 6 in. (12.7 × 24.1 × 15.2 cm)

Josiah Key (active 1692; d. 1711)

Inscription:
16|Josiah Key|92

Josiah Key (also Kay or Keye) was Blacksmith to the Office of Works from 1700 until his death in 1711. By estimation of the architect William Talman, Key was "the most ingenious man in Europe." Key's work as a locksmith can be found today at Hampton Court Palace, where Talman and Key worked alongside Christopher Wren. Key's petition to the Treasury in January 1700 states that he served his apprenticeship first under Mr. Wilkinson, master locksmith to Charles II, and then worked under Hugh May, comptroller of the King's Works. He then was in the service of Mr. Backe, the king's Blacksmith, for whom he did much of his work for sixteen years. His petition had the support of both Wren and Talman.

In his will of 1711, Key described himself simply as a locksmith of St Martin-in-the-Fields (TNA PROB 11/522/197). A relative of the armigerous Josiah Key of Milcombe, Oxfordshire—whose crest of a griffin holding a key in its beak appears on this casket—Key held land in Priors Marston, Warwickshire, and asked to be interred there as near to his parents as possible. From the profits of these lands, he set aside forty shillings a year for the benefit of the St Martin-in-the-Fields Charity School for boys and girls. He seems to have died unmarried and left all of his "working Tolls and Materials for the Smiths Trade with all may Patterns of Mettall or Wood to cast by" to his servants Thomas Churchill (a kinsman) and John Lanrock. Lanrock also received the generous bequest of one hundred pounds, and his fellow servant John Banbury fifty pounds. George Clark, Esquire, an overseer, received what Keys described as "my smallest Bunch of Keys with the Gold Locks and Case and all belonging as it now is my fine copper Box and Brass box with the Locks and keys to them." As per his wishes, Key was buried at Priors Marston on August 5, 1711.

PROVENANCE

Josiah Key, 1692, and by descent with the Key family; Christie's, London, June 7, 2007 (lot 51); bt. John H. Bryan of Lake Bluff, IL (1936–2018)

LITERATURE

Dorothy Bosomworth, "The King's Metalwork," *Apollo* 140, no. 390 (August 1994): 65–68

Christopher Wren, Arthur T. Bolton, and H. Duncan Hendry, *The Wren Ms. "Court Orders" with a Supplement of Official Papers and Manuscripts from the Public Record Office, Welbeck Abbey, &c.* (Oxford, UK: Printed for the Wren Society at the University Press, 1941), 152–53

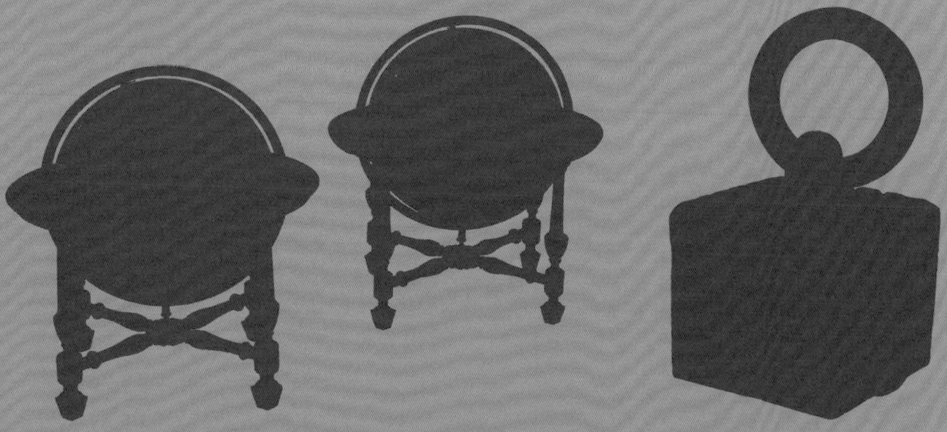

VII. MEASUREMENT & LAW

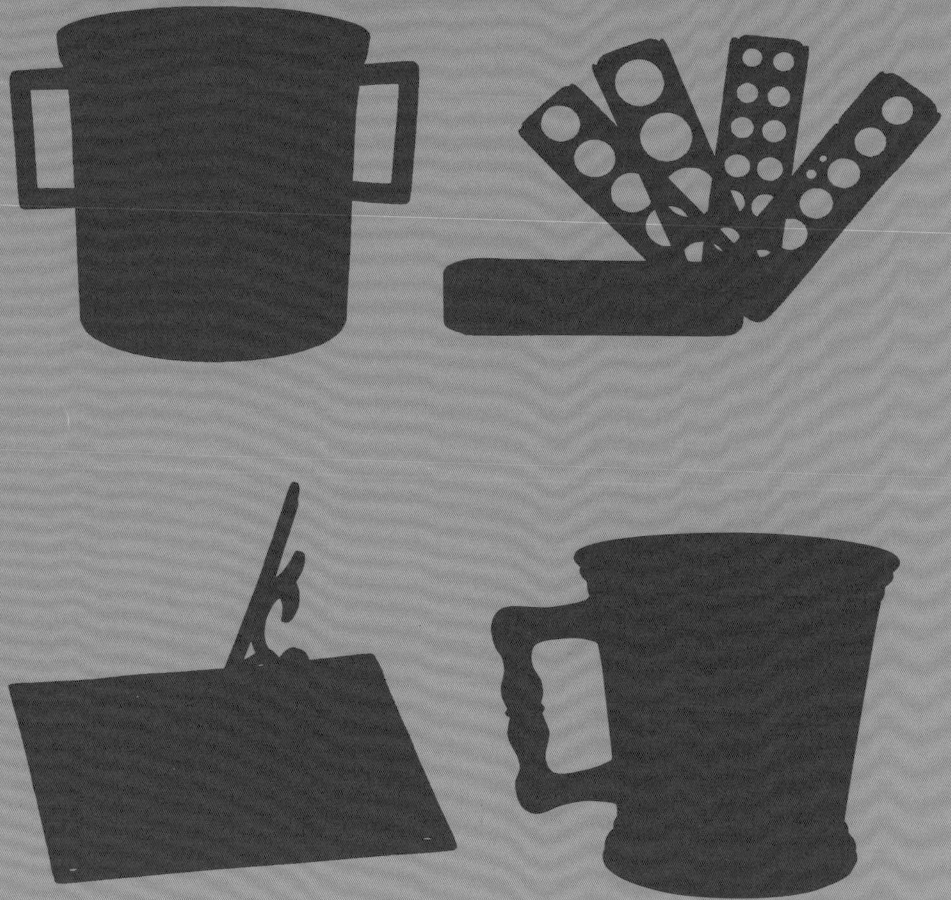

THE PURSUIT OF PRECISION AND THE CERTAINTIES AFFORDED BY exactitude intensified throughout this period, across society. Whether used to determine property boundaries or the point of entry for the surgeon's knife, instrumentation was increasingly relied upon, taking precedence over long-standing systems of received wisdom inherited from the medieval period. This shift gave rise to new professions like surveyor and mapmaker. New technologies like the barometer (pages 440–41) made it possible for the first time to predict the weather, which, within the British Isles, can be relied upon to be predictably unpredictable.

Although an accurate coin scale could protect a tradesman from being duped into accepting chipped, debased, or counterfeit currency, those at the apex of British society had the most to gain from controlling weights and measures. Thus, in the Exchequer offices in London, the Crown kept official measures from which all weights and vessel capacities were derived. The appearance of dates on these official measures imbued these objects with the authority of the act of Parliament that had issued the standards, which local magistrates used to arbitrate disputes. The measures were periodically inspected for accuracy and sometimes dated to record that event. Examples in this catalogue show that in the early modern world the fair measurement of goods was equated with divine judgment, which would measure all souls equally.

The quality of the coinage in Britain—an indicator of the health of the national coffers—reached its lowest ebb in 1551, when the precious metal content had been so badly debased that currency had to be recalled and reformed. During the chaos of the English Civil War in the following century, Charles I's mint moved a number of times, and the quality of his coinage fluctuated considerably. The fitful operations of these mints contributed to an acute lack of ready cash, which led many individuals to issue their own trade tokens. Again, those who had the most to profit—the tavern owners, taximen, and retailers who relied upon small-scale but frequent transactions—were the ones who took this remarkable step of fashioning their own coinage, which in some instances carried the same base metal content as the state-sanctioned currency (pages 442–45).

During the reign of Charles II, some of the gold coinage, including the eponymous "guinea," bore the mark of the Elephant and Castle, which indicated that the Royal Africa Company had sourced the gold from West Africa's Gold Coast (known generically in this period as Guinea). This mark was a mark of distinction, attesting to the quality of the gold content and for many years denoting a guinea that was worth a pound plus one shilling—one shilling more than the coin's nominal value. This association of the guinea with prestige and wealth continued for centuries; until recently, it was the unit of transaction employed at auctions and it endures in the names of famous horse races.

Although of no immediate exchange value, medals commemorating important moments in the nation's history were struck periodically throughout the period, often by skilled foreign medalists operating outside of Britain. Read in chronological sequence, they tell their own story of Britain's history, albeit as written by the victors.

1696

Wall clock

Brass, iron

Case: 8 × 8 in. (20.3 × 20.3 cm), with an additional 4 in. (10.2 cm) of height with the bell

Richard Savage (1663–1728)

Inscription: Will: & Jane|ffarmer; Richard Savage de Wenlock|Magna fecit 1696

Clocks of this type were the first clocks to appear in the homes of the middling sort in Britain, entering the domestic sphere in significant numbers. Previously, larger timekeeping devices could be seen and heard only on and in churches, cathedrals, towers, and bridges. With a single moving hand, wall clocks like this one told the time for thirty hours, chiming on the hour. These clocks required continual human intervention: the drive needed rewinding and resetting; the escapement and the going and striking trains needed oiling and cleaning. In the late 1650s, innovations in clock design—due to the introduction of the pendulum and improved design of the escapements—led to a dramatic reduction in the slippage of seconds over the course of a day.

Clockmakers of the seventeenth century routinely claimed authorship for their work by inscribing their names on clock dials, but it was rare for makers to date their work. At least four other wall clocks by Richard Savage with similar dated inscriptions survive: made for Edward and Millicent Pardoe, 1692; John and Susan Loe, 1693; Henry and Ursula Acton, 1694; and John and Martha Price, 1696—all of whom lived in the surrounding area of the West Midlands. All these clocks were made to commemorate recent marriages, presumably purchased as gifts for the newly married couple. A marriage between William Farmer and a woman named Jane has not been found for the year 1696, however, and the only candidates that have been identified in the vicinity are the couple who married on June 17, 1682, at Mainstone, Shropshire.

Over the course of his career, Richard Savage also made lantern, hooded, longcase, square-dial hook-and-spike, and round-dial hook-and-spike clocks. A couple of years after he made this wall clock, Savage moved from his home at Bridgcroft in Wenlock Magna (Much Wenlock, Shropshire) to the county town of Shrewsbury. There, on December 26, 1701, Thomas, son of Richard Savage of Shrewsbury, Clockmaker, was bound as an apprentice to Richard Smyth. By the time Richard Savage died in 1728, many of Shrewsbury's citizens had grown rich on the proceeds of the export trade of woolen clothes and likely could afford to purchase clocks.

PROVENANCE

With W. F. Bruce Antiques, 2015; bt. John H. Bryan of Lake Bluff, IL (1936–2018) from Howard Walwyn, Fine Antique Clocks, London, October 7, 2016

LITERATURE

Herbert Cescinsky and Malcolm R. Webster, *English Domestic Clocks* (Woodbridge, UK: Chancery House, 1976)

Douglas J. Elliot, *Shropshire Clock and Watchmakers* (London: Phillimore, 1979), 116–17

David A. Higgins, "A Brief Introduction to the Shropshire Clay Tobacco Pipe Industry," *Transactions of the Shropshire Archæological and Natural History Society* 81 (2006): 157–66

W. A. Leighton, "The Guilds of Shrewsbury," *Transactions of the Shropshire Archæological and Natural History Society* 4 (1881): 193–293, 251, 266

Brian Loomes, "The Clocks of Richard Savage," *Clocks* (November 1997): 28–31

Brian Loomes, "Richard Savage," *Horological Journal* (2001): 122–26

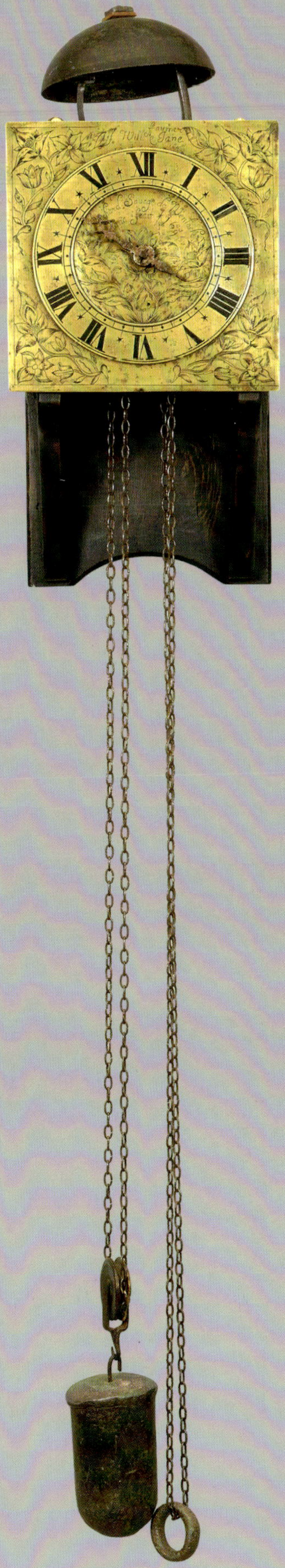

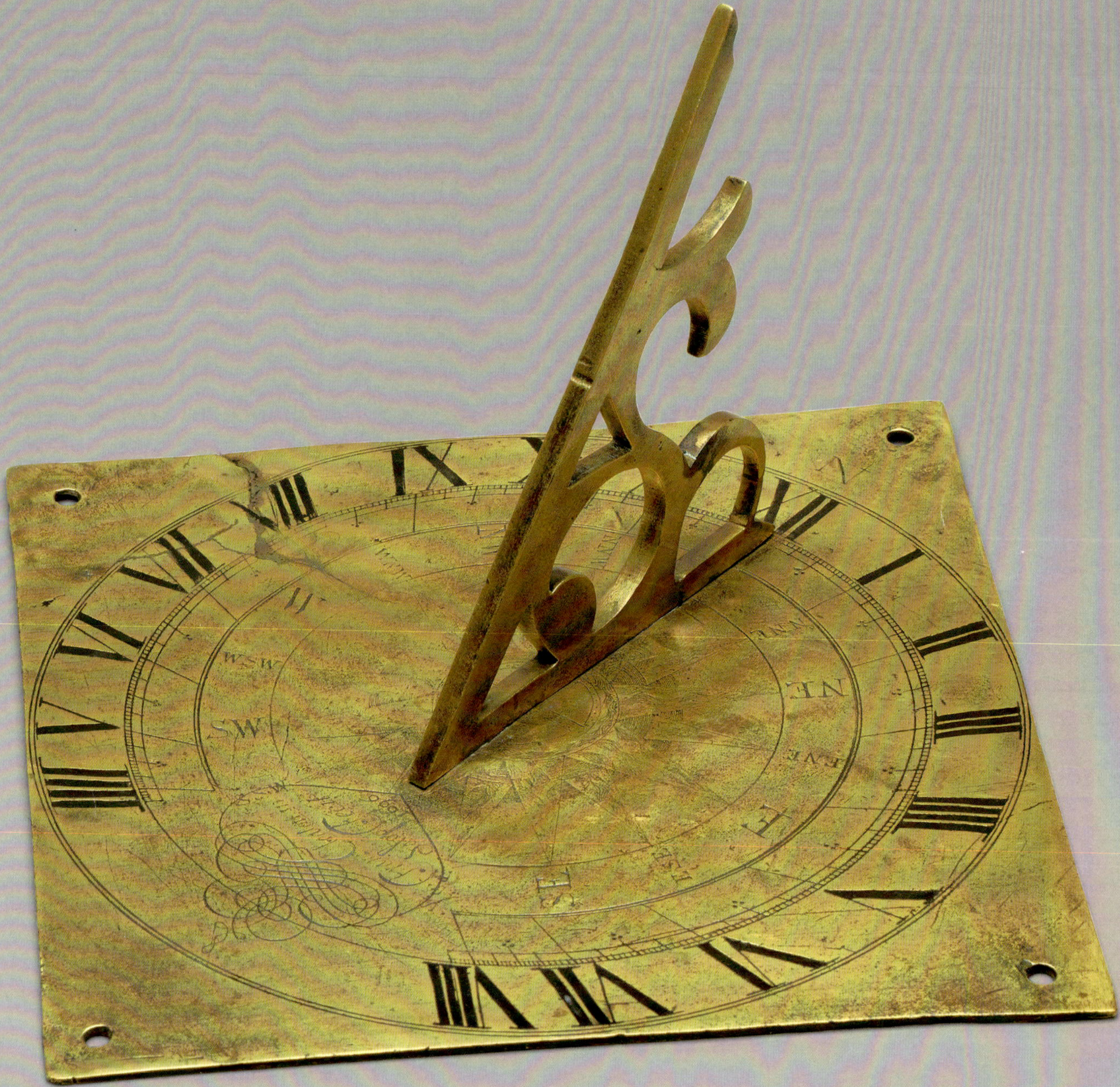

1696

Sundial

Brass

$5 \times 7\frac{1}{2} \times 7\frac{1}{2}$ in.
(12.7 × 19.1 × 19.1 cm)

Francis Cleaton (d. 1713), 1696

Inscription: Fran. Cleaton in|Salop Fecit|1696

During much of this period, owners of watches and clocks kept their timepieces accurate by setting them at noon, according to the shadow cast by the gnomon of a sundial. Noon, referred to as True Time, was the only moment at which to determine solar time. The increased use of clocks created a demand for more and better-made sundials, and their accuracy depended on calibration to their latitudinal position. So, when Francis Cleaton came to make a sundial for the churchyard of St Michael Loppington in Shropshire, it was set for Latitude 52° 52'. Like this dial, it was inscribed with a date (1695), Cleaton's name, and the distinction that he was a Salopian (from Salopesberia, the ancient name for Shropshire). There is another brass dial by Cleaton in Powys, Wales, inscribed "Fran Cleaton fecit 1710."

Cleaton—"Teacher of the Mathematicks of Shrewsbury," as his name appears among the subscribers to *The Great Historical, Geographical and Poetical Dictionary Being a Curious Miscellany of Sacred and Prophane History*, 1694—was probably the son of Elizabeth and Alexander Cleaton, tailor, who was baptized October 15, 1638, at St Alkmund, Shrewsbury. His will, dated 1713, also cites property holdings in Durham, but his links to that part of the country have not been established. During his professional life, he lived on the North East side of Mardol in Shrewsbury. The appraisal of his goods after his death records the study in which he kept his books, tables, and forms as located in the schoolhouse that was part of the property. Much like his better-known contemporary Christopher Wren (1632–1723), Cleaton would have undertaken the creation of sundials as a demonstration of his mastery of mathematics, as he provided the craftsman with the calculations and angles to fashion the dial. His work was as a land surveyor, and surveys of estates and enclosures by him date from 1694 to 1709. Yet, when his wife, Rebecca, made her will in 1733, she described herself as the widow of Francis Cleaton, "Writingmaster," which may have referred to the type of instruction he offered at his school. The cursive and highly decorative signature that ends his will with a flourish is a pared-down iteration of the inscription on the sundial.

PROVENANCE

bt. John H. Bryan of Lake Bluff, IL (1936–2018) from Tobias Jellinek Antiques, London, March 26, 2018

LITERATURE

David Herbert Somerset Cranage, *An Architectural Account of the Churches of Shropshire* (Wellington, UK: Hobson and Co., 1912), 762

Dictionary of Land Surveyors and Local Map-Makers of Great Britain and Ireland 1530–1850 (London: The British Library, 1997), 2:101

Anthony Gerbino and Stephen Johnston, *Compass and Rule: Architecture as Mathematical Practice in England* (New Haven, CT: Yale University Press, 2009), 85

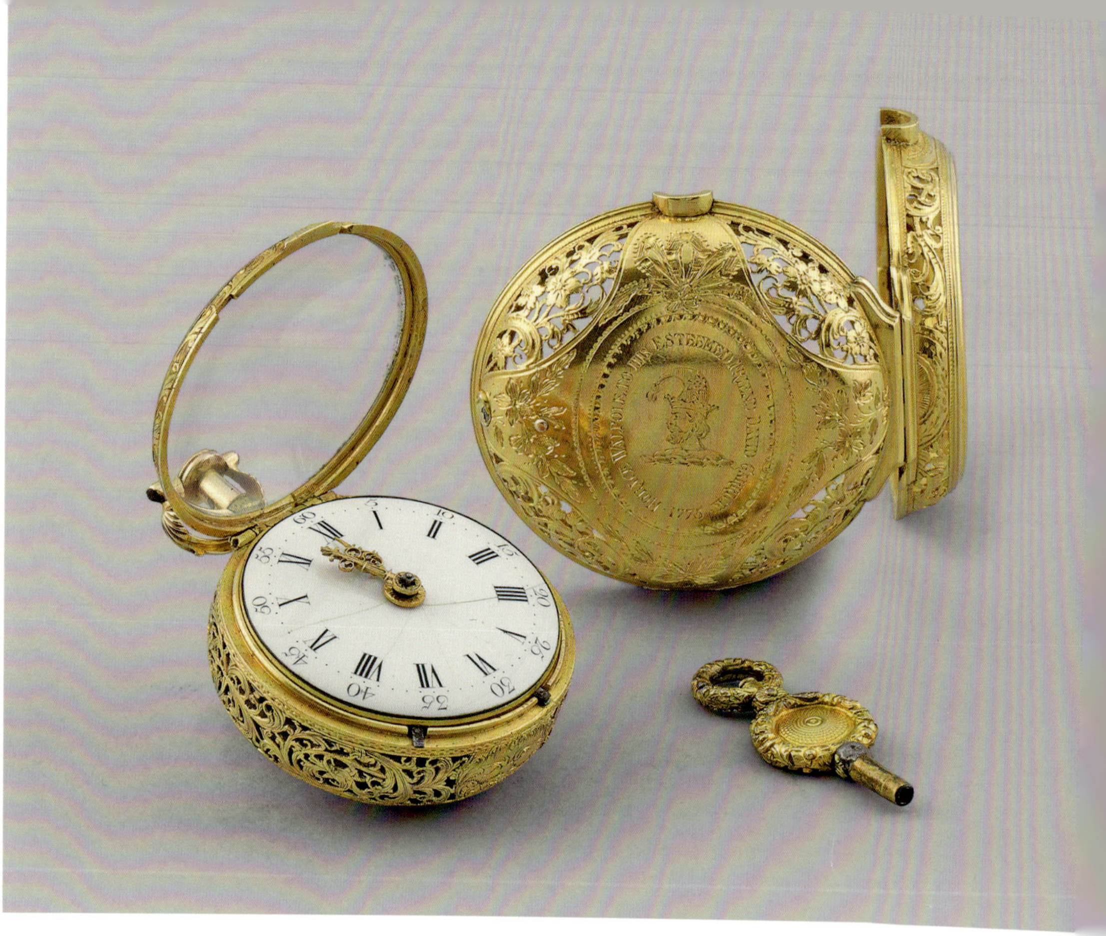

1775

Pocket watch and case

Gold

$2\frac{5}{8} \times 1\frac{7}{8} \times 1$ in. ($6.7 \times 4.8 \times 2.5$ cm)

Joseph Downes

Inscription:
HORACE WALPOLE TO HIS ESTEEMED FRIEND DAVID GARRICK 1775; Jos Downes|London| Nº 135

As the inscription describes, this gold watch was a gift from the most famous collector of the day, Horace Walpole, fourth Earl of Orford (1717–1797), to his friend, David Garrick (1717–1779), the most famous actor in Georgian England, perhaps anticipating Garrick's retirement from the stage on June 10, 1776. By the time this gift was made, Walpole and Garrick had been close neighbors and friends for more than twenty years. At Hampton House (later Garrick's Villa), Garrick built a garden folly dedicated to his hero, William Shakespeare; Walpole provided mottoes to adorn the exterior. The inscription also bears the Saracen's head, which forms part of the Walpole crest and was a recurring motif on objects Walpole commissioned.

The watch bears hallmarks for 1774 and is signed by Joseph Downes, son of Robert Downes, also a watchmaker. Joseph was born in the parish of St Sepulchre-without-Newgate in Holborn and joined St Paul's School at age ten on January 14, 1754. By 1768, Downes was active as a watchmaker and had his own apprentices. Joseph Downes the Elder, aged seventy-six, was buried on May 21, 1820, at the parish of St James Clerkenwell.

PROVENANCE

By 1891, in the possession of Horatio Walpole, fourth Earl of Orford (1813–1894); bt. John H. Bryan of Lake Bluff, IL (1936–2018) from D. S. Lavender Antiques Ltd., London, June 3, 2014

LITERATURE

Exhibition of the Royal House of Guelph (London: The New Gallery Regent Street, 1891), 159, no. 844

Dennis Moore, *British Clockmakers & Watchmakers Apprentice Records, 1710–1810* (Ashbourne, UK: Mayfield, 2003)

Michael Snodin, ed., *Horace Walpole's Strawberry Hill* (New Haven, CT: Yale University Press, 2009)

1685

Perpetual calendar

Brass

$1\frac{5}{8}$ in. diameter (4.1 cm)

William Foster

Inscription:
EASTER|1685;
W. FOSTER

No larger than a coin, this calendar contained all the information needed for its owner to navigate the forthcoming year, without having to consult a printed almanac. It noted when to observe Lent and celebrate that most movable of feasts, Easter. An identical example by the same maker in the British Museum, London (BM OA.1447), suggests that these types of calendars were made in significant numbers. At this time, the numbering of the months was made according to the legal/tax year: thus, March was deemed the first month and February the twelfth.

Beginning clockwise from the outer edge of the calendar face shown below at left, the abbreviations in the first line of the top quarter denote that, in 1685, Easter Term ("ETB") began seventeen days after Easter ("17A"), ending the Monday before Whit Sunday ("EMBW"). Trinity Term ("TB") began twelve days after Whit Sunday ("12AW").

The second line of that quarter counts nineteen days ("C19D"), indicating that Michaelmas Term ("M") begins on the twenty-third of October (which in this system is month 8) and ends ("E") the twenty-eighth of November (month 9). The third line of that quarter indicates that Hilary Term ("HB") begins on the twenty-third of January (month 11) and ends on the twelfth of February (month 12).

The other quarters on this face of the calendar provide the dates of important saints' days and feast days and also the dates of significant historical events, such as the discovery of the Gunpowder Plot, the execution of Charles I, and the birthday of Charles II.

Moving clockwise, the numbers on the outer edge of the reverse, shown below at right, tell when Easter fell each year from 1685 onward. For example, Easter 1685 was "2 19" (April 19), Easter 1686 was "2 4" (April 4), and Easter 1687 was "1 27" (March 27), and so forth. The next inner row gives the date of the southing of the moon, and the row within that the Julian Epacts, which also track movements of the moon. The numbers at center are a sunset table.

PROVENANCE

bt. John H. Bryan of Lake Bluff, IL (1936–2018) from Rupert Gentle, Pewsey, Wiltshire, December 17, 1990

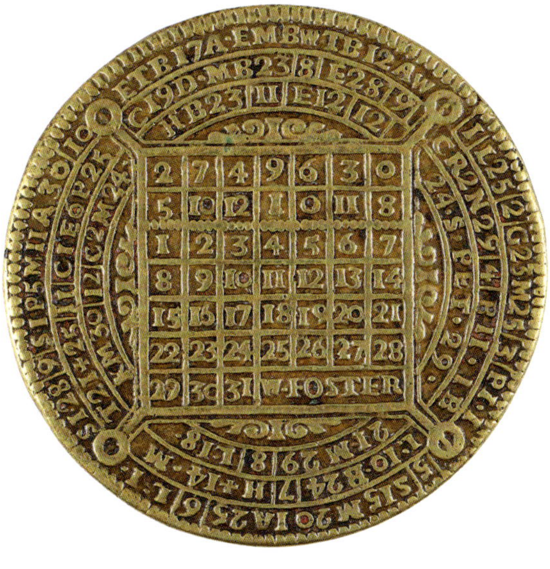

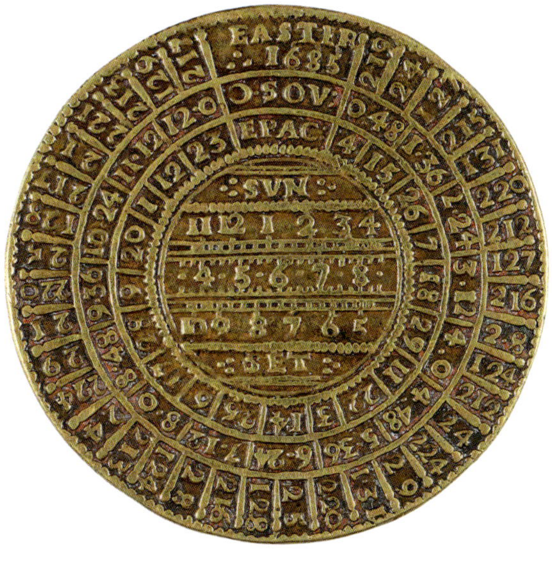

1693

Shot gauge
Brass
$\frac{1}{8} \times 2\frac{1}{4} \times \frac{1}{2}$ in. (.32 × 5.7 × 1.27 cm)

Inscription:
Nathaniel: Foster.
1693

This shot gauge has four leaves that are pierced with circles with a range of diameters, used to calibrate shot of varying sizes and weights.

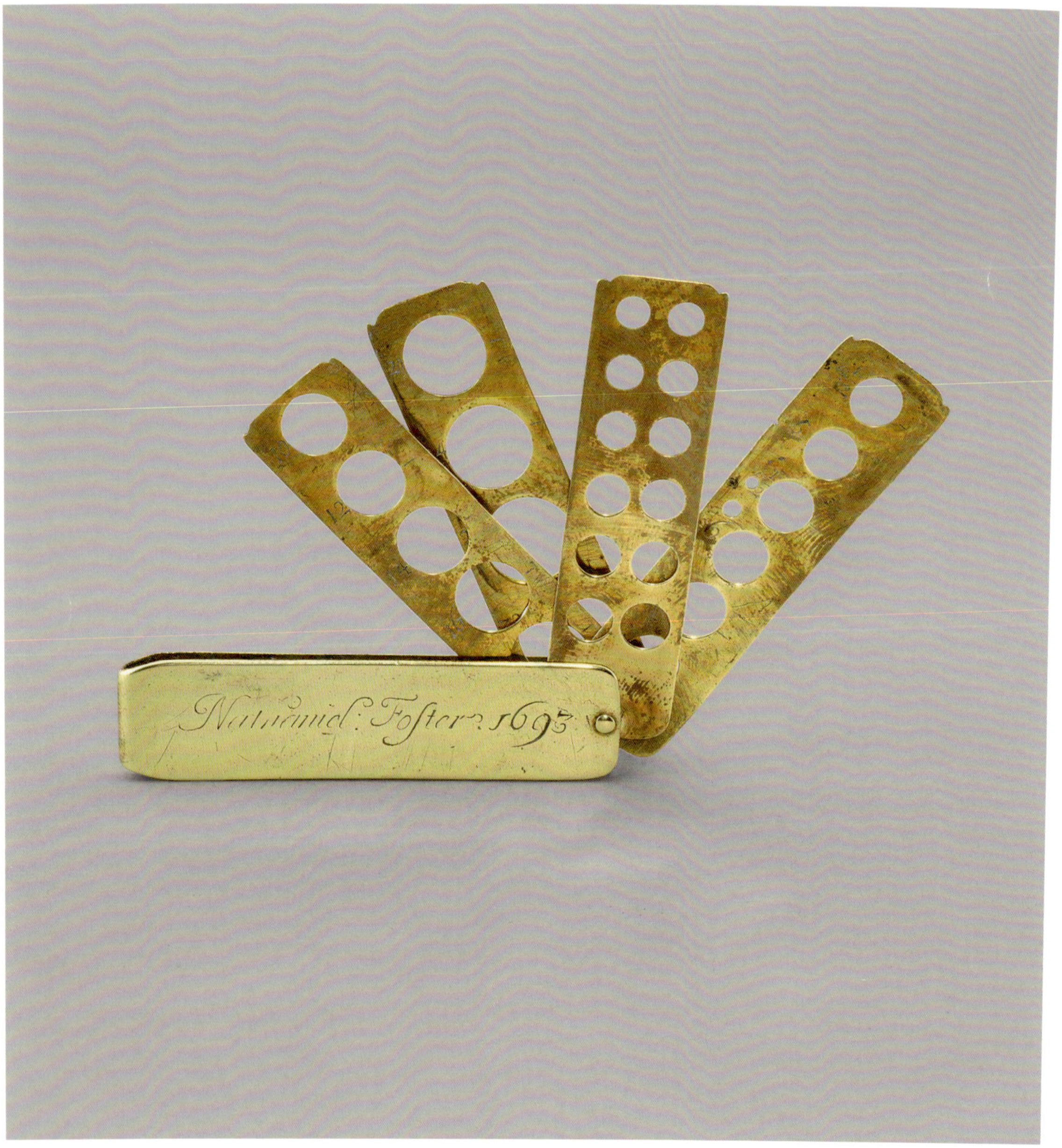

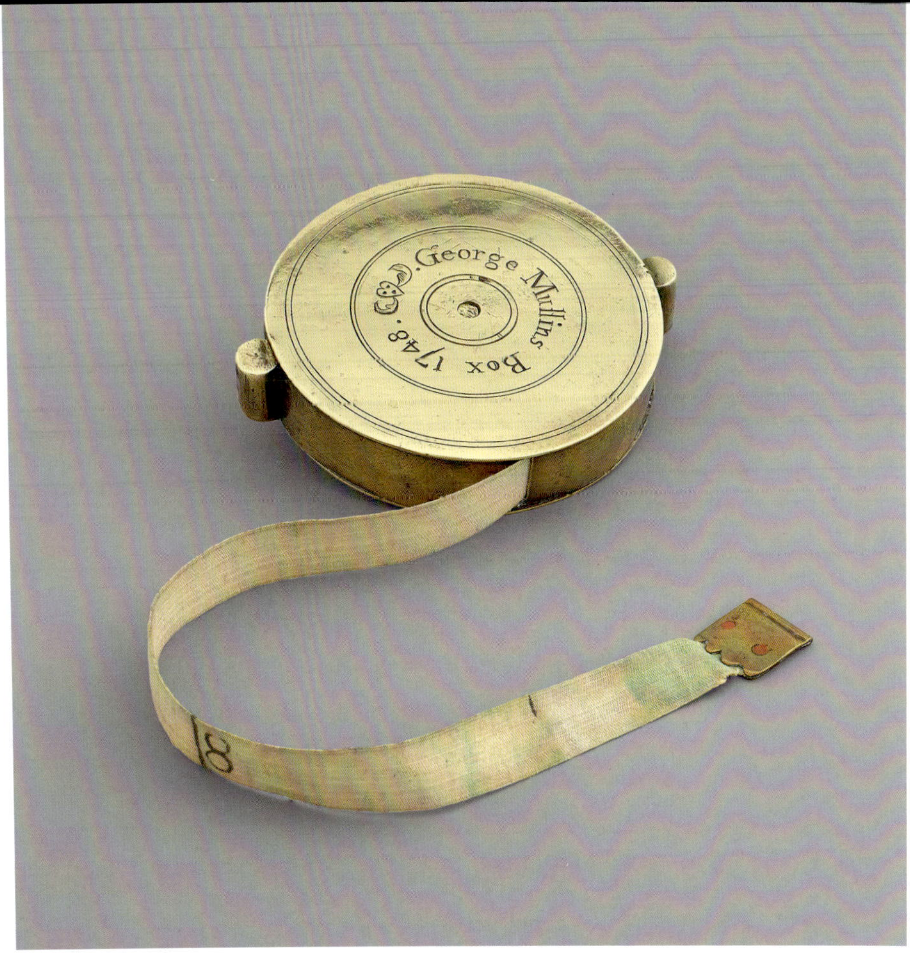

1748

Tape measure

Brass, cloth

¾ × 4 × 3⅜ in. (1.9 × 10.2 × 8.6 cm)

Inscription:
.George Mullins Box 1748.

George Mullins, the owner of this retractable tape measure, was a schoolmaster in Box, Somerset.

PROVENANCE

Michael Gillingham of London (1933–1999); bt. John H. Bryan of Lake Bluff, IL (1936–2018) from Alistair Sampson Antiques Ltd., London, December 7, 1993

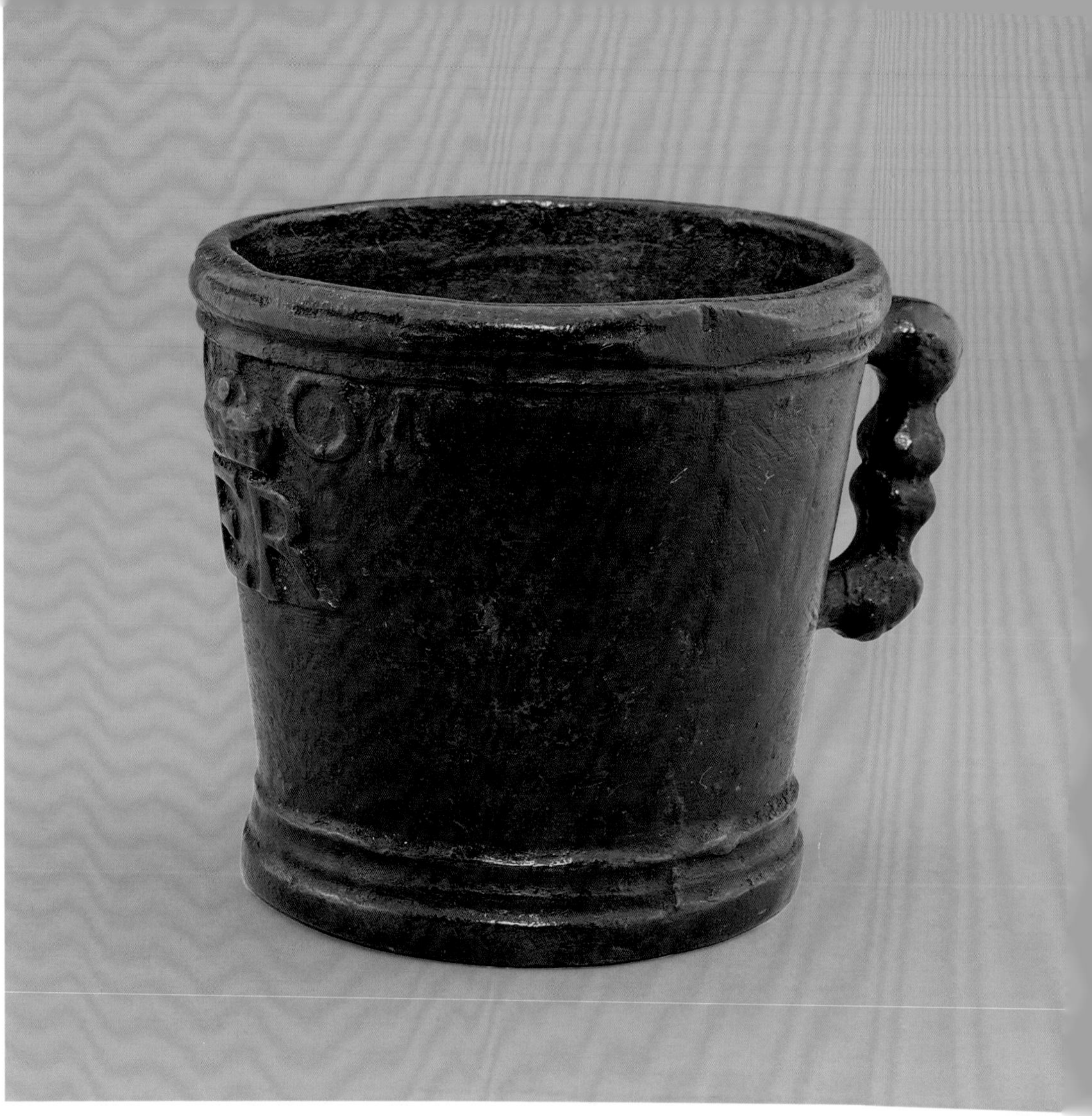

1601

Ale quart measure

Bronze

$5\frac{1}{8} \times 5\frac{1}{4} \times 6\frac{1}{2}$ in.
(13 × 13.3 × 16.5 cm)

Inscription:
16 01|ER

Standard measures were made in a heavy, durable material like bronze, as here, to prevent any damage that might alter the volume capacity. After standards had been set by the central government, copies of these measures were distributed to local officials who policed markets and retailers throughout the country. If a dispute arose about the quantity of a product bought or sold, a standard measure would determine whether the vessel in question was accurate. Queen Elizabeth I generated her Exchequer Standards of capacity from those of Henry VII, and they were issued in 1601–2, thus the date on this measure and the one on the facing page. The Exchequer still holds several of the original 1601 measures.

In addition to the inscribed date of 1601 and the initials "ER" (for "Elizabeth Regina"), the initials "HC" are stamped, upside down, on the side of the base—perhaps a founder's mark, the initials of the owner, or the initials of the inspector who verified the measure's accuracy.

PROVENANCE

bt. John H. Bryan of Lake Bluff, IL (1936–2018) from David Moulson Antique Pewter Specialist, Alcester, Warwickshire, November 13, 2003

LITERATURE

R. D. Connor, *The Weights and Measures of England* (London: Her Majesty's Stationery Office, 1987), 159–64

1601

Ale pint measure
Bronze
$4\frac{1}{4} \times 4\frac{3}{8} \times 5\frac{1}{4}$ in.
(10.8 × 11.1 × 13.3 cm)

Inscription:
16 ER 01

This 1601 measure bears no excise marks. It has a wet capacity of 550 milliliters (19 ounces) and a dry capacity of 515 milliliters (17 ounces).

PROVENANCE
bt. John H. Bryan of Lake Bluff, IL (1936–2018) from Michael Dunn Antiques, Wiscasset, ME, March 6, 1996

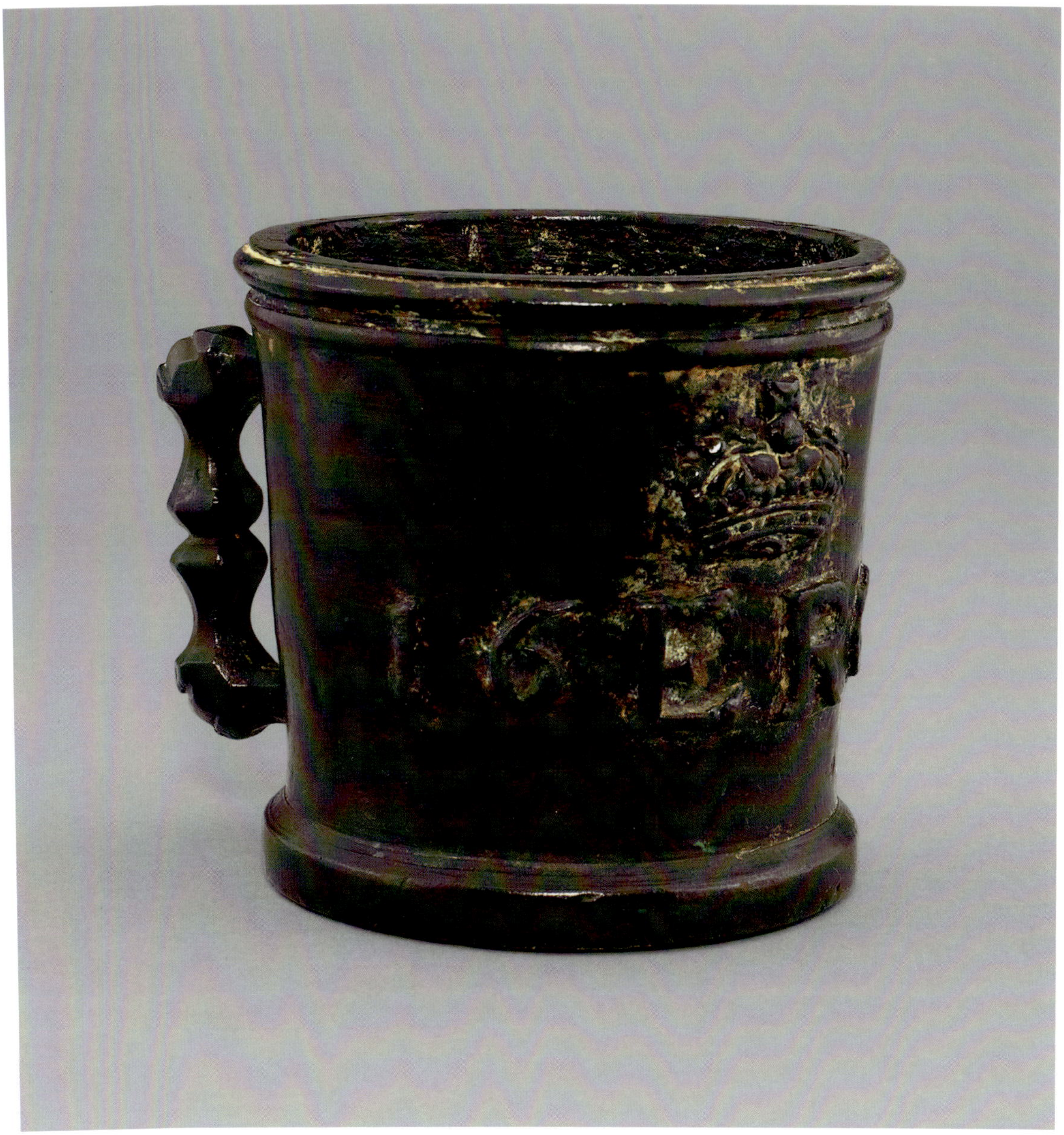

1663

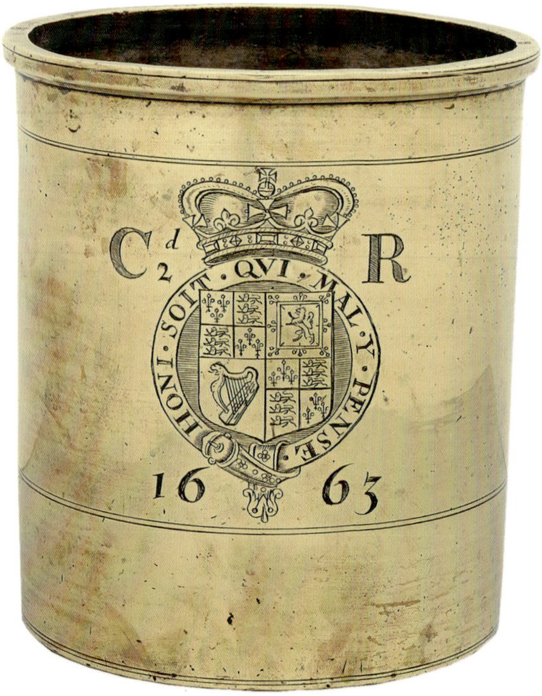

Ale gallon measure

Brass

$8\frac{1}{4} \times 7$ in. (21×17.8 cm)

Inscription:
C 2d R|16 63

The rim of this measure is stamped twice with the chequerboard symbol of the Exchequer and once with the initials "C R" within a reserve and with another faint mark.

PROVENANCE
Bonhams, London, March 15, 2017 (lot 209); bt. John H. Bryan of Lake Bluff, IL (1936–2018) from Tobias Jellinek Antiques, London, March 18, 2017

1707

Ale quart measure

Bronze

$5\frac{1}{4} \times 5\frac{1}{2} \times 6\frac{3}{4}$ in. ($13.3 \times 15 \times 17.1$ cm)

Inscription: AR; PRIMO:MAII 1707

This bronze ale quart measure, with a squared baluster handle, has a wet capacity of 1080 milliliters (37 ounces) and a dry capacity of 1075 milliliters (36 ounces). The surface of the rim is stamped with verification marks: "AR" for Queen Anne and the chequerboard of the Exchequer. The body is decorated with a Tudor rose and Scottish thistle, both crowned, and the interior has the mark "STR" impressed over the image of a reclining goat, which is the arms of the burgh of Stirling. The inscription refers to the date of the Treaty and Act of Union between England and Scotland, hence the English and Scottish emblems on the opposite surface. The treaty stipulated that English and Scottish weights and measures must be equivalent—which also standardized the quart and the Queen Anne wine gallon.

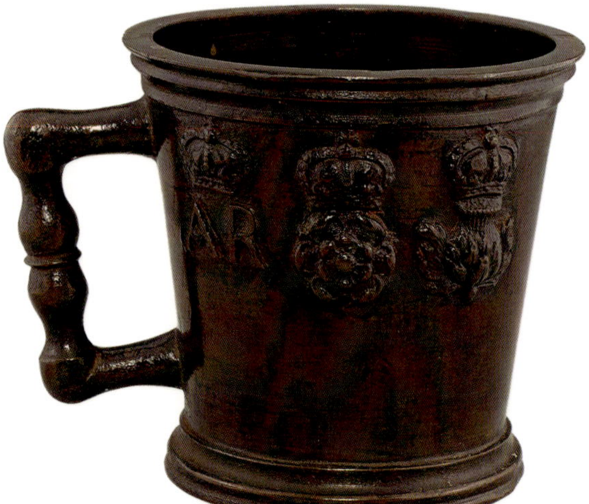

PROVENANCE
Bedford Collection; bt. John H. Bryan of Lake Bluff, IL (1936–2018) from Jack Casimir Ltd., London, April 12, 1995

1678

Wine gallon measure

Bronze

7½ × 8 × 10½ in. (19.1 × 20.3 × 26.7 cm)

Inscription:
Bakwell in Com Derb 1678

Bakewell is a small market town in Derbyshire and home of the eponymous tart and pudding. The rim of this measure is stamped with three pairs of verification marks: a crowned "C R," the crowned word "Derby," and a chequerboard.

PROVENANCE

bt. John H. Bryan of Lake Bluff, IL (1936–2018) from Tobias Jellinek Antiques, London, July 10, 2014

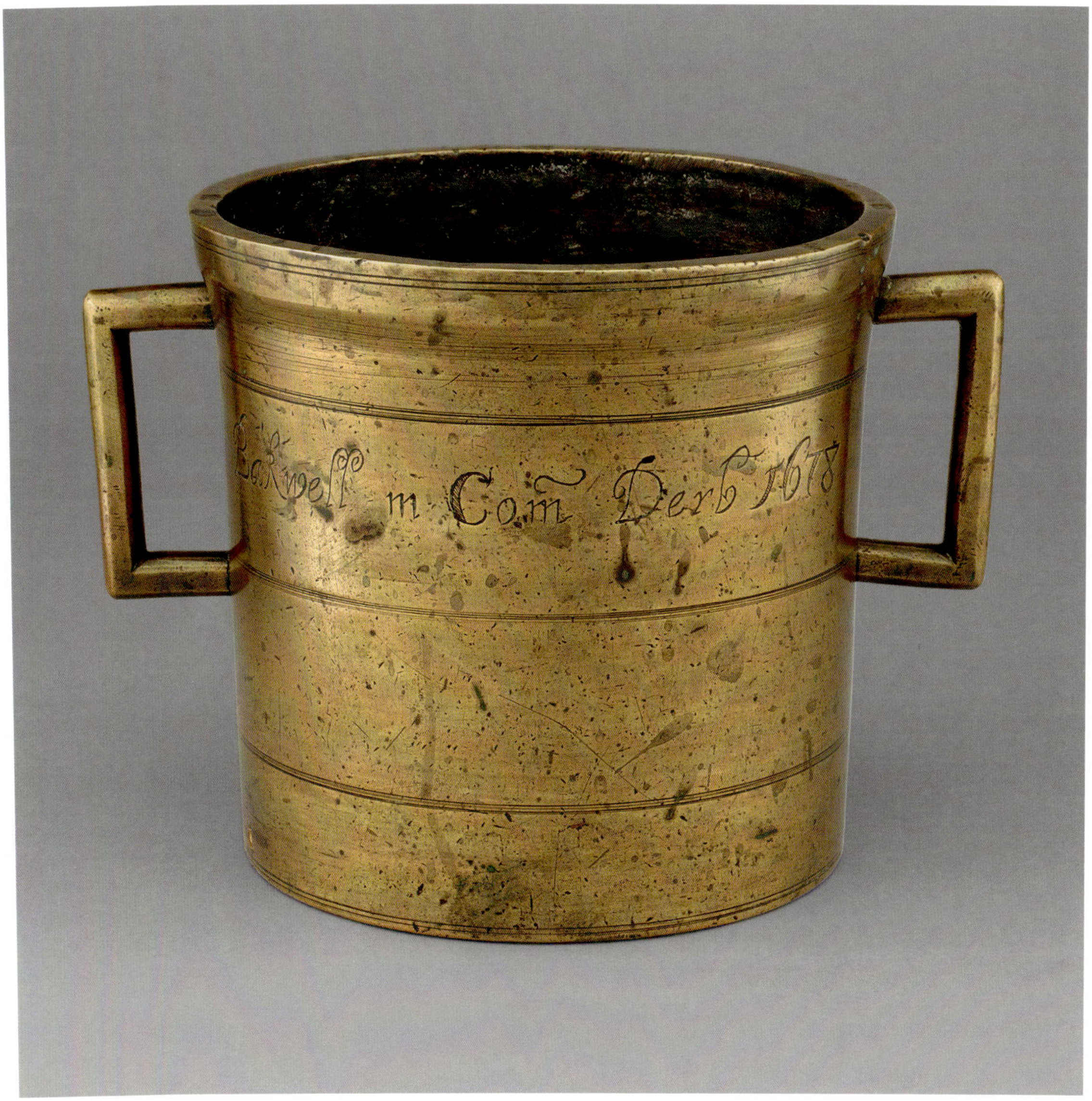

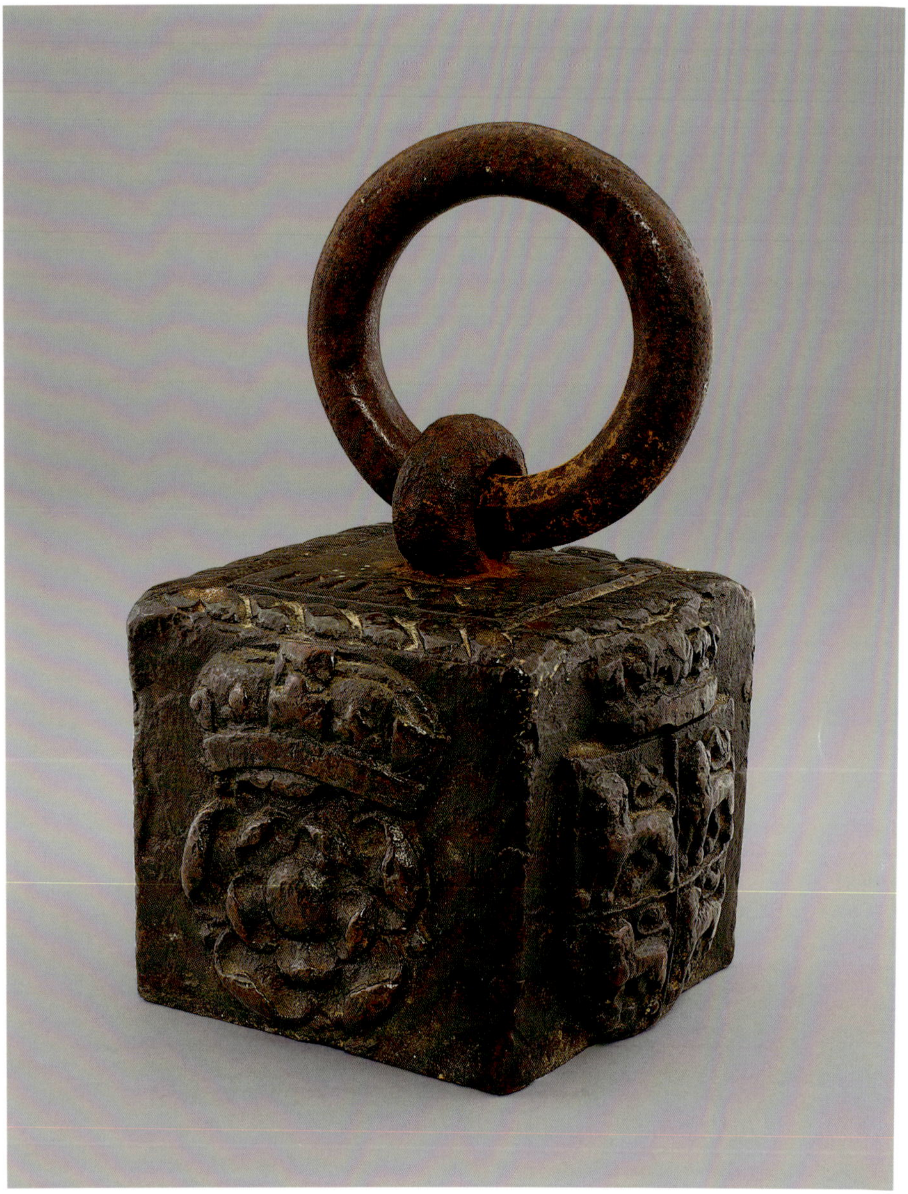

1605

Twenty-eight-pound wool weight

Bronze, lead, iron

9 × 4½ × 4½ in. (22.9 × 11.4 × 11.4 cm)

Inscription: 1605|XXIIII

This wool weight was originally twenty-four pounds, but four pounds were added to it at a later date when the ring was attached at top and lead added to the base.

PROVENANCE

Phillips, Bury St Edmunds, December 14, 1995 (lot 597); bt. John H. Bryan of Lake Bluff, IL (1936–2018) through Alistair Sampson Antiques Ltd., London

1742

Seven-pound wool weight

Bronze

7 × 4¼ × 1½ in. (17.8 × 10.8 × 3.8 cm)

Inscription: GR; DIEU ET MON DROIT; RICHARD CRESSY|1742

This weight is among the first issue of wool weights from the reign of George I. Erroneously, all had been cast with the Stuart royal arms, as had been those of Queen Anne, the last of that dynasty. When the error was noticed, the first issue was recalled, and a second issue was made using the Hanoverian royal arms.

The various small stamps are: the crown within the initials "GR" for George I; "A" for "avoirdupois"; the dagger for London, as the sword on the London Arms; and the ewer or water jug, symbol for the Founders' Company. Richard Cressy may be the name of the tronator, the official who weighed the wool.

PROVENANCE

bt. John H. Bryan of Lake Bluff, IL (1936–2018) through Alistair Sampson Antiques Ltd., London

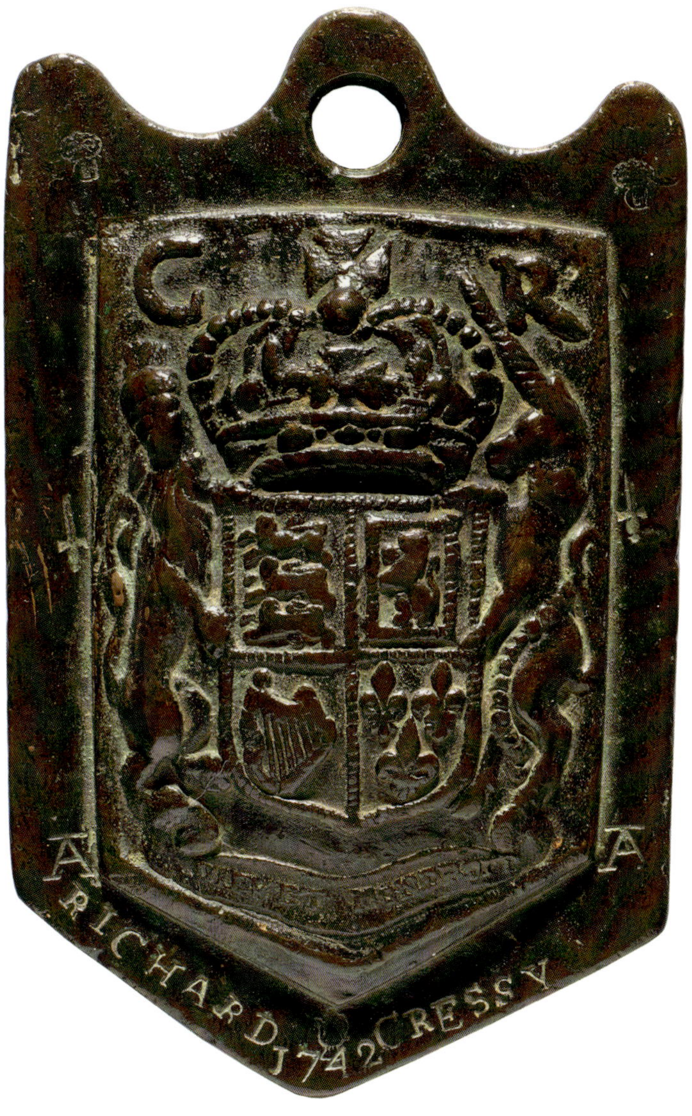

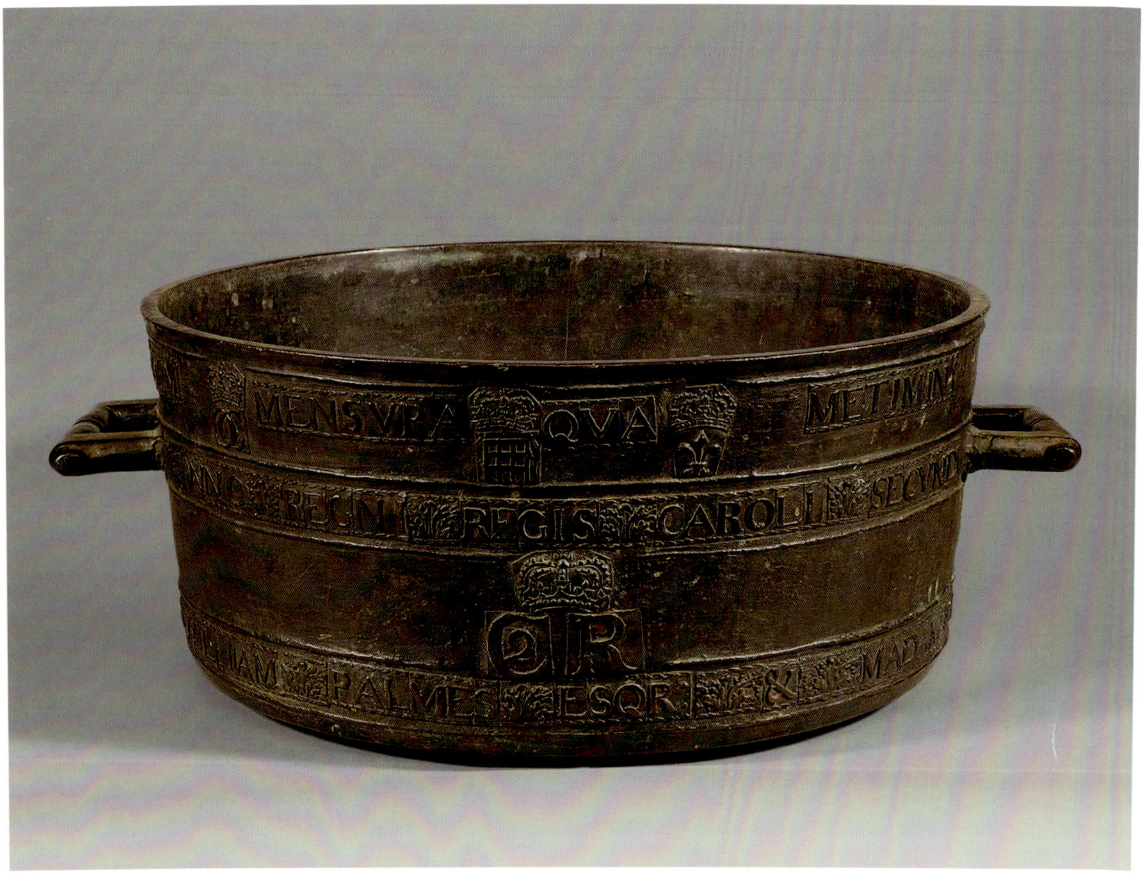

1670

Winchester corn bushel measure

Bronze

$9\frac{1}{4} \times 24\frac{1}{2} \times 20$ in. (23.5 × 62.2 × 50.8 cm)

Inscription:
EADEM MENSVRA QVA METIMINI CONTRA METIETVR VOBIS|ANNO REGNI REGIS CAROLI SECVNDI VICESIMO SECVNDO 1670|C2R|WILLIAM PALMES ESQR: & MADAM MARGARET DANBY NEW MALTON

This Winchester bushel, used for the measurement of dry goods, was made by the exasperated townspeople of New Malton in Yorkshire. Their livelihoods had suffered as a result of a long-standing dispute between two sisters: Mary and Margaret Eure (who married, respectively, William Palmes of Lindley and Thomas Danby of Farnley). The sisters were the coheirs of Lord Eure, but they could not agree on the division of the manor, so manorial meetings slipped into abeyance. By the time this measure was made, Margaret was a widow and conducting her own business affairs, whereas Mary's husband was still alive and, as the devisee of her estate, had his name cast onto the side of the measure.

PROVENANCE

Collection of the Shand-Kydd family; Boulton & Cooper; Peter Finer, Shipston-on-Stour, Gloucestershire; bt. John H. Bryan of Lake Bluff, IL (1936–2018) from Alistair Sampson Antiques Ltd., London, September 29, 1989

1709

Winchester corn bushel measure

Bronze

10 × 23 × 20 in. (25.4 × 58.4 × 50.8 cm)

Inscription: THOMAS : GOUGE : ESQR : 1709

Thomas Gouge, Esquire, was lord of the manor for Dartford, Kent, where there was a busy market every week.

PROVENANCE

bt. John H. Bryan of Lake Bluff, IL (1936–2018) from Tobias Jellinek Antiques, London, October 24, 2000

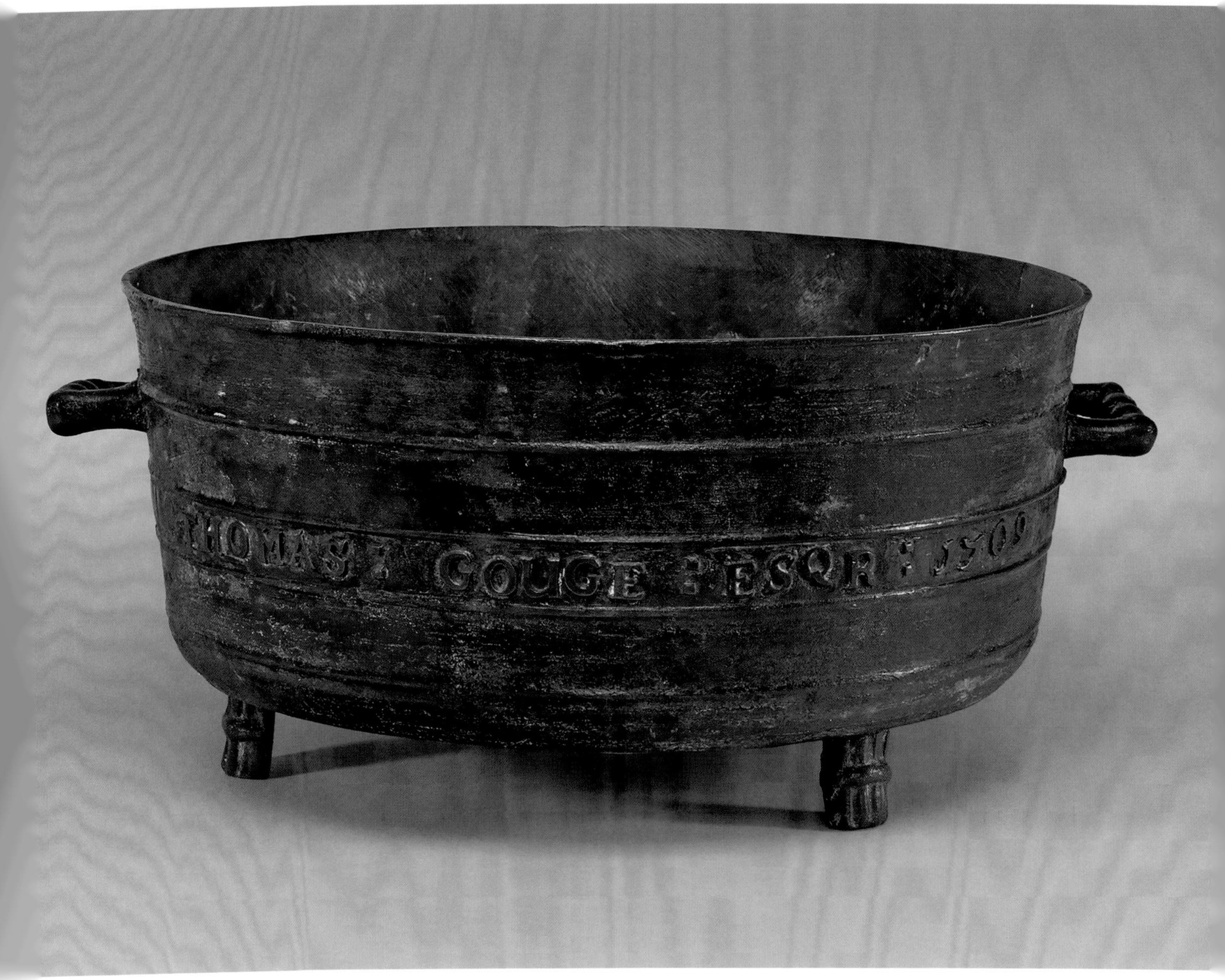

1793

Winchester corn bushel measure

Bronze

9 × 24 × 29½ in. (22.9 × 61 × 74.9 cm)

Inscription: LINCOLNSHIRE| LINDSEY .}, SPILSBY| BRACKENBURY. CLERK OF Y^E PEACE|1793

Some measures named an officeholder rather than a manorial lord, a fact indicative of the growing strength of local civic powers during this period. For example, Spilsby in Lincolnshire was a prosperous and independent town that had built a new town hall and courthouse overlooking the busy marketplace, which stands today. The power to nominate officers to Spilsby's market court rested with the lord lieutenant of the county, the Duke of Ancaster. Over time, however, leading figures in the town had appropriated these powers. In particular, the Brackenbury family had acquired control over the office of clerk of the peace, giving the family authority over the market courts. From 1787 until 1826, Joseph Brackenbury (d. 1829) held the post, and as a mark of his hereditary office, his name was permanently inscribed on the town's standard bushel when legislation required the measure's renewal in 1793.

PROVENANCE

Christie's, London, February 2, 1995 (lot 1); bt. John H. Bryan of Lake Bluff, IL (1936–2018) through Rupert Gentle Antiques, Pewsey, Wiltshire

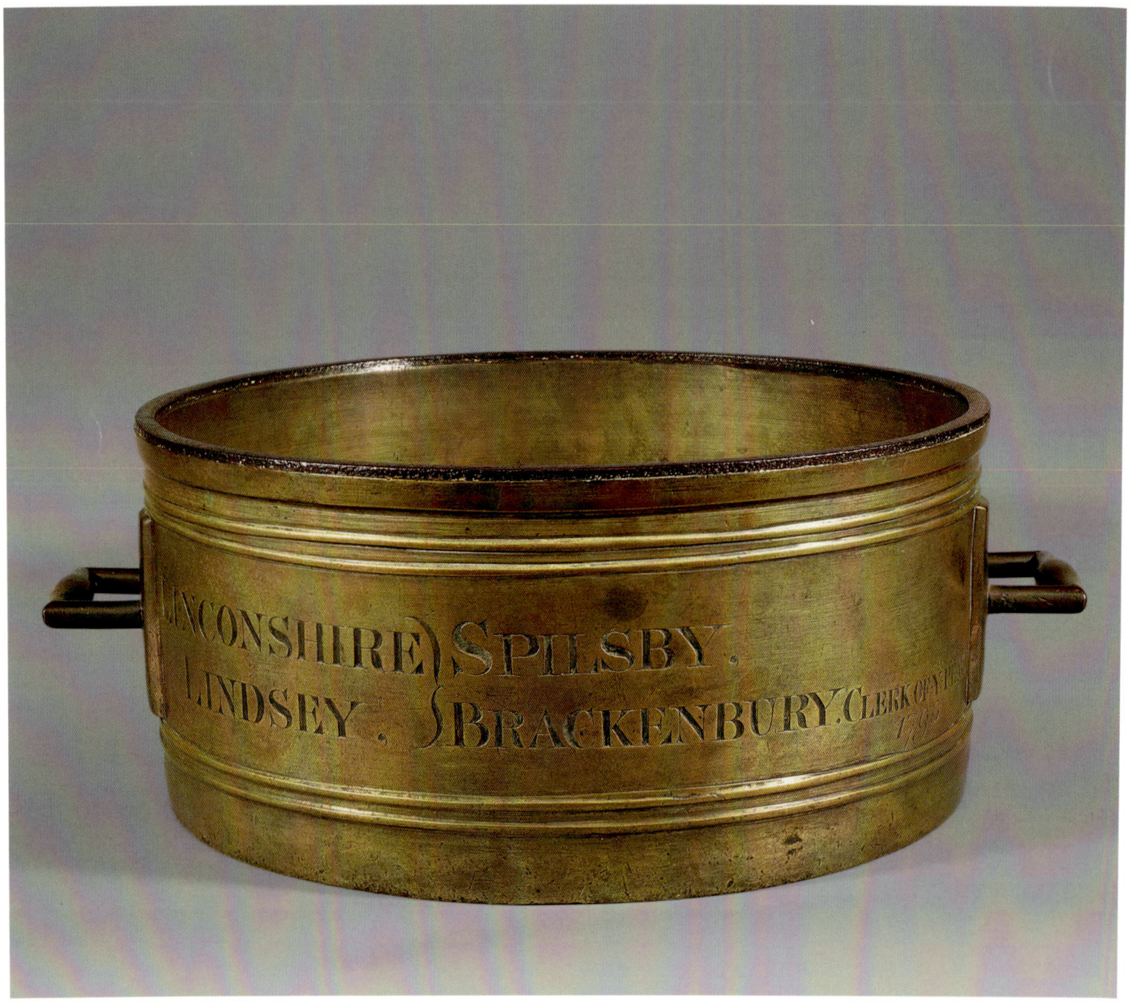

1766

Wine gill measure

Brass

$3 \times 2\frac{1}{4}$ in. (7.6 × 5.7 cm)

Inscription:
Gill Wine

Ale half-pint measure

Brass

$3\frac{7}{8} \times 3$ in. (9.8 × 7.6 cm)

Inscription: $\frac{1}{2}$ Pint Winchester

Ale pint measure

Brass

$4\frac{5}{8} \times 3\frac{3}{4}$ in. (11.7 × 9.5 cm)

Inscription: Pint Winchester

Ale quart measure

Brass

$6\frac{1}{8} \times 4\frac{3}{4} \times 6\frac{3}{4}$ in. (15.6 × 12.1 × 17.1 cm)

Inscription: Nich.^{s} Corsellis Esqr|WIVENHOE HALL.|1766.|Quart Winchester

Three of this nest of four mid-eighteenth-century brass liquid measures were for ale or beer and one was for wine. Each cylindrical tankard has a molded rim. The largest, the quart measure, has a loop handle. The inner measures include a pint Winchester, a half-pint Winchester, and a "gill wine" Winchester.

"WIVENHOE," inscribed only on the quart measure, is a town sited on the east bank of the River Colne in northeast Essex. The Corsellis family, one of the major landholders of the eighteenth century, had come to England from the Low Countries in 1570 and purchased the Wivenhoe Estate in 1657. The head of the family in 1766 was Nicholas Corsellis (1745–1826), who was at that time still a teenager.

PROVENANCE

bt. John H. Bryan of Lake Bluff, IL (1936–2018) from Tobias Jellinek Antiques, London, May 17, 1995

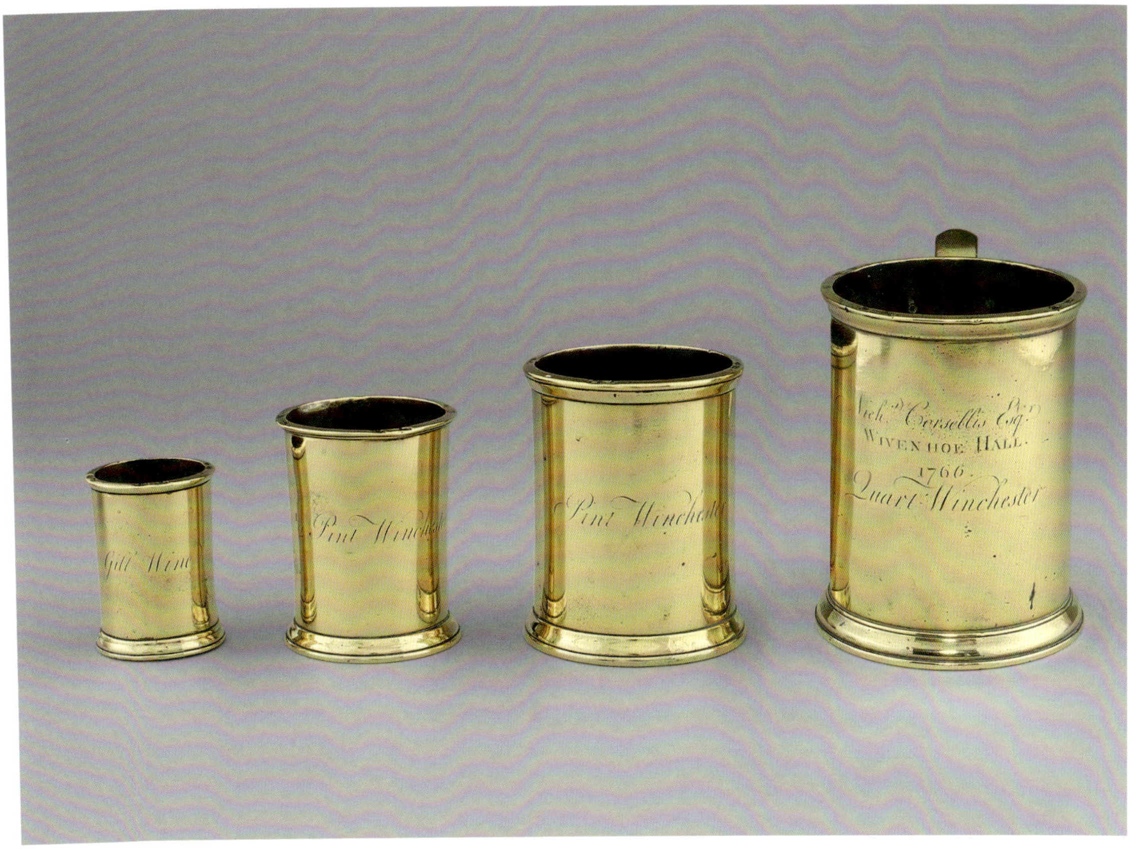

1772

Terrestrial globe

Wood, paper, brass

$5\frac{1}{8} \times 4\frac{3}{4}$ in. (13 × 12.1 cm)

Inscription: New| Terrestrial|Globe|By THarris & Son| London; fect. 1772

PROVENANCE

bt. John H. Bryan of Lake Bluff, IL (1936–2018) from Trevor Philip & Sons Ltd., London, November 16, 1999

1772

Celestial globe

Wood, paper, brass

$5\frac{1}{8} \times 4\frac{3}{4}$ in. (13 × 12.1 cm)

Inscription: New| Celestial|Globe|By THarris & Son| London

PROVENANCE

bt. John H. Bryan of Lake Bluff, IL (1936–2018) from Trevor Philip & Sons Ltd., London, November 16, 1999

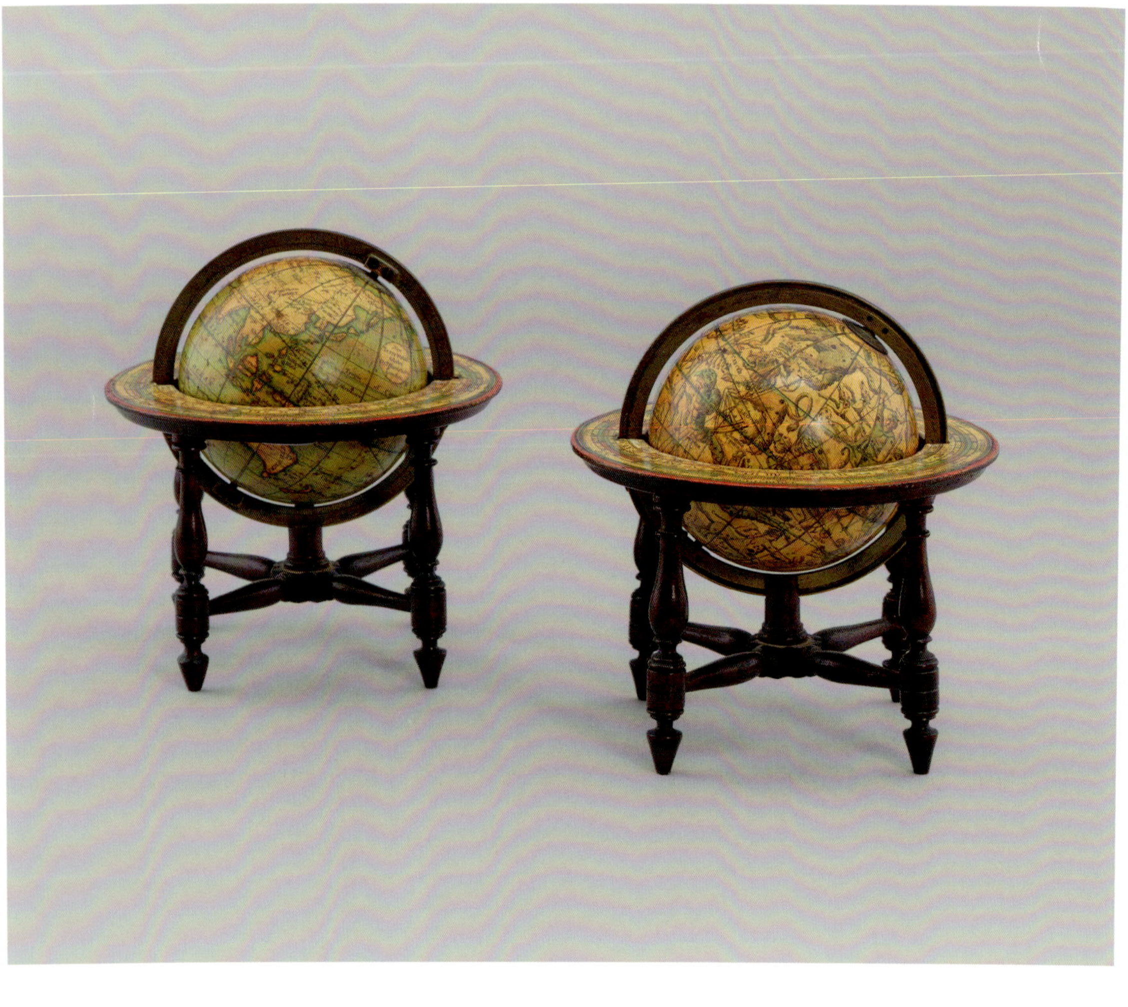

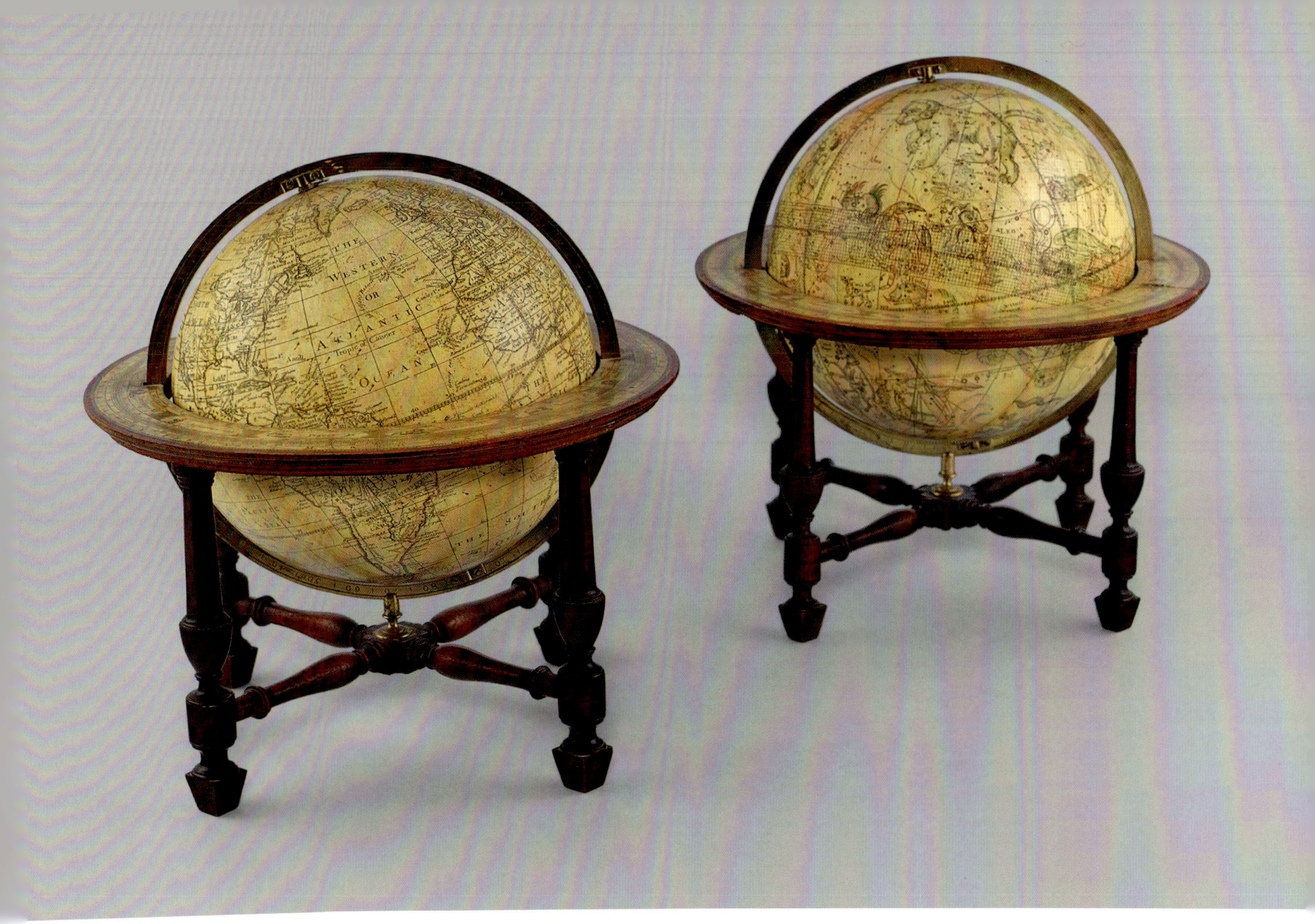

1783

Terrestrial globe

Wood, paper, brass

$13\frac{1}{2} \times 13\frac{1}{2}$ in. (34.3 × 34.3 cm)

William Bardin (d. 1798)

Inscription: A|New, Accurate, and|compleat|TERRESTRIAL GLOBE,|Accompanying the GEOGRAPHICAL MAGAZINE.|Originally laid down|By the late|Mr JAMES FERGUSON F.R.S|Improv'd by G WRIGHT,|and made by W. BARDIN,|Fleet Street, LONDON; Published as the Act directs by Harrison & Co N° 18 Paternoster Row, Augt 1st|1783

PROVENANCE

bt. John H. Bryan of Lake Bluff, IL (1936–2018) from Trevor Philip & Sons Ltd., London, November 16, 1999

1785

Celestial globe

Wood, paper, brass

$13\frac{1}{2} \times 13\frac{1}{2}$ in. (34.3 × 34.3 cm)

William Bardin (d. 1798)

Inscription: THE|CELESTIAL GLOBE,|Accompanying the|GEOGRAPHICAL MAGAZINE|Published as the Act directs by Harrison & Co N° 18 Paternoster Row, Apr 1, 1785.|made by W. BARDIN,|Fleet Street, LONDON.

PROVENANCE

bt. John H. Bryan of Lake Bluff, IL (1936–2018) from Trevor Philip & Sons Ltd., London, November 16, 1999

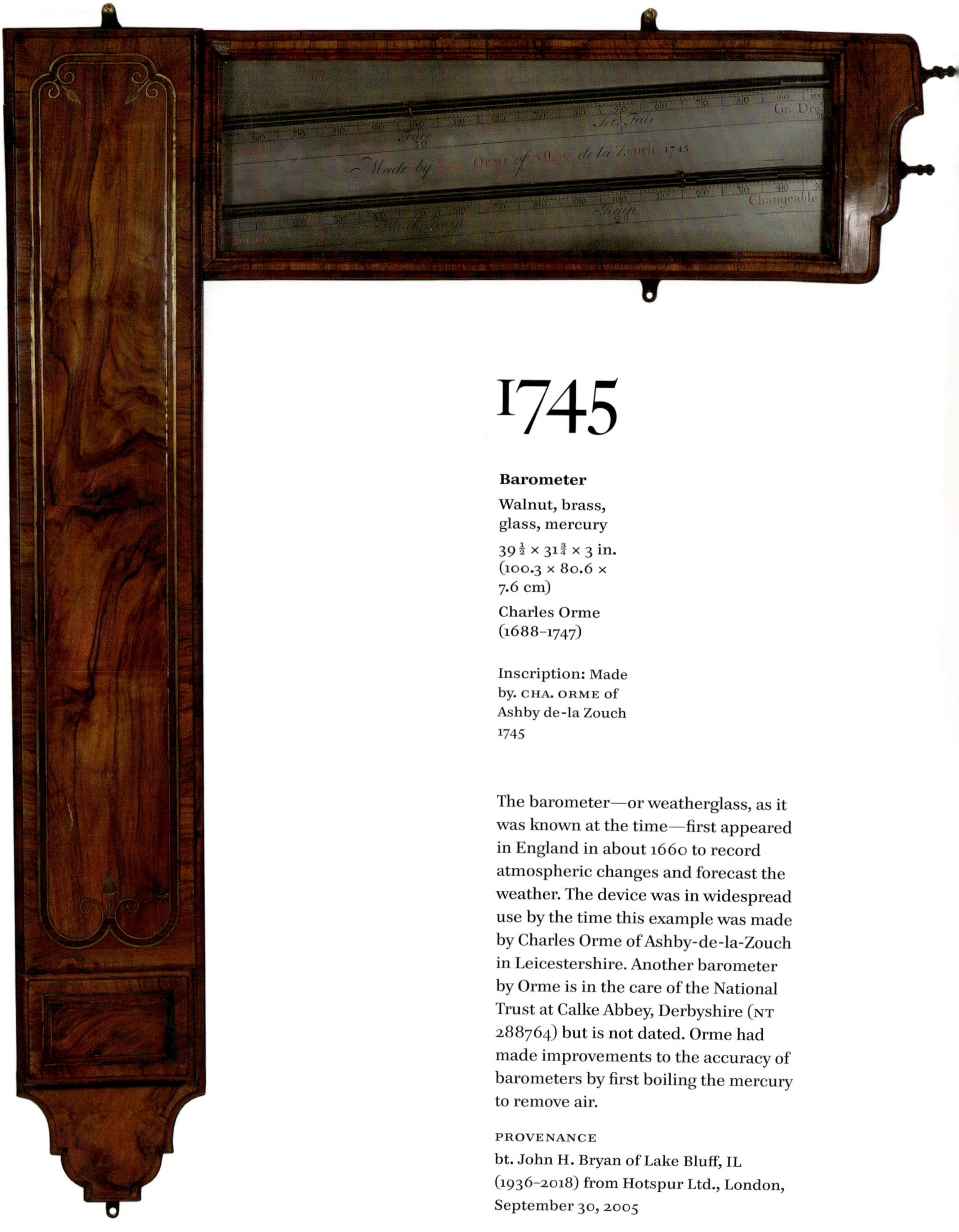

1745

Barometer

Walnut, brass, glass, mercury

$39\frac{1}{2} \times 31\frac{3}{4} \times 3$ in. (100.3 × 80.6 × 7.6 cm)

Charles Orme (1688–1747)

Inscription: Made by. CHA. ORME of Ashby de-la Zouch 1745

The barometer—or weatherglass, as it was known at the time—first appeared in England in about 1660 to record atmospheric changes and forecast the weather. The device was in widespread use by the time this example was made by Charles Orme of Ashby-de-la-Zouch in Leicestershire. Another barometer by Orme is in the care of the National Trust at Calke Abbey, Derbyshire (NT 288764) but is not dated. Orme had made improvements to the accuracy of barometers by first boiling the mercury to remove air.

PROVENANCE

bt. John H. Bryan of Lake Bluff, IL (1936–2018) from Hotspur Ltd., London, September 30, 2005

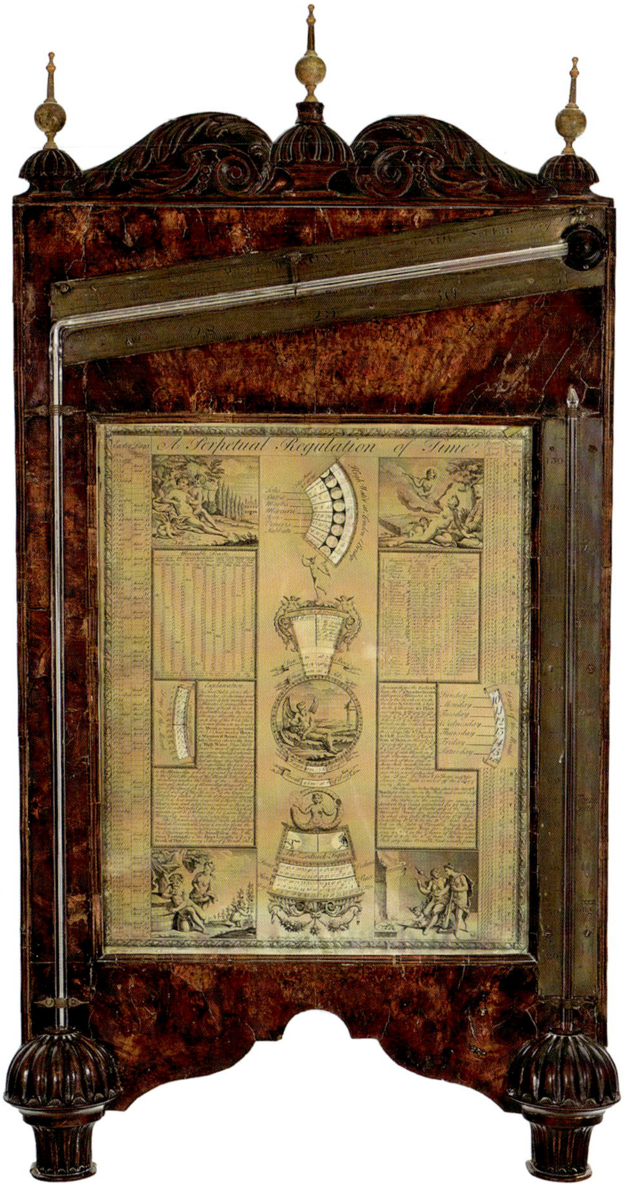

1731–1830

Calendar barometer

Walnut, brass, paper, glass, mercury

43½ × 23 × 5 in. (110.5 × 58.4 × 12.7 cm)

Inscription: A Perpetual Regulation of Time

This object is possibly the earliest recorded example of an angle barometer that incorporates a Perpetual Regulation of Time.

At center is an engraving entitled *A Perpetual Regulation of Time*. The central roundel depicts Chronos with an hourglass, scythe, and sundial. The figures in the corners represent the four seasons. Apertures in the face reveal times for "High Water at London Bridge"; dates for "The Fix'd Feasts"; "Sunrise and Sunset"; "Day Length" and "Break of Day"; "Signs of the Zodiac"; the Sun's "decline" and position; and the day of the month.

There are additional tables for the years from 1731 to 1830, "Easter Days," "Moveable Feasts," "A Regal Table" from William I to George II, and an explanation of how to use the Perpetual Regulation of Time. The movable dials are set by wooden knobs at the rear.

PROVENANCE

Sotheby's, London, The Moller Collection, November 18, 1993 (lot 14); bt. John H. Bryan of Lake Bluff, IL (1936–2018) from Hotspur Ltd., London, November 19, 1993

1664

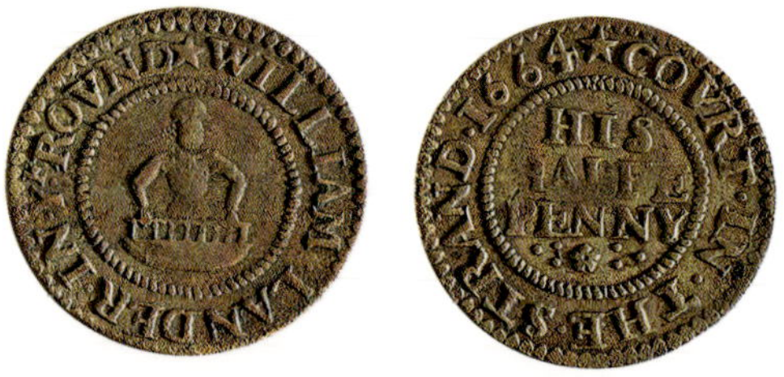

Trade token
Brass
$1\frac{1}{8}$ in. (2.9 cm) diameter

Inscription:
WILLIAM LANDER IN Y^{E} ROVND; COVRT IN THE STRAND 1664|HIS HALFE PENNY

In the absence of ready coinage in the mid- to late seventeenth century, businessmen and businesswomen issued their own trade tokens. This practice was particularly prevalent in cities, where many small transactions took place on a daily basis. This token was made for William Lander, who is believed to have been a chandler at Round Court in the Strand, in the parish of St Martin-in-the-Fields. The token was found in the River Thames.

PROVENANCE
bt. John H. Bryan of Lake Bluff, IL (1936–2018) from Timothy Millett Limited, London, October 12, 2018

1665

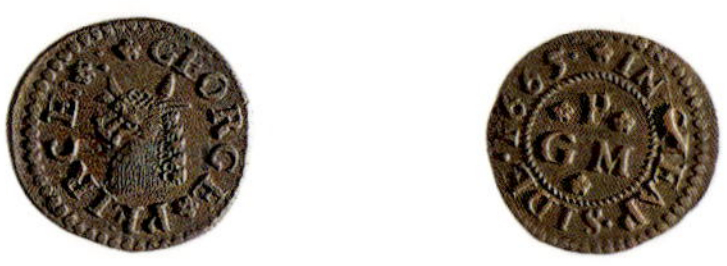

Trade token
Copper alloy
$\frac{5}{8}$ in. (1.59 cm) diameter

Inscription:
GEORGE PEIRCE; IN CHEAP SIDE 1665; G|P|M

With fourteen hearths, the Bull Head on Cheapside was one of the largest and most famous taverns of its day. Samuel Pepys (1633–1703) mentioned it on numerous occasions in his diary. The Royal Society convened at the tavern before moving to Gresham College.

The proprietors of the Bull Head were George Peirce, Vintner, and his wife, Mary, and their initials appear on this token. George was the namesake son of George Peirce, clerk of Caddington, Bedfordshire, who apprenticed to Thomas Benson in 1653. From 1663 until his death in 1670, Peirce the younger enrolled a number of his own apprentices.

PROVENANCE
bt. John H. Bryan of Lake Bluff, IL (1936–2018) from Timothy Millett Limited, London, September 20, 2018

1666

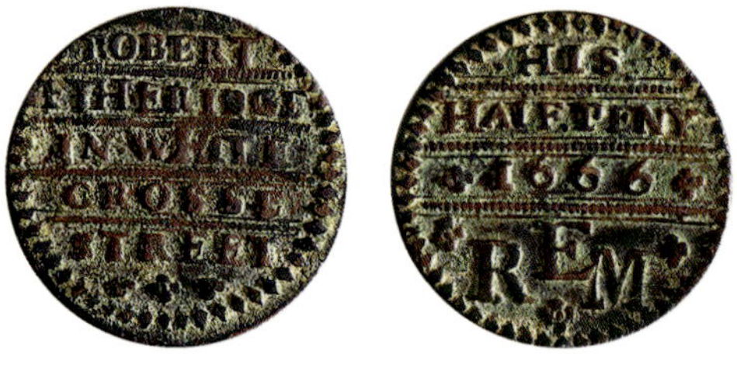

Trade token
Brass
$1\frac{1}{8}$ in. (2.9 cm) diameter

Inscription:
ROBERT|ETHERIDGE|IN WHITE|CROSSE STREET; HIS|HALF|PENY|1666|R|E|M

Like many issuers of trade tokens, Robert Etheridge was a Vintner—a wine seller, who often sold wine at taverns. He was the namesake son of a yeoman of Chingford, Essex, and apprenticed to the Vintner William Etheridge, who presumably was a relative, on February 5, 1649/50. Etheridge lived and worked with his wife, Mary, in the notoriously rough Whitecross Street in the parish of St Giles-without-Cripplegate.

PROVENANCE
bt. John H. Bryan of Lake Bluff, IL (1936–2018) from Timothy Millett Limited, London, October 12, 2018

1666

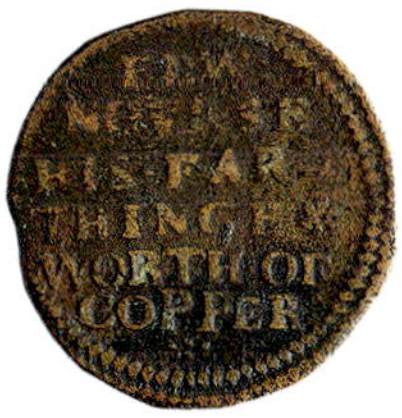

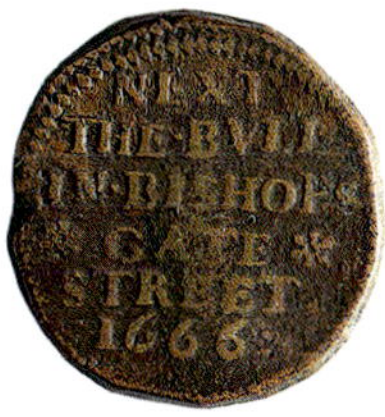

Trade token

Copper

1 in. (2.5 cm) diameter

Inscription:
EDW|NOVRSE|HIS FAR|THINGE|WORTH OF|COPPER; NEXT|THE BVLL|IN BISHOP|GATE|STREET|1666

As the inscription claims, this token carried its worth in its copper content. As such, it was an honest farthing and would have been accepted wherever it was offered.

It was commissioned by a brazier called Edward Nourse (1624–1689). He was the son of Luke Nourse (d. 1673), a Parliamentarian who served as a captain in the defense of Gloucester against Royalist forces in 1643 and then subsequently as sheriff, mayor, and member of Parliament for Gloucester. The city was a thorn in the Royalists' side and a hotbed of militant Puritan and parliamentarian zeal throughout the 1640s. Luke's involvement with the Parliamentary regime may help explain why Edward was summarily removed as Master of the Girdlers' Company in 1687 by order of James II, although he was allowed to serve in that role in the following year.

On August 23, 1649, Nourse married Mary Wells (d. 1708), daughter of Hugh Wells, the Citizen and Armorer of St Michael Cornhill, at St Olave Old Jewry. The couple returned to the city of Gloucester—where his father was issuing farthings in 1657—and there Nourse was made a freeman in 1651. In 1662, Nourse, in search of better prospects, asked to be discharged from the Gloucester Common Council and moved his family to London. This token was presumably issued by him as he and his father-in-law sought to advertise the relocation of their business to nearby Bishopsgate, which had escaped much of the devastation of the Great Fire that year. Their expertise in metallurgy allowed them to strike a coin that had comparable worth to those made at the Royal Mint. Numerous versions of Nourse's farthing survive, including at the British Museum, London (T.2259) and the Museum of London (99.66/173; NN17583; NN17584).

After Nourse's death, a strapwork cartouche containing a tablet and his coat of arms, surrounded by palm leaves and winged putti, was erected on the wall at St Michael Cornhill. The monument commemorates Luke Nourse (d. 1673), Hugh Wells (d. 1673), and Edward Nourse. Edward's will, which attests to his great wealth, was proved January 20, 1689/90 (TNA PROB 11/401/2). He was buried in the north aisle of the church on January 17, 1689/90.

PROVENANCE

Norweb Collection; Spink Auction, June 26, 2008 (lot 264); bt. John H. Bryan of Lake Bluff, IL (1936–2018) from Sovereign Rarities Ltd., London, November 10, 2017

LITERATURE

W. Dunville Smythe, *An Historical Account of the Worshipful Company of Girdlers, London* (London: Chiswick Press, 1905), 110, 256

R. H. Thompson, "Edward Nourse and a Farthing's Worth of Copper," *British Numismatic Journal* 70 (2000): 147–50

W. R. Williams, *The Parliamentary History of the County of Gloucester* (Hereford, UK: Privately printed for the author by Jakeman and Carver, 1898), 196

1666

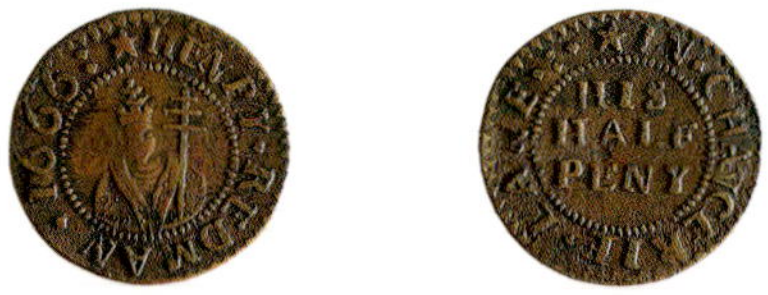

Trade token
Copper alloy
$\frac{3}{4}$ in. (1.9 cm) diameter

Inscription: HENRY REDMAN 1666; IN CHANCERIE LANE HIS|HALF|PENNY

To judge from the depiction on this trade token—the bust of a pope with a tiara and staff—in 1666 Henry Redman was the new occupant of the Old Pope's Head Tavern on Chancery Lane. In September of that year, the Great Fire had displaced Redman from the back side of St Thomas Apostle, where his previous tavern had stood. His new venture was stationed in Pope's Head Court on the south side of Jackanapes Lane. His will, proved in 1676 (TNA PROB 11/351/518), attests to his wealth and his portfolio of property in London and elsewhere. He appears also to have been the proprietor of the Sun, another tavern on Chancery Lane. After his death, his widow, Mary, had apprentices bound to her through the Vintners' Company, of which Redman had been a member.

PROVENANCE
bt. John H. Bryan of Lake Bluff, IL (1936–2018) from Timothy Millett Limited, London, September 20, 2018

LITERATURE
Jacob Henry Burn, *A Descriptive Catalogue of the London Traders, Tavern, and Coffee-House Tokens Current in the Seventeenth Century* (London: Printed for the use of the members of the Corporation of the City of London, 1855), 59

1666

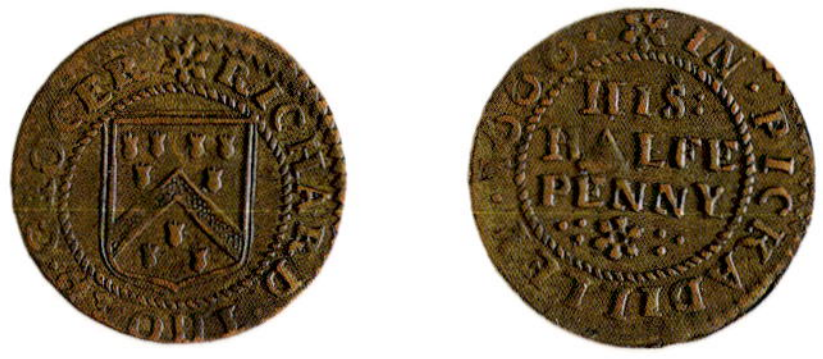

Trade token
Copper alloy
$\frac{7}{8}$ in. (2.22 cm) diameter

Inscription: RICHARD THORP GROCER; IN PICKADILLE 1666; HIS|HALFE|PENNY

Famous for the pickadils, or cut-lace ruffs and collars, that were made on the street, Piccadilly was at the heart of London's westward expansion in the Stuart period. Thorp, who appears to have operated from the Grocers' Arms in Piccadilly, has not been traced.

PROVENANCE
bt. John H. Bryan of Lake Bluff, IL (1936–2018) from Timothy Millett Limited, London, September 20, 2018

1667

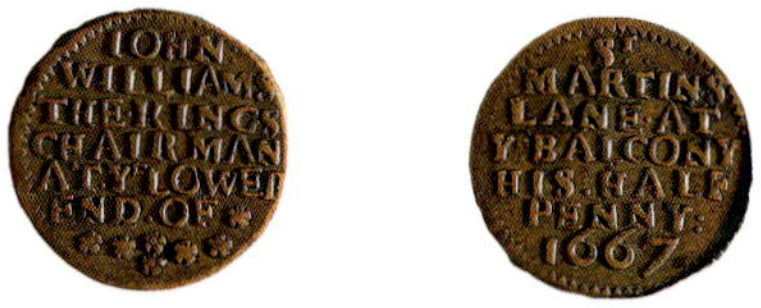

Trade token
Copper alloy
$\frac{3}{4}$ in. (1.9 cm) diameter

Inscription: IOHN| WILLIAMS|THE KINGS|CHAIRMAN| AT Y^{E} LOWER|END OF; S^{T}| MARTINS| LANE AT| Y^{E} BALCONY|HIS HALF|PENNY|1667

John Williams, of St Martin-in-the-Fields, Groom and litter man to His Majesty, made his will on June 2, 1684, and died a wealthy man (TNA PROB 11/376/507). He probably never lifted a litter, but he ran a successful business that made the most of his royal associations.

PROVENANCE
bt. John H. Bryan of Lake Bluff, IL (1936–2018) from Timothy Millett Limited, London, September 20, 2018

1667

Trade token

Brass

1⅛ in. (2.9 cm) diameter

Inscription: THO: GREEN AT THE·IN; THE CRVCHED FRYERS 67; HIS|HALFE|PENNY|T|G|I

Thomas Green operated at the Crossed Keys, near Tower Hill in the former Friary of the Holy Cross, which makes this token an appropriate sign for his business. There was a Vintner by this name who was active during the second half of the seventeenth century, so it seems possible that the Crossed Keys was a tavern. This Thomas Green was the son of Thomas Green, Gentleman, of Witham, Essex, who was apprenticed to Henry Chittie on November 12, 1650.

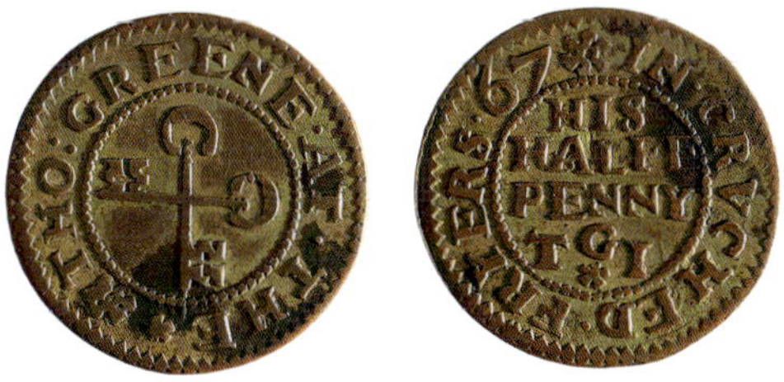

PROVENANCE

bt. John H. Bryan of Lake Bluff, IL (1936–2018) from Timothy Millett Limited, London, October 12, 2018

1669

Trade token

Copper alloy

1 in. (2.5 cm) diameter

Inscription: CHARLES KIFTELL; AT THE COFFEE HOUSE; IN|CHEAP|SIDE|1669

A hand lifts a cup into the sky, and the cup is replenished with coffee. This was the promise depicted on Charles Kiftell's token for his coffeehouse on Cheapside in 1669. Kiftell, however, has not been traced.

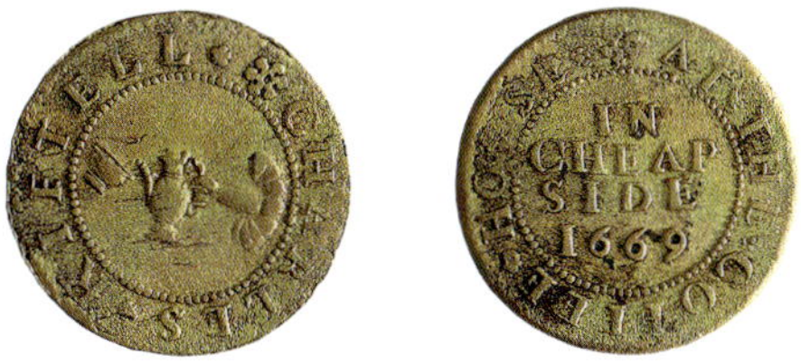

PROVENANCE

bt. John H. Bryan of Lake Bluff, IL (1936–2018) from Timothy Millett Limited, London, September 20, 2018

1752 | 1767 | 1774 | 1777

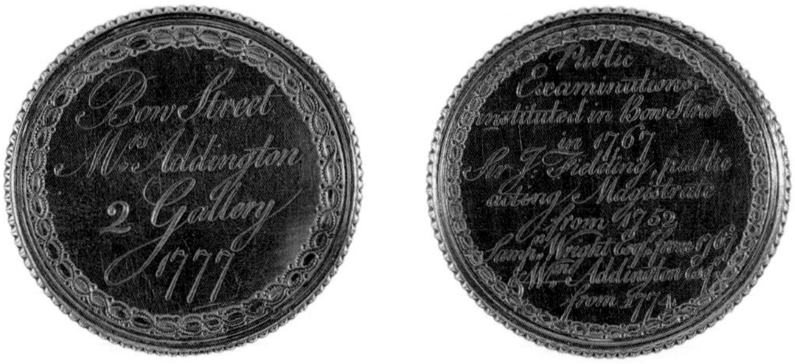

Bow Street magistrates ticket
Silver
1½ in. (3.8 cm) diameter

Inscription: Bow Street | M^{rs}, Addington | 2 Gallery | 1777.; Public | Examinations, | instituted in Bow Street | in 1767 | Sir J: Fielding, public | acting Magistrate | from 1752 | $Samp^{n}$, Wright Esqr, from 1767 | & W^{m}, Addington Esqr, | from 1774

This object is an eighteenth-century ticket for Bow Street Magistrates Court. Sir William Addington (1728–1811), son of the Reverend Charles Addington, rector of Litchborough, Northamptonshire, was appointed a magistrate in 1774. He was made chief magistrate in 1793 after the death of Sampson Wright and was knighted in 1797. Addington retired in 1800 and died in Bath in 1811. Addington's bequests in his will included a silver tea urn given to him by his friend the Duke of Cumberland and a gold box given to him by the Prince of Wales. The "Mrs. Addington" on this ticket was presumably William Addington's wife, who died in November 1788.

PROVENANCE
bt. John H. Bryan of Lake Bluff, IL (1936–2018) from Peter Cameron, London, June 28, 2017

1587

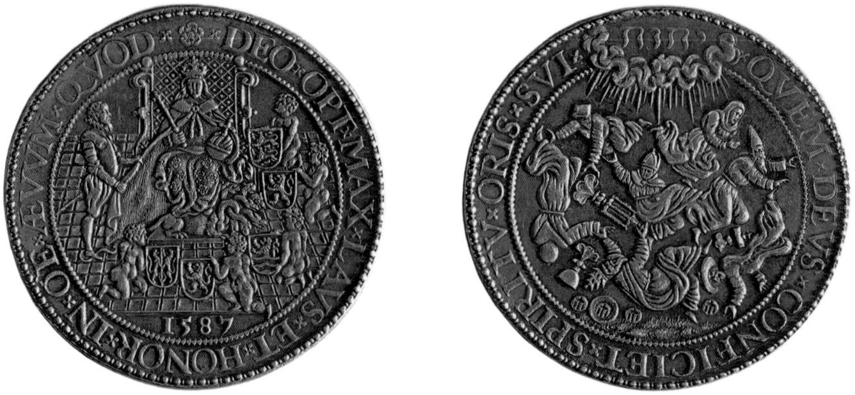

Medal
Silver
2 in. (5.1 cm) diameter

Inscription: DEO·OPT·MAX· LAVS·ET·HONOR· IN·OE·ÆVVM QVOD | 1587; QVEM·DEVS· CONFICIET·SPIRITV· ORIS·SVI

The scene of the pope and his cardinals, priests, and clergy tumbling from the sky commemorates English support for the Dutch revolt against the rule of Catholic Spain—perhaps in support of Elizabeth I's acceptance of the offer of sovereignty made to her by the Frisian clergy. Instead, Elizabeth sent as her representative Robert Dudley, first Earl of Leicester, who was made governor-general of the Netherlands. On the opposite face of this medal, Dudley attends upon the enthroned monarch alongside shield bearers who represent the states of Guilders, Holland, Zeeland, Utrecht, and Friesland.

PROVENANCE
A. H. Baldwin & Sons Ltd., London, November 2, 2015; bt. John H. Bryan III of Nashville, TN

LITERATURE
Stephen K. Scher, ed., *The Scher Collection of Commemorative Medals* (New York: The Frick Collection, 2019), 439, no. 735

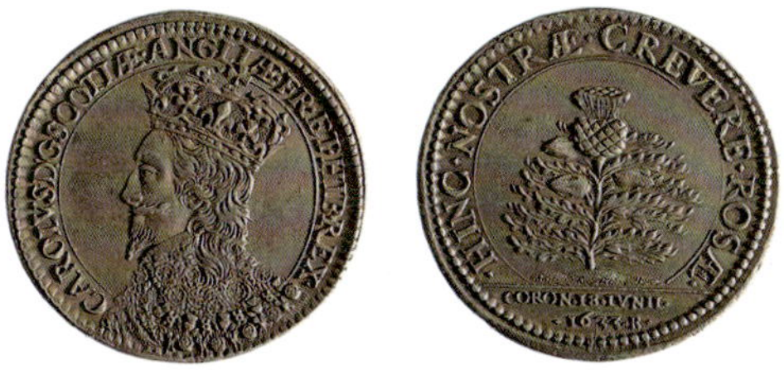

1633

Medal
Silver
$1\frac{1}{8}$ in. (2.9 cm) diameter
Nicholas Briot (1579–1646)

Inscription: CAROLVS·D:G SCOTIÆ·ANGLIÆ· FR·ET·HIB·REX·; ·HINC·NOSTRÆ· CREVERE·ROSÆ; CORON·18 IVNII·| ·1633·

This medal was struck to commemorate Charles I's Scottish coronation at Holyrood Palace on June 18, 1633. By tradition, Charles I carried the medal on his person.

PROVENANCE
Sovereign Rarities Ltd., London, June 13, 2017; bt. John H. Bryan III of Nashville, TN

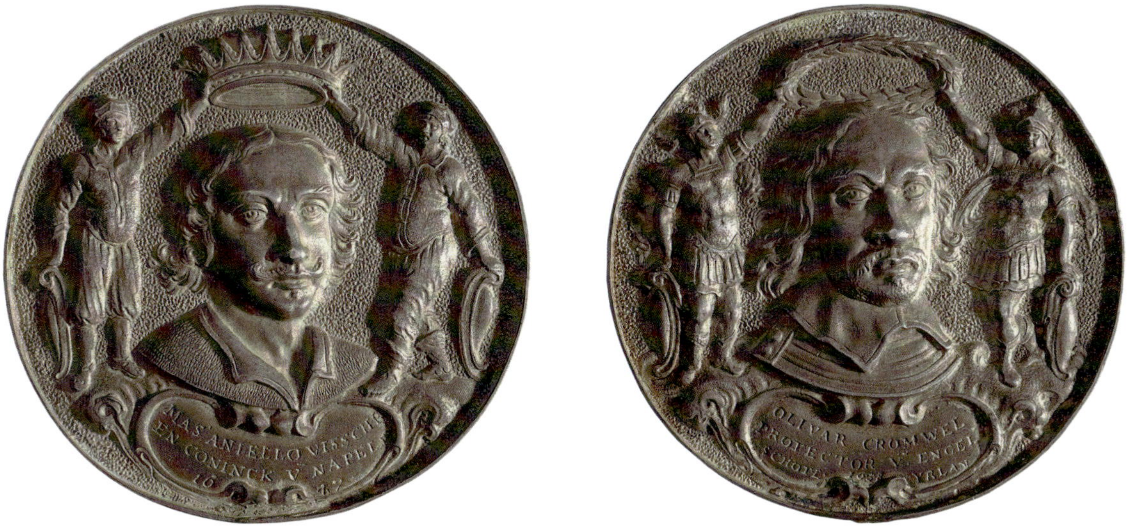

1647|1658

Medal
Silver
$2\frac{3}{4}$ in. (7 cm) diameter
Wouter Muller (1604–1673)

Inscription: MAS ANIELLO VISSCHER| EN CONINCK V'NAPEL|16 47; OLIVAR CROMWEL| PROTECTOR V'. ENGEL|SCHOTL YRLAN 1658

Celebrated in this medal are two seventeenth-century revolutionaries: Oliver Cromwell and Masaniello (Tommaso Aniello d'Amalfi). The latter, a Neapolitan fisherman, led a popular uprising against the Spanish rulers of Naples in 1647. He enjoyed some success before being assassinated at a monastery. This coin is attributed on stylistic grounds to Wouter Muller, one of the most important medalists in the Netherlands during the seventeenth century.

PROVENANCE
A. H. Baldwin & Sons Ltd., London, September 28, 2015; bt. John H. Bryan III of Nashville, TN

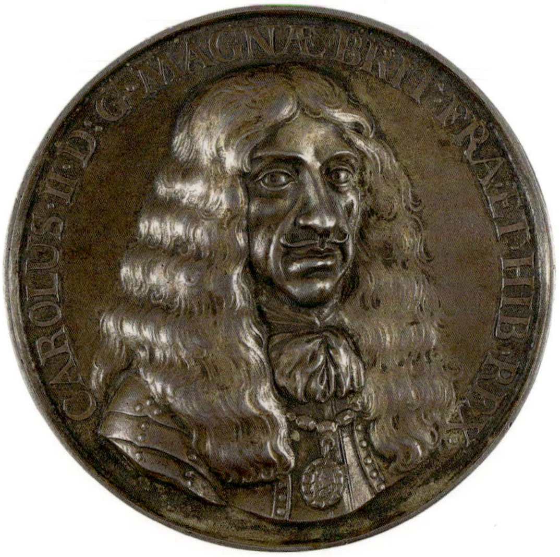

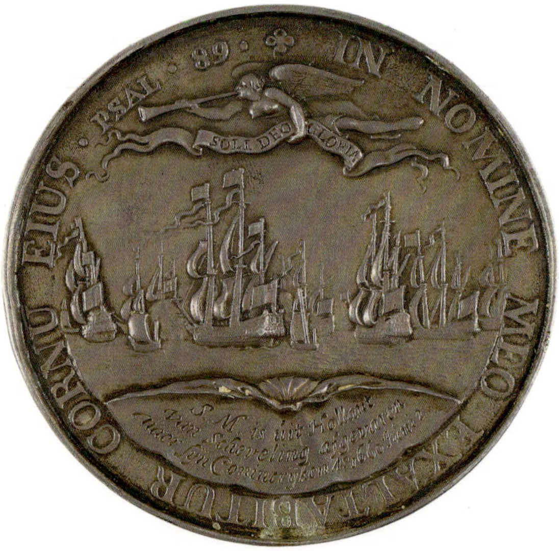

1660

Medal

Silver

$2\frac{3}{4}$ in. (7 cm) diameter

Peter van Abele

Inscription: CAROLVS·II·D:G·MAGNÆ·BRIT·FRA·ET·HIB·REX·; IN NOMINE MEO EXALTABITUR CORNU EIUS·PSAL 89·; SOLI DEO GLORIA; S.M. is úit Hollant|van Scheveling afgevaren|naer sÿn Conincrÿken A°: 1660 Júni 2.

This medal commemorates the return of Charles II to England in 1660, depicting on the reverse the fleet departing from Scheveningen, Holland. The inscription in Dutch indicates that His Majesty departed by way of Scheveningen to his own kingdom on June 2. He sailed on board the *Naseby*, the name of which was immediately changed to the *Royal Charles*. Van Abele was a Dutch medalist who joined the Silversmiths' Guild of Amsterdam in 1634. This medal was made in Holland.

PROVENANCE

A. H. Baldwin & Sons Ltd., London, July 9, 2015; bt. John H. Bryan III of Nashville, TN

LITERATURE

Edward Hawkins, *Medallic Illustrations of the History of Great Britain and Ireland to the Death of George II* (London: The Trustees of the British Museum, 1904–11), plate LXII

Stephen K. Scher, ed., *The Scher Collection of Commemorative Medals* (New York: The Frick Collection, 2019), 463, no. 776

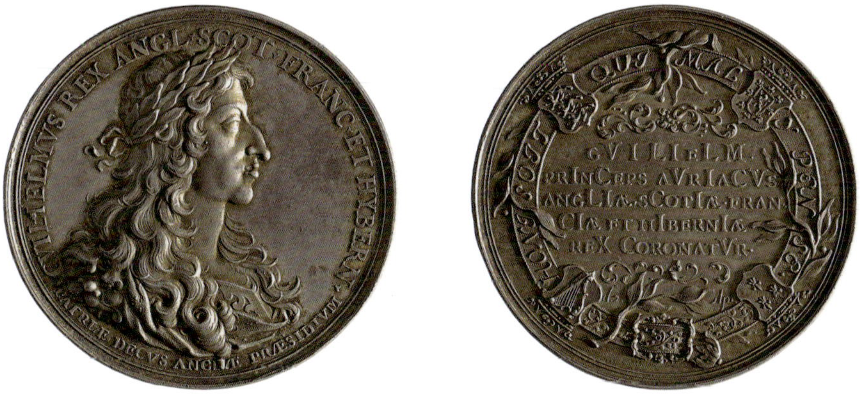

1689

Medal
Silver
2 in. (5.1 cm) diameter
Philipp Heinrich Müller (1654–1719)

Inscription: GVILIELMVS REX ANGL·SCOT· FRANC·ET HYBERN.; PATRIÆ PRÆSIDIVM; PHM; gVILIeLM.|prInCeps aVrIaCVs| angLIæ, sCotIæ·fran·|cIæ et hIbernIæ|reX CoronatVr·; d.. 21 Api.; HONI SOIT QUI MAL Y PENSE; REGIA, CREDE MIHI, RES EST, SUCCURRERE LAPSIS

Philipp Heinrich Müller was a celebrated medalist and die engraver who worked in Augsburg, Nuremberg, Salzburg, and many other mints. This medal was struck at Nuremberg. It celebrates the coronation of William III, which is explained in a chronogram. The larger characters correspond to Roman numerals for the year 1689.

PROVENANCE
Sovereign Rarities Ltd., London, November 28, 2016; bt. John H. Bryan III of Nashville, TN

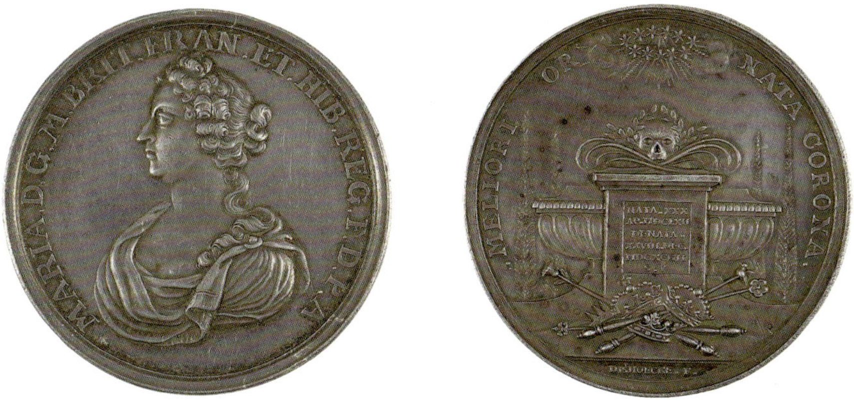

1694

Medal
Silver
2 in. (5.1 cm) diameter
Jacob van Dishoeke (1650–1723)

Inscription: MARIA. D. G. M. BRIT. FRAN. ET. HIB. REG. F. D. P. A.; NATA. XXX.| AP.MDCLXII.| DENATA.|XVIII. DEC.|MDCXCIIII| S. V; MELIORI ORNATA CORONA.; DISHOECKE. F.

A skull, sarcophagus, and cypress trees accompany the commemorative message on this coin, which was made following the death of Mary II in 1694. The medal was made by Jacob van Dishoeke, a relatively obscure sculptor who made medals for the House of Orange.

PROVENANCE
Sovereign Rarities Ltd., London, October 14, 2016; bt. John H. Bryan III of Nashville, TN

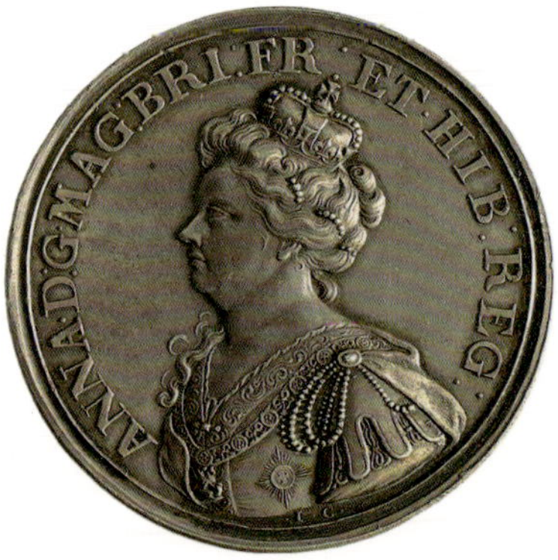

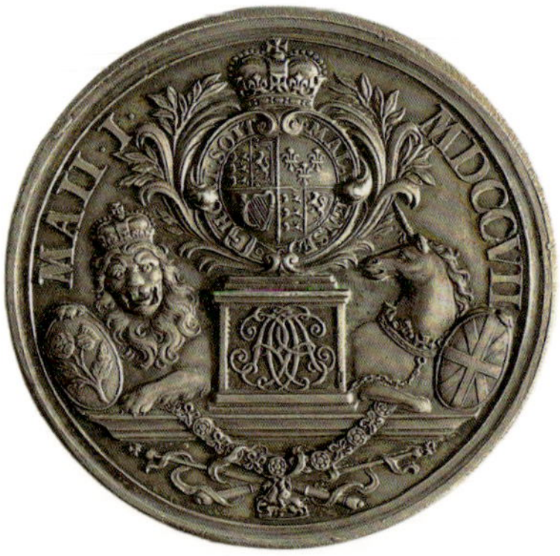

1707

Medal

Silver

$1\frac{7}{8}$ in. (4.8 cm) diameter

John Croker (1670–1741)

Inscription: ANNA D: G MAG: BRI: FR: ET HIB: REG: I C.; MAII I MDCCVII. AR [in cipher]

This medal commemorates the Act of Union of 1707, alluded to in the shield held in the paw of the crowned lion, which shows the thistle (for Scotland) and the rose (for England) growing from a single stem.

PROVENANCE

Sovereign Rarities Ltd., London, July 18, 2017; bt. John H. Bryan III of Nashville, TN

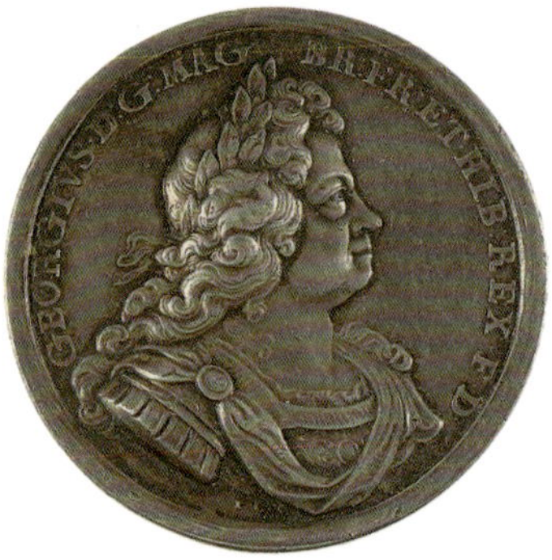

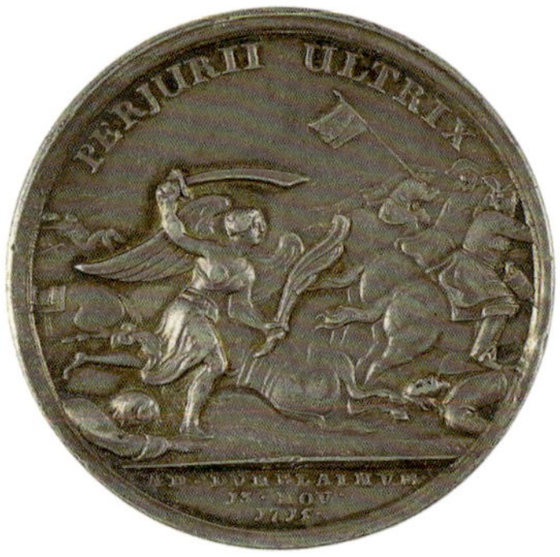

1715

Medal

Silver

$1\frac{3}{4}$ in. (4.5 cm) diameter

John Croker (1670–1741)

Inscription: GEORGIVS·D·G·MAG BR FR ET HIB: REX· F· D; PERJURII ULTRIX; AD. DVNBLAINVM | 13·NOV·1715·

The Battle of Sheriffmuir (November 13, 1715) took place at the height of the Jacobite rebellion, after the Earl of Mar, leader of the Jacobite cause in Scotland, declared James Stuart, the Old Pretender, King of the Scots. Mar had previously professed allegiance to King George I, so on this medal his "defeat" is seen as the just punishment of perjury, thus the inscription translates as "the avenger of perjury." In reality, the battle was inconclusive, and both sides claimed victory.

PROVENANCE

bt. John H. Bryan of Lake Bluff, IL (1936–2018) from Christopher Sheppard Ltd., London, June 10, 1997

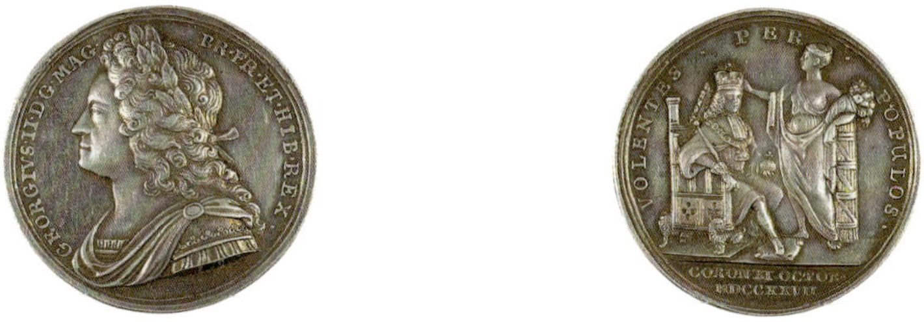

1727

Medal

Silver

$1\frac{1}{4}$ in. (3.2 cm) diameter

John Croker (1670–1741)

Inscription:
GEORGIVS·II·D G·
MAG·BR·FR·ET·HIB·
REX.; VOLONTES ·
PER · POPULOS ·;
CORON XI OCTOB·|
MDCCXXVII

This medal was made to celebrate the coronation of George II at Westminster Abbey, an event for which George Frederick Handel was commissioned to compose no fewer than four coronation anthems.

PROVENANCE

A. H. Baldwin & Sons Ltd., London, September 28, 2015; bt. John H. Bryan III of Nashville, TN

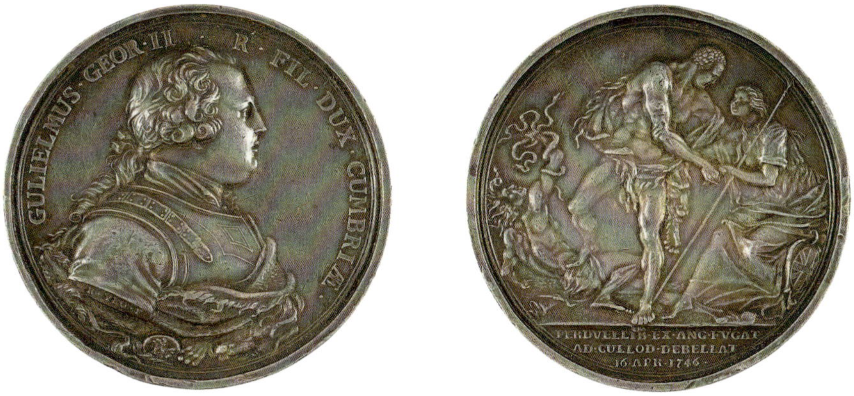

1746

Duke of Cumberland medal

Silver

2 in. (5.1 cm) diameter

Richard Yeo (ca. 1720–1779)

Inscription:
GULLELMUS·GEOR·
II · R · FIL · DUX ·
CUMBRIÆ.;
PERDVELLIB·
EX·ANG·FVGAT|AD·
CULLOD DEBEL-
LAT|16. APR·1746·

This medal commemorates the defeat of the Jacobite army by William, Duke of Cumberland, on April 16, 1764.

PROVENANCE

bt. John H. Bryan of Lake Bluff, IL (1936–2018) from Christopher Sheppard Ltd., London, June 17, 1996

1760

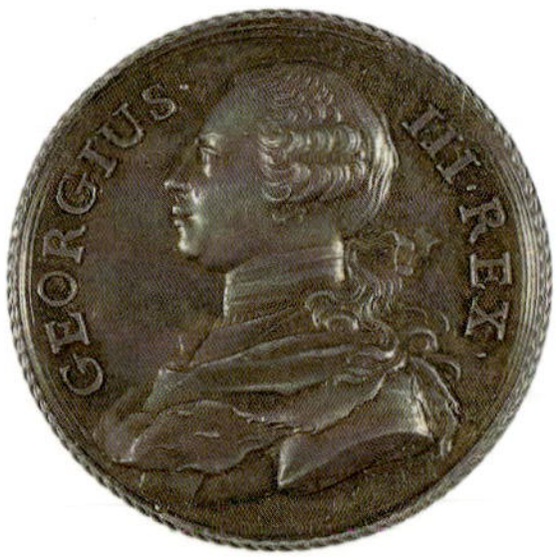

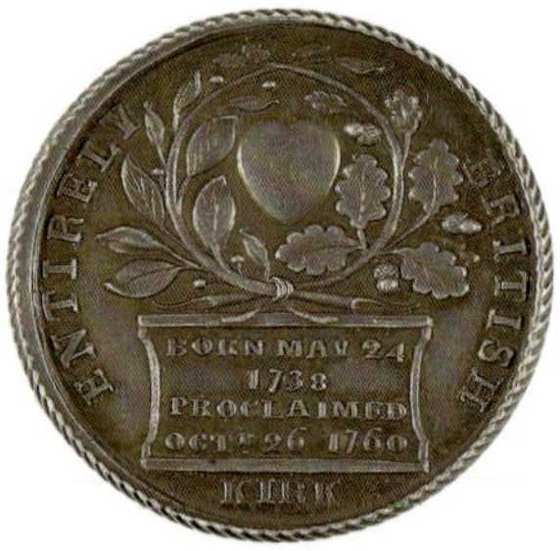

Medal

Silver

$1\frac{3}{8}$ in. (3.6 cm) diameter

John Kirk (1724–1776)

Inscription:
GEORGIUS·III·REX·
ENTIRELY BRITISH|
BORN MAY 24|1738|
PROCLAIMED|
OCT^{R}. 26 1760|KIRK

PROVENANCE

bt. John H. Bryan of Lake Bluff, IL (1936–2018) from A. H. Baldwin & Sons Ltd., London, September 17, 2015

1801

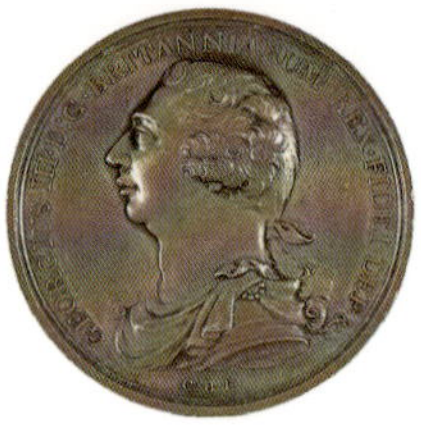

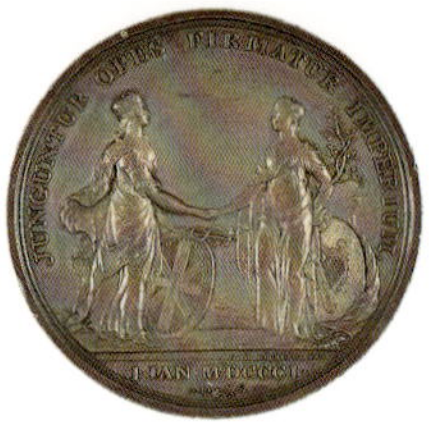

Medal

Bronze

$1\frac{1}{8}$ in. (2.9 cm) diameter

Conrad Heinrich Küchler (ca. 1740–1810)

Inscription:
GEORGIUS·III
D:G·BRITANNIARUM
REX·FIDEI DEF·&|
C·H·K; JUNGUNTUR
OPES FIRMATUR
IMPERIUM.;C·H·
KUCHLER
FC·|I·JAN· MDCCCI.

PROVENANCE

bt. John H. Bryan of Lake Bluff, IL (1936–2018) from A. H. Baldwin & Sons Ltd., London, September 17, 2015

1774

Coin scale

Brass, paper

$8\frac{1}{2} \times \frac{1}{2}$ in. (21.6 × 1.27 cm)

Inscription: John Kinsey|Shoe Maker Liverpool 1774; Scotson

This scale has an unusual form. The coins to be weighed were held in the end "claw," and the weights were held by the tassel. In 1774, John Kinsey was recorded at 12 Church Street, Liverpool.

PROVENANCE

bt. John H. Bryan of Lake Bluff, IL (1936–2018) from Tobias Jellinek Antiques, London, December 16, 1996

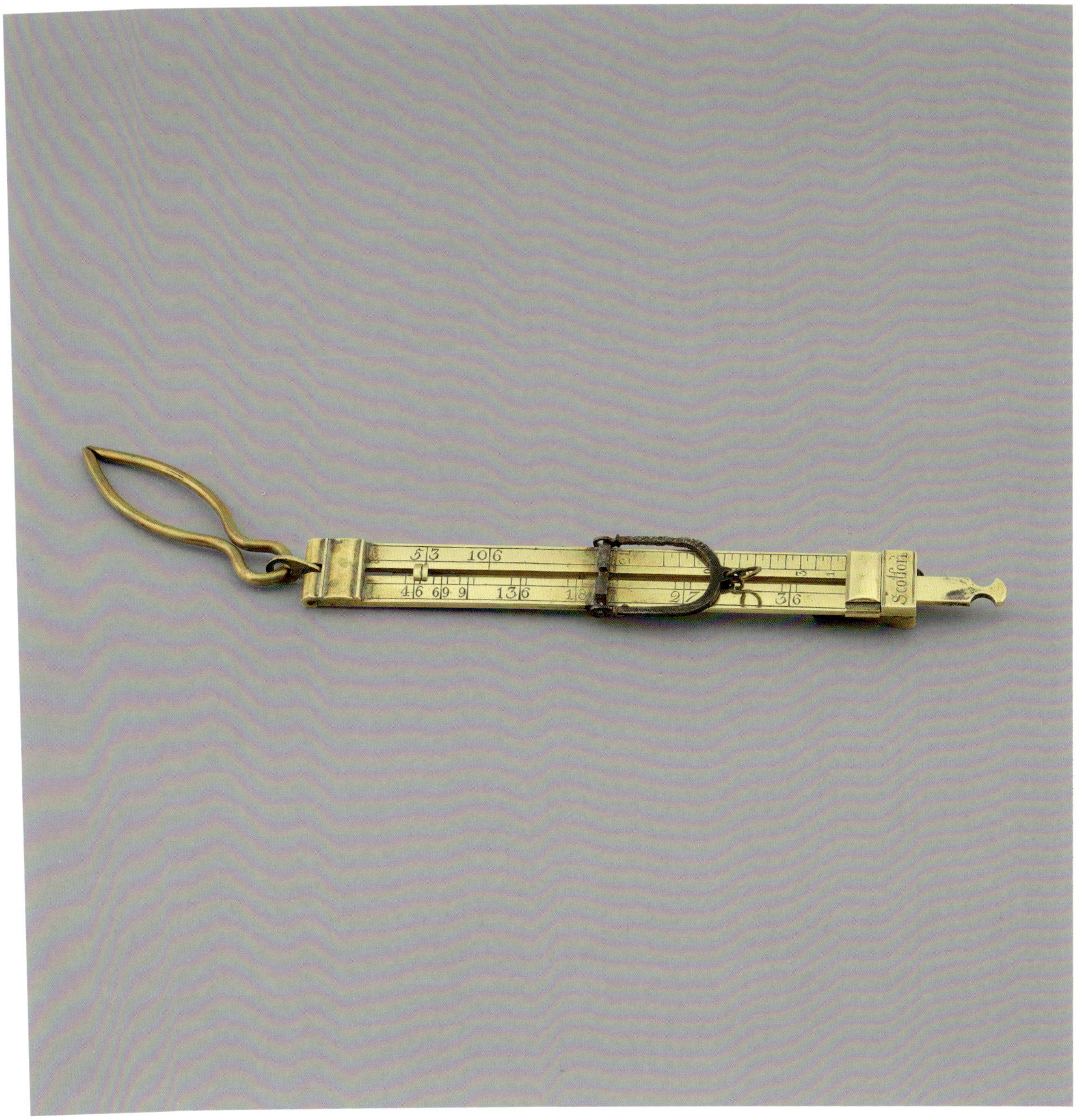

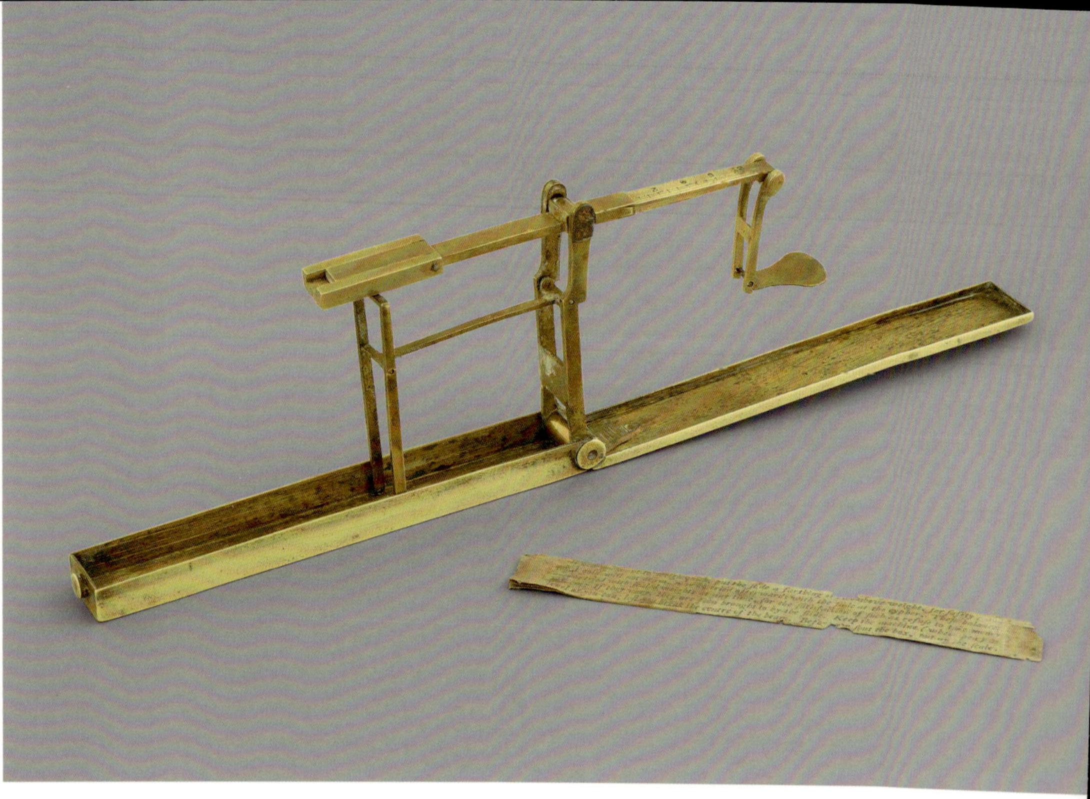

1792

Coin scale

Brass

$5 \times \frac{1}{2} \times \frac{3}{4}$ in. (12.7 × 1.27 × 1.9 cm)

Inscription:
Robert Hargreaves. Higham.1792.

This scale might have been the property of Robert Hargreaves, Gentleman, of Little Stone, Edgeware, Lancashire, who died in 1805.

PROVENANCE

bt. John H. Bryan of Lake Bluff, IL (1936–2018) from Tobias Jellinek Antiques, London, May 22, 1994

1549

Coin

Silver

$1\frac{1}{8}$ in. (2.9 cm) diameter

Inscription:
EDWARD'. VI D'. G'. AGL'. FRA'. Z. HIB'.; TIMOR DOMINI FONS VIT[AE] M D XLIX

This silver shilling of Edward VI bears the mint mark "Grapple" for the Tower of London (Royal Mint), London. The Latin inscription reads, "The fear of the Lord is the fountain of life."

PROVENANCE

A. H. Baldwin & Sons Ltd., London, August 25, 2015; bt. John H. Bryan III of Nashville, TN

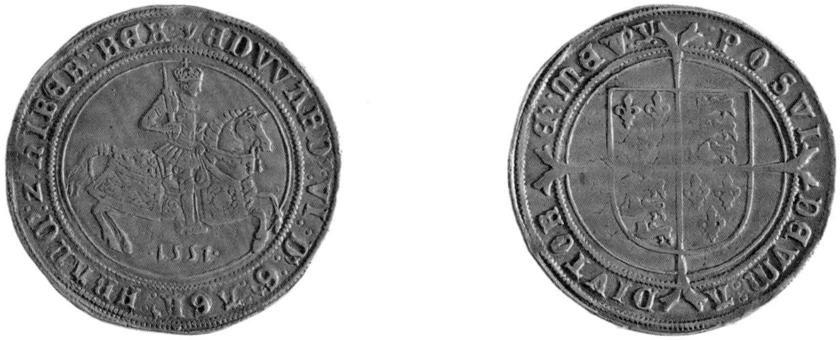

1551

Coin
Silver
$1\frac{5}{8}$ in. (4.1 cm) diameter

Inscription:
EDVVARD':VI:D':
G':AGL':FRANC':
Z:HIBER':REX|1551;.
POSVI DEVM :
ADIVTOR E' : MEV':

This five-shilling crown was made in the year when the issue of silver coins reached its lowest ebb, debasing the coinage to dangerously low levels. Retrenchment came with a fifty-percent devaluation of base silver coin, followed by this silver issue (11 oz. 3 dwt.)—which was the finest coin issued under the Tudors, albeit smaller and with less weight and precious metal.

The coin depicts the king on horseback with the date below. The Latin inscription on the reverse reads, "I have made God my helper," which is taken from Psalm 53 in the Latin Vulgate.

PROVENANCE
A. H. Baldwin & Sons Ltd., London, August 25, 2015; bt. John H. Bryan III of Nashville, TN

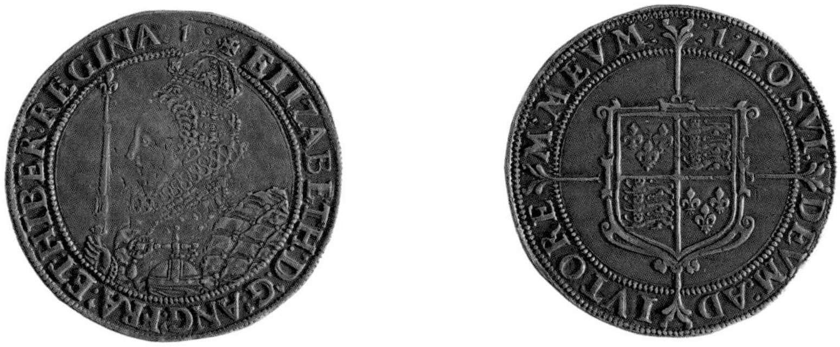

1601

Coin
Silver
$1\frac{3}{4}$ in. (4.5 cm) diameter

Inscription:
ELIZABETH:D:G:
ANG:FRA:ET
HIBER:REGINA;
POSVI DEVM:AD
IVTOREM:MEVM:

This silver crown of Elizabeth I bears the mint mark "1" for the Royal Mint at the Tower of London.

PROVENANCE
A. H. Baldwin & Sons Ltd., London, August 25, 2015; bt. John H. Bryan III of Nashville, TN

1605

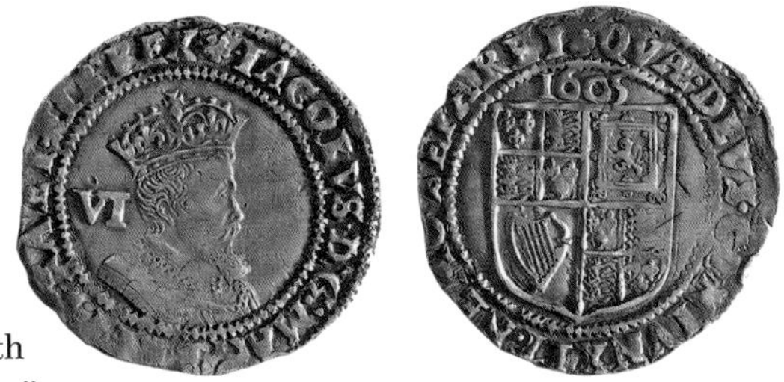

Coin
Silver
$1\frac{1}{8}$ in. (3.2 cm) diameter

Inscription:
IACOBVS·DG:
MAG [: BRIT : FR]
A·ET·HIB·REX|VI;
QVÆ·DEVS·
CONIVNXIT·NEMO·
SEPARET.|1605

This James I silver sixpence bears the mint mark "Rose" for the Tower. The Latin inscription reads, "What God hath joined together, let no man put asunder."

PROVENANCE
A. H. Baldwin & Sons Ltd., London, August 25, 2015; bt. John H. Bryan III of Nashville, TN

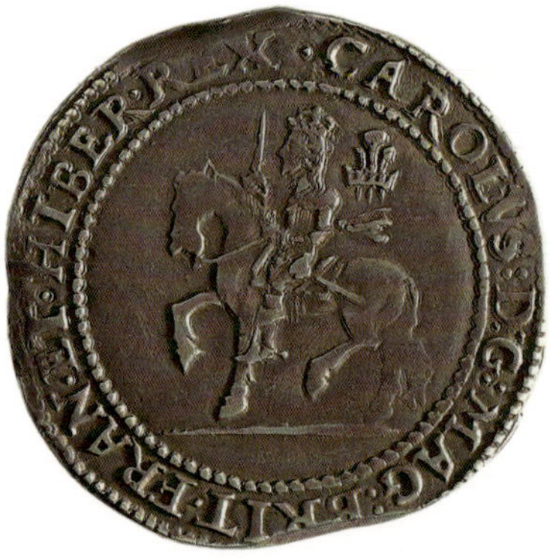

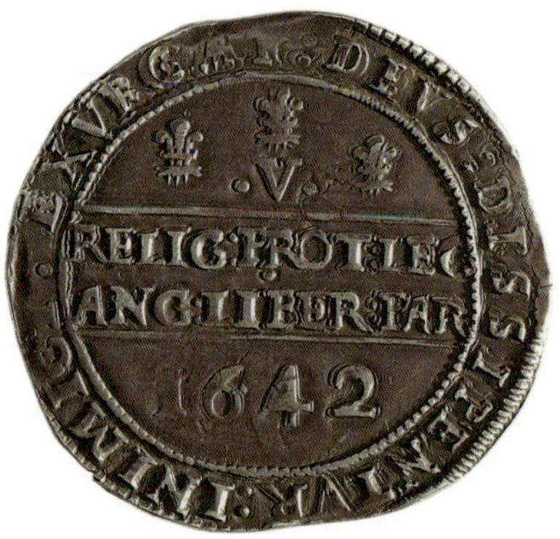

1642

Coin
Silver
$1\frac{7}{8}$ in. (4.8 cm) diameter

Inscription:
CAROLVS: D:G:MAG:BRIT· FRAN·ET·HIBER· REX; EXVRGAT: DEVS: DISSIPENTVR: INIMICI; .V. RELIG:PROT· LEG|ANG.LIBER. PAR|1642

This Charles I silver crown bears the mint mark "Plume" for Oxford but with the Shrewsbury "Horse." During the English Civil War, coins were minted up and down the country for the Royalist cause. In 1642, the mint originally at Aberystwyth, Wales, was transferred to New Inn Hall, Oxford, the Royalist capital. The Shrewsbury die on the obverse was no doubt reused there. The outside legend translates as "Let God arise and let his enemies be scattered."

PROVENANCE
Sovereign Rarities Ltd., London, September 25, 2018 (lot 55); bt. John H. Bryan III of Nashville, TN

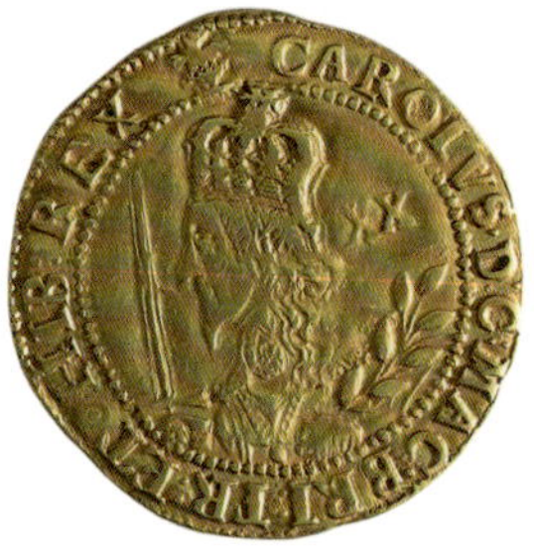

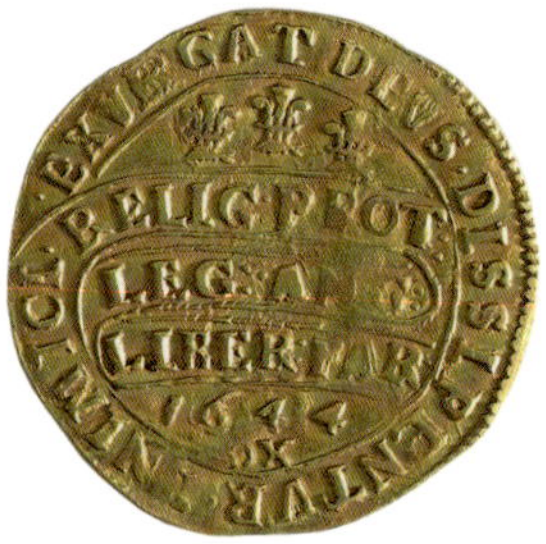

1644

Coin
Gold
$1\frac{3}{8}$ in. (3.5 cm) diameter

Inscription:
CAROLVS·D:G:MAG· BRI·FR:ET HIB: REX|XX; EXVRGAT DEVS·DISSIPENTVR· INIMICI·RELIG PROT|LEG: ANG: LIBER: PAR|1644|OX

This gold unite, or twenty-shilling coin, was minted for Charles I in Oxford. The inscription reads, "Let God arise and scatter the enemies of the religion of the Protestants, of the laws of England, of the liberty of Parliament, 1644, Oxford."

PROVENANCE
A. H. Baldwin & Sons Ltd., London, August 25, 2015; bt. John H. Bryan III of Nashville, TN

1656

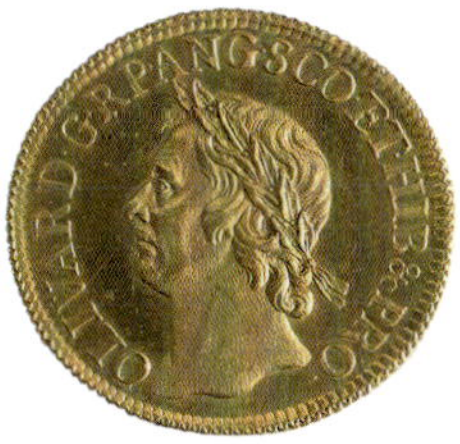

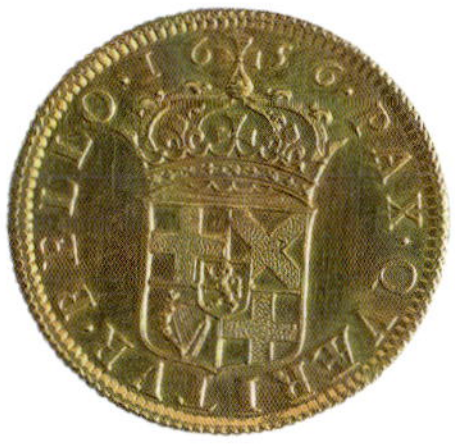

Coin
Gold
$1\frac{1}{8}$ in. (2.9 cm) diameter

Inscription:
OLIVAR·D·G·RP·ANG·SCO·ET·HIB·& PRO·; 1656·PAX·QVÆRITVR·BELLO·

A gold broad was a twenty-shilling coin issued by the Commonwealth of England in 1656. It also appeared in a piedfort edition for presentation, worth fifty shillings. It depicts Oliver Cromwell in profile in the guise of a Roman emperor within the inscription, "Oliver by Grace of God Protector of the Republic of England, Scotland and Ireland." On the reverse are the arms of the Protectorate, with the inscription "Peace is sought through war." All these coins were machine made, struck from dies by Thomas Simon (1618–1665) in the presses of Pierre Blondeau.

This coin bears no mint mark; these broads were used less frequently from the mid-seventeenth century onward as the Royal Mint, London, became the central location for issuing coins.

PROVENANCE
Sovereign Rarities Ltd., London, September 25, 2018 (lot 64); bt. John H. Bryan III of Nashville, TN

1672

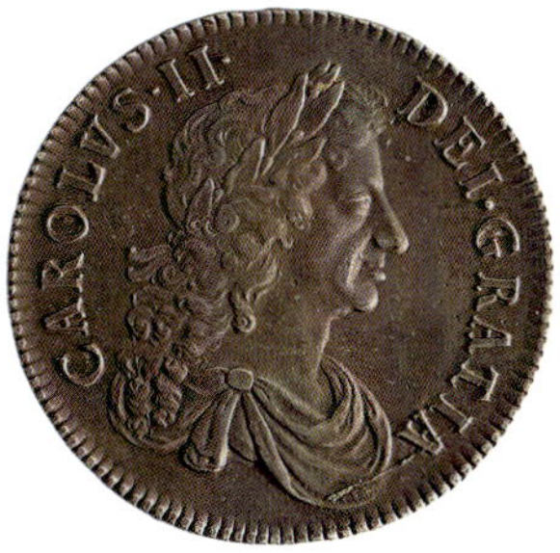

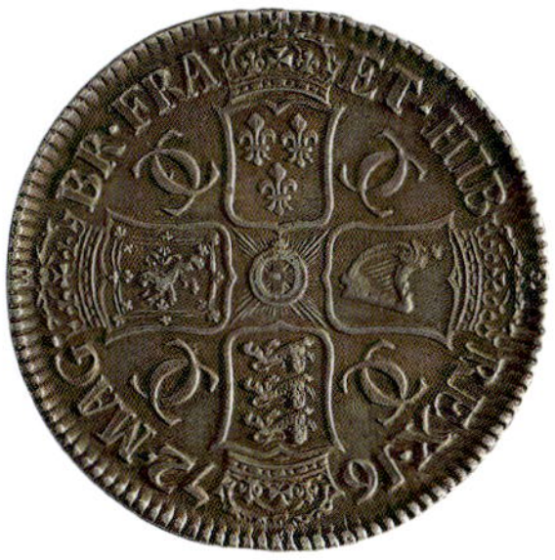

Coin
Silver
$1\frac{1}{2}$ in. (3.8 cm) diameter

Inscription:
CAROLVS·II· DEI·GRATIA; 1672·MAG BR·FRA· ET·HIB REX ·; DECVS ET TVTAMEN ANNO REGNI VICESIMO QVARTO

The cipher of Charles II appears on the reverse of this silver crown.

PROVENANCE
A. H. Baldwin & Sons Ltd., London, August 25, 2015; bt. John H. Bryan III of Nashville, TN

1677

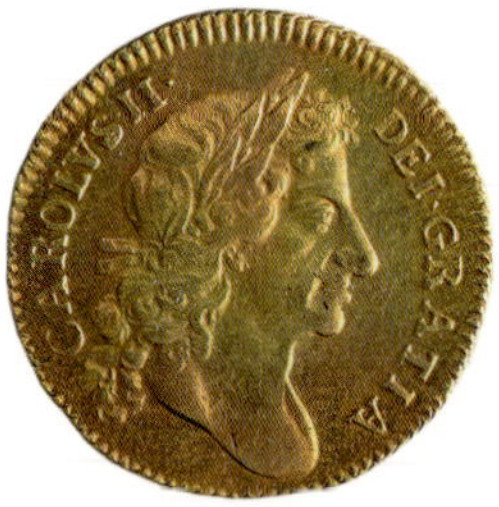
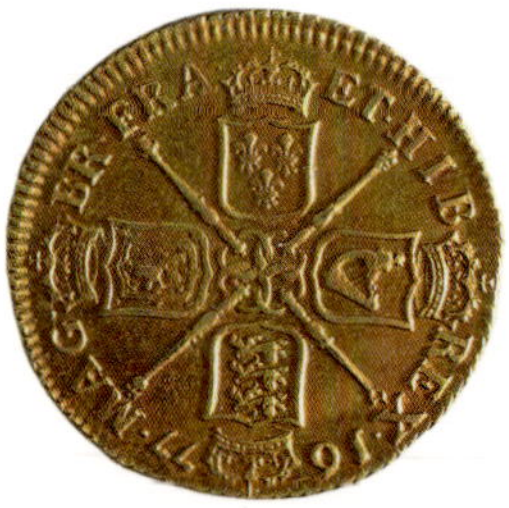

Coin
Gold
1¼ in. (3.2 cm) diameter

Inscription:
CAROLVS·II·
DEI·GRATIA; 16 77
MAG·BR·FRA ET·HIB
REX

Guinea—the generic name given to much of the west coast of Africa by English speakers—produced much of the highest-quality gold and gave its name to the twenty-shilling piece first issued by Charles II in 1663. The center of the reverse displays the cipher of four interlinked *C*'s, the initials of Charles and his queen consort, Catherine of Braganza. The one-guinea coin carried a tooth border around the edge to deter clipping—that is, the practice of removing a small part of the coin for its metal value and returning what remained of the coin to circulation.

PROVENANCE
Sovereign Rarities Ltd., London, September 25, 2018 (lot 81); bt. John H. Bryan III of Nashville, TN

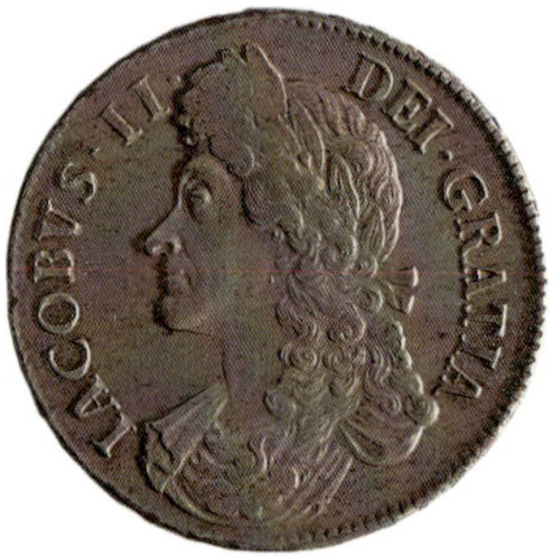
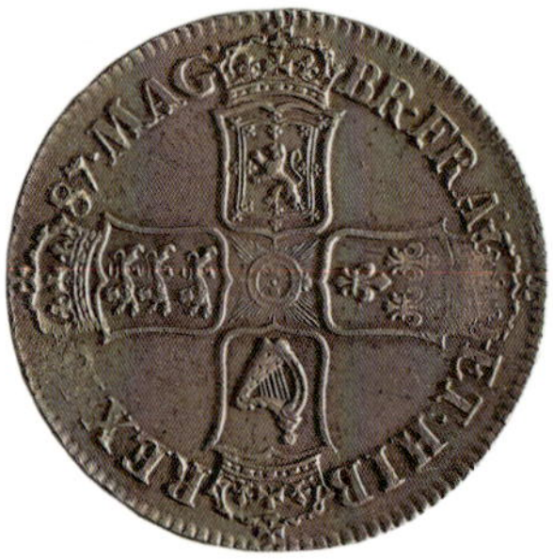

1687

Coin
Silver
1½ in. (3.8 cm) diameter

Inscription:
IACOBUS·II·
DEI·GRATIA; 87 MAG
BR·FRA·ET·HIB REX;
DECVS ET TVTAMEN
ANNO REGNI
TERTIO

This coin is a silver crown of James II.

PROVENANCE
A. H. Baldwin & Sons Ltd., London, August 25, 2015; bt. John H. Bryan III of Nashville, TN

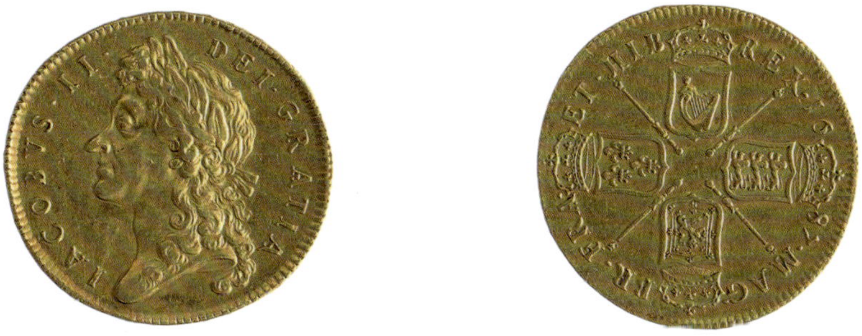

1687

Coin
Gold
$1\frac{1}{2}$ in. (3.8 cm) diameter

Inscription: IACOBVS·II·DEI· GRATIA; 16 87 MAG· BR·FRA ET·HIB REX.; DECVS·ET· TVTAMEN·ANNO· REGNI·TERTIO

Other versions of this James II five-guinea gold coin—and those issued by Charles II, James II's brother—bear the mark of the Elephant and Castle, which is the mark of the Royal Africa Company.

The company held a monopoly over the trade in West Africa between 1660 and 1689 and was also heavily involved in the transatlantic slave trade.

PROVENANCE
Sovereign Rarities Ltd., London, September 25, 2018 (lot 67); bt. John H. Bryan III of Nashville, TN

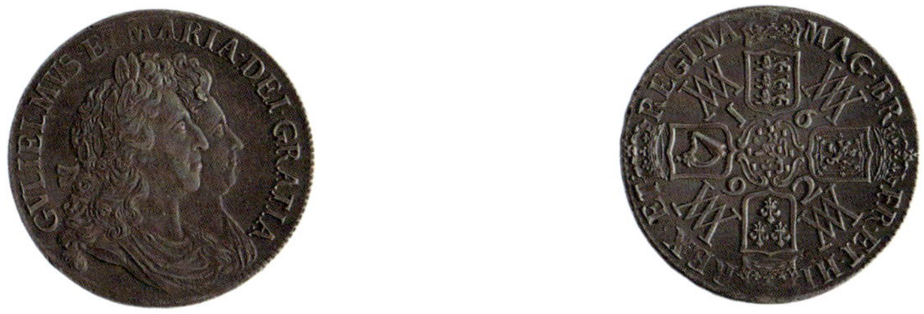

1692

Coin
Silver
$1\frac{1}{2}$ in. (3.8 cm) diameter

Inscription: GVLIELMVS·ET· MARIA·DEI·GRATIA; MAG·BR FR·ET·HI REX· ET REGINA| [W and M in cipher]; 1692.

This coin is a silver crown of William and Mary.

PROVENANCE
A. H. Baldwin & Sons Ltd., London, August 25, 2015; bt. John H. Bryan III of Nashville, TN

1694

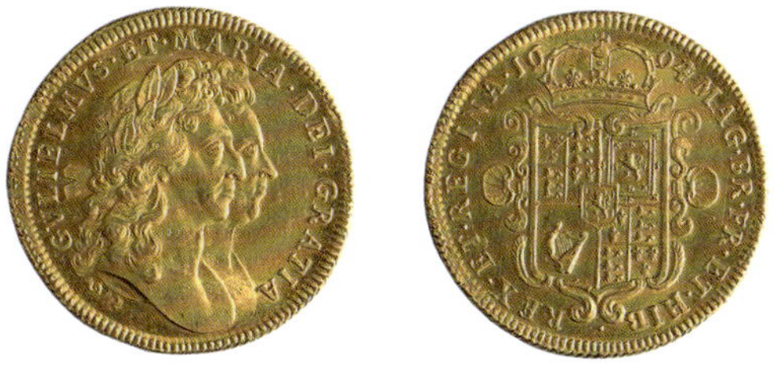

Coin
Gold
$1\frac{1}{4}$ in. (3.2 cm) diameter

Inscription: GVLIELMVS·ET· MARIA·DEI· GRATIA; 16 94 MAG·BR·FR·ET·HIB REX·ET·REGINA·

This coin is a William and Mary two-guinea gold coin dated 1694. The numeral *4* in the date "1694" is struck over the *3* from the previous year. The coat of arms includes the Gold Lion of the House of Nassau for the House of Orange-Nassau.

PROVENANCE
Sovereign Rarities Ltd., London, September 25, 2018 (lot 77); bt. John H. Bryan III of Nashville, TN

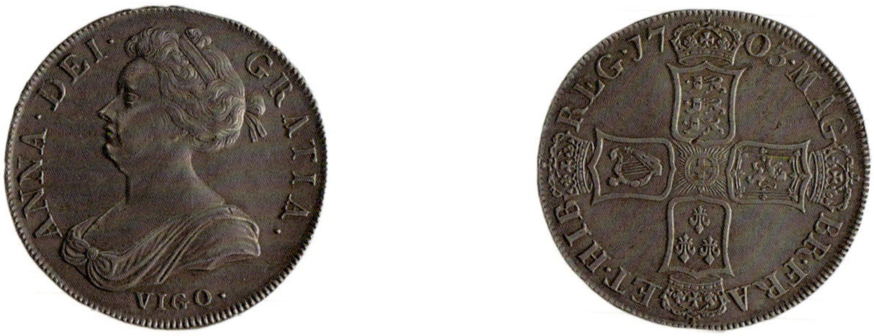

1703

Coin
Silver
$1\frac{1}{2}$ in. (3.8 cm) diameter

Inscription:
ANNA·DEI· GRATIA. VIGO.; 17 03·MAG BR· FRA ET·HIB REG·

The inscription "VIGO" below the bust on this silver crown of Queen Anne denotes that the coin was minted from silver seized from Spanish ships during the Battle of Vigo Bay during the War of the Spanish Succession.

PROVENANCE
A. H. Baldwin & Sons Ltd., London, August 25, 2015; bt. John H. Bryan III of Nashville, TN

1711

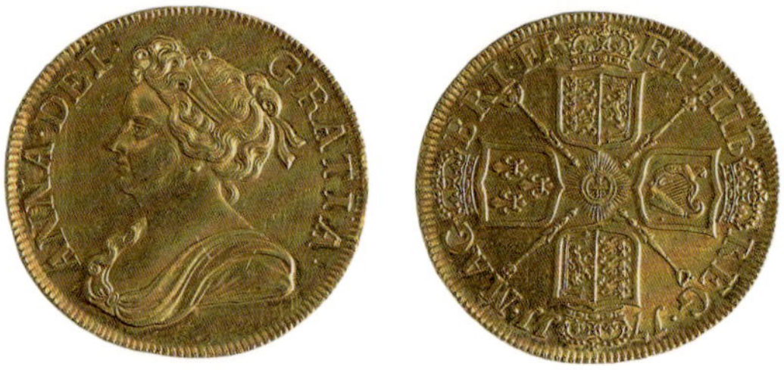

Coin
Gold
$1\frac{1}{4}$ in. (3.2 cm) diameter

Inscription:
ANNA·DEI· GRATIA·; 17 11 MAG·BR·FRA ET·HIB REX; DECVS·ET· TVTAMEN·ANNO· DECIMO

This is a Queen Anne five-guinea gold coin.

PROVENANCE
Sovereign Rarities Ltd., London, September 25, 2018 (lot 72); bt. John H. Bryan III of Nashville, TN

1716

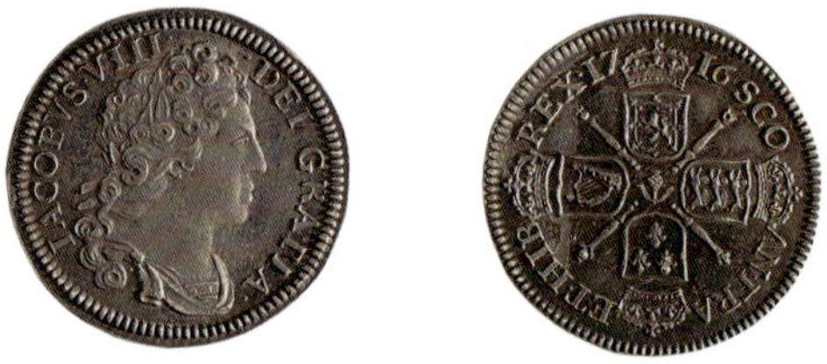

Coin
Silver
1 in. (2.5 cm) diameter
Norbert Rottiers (1665–1727)

Inscription:
IACOBVS VIII·DEI GRATIA; ·1716 SCO ANG FRA ET·HIB REX

During the Jacobite uprising of 1715, Norbert Rottiers prepared dies for a small amount of coinage, including this silver pattern guinea, for the anticipated return of James Stuart (the Old Pretender) and his proposed coronation as James VIII in 1716 at Scone Palace.

As the uprising was unsuccessful, the dies were never put into production, and it was not until their rediscovery in 1828 that Matthew Young struck a small number from them. After his short production, all the dies were destroyed. All examples in institutional and private collections are from this issue.

PROVENANCE
Noel Woolf, London; Glendinings with A. H. Baldwin's & Sons Ltd., London, November 4, 1992 (lot 99); Private Collection, Scotland; Lyon & Turnbull, Edinburgh, March 14, 2017; bt. John H. Bryan of Lake Bluff, IL (1936–2018)

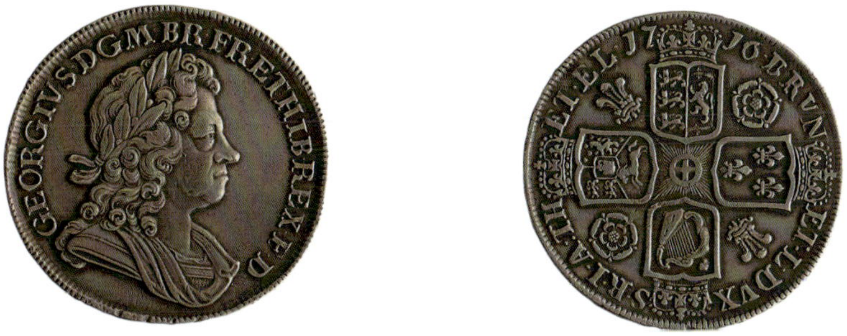

1716

Coin
Silver
$1\frac{1}{2}$ in. (3.8 cm) diameter

Inscription:
GEORGIVS DG M BR FR ET HIB REX F D; 17 16·BRVN ET·5·DVX S·R·I·A· :TH ET·EL; DECVS ET TVTAMEN ANNO REGNI SEGVNDO.

This coin is a silver crown of George I.

PROVENANCE
A. H. Baldwin & Sons Ltd., London, August 25, 2015; bt. John H. Bryan III of Nashville, TN

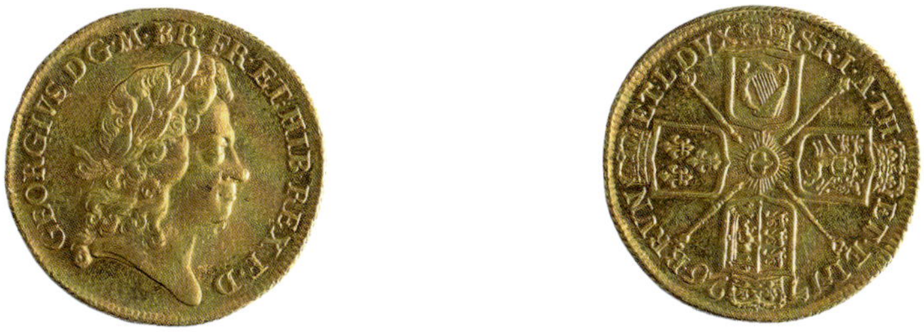

1726

Coin
Gold
$1\frac{1}{4}$ in. (3.2 cm) diameter

Inscription:
GEORGIVS DG·M· BR·FR·ET·HIB·REX· F·D; 17 26 BRVN ET L·DVX S·R·I·A·TH ET· EL

This is a George I two-guinea gold coin.

PROVENANCE
Sovereign Rarities Ltd., London, September 25, 2018 (lot 78); bt. John H. Bryan III of Nashville, TN

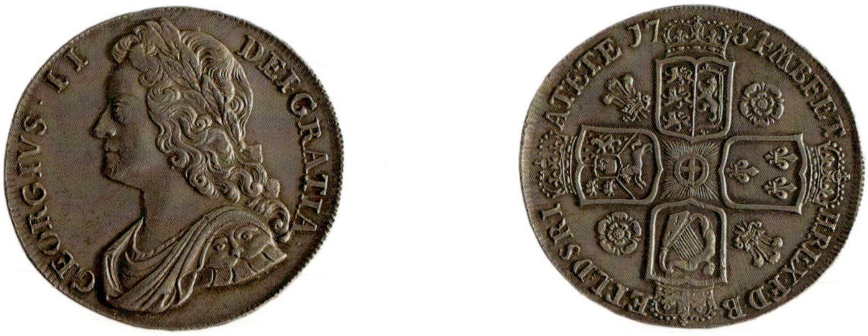

1734

Coin

Silver

$1\frac{5}{8}$ in. (4.1 cm) diameter

Inscription:
GEORGIVS · II DEI·GRATIA; 17 34 M B F ET H REX F D·B ET L D S RI A T ET E.; DECVS ET TVTAMEN ANNO REGNI SEPTIMO.

This coin is a silver crown of George II.

PROVENANCE

A. H. Baldwin & Sons Ltd., London, August 25, 2015; bt. John H. Bryan III of Nashville, TN

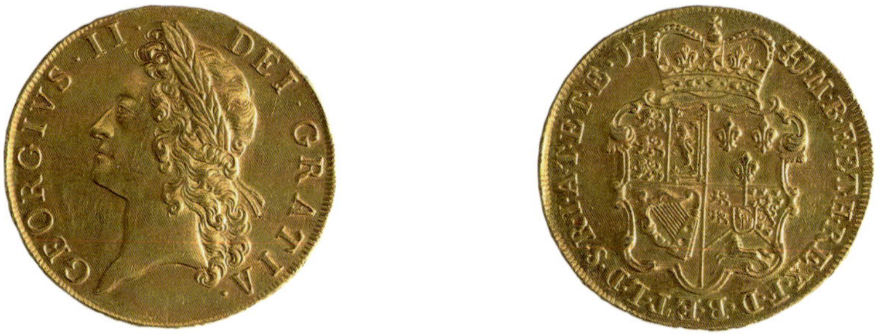

1741

Coin

Gold

$1\frac{1}{2}$ in. (3.8 cm) diameter

Inscription:
GEORGIVS·II DEI·GRATIA·; 17 41· M·B·F·ET·H·REX· F·D·B·ET·L·D·S·R· I·A·T·ET·E·; DECVS· ET·TVTAMEN·ANNO· REGNI·DECIMO· QVARTO

This is a gold two-guinea coin of George II.

PROVENANCE

Sovereign Rarities Ltd., London, September 25, 2018 (lot 73); bt. John H. Bryan III of Nashville, TN

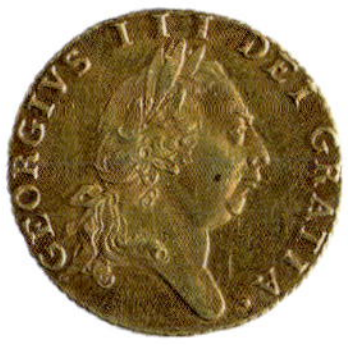

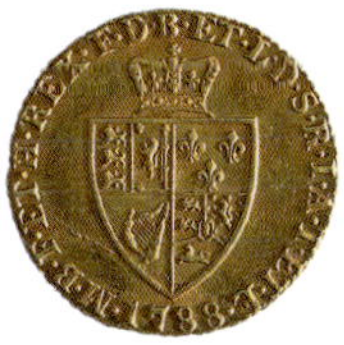

1788

Coin
Gold
$\frac{7}{8}$ in. (2.22 cm) diameter

Inscription:
GEORGIVS III DEI GRATIA.;M·B·F·ET·H·REX·F·D·B·ET ·L·D·S·R·I·A·T·ET·E·1788

This gold half guinea is struck with the likeness of George III, as is the later silver coin shown below.

PROVENANCE
Sovereign Rarities Ltd., London, September 25, 2018 (lot 87); bt. John H. Bryan III of Nashville, TN

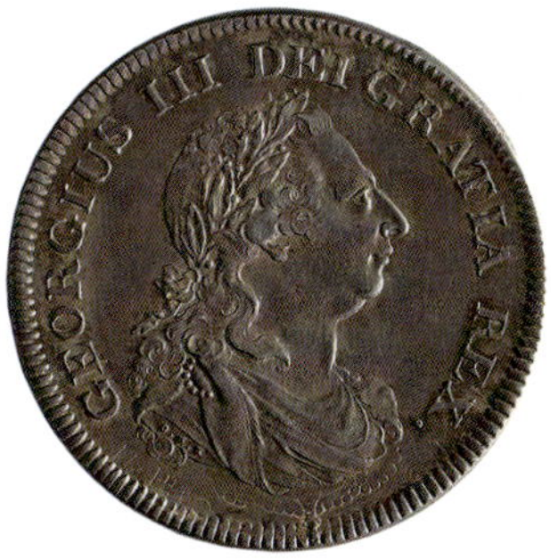

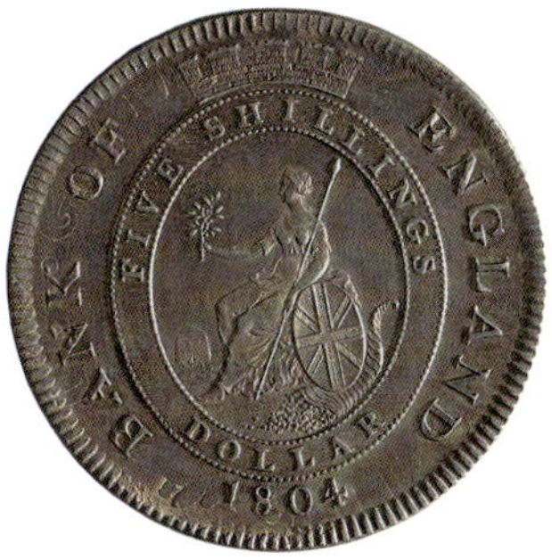

1804

Coin
Silver
1$\frac{5}{8}$ in. (4.1 cm) diameter

Inscription:
GEORGIVS III DEI GRATIA REX; BANK OF ENGLAND 1804|FIVE SHILLINGS DOLLAR

This coin is a silver dollar of five shillings of George III.

PROVENANCE
A. H. Baldwin & Sons Ltd., London, August 25, 2015; bt. John H. Bryan III of Nashville, TN

VIII. DEATH & LEGACY

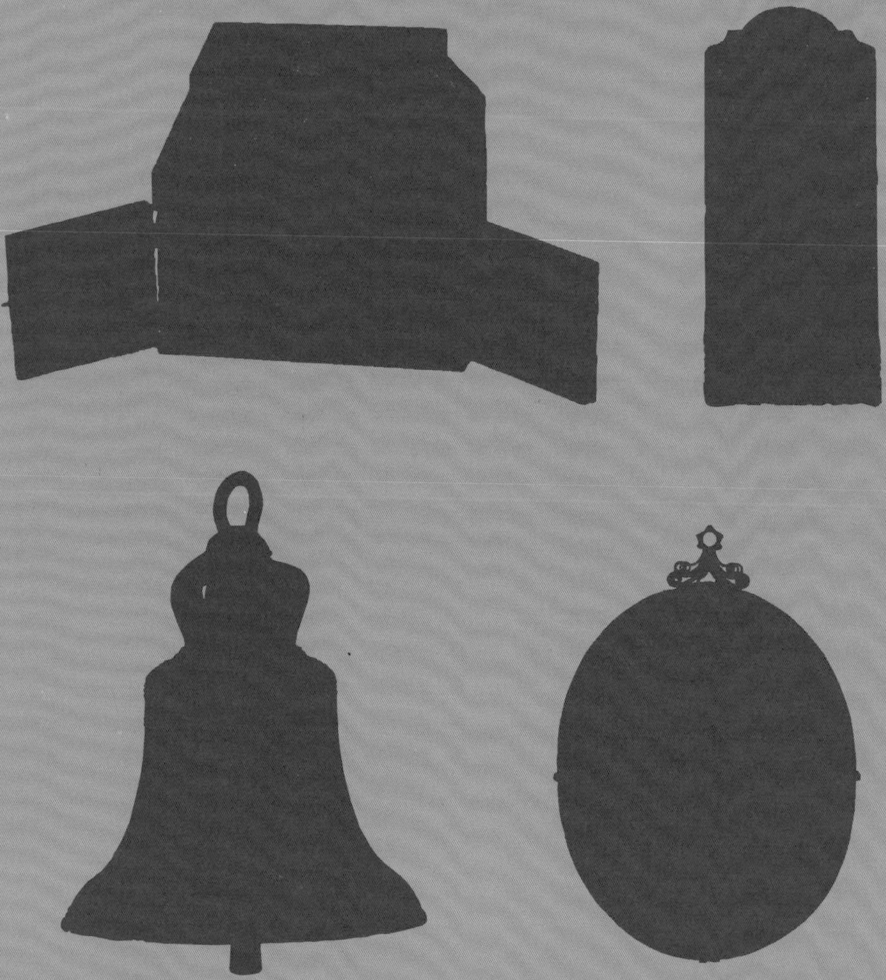

IN THE EARLY MODERN PERIOD, BRITAIN WAS A NATION PREOCCUPIED with death. War and plague brought death to every household, regardless of class or status. There was no stage of life in which the threat of death through disease did not loom large, a fact amply visualized in the imagery of skulls, bones, and hourglasses in churches and churchyards. Yet within the Christian worldview of the time, life was lived in order to die, and to die was to live eternally with God. So although death was feared, it was not unwelcome, as mortal existence was believed to be merely a period of preparation for what is to follow, which is of far greater significance. The practice of prayer at various times of the day ensured that the soul was prepared for death whenever it came and, following that, the reunion with lost friends and family members in heaven.

People lived with the dead by carrying mementos of their departed friends and family. For a Catholic, the memento might take the form of a cross, worn on the body to remind the wearer to pray for the deceased and speed that soul's passage through purgatory, where it would otherwise languish in proportion to the volume and extremity of the sins committed in life (page 466). For the majority of society members who worshipped according to the Church of England, such intercessory prayers were strictly proscribed, although "death's head" rings were frequently bequeathed in wills and worn in memory of the departed; elaborate monuments were erected to honor the achievements of those who could afford them.

No other death in this period was felt more acutely or was so subsequently venerated as that of Charles I, who was executed in early January 1649. Almost immediately after his death, objects were made to preserve his memory, helping to promote the image of the king as a martyr, a notion that only grew stronger during the second half of the seventeenth century (pages 474–78).

For those of means, bells were rung as a person lay dying, at the moment of death, and at the burial. The sound of the bell galvanized the dying person, focusing his or her mind on prayer, and encouraged others to pray for those at death's door. After death, the ringing of bells was thought to be beneficial to the deceased's soul.

The body was wrapped in linen (or, later, wool). It was then placed in a coffin and carried to the place of burial—either within the church or, more often, in the parish churchyard or burial ground. Whether the grave was marked with a tomb monument or a plain wooden marker, death brought to a close the individual's journey through time. What lived on were the material legacies of the person's life: the objects that have come down to us today.

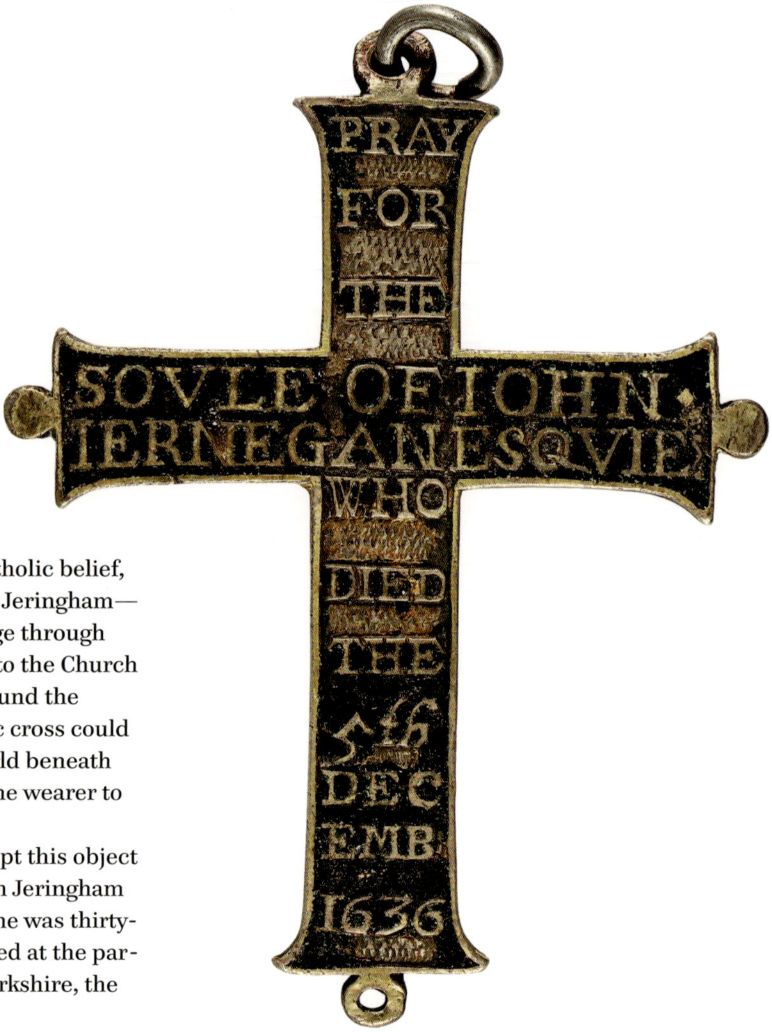

1636

Cross

Silver gilt and enamel

$3\frac{3}{4} \times 2\frac{7}{8}$ in. (9.5 × 7.3 cm)

Inscription: PRAY| FOR|THE|SOVLE OF IOHN·|IERNEGAN ESQVIE|WHO|DIED THE 5th|DEC| EMB|1636; INRI

Because it was a Roman Catholic belief, praying for the soul of John Jeringham—in order to speed his passage through purgatory—was anathema to the Church of England. When worn around the neck, however, this Catholic cross could be kept secret from the world beneath clothing while prompting the wearer to intercessory prayer.

The Jeringham family kept this object for many generations. When Jeringham died on December 5, 1636, he was thirty-eight years old; he was buried at the parish of St Mary, in Fawley, Berkshire, the next day.

Jeringham was from a family of devoted Catholics from East Anglia who could, with some confidence, trace their lineage back to before the Norman Conquest. John's great-grandfather, Sir Henry Jerningham (d. 1572), was made Vice-Chamberlain of the Household, Knight of the Bath, and Master of the Horse by Mary I, but his promising public career was curtailed by the accession of Elizabeth I. Sir Henry's namesake son—John's father—married Eleanor Throckmorton of Coughton, Warwickshire, who was from one of the country's foremost Catholic families. For this reason, John Jeringham and his family lived under constant suspicion from government officials; their movements were closely followed and their homes subject to frequent searches.

John had four siblings, one of whom, Anne, became the first abbess of the Convent at Paris. She was described on her father's funeral certificate as having a husband, heir of Sir John Arundel of Cornwall. Following his marriage in 1619 to Mary, daughter of Sir Francis Moore of Fawley, Jeringham seems to have spent most of his time in Berkshire. His son, Henry, succeeded his grandfather as the second Baronet in 1646 and may have worn this cross.

PROVENANCE

By descent in Jeringham family; gift of Simon Redburn to Irvin and Anita Schorsch, Hidden Glen Farms, Meadowbrook, PA; Sotheby's, New York, January 20, 2016 (lot 34); bt. John H. Bryan of Lake Bluff, IL (1936–2018) from Tim Gleason, February 12, 2016

1684

Slides were strung with ribbon and worn around the neck, arms, or body. These slides contain the hair of "Jerome Gatacre, Gent," who was buried January 31, 1684/85, at Willey, Shropshire, which is about ten miles northwest of Claverley. The Gatacres of Gatacre Hall were a recusant Roman Catholic family of Claverley in Shropshire. Jerome, however, appears to have been Protestant. He was born May 17, 1658, at Claverley, the third son of John Gatacre (d. 1667) and Mary Polwhele.

PROVENANCE

By descent to Galfry William Gatacre (d. 1974); with Argentum, The Leopard's Head, San Francisco, CA; Irvin and Anita Schorsch, Hidden Glen Farms, Meadowbrook, PA; Sotheby's, New York, January 22, 2016 (lot 1039); bt. John H. Bryan of Lake Bluff, IL (1936–2018) from Tim Gleason, February 12, 2016

Slides

Gold, enamel, human hair, glass

Each: $\frac{3}{4} \times \frac{5}{8} \times \frac{1}{4}$ in. (1.9 × 1.59 × .64 cm)

Inscription, on both: ob 28 Jan.|Jerrum Gatticre|1684

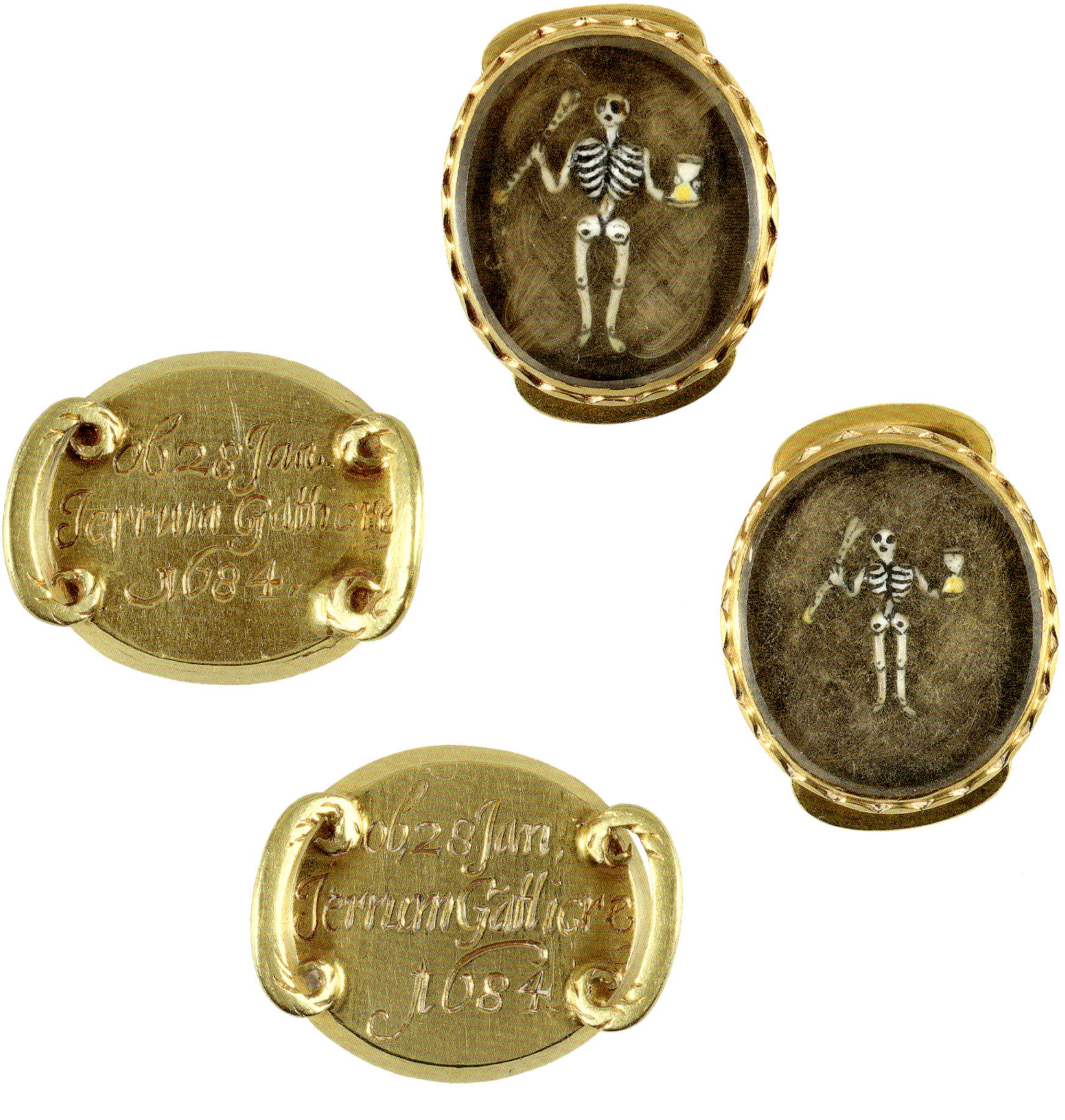

1737

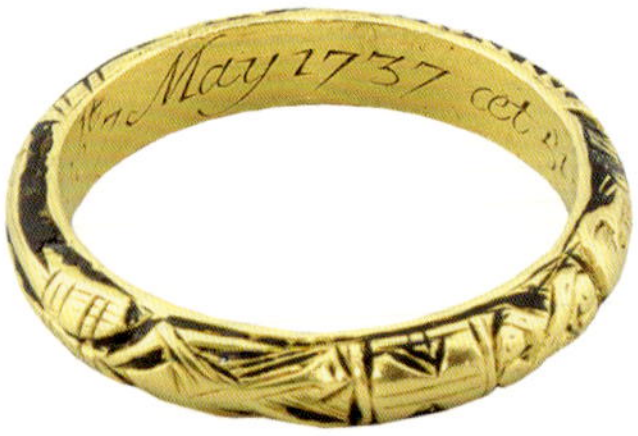

Ring
Gold, enamel
$\frac{1}{8} \times \frac{7}{8}$ in. (.32 × 2.22 cm)

Inscription:
Rach^{l} Richards ob^{t} 7 May 1737 æt 51

When a family member died, early modern funeral rituals included the distribution of gifts as a way to strengthen bonds of family and friendship among the bereaved. Anne Richards's story was typical.

Born in 1711, Anne was the only child of Rachel and Edward Richards, Esquire. She inherited the manor of Compton Beauchamp in Berkshire from her father, who died in 1729. A decade later, in 1737, Anne's mother, Rachel (b. 1686), died at age fifty-one. As was customary, Anne gifted a number of beautiful gold rings to family and close friends who came to mourn at her mother's funeral. They were wrought on the exterior with symbols of mortality and engraved inside with her mother's name and the date of her death. The rings worn by members of Anne's circle must have afforded her comfort.

Anne also commissioned a beautifully sculpted monument with a long and fulsome epitaph to her mother for the nave of her parish church. Anne died, unmarried, in 1771, and her own monument, of a similar design and perhaps commissioned at the same time, was placed next to that of her mother, according to Anne's will (TNA PROB 11/970/129).

PROVENANCE
bt. John H. Bryan of Lake Bluff, IL (1936–2018) from Les Enluminures Ltd., Chicago, IL, January 25, 2016

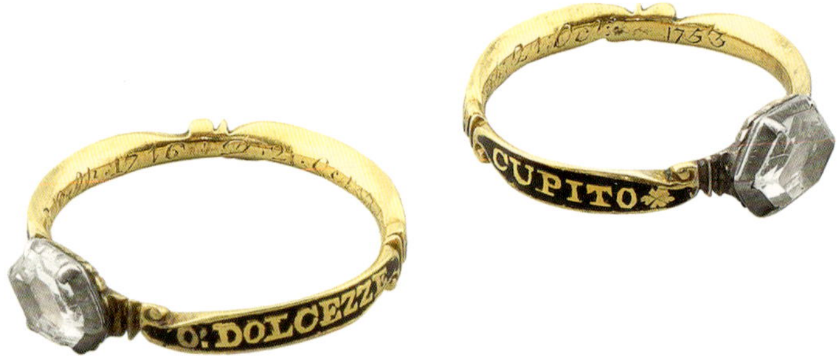

1753

Ring
Gold, rock crystal, hair
$\frac{1}{8} \times \frac{3}{4}$ in. (.32 × 1.9 cm)

Left, inscription:
O, DOLCEZZE, AMARISSIME, D'AMORE; B: 10: Ap: 1716, D: 21 Oct 1753

Right, inscription:
NIL, NIMIUM, CUPITO; B: 10: Ap: 1716, D: 21: Oct 1753

These rings bear inscriptions in Italian (left) and Latin (right), which translate, respectively, as "O, bittersweetness of love" and "Whatsoever I desire, I always have; because I desire nothing but what I have."

PROVENANCE
Sotheby's, London, March 28, 2017 (lot 390); bt. John H. Bryan of Lake Bluff, IL (1936–2018) from Les Enluminures Ltd., Chicago, IL, April 7, 2017

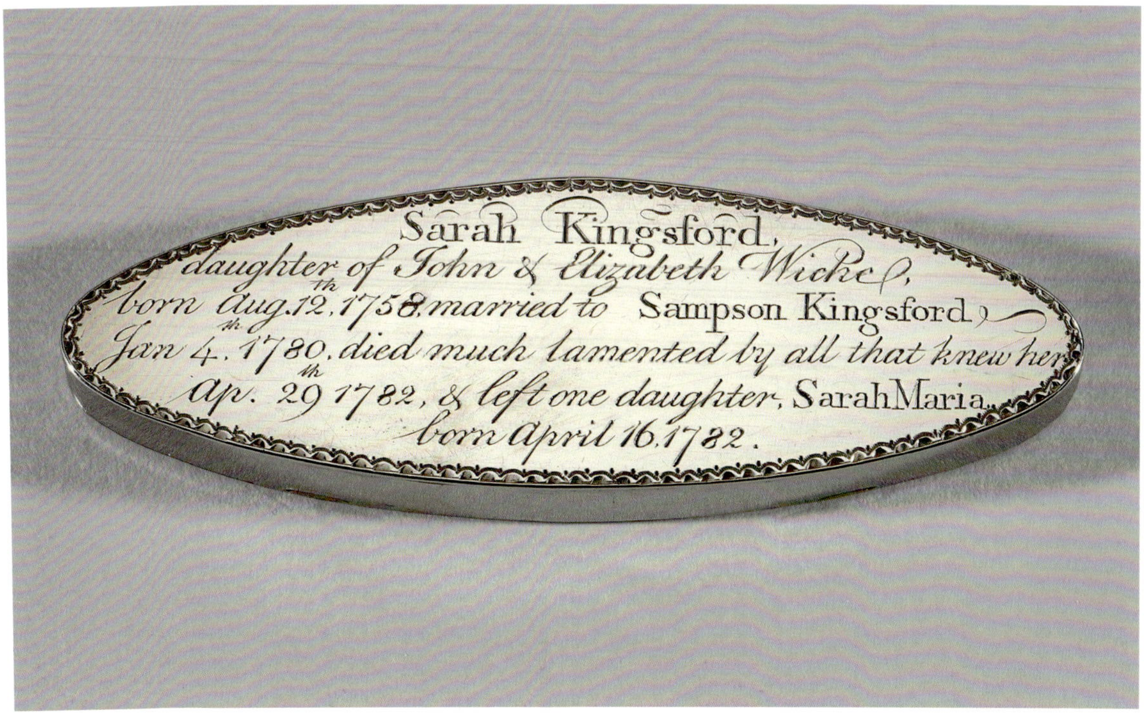

1782

Box
Silver
$\frac{3}{8} \times 4\frac{5}{8} \times 3\frac{1}{4}$ in. (.95 × 11.7 × 8.3 cm)

Inscription: Enclosed are some of the|remains of a most kind, virtuous &|affectionately loving Wife, M^{rs}: Sarah Kingsford|in whom centered every female &|desirable excellency; Sarah Kingsford|daughter of John & Elizabeth Wiche|born Aug. 12th 1758, married to Sampson Kingsford |Jan. 4th 1780, died much lamented by all that knew her|Ap. 29th 1782, & left one daughter, Sarah Maria|born April 16. 1782.

This box no longer contains the remains of Sarah Kingsford, one of the six children of John Wiche (d. 1794). Wiche was a Baptist minister from Taunton, Somerset, who moved across the country before settling in Maidstone, Kent. Sarah also married a Baptist, Sampson Kingsford, a banker of Canterbury, but she, "after a short union died, sincerely lamented by all her acquaintances, leaving a daughter."

PROVENANCE

bt. John H. Bryan of Lake Bluff, IL (1936–2018) from Peter Cameron, London, December 3, 2016

LITERATURE

Anonymous, "A Sketch of the Character of the late Rev. JOHN WICHE," *Protestant Dissenters' Magazine* 4 (1797): 121–32

1730

Mug
Salt-glazed stoneware
8 1/8 × 5 1/2 × 8 in. (20.6 × 14 × 20.3 cm)

Inscription: Drink to ye pious Memory of Good Queen Ann|John James|1730

Mugs like this one, made to commemorate Queen Anne, were used for Tory healthing (raising a glass to the health of the Tory party). Others, inscribed to the "Glorious memory of William III," were for Whigs. This mug was the product of a London pottery, probably the Vauxhall factory. It depicts a bust of Queen Anne flanked by two Yeomen of the Guard.

PROVENANCE
Jonathan Horne, London; Irvin and Anita Schorsch, Hidden Glen Farms, Meadowbrook, PA; Sotheby's, New York, January 21, 2016 (lot 578); bt. John H. Bryan of Lake Bluff, IL (1936–2018) from Tim Gleason, February 12, 2016

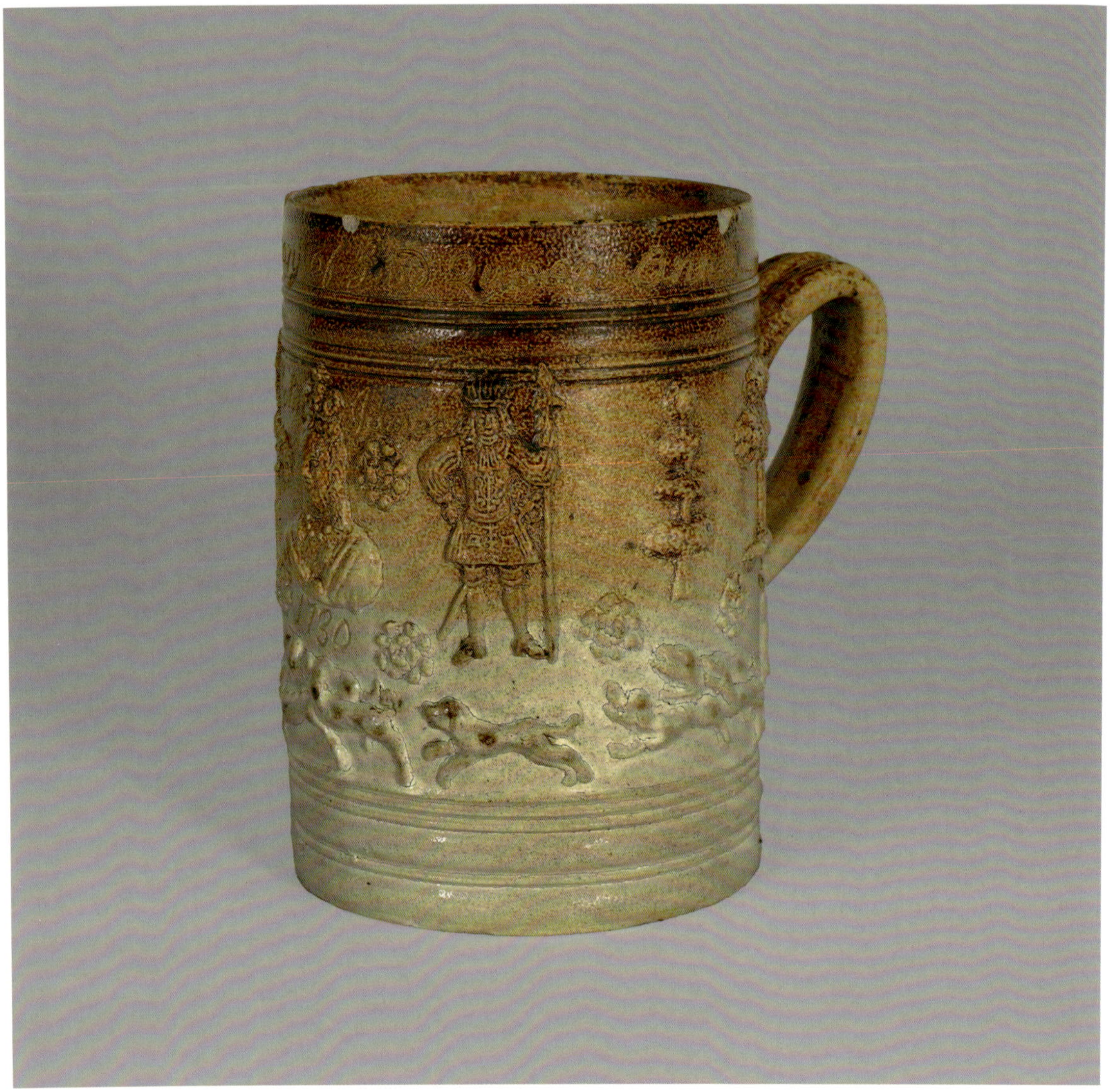

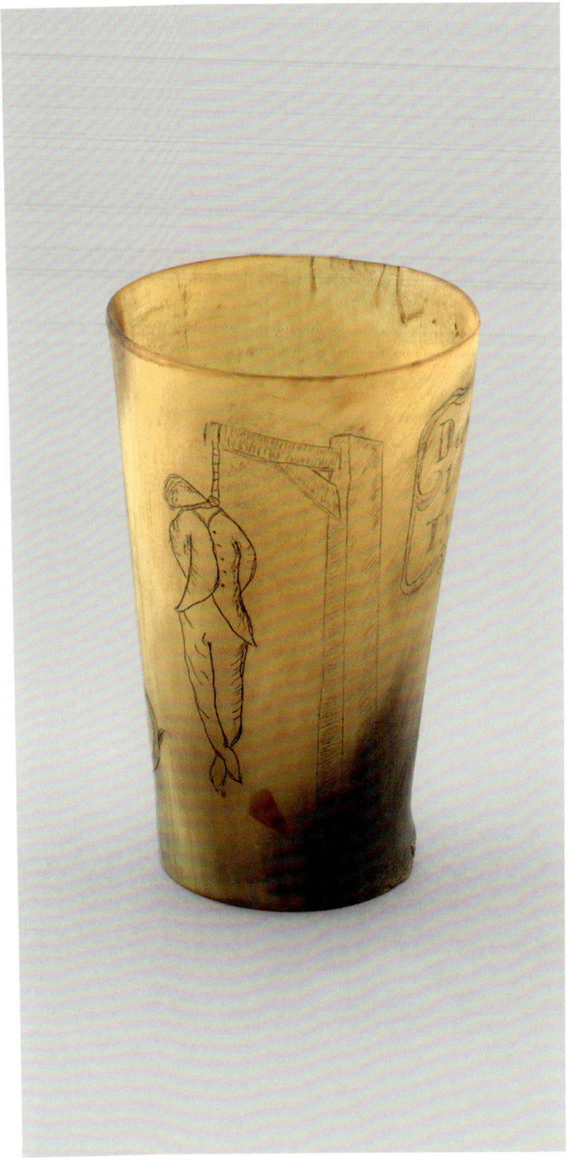

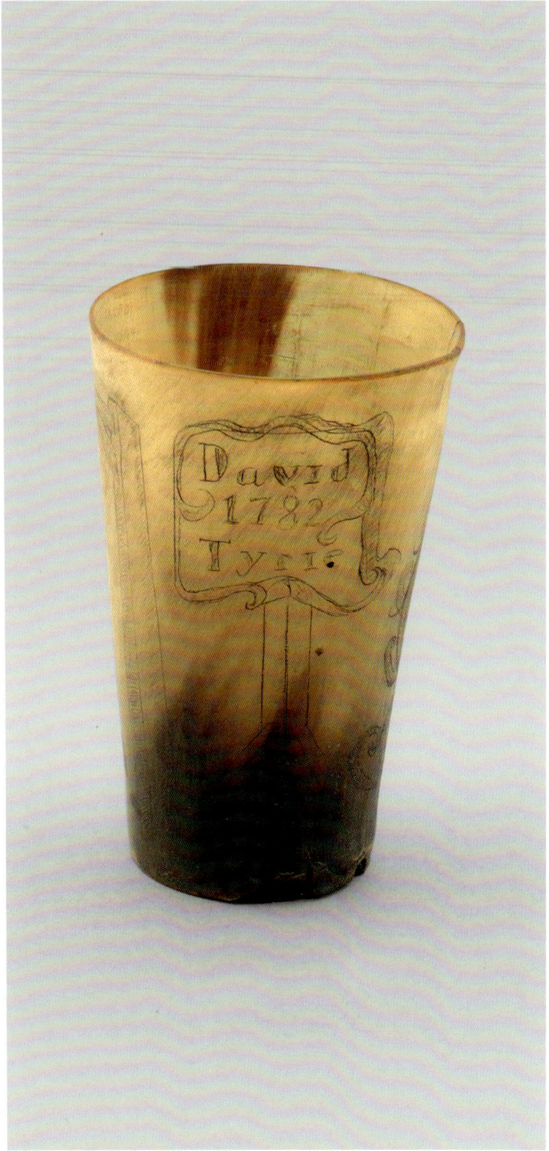

1782

Cup

Horn

$4\frac{1}{2} \times 2\frac{1}{2}$ in. (11.4 × 6.4 cm)

Inscription:
David|1782 |Tyrie;
GR [in cipher]

David Tyrie was convicted of betraying British naval secrets to the French in time of war. He was the last person in Britain to be hanged, drawn, and quartered.

The execution attracted many thousands of people, and Tyrie's head was privately exhibited for years. Popular souvenirs, like this one, served to display the event and convey the horror of disloyalty. The cup is shown here in two views.

PROVENANCE

Reputedly found in the attics of Glynn Park, County Antrim, in 2007; Bonhams, Oxford, September 30, 2015 (lot 328); bt. John H. Bryan of Lake Bluff, IL (1936–2018)

1674

Tobacco box
Horn
$\frac{3}{4} \times 3\frac{3}{8} \times 4\frac{1}{8}$ in.
(1.9 × 8.6 × 10.5 cm)

Inscription: SICVT FVMVS SIC VITA 1674

As is appropriate for a tobacco box, the Latin inscription reads, "As smoke, thus is life." It is a shortened version of a quotation from the Latin Vulgate version of Psalm 101:4.

PROVENANCE
Acanthus Collection; bt. John H. Bryan of Lake Bluff, IL (1936–2018) from Alistair Sampson Antiques Ltd., London, June 15, 1990

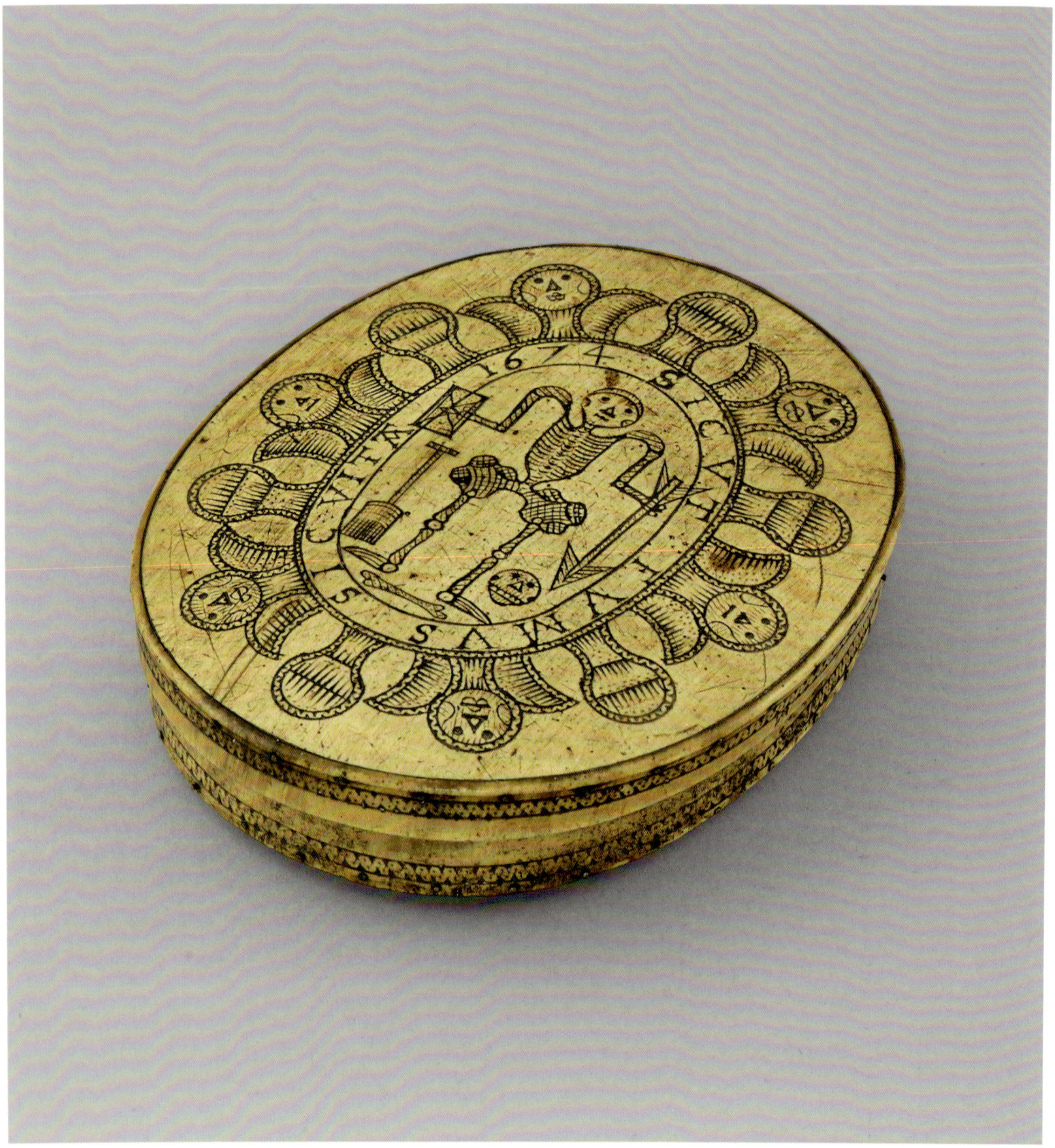

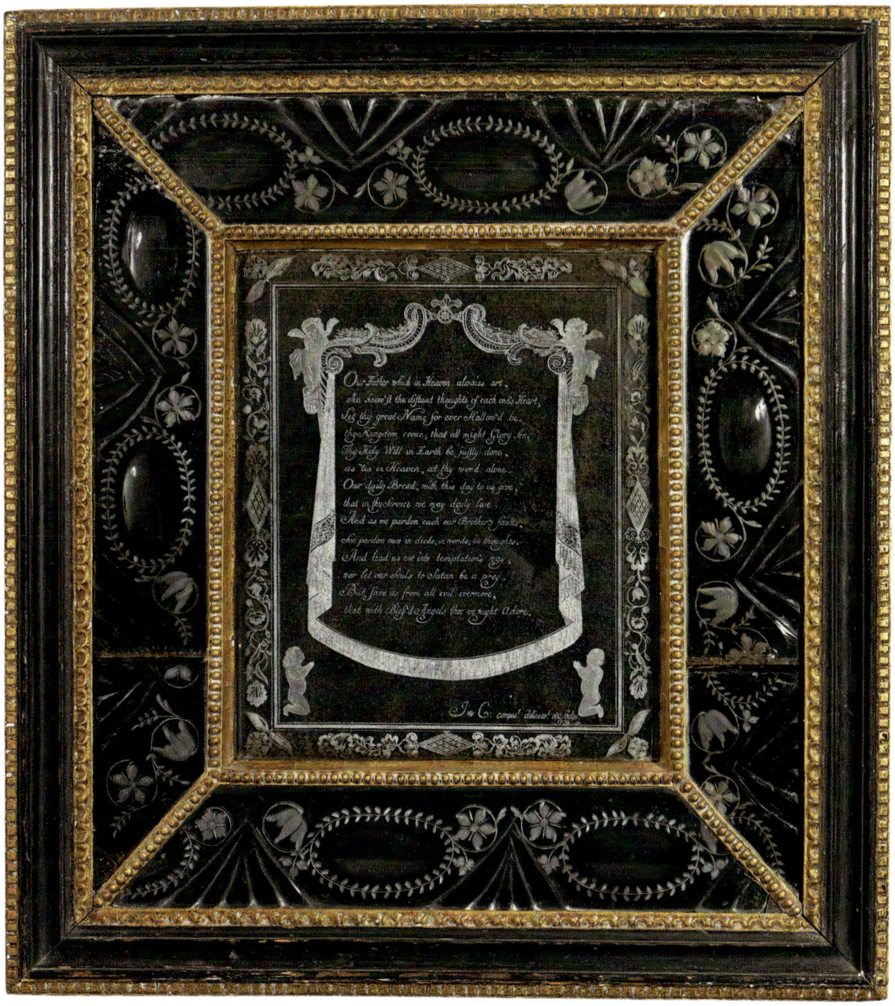

1728

Mirror

Glass

$20\frac{1}{4} \times 18 \times 1\frac{3}{8}$ in. (51.4 × 45.7 × 3.5 cm)

Inscription: Our Father which in Heaven always art,|who know'st the distant thoughts of each one's Heart,|Let thy great Name for ever Hallow'd be.|thy Kingdom come, that all might Glory See.|Thy Holy Will in Earth be justly done,|as 'tis in Heaven, at thy word alone.|Our daily Bread, with this day to us give,|that in thy Service we may daily live.|And as we pardon each our Brother's faults,|Soe pardon ours in deeds, in words, in thoughts.|And lead us not into temptation's way,|nor let our Souls to Satan be a prey.|But Save us all from all evil evermore,|that with Bless'd Angels thee we might Adore.; J C: comps.^{t} delineav^{t} et Sculps.|1728

This looking glass gave an immediate corrective to vanity by superimposing the Lord's Prayer onto the reflection of the person looking within it.

PROVENANCE

Anthony Leslie Smith of The Wadfield, Sudeley, Gloucestershire; Arthur Churchill Ltd., London; bt. John H. Bryan of Lake Bluff, IL (1936–2018) from Christopher Sheppard Ltd., London, June 10, 1997

1648

Pendant
Gold, jet
1¾ × 1 in. (4.5 × 2.5 cm)

Inscription:
IANVARY|:THE: 30:|:1648:

This pendant commemorates the day on which Charles I was beheaded outside the Banqueting House at Whitehall. The carved jet pendant shows the king in profile. It might have been worn discreetly behind a shirt in the years after Charles's death to show allegiance to the royal cause in exile.

PROVENANCE
Percival D. Griffiths of Sandridgebury, St Albans, Hertfordshire (1862–1938); Ronald A. Lee of London (1913–2000); bt. John H. Bryan of Lake Bluff, IL (1936–2018) from Georgina Gough, London, August 2013

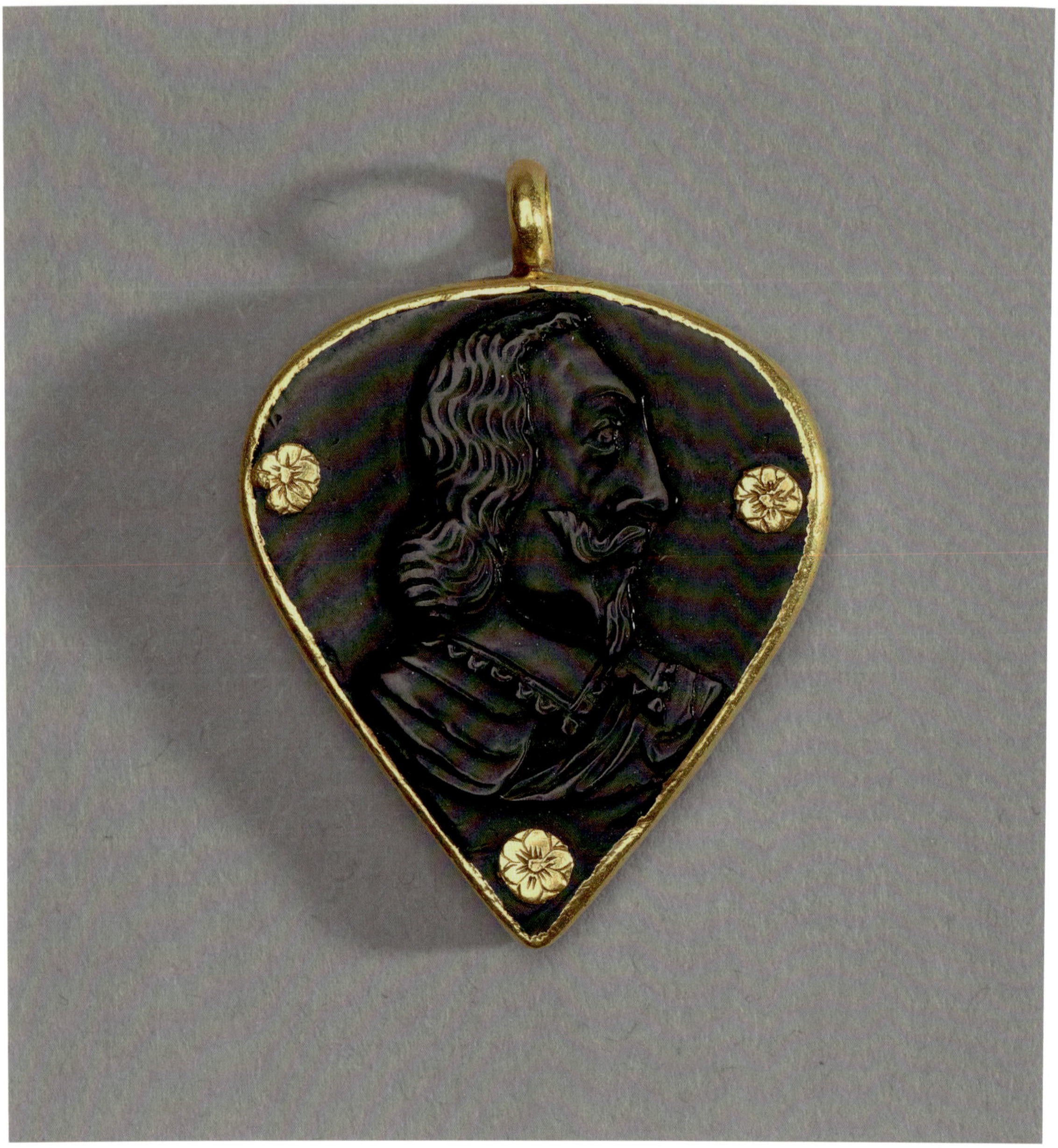

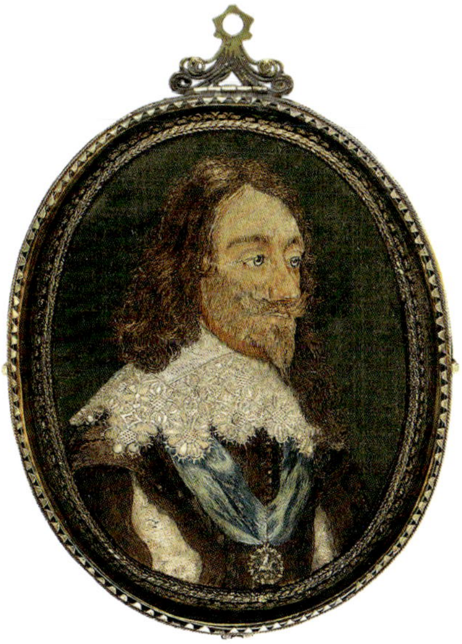

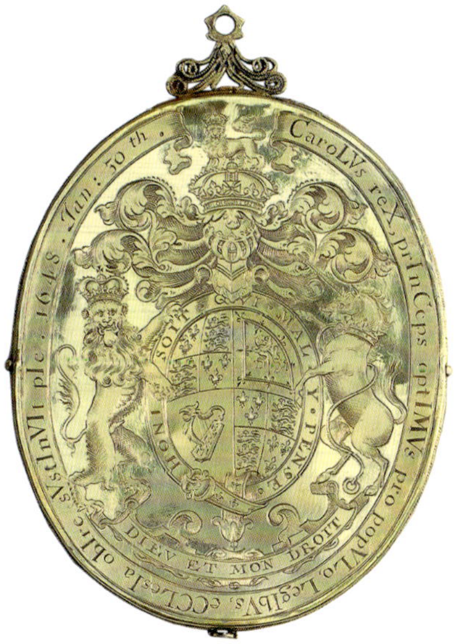

1648

Portrait

Silk, silver gilt, glass

6 ½ × 4 ½ × ½ in. (16.5 × 11.4 × 1.27 cm)

Inscription:
CaroLVs reX, prInCeps optIMVs pro popVLo, LegIbVs, eCCLesIa obIre sVstInVIt pIe

The inscription on the back of this embroidered portrait of Charles I contains a chronogram. A chronogram embeds the date into prose or verse with capitalized letters that correspond to Roman numerals: in this case, "MDCXLVIII." The date, 1648, is also expressed here in Arabic numerals.

Charles I was executed at Whitehall on January 30, 1648/9. The encryption of this date, so significant to the cult of the king as a martyr, both during and after the interregnum, speaks to the secrecy surrounding and the intensely personal nature of gestures honoring the memory of Charles I.

Eleven additional portraits of this type were identified in a study by John Nevinson published in 1965. To this list can be added another embroidered portrait sold at Mellors & Kirk, Nottingham, September 19, 2018 (lot 924), of the type best represented in the collection of the Metropolitan Museum of Art, New York (39.13.7).

All these miniatures were worked with colored silks in split, stem, long-and-short, and satin stitches on white satin backing, set on an oval background of green satin. The portraits are derived from the etching by Wenceslaus Hollar of 1641 and are thought to have all originated from the same workshop.

PROVENANCE

Hampton's, Auction Galleries, Godalming, Surrey; bt. John H. Bryan of Lake Bluff, IL (1936–2018) from D. S. Lavender (Antiques) Ltd., London, September 7, 1995

LITERATURE

John L. Nevinson, "The Embroidered Miniature Portraits of Charles I," *Apollo* 82, no. 44 (October 1965): 310–13

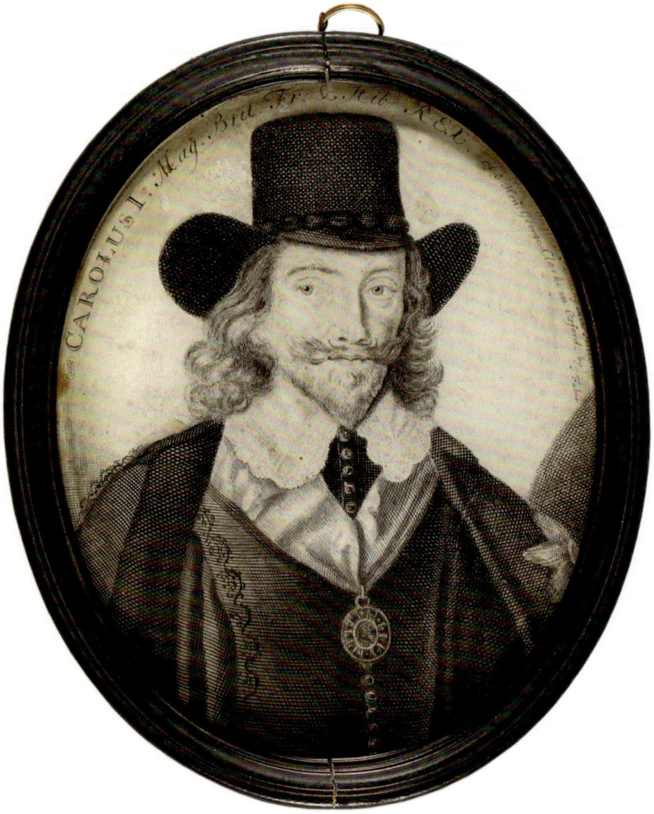

1648

Portrait

Ink on vellum

6¼ × 5 in. (15.9 × 12.7 cm)

John Faber (ca. 1660–1721)

Inscription: CAROLUS I: Mag. .Brit. Fr. & Hib: REX.|Done from y^e^ Original Painting in the Possession|of ye Hon:^ble^ George Clarke in Oxford. |by J. Faber; Portrait of|King Charles: first|as he sat on his Trial|in Westminster Hall.|January 23 1648. a Drawing with the pen.|By J. Faber.

PROVENANCE

Francis Wellesley; Sotheby's, London, June 28, 1920 (lot 286); Christie's, London, March 3, 1993 (lot 4); bt. John H. Bryan of Lake Bluff, IL (1936–2018) from D. S. Lavender (Antiques) Ltd., London, June 9, 1993; gifted, June 6, 2017, to John H. Bryan III of Nashville, TN

LITERATURE

J. J. Foster, *Samuel Cooper and the English Miniature Painters of the XVII Century*, supplement (London: Dickinsons, 1914–16), 105

G. C. Williamson, "Mr Francis Wellesley's Collection of Miniatures and Drawings, Part 2," *Connoisseur* 52 (September–December 1918): 69

1698 | 1765

Portrait

Oil on canvas

12 × $9\frac{3}{8}$ in. (30.5 × 23.8 cm)

Edwaert Collier (1642–1708)

Inscription:
Carolus Rex Primus; &:|E. Collier fecit. London 1698. The Gift of Mrs Elezth Hookes|of Gaunts in Com Dorset| 31st July 1765

This portrait of Charles I was painted nearly fifty years after his execution. A trompe l'oeil work, it is one of a series of similar portraits—resembling engravings affixed to oak or pine boards—that Collier made in the 1690s during his time in London.

The lining of the painting bears a replica of the original inscription. Elizabeth, widow of John Hookes, Esquire, of Hinton Martell, died in 1765. This portrait was presumably a bequest made on her deathbed.

PROVENANCE

Before 1765 E. Hookes, Gaunts, Dorset; Phillips, Anonymous sale, London, June 3, 1997 (lot 51); Bernadette and William M.B. Berger, Denver, CO; Sotheby's New York, January 26, 2017 (lot 225); bt. John H. Bryan III of Nashville, TN, through Gurr Johns, London

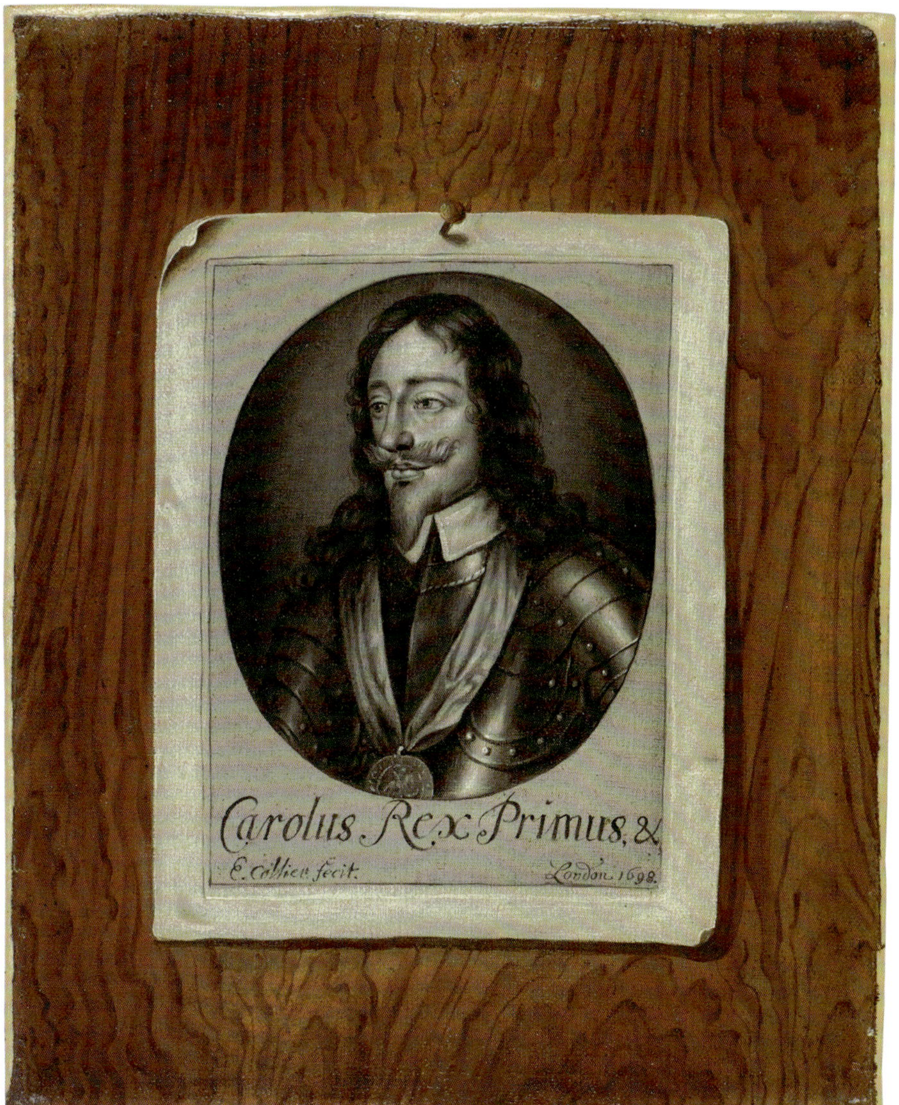

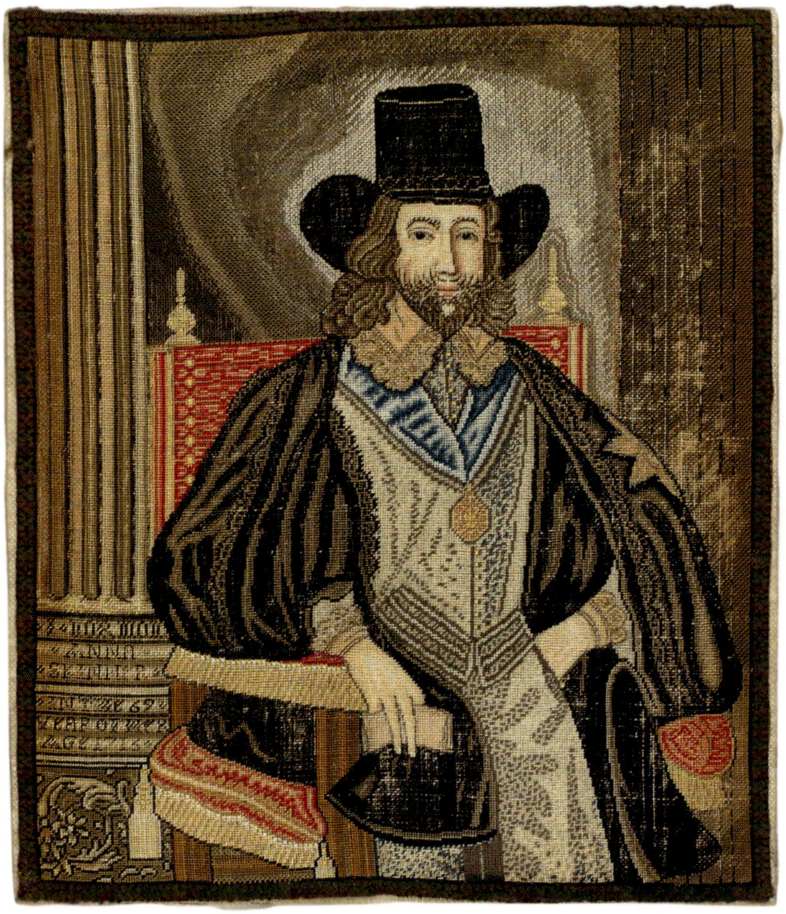

1716

Embroidered picture

Linen with silk thread

13¾ × 11½ in. (34.9 × 29.2 cm)

Anna Skinner (b. ca. 1649; active 1716)

Inscription: [____]|ANNA| SKINNER|IN THE 69|YEAR OF HER|AGE 1716.

In 1716, the sixty-ninth year of her life, Anna Skinner took it upon herself to make two embroidered pictures of Charles I. As her source, she used the mezzotint by John Faber, which in turn was based on the portrait of 1648 by Edward Bower, which depicted the haggard and obstinate monarch at trial at Westminster. The portrait was at that time in the possession of George Clark in Oxford. It was the iteration of the print made in 1713 that introduced the fluted column that appears in Anna's embroidery.

The execution of Charles I had a prolonged resonance in the collective memory of the nation, and by the time of Faber's mezzotint, the event was remembered as a martyrdom. And for Anna Skinner, who made not one but two of these embroidered pictures, this event must have had particular significance. For, as the inscription states, she had been born in about 1649, coming into the world at a major historical and cosmological moment.

The other embroidery of Charles I by Anna Skinner is in the Cecil Higgins Museum, Bedford. At this distance, it is impossible to differentiate between the various women by this name who were alive in 1716, although there is the slim possibility that this embroiderer was the woman baptized at St Margaret Westminster on April 8, 1649, who was assessed for three hearths in 1664 in the same parish. No date of death or will for this Anna Skinner has been traced.

PROVENANCE

bt. John H. Bryan of Lake Bluff, IL (1936–2018) from Mallett & Son (Antiques) Ltd., London, June 8, 1999

1677|1690

This painted elm box has a two-plank lid. The front plank is incised with initials and the date and was later painted overall in red, green, brown, and black. The lid depicts, within an oval reserve, a version of the royal coat of arms—a crowned Tudor rose flanked by the Stuart supporters (the lion and unicorn) over the royal motto "DIEV:ET:MON DROIT." On the front face of the box is a vase and flowers, flanked by red-gowned angels waving palms. One side of the box is illustrated with a winged white swan between trees, and the other with a snarling, winged green dragon flanked by two trees.

The back panel bears the name and date, "FRANCES: SARLE 1690." The name is too common to state with certainty, but this box may have belonged to Francis Searle of Awliscombe, Devonshire, who was a member of an armigerous family. He died on August 21, 1690, the same day his daughter Mary was born. The inventory of his possessions made for probate is damaged and in some parts is illegible, but it does not appear to list the box, which may not have been deemed of sufficient value to enumerate (TNA PROB 4/10347).

PROVENANCE

bt. John H. Bryan of Lake Bluff, IL (1936–2018) from Alistair Sampson Antiques Ltd., London, March 18, 1993

Bible box

Elm, painted; iron

$8\frac{1}{2} \times 21\frac{1}{4} \times 15\frac{1}{8}$ in. (21.6 × 54 × 38.4 cm)

Inscription:
F S|1677|Fear God; DIEV:ET:MON DROIT; FRANCES: SARLE|1690

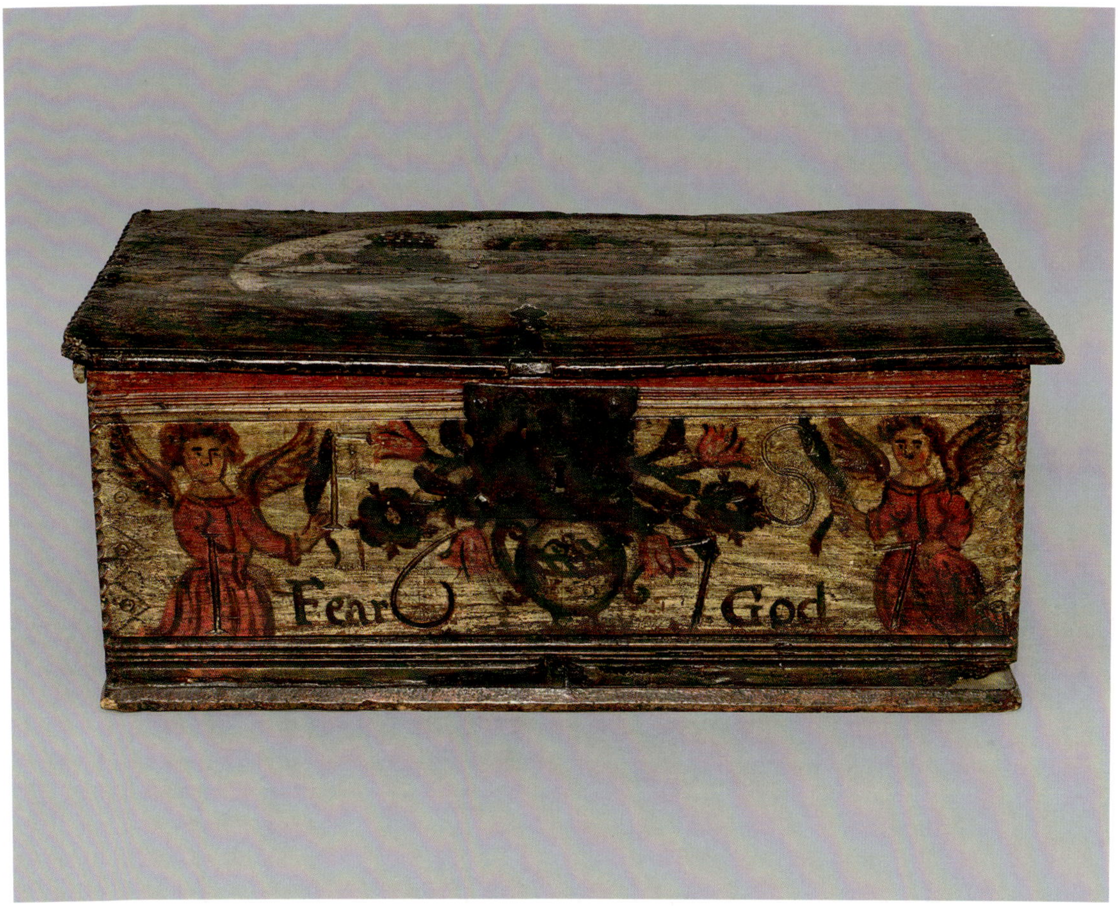

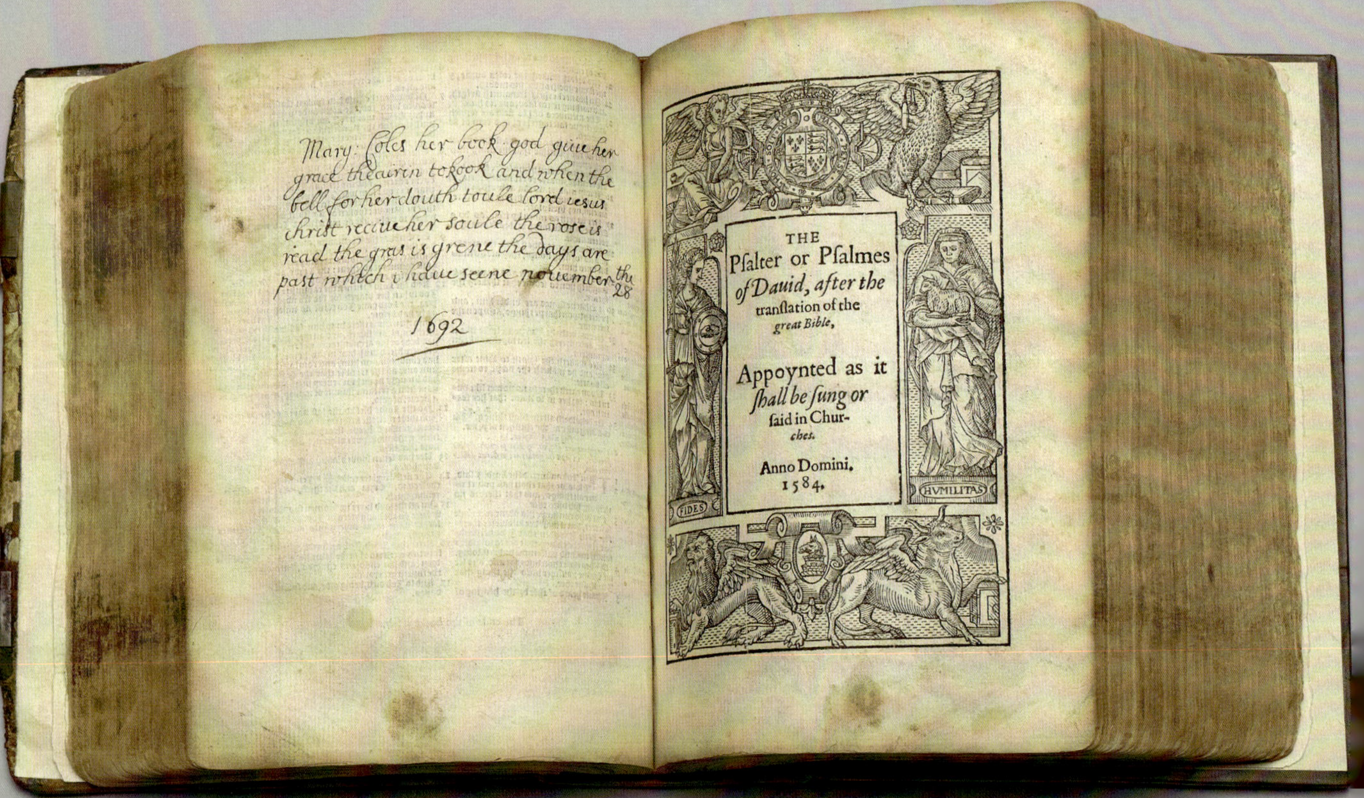
Mary: Coles her book god giue her
grace theairin tolook and when the
bell for her douth toule lord iesus
christ receiue her soule the rose is
read the gras is grene the days are
past which i haue seene nouember the 28
1692
THE
Psalter or Psalmes
of Dauid, after the
translation of the
great Bible,
Appoynted as it
shall be sung or
said in Churches.
Anno Domini,
1584.
FIDES
HVMILITAS

1584 | 1666 | 1670 | 1692

Bishop's Bible

Paper, board, leather

Closed: $2\frac{7}{8} \times 6\frac{1}{2} \times 8\frac{3}{8}$ in. (7.3 × 17.1 × 21.3 cm)

Inscription: Mary: Coles Her Boock|Novem-ber the 28: 1692; Mary: Coles her book · god give her|grace theairin to kook and when the|bell for her douth toule lord iesus|christ recive her soule the rose is|read the gras is grene the days are|past which I have seene november the 28| 1692; Com learne to love thy leson it is most plesant|having once learned it you never will forGet it|againe|Elizabeth Hawgood 1666 23 day|June; this is the holy|beble Elizabeth her [____]; Thomas Roberts|his Booke 1670|Edmond Roberts|Is My name and|with my pen I|ritt the same; Edmond Roberts|if my penn had ben bet[____] I wld|A mend it here|Every 1670; ould mab hath never a cunt; ould mab hath never a cunt.

One hears an echo of Mary Coles's father's voice in the rhyming maxims that she inscribed in her copy of the 1584 Bishop's Bible on November 28, 1692. More than two decades earlier, when her father, Robert Henderson (d. 1670), a clergyman, had come to make his will, he had appended a short note asking his friend Doctor Henry Aylworth "to bee a Friend to my wife when I am dead and gone and withall to take a Care that she may have no wrong." From an early age, his daughter would have heard such language, which relied on rhythm and rhyme to drum common truths and wisdom into the mind, certainly shaping her systems of thought. Mary was a pious woman—and not one who would have approved of the profanity later entered, not once but twice, into the pages of her Bishop's Bible.

Mary was baptized on October 27, 1645, at St Peter-in-the-East, Oxford. Her father was Curate of Goring, Oxfordshire, from 1646, and later also Rector of Checkendon. Shortly before he died, he gave his consent for the marriage of his daughter to another clergyman, Gilbert Coles (d. 1676), a widower and probably about twice as old as his twenty-four-year-old bride. Coles had matriculated at New College, Oxford, and become a fellow of Winchester College in 1648, from which he was expelled by Parliamentary visitors in the same year. There seems to have been some disagreement about the dowry due to Coles at the time of Henderson's death, the latter insistent that his son-in-law was "not to demand anything of my Wife."

After her marriage, Mary had to contend with a large number of children (Edward, Mary, Anne, Cicely, Martha, and Gilbert) who were not her own, some of whom were only a decade younger than she. Mary and her husband had their own son, Robert, referred to in his father's will of 1676 as "our deare Infant Robert."

When she came to make her own will, Mary bequeathed her "Soul into the hands of my faithfull Creator and Redeemer and my body I bequeath to the dust in hopes of the Resurrection unto eternall life." She gave money to the poor of Lockerly and gave to her niece, Mary Henderson, her "best Diamond Ring, a little wrought Box, a damask Table Cloth," along with twelve napkins, half of her household goods, and apparel. She also gave specific instructions as to her burial, which was to be under her pew in Lockerly Chapel. The pallbearers were to be six clergymen, all of whom were given rings for the occasion. She was buried at Lockerley, on the Wiltshire border of Hampshire, on May 10, 1710.

PROVENANCE

bt. John H. Bryan of Lake Bluff, IL (1936–2018) from David C. Lachman, Wyncote, PA, July 7, 2017

1673–1863

Cabinet

Wood, paper, mica, woven silk, silk and metal thread, silk wrapped wire, plush, brass

Contents: paper, cotton, silk, metal and linen thread, glass beads, wire, woven silk, velvet, brass, silk ribbon, ivory, unidentified metal, pewter, glass, moss, spun glass, mirror, engraving on paper, spangles, human hair, dried flowers

11 ¾ × 10 ¾ × 8 ½ in. (29.8 × 27.3 × 21.6 cm)

Inscription on cabinet: for a glass and cushins

This cabinet, replete with a remarkable assemblage of contents, is an unusual survival and something of a puzzle. Its contents associate it to one family—the Wesleys—whose members were responsible for founding the Methodist Church, but as there is no provenance for the object, it is unclear to whom it belonged. The assorted fragments and keepsakes suggest that the cabinet served as a miniature museum in the family home, holding religious relics (some of which seem to have been sent as gifts) and other objects with no demonstrable link to Methodism but rather saved from moments of particular national significance. These range from a piece of cake from the royal wedding of Prince Edward and Alexandra on March 10, 1863, to a sculptural fragment brought from Greece to England by Lord Elgin.

Many of these objects are wrapped in paper, and the drawers contain cards with notes describing the contents. None of this information should be taken at face value, however. For example, it is impossible that the lace mounted onto pink paper was worked by Sarah Wesley (1759–1828), daughter of Charles (1707–1788), because most of the lacework predates her lifetime, and the French lace of about 1750–60 looks professionally made. Sarah was the niece of John Wesley (1703–1791), brother of Charles. Both men were major figures in the establishment of Methodism.

Likewise, the silk-and-glass-bead ornamental sprigs of carnation and barley on wire are not, as the note suggests, the work of French nuns. They are English and contemporaneous with the cabinet. In fact, the cabinet contains an example of a *colifichet*, a form of double-sided needlework popular in France and taught in convent schools, so it seems that the memory of this link to France persisted with the owners but was, at some point, graphed onto the wrong objects—a pertinent reminder to us today of the fallibility of memory.

What follows is a transcription of the paper notes that describe the full content of the cabinet. Due to the small size of many of these objects, such as the strand of hair, not all of these are illustrated. The inscriptions are keyed to the objects shown in the photograph on pages 484–85.

PROVENANCE

With the shipping company C. R. Fenton & Co. Ltd., Beachy Road, Old Ford, London; bt. John H. Bryan of Lake Bluff, IL (1936–2018) from Cora Ginsburg, New York, September 14, 1988

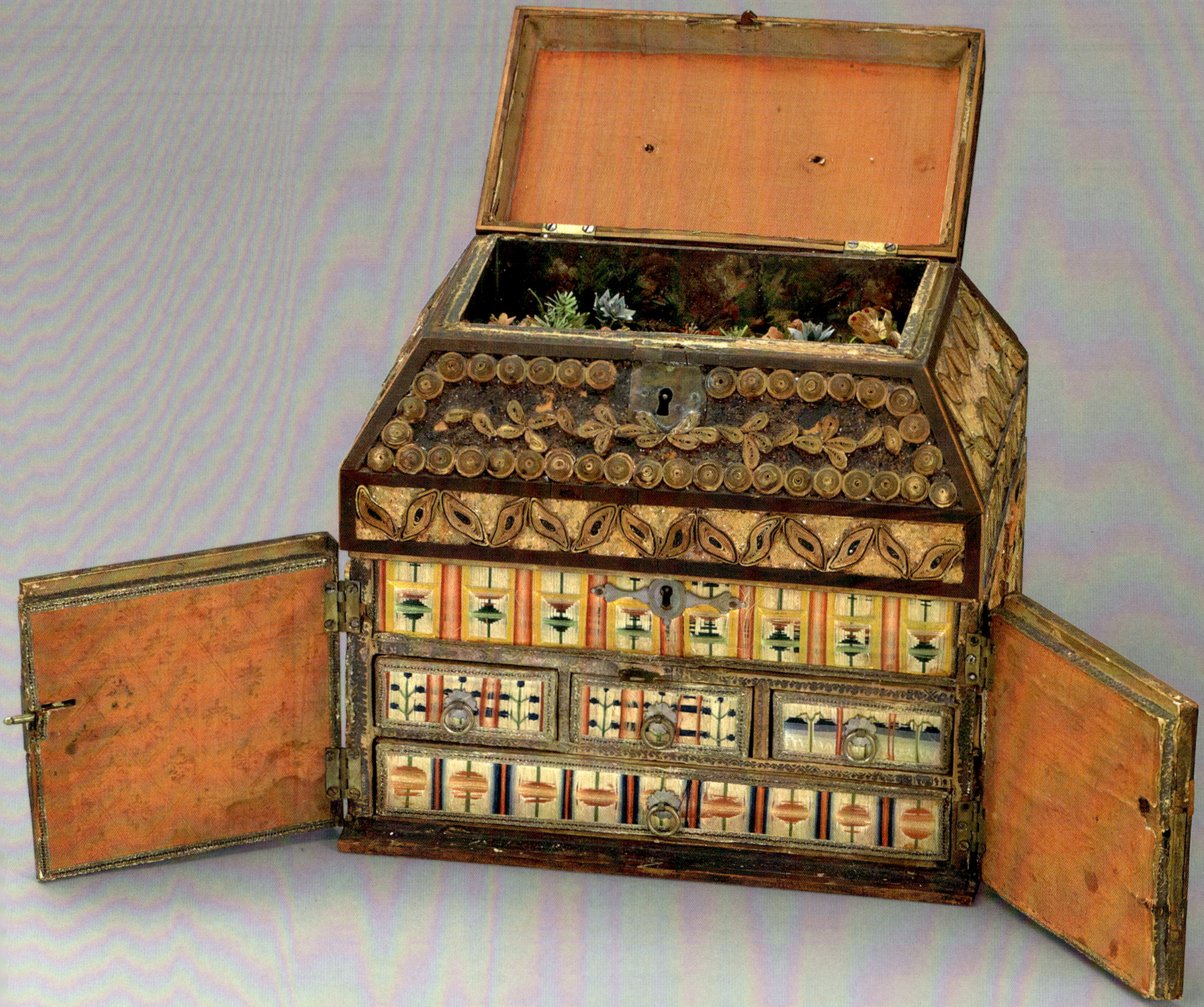

Inscriptions on and descriptions of contents:

A: Lace - worked by Sarah Wesley| daughter of Charles + niece of|John Wesley|founder of Methodism.| From Cabinet.

B: Squirrel|running up an Ear of Corn|The work of French|Nuns.| (Curious idea of Natural History)| From Cabinet.

C: THE REV. IOHN WESLEY. A.M; FOLLOW THOU ME; IOANNES WESLEY OBIIT|SECUNDO DIE MARTIS 1791|ANNO SUÆ ÆTATIS OCTO|GESIMO OCTAVO.; HAVING DONE ALL IN|MY POWER I ACCOUNT MY|SELF AN UNPROFITABLE| SERVANT.

D: THE REV. IOHN WESLEY. A.M; FOLLOW THOU ME; IOANNES WESLEY OBIIT|SECUNDO DIE MARTIS 1791|ANNO SUÆ ÆTATIS OCTO|GESIMO OCTAVO.; HAVING DONE ALL IN|MY POWER I ACCOUNT MY|SELF AN UNPROFITABLE| SERVANT.

E: THE REWARD OF ATTENTION & INDUSTRY: LEARN TO LIVE; REMEMBER THY CREATOR IN THE DAYS OF THY YOUTH

F: Bead Chain (Intended to be hung round old fashioned|circular Brazier for airing linen) with| curious inscription|"my ring is round and hath no end|so is my love unto my friend"|"Mary Blackmore" "anna dominy 1673." (N.B. Ladies' Latin); lent by Mrs A. Harper, Charlton Adam.

G: Baskets of|Primroses &|Horn of Flowers|Cut out by Miss Sarah Wesley

Part of a Bead Robe|worn by|TIPPOO SAIB|an Indian Prince (b. 1749)|who was defeated and|slain by the British at|Seringapatom.|From Cabinet.

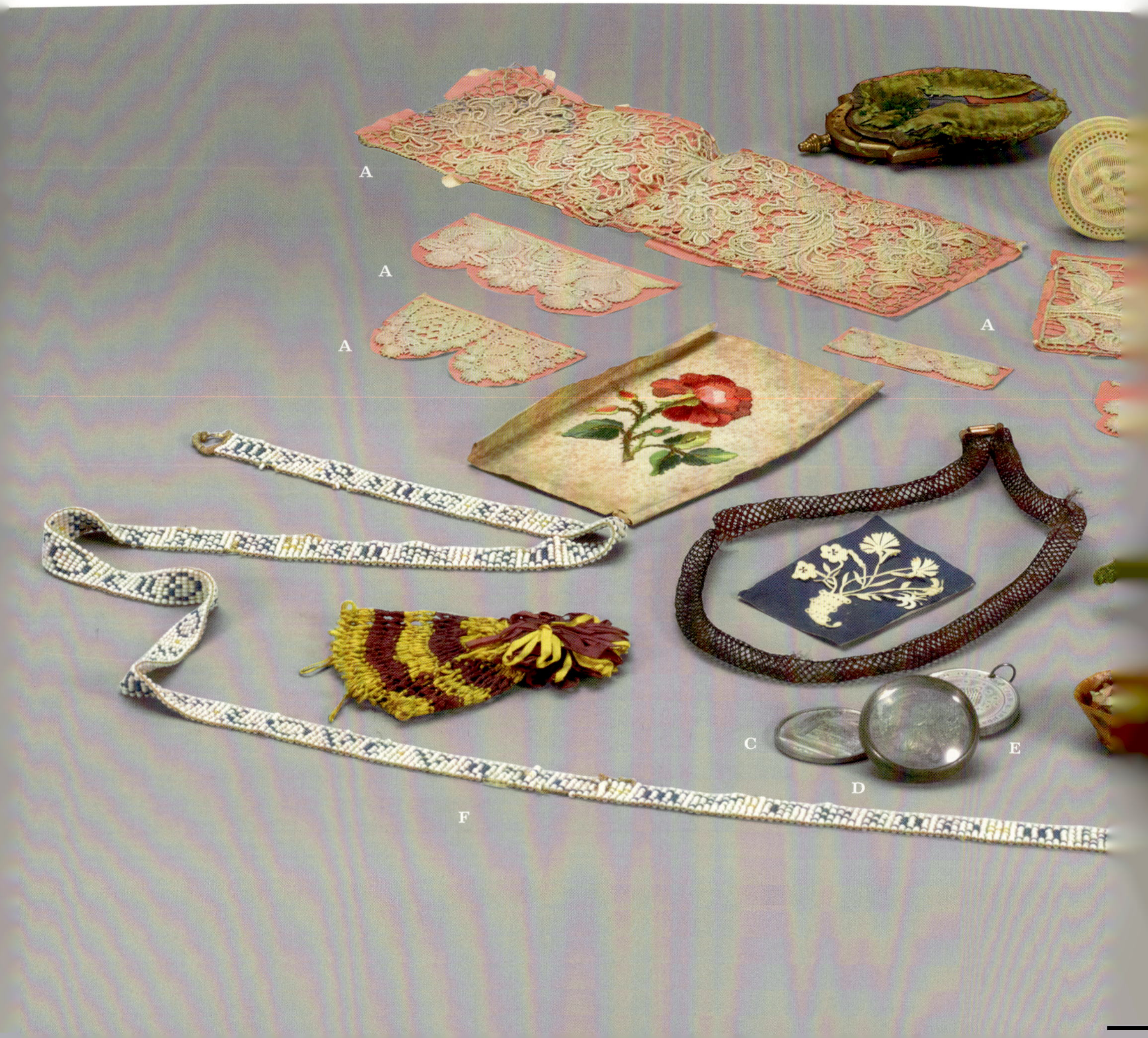

H: Ornaments|(for the hair)| Made of Spun glass|Worn at the Thanksgiving| Service in S. Pauls Cathedral|on the occasion of the|restoration to reason of King George III|April 23rd1789.|From Cabinet

I: Portrait of George iii|Such as was worn on the|heart by gentlemen at|-tending the Thanksgiving| Service for his Majesties|Restoration to Reason|held at St Pauls Cathedral, April 23rd 1789.|From Cabinet.

S Bradburn|Revd John Wesley's| Hairs

A piece of a Palm tree|two hundred & forty years|old brought from Haarlem|where it is now growing| but is not considered to have|arrived at full growth till a| thousand years

Wellington Tree; Piece of the tree under|which Wellington|stood when writing his|despatches at the|battle of Wartloo 1815|From Cabinet [information repeated]

Peter's Vine; A piece of the Vine that|grew over the house at| Zaandam in which|Peter the Great dwelt|while studying [   ]|the art of ship building

Royal Cake, March 10th 1863, Prince and Princess of Wales

Phidias; Brought from Greece by Lord Elgin. A piece from one of Phidias's statues [information repeated on second label]

Flower from the wreath placed by the|Queen Victoria on Prince Alberts|grave in St George's Chapel|Windsor [information repeated on second label]

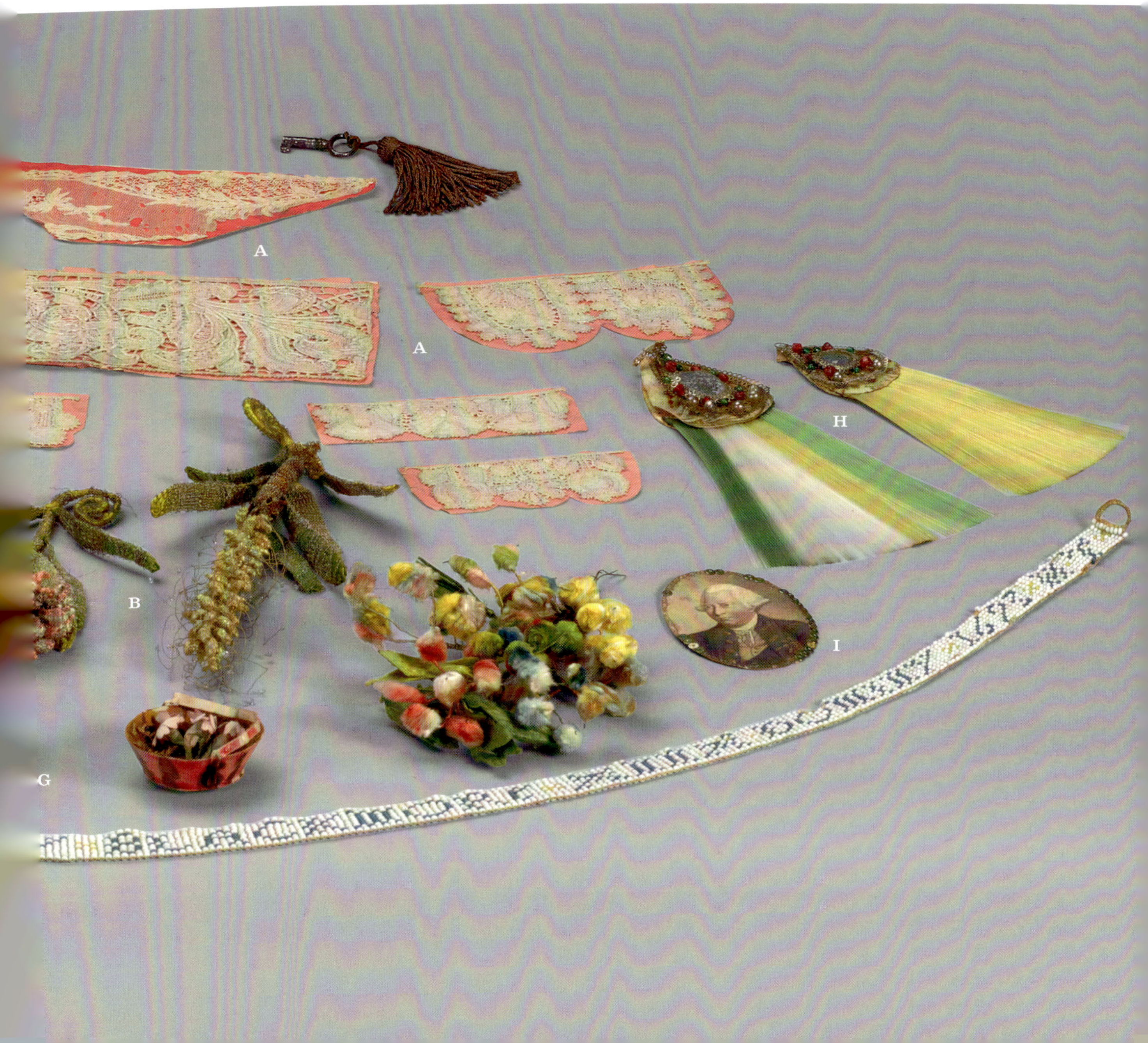

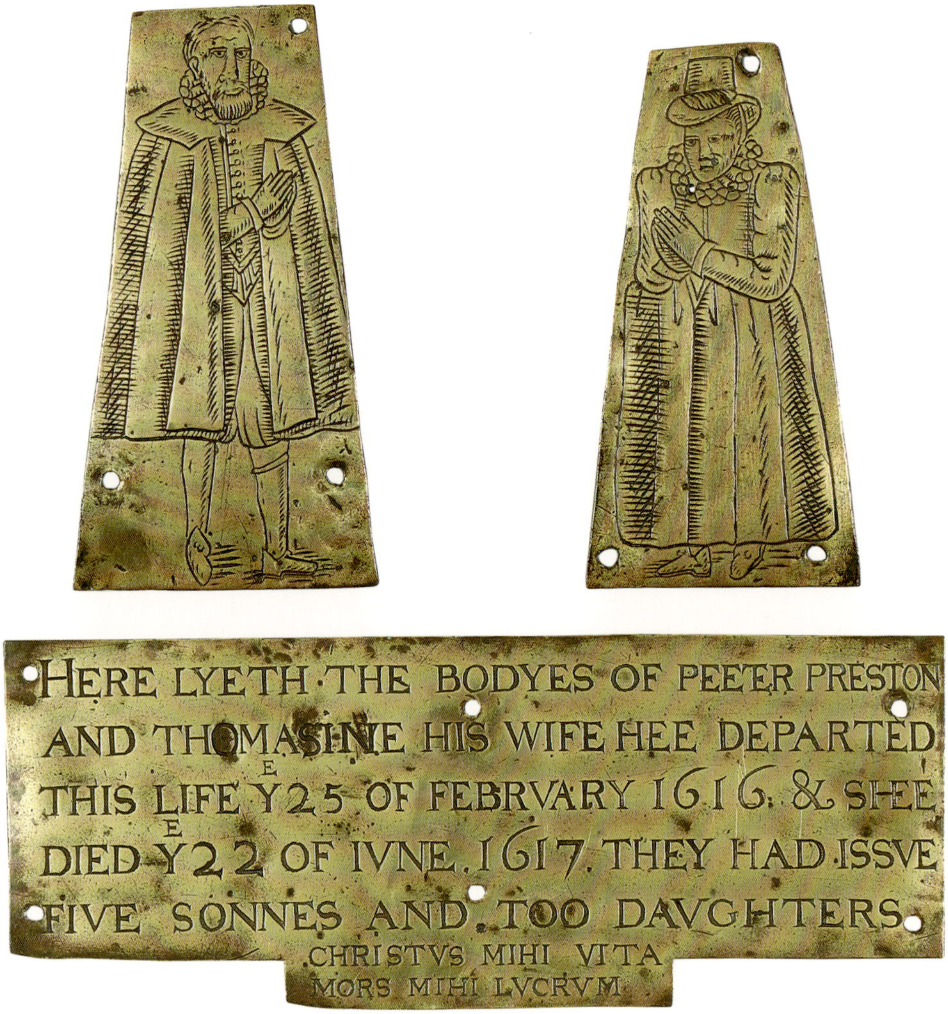

1616|1617

Memorial plates

Brass

$8\frac{3}{4} \times 4\frac{1}{2}$ in. (22.2 × 11.4 cm); $8\frac{1}{4} \times 4\frac{1}{2}$ in. (21 × 11.4 cm); $14\frac{3}{4} \times 6$ in. (37.5 × 15.2 cm)

Inscription:
HERE LYETH·THE BODYES OF PEETER PRESTON|AND THOMASINE HIS WIFE HEE DEPARTED|THIS LIFE Y^E 25 OF FEBRUARY 1616 & SHEE|DIED Y^E 22 OF IVNE . 1617. THEY HAD ISSVE|FIVE SONNES AND TOO DAUGHTERS; CHRISTVS MIHI VITA|MORS MIHI LVCRVM

Peter and Thomasine Preston, husband and wife, died within three months of each other. As the parish register of St Mary and St Lambert, Mickfield, Suffolk, records, they were buried on February 29, 1616/7, and June 24, 1617, respectively. The Prestons were members of the Suffolk gentry and owned lands in Framlingham and elsewhere in the country. The Latin epitaph is from St. Paul's Epistle to the Philippians and translates as "For me to live is Christ, and to die is gain."

PROVENANCE

bt. John H. Bryan of Lake Bluff, IL (1936–2018) from Tobias Jellinek Antiques, London, February 10, 1994

LITERATURE

Rev. Edmund Farrer, *A List of Monumental Brasses Remaining in the County of Suffolk* (Norwich, UK: Goose, 1903), 44

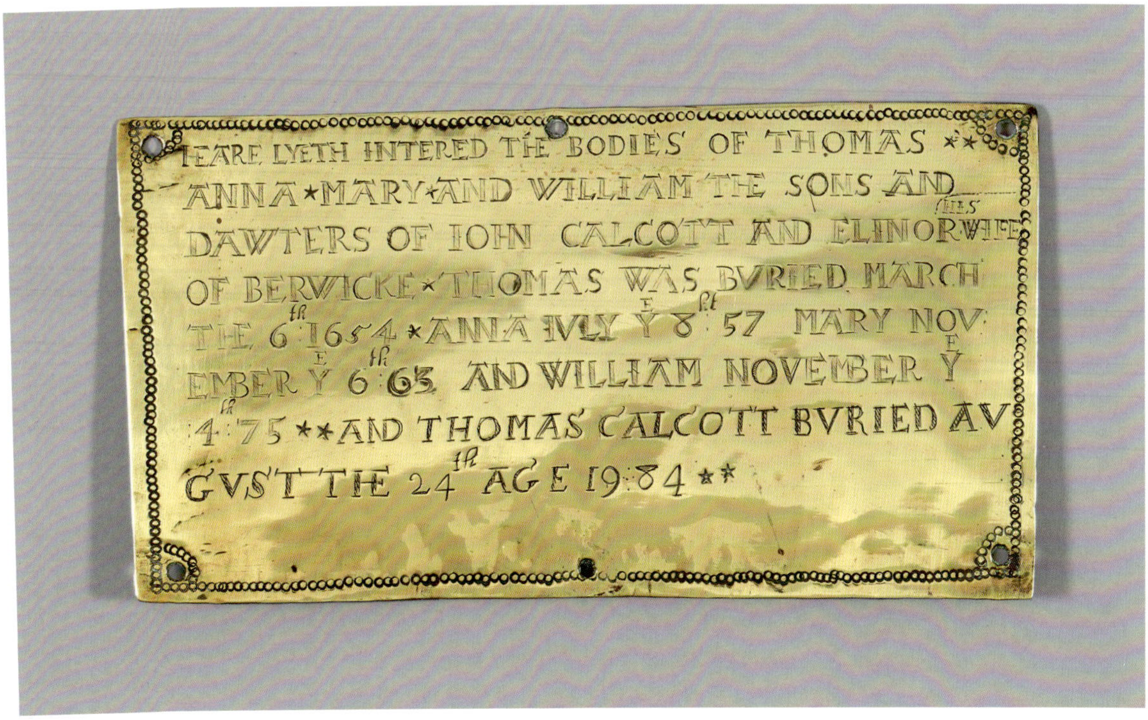

1654 | 1657 | 1663 | 1675 | 1684

Memorial plate

Brass

4¾ × 8¾ in. (12.1 × 22.2 cm)

Inscription:
HEARE LYETH INTERED THE BODIES OF THOMAS ·|ANNA·MARY·AND WILLIAM THE SONS AND|DAWTERS OF IOHN CALCOTT AND ELINOR HIS WIFE|OF BER-WICKE· THOMAS WAS BVRIED MARCH|THE 6th 1654· ANNA JULY YE 8 th : 57 MARY NOV|EMBER YE 6th : 63 AND WILLIAM NOVEMBER YE |4 th : 75 ·· AND THOMAS CALCOTT BURIED AV|GVST THE 24th AGE 19 : 84··

This plaque records the succession of bereavements suffered by John and Eleanor Calcott of the manor of Berwick, a township about four miles from Shrewsbury in Shropshire and part of the parish of Atcham. Their son Thomas died at age nineteen, daughter Anne died as an infant, and son William at age seventeen. Their children Susanna, John, Martha, and Richard all seem to have lived at least until early adulthood. The parish register, which is missing thirteen years, belies the frequent slippage in the paper record that occurred between an event and its entry into an official record. For example, the register shows Mary was buried on September 29, 1663 (which does not accord with the memorial plate); William on November 4, 1675 (which does); and Thomas on August 23, 1684 (which is off by a day).

The family occupied a moated dwelling, which is now the Attingham Home Farm, where the name "Calcot" is inscribed into one of the windows. This brass plate appears to have been on the west end of the south choir stalls of the twelfth-century parish church—the only church in the country to be dedicated to St. Eata, Bishop of Hexham.

PROVENANCE

bt. John H. Bryan of Lake Bluff, IL (1936–2018) from Lindsay Antiques, London, December 12, 1994

LITERATURE

"The Register of Atcham," in *Shropshire Parish Registers, Diocese of Lichfield*, Vol. XIV (Privately printed for the Shropshire Parish Register Society, 1921)

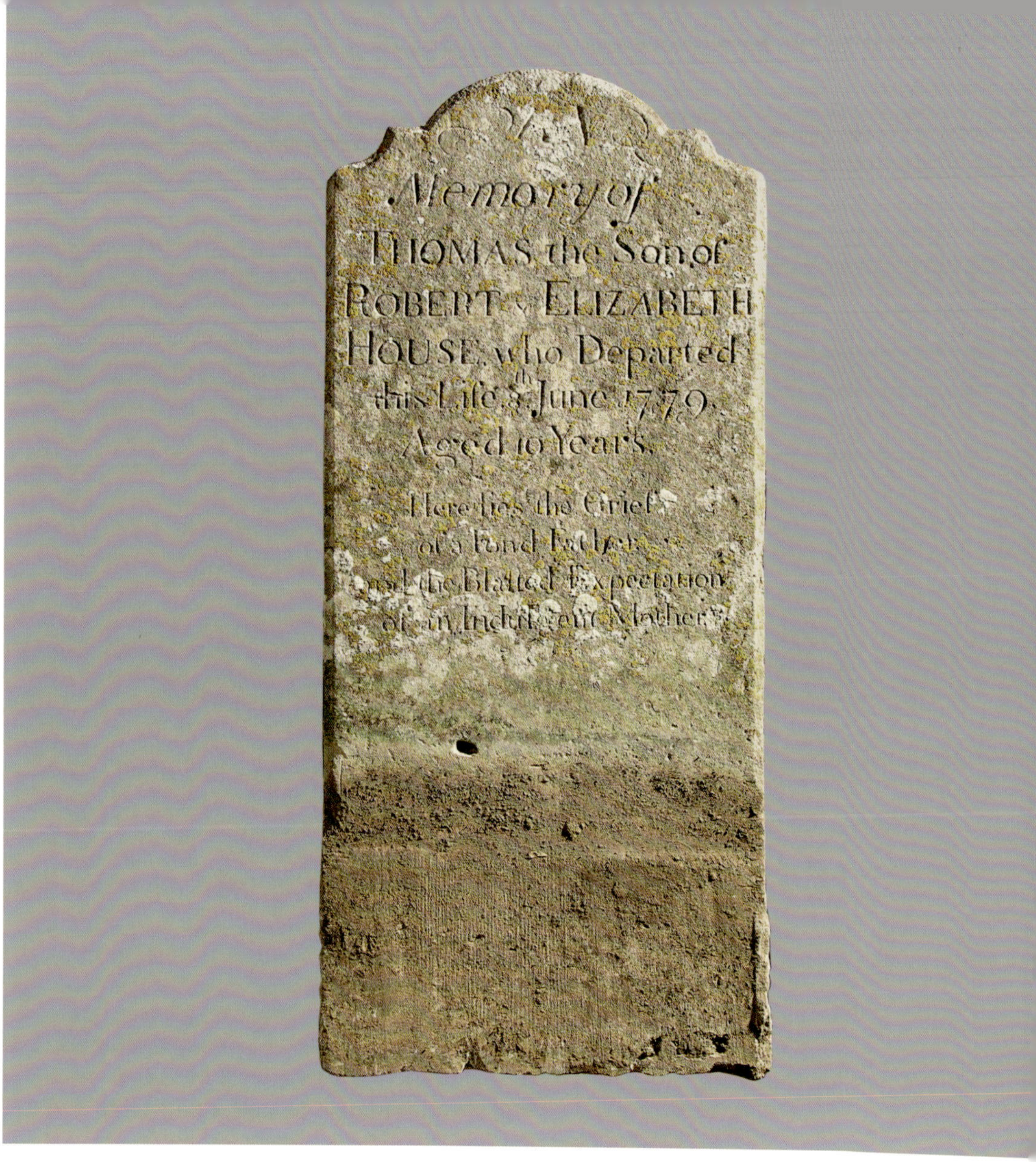

1779

Tombstone

Stone

53 × 23½ × 4½ in. (134.6 × 59.7 × 10.2 cm)

Inscription: IN Memory of | THOMAS the Son of | ROBERT & ELIZABETH | HOUSE who Departed | this Life 8. June 1779 | Aged 10 Years. | Here lies the Grief | of a Fond Father | and the Blasted Expectation | of an Indulgent Mother

Thomas House died at age ten. His parents appear to have been semi-itinerant farm workers or laborers who spent their adult lives moving within a twenty-mile radius of rural Dorset. Other members of this family were "copicers," that is, laborers who kept the woods pruned, or coppiced. Robert House married Elizabeth on September 13, 1755, at Affpuddle, Dorset, about thirteen miles northeast of Broadwey. Children baptized at this parish were Christian (1756), Robert (1757), Elizabeth (1759), Charity (1761), and Benjamin (1763). When on October 6, 1772, their daughter Anne was baptized at Buckland Newton, Robert and Eizabeth were said to be of Clinger Farm, part of the Middlemarsh estate. Anne was buried January 12, 1773.

Thomas, the child commemorated on this tombstone, was born in Tincleton and baptized on April 7, 1769. He was buried on June 12, 1779, at St Nicholas Church, Broadwey. Robert House died at age eighty-six in 1819.

PROVENANCE

St Nicholas Church, Broadwey, Dorset

1612

Church bell
Leaded bronze
22 × 19½ in.
(55.9 × 49.5 cm)

Inscription: GIVE THANKES TO GOD 1612 I S

Bells played an integral role in everyday life, signaling to all those within earshot a call to action. Varying considerably in size, sound, and cadence, bells rang on every quarter of every hour and at times of daily prayer. Church bells like this one would also announce births, marriages, and deaths as well as national events of calamity or celebration. Bell tones announcing deaths often denoted the age or sex of the deceased; the number of knocks also conferred information about the person.

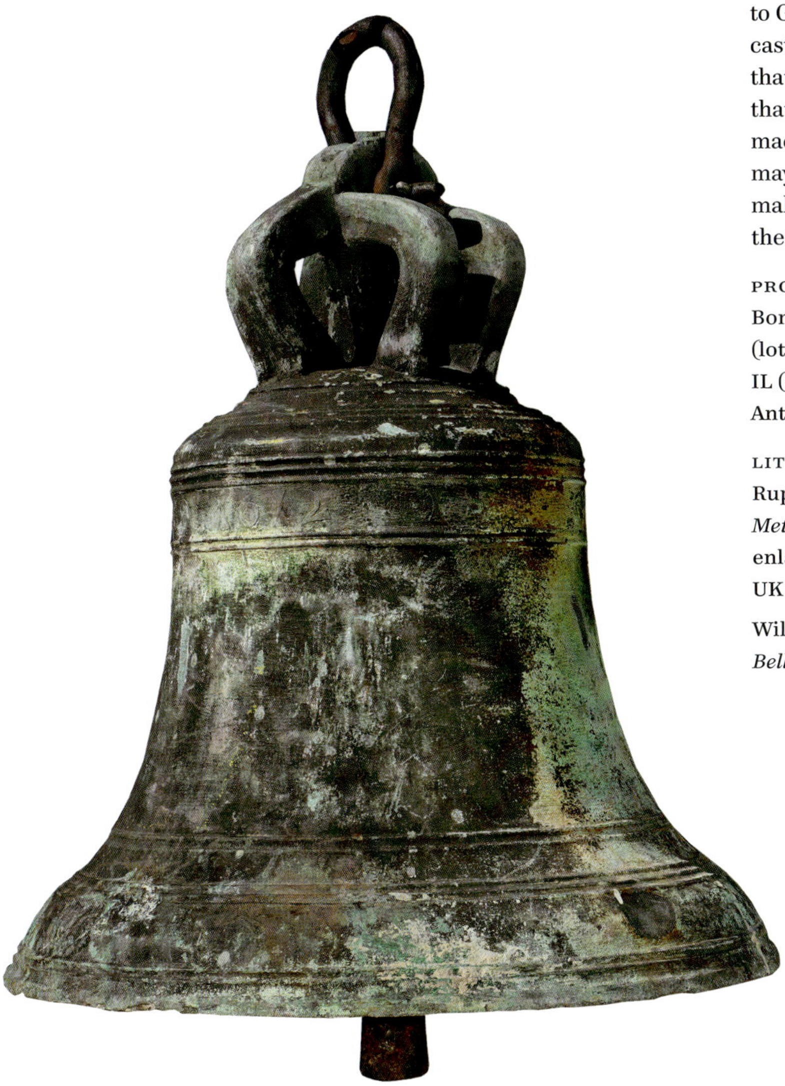

During this period, it was widely believed that ringing bells could dispel thunder and the dangerous lightning that might strike a church steeple and melt the bell within it. Prior to, and even after the Reformation, bells were blessed and baptized.

Casting the year onto a church bell has a long tradition, as does including a dedication to the donors or churchwardens. Bells with similar inscriptions can be found at Ilchester in Somerset, Coombe Bissett in Wiltshire ("O give thanks to God I. W. 1617"), and Lower Clatford in Hampshire ("Give thanks to God I. W. 1622"). That these bells are cast with different initials than those that appear on this example suggests that, although the bells may have been made at the same foundry, the initials may record the commissioners, normally the churchwardens, rather than the makers.

PROVENANCE

Bonhams, London, September 20, 2017 (lot 404); bt. John H. Bryan of Lake Bluff, IL (1936–2018) through Tobias Jellinek Antiques, London

LITERATURE

Rupert Gentle and Rachel Field, *Domestic Metalwork: 1640–1820*, revised and enlarged by Belinda Gentle (Woodbridge, UK: Antiques Collectors' Club, 1994)

William C. Lukis, *An Account of Church Bells* (London: J. H. Parker, 1857), 77, 104

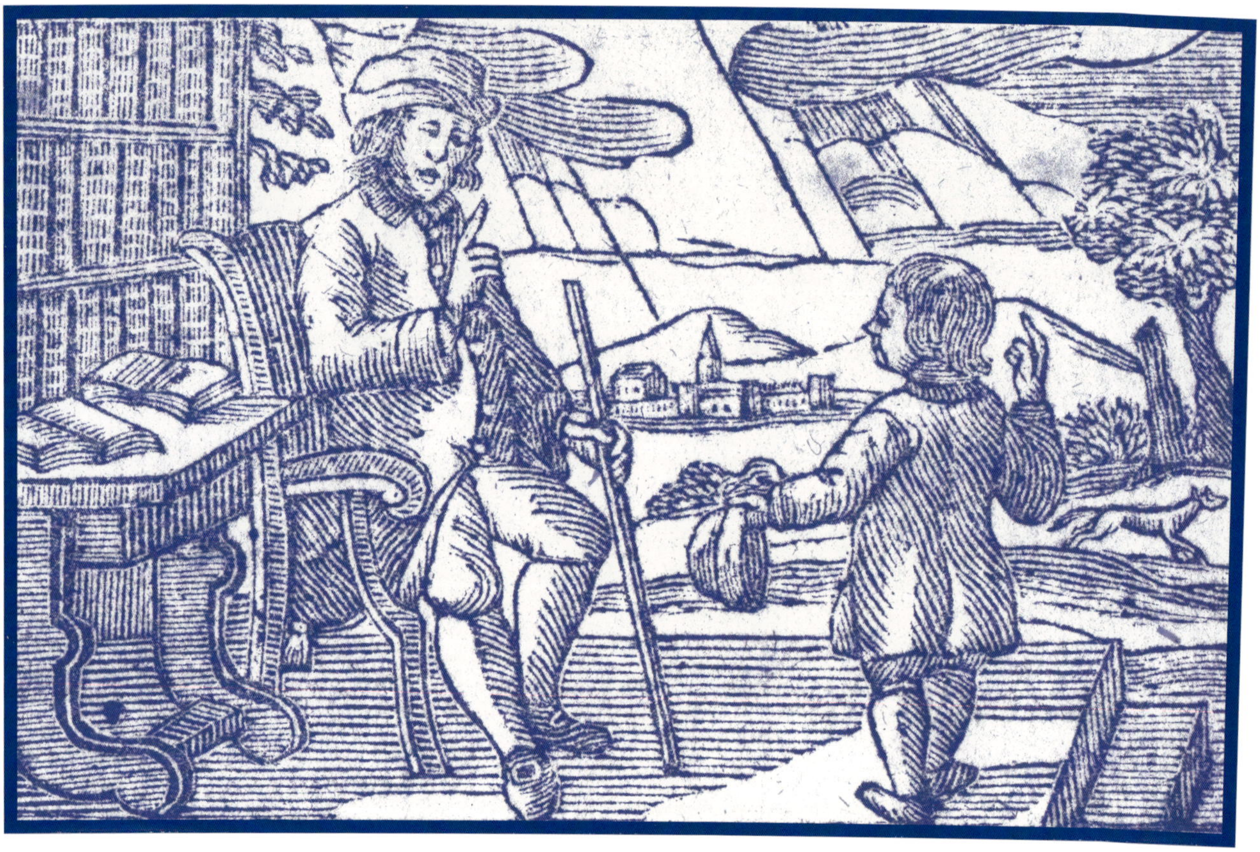

Afterword

A Reflection on the Collector

JENNY SAUNT

HAVING WORKED WITH JOHN H. BRYAN ON A NUMBER OF PROJECTS, as a member of his team, I was fortunate enough to play a small part in acquiring and researching the incredible collection of objects he gathered at Crab Tree Farm, his estate in Lake Bluff, Illinois. Many of these objects are included in this book, each one speaking to Mr. Bryan's boundless enthusiasm for acquisition and study. Yet, since his passing in October 2018, I and many others have felt his absence acutely. There is no doubt that, had he lived longer, Mr. Bryan would have had an opinion on every decision that went into the making of this book. As such, we knew it was important to convey the sense of playfulness and theatricality that went hand in hand with the attention to detail that Mr. Bryan brought to the study of early modern Britain—and indeed to many other subjects dear to his heart. To that end, I would like to explore three ideas here: his gift, his idea of cliché, and his process.

To experience John Bryan's collections at Crab Tree Farm—in the way he intended visitors to experience them—was to receive a gift: the chance to be transported to another world, to escape the present, and to enter an alternate reality, one that had a magical, storybook quality. Beyond that was another rich gift: the opportunity to engage with precious and rare historic objects in a completely different way, to handle directly things that were usually only viewed behind glass in museums.

In John Bryan's world, all objects have their place and are best appreciated through physical, sensory relationships, as they were intended to be: brass, glass, and ceramic against oak or walnut; paneled rooms with flagstone, rush, or boarded floors; historic textiles, dressing beds, tables, and seats. On occasion, it might be candlelight that heightens the glint and glitter of a sixteenth-century gilded goblet, set upon an ancient Turkish carpet draped over a fifteenth-century oak table. More often a beam of sunlight enhances the joy of turning a treasure in your hands, feeling its material characteristics, judging the balance of its weight—and thereby learning something about the object's original function and purpose. When at Crab Tree Farm with Mr. Bryan as your guide, you felt as if time stood still.

This gift that John Bryan gave to others gave back to him, too. It drew many people to him: his family and friends, museum folk, academics, antique dealers, and specialists from throughout the world. He used the collection he had designed and made both to learn and to teach. A natural and generous teacher, he loved to share his gift with anyone who expressed interest.

The collection also presented John Bryan with an opportunity to play with the idea of cliché, consciously, with a sense of knowingness and a well-practiced lightness of touch. He understood that a cliché is a cliché for a reason: if it sells an idea, it works, so use it. The Hermit's Cottage is the perfect example. With its thatched roof and tiny windows, the Cottage appears as if out of a fairy tale. Yet, this cliché is playfully used, just sufficiently to somewhat suspend reality, to signal to the visitor that enchantment plays an important part in gaining access to this other world, which was created to provide a means of study but also a means of escapism.

And with a sense of enchantment is how I think the collections at Crab Tree Farm began. As a young man, recently married, John Bryan traveled for the first time to the United Kingdom. He was enchanted by the world he found there and wanted to capture something of what he was experiencing. He often loved to tell a story about that trip, about getting lost on narrow and winding English country lanes, of hurtling happily along (on the wrong side of the road). This journey is when his collecting started. As he and his wife, Neville, made their way through the UK, he bought things that caught his eye: "country" furniture, Windsor chairs, plenty of oak tables and chairs. He arranged for it all to be shipped to the United States. Some months after the couple returned home, he was summoned by his furious father, who demanded to know about the container load of "junk" that had arrived. As John Bryan liked to muse, concluding his story, his father was right, of course. He discovered in years to come that the furniture he bought was, indeed, not the finest. So not only was this trip the birth of a collector, it was also the genesis of John Bryan's process and his determination to acquire nothing but the best. The desire to capture something of the places he had traveled to was perhaps a standard, even clichéd, response of any young person on an exciting overseas trip, but what John Bryan eventually made of that response was unique.

In terms of process, he developed his collections as a counterbalance to an extraordinary business life, so the collections were always about escapism, about offering a fantasy world. After working with him for a number of years, I started to see John Bryan as a storyteller. He was creating histories with a small *h*, which were, quite literally, "his stories." It was not that he was not serious about history, but he did not want to overstate his own credentials as an historian. Every scene was crafted and narrated perfectly—and always imbued with a sense of great urgency. There were always multiple projects running at speed: the creation of Arts and Crafts interiors or a sixteenth-century room, the restoration of an early American cabin or a relandscaping scheme. All these projects ran concurrently and alongside his numerous publication and exhibition plans. John Bryan was prolific, and the pace of his projects could never be fast enough for him, so however large or experienced was the team at work, they were always struggling to keep up with him. In light of his great business achievements, the ferocious creative energy that shaped all his projects makes perfect sense. Crucially, though, his collecting, his creation of displays, his work with museums, the exhibitions and publications were all about education—which takes us deeper into a consideration of his process.

In terms of his acquisitions, there was no such thing as enough. There was always something more to acquire. It is interesting to flip that fact and ask the question, why acquire so much? My answer is that his process as a collector was really a process of education. His learning process was rooted in objects, through which the people of the past spoke to him and told him about their lives. But his process was also about the exchange of information. Many people, from throughout the world, visited him and his collections, frequently. Every visit provided him with an opportunity to teach and to learn. These visits led to conversations, and those exchanges of ideas led to the commissioning of research, books, and exhibitions. For John Bryan, there always had to be an outcome that was bigger than the original thought, that would give people broader access to objects and information.

After one of these visits, he would set about rearranging his display spaces, rewriting his ideas about the past—rewriting through his collections—to reflect his new understandings or perspectives. In this light, John Bryan's process involved furnishing histories, or narratives, as most clearly expressed in the Hermit's Cottage: The imagined resident, rarely at home to visitors, sleeps on the small, simple bed, surrounded by traps and other devices for ensnaring the vermin that comprise his diet. By the window stands a stack of walking canes ("pizzle sticks"), fashioned from bull penises (a knowing nod to the Bryan family's meat distribution business during the first half of the twentieth century). Atop the roughly hewn fireplace mantel is a photograph of a lost friend, the furniture historian Victor Chinnery, with whom Mr. Bryan created the hermitage—the two men relying on historic materials and methods, which included erecting brick-nogged walls and bringing expert thatchers from England to lay thatch. Even this most whimsical space at Crab Tree Farm was built with the same blend of theatricality and precision that characterized all the building projects Mr. Bryan undertook during his lifetime.

It was only when Mr. Bryan had gone that I realized that each of these spaces, or stories, was created for a particular character. That character was John Bryan himself. He had created his magical kingdom as a way to inhabit his favorite places and to travel through time.

Rigorous in its approach and ambitious in its scale, the collaboration between Crab Tree Farm and the Yale Center for British Art to produce the publication *Marking Time* is, in many ways, the perfect expression of John Bryan's process. Although his legacies are legion, this book is a fitting testament to a great collector, his final mark in time.

Chronology of Objects & Historical Timeline

1500

Object	Year	Historical event
Portrait (of Edward VI) (p. 350) **Silver shilling** (p. 454)	1549	Book of Common Prayer issued in England and Wales; Prayer Book Rebellion
Silver five-shilling crown (p. 455)	1551	Bad harvests, high prices, and a decline in wool exports; precious-metal content in coinage falls to its lowest point
Bishop's Bible (p. 481)	1584 | 1666 1670 | 1692	Charter granted to Walter Raleigh to colonize parts of North America; James VI becomes Head of the Scottish Church, ending the Presbyterian system
Medal (p. 446)	1587	
Posnet (p. 204) **Skillet** (p. 205)	1592	Plague year in London; Pope Clement VIII elected
Bellarmine bottles (p. 363)	1594	
Shoehorn (p. 326)	1596	
Shoehorn (p. 328)	1597	Anthony Sherley attacks the Spanish territory of Jamaica; James VI publishes his study *Dæmonologie*
Shoehorn (p. 328)	1598	

1600

Objects	Year	Events
Ale pint measure (p. 429) **Ale quart measure** (p. 428) **Silver crown** (p. 455) **Walking stick** (p. 332)	1601	
Mortar (p. 199)	1602	The United East India Company (Dutch East Asia India Company) established; James Lancaster lands in Aceh, Sumatra, to trade British goods for black pepper
Silver sixpence (p. 455) **Twenty-eight-pound wool weight** (p. 432)	1605	Discovery of the Gunpowder Plot; execution of Guy Fawkes and co-conspirators
Chest (p. 198)	1609	News of a "monstrous child" born in Old Sandwich, Kent
Tobacco box (p. 305)	1610 | 1661	
Church bell (p. 489) **Powder horn** (p. 291)	1612	
Spoon (p. 117) **Warming pan** (p. 216)	1615	Reports of the births of two monsters by two women and an earthquake on July 25, near Faversham in Kent
Memorial plates (p. 486)	1616 | 1617	Walter Raleigh, imprisoned since 1603 for his attempt to depose James I, released from the Tower of London to seek lost city of El Dorado in Guiana, South America (now Guyana)
Mortar (p. 200)	1617	
Portrait (of Thomas Trayton) (p. 414)	1618	Walter Raleigh returns to England and is executed for treason
Armchair (p. 196)	1621	
Counter box (p. 298) **Sweet bag** (p. 341)	1625	Death of James I and succession of Charles I; royal proclamation makes further attempt to prohibit the importation of tobacco produced by other nations; plague in London kills approximately 10 percent of population

Wine bottle (p. 296)	1628	Enslaved Africans transported to Manhattan Island in the Dutch colony of New Netherland (now New York)
Armchair (p. 197)	1631	Clockmakers' Company of London receives Charter of Incorporation
Mortar (p. 201)	1632	British colonists settle in Antigua, West Indies
Medal (p. 447)	1633	
Cross (p. 466)	1636	
Bible (p. 242)	1636 | 1639 | 1641 | 1693 | 1872	British East India Company establishes Fort St George, first settlement of British India
Mortar (p. 202)	1637	
Bible box (p. 240)	1639	
Fuddling cup (p. 285)		
Silver crown (p. 456)	1642	Outbreak of the English Civil War; Battle of Edgehill
Gold unite (twenty shilling) (p. 456)	1644	
Medal (p. 447)	1647 | 1658	News of a monster born near Edinburgh; publication of *The Discovery of Witches* by the self-titled Witch Finder General, Matthew Hopkins
Pendant (p. 474)	1648	
Portrait (of Charles I) (p. 475)		
Portrait (of Charles I) (p. 476)		
Wine serving bottle (p. 275)		
Wine serving bottle (p. 275)	1649	Execution of Charles I; Charles II proclaimed king in Edinburgh; England declared a commonwealth republic
Case (p. 298)	1652	First coffeehouse opens in London
Penner (p. 336)	1653	
Memorial plate (p. 487)	1654 | 1657 | 1663 | 1675 | 1684	

Objects	Year	Events
Chalice (p. 354) **Sealing wax case** (p. 337)	1655	British capture of Jamaica from Spain; Oliver Cromwell prohibits Anglican services
Gold broad (twenty shilling) (p. 457) **Pastry jigger** (p. 207)	1656	
Cauldron (p. 206) **Spoon** (p. 210)	1658	Death of Oliver Cromwell; publication of *Orbis sensualium pictus* by John Amos Comenius; Robert Hooke invents the balance spring for watches
Wine costrel (p. 280)	1659	News of two wonders: a "huge and mighty fish much like unto a great Black Bull" and a mermaid, both sighted in the Thames near Blackwall
Cabinet (p. 124) **Cup** (p. 368) **Medal** (p. 448)	1660	Restoration of Charles II; Company of Royal Adventurers Trading to Africa Company founded to trade with West Africa for gold, silver, and slaves; Samuel Pepys begins his diary and records, inter alia, his first cup of tea
Dish (p. 360) **Plate** (p. 223)	1661	
Ale jug (p. 281) **Tobacco box** (p. 306) **Wine serving bottle** (p. 275)	1662	Portugal cedes Bombay, India, and Tangier, Morocco, to England
Ale gallon measure (p. 430) **Fireback** (p. 213) **Spoon rack** (p. 224)	1663	The first gold guinea minted
Pipe tamper (p. 302) **Trade token** (p. 442)	1664	
Candlesticks (p. 293) **Tobacco box** (p. 309) **Trade token** (p. 442)	1665	The Great Plague of London, the last widespread outbreak of bubonic plague in Great Britain
Mortar (p. 365) **Trade tokens** (pp. 442, 443, 444)	1666	Great Fire of London; passage of the first Burying in Woollen Act
Cup (p. 188)	1666 | 1783 | 1791	

Mortar (p. 203) — **1667**

Trade tokens (pp. 444, 445)

Posset pot (p. 175) — **1668** — East India Company takes over the running of Bombay; Isaac Newton invents first reflecting telescope

Comb and case (p. 322) — **1669** — News of a flying serpent, spotted near Saffron Walden, Essex

Trade token (p. 445)

Bed hangings (p. 152) — **1670** — Introduction of minute hands on watches

Mirror (p. 158)

Winchester corn bushel measure (p. 434)

Goblet (p. 276) — **1670 | 1671**

Wet drug jars (p. 402)

Silver crown (p. 457) — **1672** — England declares war on the Dutch Republic; Royal African Company established as new charter of the collapsed Company of Royal Adventurers Trading to Africa Company

Snuff flask (p. 342)

Charger (p. 369) — **1673**

Cradle (p. 112)

Hatbox (p. 324)

Cabinet (p. 482) — **1673–1863**

Ale bottle (p. 252) — **1674** — News of a woman from Shoreditch, in London, who sold her soul to the Devil six years prior

Tobacco box (p. 472)

Calligraphic miniature (p. 409) — **1674 | 1718/9**

Tobacco box (p. 309) — **1676** — The Lord's Day Act passed in Parliament, enforcing the observance of the Sabbath

Gold one guinea (twenty shilling) (p. 458) — **1677**

Spoon (p. 225)

Wet drug jar (p. 402)

Bible box (p. 479) — **1677 | 1690**

Winc gallon measure (p. 431)	**1678**	Discovery of the Popish Plot; the second Burying in Woollen Act, more rigidly enforced than the first
Dry drug jar (p. 404) **Jar** (p. 277) **Spoon** (p. 226)	**1681**	
Plate (p. 159)	**1682**	Edmond Halley sights the comet he will later accurately predict will return; Bideford witch trials
Alms dish (p. 361) **Skimmer** (p. 210) **Tobacco box** (p. 309)	**1683**	
Box (p. 299)	**1683/4**	
Ale or wine bottle (p. 253) **Slides** (p. 467)	**1684**	Frost fair held on the frozen River Thames; East India Company builds trading post in Canton, China (now Guangzhou)
Brush (p. 330) **Perpetual calendar** (p. 425) **Wassail bowl-and-cup set** (p. 194)	**1685**	Death of Charles II and accession of his brother James II; failure of the Monmouth Rebellion; execution of the Duke of Monmouth
Fire tongs (p. 212) **Gold five guinea** (p. 459) **Silver crown** (p. 458)	**1687**	
Armchair (p. 379)	**1687/8**	
Comb and case (p. 323) **Cup** (p. 176) **Hand warmer** (p. 345) **Plate** (p. 160)	**1688**	Glorious Revolution begins
Spoon (p. 227)	**1688/9**	
Garters (p. 380) **Medal** (p. 449)	**1689**	William III and Mary II proclaimed co-rulers of England, Scotland, and Ireland; Jacobite uprising
Tobacco box or pounce pot (p. 318)	**1689 | 1739**	

Friendship book (p. 140) — **1689 | 1766 | 1797 | 1810**

Nutcracker (p. 233) — **1690**

Fountainhead (p. 412) — **1691**

Plate (p. 370)

Wine bottle (p. 254)

Casket (p. 417) — **1692** — Earthquake in Jamaica decimates Port Royal

Silver crown (p. 459)

Water cistern (p. 109)

Cabinet (p. 272) — **1693**

Shot gauge (p. 426)

Cutlery set (p. 138) — **1694** — Mary II dies of smallpox; voyage of the slaver *Hannibal*; Bank of England established

Gold two guinea (p. 459)

Medal (p. 449)

Nutcracker (p. 234)

Portrait (of James Francis Edward Stuart) (p. 376)

Shoehorn (p. 329)

Communion flagons (p. 356) — **1695** — Quakers Act, releasing Quakers from the legal obligation to swear an oath, a violation of their religious beliefs; Window Tax results in bricking up of windows

Plate (p. 146)

Hunting horn (p. 288) — **1696** — The Jamaica Slave Act provides code of law to slave holders in that colony; Great Recoinage Act passed to curtail devaluation and deter counterfeiting

Plate (p. 160)

Porringer (p. 114)

Sundial (p. 423)

Wall clock (p. 420)

Spit jack (p. 214) — **1697**

Brush (p. 330) — **1698** — The Palace of Whitehall burns to the ground

Cradle (p. 115)

Ember tongs (p. 212)

Girdle (p. 381)

Portrait (of Charles I) (p. 477) — **1698 | 1765**

1700

Sampler (p. 126)	1700	
Standing cup (p. 343)		
Punch bowl (p. 387)	1701	
Skimmer or salamander (p. 211)		
Bellows (p. 215)	1701 | 1781	
Candlesticks (p. 236)	1702	Death of William III; Battle of Vigo Bay; silver captured from Spanish ships brought to England and made into coins
Fountain pen (p. 338)		
Silver crown (p. 460)	1703	Great Storm ravages southern England; Isaac Newton elected president of the Royal Society; death of Samuel Pepys
Plaque (portrait of Queen Anne) (p. 366)	1704	
Plate (p. 162)		
Portrait (of James Francis Edward Stuart) (p. 377)		
Ale bottle (p. 255)	1705	
Communion paten (p. 359)		
Apple corer (p. 232)	1706	The Treaty of Union unites England and Scotland into one kingdom named Great Britain
Spoon (p. 116)		
Ale quart measure (p. 430)	1707	
Medal (p. 450)		
Mug (p. 181)	1707 | 1755	
Ale pistol flask (p. 282)	1708	
Buckle (p. 325)		
Candlestick (p. 294)	1709	The Great Frost, the coldest European winter in five hundred years
Goblets (p. 157)		
Winchester corn bushel measure (p. 435)		
Alms dish (p. 362)	1710	
Fork (p. 232)		

Gold five guinea (p. 460)	1711	
Ale or wine bottle (p. 256) **Plate** (p. 163) **Shadowbox** (p. 130)	1712	New York Slave Revolt
Blunderbuss pistols (p. 388)	1714	Death of Queen Anne; Parliament offers Longitude Prize for method of establishing longitude at sea
Medal (p. 450) **Stay busk** (p. 169)	1715	Jacobite uprising
Embroidered picture (portrait of Charles I) (p. 478) **Glass** (p. 364) **Silver crown** (p. 461) **Silver pattern guinea** (p. 460)	1716	James Stuart, the Old Pretender, returns to France after the failed Jacobite uprising
Parish chest (p. 352) **Plate** (p. 371)	1717	
Nutcracker (p. 234) **Posy ring** (p. 178)	1719	
Model dogs (p. 245) **Plate** (p. 180)	1720	South Sea Company purchases much of Britain's national debt, leading to the scandalous "South Sea Bubble" and market collapse; execution of the pirate John Rackham (alias Calico Jack) at Port Royal, Jamaica
Tobacco box (p. 314)	1720 | 1745	
Dog bowl (p. 246) **Dog collar** (p. 247)	1721	
Tankard (p. 355)	1722	First Caslon serif typeface set by William Caslon in London
Ale or wine bottle (p. 257) **Beaker** (p. 182) **Goblet** (p. 278) **Wall hanging** (p. 195) **Wine bottle** (p. 258)	1723	Deaths of Godfrey Kneller and Christopher Wren

Objects	Year	Events
Pocket watch and case (p. 347)	1723 | 1740 | 1744	
Sampler (p. 129) **Stay busk** (p. 170) **Tobacco box** (p. 318) **Wine bottle** (p. 279)	1724	
Ale or wine bottle (p. 260) **Tobacco box** (p. 310)	1725	"Peter the Wild Boy," a feral child, found in woods near Hamelin, Hanover, and brought to George I at Kensington Palace where he lived as a court "pet" and public curiosity
Candlestick (p. 406) **Gold two guinea** (p. 461) **Tobacco box** (p. 311)	1726	Clockmaker John Harrison invents the gridiron pendulum, improving clock precision; Mary Toft, of Godalming, Surrey, claims to have given birth to a monster and a litter of rabbits
Architectural fragment (p. 413) **Medal** (p. 451)	1727	Death of George I, succession of George II; death of Isaac Newton
Embroidered picture (p. 119) **Embroidered picture** (p. 367) **Mirror** (p. 473) **Sampler** (p. 131)	1728	
Hammer (p. 144) **Punch bowl** (p. 292) **Walking stick** (p. 333)	1729	Publication of *The Sea-surgeon, or The Guinea man's vade mecum* by Thomas Aubrey
Mug (p. 470) **Plate** (p. 120) **Tobacco box** (p. 319)	1730	Tobacco Inspection Act passed to improve the quality of tobacco exported from Colonial Virginia
Ale bottle (p. 261)	1731	
Calendar barometer (p. 441)	1731–1830	
Fork (p. 211) **Plate** (p. 120)	1732	

Ankle iron and key (p. 384) —— **1733** —— Sugar and Molasses Act is passed by Parliament to tax British colonists in North America; John Kay patents his "shuttle" engine for twisting reed twine

Candlesticks (p. 393) —— **1734**

Cribbage board (p. 300)

Plate (p. 120)

Portrait (of Thomas Western) (p. 142)

Silver crown (p. 462)

Mug (p. 177) —— **1735**

Sauceboat (p. 174) —— **1736** —— "India rubber" comes to England; the Gin Act of 1735 establishes a retail tax on gin and annual licenses for sellers

Flower brick (p. 244) —— **1737**

Ring (p. 468)

Workbag (p. 231)

Plate (p. 372) —— **1738** —— John Wesley launches the Methodist movement

Sampler (p. 132)

Tea caddy (p. 172)

Screen (p. 192) —— **1738 | 1742**

Model shoe (p. 166) —— **1739**

Plate (p. 374)

Wine bottle (p. 262)

Mug (p. 286) —— **1740** —— Benjamin Huntsman improves crucible process for smelting steel

Gold two guinea (p. 462) —— **1741**

Plate (p. 375)

Plates (p. 219) —— **1742** —— Death of Edmond Halley; cotton factories established in Birmingham and Northampton

Seven-pound wool weight (p. 433)

Ale bottle (p. 263) —— **1745**

Barometer (p. 440)

Candlesticks (p. 407)

Cup (p. 289) **Duke of Cumberland Medal** (p. 451) **Plate** (p. 373) **Portrait (of Samuel Irton)** (p. 304) **Walking stick** (p. 333)	1746	After the last great Jacobite rising and defeat at the Battle of Culloden in Scotland, Charles Edward Stuart, the Young Pretender, flees to France and lives life in exile; the Dress Act prohibits the wearing of the kilt and other Highland dress
Plaque (p. 113)	1747	Liverpool overtakes Bristol as the busiest slaving port in Britain
Sampler (p. 133) **Tape measure** (p. 427)	1748	
Plates (p. 218)	1750	
Candle box (p. 179) **Jar** (p. 265) **Pipe tamper** (p. 303) **Wine bottle** (p. 264)	1751	Calendar Act makes January 1 the first day of the year; death of Frederick, Prince of Wales
Stirrup cup or powder flask (p. 282)	1752	Great Britain adopts Gregorian calendar on September 14
Bow Street magistrates ticket (p. 446)	1752 | 1767 | 1774 | 1777	
Plate (p. 147) **Ring** (p. 468) **Tobacco box** (p. 312) **Wall pocket** (p. 222)	1753	Marriage Act seeks to end clandestine marriages
Mug (p. 287) **Plate** (p. 122) **Plate** (p. 394) **Pocket globe and case** (p. 344)	1754	Quaker abolitionist John Woolman publishes his essay "Some Considerations on the Keeping of Negroes"
Staff (p. 390) **Teapot** (p. 237) **Tobacco box or boot-powderer** (p. 313)	1756	Outbreak of the Seven Years' War; British troops imprisoned in the event known as the Black Hole of Calcutta
Candlesticks (p. 408) **Mug** (p. 405)	1757	

Portrait (of Prince William Augustus, Duke of Cumberland) (p. 378) — **1758** — Eponymous comet Edmond Halley predicted would return returns

Box (p. 320) — **1759** — John Harrison invents the first successful marine chronometer, solving the problem of calculating longitude at sea

Medal (p. 452) — **1760** — Death of George II; succession of George III

Model shoe (p. 166)

Cribbage board (p. 300) — **1761**

Flower brick (p. 410)

Horse-racing pass (p. 345)

Ink stand (p. 334)

Mug (p. 238)

Plate (p. 395)

Sealing wax case (p. 339)

Chandelier (p. 353) — **1762**

Swear box (p. 273)

Bowl (p. 149) — **1763**

Brick fragment (p. 411)

Teapot (p. 173)

Knitting sheath (p. 228) — **1764** — Sugar Act passed to end colonists' smuggling of sugar and molasses from French and Dutch West Indies; Currency Act prohibits colonies issuing paper money; James Hargreaves invents the "spinning jenny," initiating machine production of cotton and wool thread

Mug (p. 184)

Plate (p. 164)

Portrait (of Parnell Villiers) (p. 154)

Portrait (of Thomas Abney) (p. 155)

Candle box (p. 235) — **1765** — Stamp Act levies taxes on American colonies

Inkstand (p. 334)

Porter's tally (p. 400)

Rastrum (p. 340)

Tailor's smoothing iron (p. 145)

Communion flagon (p. 357) — **1765 | 1968**

Liquid measures (p. 437) — **1766** — Repeal of the Stamp Act; death of James Stuart, the Old Pretender

Mug (p. 243)
Plate (p. 396)
Quill holder (p. 335)
Tobacco boxes (p. 314) — **1767** — Townshend Act taxes imports of certain products in American colonies

Carving steel (p. 209)
Plate (p. 165) — **1768** — James Cook embarks on first voyage to the Pacific Ocean

Ale bottle (p. 268)
Dog or monkey collar (p. 248)
Tipstaff (p. 392) — **1769** — Wig maker Richard Arkwright patents the spinning frame, developed with clockmaker John Kay, making it possible to mass-produce stronger thread than the jenny

Plate (p. 397)
Wine bottle (p. 267) — **1770**

Cribbage board (p. 300) — **1771** — Spain cedes Falkland Islands to Britain

Apple corer (p. 233)
Celestial globe (p. 438)
Terrestrial globe (p. 438)
Tobacco box (p. 316) — **1772** — British court ruling prevents forced return of former slave James Somerset to West Indies to be resold

Dog collar (p. 249)
Magnifying glass (p. 346)
Tobacco box (p. 317) — **1773** — Tea Act grants East India Company monopoly on North American tea trade, leading to colonists' protest, later known as the Boston Tea Party

Brush (p. 330)
Coin scale (p. 453)
Plate (p. 220) — **1774**

Pocket watch and case (p. 424)
Tea canisters (p. 237) — **1775** — First battles of the American Revolution; influenza outbreak in England and parts of Europe

Fishing reel (p. 290)
Plate (p. 283)
Sampler (p. 134)
Shaker or muffineer (p. 208)
Stay busk (p. 171) — **1776** — American Congress carries Declaration of Independence from Britain

Sampler (p. 135) — **1777**

Tobacco box (p. 145)

Etui (p. 321) — **1778** — Britain declares war on France for aiding the Americans

Jug (p. 239)

Coffer (p. 401) — **1779**

Tombstone (p. 488)

Box (p. 469) — **1782**

Candlesticks (p. 389)

Cup (p. 471)

Knitting sheath (p. 230)

Mug (p. 185) — **1783** — Peace of Versailles, in which Britain recognizes independence of United States of America

Sampler (p. 136)

Terrestrial globe (p. 439)

Telescope (p. 399) — **1784**

Celestial globe (p. 439) — **1785** — James Watt and Matthew Boulton install a steam engine in a cotton-spinning factory in Nottinghamshire

Sampler (p. 137)

Tobacco box (p. 319)

Wine bottle (p. 269)

Candlesticks (p. 358) — **1786**

Knitting sheath (p. 229)

Pincushion (p. 118) — **1787** — Founding, in London, of the Society for the Effecting of Abolition of the Slave Trade; Britain establishes Freetown, Sierra Leone, as a colony for freed slaves

Cups (pp. 186, 187) — **1788**

Gold half guinea (p. 463)

Puzzle (p. 123)

Staff (p. 391) — **1789** — Aided by John Wesley and others, British abolitionist and former slave Olaudah Equiano publishes his widely read account of the Middle Passage from Nigeria to Barbados and enslavement in Virginia

Bloodletting knife (p. 148) — 1790

Knitting sheath (p. 229)

Bird feeder (p. 168) — 1792

Coin scale (p. 454)

Ale bottle (p. 270) — 1793

Box (p. 398)

Plate (p. 221)

Winchester corn bushel measure (p. 436)

1797 — Parliament passes the Duties on Clocks and Watches Act, leading to drastic decline in the manufacture and purchase of the timepieces and, as a result, repeal of the tax the next year

Flask (p. 284) — 1799

Knitting sheath (p. 230)

1800

Medal (p. 145) — 1800 — Act of Union of Great Britain and Ireland; Eli Whitney makes muskets with interchangeable parts near New Haven, Connecticut

Medal (p. 452) — 1801 — Union Jack becomes the official flag of the United Kingdom; the first census conducted in Britain

Wine bottle (p. 271)

1803 — After people in Hammersmith, London, claim attacks by ghost of a suicide victim, bricklayer in white clothing shot, mistaken for the apparition

Silver dollar of five shillings (p. 463) — 1804 — Hammersmith Ghost murder conviction opens debate as to whether mistaken belief is valid criminal defense, with verdict overturned in 1983 and British law amended in 2008

Index of Objects

Index of Known Makers, Skills & Trades

Known Makers

Note: Makers' skills and trades are specified within parentheses. Guild members' trades are capitalized.

Skills & Trades

Acknowledgments

The editors would like to thank the following specialists who contributed to the Crab Tree Farm dated project and, by extension, to the exhibition and publication *Marking Time*. Their areas of expertise are noted within parentheses.

Michael Archer (ceramics)
Stephanie Aspin (general research)
Garry Atkins (ceramics)

Christopher Bangs (metals)
Joe Barabe, McCrone Associates Inc. (paint analysis)
Charlotte Bolland (sixteenth-century paintings)
Adam Bowett (furniture)
David Brown (sundials)
Francesca Butcher (general research)
Roderick Butler (cookware)
Graham Byfield (coins and medals)

Peter Cameron (silver)
Victor Chinnery (furniture)
Sophie Cope (general research)
Dennis Cowan (sundials)
Oliver Cox (racing)

Christopher Daniel (sundials)
Jessica David (sixteenth-century paintings)

Michael Finlay (mortars, pastry jiggers, weights, measures)

Titi Halle (textiles)
Peter Harrington (books)
Catherine Hassell (paint analysis)
Steve Hill (coins and medals)
Sandra Hindman (jewelry)
Diana Hitchens (scales)
Jeremy Hodgkinson (firebacks)
Jo Hormuth (paint analysis)
Sylvia Houghteling (textiles)
John Howard (books)
Robert Hunter (ceramics)

Richard Ireland (paint analysis)

Toby Jellinek (furniture)

Jasmine Kilburn-Toppin (general research)
Eloy Kolderweij (candlesticks)

David Lachman (books)
Dwight Lanmon (glass)
Santina Levey (textiles)

Jim Marshall (writing materials)
Timothy Millett (coins and tokens)
Eve Mongurtan (general research)
Kitty Morris (textiles)
David Moulson (pewter)
Tessa Murdoch (silver)

Angus Patterson (brass, pewter, copper)

Roxanne Ravenhill (general research)
Peter Robinson (canes)
Isabella Rosner (samplers)
Emma Rutherford (miniatures)

Annie Sheen (general research)
Harry Smith (paintings)
Hannah Stockton (general research)
Dominick Strickland (canes)
Danai Synadinou (general research)

Ian Tyers (dendrochronology)

Melinda Watt (textiles)
Annabel Westman (textiles)
Tony Weston (spit jack)
Tony Willis (firearms)
Jeremy Wood (identification of print sources)
Helen Wyld (samplers)

Dario Zorza (Latin translations)

Photography & Illustration Credits

Photographs

Unless otherwise noted, all images are courtesy of The Bryan Collection, Lake Bluff, IL, and all photography is by Jamie Stukenberg, Professional Graphics Inc.

© 2020 Copyright The Chapter of St Paul's Cathedral: fig. 24

© 2020 National Museums Liverpool: figs. 11, 13

© 2020 Trustees of the British Museum: fig. 17

Beinecke Rare Book & Manuscript Library: figs. 3, 6

Bodleian Library, John Johnson Collection, University of Oxford: fig. 2

By permission of the Pepys Library, Magdalene College, Cambridge: fig. 16

© National Maritime Museum, Greenwich, London: fig. 8

© National Portrait Gallery, London: figs. 15, 25, 26

Parliamentary Archives, London, HL/PO/JO/10/1/297A: fig. 27

Photograph © 2020 Museum of Fine Arts, Boston: figs. 18, 21

Reproduced by courtesy of Abbot Hall Art Gallery, Lakeland Arts Trust, England: fig. 23

© Tate, London 2020: figs. 9, 12

The Art Institute of Chicago/Art Resource, NY: pages 156, 276, 364

© The British Library Board: figs. 29, 30

The Lewis Walpole Library, Topographical Prints Collection, Yale University: fig. 5

Wellcome Collection, London: fig. 14

Yale Center for British Art, photography by Richard Caspole: fig. 19

Yale Center for British Art, Gift of Andrew Cavendish, eleventh Duke of Devonshire, photography by Richard Caspole: fig. 4

Yale Center for British Art, Gift of Susan M. Yecies, BA. 1971, photography by Richard Caspole: fig. 28

Yale Center for British Art, Paul Mellon Fund, photography by Richard Caspole: figs. 22, 31

Yale University Art Gallery, Mabel Brady Garvan Collection: fig. 20

Illustrations

All the woodcut prints that open the prose sections of the book are from John Amos Comenius, *Orbis sensualium pictus... Or, A Nomenclature, and the Pictures, of All the Chief Things That Are in the World* (London: Printed for S. Leacroft, 1777). Courtesy of the Beinecke Rare Book & Manuscript Library

page 8: "The Seven Ages of Man," plate 39

page 18: "Society betwixt Man and Wife," plate 202

page 34: "Dials," plate 80

page 42: "Ship-wreck," plate 93

page 56: "Linen Cloths," plate 63

page 68: "Measures and Weights," plate 129

page 82: "The Box-maker and the Turner," plate 72

page 90: "A Burial," plate 133

page 490: "The Close," plate 153

Published by
Yale Center for British Art
P.O. Box 208280, New Haven, CT 06520-8280
britishart.yale.edu

Distributed by
Yale University Press
P.O. Box 209040, New Haven, CT 06520-9040
47 Bedford Square, London WC1B 3DP
yalebooks.com / yalebooks.co.uk

Library of Congress Cataloging-in-Publication Data
Names: Town, Edward, editor. | McShane, Angela, 1959– editor. | Adamson, Glenn. Running like clockwork? | Yale Center for British Art
Title: Marking time : objects, people, and their lives, 1500–1800 / Edward Town, Angela McShane.
Other titles: Marking time (Yale Center for British Art)
Description: New Haven : Yale Center for British Art, [2020] | Includes bibliographical references and index.
Identifiers: LCCN 2020003822 | ISBN 9780300254105 (board)
Subjects: LCSH: Time perception. | Time and art.
Classification: LCC BF468 .M35 2020 | DDC 304.2/370941—dc23
LC record available at https://lccn.loc.gov/2020003822

Produced by Lucia | Marquand
luciamarquand.com

Book and cover design by Ryan Polich
Map design by Jeremy Linden
Indexed by Kathleen Friello
Proofread by Livia Tenzer
Typeset in Big Caslon, Paperback, and Surveyor by Tina Henderson
Project management by Melissa Duffes
Color management by Prographics
Printed and bound in China by Artron

Front cover: Richard Savage, wall clock (detail), 1696, brass, iron; case: 8 × 8 in. (20.3 × 20.3 cm). The Bryan Collection, Lake Bluff, IL